Millstone workshop, Olho Marinho, Portugal

UNIVERSITY OF Southampton
Series In Archaeology No. 3

Bread for the People: The Archaeology of Mills and Milling

Proceedings of a colloquium held in the British School at Rome 4th - 7th November 2009

Edited by

David Williams
David Peacock

BAR International Series 2274
2011

Published in 2016 by
BAR Publishing, Oxford

BAR International Series 2274

University of Southampton Series in Archaeology Monographs No. 3
Bread for the People: The Archaeology of Mills and Milling

ISBN 978 1 4073 0848 7

BAR Publishing is the trading name of British Archaeological Reports (Oxford) Ltd. British Archaeological Reports was first incorporated in 1974 to publish the BAR Series, International and British. In 1992 Hadrian Books Ltd became part of the BAR group. This volume was originally published by Archaeopress in conjunction with British Archaeological Reports (Oxford) Ltd / Hadrian Books Ltd, the Series principal publisher, in 2011. This present volume is published by BAR Publishing, 2016.

Printed in England

BAR
PUBLISHING

BAR titles are available from:

BAR Publishing
122 Banbury Rd, Oxford, OX2 7BP, UK
EMAIL info@barpublishing.com
PHONE +44 (0)1865 310431
FAX +44 (0)1865 316916
www.barpublishing.com

Contents

List of contributors

Natàlia Alonso
Universitat de Lleida
Spain.
nalonso@historia.udl.cat

Timothy J. Anderson
University of Pierre Mendès
Grenoble
France.
timanderson.granada@gmail.com

Martina Andreoli
University of Trento
Italy.
martina.andreoli@unitn.it

Paul Arthur
University of Salento
Italy.
paul.arthur@unisalento.it

Redha Attoui
University of Trento
Italy.
redha.attoui@unitn.it

Meritxell Aulinas
Universitat de Barcelona
Spain.

Marie-Christine Bailly-Maître
Laboratoire d'Archéologie Médiévale Méditerranéenne
France.
baillymaitre@wanadoo.fr

Irene Baug
University of Bergen
Norway.
irene.baug@ahkr.uib.no

Alain Belmont
University of Pierre Mendès
Grenoble
France.
alain.belmont@club-internet.fr

Elizabeth Bloxam
University College London
UK.
e.bloxam@ucl.ac.uk
and
Monash University, Melbourne
Australia.
Elizabeth.Bloxam@monash.edu

André Buisson
Université Lyon III
France.
andre.buisson@univ-lyon3.fr

Carolyn Coulter
University of Southampton
UK.
cc1104@soton.ac.uk

Lyn Cutler
University of Southampton
UK.
lhc200@soton.ac.uk

John Paul Duc
Conservatoire des meules et pavés d'Epernon
France.
jeanpaul.duc@wanadoo.fr

Jean-Yves Dufour
Institut national de recherches archéologiques préventives
France.
Jean-yves.dufour@inrap.fr

Juan Manuel Fernández Soler
Universidad de Granada
Spain.
jmfsoler@ugr.es

Paola Galetti
Università di Bologna
Italy.
paola.galetti@tin.it

Maria Teresa Garcia
Universitat de Barcelona
Spain.

Pedro García Ruiz
Logroño (La Rioja)
Spain.
pilarpascual4@gmail.com

Antonio Granero Gallegos
Universidad de Murcia
Spain.
agranero@um.es

Chris Green
Freelance archaeologist
St Albans
UK.
cmg@waitrose.com

Tor Grenne
Geological Survey of Norway (NGU)
Norway.
Tor.Grenne@ngu.no

Caroline Hamon
CNRS, UMR 7041
France.
caroline.hamon@mae.u-paris10.fr

Tom Heldal
Geological Survey of Norway (NGU)
Norway.
Tom.Heldal@ngu.no

Charles D. Hockensmith
Archaeological Consultant
Kentucky
USA.
charleshockensmith@fewpb.net

Luc Jaccottey
Institut national de recherches archéologiques préventives
France.
luc.jaccottey@inrap.fr

Florent Jodry
Institut national de recherches archéologiques préventives
France.
florent.jodry@inrap.fr

Valerie Le Gall
Institut national de recherches archéologiques préventives
France.
valerie.legall@inrap.fr

Achim Lehmkuhl
Staatliches Museum für Naturkunde
Stuttgart
Germany.
achim.lehmkuhl@smns-bw.de

Torbjørn Løland
Hyllestad
Norway.
torbjorn@aafjordstein.no

Samuel Longepierre
Flaux
France.
samuel.longepierre@wanadoo.fr

Félix Martín
Master stonemason
Lleida
Spain.

Francisco José Martínez López
Universidad de Murcia
Spain.
fjmartinez@um.es

Gurli Birgitte Meyer
Geological Survey of Norway (NGU)
Norway.
Gurli.Meyer@ngu.no

Nicolas Minvielle Larousse
Université Pierre Mendès-France
Grenoble
France.
minvielle.nicolas@gmail.com

Holly Parton
Freelance archaeologist
UK.
h_ptn@yahoo.co.uk

Pilar Pascual Mayoral
Universidad de Zaragoza
Spain.
pilarpascual4@gmail.com

David Peacock
University of Southampton
UK.
dpp@soton.ac.uk

Meinrad Pohl
Deutsches Bergbau-Museum Bochum
Bochum
Germany.
Meinrad.Pohl@bergbaumuseum.de

Georgina Prats
Universitat de Lleida
Spain.

Fiona Roe
Freelance stone specialist
UK.
fiona.roe@virgin.net

Joaquin Sanchez Navarro
Freelance archaeologist
Menorca
Spain.
jsanchez@silme.es

Jane H. Scarrow
University of Granada
Spain.
jscarrow@ugr.es

Ruth Shaffrey
Oxford Archaeology South
Oxford
UK.
ruth.shaffrey@oxfordarch.co.uk

Juan Manuel Fernández Soler
Universidad de Granada
Spain.
jmfsoler@ugr.es

Sílvia Vila
Universitat de Lleida
Spain.

Mariette de Vos
University of Trento
Italy.
mariette.devos@unitn.it

Martin Watts
Traditional Millwright
and Historic Milling Consultant
UK.
martin.watts@virgin.net

Susan Watts
University of Exeter
UK.
S.R.Watts@exeter.ac.uk

Stefanie Wefers
Römisch-Germanisches Zentralmuseum
Mainz
Germany.
wefers@rgzm.de

David Williams
University of Southampton
UK.
dfw@soton.ac.uk

Foreword

To the memory of Cecil Curwen, a pioneer in the study of querns and mills.

The study of mills and querns essentially developed during the 20th century. At the very end of the 19th century two seminal works set the stage. The first was Bennett and Elton's *History of corn milling* published in 1898 followed in 1899 by an article by Lindet published in *Revue archéologique.* Both were aware of the scantiness of available information as is apparent from their reference to earlier 19th century works including those of Mongez published in 1818, Keller on the Swiss lake dwellings which appeared in 1866, Hume and Stafford in 1868 and Blümner's *Technologie und Terminologie der Gewerbe und Künste bei Greichen und Römern* (1875-87). An outstanding if little known contribution was Diviš-Čistecký's (1893) account of the querns from Kunětické Hořa in the Czech Republic. Here he identified quarries and illustrated typology with both cross sections and perspective views in a work well ahead of its time. However, it seems that Bennett and Elton and Lindet were the first to draw attention to the importance and interest of the subject despite the paucity of the evidence at their disposal.

Despite this propitious start, little new work seems to have been generated and one of the most influential figures in our study, Cecil Curwen, was a babe in arms at this time. Nearly forty years later he was still able to write:
The development of the quern presents an immense field for study, and represents a gap in our knowledge which is crying out to be filled (Curwen 1937, 151).

Without doubt Eliot Cecil Curwen (1895-1967), was a pioneer of quern and mill studies and for this reason we dedicate this volume to his memory. Like his father, Eliot Curwen, he was a doctor by profession, but he nevertheless had an enduring passion for archaeology. Cecil Curwen was born in China, while his father was a surgeon at Peking Hospital with the London Missionary Society, before the family returned to England at the onset of the Boxer Rebellion. They settled in Hove, where Cecil Curwen began a medical practice and in his spare time became an amateur archaeologist, at a time when there were few professionals around. He was introduced to archaeology at an early age and published his first academic paper while still a boy at Rugby School in 1912. In the years following, father and son carried out a series of excavations and prehistoric field surveys on a number of sites across the South Downs and also collaborated on a wide range of archaeological papers for the *Sussex Archaeological Collections*. Such was the call of archaeology that ECC spent his honeymoon surveying on Dartmoor in the 1920's. His book on *Prehistoric Sussex* was published in 1929 to be followed by *The Archaeology of Sussex* in 1937 and *Plough and Pasture: the early history of farming* in1953. In 1955 he received the OBE for services to archaeology.

However, it was his pioneering work on querns and mills in influential papers published in the journal *Antiquity* that especially concern us here. In particular, his work on the development of the rotary quern in Britain, dealt with in his 1937 and 1941 papers is particularly relevant. Curwen was one of the first to appreciate the important role that querns provided in a subsistence economy, converting corn into flour and hence bread; while many of his contemporaries tended to overlook stone finds in the interpretation of archaeological sites and rarely looked upon them as objects of trade.

Moreover, he was not just an academic when it came to the study of querns but was also interested in their practical performance. He was one of the first to do ethnographic field-work on milling and visited the Isles of Scotland in the late 1930's, especially Shetland and the Hebrides. He considered these a cultural backwater in which Iron Age lifestyles persisted, where he not only saw rotary querns being used in the household but also being shaped and dressed. Perhaps this inspired him 'in 1937, to reconstruct some querns, make flour from them and treat members of the Sussex Archaeological Society to scones made from the flour' (Penn 1966). It may have been his visit to the north that stimulated his interest in watermills resulting in a paper debating their diffusion and development (Curwen 1944).

Later, in 1956, he visited the *Grottes d'Hercules* 14 km west of Tangiers. Today it is a tourist attraction dedicated to quern quarrying, but he actually witnessed workers extracting querns by candle light.

He was especially interested in the dating of querns and his work would lead to their recording on archaeological sites in a more systematic manner than had been usual up to that time. Allied to this, he proposed 'a tentative (typological) scheme as a basis of study', in which he classified a selection of British rotary querns based on a progression over time, from the 'bee-hive' shaped Iron Age types through the flatter and thinner Roman forms to the 'pot-quern' of the Medieval period (Curwen 1937, 133). Yet his work was not without its shortcomings. Recent evidence has shown that Curwen's notion that the use of the saddle quern finished with the introduction of the rotary quern does not stand up, as we can see from a number of papers in this volume, and his view that there is little correlation between form and lithology has been shown to be wrong (Peacock 1987, 61). Equally it is unfortunate that he differentiated what he called the 'vertical mill' from the 'Vitruvian' as most people nowadays would regard the Vitruvian as vertical and the simpler ungeared variety as horizontal. However, it says much for Curwen's vision and industry that although there have been a number of recent detailed regional studies on British querns, no general reassessment of his scheme has yet been undertaken.

Curwen's purpose in his quern and mill papers was to '*stimulate research by suggesting a rough classification of some leading types in Britain only. These need checking, correcting and amplifying, and the whole of the Continental and Irish material needs studying, in order to place our own in its proper setting*' (Curwen 1937, 151). It is clear from this that he saw his work merely as a first step along a long road of future research into quern and mill studies. Unfortunately, that journey has been one of stops and starts and it has only really been in the last few years that the detailed research into the subject that Curwen felt was needed has really come about. It is to be hoped that this volume takes us a little further on our way.

Acknowledgements

Many people helped us to organize the conference and we should like to thank the Institution and staff of the British School at Rome, for not only for putting their organization skills and facilities at our disposal but also for the generous hospitality that was extended to the group as a whole. Special mention must be made to the Director, Prof. Christopher Smith, and his predecessor Dr Andrew Wallace-Hadrill for their support and encouragement, Dr. Susan Russell, Assistant Director, who helped enormously in all manner of ways, Geraldine Wellington, the Residence Manager, Isabella Gelosia and Alvise di Giulio who handled the accounts splendidly and Giuseppe Pellegrino, who always had the perfect solution when presented with technical difficulties over power points and handouts.

We very much look forward to meeting friends and colleagues, old and new, at the next venue for this series of colloquia which is to be held at Bergen in October, 2011.

David Peacock
David Williams
Department of Archaeology,
University of Southampton.

References

Bennett, R., and Elton, J., 1898. *History of corn milling.* London.

Blümner, H., 1975-87. *Technologie und Terminologie der Gewerbe und Künste bei Greichen und Römern.* Leipzig.

Curwen, E.C., 1937. Querns. *Antiquity*, 11, 133-151.

Curwen, E.C., 1941. More about querns. *Antiquity*, 15, 15-32.

Curwen, E.C., 1944. The problem of early water-mills. *Antiquity*, 18, 130-146.

Curwen, E.C., 1956. A Tangier quern quarry. *Antiquity*, 30, 174.

Diviš-Čistecký, V., 1893. Starobylá dilna ručnich mlýnků na Hoře Kunětické. *Český Lid,* 2, 708-11.

Hume, D. and Stafford, W.C., 1868. *The History of England, from the earliest period to the present time. Compiled from the most authentic sources.* London.

Keller, F., 1866. *The lake dwellings of Switzerland and other parts of Europe.* London.

Lindet, L., 1899. Les origines du moulin a grains. *Revue Archéologique*, 35, 413-27.

Mongez l'Aîné, A., 1818. Mémoire sur les meules de moulins employées par les anciens et les modernes, et sur les meules à bras, antiques, trouvées près d'Abbeville. *Histoire et mémoires de l'institut royal de France, classe d'histoire et de littérature ancienne*, 3, 479.

Peacock, D.P.S., 1987. Iron Age and Roman quern production at Lodsworth, West Sussex. *Antiquaries Journal*, 67, 61-85.

Penn, W.S., 1966. A quern survey in Kent. *Kent Archaeological Review*, 3.

Introduction

David Peacock and David Williams

Since the Neolithic, and occasionally before, grain has often been ground to flour before consumption. While it can be consumed without grinding, as in the medieval dish *frumenty*, flour can be stored for longer periods without germination and is more easily transportable as bulk is reduced. This can then be mixed to make bread which has long been a staple of the European diet. Because of this a Europe without stone querns and mills is unthinkable, as up to the nineteenth century they were needed to produce flour and they must have comprised a major element in early economies. Indeed they continue to be made today and until recently the villages of Olho Marinho and Alveite, near Vila Nova de Poiares, in central Portugal were major centres of production, exploiting the excellent hard arenite of the nearby Serra de São Pedro Dias. Today just one millstone maker continues in production, Eduado Manuel Fernandes of Olho Marinho. He began work at the age of 13, learning the craft from his parents. Smaller millstones up to 60 cm in diameter are still made, but for use in electric mills, favoured by those preferring stone ground flour, although many of the larger ones, once destined for windmills or *azenhas* (watermills), are used nowadays as garden ornaments. Production has also expanded into other decorative items such as shrines, fountains, basins and crosses. They are marketed through fairs and by local custom. It is probable that Eduado Manuel Fernandes is one of the last millstone makers in Western Europe: the end of a great tradition once vital for survival. For this reason we have chosen to feature the work of this man on our cover in our frontispiece.

Millstone production and trade were of crucial economic importance and a vital element in human subsistence. The efforts made to quarry millstones were considerable, ranging from the Neolithic to recent times and sometimes involving underground operations. The economic value of good quality mill rocks is reflected, in wide trade patterns so often encountered. To the archaeologist querns and mills are important artefacts with potential to inform on trade, exchange and technology while millstone quarries are important monuments which deserve protection and evaluation.

This volume contains a series of papers that resulted from the proceedings of *Bread for the People: a Colloquium on the Archaeology of Mills and Milling*, which was held between the 4th and 7th November 2009 at the British School at Rome. This was the third in a series of colloquia whose central topic was current research on ancient, medieval and recent millstones and quernstones. The previous two being held at La Ferté-sous-Jouarre (2002) and Grenoble (2005). The broad themes of the Rome conference encompassed: the study of quarries of all periods; quality, production and trade in querns and millstones; archaeometrical studies; ethnographic studies, to include agriculture, ore processing and glass making; protection and evaluation of millstone quarries; and a poster session. We are especially grateful to Professor Alain Belmont for kindly agreeing to give the keynote paper at the start of the conference which admirably set the tone for the rest of the proceedings. However, we have taken this opportunity to include papers not delivered at the conference as some were prevented from attending for a variety of reasons.

Millstones and querns have for far too long been regarded as little more than tedious lumps of stone with little significance by all too many archaeologists. However, as this volume indicates such attitudes are unjustified and outmoded. It is clear from the wide range of papers here, both geographically and in terms of topic, that although trade in these objects is a valuable source of evidence, they also provide coded information which can shed valuable light on a range of social practices in the preparation and consumption of certain foods across regions and provinces. Looking to the future, research on food residues and phytoliths in the vesicles of mills and querns will doubtless add to that social understanding.

Why dig a millstone quarry? The case of Claix in the south-west of France (5th-19th centuries)

Alain Belmont

Translated by Timothy Anderson

Millstone quarries: a neglected subject of research in archaeology

For thousands of years millstones for grinding cereals have played a fundamental role in the sequence of operations for nourishing populations. We would therefore suppose that archaeologists would have a penchant for digging and studying the source of these stones. This, however, is not the case. Excavations of these industrial sites are extremely rare in Europe and in the world. Yet the first excavations took place very early, such as at Würenlos in the north of Switzerland, where Haberbosch unearthed in 1938 a quern quarry dating to Roman times (Haberbosch 1938; Doswald 1994). A few years later, in 1943, construction work at Chavannes-le-Chêne in western Switzerland resulted in the excavation of a second Roman quern quarry, as well as a small cemetery from the early Middle Ages (Bosset 1943; Weidmann 2002) (Fig. 1). A third more recent site in Switzerland, Châbles (1996-1997), was explored during construction of the Lausanne-Bern motorway. This model rescue operation unearthed a quern quarry dating to the first two centuries AD (Anderson *et al.* 2003). Elsewhere in Europe (Fig. 2), three millstone quarries were also dug in Norway at Selbu (Alsvik *et al.* 1981), Saltdal (Helberg 2007; Grenne *et al.* 2008) and Hyllestad (Baug 2006). These sites were exploited since the early Middle Ages and two were still in operation as recently as in the 20th century. These Scandinavian sites are presented in more detail elsewhere in these proceedings. Germany also is a pioneer in this field. Early amateur explorations were succeeded by the formal intervention of the Römisch-Germanisches Zentralmuseum (RGZM) of Mainz, which excavated the sites of Hochstein and Mayen under the direction of Fritz Mangartz (Mangartz 1993; 2002). Although France has a long tradition of digging even the smallest Roman theatre or villa, they forgot that their past cannot be summarised simply as the 'Latin way of life' and have, until recently, totally disregarded their millstone exploitations. This changed toward the end of the 1990's with a sudden increase in the number of excavations starting with that of Vioménil, a grinding stone site in the Vosges dating from the 17th century (Bertin 1998). Also in the Vosges, at La Salle, Virginie Farget directed several interventions dating from the 6th century BC to the 4th century AD (Farget 2006). In the Jura mountains, Luc Jaccottey explored the Neolithic quern quarry of Malange, as well as a medieval site of Offlanges (Jaccottey 2007; 2008). In the Massif Central, Ulysse Cabezuelo dug a site dating from the 17th – 18th centuries at Vic-le-Comte (Cabezuelo 2001). In the Alps, the University of Grenoble excavated the 13th-15th century sites of Ecouges, as well as Quaix-en-Chartreuse dating from the 15th to the 19th century (Belmont 2001; 2006a). Finally, in the southwest of France in 2008, the same institution undertook a programmed excavation of the millstone quarry of Claix, the subject of this paper.

Figure 1. The quern quarry of Chavannes-le-Chêne, Switzerland, excavated in 1943.

Figure 2. Location of the millstone quarries excavated in Europe up to 2009.

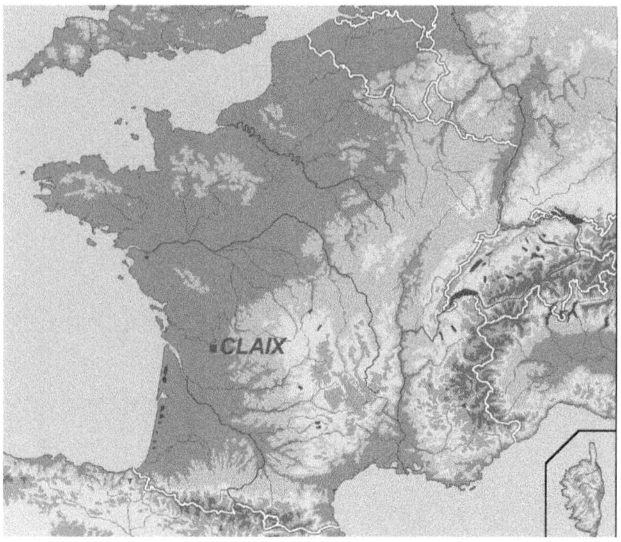

Figure 3. Location of the millstone quarry of Claix.

Therefore, if we have not omitted or forgotten any sites, up to 2009, not more than 16 millstone quarry sites have been excavated in Europe. This is a very modest count compared with the hundreds of excavations of mines. For example, Marie-Christine Bailly-Maître, co-author of an article appearing in these proceedings, alone has dug more than thirty. So what are the reasons for this neglect? Europe is far from lacking in millstone quarries. There are probably several thousand major sites spread across the continent that distributed their products over a radius of several hundred kilometres. In France alone there are more than 300. Moreover, the smaller, more modest exploitations, which furnished stones only as far as the neighbouring villages, must be counted in the tens of thousands. Even the most celebrated European sites, such as Melos in Greece, La Ferté-sous-Jouarre in France, Lugnås in Sweden, Jonsdorf in Germany, the Peak District of Derbyshire in England, Orvieto in Italy, etc., are still waiting for the archaeologist. When we inquire of colleagues and local authorities the reason behind this aversion, they invariably respond that this type of field work is too dangerous, too expensive, devoid of interest, unattractive, or even impossible to carry out. The case of Claix, however, refutes these pessimistic assertions and proves, we hope to show, the interest and necessity of multiplying the number of these excavations.

Claix: one of the largest and oldest millstone quarries in France

In France there are two communes (municipalities) called Claix. The first, devoid of millstone production, is in the Alps near Grenoble. The second, with an important millstone quarry, is in the southwest of France, 100 km north of Bordeaux and 15 km to the south of Angoulême (Fig. 3). The name derives from the latin cos or cotis meaning rock outcrop or rocky place. This is exactly the case of Claix that, like the White Cliffs of Dover, albeit less spectacular, dominates a landscape of gently sloping hills covered with vineyards (cognac is manufactured nearby). These cliffs form a plateau called 'Les Meulières' (from the French term 'millstone quarries') and designate three groups of millstone exploitations located on the summit of the plateau that cover a surface of 20 acres (Fig.

Figure 4. Aerial view of the Meulières Plateau (Claix).

Figure 5. The present scenery of the Meulières Plateau recalls a southern European Mediterranean landscape.

4). The topsoil of the plateau has been so altered by the stone cutters that the original farmlands and vineyards dating from the Middle Ages and the 17th century gave way to a flora now characterised by dry grasslands and thorny shrubs typical of Mediterranean landscapes (Fig. 5). So, although Claix is near the Atlantic Ocean and at the same latitude as Quebec, its landscape is more reminiscent of Sicily or the South of Spain. Several extremely rare plants prosper in these conditions, such as the Globularia valentina (globulaire de Valence in French), a purple-coloured flower that in France grows only here and in the surrounding hills. This peculiar landscape thus benefits from the highest legal status known for natural reserves and wildlife and is protected by the Conservatoire Régional d'Espaces Naturels Poitou-Charentes (CREN). This is the agency that was behind the initiative formulated in 2005 to carry out an archaeological excavation, fostered by their desire to learn more of the history of the industries that originated this singular flora. In 2005 they solicited the University of Grenoble and the excavation took place three years later with the financial backing of 12 public institutions and private enterprises. This operation was led by the LARHRA (Laboratoire de Recherches Historiques Rhône-Alpes, UMR CNRS 5190) in partnership with the CREN of Poitou-Charentes. Financial support came from the following entities: the municipalities of Claix and Roullet-Saint-Estèphe, the federation of municipalities of Blanzacais and Charente-Boëme-Charraud, the Pays Sud-Charente, the Conseil Général of Charente, and the cement company Ciments Lafarge. We sincerely thank them for their assistance. The investigation, stretched out over 2008, consisted of exploring 250 m^2 distributed over five distinct zones (one in the municipality of Roullet-Saint-Estèphe and four in Claix). The field work was complemented by petrographical analyses, and research in archives dating from 14th to the 19th century. The field work was undertaken by a group of students of history and archaeology of the University of Grenoble 2. Timothy Anderson was responsible for zone I, Séverine Penon for the drawings, and Nicolas Minvielle for zone IV, while Guillaume Guidon was a temporary worker. The petrographical analyses are the work of Jacques Gaillard and Jean-Pierre Mercier (UMR LIENSs, Université de la Rochelle), and of Jacques Morel (CRPG, Nancy). Research in the historical archives was undertaken by the director of the operation assisted by Nicolas Minvielle and Amandine Vayr.

One of the first goals consisted in establishing the chronology of the exploitation. From the archives we learned that production had been abandoned between 1810 and 1840 due to the fierce competition of the excellent siliceous millstones from the Périgord and Parisian Basin regions. The specific production of millstones was then definitely halted giving way to the extraction of building blocks. The same records show that at the beginning of the 19th century, the site belonged to a family of millstone merchants named Parenteau. These members of the countryside bourgeoisie lived in a manor and married the daughters of the local gentry. They exploited several large millstone quarries in the area of Angoulême from at least the end of the 16th century and in 1702 acquired the perpetual concession of the Claix quarries from the village Lord (AD 16, 2 E 4895, Registry of Master Tiffon, Notary of Roullet, perpetual concession of the millstone quarries of Claix by Pierre de Gallard in favour of Jean Parenteau, 5/7/1702). Before 1702 and throughout all the 17th century, the Parenteau family rented not only the quarries from the village Lord, but one of his fortified houses and the administration of his feudal rights. Records earlier than the 17th century bear witness that local nobles had constant control of the quarries throughout the early Modern period and from the end of the Middle Ages. The oldest conserved manuscript, a vassalic tribute dating to 1306, reveals that the *Meulières* plateau was under the possession of Lord Alain Delisle (AD 16, G 131/15, Avowal and Census of the Lordship of Rocheraud, 1306). Even earlier, prior to the 14th century, the site probably belonged to the Church. This is based on the fact that Alain Delisle declared in 1306 that he obtained his concession through the Lordship of the Bishop of Angoulême while the other nearby quarries remained under the control of a nearby priory. Since the exploitation was large enough to be divided between notables of the Church and the nobility (Delisle and the priory), its origin cannot be recent. The ownership must have its roots deeper in the past. However no document prior to the 14th century has surfaced to reveal information about the first phases of the exploitation.

Figure 6. Zone III, before the excavation.

The physical features of the quarry today fill in pieces of information that the archives are silent about. A hike from the eastern to western edge of the plateau clearly reveals a gradual decrease in the diameter of the millstone extractions. The first group of extractions measure from 1.70 m to 1.50 m. These then give way to diameters of 1.20 m and 1 m and then end up at 0.50 to 0.60 m. Hence, before the final, latest production of huge millstones – of a diameter equivalent that of the height of a man – there existed the production of small, hand-operated, rotary querns. In France, however, the use of these handmills was prohibited around the 11th-13th centuries during the establishment of the feudal system that obligated the peasantry to grind their cereals at the mill of the Lord. The oldest quarry faces of Claix with rotary quern extractions are therefore older than the 11th-13th centuries, presumably from early Middle Ages or even late Antiquity. Hence, production on the whole of the plateau lasted a very long time, spanning at least 10 to 15 centuries. The aspect of an extended production is also one of the reasons for the interest in this excavation. To date, all of the excavations of millstone production sites in Europe, and in France in particular, concern sites that were short-lived. The five excavation zones of Claix were chosen intentionally to represent the chronology of the exploitation, from the oldest hand quern extractions on the western edge of the plateau to the most recent scorings on the eastern side near the farmlands.

Figure 7. Zone III, after the excavation.

Elisabeth Bloxam's exquisite phrase, 'to make the invisible visible' defines precisely the second objective of our investigation. In general millstone quarries are not easy to comprehend and sometimes even almost impossible to perceive in the field. This is due to the accumulation over the centuries of the discarded overburden and working debris (in French *haldes*) produced by the stone cutters themselves. The debris of each new extraction was placed in the cavities left by older nearby scorings. At times the debris reaches such a large scale that even a specialist accustomed to studying this type of industrial site has difficulty visualising its aspect when production was in full swing. Moreover, in these cases an inexperienced archaeologist or amateur is totally lost. Therefore, without an archaeological excavation it is practically impossible to understand the magnitude of a site, its phases of exploitation, the volume of its production, and the extraction techniques put to use by the early craftsmen. An archaeological excavation changes all this, as illustrated by the before (Fig. 6) and after (Fig. 7) photographs of several areas from Claix taken from the same location with approximately the same angle. Hence the invisible becomes visible.

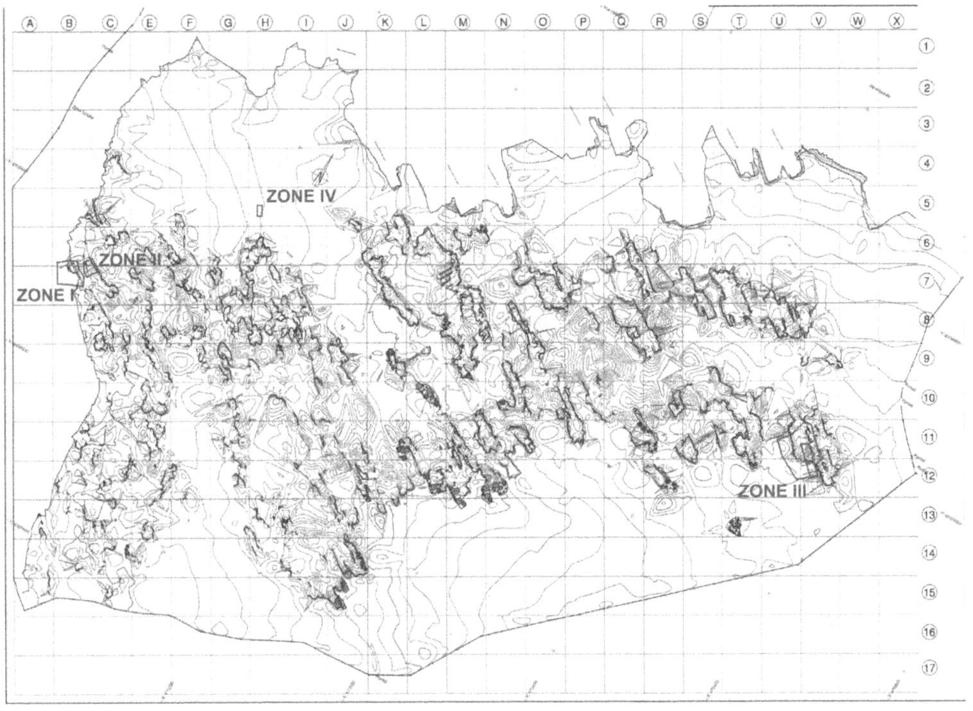

Figure 8. Topographical map of the Claix millstone quarry. Each square = 25 m (Measurements and CAD by M. Boissard, Topo 16, Co.).

Figure 9. The rutted road (carved into the bedrock) servicing the Claix quarries.

The huge pit of Fig. 7, zone III, was exploited during the 18th century. The surface explored by the students of the University of Grenoble was approximately 130 m^2, corresponding to about a quarter of the total surface (50 x 30 x 6 m) for this sector of the quarry. In the excavated area the production consisted of 200 monolithic large millstones measuring 1.73 m in diameter. On a larger scale, in order to evaluate the production of the whole of the quarry, a surveying company (Topo16) was hired to map the whole of the plateau. Four months of work consisting of processing 20,000 topographical points were necessary to create a detailed map of the site's 13 acres. Although very costly, this operation was absolutely necessary since a site of these proportions cannot be interpreted from the ground and quickly becomes a labyrinth in which one gets lost. Aerial photography, as seen in the comparison of Fig. 4 and Fig. 8, although providing a better vantage point, also does not convey the complexity of the site. A detailed topographical map is therefore a necessity for fathoming the features of this type of site.

The 130 m^2 excavated in zone III (bottom right of fig. 8) represents only a minute fraction of the site. Over the centuries, 190 similar elongated pits were hewn into the bedrock, a volume corresponding to between 50,000 and 100,000 millstones. This profusion of pits adheres to an orderly schema, far from the anarchic labyrinth of the other phases. They are lined up following an orderly parallel axis, like soldiers of the English military. In fact, their orientation adhered to a series of natural geological joints oriented SE-NW, which the millstone makers called '*fins*' (ends), that defined the lateral boundaries of each pit.

The topographical map also reveals in the heart of the plateau several unworked areas. One of these 'empty' spaces (zone IV) was investigated and revealed rutted tracks cut into the bedrock (Fig. 9), a type of roadway systematically found on the large millstone quarries. Any jolt of a cart might fracture the precious and fragile millstone cargo. The need to protect the millstones from unexpected shocks explains why these roadways were built either with paving stones or, as in this case, by cutting ruts directly into the bedrock. Each empty area on the map corresponds to this type of roadway; several secondary tracks lead to the exploitations up to the centre of the quarry and joined two more important roadways to the north and to the south of the quarry. The millstones transported along these major roadways led to either river ports or even larger regional thoroughfares that probably date to either Antiquity or Medieval times. To ensure that these areas were not encroached upon by greedy millstone makers, stone markers were erected to mark the limits of the pits, in the same manner as strips of unfarmed land the along earlier Roman aqueducts. Finally, the topographical map reveals a more chaotic outline of the excavated areas in the older scorings on the western area of the plateau. To the east, in the more recent work, the pits become larger and more structured. Progressively they pass from 4 or 5 m wide to 15, 20 even 40 m wide and 50 m long. These variations in size are accompanied also by changes in extraction techniques.

Changing techniques

Extraction techniques in millstone quarries have often been described as immutable and have practically not evolved from Antiquity to the 19th century. Even the tool kit of the millstone maker, with few exceptions, did not vary over the centuries. The excavations of Claix yield a much more nuanced vision of this aspect.

The 18th century exploitation

The pit explored in zone III, on the eastern area of the plateau of the *Meulières*, was the next-to-last exploitation of the site and dates to the 18th and beginning of the 19th centuries. The excavation first revealed the technique to evacuate the overburden, a layer about 30 cm thick of arable topsoil and clay that covered a second layer of weathered bedrock 50 cm thick not apt for scoring millstones. The earth was removed with picks from areas of about 20 m2, just enough space to score two or three rows of new millstones. At times, during this stage, this weathered rock was cut with a pick leaving arc-shaped excavations about

Figure 10. The work of removing the overburden of the Claix quarry, before the quarry face (zone III).

2 m in diameter that sketched the outline of the future cylinders (Fig. 10). The 18th century craftsmen called this phase '*décroûter*,' meaning 'removing the crust.' The stone cutters not only avoided the natural fissures but put them to use as the edge of future cylinders so as to spare themselves some of the work of cutting the trench.

Once the overburden was discarded, the craftsman then traced the circumference of the future cylinders using a compass with one point placed in a small cavity pecked in the centre, arranging the markings in a staggered fashion so as to obtain a schema resembling a honeycomb (Fig. 11). This planning aimed at optimising production to extract the highest number of millstones with the minimum of debris. This strategy was apparently implanted in millstone

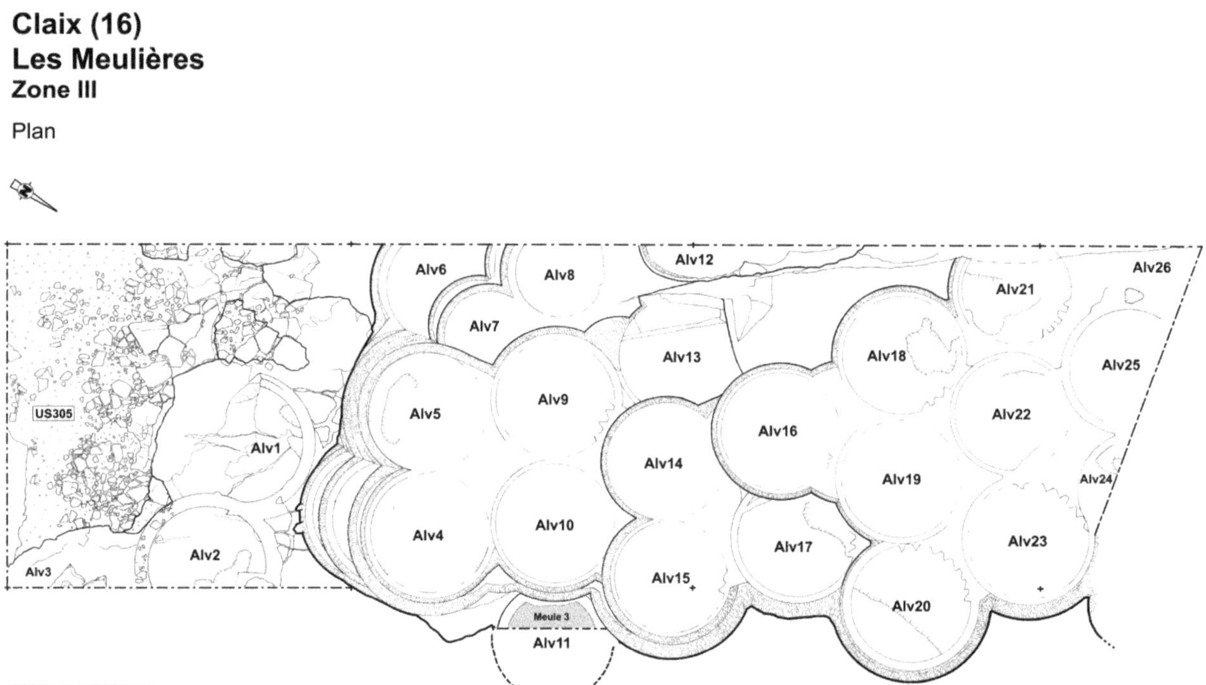

Figure 11. Plan of zone III: the honeycombed-shaped layout is the result of the 'economic' strategy of exploitation of Claix (drawing by S. Penon).

Figure 12. A circular channel in zone III.

quarries of the southwest of France in the middle of the 18th century. Indeed, a document from 1766 describing a lease of a quarry near Claix, and formerly held by the Parenteau family, stipulated that the millstones should no longer be hewn in a disorderly fashion. From that moment they should be scored 'in the most regular, useful and economic mode.' (AD 16, 2 E 155, Registry of Master Bernard, Notary at Angoulême, of the sale of the millstone quarry of Crages at Angoulême, 7/6/1766, and then the hiring of millstone cutters, 15/6/1766. The following original phrase, as well as other phrases appearing later in this paper, are in their original old French. '*dune maniere regulliere la plus utille et autant oeconomique quil sera possible.*') This 'economic' mandate was thus imposed as an absolute rule on both the owners and their craftsmen.

The next phase of extraction, after the tracing of the circumference, consisted of cutting a circular flat-based channel (U section) 19 to 22 cm wide and about 30 cm deep around the future roughout. The well-preserved tool marks on the Claix bedrock indicate that the channels were carved after four linear 'passes' with the pick. In passing let us affirm - and this is another advantage of the archaeological excavation - that even the finest tool marks, sheltered under the layers of sediment, are much better preserved than the weathered, often invisible imprints, of the exposed areas of the sites (Fig. 12). But returning to the extractive channel, the first 'pass' of the pick, seen as a curvilinear groove made up of repeated impacts, is placed along the perimeter of the cylinder. The three ensuing passes are spaced concentrically at intervals between 3

Figure 13. The 'crown' of wedge holes at the base of a millstone in zone III.

and 6 cm and placed along the outer edge of the channel. The diagonal working plane at the bottom of the trench in this case follows an angle of 28°. Several well-conserved grooves at the base of the channel, measuring from 20 to 24 cm, are indicative of the rhythm or interval of change in the working position of the cutter as he advanced his trench. The individual impacts visible here also reveal the use of a pick with a point 0.8 cm wide.

The next step of work consisted of splitting the roughout from the bedrock. This process consisted of carving from three to nine wedge holes ('*emboîtures*' in French) with a pick at the base of the roughout. These cavities were found spread out over about one half of the circumference and followed the bedrock's plane of sedimentation (Fig. 13). They were either slightly rectangular or triangular and measured between 11 and 13 cm wide, 7 to 11 cm deep, 5 to 7 cm thick at their start and 2.5 cm thick at their deepest point. Iron wedges were then inserted in each cavity between two metal shims, like sandwiches. These shims, measuring 6 x 5 cm, have also left their imprint on the rock. Then one, two, or three men began to strike

Figure 14. The splitting of the millstone leaves a 'stump' at the floor level that is indented with the marks of wedge holes (zone III).

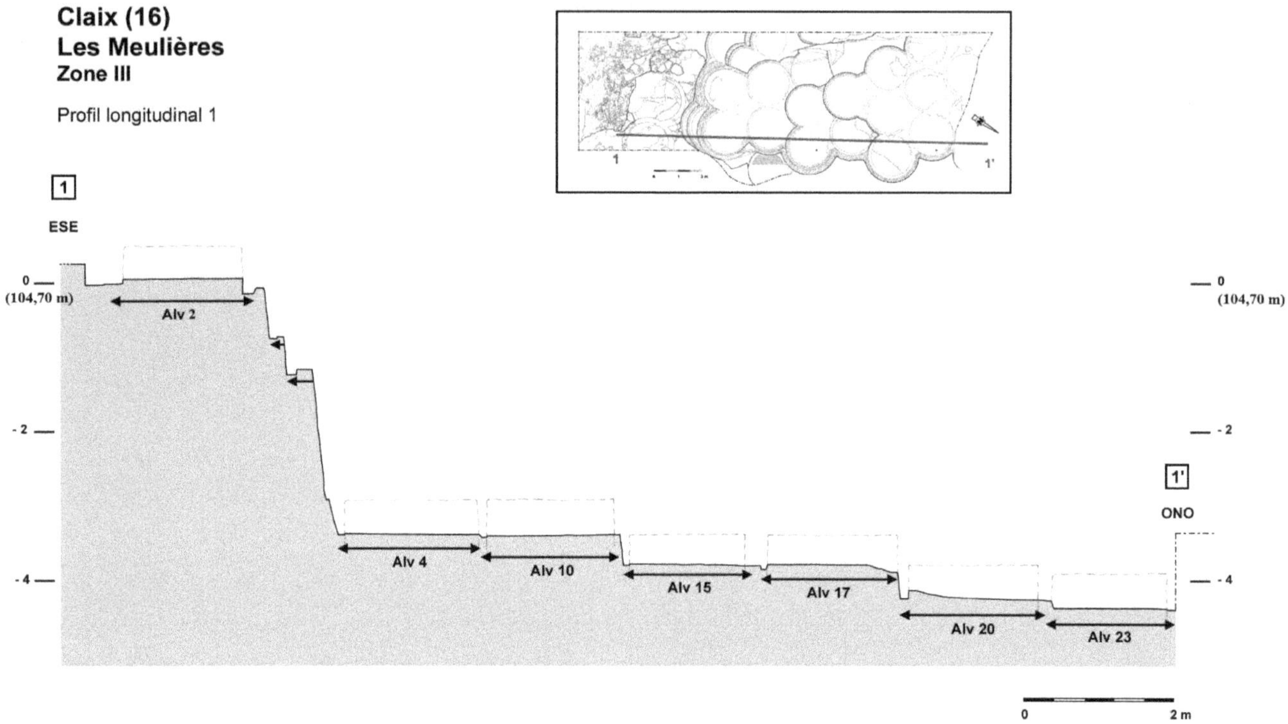

Figure 15. Transversal section of zone III revealing the tiered quarry floor (drawing by S. Penon).

the heads of the wedges with hammers. The natural rock bedding facilitated the progression of the fracture toward the interior of the cylinder. Since the wedge holes were placed only on one side of the cylinder, the splitting did not necessarily follow a perfectly horizontal plane and resulted in a slight bulge on the surface of the roughout at the opposite side of the wedges, as well as a 'stump' on the quarry floor that needed to be levelled once the roughout was removed (Fig. 14). Once split from the bedrock, the millstone roughout was removed from the circular socket (*alvéole* in French) by means of long iron levers ('*pinces*' in French) slipped into rectangular cuttings placed in the periphery of the channel. Then the workers lifted the stone and placed it on the ground behind the quarry face or outside of the pit where a second team proceeded to the next stage of fashioning or putting on the finishing touches. This consisted of smoothing the working faces and piercing the central eye. This phase of work produced minute stone flakes, some the size of grains of rice, that contrasted with the larger debris from the channel-cutting and formed a 'crown' at floor level around the millstone. This process over time resulted in the formation of clearly visible circular deposits of sediments.

Figure 16. A vertical extraction 'tour' (tower) in zone III.

The fashioning of the millstone phase of work outside of the pit freed the quarry face so that other workers could pursue scoring new roughouts adjacent to those previously extracted. Several teams worked simultaneously at the base of the pit. We can deduce this from the slight differences of tool marks from one extraction to another and by the arrangement of the different staggered tiers of the quarry floor (Fig. 15). Hence, three activities took place simultaneously: 1) removal of the overburden (earth and altered rock), 2) carving of cylindrical roughouts by several teams of workers, and 3) fashioning of the extracted roughout. This almost 'scientific' management of tasks, precursor of the method heralded by Frederick Taylor, allowed an important increase in the rhythm of production. The standardisation of the diameter of the 18th century millstones of Claix, between 1.49 and 1.76 m, with a preference for models between 1.70 and 1.73 m (2/3 of

Figure 17. Darney (Vosges, France): a grindstone quarry towards 1900 at work. Notice the mounds of working debris (collection of A. Belmont).

Figure 18. General view of zone I dating from the 5th -8th centuries AD. Photograph by T. Anderson.

the production), also facilitated the gain of production. In passing, it is necessary to clarify that in the 18th century, these millstones were destined exclusively for both watermills and windmills; no hand operated rotary quern was scored at this time. The consistent diameters and the fact that the cylinders were hewn in succession, one below the other, resulted in the formation of vertical tube-shaped imprints along the quarry faces that the craftsmen called '*tours*' (towers) that served as units of counting. A millstone maker could lease, for example, six '*tours*' or hire workers to exploit three '*tours*.' The tubes of zone III reached a height of 4.7 m. Other tubes in pits near this area reached up to 7 m (Fig. 16) in height.

As work progressed, the quarrymen piled the huge mounds of overburden and stone debris behind them. Due to the invasive nature, the mounds ('*haldes*') were a recurring problem in all mining and quarry activity. The debris at Claix was originally stored on the upper edges of the pits in mounds 2 m high and 3 to 4 m wide. But when the pits became too deep, the workers began to move the debris further away and backfill earlier finished pits, a practice that also facilitated the transport access to the working areas. Therefore, to gain as much working space as possible, each abandoned 'tour' was filled in with a few square metres of spoil. To restrain the debris in these finished workspaces, simple retaining walls were assembled with blocks. Claix at that time thus probably resembled a photograph taken towards 1900 of a grindstone quarry at Darney in the Vosges where we see the workers working between the mounds of spoil held back by retaining walls (Fig. 17). Finally, in the modern phases at Claix, when work in one pit was completed, a new one was initiated just beside it, and the former was used to store debris moved with baskets, stretchers, and carts.

The first quarries from the end of Antiquity and the early Middle Ages

Two other sectors located about 500 m to the west of zone III were opened. Zone I (40 m^2) brought to light half of a tiered, horseshoe-shaped quarry, 16 m long and 6 m wide (Fig. 18). The adjacent zone II (30 m^2) explored by means of a transversal section a crescent-shaped pit 44 m long and 5 to 6 m wide. This specific site, near the western edge of the plateau, was the first exploited by the millstone makers of Claix (Fig. 19). Both of these sites were relatively short-lived. This is inferred from the debris found in the backfill of zone I, which originated in zone II. The chronology is based on the morphometric analyses of the products. As opposed to the large, 17th and 18th century roughouts (1.70 m in diameter) of zone III, these quarries yielded, on the one hand, much more modest millstones measuring between 0.80 and 1.40 m in diameter (with an average between 1.05 and 1.15 m), as well as hand operated rotary querns measuring between 0.44 and 0.60 m. The typology of the millstones, hardly over one metre, combined with the knowledge that the smaller hand querns were prohibited during the 11th to the 13th centuries, suggests a date from the early Middle Ages, or possibly even late Antiquity, for these exploitations. This early date is substantiated by the result (AD 630 $\pm$ 170) of a thermoluminescence analysis of a group of common grey-coloured potsherds collected in a layer of working debris covering the quarry floors. The firing of this pottery is calibrated to between AD 460 and 800, suggesting a range between the 5th and 8th centuries, 1000 to 1400 years earlier than that of zone III (Thermoluminescence analysis by Archéolabs A-08-39-05 TL, potsherd, US 203).

It is generally admitted that the extraction techniques put to use by millstone quarrymen did not evolve very much over a long period. The case of Claix, on the contrary, reveals great differences over a period of 1000 to 1400 years from one end of the plateau to the other. The only common factor is the use of the pointed pick to carve the channel around the future cylinder that will be split from the bedrock along the horizontal natural rock bedding. The first difference is the sheer volume of the productions. The recent work at zone III covered a surface 10 to 20 times greater than that of zone I and II, and its working faces are 5 to 7 m high while those of zones I and II never surpass 2 m. The organisation of the older sites also is much less rigorous than the later 18th century site. Even if the quarrymen of the older sites attempted as best as possible to make the edges of the extractions coincide with the rock's natural faults, their floors of these older sites reveal a strategy far from the ordered 'honeycomb' 18th century imprints (Fig. 19). And while the later succession of extractions result in perfectly vertical tube-shaped tubes, those of Roman or Merovingian times recall the Tower of Pisa (Fig. 20). Hence the productivity of the older quarries was far from that of the later zone III. Instead of following

Figure 20. General view of zone II.

an 'economic' strategy, the older quarrymen squandered part of the mineral resource, as seen by the fact that the volume of working debris of the older quarries reaches is equivalent to half of the excavated material while that of the more recent pits represents only a quarter.

The technique of the cutting the channel around the cylinder also differs from the older to the newer quarries. Although sharing a circular shape, the older trenches are from 20 to 40 cm wide compared to 20 cm in zone III. In addition the earlier quarrymen used four passes of the pick in the upper part of the trench and one or two at the base, resulting in V-shaped sections that lean to the outside. These older channel sections contrast radically with the perfectly U-shaped channels of the 18th century. In the case of the smaller hand querns of the older quarry,

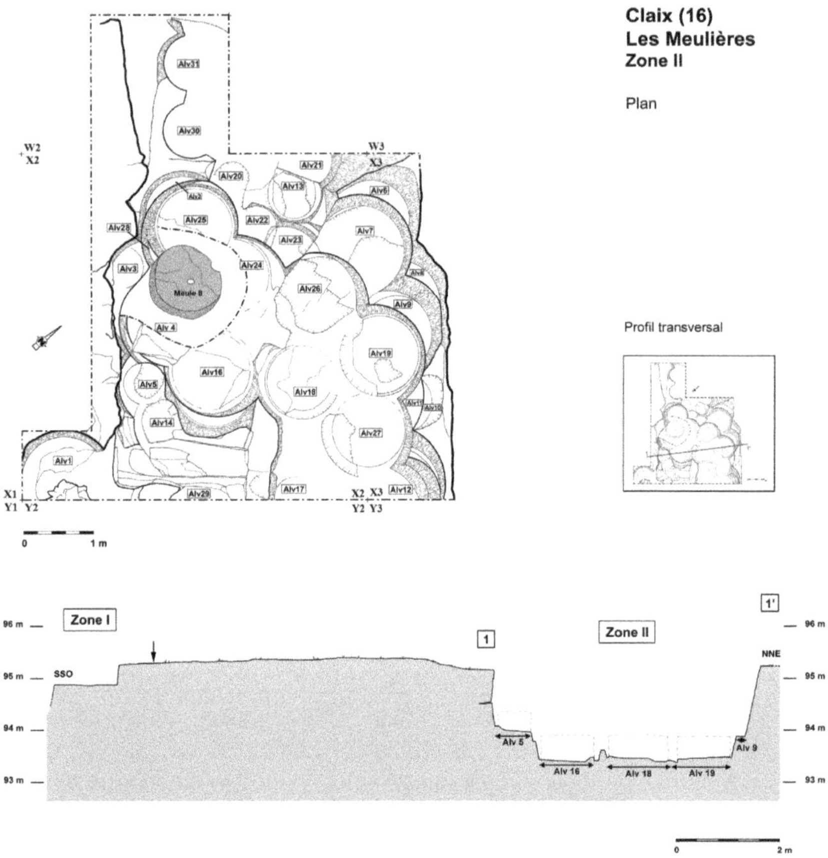

Figure 19. Plan and section of zone II (drawing: S. Penon).

Figure 21. Zone I: large wedge hole and defective roughout (photograph by T. Anderson).

the channel is only 8 to 10 cm wide and cut with two passes of the pick. The difference also is seen in the mode of splitting the cylinders from the bedrock. While in the 18th century splitting was performed with iron wedges, the earlier craftsmen most often used wooden wedges, and instead of cutting a series of small wedge holes resembling a 'crown,' the earlier quarrymen fashioned one large rectangular cavity (*encoignure*) for both types of extractions. For the querns, the cavity was 25 cm long, about half the size of the circumference, and for the large water millstones, 86 cm on an average, about one third of the cylinder's diameter. According to the fine WW-shaped tool marks visible at their base, the cavities were fashioned with a small pick (*mortaisoir*) fitted with a short cutting blade about 2 to 3 cm wide (Fig. 21). Each of these cavities is preceded by a rectangular corridor or space that is only about 20 cm long in the case of the hand querns, but 1.10 by 0.60 m in the case of the larger hydraulic millstones. These corridors probably facilitated placing the wooden wedges and created an access of long iron levers to lift the stone.

The team of diggers of realised that the mode of splitting the cylinders in zone II was very different than that of zone I, which is indicative how rapidly extraction techniques can change from one group of millstone makers to another. Instead of applying large cavities (*encoignures*), the craftsmen of zone II used three types of cavities for their wooden wedges: 1) one or two W-shaped cavities,

Figure 22. Zone II: large wedge cavity in zone II.

2) trapezoidal, U-shaped cavities for hand querns, and 3) one large linear cavity for the larger millstones, up to 1.14 m long (Fig. 22). In these cases, the base of the cavity was meticulously flattened with a chisel and not a small pick. This factor, coupled with the fact that the slope of the cavity was systematically inclined at 20° toward the heart of the cylinder and that the outer corridor surpasses the circumference about 15 cm, suggests the use of wooden wedges snugly inserted with a mallet. To make them swell and fracture the rock they were drowned in water. In this sense the long cavities served as a basin to contain the water.

Unfortunately, no tool was recovered during the excavation that might confirm this interpretation of the tool marks. In zone II, once the millstone was split, the phase of fashioning the roughout took place both inside the quarry pit and along its edges. We can infer this from the fine working debris observed in these areas, but particularly in the quarry pit where there is a white layer, about 15 cm thick, of dust and crushed rock the size of rice grains. To complete the sketch of the techniques put to use at Claix, it would be necessary to describe in more detail the differences from one quarry area to another, the ramps erected to evacuate the product, and the workshops for the finishing phase, as well as to determine the size of the different tools and measure the frontal and lateral impacts of the tools. These are all factors, if observed systematically in Europe and in a quantity sufficiently large to establish regional nuances, that might permit us one day to establish a chronological basis or reference to date other quarries through observations limited only to the features apparent on the surface.

The sphere of distribution

Besides the study of techniques, research in 2008 by the University of Grenoble planned to examine the area of distribution over time of the products extracted at the quarry of Claix. The reconstruction of the distribution during late Antiquity and the early Middle Ages, not yet completed, requires the study of the querns and millstones stored in the excavation depositories and collections of museums of the southwest of France. Another means of approaching the question of distribution is through geological analyses. The counting and dispersion of the fine terreginous inclusions (measuring about 50 µm) in the Claix limestone with a scanning electron microscope has permitted Jacques Gaillard and Jean-Claude Mercier of the University of La Rochelle to create a physical and chemical reference that will permit one to determine with a very high margin of certainty whether the plateau of Claix is the source of millstones unearthed in future excavations of Antique or Medieval mills and settlements. But for the present, the sketch of the commercial contours of the products of the Claix quarry is based on 163 handwritten registers from the 17th and 18th centuries conserved in the National Archives at Paris, in the Departmental Archives of Charente at Angoulême, and in the Departmental Archives of Charente-Maritimes at La Rochelle.

In the 17th and 18th centuries, the owners of Claix managed their own accounting and noted all the details of their commercial transactions. These records are cited explicitly in the inventories of their possessions after their death. The following are two examples: '. . . *an inventory of all the millstones extracted from the millstone quarries . . . .*' and '. . . *daily records bounded in a carton and parchment cover for the master of the millstone quarry to inscribe all his accounts and that of the farmers who work for him and other related transactions, containing 93 sheets written by hand of Master de La Meulière.*' (AD 16, 2 E 4365, Registry of Master Delafont, Notary of Nersac, inventory of the possessions of J.B. Parenteau after his death, 11/12/1780: '. . . *memoire de touttes les meulles qui sont tirées auxdictes meullieres*' and ' *livre journal relié dont la couverture est de carton et parchemin servant audit sieur de la Meulliere pour y inscrire tous les comptes de ses metayers et autres affaires relatives a sa maison, ledit livre contenant 93 feuillets coté et paraphé de la main du sieur de la Meulliere.*'). Contrary to the case of Selbu in Norway, where the records of commerce of the millstone merchant Frederik Birch (1852-1912) make up a formidable set of archives (Archives of the municipality of Selbu, Norway, B1 to Y6. These exceptionally rich records consist of commercial transactions (purchases and sales of millstones, maps . . .) as well as private correspondence. These documents to date have yet to be studied), the original commercial transactions alluded to in the text unfortunately are not preserved. Therefore, most of the data we have gathered (consisting of mill descriptions, inventories of possessions after the decease of the quarry owner, contracts of quarry concessions, invoices and receipts of millstone sales) are provided by old records of notaries. For example, the 13th of November, 1612, a miller called Jehan Mérichon from the region of Saintes, '. . . *confesses to have perceived in the presence of Master Anthoine Chabaniel, attorney of the Lord of Gons, the sum of sixteen pounds to purchase five white millstones from Claix near Angoulême for the mills of the Lord of Gons, the sum that I paid to the merchants from whom I acquired the millstones.*' (AD 17, 3 E 88/20, Registry of Master Verjat, Notary at Saintes: ' . . . *confesse avoir heu et receu cy devant de maitre Anthoine Chabaniel procureur de la seigneurie des Gons, la somme de seize livres pour aller achepter cinq meulles blanches au lieu de Claix en Angoulmois, pour icelles estre mises aux moullins du seigneur desdits Gons, laquelle somme jay baillé et deslivré aux marchands desquels jay acheрté lesdictes meulles.*'). Five years later, in 1617, the same Antoine Chabaniel travelled to Claix where he made a purchase from the owner of the quarries, Jehan Parenteau: '. . . *eight white millstones for grinding wheat, that is, four from the millstone quarry of La Roche Andry and four from the millstone quarry of Claix,* [which must be] *delivered and taken to the mill of Gong called la Debouyse* [within] *the next three weeks*' (AD 16, 2 E 4886, Registry of Master Pastureau, Notary at Roullet, act of 20/4/1617: '. . . *le nombre de huit meulles blanches a usage de mouldre le bled, scavoir quatre de la meulliere de la Roche Andry et quatre de la meulliere de Claix, rendues et mené au moulin du Gong appelle la Deboyse* [d'ici à] *troys semaynes prochaines venents.*') (Fig. 23).

During the reigns of Henry IV, Louis XIV, and Louis XVI, the stones of Master Parenteau outfitted the mills of Claix and surrounding villages such as Roullet (AD 16, 2 E 4380, Registry of Master Viaud, Notary at Mouthiers, lease of 5/5/168) (Fig. 24). They probably circulated in the areas

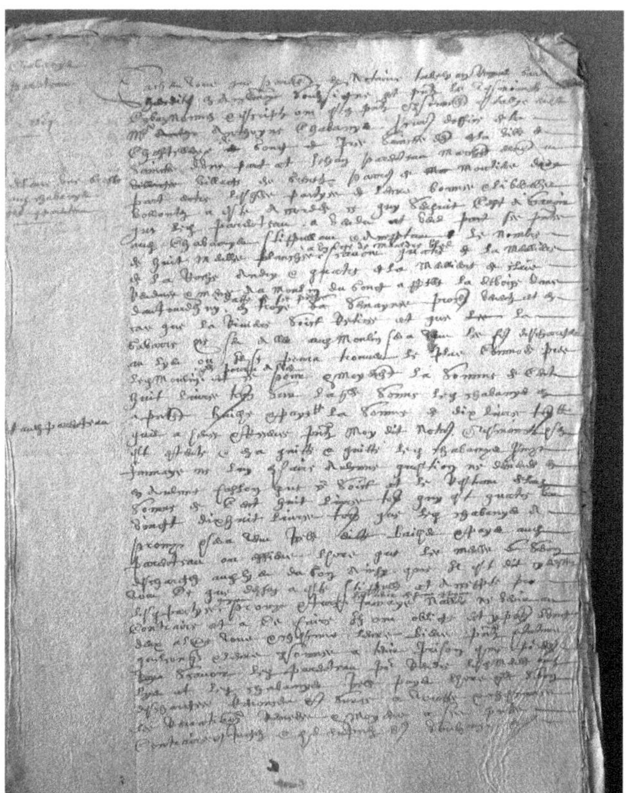

Figure 23. Record of a sale of a Claix millstone dating from 1617.

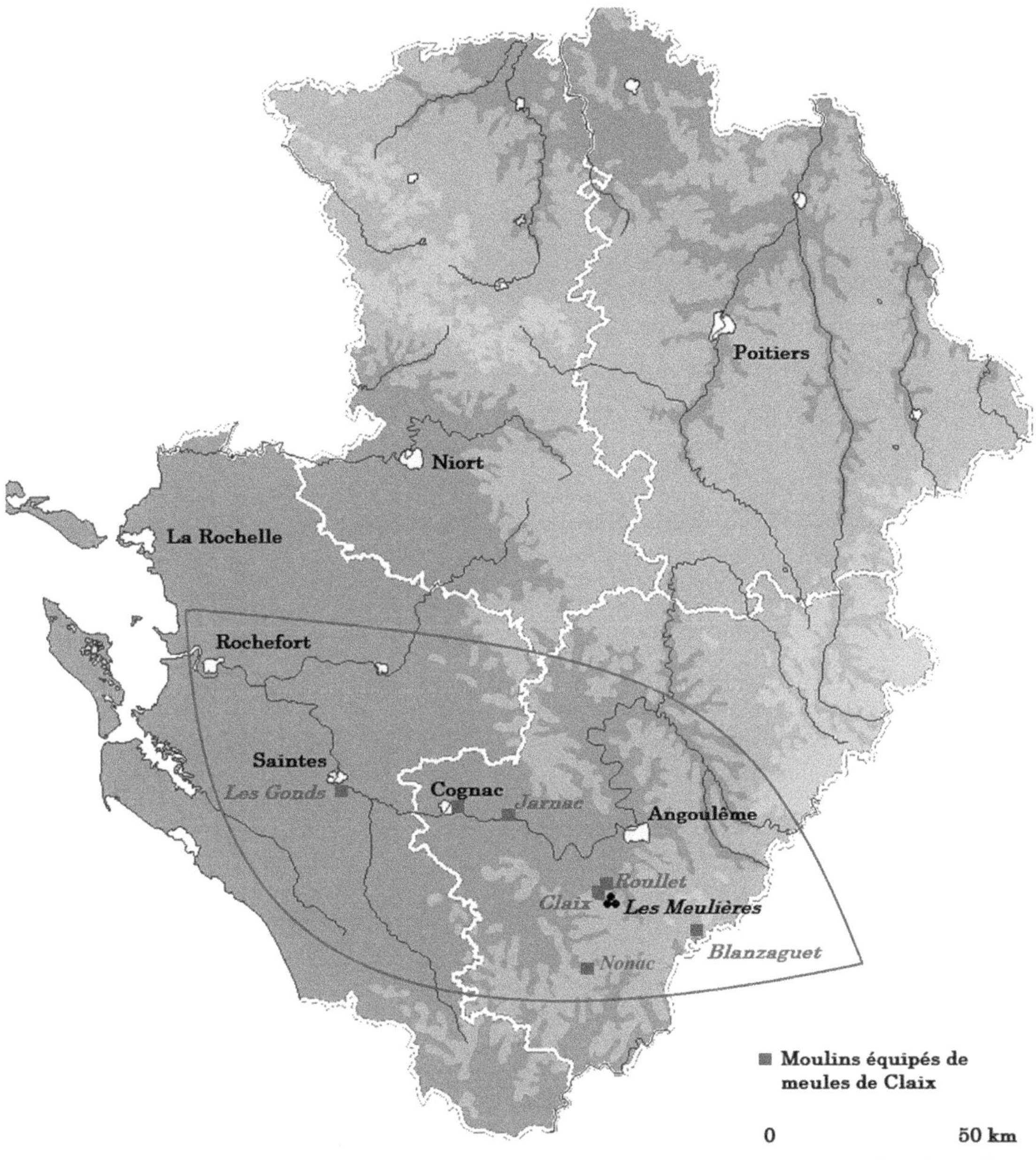

Figure 24. The sphere of distribution of Claix millstones in the 17th and 18th centuries.

of Angoulême and Cognac, in spite of the fact that there are no specific finds on the map of Fig 24. This is probably related to the fact that in mills, inventories for the most part do not cite the geographical source of the millstones but describe them simply as 'white millstones.' We know they reach the city of Saintes and its surroundings due to the sale of a dozen to the mills of Gonds between 1612 and 1617. According to an 18th century geographer, they attain a great success among the millers of this region: '. . . *the dry hills that are seen to the left of the road* [leading to Claix] *yield building stones and millstones that are taken to Saintonge*' (Munier 1779, 235). We also know that they equipped the mills of Nonac, between Blanzac and Montmoreau, as well as those of Blanzaguet, half way between Angoulême and Brantôme, bordering the Périgord region (Nonac: AD 16, 2 E 4399, Coutord Mill, 1692: a lower stone of *'le pierre de Claix'* ('stone from Claix'). Document furnished by Mme Michèle Aillot of Blanzaguet: AD 16, 2 E 4895, Registry of Master Tiffon, Notary at Roullet, records of 5/7/1702). Even if the millstones of the Parenteau family did not travel as far as La Rochelle, or even Nantes, where their competitors from Caunay (Deux-Sèvres), L'Hermenault (Vendée), and the forest of Moulière (Vienne) maintained the monopoly (Coutant 1986), the area of distribution of the Claix millstones far surpasses the local sphere and must be considered as regional since the distance between Saintes and Blanzaguet, as the crow flies, is 80 kilometres.

This 'almond-shaped' commercial sphere of distribution is the result, on the one hand, of the thoroughfares passing the *Meulières* plateau of Claix, notably the '*chaussées des*

Figure 25. The white vesicular limestone of Claix.

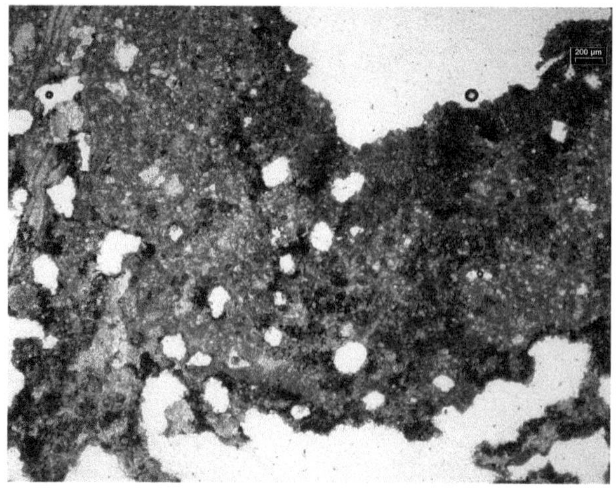

Figure 26. Thin section of the Claix limestone (photograph by C. Malacour).

meules' ('millstone roads') of which at least one dates presumably to Roman times (Vernou-Magister 1993, 27). But it is the navigable routes that were the principal means of transport, starting with the Charente River, which joins the Atlantic just after Rochefort and is known to have depositories of millstones. In 1617, sixteen millstones purchased from master Jehan Parenteau were taken down the Charente on a large sailing boat (*gabarre*) and unloaded as near as possible to the mills of the clients. The purchase of the stones and their transport, first on ox carts and then on a sailboat cost £8 each for a total of £108, a sum equivalent at the time, to a small fortune, the price of two or three houses.

White millstones for white bread

The reason for the popularity of the millstone quarries of Claix resides in the quality of the rock. It is a type of Late Turonian limestone very similar to a very large number of outcrops spread out across western France. But contrary to other limestone, it is pocked by many minute vesicles (level of porosity: 26 to 30%) that gives it a sponge-like, abrasive quality essential for the grinding cereals (Figs. 25 and 26). Another important property is the ease with which it is carved. When freshly cut in the quarry it is as white as chalk and so tender that it can be sculpted with a trowel. But once in contact with air, it becomes rapidly very hard due to the migration of salts of calcium towards its surface. This gain in hardness is nonetheless not very important and is limited mostly to the surface. A series of mechanical tests conducted in a laboratory show that the Claix limestone resisted a pressure of 9 megapascals, that is 90 kg/cm^2, and a tensile strength of 0.8 megapascals, that is, 8 kilos/cm^2 (Fig. 27). In comparison, most sandstones and conglomerates exploited in the past for millstones average around 20 to 50 megapascals, that is 200 to 500 kg/cm^2 (Belmont 2006a, 175-181; The mechanical tests are the work of Yves Orengo, LIRIGM, University of Grenoble 1).

The quality of hardness is therefore not an essential property of this stone. Once installed in the mill and submitted to 60 to 120 rotations per minute, a Claix millstone wears very quickly. In 1809 the Prefect of Charente declared that '*the millstones are not the same everywhere. Those that come from departmental quarries must be dressed every two days (. . .); this operation takes 5 to 6 hours.*' (AN F/20/295, Census of the mills of France, response of the Prefect of Charente, 1809: '. . . *les meules ne sont pas partout les mêmes. Celles provenant des carrières du département veulent être repiquées tous les deux jours* (. . .); *cette opération fait perdre 5 à 6 heures de tems.* '). At this rhythm the Claix millstones were transformed into 'pancakes' several centimetres thick in a matter of three to four years. Their density compensates in part for this defect. At 1.9 tons per cubic metre, the relation of weight to volume is very near 2 to 2.6 tons per cubic metre, normally used for millstones, yielding a density of 0.8 to 1.2 tons for the smaller hydraulic millstones scored from zones I and II and 2.5 to 3.5 tons for the 'monsters' measuring 1.72 m in zone III. In any case, these figures indicate that the density of the stone was more than sufficient to grind grain.

Sample	Density (t/m3)	Porosity (%)	Vp mean (m/s)	Vs (m/s)	Piston cylinder (Gpa)	Compression resistance (Mpa)	Tensile strength resistance (Mpa)
Claix 1	1.972	26.2	3109	2318	19.1		
Claix 2	1.866	30.1	2668		13.3		
Claix 3	1.943	27.2				9.1	0.8

Figure 27. Results of mechanical tests done on the vesicular limestone of Claix

COMPONANT	CLAIX (%)	*La Motte d'Aveillans (%)*
Silica	0.41	*67.9*
Aluminium	0.46	*16.93*
Iron	0.11	*6.38*
Manganese	<ld	*0.03*
Magnesium	0.2	*1.19*
Calcium	54.17	*<ld*
Sodium	<ld	*0.25*
Potassium	<ld	*3.5*
Titanium	<ld	*1.02*
Phosphorus	<ld	*0.06*
Loss at ignition of 1000	43.33	*3.71*
Total	98.68%	*100.97%*

Figure 28. Results of chemical analyses of the porous limestone of Claix (<ld indicates a quantity inferior to the limit of what the spectrometer can detect)

It is, however, the exceptional purity of the Claix limestone that raises it above other stones. Chemical analyses reveal that the combination of a high proportion of calcium (54.17%) and other elements lost on ignition (43.33%) make up 97% of the rock (The chemical analyses are the work of the CRPG of Nancy) (Fig. 28). The other components are less than 1% and often below the limit of detection by the spectrometer. This is the case with aluminium oxide (0.46 %) and iron (0.11 %). These constituents do not attain a sufficiently important proportion to colour the rock with a greyish or reddish tint. As an example of comparison, the table of Fig. 28 presents the chemical analyses of a breccia-schist exploited at a mediaeval millstone quarry at La Motte d'Aveillans in the southeast of France. The 17% of aluminium and 6% of iron in this stone result in the grey, blue, red, and black tints, the opposite of the snow-white coloured rock of the Claix limestone.

This is the principal property of the Claix limestone. It is not only abrasive, contrary to other limestones, but it possesses an immaculate whiteness. It was therefore ideal to equip the 'white mills' known in France and in other countries of the south of Europe, as these were specifically designed to grind wheat. By contrast, the 'dark mills' ground rye, oats, and other grains of secondary quality. During the grinding process, only white mineral particles of the stone, invisibles to the eye, would end up in the white wheat flour. It is this immaculate virtue that is cited in the records of sales from the 17th and 18th centuries. In 1612, for example, the boats of Master Parenteau transported to the mills of Gonds '*millstones for the white mills*,' and in 1617, the master millstone maker delivered '*eight white millstones for grinding wheat*.' (AD 17, 3 E 88/20, Verjat, Notary at Saintes, records of 29/10/1612 and 13/11/1612. AD 16, 2 E 4886, Pastureau, Notary at Roullet, sale dating to 20/4/1617: '*. . . les meulles pour ledict moullin blanc*;' '*huit meulles blanches a usage de mouldre le bled.*' The term 'bled' meant wheat during the Ancien Régime).Other coloured stones, by contrast, such as sandstone, granite, or conglomerate would have shaded the wheat flour with an unappetising tint and dirtied it with sand and minute red, black, green, and yellow particles. Three of the most celebrated French agronomists and physicians of the 16th century, Jean Bruyérin, Charles Estienne, and Jean Liébault stated: '*the millstones are of a soft and sandy stone (. . .) and always release a little grit that ends up in the flour and removes all the charm and taste of the bread.*' (Estienne, 1578, 311: '*. . . les meules qui sont de pierre tendre et mollastre (. . .) delaissent tousjours quelques gravois en tournant, lesquels meslés parmi la farine ostent toute la grace et saveur du pain.*'). By contrast, according to the ancient millers, white stone is the only stone capable of producing white flour. '*The millstones of those quarries are white,*' states an anonymous author in 1796 about the quarries of Molières near Paris. '*Their whiteness comes, in fact, that they do not contain, or hardly contain, iron* (. . .). *In the end, while they wear down very quickly, they are lighter, do not heat up, and yield, it is said, a whiter flour.*' (Anonymous, 1796, 32-33. '*. . . les meules de ces carrières sont blanches (*à propos des meulières des Molières, près de Paris*). Leur blancheur vient en effet, de ce qu'il ne se trouve point, ou presque point, de fer parmi leurs parties constituantes* (. . .). *Il en résulte qu'elles s'usent assez vite; mais d'un autre côté, elles sont plus légères, s'échauffent moins, et donnent dit-on, de la farine plus blanche.*'). A century later, in 1885, a geologist makes the same observation about the millstones from La Ferté-sous-Jouarre, famous all over the world and produced in a number surpassing that of a million: '*. . . the chert of La Ferté (. . .) has the merit of being easily dressed with the hammer and conserving its abrasive quality for a long time, and especially the fact that it yields a very white flour.*' (Jagnaux, 1885, 277. '*. . . Le silex de La Ferté (. . .) a le mérite de prendre bien le marteau, de conserver longtemps la rhabillure* (son abrasivité) *et surtout de faire des farines très blanches.*').

The limestone of Claix therefore played an essential rôle in obtaining a snow white bread, a bread that was

Figure 29. The millstone quarry of Sarospatak in Hungary located near the border of Ukraine.

synonymous with refinement, power and good taste, of which the elite of the 13th to 15th centuries were so fond. And through a process of social imitation, even the townsfolk of certain regions not necessarily reputed to be rich, such as the Dauphiné principality, also demanded this bread (Belmont 2006a, I, 96-105, and II, 148-153; Belmont 2006b, 81-92). It was this concern, in the cultural and social realm as well as in relishing the flavour, that explains the immensity of the Claix quarries. Evidence of a relation between the nutritional concerns and a vast millstone production industry is nonetheless not new or strictly limited to the southwest of France. It appeared a dozen years earlier at the medieval site of Lhuis-en-Bugey in the east of France, where millstones were scored from quarry faces of Urgonian limestone over a kilometre long (Belmont 2006a, 99-100) and from the six acres of La Livinière, near Toulouse, where millstones of sandy limestone starting at least in the 11th century were scored (Belmont 2007). The same applies also to the labyrinth of subterranean galleries on the Cycladic island of Melos (Vrettou-Souli 2002), as well as the astounding crater of Sarospatak, in Hungary, 300 m in diameter and 70 m deep, where was exploited a vesicular volcanic rock white as snow, like that of Melos (For these sites, see the Atlas of European Millstone Quarries (http://meuliere.ish-lyon.cnrs.fr) (Fig. 29). We could pursue the list with examples from Italy, Portugal, Spain, Slovenia, etc.

In this sense, the novel aspect of the site of Claix is not so much the pre-occupation of the colour, but the early, precocious, manifestation of this concern. The millers and inhabitants of this sector of the southwest of France had access to products that were more abrasive, harder, and better suited to grind grains than the limestone of Claix. They could have acquired sandstone millstones from the outcrops of Aignes-et-Puypéroux, only 13 kilometres away, known to have been exploited for millstones since the Middle Ages. They also could have purchased stones from the excellent arkose quarries of Saint-Crépin-de-Richemont (spread out over 260 acres) only 45 kilometres away, exploited since Neolithic times. But the Aignes rock is grey and the rock at Saint-Crépin is at times grey and at times yellow – both incapable of yielding white flour. We can therefore gather from the study of Claix, dating either to late Antiquity or Merovingian times, that there existed a very early desire - older, at least, than what is documented to date - of the population for a bread of a particular colour. Thanks to a large number of archaeological finds in the last 20 to 30 years, we now know that the 'dark times' of the beginning of the Middle Ages were certainly not as sombre and sinister as described by our parents. The same conclusion seems to apply to the daily bread of at least a part of this past population.

References

Alsvik, E., Sognnes, K. and Stalsberg, A., 1981. Kulturhistoriske undersøkelser ved Store Kvernfjellvatn, Selbu, Sør-Trøndelag, 1978. *Rapport arkeologisk serie.* 1.

Anderson, T. J., Augustoni, C., Duvauchelle, A., Serneels, V. and Castella, D., 2003. Des artisans à la campagne. Carrière de meules, forge et voie gallo-romaine à Châbles. *Archéologie Fribourgeoise*, 19, Fribourg.

Anonyme (Coquebert-Montbret ?), 1796. Description des carrières de pierres à meule qui existent dans la commune des Molières, département de Seine-et-Oise. *Journal des Mines*, Messidor an IV, 32-3.

Baug, I., 2006. The quarries in Hyllestad – production of quernstones and millstones in Western Norway. In A. Belmont et F. Mangartz (eds.),

Mühlsteinbrüche. Erforschung, Schutz und Inwertsetzung eines Kulturerbes europäischer Industrie. Les meulières. Recherche, protection et valorisation d'un patrimoine industriel européen. Actes du colloque de Grenoble, Mayen, 55-9.

Belmont, A., 2001. La pierre et le pain. Les carrières de meules de moulins de Quaix-en-Chartreuse, XVIe-XVIIIe s. *Histoire et sociétés rurales*, 16, 45-79.

Belmont, A., 2006a. *La Pierre à pain. Les carrières de meules de moulins en France, du Moyen Age à la révolution industrielle*, Grenoble, 2 vols.

Belmont, A., 2006b. Les meulières médiévales. Résultats d'une moisson dauphinoise. In Belmont and Mangartz 2006, 81-92.

Belmont, A. and Mangartz, F., 2006. *Mühlsteinbrüche. Erforschung, Schutz und Inwertsetzung eines Kulturerbes europäischer Industrie. Les meulières. Recherche, protection et valorisation d'un patrimoine industriel européen.* Actes du colloque de Grenoble, Mayen.

Belmont, A,2007. Les carrières de meules de moulins. In M. E. Gardel et M.-C. Bailly-Maître (éds.), *La pierre, le métal, l'eau et le feu. L'économie castrale en territoire audois, Carcassonne*, Bulletin de la SESA, 39-50.

Belmont, A. and Badin de Montjoye, A. 2008. Aux origines de l'ordre des Chartreux : le monastère des Ecouges et son domaine. *Archéologia*, 458, 56-64.

Bertin, O., Fetet, P. and Gaffiot, J.-J., 1998. *Le vallon Saint-Martin, Vosges*. Epinal.

Bosset, L., 1943. Chavannes-le-Chêne: une nécropole burgonde dans une ancienne carrière romaine. *La Suisse primitive*, 7, 34-41.

Cabezuelo, U., Connier, Y. and Gauthier, F., 2001. Les carrières de meules de moulins dans la région de Vic-le-Comte, premier état de la recherche. *Revue d'Auvergne*, 184-200.

Coutant, B., 1986. *Moulins à vent d'Aunis dans La Rochelle et la banlieue, sl*, 94.

Doswald, C., 1994. Herkunft und Verbreitung der römerzeitlichen Mühlsteine im Kanton Aargau. *Minaria Helvetica*, 14a, 22-33.

Estienne, C. and Liébault, J., 1578. *L'agriculture et maison rustique*. Paris.

Farget, V. 2006. *Carrières, matériau et façonnages des productions en rhyolite des 'Fossottes' de LA SALLE (Vosges) entre le VIe av. J.-C et le Ve siècle ap. J.-C.* Mémoire de Master 1 sous la direction de Françoise Dumasy, Universtié Panthéon-Sorbonne Paris-I.

Gren, T., Heldal, T., Meyer, G.B. and Bloxam, E.G., 2008. From Hyllestad to Selbu: Norwegian millstone quarrying through 1300 years. In T. Slagstad (ed.), *Geology for Society*, Geological Survey of Norway Special Publication, 11, 47-66.

Haberbosch, P., 1938. Römischer Steinbruch bei Würenlos. *Badener Neujahrsblätter*, 57-61.

Helberg, B. H., 2007. *Rapport vedrørende overvåking av inngrep i kvernsteinsbrudd i damområdet for Saksenvik kraftverk, Saksenvik I Saltdal Kommune, Nordland.* Unpublished report, Tromsø.

Jaccottey, L. and Milleville, A., 2007. Première découverte de zones d'extraction de moulins de type 'va-et-vient' Jura. *Bulletin de la Société Préhistorique Française*, 104.4, 827-31.

Jaccottey, L. and Milleville, A., 2008. Les carrières du massif de la Serre (Jura). Sept millénaires d'exploitation meulière. *Archéopages*, 22, 16-23.

Jagnaux, R., 1885. *Traité de minéralogie appliquée aux arts, à l'industrie, au commerce et à l'agriculture*, Paris.

Mangartz, F., 1993. Mittelalterliche Mühlsteingewinnung am Hochstein, Der Hochstein. Führer zu einem Vulkan der Osteifel. Mendig. *Schriften des Förderverein Kultur und Heimatmuseum*, 23-33.

Mangartz, F., 1998. Bernhard Keibs Lay. Neueu Erkenntnisse zur spätmittelalterlichen Mühlsteinproduktion im oberen Niedermendiger Lavastrom. *Pellenz-Museum*, 7, 101-37.

Munier, E., 1779 (reprint 1981). *Essai d'une méthode générale propre à étendre le connaissances des voyageurs*. réimpr. L'Angoumois à la fin de l'Ancien Régime, Paris, 235.

Vernou-Magister, C. and Gomez de Soto, J., 1993. *Carte archéologique de la Gaule. La Charente*. Paris.

Vrettou-Souli, M., 2002. *H Milopetra tis Milou. Apo tin exorixi stin emporiki a diakinisi* (La meule de Milos, de l'extraction jusqu'au commerce). Athens.

Wefers, S., 2006. Ein Latènezeitlicher (?) drehmühlensteinbruch im Stadtwald von Borken (Schwalm-Eder-Kreis). *Archäologisches Korrespondenzblatt*, 36, 1, 65-74.

Weidmann, D., 2002. Chronique Archéologique 2001, Chavannes-le-Chêne. *Revue Historique vaudoise*, 115-6.

Les meules en pays Minyanka (Mali): étude des carrières et techniques de production actuelles

Caroline Hamon et Valérie Le Gall

L'histoire des techniques a longtemps considéré les meules comme un indicateur peu significatif de l'organisation économique et sociale des sociétés. Pourtant, les meules sont un élément important de la culture matérielle des premières populations agricoles du Néolithique. D'un point de vue archéologique, l'analyse technique et fonctionnelle des meules a fourni depuis une douzaine d'années des éléments significatifs pour aborder les économies de subsistance et l'organisation sociale des populations préhistoriques (Adams 1999; De Beaune 2000; Dubreuil 2004; Fullagar et Field 1997; Gonzalez et Ibanez 2002; Hamon 2008; Mauldin 1993; Risch 2002; Wright 1994; 2000). A la lumière de nombreux exemples ethnographiques, les meules constituent un point de ralliement des communautés pastorales et symbolisent le cœur de la maison chez les communautés sédentaires (Gast 1968; Gelbert 2005; Jackson 1991; Baudais et al. 2002; Roux 1985). Ils montrent que la mouture des céréales constitue une des activités les plus importantes de la vie quotidienne des populations traditionnelles d'agriculteurs. Elle fournit la nourriture de base, et joue un rôle clef dans les activités domestiques et dans la répartition des tâches entre les femmes (David 1998). Les meules sont un élément de stabilité de la communauté et contribuent à la pérennité du lien social. Dans les sociétés d'agriculteurs dits traditionnels, les activités de mouture participent aux rencontres quotidiennes entre les femmes d'une même famille, d'un même quartier ou d'un même village. Les meules rappellent enfin le rôle essentiel joué par l'agriculture et les céréales pour la subsistance de la communauté.

Contexte d'utilisation des meules va-et-vient

Malgré l'utilisation de plus en plus répandue des moulins mécaniques dans toute l'Afrique de l'Ouest, la fabrication de meules va-et-vient et leur utilisation à des fins alimentaires sont toujours d'actualité en pays Minyanka (Mali). Notre étude de l'utilisation des meules va-et-vient a porté sur une trentaine de familles de trois villages localisés à une vingtaine de kilomètres à l'est de Koutiala: Nampossela, Nintabougouro et Karangasso (Fig.1). L'utilisation quotidienne de près de 145 meules (et autant de molettes) a été documentée en parallèle de l'utilisation des couples de mortier/pilon.

Les Minyankas sont des agriculteurs de langue Senoufo,

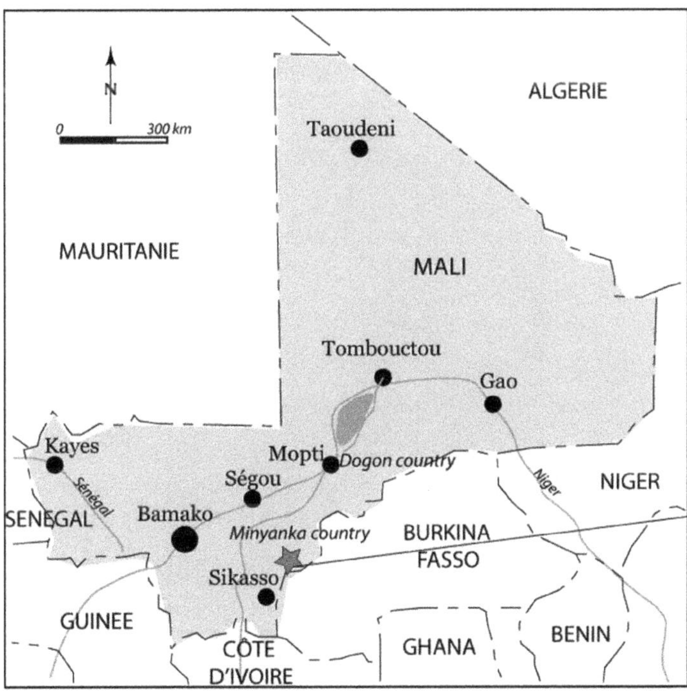

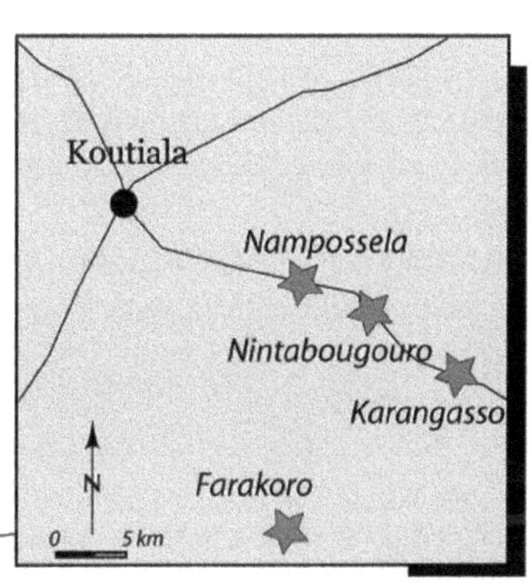

Figure 1. Carte de localisation du village de Farakoro et des trois autres villages Minyanka visités (DAO : C. Hamon).

Figure 2. La préparation du tô quotidien à base de farine de mil (cliché : C. Hamon et V. Le Gall).

Figure 3. Grande meule utilisée pour la mouture du mil (cliché : C. Hamon et V. Le Gall).

cultivant principalement le mil et le sorgho pour leur alimentation quotidienne (Jonckers 1987; Rondeau 1994). Le mil est consommé à chacun des repas de la journée: il est utilisé pour préparer la crème de mil, le tô consommé frais ou séché, ou encore le couscous. A chacun de ces plats correspond un type de préparation et une farine plus ou moins fine. La préparation des repas rythme l'organisation de toutes les activités de la journée: le bruit du pilon est l'un des premiers résonnant dans le village au lever du jour et la préparation des repas s'étale tout au long de la matinée et une bonne partie de la soirée (Fig. 2).

Dans les habitats Mynianka, les meules et les mortiers de grandes dimensions sont plutôt utilisés pour la préparation du mil tandis que les meules et les mortiers plus petits sont employés pour broyer les différents ingrédients composants les sauces d'accompagnement (feuilles d'hibiscus ou de baobab, graines, arachides, poissons séchés, soumbala, oignons, poivre, etc.) (Fig. 3). Les mortiers et les meules sont utilisés à différentes étapes de la préparation du mil (Fig. 4):

- le bris des épis de mil se fait souvent au petit matin dans de grands mortiers à fond plat disposés à l'extérieur des maisons voire du village.
- après vannage et nettoyage, les grains sont décortiqués dans des mortiers profonds situés dans l'espace domestique, dans lesquels un peu d'eau est versé pour faire gonfler les grains.
- après lavage des grains de ses impuretés, on procède à l'étape de mouture proprement dite: l'usage du mortier permet de produire une farine plus fine tandis que la meule permet d'obtenir plus rapidement une grande quantité de farine.

Elément central de la dot du mariage et de la fondation du foyer, chaque femme possède sa propre meule offerte par la famille du mari à la future épouse. Les meules utilisées par leur seule propriétaire sont généralement rangées avec leur molette (partie mobile) contre le mur intérieur ou extérieur de sa chambre (Fig. 5). Lorsque la meule est mise

Figure 4. Quelques étapes de la chaîne opératoire de transformation du mil a. bris des épis, b. vannage ; c. décorticage avec ajout d'eau (cliché : C. Hamon et V. Le Gall).

en commun entre toutes les femmes de la famille, elle est rangée dans un espace commun à toutes les meules, dans la cour de la maison ou dans la réserve. On observe également une complémentarité entre les meules individuelles liées à l'espace strictement domestique (voire personnel) et les mortiers utilisés en extérieur dans des espaces collectifs et socialement partagés. En cela, la distribution des meules dans l'espace domestique nous informe sur l'organisation de la préparation des repas et la structure familiale. Au delà de son aspect purement fonctionnel, la meule revêt donc une forte connotation sociale qui justifie aujourd'hui encore leur ancrage dans les jeunes foyers des villages de la région et l'existence de villages spécialisés dans la production de ces meules va-et-vient.

Figure 5. Rangement des meules à l'extérieur de la réserve collective (cliché : C. Hamon et V. Le Gall).

Extraction et carrière de meules

Ces meules sont produites dans le village de Farakoro à une vingtaine de kilomètres au sud-ouest des villages étudiés. Son nom signifie 'près (koro) de la pierre (fara)' en Minyanka: ce village est en effet installé à proximité d'une série de petits affleurements de grès quaternaires exploités pour la production de meules, et localisés dans un rayon de trois à quatre kilomètres autour. Il s'agit du seul village producteur de meules du cercle de Koutiala.

Dans ce village, la production des meules est réalisée par les hommes du village dont l'activité principale reste l'agriculture et l'élevage. La production des meules constitue donc une activité secondaire bien que spécialisée. Elle est saisonnière, et pratiquée surtout durant les périodes creuses pour l'exploitation agricole, le plus souvent à la saison chaude. Ils travaillent à la production de meules six jours sur sept en saison sèche et seulement deux jours pendant l'hivernage. L'extraction est prise en charge par des carriers spécialisés. Treize équipes de carriers sont recensées dans le village de Farakoro et les hameaux qui en dépendent: chacune d'elle est généralement constituée d'un maître-carrier travaillant avec deux à six apprentis ou ouvriers. Deux à trois années sont nécessaires pour finaliser cet apprentissage. Chaque maître-carrier sélectionne l'affleurement qu'il souhaite exploiter et le 'baptise' avant de démarrer l'exploitation. Le gisement devient alors sa propriété. Deux types d'exploitation sont réalisées à Farakoro: à ciel ouvert en front de taille (Fig. 6), et en sape dans les niveaux inférieurs des affleurements gréseux, mieux conservés. La roche est testée au son avant d'être exploitée: un son sec signale une roche de bonne

Figure 6. Ancien front d'extraction des meules sur affleurement de grès (cliché : C. Hamon et V. Le Gall).

qualité tandis qu'un son sourd correspond à une roche trop tendre ou fissurée impropre à la production de meules.

L'outillage principal des carriers se compose de barres à mine, de masses, de massettes et de coins. L'ensemble est confectionné par le forgeron du village; la réfection des manches de ces outils est continuelle, même durant l'opération d'extraction elle-même. Lorsqu'un outil d'extraction casse, l'équipe met de côté les meules nécessaires pour payer le forgeron.

La technique d'extraction utilisée serait à l'origine une tradition Bobo (peuple vivant entre le Mali et le Burkina Faso), et se rapproche des techniques relativement répandues dans les carrières de pierre traditionnelles (Nelson 1987, 123). Les carriers utilisent parfois le feu pour révéler préalablement les pans de fractures naturels et faire éclater la roche (Fig. 7). L'extraction des blocs à ciel ouvert se fait en 'escalier'. L'extraction des blocs est obtenue par insertion de coins métalliques d'une dizaine de centimètres à l'aide d'une masse en fer (Fig. 8).

Une série d'encoche est d'abord réalisée dans la roche à l'aide de coins en fer. Les carriers se relayent pour l'insertion des coins, espacés les uns des autres d'une trentaine de centimètres. Trois coups de masse permettent de les enfoncer superficiellement et de créer une cavité dans laquelle est versée un peu d'eau. Les coins sont ensuite réintroduits et enfoncés jusqu'à fissuration du bloc. Les coins sont d'abord insérés à la base des blocs, jusqu'à ce que la fissuration soit totale. Puis de nouveaux coins sont insérés sur le dessus et le côté, afin de dégager de nouveaux plans de fissures délimitant les contours du bloc recherché. Le détourage des blocs permet de révéler et d'exploiter les fissures naturelles du gisement (Fig. 9). Les blocs ainsi détachés sont ensuite extraits par effet de levier à l'aide d'une barre à mine (Fig. 10). Dans tous les cas, le volume extrait est largement supérieur au volume réellement exploité. Les déchets d'extraction couvrent la pente de la carrière sur plusieurs dizaines de centimètres (Fig. 11). Une heure quinze a été nécessaire au total pour dégager et extraire le bloc de la figure 10 qui permettra de produire environ cinq meules.

Figure 7. Tronc préparé pour la mise au feu visant à faire éclater la roche et à souligner les pans de fissuration (cliché : C. Hamon et V. Le Gall).

Figure 9. Détourage du bloc avec des coins métalliques avant extraction (cliché : C. Hamon et V. Le Gall).

Figure 8. Espacement, disposition et enfoncement des coins métalliques à la base de la roche (cliché : C. Hamon et V. Le Gall).

Figure 10. Extraction du bloc à la barre à mine(cliché : C. Hamon et V. Le Gall).

Figure 11. Déchets d'extraction et de mise en forme autour de la carrière (cliché : C. Hamon et V. Le Gall).

Les blocs extraits (1 m^3 environ) sont ensuite déplacés en contrebas de la carrière. A mi-pente, une zone est réservée à la fracturation des blocs quadrangulaires de plus petites dimensions. Le plan de litage naturellement horizontal est utilisé: le bloc est alors fracturé en deux dans son épaisseur par deux coups de marteau sur un coin métallique positionné sur le flanc du bloc (Fig. 12). Un premier épannelage confère sa forme quasi définitive à l'outil; il est réalisé à l'aide d'un marteau sur la meule posée à terre légèrement en porte-à-faux (Fig. 13). Cette première série d'enlèvements permet d'abattre les angles.

Entre dix et vingt blocs préformés sont ainsi produits quotidiennement sur la carrière. Une fois les blocs extraits, ils sont répartis entre les carriers selon leur ancienneté (le maître choisi en premier puis l'apprenti le plus expérimenté etc.). Chacun est alors libre de finir sa meule lui-même dans un 'atelier', ou de l'envoyer voire de la vendre à l'un des ateliers situé dans le village où la meule sera façonnée puis revendue. Tous les carriers peuvent tailler, mais les tailleurs, eux, ne peuvent pas prendre en charge l'extraction qui constitue le savoir faire le plus difficile à acquérir.

Des ateliers pour la fabrication des meules

Les ébauches issues de la carrière sont ensuite façonnées dans les différents ateliers installés dans le village ou dans

|Figure 12. Fracturation d'un bloc pour obtention (cliché: C. Hamon et V. Le Gall).

Figure 13. Epannelage du bloc et mise en forme grossière des flancs de la meule (cliché : C. Hamon et V. Le Gall).

les hameaux environnants (Fig. 14). Deux types d'ateliers se distinguent, d'une part ceux composés de carriers-tailleurs qui façonnent les ébauches qu'ils ont eux même extrait des carrières. D'autre part les ateliers de tailleurs, qui ne maîtrisent pas les techniques d'extraction, ils se fournissent en ébauches auprès des carriers; ils y travaillent de manière individuelle ou collective. La technique de façonnage des meules reste identique quelque soit le type d'atelier. Ce type de différentiation dans l'espace des étapes de la fabrication des meules rappelle entre autre l'organisation de la production des meules dans les Hautes Terres mayas (Nelson 1987, 120).

Figure 14. Atelier de tailleur à l'entrée du village (cliché : C. Hamon et V. Le Gall).

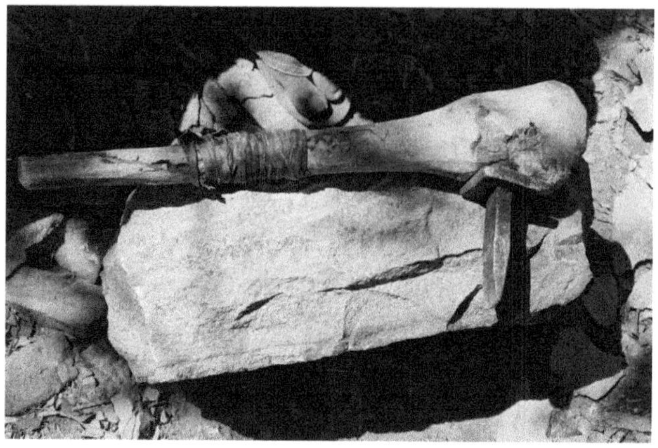

Figure 15. Outil de mise en forme et de régularisation de la meule (cliché : C. Hamon et V. Le Gall).

Nous présentons ici l'une des séquences de façonnage exécutée par un tailleur de Farakoro. L'étape de façonnage débute par l'entame d'un des flancs de la meule à partir de la future surface active. La meule ou la molette sont alors posées au sol. Cette étape est réalisée à l'aide d'un outil métallique plat de section quadrangulaire, dont la lame est large de 3 à 4 cm; il servira jusqu'aux dernières étapes de régularisation de la meule (Fig. 15). Une fois le bord épannelé, la meule est retournée face contre terre et une série d'enlèvements est alors détachée sur le bord opposé à partir du dos de la meule. Les flancs sont ainsi progressivement rétrécis par le détachement de gros éclats. Le bourrelet dorsal ainsi créé est ensuite éliminé par une série d'enlèvements détachés par percussion à partir d'une des extrémités de la meule posée verticalement (Fig. 16). Les arêtes de la surface active sont rectifiées par une série de retouches régulières. Comme cela est également observé dans d'autres contextes (Hayden 1987, 43), la séquence de façonnage de la meule est quasiment identique à celle suivie ensuite pour la molette. Dix à quinze minutes sont nécessaires pour le façonnage d'une meule (Fig. 17), cinq à dix minutes pour le façonnage d'une molette, avivage de la surface active non compris (Fig. 18).

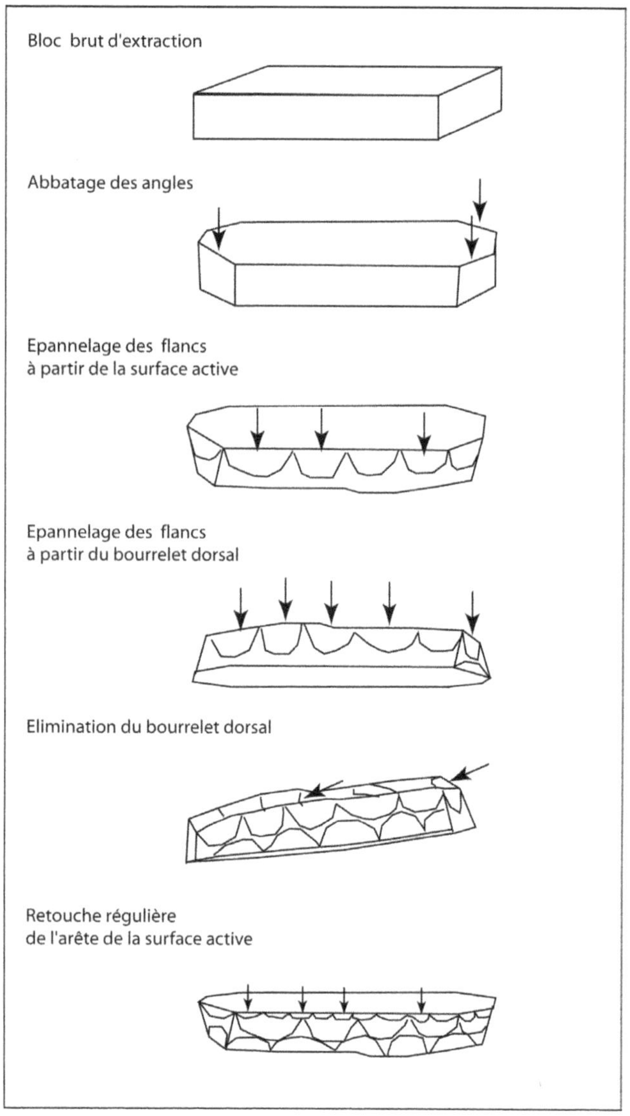

Figure 16. Etapes de façonnage d'une molette (dessin : C. Hamon).

Figure 17. Meule au sortir de la carrière (noter la marque du carrier) (cliché : C. Hamon et V. Le Gall).

Figure 18. Ebauche de molette (cliché : C. Hamon et V. Le Gall).

Figure 19. Régularisation progressive des flancs d'une meule (cliché : C. Hamon et V. Le Gall).

Figure 20. Régularisation des flancs d'une molette (cliché : C. Hamon et V. Le Gall).

Figure 21. Traces de régularisation des flancs et d'avivage de la surface active sur une meule en cours de finition (cliché : C. Hamon et V. Le Gall).

Figure 22. Meule et molette en cours d'appariement avant finition (cliché : C. Hamon et V. Le Gall).

Figure 23. Production d'un atelier (cliché : C. Hamon et V. Le Gall).

Le dos des meules et molettes est généralement laissé brut d'extraction ou de façonnage. La finition des meules et molettes consiste en une régularisation des flancs et un avivage de la surface active (Fig. 19 et 20). La régularisation des flancs s'effectue de manière régulière, avec des coups parallèles tous orientés obliquement. Pour le premier avivage de la surface active, les coups de marteaux sont également parallèles et orientés cette fois longitudinalement suivant le sens d'utilisation de l'outil (Fig. 21). La taille de la lame des outils est proportionnelle à la taille des objets produits: grande pour les meules et plus petites pour la molette. Elle est aussi liée au degré de finesse du résultat souhaité.

Si nécessaire, les blocs peuvent être mouillés pour faciliter le travail. Une fois la molette façonnée, une pierre est utilisée pour égriser l'ensemble des surfaces et adoucir les arêtes sur toutes les faces de la molette.

L'appariement entre meule et molette se fait après la phase d'ébauchage et avant la phase de finition et d'avivage des surfaces actives (Fig. 22). Le critère principal d'appariement réside dans les dimensions respectives des meules et molettes. La fabrication d'une molette se révèle plus difficile que celle d'une meule du fait du calibrage lié à la prise en main et du plus grand risque de fissuration de la molette lors du façonnage. Quatre meules neuves peuvent être avivées par un seul tailleur sur une journée (Fig. 23). L'apprentissage démarre à 15 ans pour la fabrication des molettes et à 20 ans pour la fabrication des meules.

Vente et diffusion d'une production standardisée

Une fois terminées, les meules sont relativement standardisées dans leur forme, un peu moins dans leurs dimensions. Il s'agit de meules quadrangulaires et épaisses associées à des molettes dont la longueur dépasse la largeur de la meule (Fig. 24). Ce type de molette est propre aux productions de Farakoro, car d'autres carrières de meules Minyanka produisent des molettes plus courtes (à San notamment). Leur dos est plat, leurs flancs sont droits à légèrement bombés, leurs surfaces sont plano-convexes transversalement et plano-concaves longitudinalement. La

Figure 24. Meule et sa molette prête à la vente (cliché : C. Hamon et V. Le Gall).

longueur des meules est de 30 à 50 cm et est proportionnelle à leur largeur comprise entre 20 et 35 cm. Les molettes ont une épaisseur de 3 à 7 cm, tandis que leurs extrémités débordantes sont longues de 3 et 5 cm. La meule est désignée par le terme de « mère » et la molette par celui d'« enfant ».

Les carriers et les tailleurs peuvent vendre eux-mêmes les meules ou les confier à des intermédiaires. Les meules peuvent donc être achetées directement au sortir de la carrière ou auprès des ateliers du village. Les meules sont le plus souvent vendues à des particuliers, à la commande pour des mariages, ou sur les marchés de la région dans un rayon pouvant atteindre plusieurs dizaines de kilomètres (marché de Sikasso). L'achat de nouvelles meules dans une famille constitue encore aujourd'hui un signe de richesse au même titre qu'une tête de bétail. Si une ébauche de meules vaut aux environs de 1000 CFA (soit l'équivalent de 15 €), le prix des meules finies varient selon leur taille; il est généralement compris entre 2000 et 5000 CFA. Les meules une fois livrées ne pourront pas être utilisées directement: leur propriétaire devra réaliser un avivage soigné de la surface active avant la première utilisation de la meule au sein de l'espace domestique.

Conclusion

Cette étude de la production de meules en pays Minyanka nous a permis de documenter toute la chaîne opératoire de fabrication de ces meules, depuis l'extraction de blocs de grès de plusieurs tonnes en carrière jusqu'à leur finition dans les ateliers du village de Farakoro. Cette production spécialisée est en concurrence toujours croissante avec le moulin mécanique, dont le coût d'utilisation est compensé par un rendement efficace. L'usage des meules va-et-vient est aujourd'hui plus une contrainte, liée aux maigres revenus de beaucoup de famille, qu'un choix, même si la volonté de perpétuer les traditions explique le succès toujours important de la production de meules notamment en tant qu'élément indispensable de la dot de la future épouse.

Remerciements

Ce projet a été financé grâce à une bourse de recherche post-doctorale de la fondation Singer-Polignac. Nous souhaitons remercier l'ensemble des habitants des villages de Nintabougouro, Karangasso, Nampossela et bien sûr Farakoro, pour leur aide et accueil sans lesquels cette étude n'aurait pu être réalisée. Nos sincères remerciements en particulier à nos amis et 'guides' M'Pougo Dissa, Joseph Dembélé, Felix Dembélé et Jean-de-Dieu Dembélé, et aux carriers de Farakoro, en particulier MM. Yaya Dembélé et Abdoulaye Sanogo.

Bibliographie

Adams, J., 1999. Refocusing the role of food-grinding tools as correlates for subsistence strategies in the US. Southwest. *American Antiquity*, 64, 475-498.

Baudais, D., et Lundström-Baudais, K., 2002. Enquête ethno-archéologique dans un village du Nord-Ouest du Népal : les instruments de mouture et de broyage. In H. Procopiou et R. Treuil (eds.), *Moudre et broyer. L'interprétation fonctionnelle de l'outillage de mouture et de broyage dans la Préhistoire et l'Antiquité : actes de la table ronde internationale, Clermont-Ferrand, 30 nov. - 2 déc. 1995. I – Méthodes*. Paris, 181-209.

David, N., 1998. The ethnoarcheology of grinding at Sukur, Adamawa state, Nigeria. *African review*, 15, 13-63.

De Beaune, S. A., 2000. *Pour une archéologie du geste. Broyer, moudre, piler. Des premiers chasseurs aux premiers agriculteurs*. CNRS éditions, Paris.

Dubreuil, L., 2004, Long-term trends in Natufian subsistence: a use-wear analysis of ground stone tools. *Journal of Archaeological Science*, 31, 1613-1629.

Fullagar, R., et Field, F., 1997. Pleistocene seed-grinding implements from the Australian arid zone. *Antiquity*, 71, 300-307.

Gast, M., 1968. *Alimentation des populations de l'Ahaggar. Etude ethnologique*. Mémoires du centre de recherche anthropologique, préhistorique et ethnologique VII, Paris.

Gelbert, A., 2005. Evolutions du matériel de broyage dans la communauté Dii de Djaba (Nord-Cameroun) durant les deux derniers siècles (Etudes ethno-archéologique et archéologique). In C. Raimond, O. Langlois et E. Garine (eds.), *Ressources vivrières et choix alimentaires dans le bassin du lac Tchad*. Paris, 319-347.

Gonzalez, J.E., et Ibanez, J.J., 2002. The use of pebbles in Eastern Vizcaya between 12000 and 10000 B.P. In H. Procopiou et R. Treuil (eds.), *Moudre et broyer. L'interprétation fonctionnelle de l'outillage de mouture et de broyage dans la Préhistoire et l'Antiquité : actes de la table ronde internationale, Clermont-Ferrand, 30 nov. - 2 déc. 1995. I- Méthodes*. Paris, 69-80.

Hamon, C., 2008. Functional analysis of stone grinding and polishing tools from the earliest Neolithic of north-western Europe. *Journal of Archaeological Science*, 35, 1502-1520.

Hayden, B., 1987. Traditional metate manufacturing in Guatemala using chipped stone tools. In B. Hayden (ed.) *Lithic studies among contemporary Highland Maya*. Tucson, 8-119.

Jackson, T., 1991. Pounding acorn: women's production as social and economic focus. In J.M. Gero and M.W. Conkey (eds.), *Engendering archaeology: women and prehistory*. Oxford, 301-325.

Jonckers, D., 1987. La société minyanka du Mali: traditions communautaires et développement cotonnier. *Collection "Connaissance des Hommes"*. L'Harmattan, Paris.

Mauldin, R., 1993. The relationship between ground stone and agriculture intensification in western New Mexico. *KIVA*, 58.3, 317-330.

Nelson, M., 1987. Site content and structure: quarries and workshop in the Maya Highlands. In Hayden, B. (ed.) *Lithic studies among contemporary Highland Maya*. Tucson, 120-147.

Rondeau, C., 1994. *Les Paysannes du Mali. Espaces de liberté et changements*. Karthala.

Roux, V., 1985. *Le matériel de broyage. Etude ethnoarchéologique à Tichitt (R.I) Mauritanie*. Edition Recherches sur les civilisations, mémoire n° 58, Paris.

Risch, R., 2002. *Recursos naturales, medios de produccion y explotacion social. Un analisis economico de la industria litica de Fuente Alamo (Almeria), 2250-1400 antes de nuestra era*. Mainz (Iberia archaeologica 3).

Wright, K., 1994. Ground-stone tools and hunter-gatherer subsistance in southwest Asia: implications for the transition to farming. *American Antiquity*, 59, 238-262.

Wright, K., 2000. The social origins of cooking and dining in early villages of Western Asia. *Proceedings of the Prehistoric Society*, 66, 89-121.

The hand-mills of Olymbos: an ethnographical study of their form, function and role in a Greek village

Holly Parton

Introduction

The village of Olymbos is situated in the north-west of the Dodecanese island of Karpathos, between Crete and Rhodes in the south-east Aegean, Greece (Fig. 1). Karpathos and Olymbos in particular, has retained a number of traditions due to its remote location. These traditions include; music and dance, dialect, social customs, dress, religious festivals and most specifically technology in relation to milling, and agricultural methods. In Olymbos the technological development in milling can be traced historically from the hand-mill, through animal, wind and water powered mills, to diesel operation and electric. Whilst electric powered mills have largely replaced the others, two windmills remain in use as a tourist attraction and, of greater significance, a few people continue to operate hand-mills.

The majority of the village mills were and continue to be used for the processing of flour and olives. This paper will provide an overview of the flour mills, an outline of daily bread production, and a detailed examination of three key aspects of the hand-mills; their form, function and role.

Form: The physical design of the mill stones and the setting in which they were used.
Function: Raw materials and produce.
Role: The place of hand-mills within the overall milling industry in the village, highlighting why they continued in use at the height of the self-sufficient agricultural period when more efficient mills were available and why they endure up to the present day.

The methodological approach employed in this study has been qualitative, compiling information from local sources

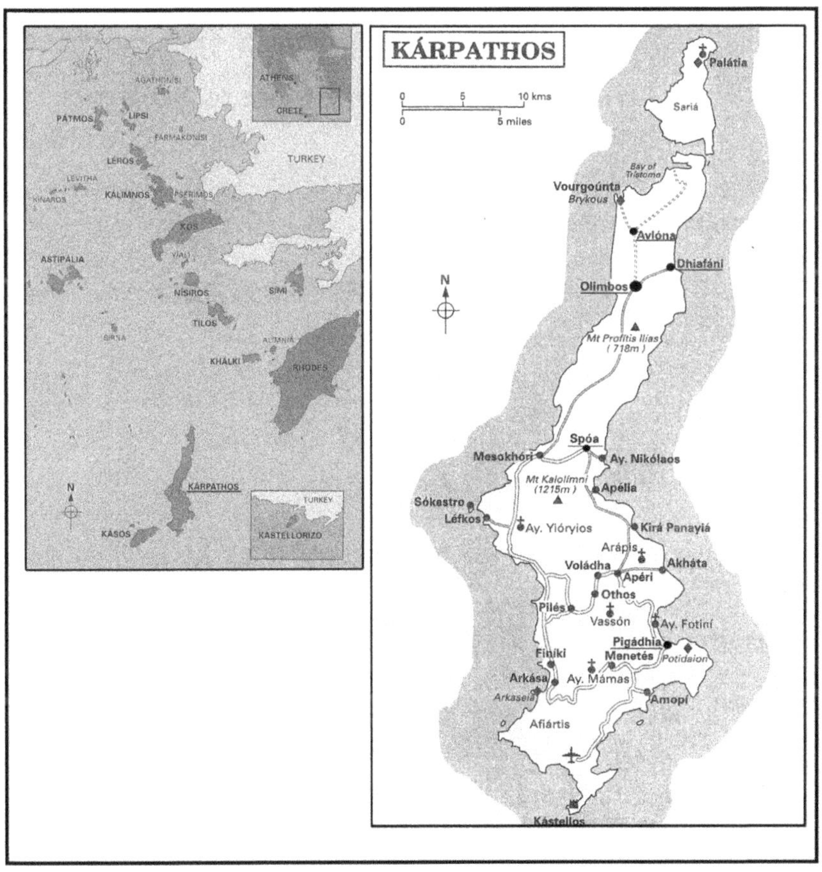

Figure 1. Map of Greece with detail of the Dodecanese Islands and map of Karpathos with main places mentioned in the text highlighted. (Adapted from Barber 1997, frontispiece and 119).

Figure 2. Olymbos from the north with a row of windmills along the crest of the hill highlighted in the setting sun.

via questionnaires to establish the wider socio-ethnographic implications of mills within the local landscape.

The island of Karpathos

Karpathos is in the Dodecanese island group which are scattered in a roughly north-south alignment along the western coast of Turkey. The term Dodecanese can be linked to the Greek words for twelve (δώδεκα – do-deka) and island (νηςή – ni-si), and whilst the whole group includes over one hundred islets the main populated islands are the following: Patmos, Leros, Kalymnos, Kos, Astypalia, Nisyros, Symi, Chalki, Rhodes, Karpathos and Kasos. In his study of the Dodecanese Kasperson (1966, 10) also includes the islands of Leipsos and Kastellorizio which grew in importance during the medieval period.

Karpathos is situated together with Kasos, just to its south, on the south-western edge of the Dodecanese. It lies on a north-south alignment and is 48 km in length, 12 km wide at its maximum, 1 km wide at its minimum and has a total area of 288 square kilometres (Melas 1985, 17). Just to the north, across a 100 m wide strait is the small island of Saria, which is seen almost as an extension of Karpathos (Philippides 1973, 24). This island is 8 km long, 4 km wide and is 17 ½ square kilometres in area (Melas 1985, 19). Karpathos is a rugged mountainous island with a distinctive relief running along its spine, where the following peaks, listed from south to north, are located: Chomali 685 m, Kalolimni 1,215 m (the highest in the Dodecanese), Kymaras 692 m, Profitis Ilias 718 m, Koryphi 590 m and Malo 713 m (Philippides 1973, 24). The coast line varies from steep cliffs to undulating hills that gradually fall to sea level and there are a few areas, mainly in the south, west and at the northern tip, where the land levels out. Saria also has steep cliffs, especially on the east side, and one inland, upland plain in the north-west.

The climate is typical of the Mediterranean with a temperature ranging from 32 to 100 degrees Fahrenheit (*ibid.*, 26). The summer months of May to October are dry with July and August being the hottest. Winters are mild with rainfall from as early as October, but mainly in the months of November onwards. Spring includes the months of March to May when the temperatures can be variable, and rainfall scattered.

A particular feature of the climate are the winds, of which Melas gives a very detailed account (Melas 1985, 20-21. The most important are the annual etesians (*meltemi*) that blow in the summer throughout the Mediterranean from the north and north-west. These usually pick up in June but can start in May and blow almost constantly until September. There are days when the wind will drop and a total calm pervade and other days when the wind changes coming from the south to south-east. This is the sirocco, an almost unbearable hot furnace blast that brings with it the sand of the deserts of North Africa. In the spring, autumn and winter the wind is more variable and comes mainly from the west to south-west and north to north-east. This annual wind pattern has had a defining influence on the type of windmill people have chosen to build with the uni-directional mill dominating over the tower mill, the former being powered by the summer etesians.

There are no perennial rivers but river beds are fed in winter from the rains and storms. The main water source is from springs and/or wells. Olymbos itself has a spring on the southern outskirts of the village. In addition, the houses have cisterns to collect the rain water from the roof. As with the wind, the water supply has affected the type of watermills built in Olymbos. They are situated in the river bed as they rely on the winter rains and are placed in a stepped series to take advantage of the terrain and have drop-towers so that the maximum power can be gained from the water.

Karpathos is fairly well forested, especially in the north, even today and despite recent forest fires. The predominant species is the subtropical pine. Other trees include oak,

Figure 3. The valley on the east side showing gardens, the river bed and some windmills.

Figure 4. Avlona from the south.

plane, cypress, olive, almond, carob, strawberry, myrtle and bay. Melas also cites cedar trees in the area of Afiartis in the south (Melas 1985, 18). Being a very mountainous island the main agricultural land is the coastal lowlands of Pigadia, Arkassa and Afiartis in the south and Lefkos on the west. Travelling inland you find the upland valleys are cultivated and many of the hillsides surrounding the villages are terraced, showing that every available piece of land has been utilised. Kasperson (1966, 111-2) identifies three economic types within the Dodecanese: the 'diminutive maritime', the 'undiversified agricultural' and the 'modified multi-resource'. Karpathos falls into the middle category as somewhere that 'remained primarily subsistence-orientated throughout their history' (Philippides 1973, 17). Being quite fertile the island could support agriculture, but not to the extent that it became a major commercial venture bringing in revenue although at times the island did export grain crops, for example in the 19th century to Kasos (*ibid.*, 18). Farming used to be carried out throughout the island, though today this is no longer the case. Smaller gardens are still maintained, especially in areas with springs where the water is channelled to the plots by intricate irrigation networks. Olymbos is one region where farming is still carried out on a larger scale. The staple crops are cereals and various legumes. Vines and olives are also cultivated as well as vegetables and fruit. Because of the climate a system of 'dry-farming' prevails.

Olymbos lies on the northern slopes of Mount Profitis Ilias (Fig. 2). To the west the land falls away steeply to the sea. To the east the land slopes into an enclosed valley (Fig. 3). Most of the houses have been built on this side and are thus protected from the weather that comes in off the sea and the strong summer winds blowing from the north-northwest. The lower terraces and the slopes opposite the village have small garden plots. The riverbed in which the watermills are situated runs through the valley around the foot of the village and turns west continuing down to the sea. North of Olymbos, about an hour and a half by foot or ten minutes drive, is Avlona, a north-south aligned upland plain bordered by terraced hills (Fig. 4). This is the main agricultural area today. On the eastern slopes is a small settlement of houses used for agricultural purposes. The majority are abandoned but each family in Olymbos would have owned one. East of Olymbos, approximately two hours by foot or fifteen minutes by car is the small port of Diafani.

Olymbos has always been and continues to be remote. In the past before the construction of roads, it would have taken almost a day to walk to the other villages in the south of the island, along the old stone paths or *kalderimia*. Today Karpathos owns an airport in the south of the island and is serviced by ferries which run a few days a week. Of these, only one stops at both Pigadia, the capital in the south and Diafani. The other ferry only stops at Pigadia before continuing to Rhodes. Of course, Olymbos is also accessible by car but there is no bus service as the road north from Spoa is not asphalt and is currently being upgraded. The drive from Spoa to Olymbos is a little less than an hour. Alternatively, there is a small boat that runs between Diafani and Pigadia a few times a week, journey time about 2 hours. This remoteness is a key reason why traditions including the use of mills and some agricultural methods have continued in Olymbos until today, where elsewhere in the island and in Greece they have been replaced by more modern technology and no longer play a part in daily life. Much has been written about the customs of Karpathos but some literature is of a more general nature, covering different aspects including the landscape, history, archaeology and even the mills.

A sample of comparative literature

Theodore Bent who wrote about the islands of Telos and Karpathos in 1885 visited Olymbos with the purpose of studying the local dialect to see how similar it was to ancient Greek and describes the people as 'a wild, uncultured race of shepherds, and their customs of great value to the student of folk lore and comparative mythology' (Bent 1885, 240). He was very interested in the local pastoralism and many of the words he highlights are those associated with animals and farming. In 1903 the archaeologist Dawkins visited Karpathos and gives a detailed account of several aspects including the mention of the island in ancient sources, recent travellers accounts, the geography and settlement pattern of the island, archaeological remains, general observations on agricultural practises he witnessed such as the harvesting, on women's dress, inheritance patterns and even a section on wooden door locks he saw in Olymbos. He also writes about an olive mill in Olymbos which caught his attention because of the wooden machinery and single, re-used column drum used for crushing the olives (Dawkins 1902-1903, 178-210). In 1962 Simpson and Lazenby published their explorations in the Dodecanese. The focus of their report is archaeology and they explored Karpathos and Saria searching for remains. Fortunately they also comment on travel distances between places, modern villages, agriculture, water sources and even

Figure 5. Looking across the northern end of the village. The Uni-Directional Taverna windmill with sails and a tower mill to its left.

Figure 6. Watermill showing ring built drop tower above and exit hole for water in front wall. Scale 1m.

a watermill at Limanáki (Hope Simpson and Lazenby 1962, 166). Finally a more recent publication is that of Halstead and Jones (1989) who carried out a study of crop-processing and farming in Olymbos and on the island of Amorgos. They give much valuable detail on the species grown and on the processing and harvesting methods. Of particular note is their reference to the fact that at the time of their fieldwork only one windmill was operational in Olymbos along with petrol-driven mills (Halstead and Jones 1989, 42).

In addition to material specifically about Karpathos, publications for other areas in Greece provide good comparative reading for building up a general picture of the island. Archaeological surveys are valuable as one of their main aims is the examination of changes in settlement patterns across time and for the later historical period they are able to build a much more detailed picture through the use of written and oral evidence in addition to the recording of physical structures on the ground. An example is that for the island of Keos. Whitelaw (1991, 403-54) not only describes various elements within the landscape itself such as field terraces, walls, roads, cisterns, threshing floors, houses, animal shelters and wine presses, he discusses agriculture generally looking at farming techniques, what crops are grown, and how land use has changed over time in addition to the changes in settlement patterns. He makes specific reference to how oral and written data can 'set the material evidence for the changes in the agricultural exploitation of the area into a wider social context' (Whitelaw 1991, 405).

Nevertheless, such surveys are still essentially archaeological. The ethnographic approach and methodology itself is not elaborated on, interviews are not generally referred to in detail and although in some cases reference is made to social structures such as inheritance systems of land and property (Whitelaw 1991, 454), no detailed analysis is made of exactly how these affected the interpretation of the very landscape under study. This would involve a more anthropological approach, as seen in Forbes' (2007) work on the Methana peninsular in which the two disciplines are combined. Forbes records the material landscape and agricultural system and then, through extensive ethnographic research, examines what they meant to the local people themselves. By looking at various social constructs within the society, especially kinship relationships and how people are connected with each other on a familial, village and wider community level, he is able to present a much more intimate view of the landscape, almost that of an insiders' viewpoint.

Then there is the literature that is purely anthropological in approach some of which provide valuable comparative detail for the rural, social and economic setting by describing rural daily life, agriculture, land ownership, family relationships and the idea of self-sufficiency within a community where one has to be involved in relationships with fellow villagers. Examples include du Boulay's (1994) book describing life in a village on the island of Evia, Kenna's The Idiom of Family (1976) and her article on family and economics, which also makes specific reference to mills and flour production and how the latter decreased when a bakery was established (1990, 155-6).

What stands out amongst all of the above references is how, despite the emphasis given to the agricultural cycle with crops, harvesting methods, land ownership and production quotas being analysed, hardly any reference is made of the mills that would have been the main method of actual flour production. In addition, whilst much has been written specifically on mills, both in Greece and generally, the majority concentrate on the technical and architectural characteristics. Examples include: Blom (1999) and Limona-Trebela (1983), who have written comparative overviews of windmills in the Greek islands; Calvert's (1971-1972, 1972-1973, 1975, 1978) writing on such details as the aerodynamics of wind power and Moog's (1994) work on the horizontal watermill. Although these publications focus on mills they contain little or no reference to how mills fit within the social and economic contexts of everyday life, neither do they look at

Figure 7. Watermill showing rubble masonry drop tower with water channel coming in on the left and exit hole in front wall. Scale 1m.

Figure 8. Women harvesting barley in Avlona.

how different types of mills compliment each other and work together within a community so that the facility to grind flour is available year-round. Another factor is that little mention is made of hand-mills. In the remainder of this paper I would like to address some of these issues.

The flour mills – an overview

The main mills used for flour production in Olymbos in the past were windmills and watermills. There were also two diesel mills.

The windmills

I have counted 54 windmills in Olymbos. Of these, 51 are the horseshoe-shaped, uni-directional type, facing north-northwest so that they catch the summer winds. Three are tower mills that have revolving caps to enable them to operate with wind from different directions (Fig. 5). The largest group of windmills is situated along the crest of the ridge upon which Olymbos is built and is one of the most striking aspects of the village (see Fig. 2).

Figure 9. A threshing floor in Avlona with motor powered threshing machine and wooden pitch forks for winnowing.

Figure 10. A family sized model of an electric mill of the Zanon manufacturer, capable of a production rate of 50-300 kg of flour an hour.

The watermills

There are fourteen watermills in total. They are all of the Arubah Penstock type having a drop tower and a horizontal water-wheel. There are two forms of tower construction. Most are built of stacked limestone rings sometimes with an additional external skin of stone rubble for support (Fig. 6). A few are built entirely of regular, rubble stone masonry lined with a waterproof mortar (Fig. 7). They vary in size but all, apart from one square example, taper towards the top. In some cases both the stacked ring and rubble construction are found. The mills are placed in a stepped series along the river bed. A water channel can

Figure 11. A miller inside his Uni-Directional windmill.

Figure 12. A stone bread oven with an unbaked loaf being placed inside.

be traced between them though it does not appear to have connected all the mills. If several consecutive mills were running then the millers downstream could utilise the water running from a mill up-stream. In addition, each mill had a separate by-pass channel enabling them to be operated independently.

The condition of the different mills varies from ruinous with just foundations or a few walls intact to fully operational. The watermills stopped working about 30 years ago. Today there are two windmills in Olymbos that work in the summer months mainly for the benefit of tourists. However they are also used by the few families who still grow their own wheat and barley.

Flour today

About 15 families still grow their own wheat and barley up in Avlona. The main reason is because they have animals – sheep, goats and donkeys, and they need the fodder. For some though it is simply continuing a way of life that they have grown up with and do not want to change and also because they prefer their own produce as they do not use fertilizers and they know where and how it has been grown. The crop is harvested by hand, cut with sickle blades, and threshed using a motor but then winnowed by hand using a wooden pitchfork (Figs. 8 and 9). The resulting grain is ground mainly with small, individually owned electric mills but also at the working windmills (Fig. 10). The owners of the Taverna Windmill run a restaurant in the summer and the mill is open for the tourists but they also grow their own grain to be ground in the mill. The other windmill is run more as a hobby by a man who has returned to Olymbos in his retirement (Fig. 11). Of the people who don't grow their own grain some buy it and then grind it as above, others buy sacks of flour, usually of 50 kg in weight.

The reason why people buy either grain or flour is because they still prefer to bake their own bread; for this purpose large wood-burning stone ovens recur throughout the village (Fig. 12). In the past this was a weekly activity but the facility to freeze has altered this pattern. However, in the summer months when the tavernas need a constant supply of bread, there are usually one or two ovens being fired up somewhere on a regular basis. Bread can be also bought in the village shop. It is delivered from Spoa, nearly an hour's drive away. This bread is bought by tourists, workers from outside the village and also those locals who are unable to make their own bread, for reasons of old age or illness for example. The reason why most villagers still go to all the effort of collecting firewood, firing the oven, making four or five, and sometimes more, bowls full of hand-kneaded dough to then shape into the large round loaves is very simple indeed – the local, home-made bread tastes better than anything else. Having eaten both the local and shop bread I can testify to that, the former being especially delicious when combined with local olive oil and fried eggs.

Figure 13. Hand-mill with stone collar. Scale c. 23cm.

Figure 14. Hand-mill with cement collar. Scale c. 23cm.

The hand-mills

In the past this would have been the most common type of mill as every household owned one or even two. I have seen ten single stones and 17 pairs, all of which are rotary querns. Of the latter four are no longer used, six are still used regularly, two occasionally, and five possibly. Materials used that I can identify are sandstone, vesicular basalt and limestone. I have not located any quarries and nobody has been able to confirm where their quern was made as most have been passed down in dowries through families. The average dimensions are a diameter of between 38 and 44 cm with the largest being 54 cm, and a thickness of 5-6 cm. Although the number of mills examined is small there is a

Figure 15. Hand-mill with collar made from mud and straw and two horizontal handle sockets; a shallow groove and a cut with handle.

Figure 16. Hand-mill with cement covering the upper face and a shallow hollow around the perforation.

Figure 17. Hand-mill with a domed cemented upper face and a metal rod for a handle. Scale c. 23cm.

Figure 18. Hand-mill with horizontal handle socket, wooden handle and a wooden rynd cemented in place. Scale c. 23cm.

Figure 19. Hand-mill with horizontal handle socket and rynd cemented in place.

surprising variety amongst the sample. I will consider first the form of these mills and then their function and role. Unlike the two working windmills, the few hand-mills that are still used today are not done so for tourism purposes.

Form: the upper grinding stones

Thirteen of the upper stones have a raised collar around the central perforation on the upper face. Some are cut out of the stone itself, some are constructed by the owners from cement and one is made out of mud and straw (Figs. 13, 14 and 15). The purpose of the collar is to help stop the substance being ground from spilling out as the quern is turned. Of those that do not have a collar four have an upper face covered with cement which on one is shaped into a shallow hollow around the perforation and on the other three is shaped into a high dome providing a much deeper perforation (Figs. 16 and 17). These would all have been effective for holding the produce. Two other querns have traces of cement and of the remaining four it can be presumed that they may well have had some form of collar

Figure 20. Hand-mill with one broken horizontal handle socket and one intact socket. Scale c. 23cm.

Figure 21. Hand-mill with 'key-hole' design handle socket. Scale c. 23cm.

or hollow when they were still in use.

There are several different methods for attaching the handles. The most common is a shallow horizontal groove cut into the upper face on the edge. A wooden fitting with a hollow or hole for a vertical handle is slotted into the groove (Fig. 18). In some examples the fitting is cemented into place otherwise it is cut to fit the groove and is wedged in firmly (Fig. 19). Three querns have vertical hollows cut into their upper faces. One actually has an unfinished cut that was set too close to the edge causing part of it to chip off. A second hollow has been cut opposite and this holds the vertical handle (Fig. 20). The other two are

Figure 22. Hand-mill with two unused handle sockets and a tap used as a handle instead.

Figure 23. Hand-mill showing 'bow-tie' shaped rynd cut without in situ rynd.

more elaborate being of the 'key-hole' design (Fig. 21). They both have stone collars around the perforation and the surface of the upper face is further cut in relief to define the area where the vertical hollow for the handle is placed. Two have a horizontal cut in the edge, as if a bite has been taken. One also has a groove opposite but the handle fitting is wedged in the cut today (Fig. 15). On another example the handle grooves are again no longer used, rather the owner has cemented an old tap onto the upper face to use as a handle (Fig. 22). A similar case is seen with one of the domed cemented querns where a metal rod is fixed into the cement (Fig. 17).

Figure 24. Hand-mill with 'bow-tie' shaped rynd cut without in situ rynd. Scale c. 23cm.

Figure 25. Hand-mill with cross shaped rynd cut. Scale c. 23cm.

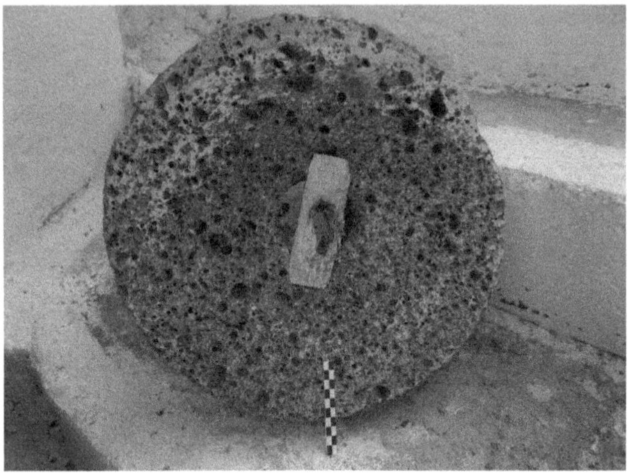

Figure 26. Hand-mill with a wooden ryndfixed in the rynd cut. Scale 10cm.

Figure 27. Hand-mill showing perforation with rynd in situ and no rynd cut. Scale c. 23cm.

Figure 28. Hand-mill showing re-used upper stone used as a lower stone with wooden plug secured by leather, handle socket and rynd cut.

Turning now to the rynds, these are made of wood or metal with the former being more popular having 11 examples whilst only five are made of iron. There is no set method for fixing them. Of those querns where I was able to examine the grinding face 11 of the perforations had a

Figure 29. Hand-mill showing reused upper stone used as a lower stone with wooden plug fixed in cement and handle socket. Scale c. 23cm.

rynd cut, one of which went right through to the upper face. All, apart from one, are of the 'bow-tie' style varying only in size and shape, from rounded to more rectangular or flaring (Figs. 23 and 24). One cut is shaped like a cross with four arms (Fig. 25). Only four had a rynd *in situ* but of these only one was actually fixed in the cut itself (Fig. 26). The remainder were fixed in the cement collar or dome on the upper face (Figs. 18 and 19). Seven querns have no rynd cut and of these six have *in situ* rynds, five fixed in cement and one wedged in the perforation itself (Fig. 27). I was unable to examine the grinding face of six querns only one of which did not have a rynd. Three rynds are fixed in cement and one is fixed in the collar of mud and straw. These may or may not have had a rynd cut. The last has a stone collar, the rynd could just be wedged in the perforation or it may have a cut.

Form: the lower grinding stones

Of those that I was able to look at in detail all except one had a wooden plug in the perforation most of which fitted snugly whilst one was secured tightly by a piece of leather on one side and one was cemented in place on the lower face. One quern had a plug made of cement (Figs. 28, 29 and 30). All of the spindles are iron. The grinding faces are flat to level or slightly convex. An interesting detail is that four examples are re-used upper grinding stones.

Figure 30. Hand-mill showing lower stone with cement plug. Scale c. 23cm.

Figure 31. Hand-mill in use with cloth for collecting the produce.

Figure 32. Hand-mill in use with sack for collecting the produce.

Figure 33. Hand-mill on platform Scale c. 23cm.

They have handle sockets cut in the lower faces and two also have rynd cuts (Figs. 28 and 29). These latter two may have been intended to be used with the face with the rynd cut as the grinding face but then had handle sockets cut and the opposite face used for grinding before finally being used as lower stones.

The physical setting

Most of the querns are portable and kept in a cupboard or storage room. They are used on the floor or ground, both outside and inside and a cloth is placed under them to collect the ground substance (Figs. 31 and 32). Five are placed on concrete platforms built especially for the mill. Most are constructed from wooden planks placed across a corner space, sometimes with additional wooden leg supports and with wooden sides to contain the concrete (Figs. 33 and 34). One is a flat concrete slab supported on a concrete pillar (Fig. 35). Another built entirely of concrete has additional storage space below and a concrete edge which is cut away on one side to enable the ground substance to be easily collected from the platform (Fig.

Figure 34. Hand-mill on platform with kondalethos type handle. Scale 1m.

Figure 35. Hand-mill placed on platform. Scale c. 23cm.

36). In four cases the lower grinding stone is securely fixed in the concrete. One has a piece of sacking placed under the quern to collect the produce (Fig. 37). Another interesting design feature of these platform mills is the handle. In most cases today a short handle is used but two examples have a long wooden handle which has one end in the socket and one fixed in a hole drilled through a wooden plank placed above the platform (Figs. 34 and 37). This handle is called a kondalethos locally and it enables the mill to be turned more easily and with less force thus making grinding less arduous.

Figure 36. Concrete platform for hand-mill with lip for collecting produce.

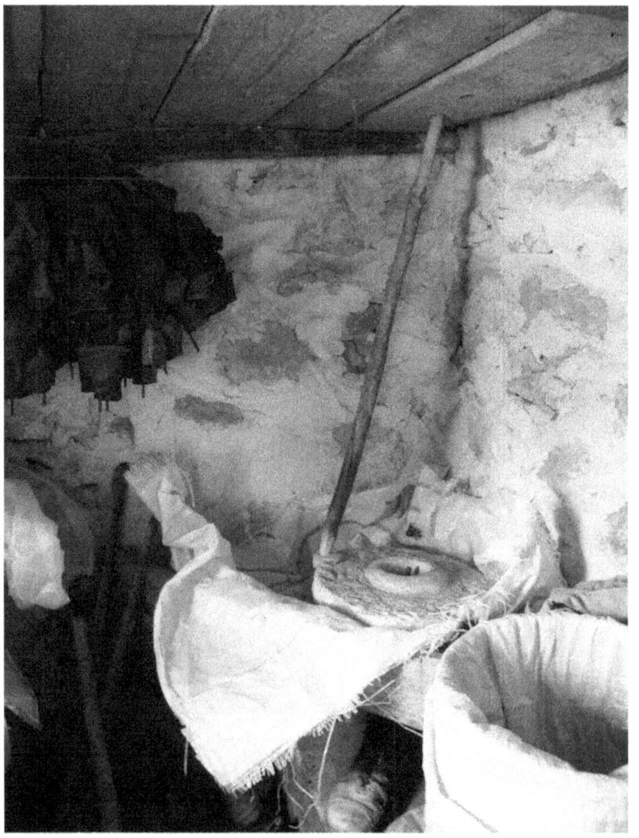

Figure 37. Hand-mill on platform with sacking for colleting produce and long kondalethos type handle.

Figure 38. Close up of fava bean being ground with the purple shell and yellow bean.

Function and role

Function and role are intrinsically linked. Function will consider specific use and role will address two questions; why were hand-mills not ousted by the larger, more efficient mills and why do they continue to be used today? Regarding function, flour for bread was only processed occasionally as it is very time consuming to produce fine flour with a hand-mill and much quicker to use the larger wind, water or diesel mills. Wheat and barley were coarsely ground to make a substance called *chondro* which was used to thicken soups. You can still buy this today and it is served in some tavernas. The main substance ground was the fava bean, a yellow split pea with a hard outer purple shell (Fig. 38). This bean is made into a dip, usually served with olive oil and raw onion. It is a common dish eaten in Greece today.

The substances being ground provide the key to the first question being addressed under role. The reason why hand-mills continued to be used even when the larger mills were operating is twofold. Firstly, they were the only mills used to grind *chondro* and fava, two staple items of peoples' diet. Secondly they actually provided a back-up for times when the larger mills were not working, which could happen under certain conditions. For example, the majority of windmills were of fixed direction working only with the summer seasonal winds. In the winter when the wind came from different directions the few tower mills could be turned into the wind but it was often too strong for them to operate safely. The water mills provided an alternative for the winter months but would only work if there was adequate rainfall which was not always the case. The diesel mills only operated a few days a week as diesel was not easily available, even today the nearest petrol station is nearly an hour's drive away. So in the winter months it was often the case that no larger mills were working. One way in which the people prepared for this was to grind surplus flour in the summer to store for the winter, but they would also use their hand-mills if they needed extra flour.

Another case where the hand-mills took on a vital role was during World War II. People relate that the Germans and Italians forbade the use of hand-mills, confiscating and in some cases breaking them. The villagers were only supposed to use the larger mills where the flour production could be monitored and a quantity of flour, sometimes as much as 80% according to memory, was taken. In response to this people hid their hand-mills and ground flour secretly at night or away from the village. The mills became a means of survival during these hard and difficult times.

Regarding the second role issue, of why some older people still use hand-mills today a variety of factors come into play. One reason is because they grow their own produce and whilst you can buy fava and *chondro* in the local shop it is easier and cheaper to grind it at home. There is also the question of flavour – the *chondro* tastes better. This is also why even a little flour is still hand-milled occasionally, because these mills produce the best flavour as the least heat is generated during the grinding process. There is also a sentimental value; hand-mills, though a basic commodity that everyone owned, were of high value, passed down in dowries and thus with strong memories associated with them. Finally they are also still a security; the memory of the war is strong, electricity was only installed in the 1980's and power cuts often occur. Even people who no longer use a hand-mill have still kept them, ready to be put into action again.

Conclusion

My aim in this paper was to try and understand hand-mills from the perspective of the people who owned and used them rather than simply as objects. To do this it has been necessary to consider them not in isolation but in relation to the milling industry as a whole as well as the agricultural system. Yet the technological aspect has not been ignored for this is also pertinent to placing the hand-mills within a social and economic context. I chose the village of Olymbos for my study area because here elements of an older way of life with its accompanying technology are still to be found, though less and less. However, the advent of tourism, attracted to the village because of its traditions has brought in new income, and the younger generation have to find a way to sustain their traditions whilst modernising the village. This combination makes Olymbos an ideal place to study milling and its role within the life of a community and how such a community adapts to a changing world around it.

Acknowledgements

This paper would not have been possible without the generous and patient help of the people of Olymbos in allowing me to study their mills and answering numerous questions. I am also grateful to The British School at Athens and Leicester University for their support and to Dr C. Hebron for her comments and feedback.

References

Barber, R., 1997. *Blue Guide: Rhodes and The Dodecanese.* London.

Bent, T., 1885. The Islands of Telos and Karpathos. *Journal of Hellenic Studies,* 6, 233-42.

Blom, L., 1999 *The Windmills of the Greek Islands.* The Hague, Netherlands, TIMS.

Calvert, N.G., 1971-1972. Windpower in Eastern Crete. *Transactions of the Newcomen Society*, 44, 137-44.

Calvert, N.G., 1972-1973. On Water Mills in Central Crete. *Transactions of the Newcomen Society*, 45, 217-22.

Calvert, N.G., 1975. The Mono-Kairos Windmills of Lasithi. *Annual of the British School at Athens*, 70, 51-7.

Calvert, N.G., 1978. The Characteristics of a Sail Mill. *Journal of Industrial Aerodynamics*, Amsterdam, 3, 79-84.

Dawkins, R., 1902-1903. Notes from Karpathos. *Annual of the British School at Athens,* 9, 176-210.

Du Boulay, J., 1994. *Portrait of a Greek Mountain Village.* Denise Harvey, Evia, Greece.

Forbes, H., 2007. *Meaning and Identity in a Greek Landscape: an Archaeological Ethnography.* Cambridge.

Halstead, P. and Jones, G., 1989. Agrarian Ecology in the Greek Islands: Time Stress, Scale and Risk. *Journal of Hellenic Studies,* 109, 41-55.

Hope Simpson, R. and Lazenby, J.F., 1962. Notes from the Dodecanese. *Annual of the British School at Athens,* 57, 154-75.

Kasperson, R.E., 1966. *The Dodecanese; Diversity and Unity in Island Politics.* Chicago.

Kenna, M., 1976. The Idiom of Family. In J.G. Peristany (ed.) *Mediterranean Family Structures.* Cambridge, 347-62.

Kenna, M., 1990. Family and Economic Life in a Greek Island Community. In C.C. Harris (ed.) *Family, Economy and Community.* Cardiff, 143-63.

Limona-Trebela, E., 1983. *Windmills of the Aegean Sea.* TIMS, 7, The Hague, Netherlands, TIMS.

Melas, E.M., 1985. *The Islands of Karpathos, Saros, and Kasos in the Neolithic and Bronze Age.* Studies in Mediterranean Archaeology, 68, Göteborg.

Moog, B., 1994. *The Horizontal Watermill.* The Hague, Netherlands, TIMS

Philippides, A., 1973. *The Vernacular Design Setting of Elymbos: A Rural Spatial System in Greece.* Michigan.

Whitelaw, T.M., 1991. The Ethnoarchaeology of Recent Rural Settlement and Land Use in Northwest Keos. In J.F. Cherry, J.L.Davis, and E. Mantzourani, (eds). *Landscape Archaeology as Long-Term History Northern Keos in the Cycladic Islands from Earliest Settlement until Modern Times.* Los Angeles, 403-54.

Visualising the Invisible: re-discovering the ancient grinding stone quarries of the Aswan West Bank, Egypt

Elizabeth Bloxam

Introduction

As this volume of papers demonstrates, many important advances have been made in the study of European millstone quarries, particularly in terms of enhancing our understanding of how major transitions in food production impacted on millstone production over time. By comparison, our understanding of the procurement and distribution of grinding stones (millstones) in ancient Egypt lags significantly behind, such studies being a low research priority in preference to stone procurement for 'elite' objects and royal monuments. Questions relating to the interplay between procurement of high-quality hard stones for ornamental purposes, as well as for domestic utilitarian use, tend to be overlooked. As a consequence, studies of ancient quarries rarely investigated or even recognised the extent to which prestige stone sources had long histories of appropriation, often associated with a primary utilitarian use of the resource for the processing of food.

Covering a 12 kilometre stretch of hyperarid desert from the Wadi Kubbaniya in the north to the old Aswan Dam in the south, the large deposits of silicified sandstone (often termed 'quartzite') that span the Aswan West Bank, were first noted as a place of *only* pharaonic period quarrying (3rd – 1st millennium BC) for ornamental objects by De

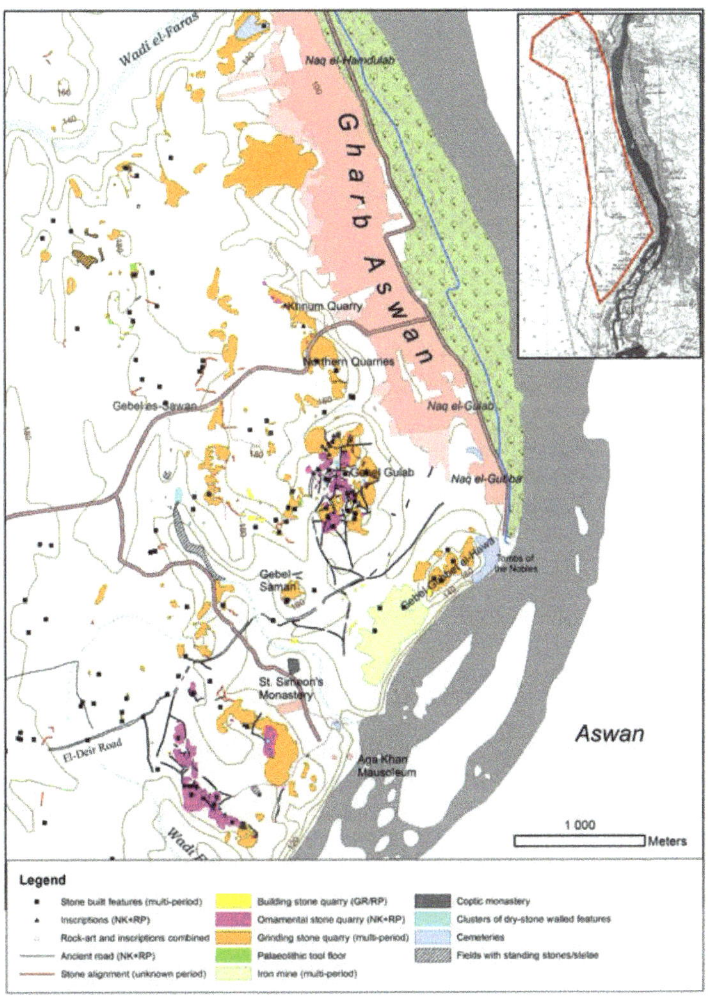

Figure 1. Map of the Aswan West Bank showing major areas of grinding stone and ornamental stone quarries surveyed during the 'QuarryScapes' project. Inset map shows the full extent of the quarry landscape from Wadi Kubbaniya in the north to the Old Aswan Dam in the south. Map produced by Tom Heldal and Per Storemyr.

Figure 2. View towards the Aswan West Bank ancient quarry landscape; centre foreground Elephantine Island; background left the ornamental and grinding stone quarries at Gebel Gulab.

Morgan (*et al.* 1894, 125-7) in the late 19th century. Even into the 1960s and 1980s, research of the quarries only recognised the Aswan West Bank as a place of essentially New Kingdom (1550-1069 BC) and Roman Period (30 BC – AD 395) ornamental quarries, centred on two main gebels (hills) at Gebel Gulab and Gebel Tingar (Habachi 1960; Klemm and Klemm 1993, 289-303; 2008, 219-31) (Fig. 1). Easily visible by large amounts of quarry waste that cascades down the sides of the gebels and ramps constructed to transport the stone down to the Nile, the Aswan West Bank was established as a major source of 'quartzite' for royal and elite ornamental objects. These quarries being second only to that of Gebel Ahmar, 800 kilometres north near Cairo.

During the 'QuarryScapes' project (2005-2008), a multi-disciplinary team of researchers returned to the Aswan West Bank quarries with the objective of building on this earlier work by undertaking a more thorough investigation of the ornamental stone quarries and to characterise all the material culture associated with them (Bloxam *et al.* 2007). During these investigations it became clear that ornamental stone quarrying was a minor activity in comparison to 16,000 years of grinding stone production, from the Late Palaeolithic (18,400 – 18,100 BP) into the early Roman period (30 BC). This discovery has led to a complete revision of how we visualise archaeologically grinding stone quarries and their implications to understanding the early history and longevity of quarrying along the Aswan West Bank. Findings from this research are discussed below.

The silicified sandstone of the Aswan West Bank: a brief history of consumption

The properties of silicified sandstone (quartzite) in terms of hardness, durability, abrasion and aesthetics were all qualities that made this one of Egypt's most sought-after

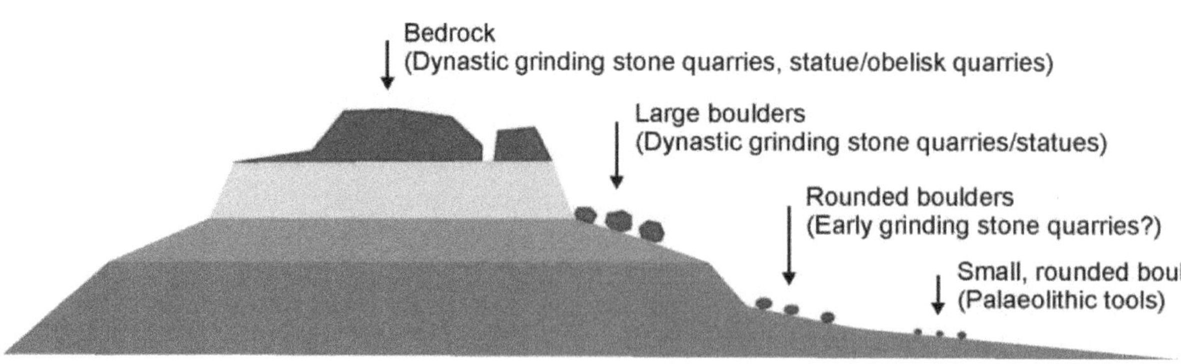

Figure 3. Schematic cross-section of a typical Aswan West Bank gebel (hill) showing the distribution of silicified sandstone (top layer and boulders) and areas of exploitation through time. Produced by Tom Heldal.

Figure 4. Saddle quern (silicified sandstone) found in Old Kingdom settlement levels on Elephantine Island.

Figure 5. Head of Nefertiti in the gold variety of silicified sandstone (quartzite), Egyptian Museum, Cairo.

stones in antiquity. In simplified terms, silicified sandstone is sandstone that has been cemented with silica, this giving the stone its hardness, extra abrasive qualities as well as its 'glittering' appearance (Heldal *et al.* 2007, 51-67; Bloxam 2007, 5-6; Heldal 2009, 134). The most durable and harder of the silicified sandstones are the brightest in colour and so the pure gold/yellow and purple varieties were the most sought-after for ornamental objects. As these hardest varieties cap the sandstone and mudstone gebels, this is where we find almost all of the ornamental stone quarries, together with their networks of roads (Bloxam *et al.* 2007; Heldal and Storemyr 2007, 102-22) (Figs. 1 and 2). Weathering of these caps has caused blocks to also fall down the sides of the gebels and it is these loose blocks that were utilised by the earliest Lower Palaeolithic tool makers and later grinding stone producers (Heldal 2009, 136-42) (Fig. 3).

The Aswan West Bank still provides us with a relatively well-preserved window into several episodes of quarrying for a range of products, covering a time depth from the Lower Palaeolithic into the Roman period. The key

Figure 6. Foreground: part of the Late Palaeolithic grinding stone quarries/workshop between low gebels of silicified sandstone. Background: the Wadi Kubbaniya.

properties of silicified sandstone such as durability and hardness made this the resource of choice for mobile groups of hunters in the Lower Palaeolithic. Evidence of Acheulean tool making industries for producing hand-axes and cleavers were located in several places and so establishing a time-depth to the first exploitation of the resource to at least 350,000 years ago with *Homo erectus/ Homo ergaster* (Bloxam *et al.* 2007; Heldal and Storemyr 2007, 72-77, 132-34; Bloxam and Moloney 2009). The stone's abrasive properties were clearly recognised by some of the earliest semi-permanent settlers in the Nile Valley for the grinding of early cultigens from at least the Late Palaeolithic, 18,400 – 18,100 BP (Wendorf and Schild 1989; Roubet 1989). Its use for grinding continuing into the early Roman period (30 BC) (Fig. 4).

Aesthetically, the stone's 'glittering' properties and its spectrum of colours from gold to purple were recognised by early elites, when cut and polished, as providing a beautiful medium for ornamental objects (Fig. 5). By the 3rd millennium BC (Old Kingdom, 4th Dynasty 2613 BC) silicified sandstone became one of Egypt's most sought-after stones for prestige royal objects and became symbolically associated with the rise of the solar gods (Quirke 2001, 76; Bloxam *et al.* 2007, 38-9, 42-3). This connection peaked during the early New Kingdom (mid 18th Dynasty 1390 BC) when there was a re-focussing of religious ideas to solarise the major cults of Egypt (Kozloff *et al.* 1992, 76, 110). Hence, the New Kingdom forms the longest and most intensive period of ornamental stone production, particularly between the reign of Amenhotep III and Seti I (1390 – 1279 BC). Its appeal for ornamental products gradually declined by the end of the New Kingdom, with sporadic consumption occurring into the Roman period.

Although today the non-silicified sandstone deposits are mainly used to build nearby Nubian houses, it is important to recognise that human exploitation of the Aswan West Bank's stone resources has been an almost continuous activity since the Lower Palaeolithic. Even during the resources' consumption for ornamental objects by the pharaonic era (3rd millennium BC), its use for grinding stones was always maintained into the early Roman period. Perhaps these earliest uses of the stone in Prehistory and its key role in the processing of food may have remained integral to the stone's name in antiquity, as written sources and hieroglyphs inscribed on some New Kingdom silicified sandstone objects refer to the stone as *inr n bnwt* - the primary meaning of *bnwt* being 'millstone' (Harris 1961, 75-6; Bloxam *et al.* 2007, 38-9).

Figure 7. Left: base silicified sandstone grinding stone with smooth concave surface; right: hand-held rubber (silicified sandstone) used for grinding. Late Palaeolithic settlement, Wadi Kubbaniya.

Unravelling the history of grinding stone production along the Aswan West Bank

Late Palaeolithic (18,400 - 18,100 BP)

Although the discovery of extensive grinding stone production along the Aswan West Bank was made during the 'QuarryScapes' project in 2005, the Late Palaeolithic (18,400 and 18,100 BP) precursor to this 'industry' had been identified 20 years previously (1980s) in the Wadi Kubbaniya (Roubet 1989, 589-608 in Wendorf and Schild 1989) (Fig. 1). Situated in the northernmost reaches of the Aswan West Bank, the grinding stone quarries or 'workshops' are located at the nearest source of silicified sandstone between a series of low-rise gebels, 200 metres north of the Late Palaeolithic settlements (Site E-82-1, Roubet 1989) (Fig. 6). Here, the stone occurs as loose boulders that were split to make thick slabs and then trimmed by re-touching with locally sourced hammerstones. Final shaping occurring in the settlements (*op. cit.*).

Representing one of the first occurrences of grinding stones in the Nile Valley, the procurement and use of these objects is directly related to fundamental changes in subsistence patterns during the Late Palaeolithic. Such changes have been related to a broadening of subsistence from just hunting of large mammals, to include a range of local (wild) floral resources such as wheat, barley and wetland nut-grass tubers that clearly required grinding (Wendorf and Schild 1989, 820-1). The grinding stones occur in both the settlement and quarries in a variety of shapes: oval/suboval, rectangular, triangular and other irregular shapes, which were primarily used as base-stones (site E-78-3, *op. cit.*). A small roughly squared hand-pebble of silicified sandstone or other locally sourced hard stone was then used to grind the material, causing many of the base-stones to have a smooth concave surface (Fig. 7). This is when the 'oval' or boat-shaped grinding stone emerges as probably the precursor to the later 'saddle quern' (Banks 1980, 241; Roubet 1989, 427-73). Even though this discovery was extremely important, these researchers did not realise that grinding stone production in the Wadi Kubbaniya was just

Figure 8. Predynastic grinding stone quarry represented by a sand-filled depression (marked with the scale) and surrounded with workings. Gebel es Sawan.

the 'tip of the iceberg' to an 'industry' that spanned 16,000 years and covered an area of almost 60 square kilometres.

Predynastic (4th millennium BC)

The grinding stone quarries discovered by Roubet (1989) in the Wadi Kubbaniya more or less set the standard in terms of how these objects were produced for several thousand years, into the Predynastic (4th millennium BC). Although by the Predynastic the centre of grinding stone production moves south-west and becomes concentrated at Gebel es-Sawan (Bloxam 2007, 9-10; Heldal and Storemyr 2007) (Fig. 1). 'Quarries' may be a rather over-stated term for these scattered areas of working loose silicified sandstone boulders, which are consequently some of the most difficult sites to visualise. These workings, or quarries, usually occur as small pits or sand-filled depressions surrounded by small amounts of waste, broken tools and oval shaped grinding stones (Fig. 8). Using similar production techniques as found in the Wadi Kubbaniya, a boulder measuring about one metre across was first worked by peeling the weathered exterior by hammering with locally sourced tools to produce a squared 'core'. This was then split along the sandstone bedding planes into one or a few rough-outs (Heldal and Storemyr 2007, 86-7). Similar to the Wadi Kubbaniya quarries, final shaping of the grinding stone would have occurred in the settlements.

Dating these quarries more accurately is difficult because unlike the Wadi Kubbaniya, there are no known settlements in the area that can be directly connected with them. There is also an absence of datable organic material in the quarries and ceramics are negligible. A Predynastic date is proposed from indirect evidence related to technology of production, grinding stone shape and from the amount of weathering (Heldal and Storemyr 2007, 97-102). In essence, we see a refining of the boat-shaped or oval form by the Predynastic, given their occurrence in settlement contexts at Hierakonpolis dating to the Naqada period (4th millennium BC) (Fairservis 1972, 11-12, Fig. 6; Hoffman 1972, 52). Although their provenance is unknown, this shape remains a well standardised form, which into the dynastic period became the 'rider' for the saddle quern (Heldal and Storemyr 2007, 100-2). In several instances as discussed below, rock art typical of the Predynastic is closely associated with these quarries. We can also observe that exploitation of loose and more accessible boulders in these 'low intensity' quarries did not significantly change until the dynastic period, when there is a transition to bedrock quarrying.

Dynastic to Early Roman Period (3rd millennium BC – 30 BC)

Production of grinding stones by exploiting massive bedrock and dividing of much larger blocks, rather than just the more accessible loose boulders, certainly occurred during the dynastic period (3rd millennium BC onwards). Determining exactly when this came about is difficult, although we can suggest this might be connected with the onset of silicified sandstone quarrying for prestige 'royal' objects that began by the Early Dynastic (*c.* 3000 BC). As ornamental quarrying for large objects intensified by the 4th Dynasty (2613 BC) and required sound, massive and good quality silicified sandstone, we can observe a shift in the location of quarrying onto the large gebels of Gebel Gulab and Gebel Tingar, one kilometre west of the Nile (Fig. 1).

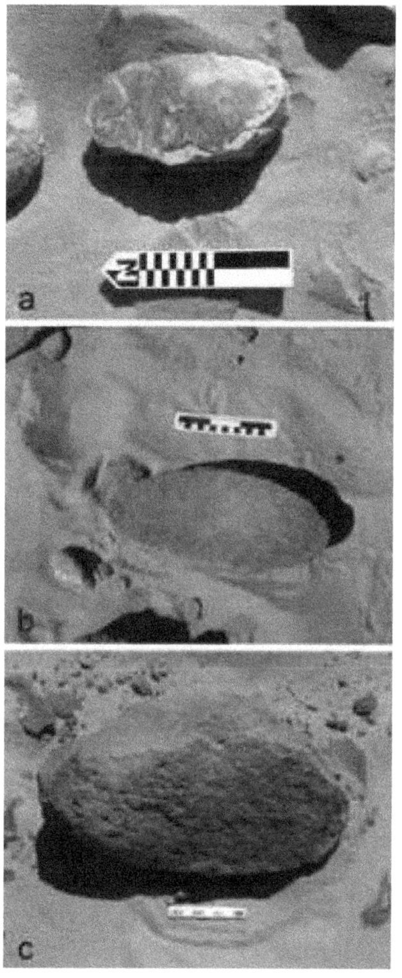

Figure 9. Grinding stone rough-outs: (a) Late Palaeolithic; (b) New Kingdom; (c) Ptolemaic – early Roman period.

The use of fire-setting to extract large blocks from massive bedrock was a feature of large-scale ornamental quarrying, a technique that was also used in grinding stone production, certainly by the dynastic period (Heldal *et al.* 2005; Heldal and Storemyr 2007, 102-16; Heldal 2009, 136-42). Given that bedrock quarrying probably required better quality tools, a much larger proportion of non-local stone tools appear in the quarries, these probably brought across from the East Bank granite quarries (Heldal and Storemyr 2007, 98; Heldal 2009, 141). Yet, despite the changes in method of primary extraction, secondary shaping and trimming of

Figure 10. Foreground: part excavated grinding stone quarry into bedrock; background: spoil heaps associated with other bedrock grinding stone quarries. Gebel Gulab.

Figure 11. Broken rough-outs of granite rotary querns (Roman Period) found in the East Bank Aswan granite quarries. .

the flakes with pounders into the traditional 'oval' shaped rough-outs transformed little from its earliest antecedents in the Late Palaeolithic (Heldal and Storemyr 2007, 98; Bloxam 2009, 172-5). Hence producing a diagnostically similar object over several thousand years (Fig. 9).

Bedrock quarrying heralded a significant change in terms of the size of grinding stone quarries, which at Gebel Gulab are deeper and so much more visible than those of the Predynastic. The landscape surrounding these quarries is dominated by large spoil heaps in amongst which numerous oval grinding stone rough-outs occur, often broken (Fig. 10). Although remembering that we see this landscape now as a 'totality' of grinding stone procurement spanning hundreds of years, nevertheless, the impression is of a considerable intensity in production by the pharaonic era. This increasing demand probably being related to the establishment of permanent settlements in Aswan and Elephantine Island. Silicified sandstone 'saddle querns' with the oval rider are known in these settlements since the Old Kingdom, but whether production exceeded local demand into wider networks of distribution remains uncertain, as discussed below.

Distribution and cessation of grinding stone production on the Aswan West Bank

Many important questions remain as to the extent to which grinding stones from the Aswan West Bank were distributed and traded to other places in Egypt. As opposed to research of millstone distribution across Europe, that can be more easily assessed through provenance studies (Carelli and Kresten 1997; Hansen 1997; Baug 2002; Belmont 2006; Grenne *et al.* 2008; Peacock and Cutler 2009), the numerous occurrence of silicified sandstone in Egypt hinders this task quite considerably. However, a petrological study of several grinding stones found in Predynastic settlements dating to Naqada I and II (3500 BC) at Hu-Semaineh in Upper Egypt were found to be imported from sources upwards of 150 km away from the Wadi Hammamat and Aswan regions (Mahmound and Bard 1993, 244-5). Even though local limestone was available these good quality grinding stones, produced from a range of igneous and metamorphic rocks, were clearly valued items that may have been incorporated into extensive long-distance trade and exchange networks by the Predynastic (*op. cit.*). Thus, we cannot discount distribution of Aswan West Bank grinding stones to other settlements in Egypt, such networks utilising the strategic location of the area not only via its proximity to the Nile, but also as the embarkation point of several desert routes into Nubia and the Western Desert oases (Goedicke 1981; Jaritz 1981; Storemyr 2007, 173-8; Storemyr *et al.* in press).

It is difficult to know exactly when the Aswan West Bank ceased to be at the centre of grinding stone production in the region. Evidence from the settlement excavations at Elephantine Island has revealed saddle querns with the typical oval-shaped riders from the Aswan West Bank in contexts from the Old Kingdom into the Ptolemaic Period (Fig. 4). With the advent of the rotary quern by the Roman Period, rough-outs of which can be seen in the East Bank granite quarries, it seems as if this transition led to the eventual phasing out of the silicified sandstone saddle quern and so their production on the Aswan West Bank (Fig. 11).

Figure 12. Typical shelter showing low-level dry-stone walls butted up against a rock outcrop, inside fragments of pottery. Gebel Gulab.

Who were the grinding stone quarriers?

In the absence of written records, determining the social context of grinding stone production in terms of labour force size and organisation can only be hypothesised through aspects of the social infrastructure in the quarries such as settlements, ceramics and other elements of material culture. Given the long time depth over which quarrying occurred, some periods are more represented by such evidence than others, as summarised below.

Settlements

Settlements directly associated with grinding stone quarries, apart from the direct association between the Late Palaeolithic semi-permanent settlements in the Wadi Kubbaniya, remain unknown. From the Predynastic onwards, the only features found in the quarries that could be termed temporary dwellings are small enclosures, often made up of three to four courses of dry-stone walls, usually butted up against a rock outcrop (Bloxam and Kelany 2007, 194-9) (Fig. 12). These occur most numerously on Gebel Gulab and Gebel Tingar, thus connecting them with both ornamental and grinding stone quarrying. Pottery is usually associated with these features and in many instances later re-use in the Roman Period means that predominantly New Kingdom pottery is discarded around the outside, with the later pottery found inside (Bloxam and Kelany 2007, 203-6; El-Senussi 2007, 254-75).

The function of these scattered shelters is difficult to determine, but given the ceramics associated with these places are minimal, and with a notable absence of cooking vessels and bread moulds, we can suggest that these were *not* dwelling places for any length of time (Bloxam and Kelany 2007, 194-9). It seems feasible to suggest that from the Early Dynastic onwards, the quarry workers travelled to the quarries on a daily basis from the permanent settlements at Elephantine Island and Aswan. Although given the little ceramic evidence and size/occurrence of the shelters, numbers involved in grinding stone quarrying were probably limited to small (less than 50?) local groups (Bloxam and Kelany 2007, 222-3; Bloxam 2007,12-15).

Whether numbers of quarry workers and the social organisation of production changed during periods of ornamental quarrying, particularly in the New Kingdom, is difficult to determine from the archaeological record. Although if such short-lived periods (in the New Kingdom 80 years) required larger numbers of people (into hundreds) and higher levels of social organisation, this is not attested in the quarries (Bloxam and Kelany 2007, 223-4). So perhaps we can argue that the social infrastructure we see today is related to grinding stone production more than anything else.

The inscribed landscape

Although the settlement and ceramic record is scant, inscribing of the landscape by successive generations of quarriers is the one continuous link that connects people with the Aswan West Bank landscape and its strategic resources (Bloxam 2007, 7-11; Bloxam 2009, 172-5). Studies of prehistoric quarries in European and Australian contexts have stressed the importance of ancestry and social networks being linked with the exploitation of key resources (McBryde 1997; Edmonds 1999; Cooney 1999; Bradley 2000). As meeting places and where social

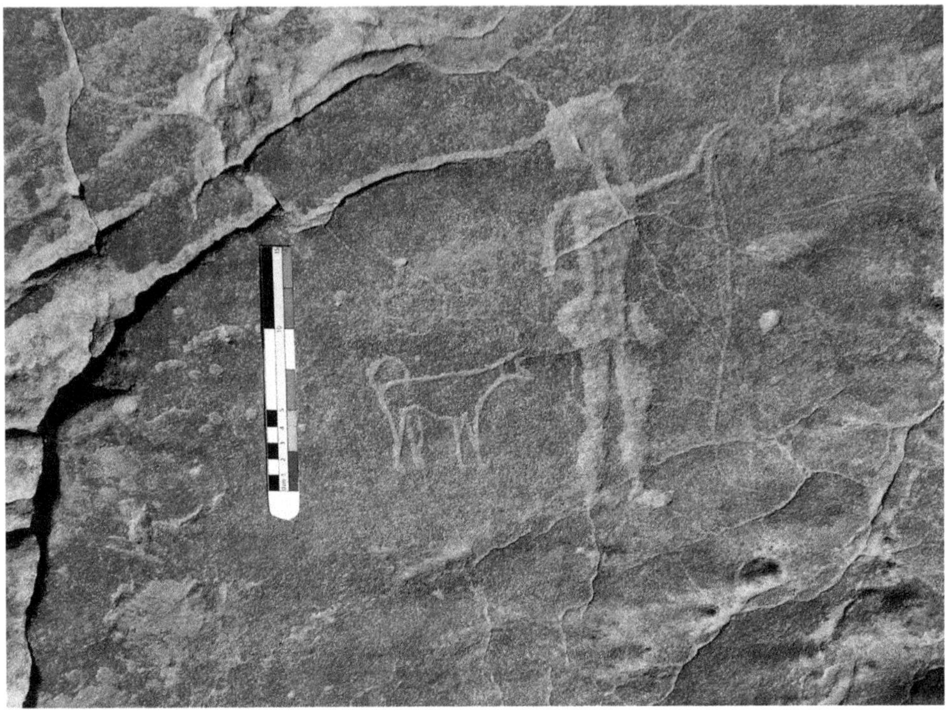

Figure 13. Rock art of Nubian A-group man (4th millennium BC) found close to Predynastic grinding stone quarries. Gebel es Sawan.

relations were reinforced, the role that quarrying played in this has only recently been recognised (Taçon 1991; Fullager and Head 1999; Boivin 2004; Bloxam and Heldal 2007; Bloxam 2009). Hence, these markers of presence by the inscribing of the landscape and its connection with grinding stone quarrying along the Aswan West Bank is beginning to emerge. This tradition may be related to the earliest grinding stone quarrying in the Wadi Kubbaniya, where there is evidence of highly patinated roughly pecked forms that may represent the earliest examples of this activity. Sound may also have been important, given the strategic placing of a stone slab on top of an in-situ outcrop that acts as a drum, emitting the sound directly into the grinding stone quarries.

Hundreds of previously undocumented panels of rock art were recorded during the 'QuarryScapes' survey (Bloxam *et al.* 2007; Storemyr 2007). Panels consisting of geometric designs, animals and people dating from at least the 5th millennium BC are often found close to grinding stone quarries (Storemyr 2007, 164-9; Bloxam 2007, 8-10; Storemyr 2008; Bloxam 2009, 173-4). In some instances, the representations of people are typical of Nubian A-group (4th millennium BC) and C-group (2nd millennium BC) (Fig. 13). As the Aswan West Bank lies on the cultural crossroads between Egypt and Nubia, then a Nubian ethnicity to the early quarry workers should not be ruled out. Recent excavations of a Predynastic settlement and burials at Nag el-Fugani, approximately 2 kilometres south-east of the Wadi Kubbaniya in the Nile floodplain, attest to a strong Nubian presence in the area (Gatto and Giuliani 2007; Gatto 2009). This Nubian identity may persist into the later Meroitic representations of the late 1st millennium BC attested at Nag el-Fugani (Bloxam and Kelany 2007, 212-4). Moreover, a Demotic inscription on an early Roman period *ostrakon* from Aswan describes the 'stone carriers' as 'people from Syene', 'people from the cataract' and most interesting is that the 60 names listed are all of Nubian *not* Egyptian origin (Rachel Mairs, pers. comm. 2007). Today, the Aswan West Bank still remains the domain of Nubians and one may speculate whether an ancestry to the earliest stone workers of the Palaeolithic can be traced.

In essence, we can hypothesise that traditions such as inscribing of the landscape and its link with early grinding stone production was also about linking local groups, through kinship ties, with the landscape as central in its socialisation over time (Bloxam 2009, 172-5; in press). Even into the pharaonic era inscribing of the landscape continues and notably retains a local flavour with depictions of local gods such as Khnum and Satet. Graffiti figures of people (in 'dynastic' style) and depictions of animals such as the baboon, which has antecedents to solar worship since the Predynastic (Pinch 2004, 113-4), were recorded in several New Kingdom quarries (Bloxam and Kelany 2007, 207-15).

Cross-culturally in prehistoric European and Australian contexts, inscribing of the landscape is attributed to the special place that exploitation of strategic resources played in linking past generations of quarriers with the present through ancestry (Taçon 1994; Bradley 2000). As well as quarries being nodes of social interaction, they may also be where transmission of technologies were passed down between local groups. Such practices could explain why we see little diagnostic or technological change in the production of grinding stones over a period of 16,000 years across the Aswan West Bank. Given that ornamental quarrying was only a minor activity, less than

11% within this long history of grinding stone production, then it is highly probable that quarrying for large objects was undertaken by the same groups who had an intimate knowledge of the resource and skill in how to extract it, transmitted over many generations (Bloxam in press).

Conclusion

Re-discovering the grinding stone quarries of the Aswan West Bank has been a major development in how we understand the exploitation of a key resource that had previously only been connected with ornamental quarrying in the pharaonic period. As the first survey undertaken of a grinding stone quarry landscape, this work has provided an important foundation for how we research ancient quarries in Egypt. In particular, the need to see ancient quarry landscapes in a broader perspective outside of procurement for just ornamental objects. Recognising that most strategic resources appropriated by elites usually have prehistoric origins, has important implications for future research of quarries that also needs to be looking for procurement of utilitarian objects.

This research of the Aswan West Bank and its transformation by quarrying for grinding stones can now be seen as having important implications in how we relate transitions in the processing of food, since the Late Palaeolithic, with the procurement of silicified sandstone. We have been able to trace how little grinding stone production changed over a period of 16,000 years, suggesting that technological transmission made have emerged through successive generations of local skilled kin groups who had an intimate knowledge of the resource. Even with the advent of ornamental quarrying by the 3rd millennium BC and quarrying into bedrock, this did not significantly change methods of secondary production. The end to this longevity of grinding stone production for saddle querns only seems to have occurred with the transition to the rotary quern, probably by the early Roman period.

Outcomes from this research of the Aswan West Bank grinding stone quarries can also give us a useful comparative context for current studies of Norwegian millstone quarries. Of particular interest in the study of Norwegian millstone landscapes, is exploring the extent to which exploitation of a resource for utilitarian objects may also make it desirable for other objects, in particular the ornamental that can often run side-by-side. Arguably, the original use of the resource and its importance in early food production may be an important symbolic factor in why it was a sought-after stone by 'elites'. Certainly there is a case to be argued in terms of silicified sandstone consumption in Egypt. Hence, cross-cultural comparison can be an important analytical tool in terms of how we fill data gaps, access the micro-level social context and develop strategies in the investigation of ancient quarries. In particular, how we go about articulating the significance and value of these difficult, hard-to-see, non-monumental landscapes in terms of their historical importance in understanding the social, political and economic transformations that were linked with the production of food.

Acknowledgments

I would like to thank the Supreme Council of Antiquities (SCA) in providing us with the opportunity to carry out this work in Egypt, as partners in the EU-funded 'QuarryScapes project' co-ordinated by the Geological Survey of Norway (NGU). Special thanks to Zahi Hawass and the Permanent Committee of the SCA, Mohamed el-Biely, Director of SCA Aswan, Mohi ed-Din Mustapha, Assistant Director of SCA Aswan and Magdy el-Ghandour, Director of Foreign Missions, SCA Cairo. Thanks are also extended to those colleagues who made enormous contributions during fieldwork in Egypt and whose work is integral to this paper: Tom Heldal, Per Storemyr and Reidulf Bøe (NGU); Adel Kelany (SCA documentation centre Aswan); Patrick Degryse (KUL Leuven) and Ashraf el-Senussi (SCA). To the Sixth Framework Programme (FP6) of the European Union for giving financial support to the project and also to the Leverhulme Trust for funding some aspects of the Egypt fieldwork. Thanks also to my colleagues involved in the 'Norwegian Millstone Landscape' project (funded by the Norwegian Research Council): Gurli Meyer, Tor Grenne, Tom Heldal, Irene Baug, Torbjørn Løland, Øystein Jansen and Tim Anderson.

References

Banks, K.M., 1980. The Grinding Implements of Wadi Kubbaniya. In A. Close (ed) *Loaves and Fishes: The Prehistory of the Wadi Kubbaniya* (assembled by F. Wendorf and R. Schild). Dallas, 3, 239-44.

Baug, I., 2002. Kvernstensbrota i Hyllestad. Arkeologiske punktundersøkingar i steinbrotsområdet i Hyllestad i Sogn og Fjordane. *Norsk Bergverksmuseums skrifserie*, 22, 113.

Belmont, A., 2006. *La Pierre à pain. Les Carrières de meules de moulins en France du Moyen Âge à révolution industrielle.* Grenoble.

Bloxam, E., 2007. *The Assessment of Significance of Ancient Quarry Landscapes – Problems and Possible Solutions. The case of the Aswan West Bank.* QuarryScapes report, Trondheim, Geological Survey of Norway (downloadable from www.quarryscapes.no).

Bloxam, E.G., 2009. New directions in identifying the significance of ancient quarry landscapes: four concepts of landscape. In N. Abu-Jaber, E. G. Bloxam, P. Degryse and T. Heldal (eds), *QuarryScapes: Ancient Stone Quarry Landscapes in the Eastern Mediterranean.* Norway, Geological Survey of Norway Special Publication, 12, 163-81.

Bloxam, E. (in press). Ancient quarries in mind: pathways to a more accessible significance. *World Archaeology* 43, 2.

Bloxam, E. and Heldal, T., 2007. The industrial landscape of the Northern Faiyum Desert as a world heritage site: modelling 'outstanding universal value' of 3rd millennium BC stone quarrying in Egypt. *World Archaeology,* 39.3, 305-23.

Bloxam, E. and Kelany, A., 2007. The material culture of the West Bank quarry landscape: constructing the social context. In E. Bloxam, T. Heldal and P. Storemyr (eds) *Characterisation Of Complex Quarry Landscapes: An Example From The West Bank Quarries, Aswan.* QuarryScapes report, Trondheim, Geological Survey of Norway (downloadable from www.quarryscapes.no), 183-226.

Bloxam, E. and Moloney, N., 2009. *Aswan West Bank Ancient Quarry Survey: Palaeolithic Tool Working Analysis.* Report to the Supreme Council of Antiquities, Egypt.

Bloxam, E., Heldal, T. and Storemyr, P., (eds) 2007. *Characterisation of Complex Quarry Landscapes: An Example From The West Bank Quarries, Aswan.* QuarryScapes report, Trondheim, Geological Survey of Norway (downloadable from www.quarryscapes.no).

Boivin, N., 2004. Rock art and rock music: petroglyphs of the south Indian Neolithic. *Antiquity,* 78, 38-53.

Bradley, R., 2000. *An Archaeology of Natural Places.* London.

Carelli, P. and Kresten, P., 1997. Give us this day our daily bread. A study of Late Viking Age and Medieval Quernstones in South Scandinavia. *Acta Archaeologica,* 68, 109–37.

Cooney, G., 1999. Social landscapes in Irish prehistory. In P. J. Ucko and R. Layton (eds) *Archaeology and Anthropology of Landscape: Shaping Your Landscape.* London, 46-64.

De Morgan, J., Bouriant, U., Legrain, G., Jéquier, G. and Barsanti, A., 1894. *Catalogue des monuments et inscriptions de l'Égypte antique, I, de la frontiére de Nubie a Kom Ombos.* Vienna.

Edmonds, M., 1999. *Ancestral Geographies of the Neolithic: Landscapes, Monuments and Memory.* London.

El-Senussi, A., 2007. Reports of pottery found at the Aswan West Bank, 2004 – 2007. In E. Bloxam, T. Heldal and P. Storemyr (eds) *Characterisation Of Complex Quarry Landscapes: An Example From The West Bank Quarries, Aswan.* QuarryScapes report, Trondheim, Geological Survey of Norway (downloadable from www.quarryscapes.no), 253-75.

Fairservis, W.A., 1972. Peliminary Report on the First Two Seasons at Hierakonpolis: I. *Journal of the American Research Center in Egypt,* 9, 7-28.

Fullagar, R. and Head, L., 1999. Exploring the prehistory of hunter-gatherer attachments to place: an example from the Keep River Area, Northern Territory, Australia. In P. J. Ucko and R. Layton (eds) *Archaeology and Anthropology of Landscape: Shaping Your Landscape.* London, 322-35.

Gatto, M.C., 2009. The Aswan area at the dawn of Egyptian history. *Egyptian Archaeology,* 35, 12-15.

Gatto, M.C. and Giuliani, S., 2007. Survey between Aswan and Kom Ombo. *Egyptian Archaeology,* 30, 6-9.

Goedicke, H., 1981. Harkhuf's Travels. *Journal of Near Eastern Studies,* 40.1, 1-20.

Grenne, T., Heldal, T., Meyer, G.B. and Bloxam, E., 2008. From Hyllestad to Selbu: Norwegian millstone quarrying through 1300 years. In T. Slagstad (ed) *Geology for Society,* Geological Survey of Norway Special Publication, 11, 47-66.

Habachi, L., 1960. Notes on the unfinished obelisk of Aswân and another smaller one in Gharb Aswân. In V. V. Struve and V. S. Golenishchev (eds) *Drevnii Egipet, Sbornik Statei.* Moscow, 216-35.

Hansen, A.M., 1997. Maritime perspektiv på kvernsteinsproduksjonen i Hyllestad. In H. Sørheim (ed) *Arkeologi og kystkultur, 25 and 26 October 1997.* Ålesund, 58-63.

Harris, J.R., 1961. *Lexicographical Studies in Ancient Egyptian Minerals.* 54, Berlin.

Heldal, T., 2009. Constructing a quarry landscape from empirical data. General perspectives and a case study at the Aswan West Bank, Egypt. In N. Abu-Jaber, E. G. Bloxam, P. Degryse and T. Heldal (eds), *QuarryScapes: Ancient Stone Quarry Landscapes in the Eastern Mediterranean.* Norway, Geological Survey of Norway Special Publication 12, 125-153.

Heldal, T., and Storemyr, P., 2007. The quarries at the Aswan West Bank. In E. Bloxam, T. Heldal and P. Storemyr (eds) *Characterisation Of Complex Quarry Landscapes: An Example From The West Bank Quarries, Aswan.* QuarryScapes report, Trondheim, Geological Survey of Norway (downloadable from www.quarryscapes.no), 69-140.

Heldal, T., Bøe, R. and Müller, A., 2007. Geology and stone resources of the Aswan West Bank. In E. Bloxam, T. Heldal and P. Storemyr (eds) *Characterisation Of Complex Quarry Landscapes: An Example From The West Bank Quarries, Aswan.* QuarryScapes report, Trondheim, Geological Survey of Norway (downloadable from www.quarryscapes.no), 51-67.

Heldal, T., Bloxam, E., Storemyr, P. and Kelany, A., 2005. The geology and archaeology of the ancient silicified sandstone quarries at Gebel Gulab and Gebel Tingar, Aswan, Egypt. *Marmora: International Journal for Archaeology, History and Archaeometry of Marbles and Stones* 1, 11-35.

Hoffman, M.A., 1972. Peliminary Report on the First Two Seasons at Hierakonpolis: Part IV, Test Excavations at Locality 14. *Journal of the American Research Center in Egypt,* 9, 49-66.

Jaritz, H., 1981. Zum Heiligtum am Gebel Tingar. *Mitteilungen des Deutschen Archaologischen Instituts Abteilung Kairo,* 37, 241-6.

Klemm, R. and Klemm, D., 1993. *Steine und Steinbrüche im Alten Ägypten.* Berlin and Heidelberg.

Klemm, D. and Klemm, R., 2008. *Stones and Quarries in Ancient Egypt.* London.

Kozloff, A.P., Bryan, B.M. and Berman, L.M., 1992. *Egypt's Dazzling Sun: Amenhotep III and his World.* Cleveland.

Mahmoud, A. and Bard, K., 1993. Sources of the Predynastic Grinding Stones in the HuSemaineh Region, Upper Egypt, and their Cultural Context. *Geoarchaeology,* 8.2, 241-5.

McBryde, I., 1997. The landscape is a series of stories. Grindstones, quarries and exchange in Aboriginal Australia: a Lake Eyre case study. In A. Ramos-Millán and M. A. Bustillo (eds) *Siliceous Rocks and Culture.* Granada, 587-607.

Peacock, D. and Cutler, L., 2010. A Neolithic Voyage. *The International Journal of Nautical Archaeology,* 39.1, 116-24.

Pinch, G., 2004. *Egyptian mythology : a guide to the gods, goddesses, and traditions of ancient Egypt.* Oxford.

Quirke, S., 2001. *The Cult of Ra: Sun-worship in Ancient Egypt.* London.

Roubet, C., 1989. Report on Site E-82-1: A Workshop for the Manufacture of Grinding Stones at Wadi Kubbaniya. In F. Wendorf and R. Schild (eds) *The Prehistory of Wadi Kubbaniya,* 3, 589-608. Dallas.

Storemyr, P., 2007. Overview of rock art, stone alignments, desert routes and a possible hermitage at the West Bank of Aswan. In E. Bloxam, T. Heldal and P. Storemyr (eds) *Characterisation Of Complex Quarry Landscapes: An Example From The West Bank Quarries, Aswan.* QuarryScapes report, Trondheim, Geological Survey of Norway (downloadable from www.quarryscapes.no), 163-81.

Storemyr, P., 2008. Prehistoric geometric rock art at Gharb Aswan, Upper Egypt. *Sahara,* 19, 61-76.

Storemyr, P., Bloxam, E., Heldal, T., and Kelany, A., (in press). Ancient desert and quarry roads on the West Bank of the Nile in the First Cataract region. In H. Riemer (ed.) *Desert Road Archaeology.* University of Cologne.

Taçon, P.S.C., 1991. The power of stone: symbolic aspects of stone use and tool development in western Arnhem Land, Australia. *Antiquity,* 65, 192-207.

Taçon, P.S.C., 1994. Socialising landscapes: the long-term implications of signs, symbols and marks on the land. *Archaeology in Oceania,* 29, 117-29.

Wendorf, F. and Schild, R., (eds) 1989. *The Prehistory of Wadi Kubbaniya,* 1-3. Dallas.

Manufacturing rotary querns in the 4th century BC fortified settlement of Els Vilars (Arbeca, Catalonia, Spain)

Natàlia Alonso, Meritxell Aulinas, Maria Teresa Garcia, Félix Martín, Georgina Prats and Sílvia Vila

Transalted by Timothy Anderson

A group of thirty discarded rotary querns uncovered during excavations at the Iron Age site of Els Vilars provide clues as to the sequence of production of this type of early grain mill. This paper presents the typology of these querns, the source of their petrography, and the different steps in their production sequence at Els Vilars, a fortified settlement located on the Western Catalonian Plain. Although the first phase of occupation of this site dates to the 8th century BC, the production of these hand-operated querns and larger mills, possibly driven by two individuals, dates to its last phase of occupation during the 4th century BC.

1. The fortified Iron Age site of Els Vilars

"Els Vilars d'Arbeca" is located in western Catalonia to the north-east of the present village of Arbeca (Lleida, Catalonia). The site, about 300 m above sea level, is established on an extensive alluvial plain formed by deposits of the different various tributaries of the Segre River (Fig. 1). The archaeological excavation is a research project conducted by the University of Lleida, that began in the late eighties. It is a high profile project due to the spectacular results and the many interesting conclusions that have been reached. In the framework of archaeological research in Catalonia and, in particular, the western Catalonian plain, this site reveals facets that are typical of the model of socio-economic transformation observed in this area throughout the Iron Age. Additionally, the defensive features of Els Vilars are unusual and there are no known parallels among other contemporary sites both in the Iberian Peninsula and elsewhere in Europe. The fortified features make up 80% of the constructed surface and consist of five major defensive elements: a wall, 14 wall towers, a *chevaux-de-frise* (a barrier of sharpened rocks), a moat, and a fortified gate (Fig. 2) (GIP 2003; Alonso *et al.* 2010).

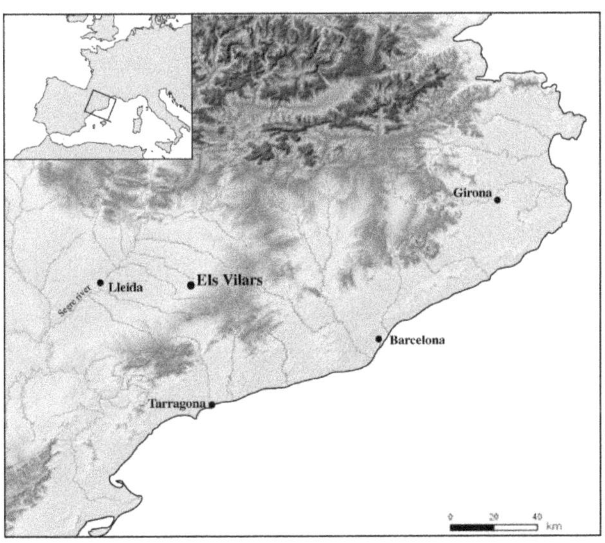

Figure 1. Location of the fortress of Els Vilars.

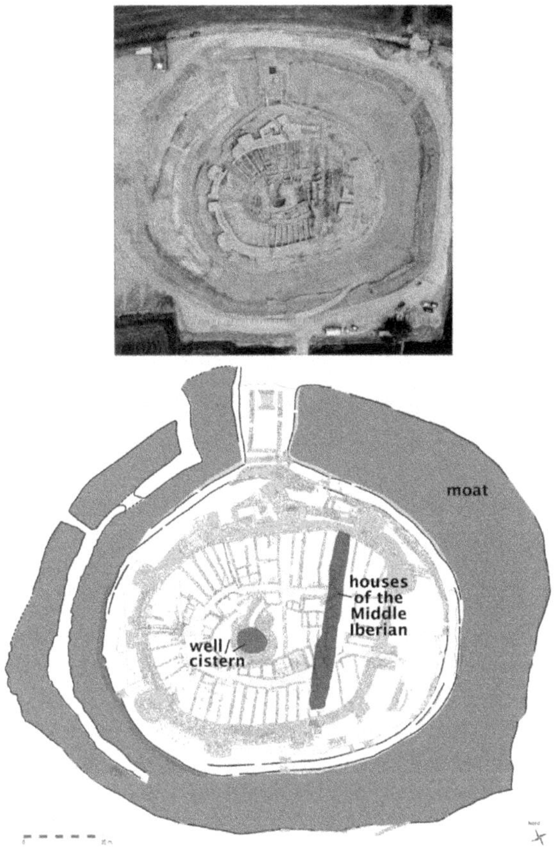

Figure 2. Aerial photograph and plan of the Els Vilars fortification. The plan highlights the features and areas of the Vilars III and IV phases where the rotary querns presented in this article were conserved.

Figure 3. Rotary querns reused as construction material (a) in the wall of the moat and (b) in a wall of a house at Els Vilars.

The occupation of the site spans a large part of the First Iron Age and the Iberian Culture, between 800/775 cal BC and 350/325 BC. Its stratigraphy reveals five major phases:

- Vilars 0 (800/775 cal. BC - 700/675 cal. BC) and Vilars I (700/675 cal. BC – 550/525 BC) belong to the First Iron Age.
- Vilars II (550/525-450/425 BC) belongs to the early Iberian Culture.
- Vilars III (450/425-375 BC) and Vilars IV (375-350/325 BC) belong to the middle Iberian Culture.

The Vilars III and IV phases saw a significant remodelling in the urban features, especially the cistern/well and the defensive system, with the construction of a massive moat (13 to 25 m wide and 3.60 to 4.80 m deep) filled with water. Its slopes were coated with stones and the access to cross it to reach the northern gate was fortified (GIP 2003; Alonso *et al.* 2010). Yet, these two phases of the middle Iberian Culture – which include the rotary querns under study - are only conserved in certain areas of the site due to recent farm work that has destroyed the upper levels. These areas are limited to (1) a narrow surface (oriented N-S) located toward the middle of the eastern sector and (2) the central area of the site, where a wide underground cistern/well was built presumably on the remnants of an earlier cistern, and (3) in the moat (Fig. 2).

The fortress is representative of the centre of power, exalting the might and prestige of an aristocratic lineage. It was also an instrument of military domination of the territory and a means of ideological coercion of its own inhabitants, as well as a force for dissuasion for potential enemies among neighbouring settlements.

The principal lines of research of the Els Vilars project are the following: the depiction of the historical-archaeological development of the fortress and its inhabitants over a period of almost 400 years (Alonso *et al.* 1998; 1999; 2010; GIP 2003; Junyent *et al.* 2009); the reconstruction of the early ecology and economy (Alonso 1999; Alonso *et al.* 2008; Ferrio *et al.* 2007; Nieto *et al.* 2010), the restoration of the structure, and the dissemination of information to the public (Alonso *et al.* 2000; Vidal *et al.* 2007).

2. The querns: typology and petrography

Research on the site of Els Vilars therefore permits us to observe the development of these grinding mechanisms between the 8th and the 4th century BC. Regrettably, the number of querns is unequal from one phase to another and most were unearthed in secondary positions such as in walls, construction backfills or surface finds (Fig. 3).

Mills from the phases of the site corresponding to the first Iron Age (Vilars 0 and Vilars I phases) are exclusively saddle querns (Fig. 4). This type also dominates the early Iberian

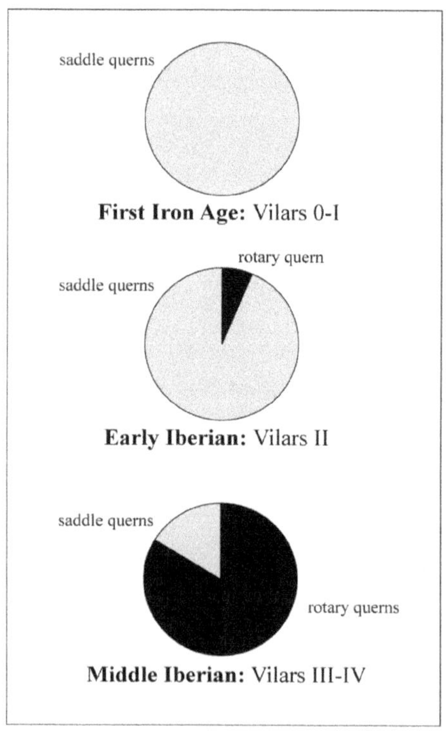

Figure 4. Comparison of the proportions of saddle querns and rotary querns in the different chronological phases of Els Vilars.

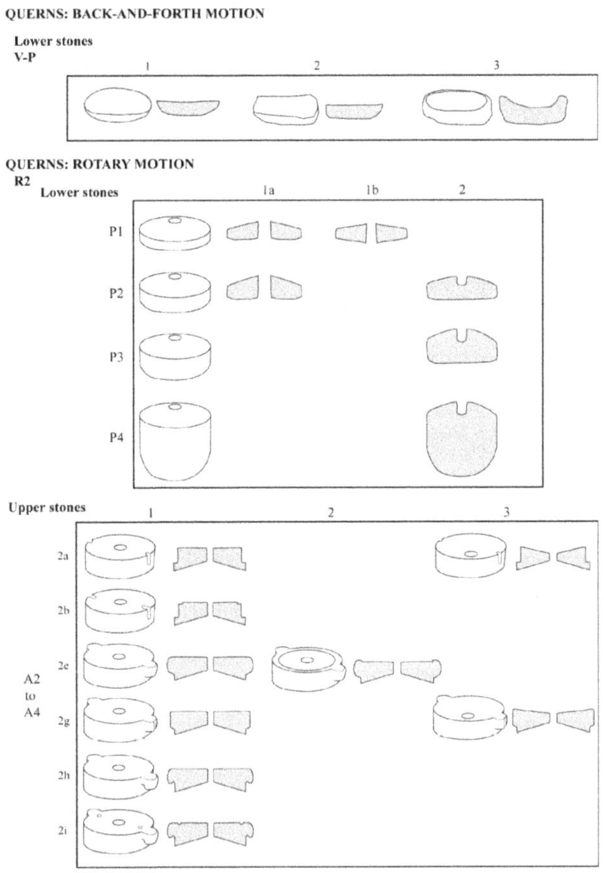

Figure 5. Typological diagram of the back-and-forth (saddle querns) and rotary querns of Els Vilars.

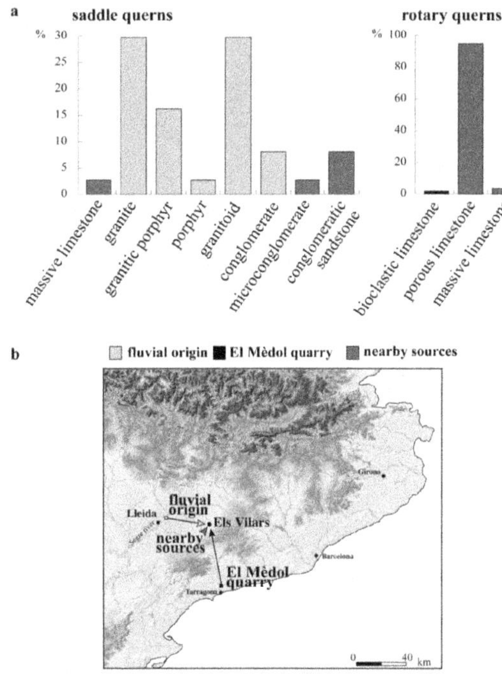

Figure 6. (a) Percentage of the different lithological types of saddle querns and rotary querns of Els Vilars (b) and sources of the different raw materials for their production.

Culture (Vilars II). It is during this phase, in the middle of the 5th century BC, that the rotary quern – represented by one single stone – first appears on this site. It is only later, during the middle Iberian Culture, that hand-operated rotary querns became the predominant milling tool. The earlier back-and-forth type, nonetheless, does not vanish (Alonso 1999). Most of the saddle-shaped finds consist of the larger lower stones of this mechanism. These fall into any of three morphological types (VP-1, VP-2, or VP-3). VP-2 is by far the most common, and VP-3 with a deeper, hollowed-out, working surface is a single find from the Vilars III phase (Fig. 5). The petrography of these saddle querns is heterogeneous. Those of igneous composition are the type most often represented (granites and other granitoids as well as granitic porphyr and other porphyrs of different composition). Some were hewn from detrital sedimentary conglomerates and micro-conglomerates, while only a few were hewn from sandstone conglomerates and massive limestones (Fig. 6).

The source of most of the Els Vilars saddle querns (e.g. igneous rocks) is most likely to be found along the terraces and the shores of the Segre River, a tributary of the Ebro River about 30 km away proceeding from the Pyrenees. These stones were either collected by the inhabitants of Els Vilars themselves or delivered through intermediaries.

The oldest case of a quern driven with the rotary movement on the site is the single find mentioned previously coming from a layer dating to the middle of the 5th century BC (VI-21, Fig. 7). It is a remarkably well-crafted lower stone bearing a very particular bi-conical morphology. This type of section could either indicate reversible grinding surfaces or simply that the original base was transformed into a working surface when the first upper working surface was worn. For the moment the only known parallel is a quern from Bad Nauheim (Wetteraukreis, Hessen) in Germany (Weffers in press). In any case, the Catalonian stone is carved from a fossiliferous limestone typical of outcrops from the Camp de Tarragona located about 70 km away from Els Vilars, in a region where several Roman, or possibly earlier, quarries (Mèdol and Lorito) have been identified (Fig. 6).

As we have noted earlier, the use of rotary querns in the 4th century BC became widespread. According to their aspect and morphology, however, they appear to be less well-crafted than the early examples. The majority (96%) were hewn from porous limestone layers while only 4% were scored from a more compact, massive limestone (Fig. 6). Later in this paper, we will examine the sources of all of these stones, most likely extracted from nearby outcrops. Since a large proportion of the finds were left incomplete, they do not bear all of the different handle and driving fittings, the criteria used to determine their typology. However, several types have been established for the later middle Iberian Culture. These are reflected in Fig. 5, a revised typology (including the most recent Vilars finds) adapted from previous quern research in southern

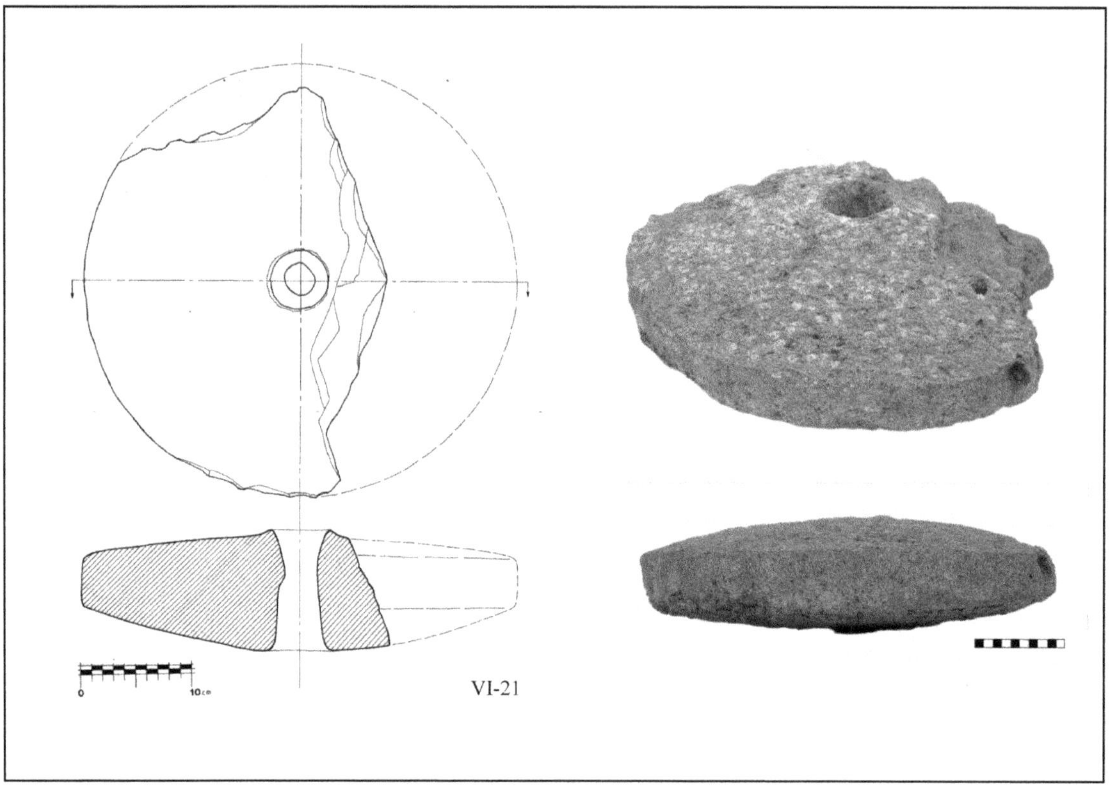

Figure 7. Rotary quern VI-21 from the middle of the 5th c. BC d (Vilars II phase). Drawing: Antoni Llussà i Guasch.

France and north-eastern Spain (Py 1992; Alonso 1999; Equip Alorda Park 2000; Portillo 2005). However, we will not delve into the details of this subject since this is not the objective of this paper.

In general, grinding surfaces show an inclination superior to that of 10° (R2). Lower stones are on the whole thick (P3: thickness/diameter x 100 between 35 and 70) to very thick (P4: thickness/diameter x 100 > 70) and bear a blind (partially perforated) eye (P3-2 and P4-2). There are nonetheless cases of thin lower stones bearing completely perforated eyes (P1-1) and cases of stones of medium thickness with both types of eyes (P2-1 and P2-2).

Upper stones are either medium thick (A2: thickness/diameter x 100 between 25 and 35), thick (A3: thickness/diameter x 100 between 35 and 70), and very thick (A4: thickness/diameter x 100 > 70). Their upper surface is usually flat (1); one case has a peripheral rim (2); a few cases have upper surfaces inclined towards the eye (3). They always bear either two opposite protruding handle lugs or handle cuttings. The fittings (probably for wooden handles) are classified as follows (Fig. 5): (2a) vertical cuttings; (2b) horizontal cuttings (there is only one of this type, so the cutting could be a repair); (2e) protruding handles (lugs) fashioned directly into the stone; (2g) lugs bearing a vertical cutting; (2h) lugs bearing a horizontal cutting at their base (possibly to adjust a cord around the handle); (2i) lugs bearing both a horizontal cutting at their base and shallow vertical slots on their upper surface.

3. Evidence of the manufacture of rotary querns

The presence among the grinding stones of an important number of unfinished and discarded rotary querns in different stages of production, in addition to that of rough, unworked cylindrical blocks, proves without a doubt that rotary querns were produced at Els Vilars during the 4th century BC. The rejected models were sometimes recycled as construction material in both the walls of the houses of the Vilars III phase (450/425-375 BC) or in the coating material of the moat, both in the phases of Vilars III and IV (375-350/325 BC) (Fig. 3). They are also found in the backfills of the cistern and the moat in the Vilars IV phases. Since it is not possible to observe significant typological differences from one phase of the middle Iberian Culture to another, they are treated as a whole.

3.1. Raw materials and the location of the stone outcrops

Based on the lithology of the querns (Fig. 6) and the availability of the raw material, five major stone types have been determined:

(1) porous limestone
(2) massive limestone
(3) micro-conglomerate
(4) conglomeratic sandstone
(5) fluvial rocks

The last group (5) corresponds to different rock types (including rocks of igneous composition) associated with deposits along the Segre River. The first four groups, on the other hand, are found as outcrops near the site and, therefore, require a more detailed description (Fig. 8).

From the geological point of view, Els Vilars is located in an area dominated by Pleistocene deposits (Fig. 8, Qvpu). In the surrounding areas there are also Oligocene and Holocene outcrops which include lutites, sandstones, marls, limestones, as well as dolostones and gypsum). The extraction of dolomite and gypsum sedimentary rocks by the settlers of Els Vilars was not only destined for quern production, but for plaster for walls or for other defensive features. Nonetheless, although no extractive ancient quarry has been identified in the surrounding outcrops, we presume that some were exploited for the quern production:

(1) *Porous limestone* (POmc4)
This type of limestone is a biomicrite with calcareous organism remains (gastropods and ostracods). According to the classification of Dunham (1962), carbonated sedimentary rocks of this type may be classified as grainstones and wackestones. In general, this facies exhibits an intense bioturbation and moldic porosity related to dissolution by digenetic processes (see, for example, the detail in Fig. 9). The resulting lack of uniformity of the degree of porosity explains why some querns of the same facies seem to be more porous than others. On the other hand, the porosity has been altered and fully or partially silted due to its condition of being buried underground. The potential outcrops of this type of porous limestone are found to the south and the south-east of Els Vilars. Details of this lithology can be observed in Fig. 8.

(2) *Massive limestone* (POmg3)
This rock is a micritic limestone with a low fossil and quartz content that is materialised on the surface in the form of easily detachable, uniform layers less than a metre thick. These properties explain its use in the construction

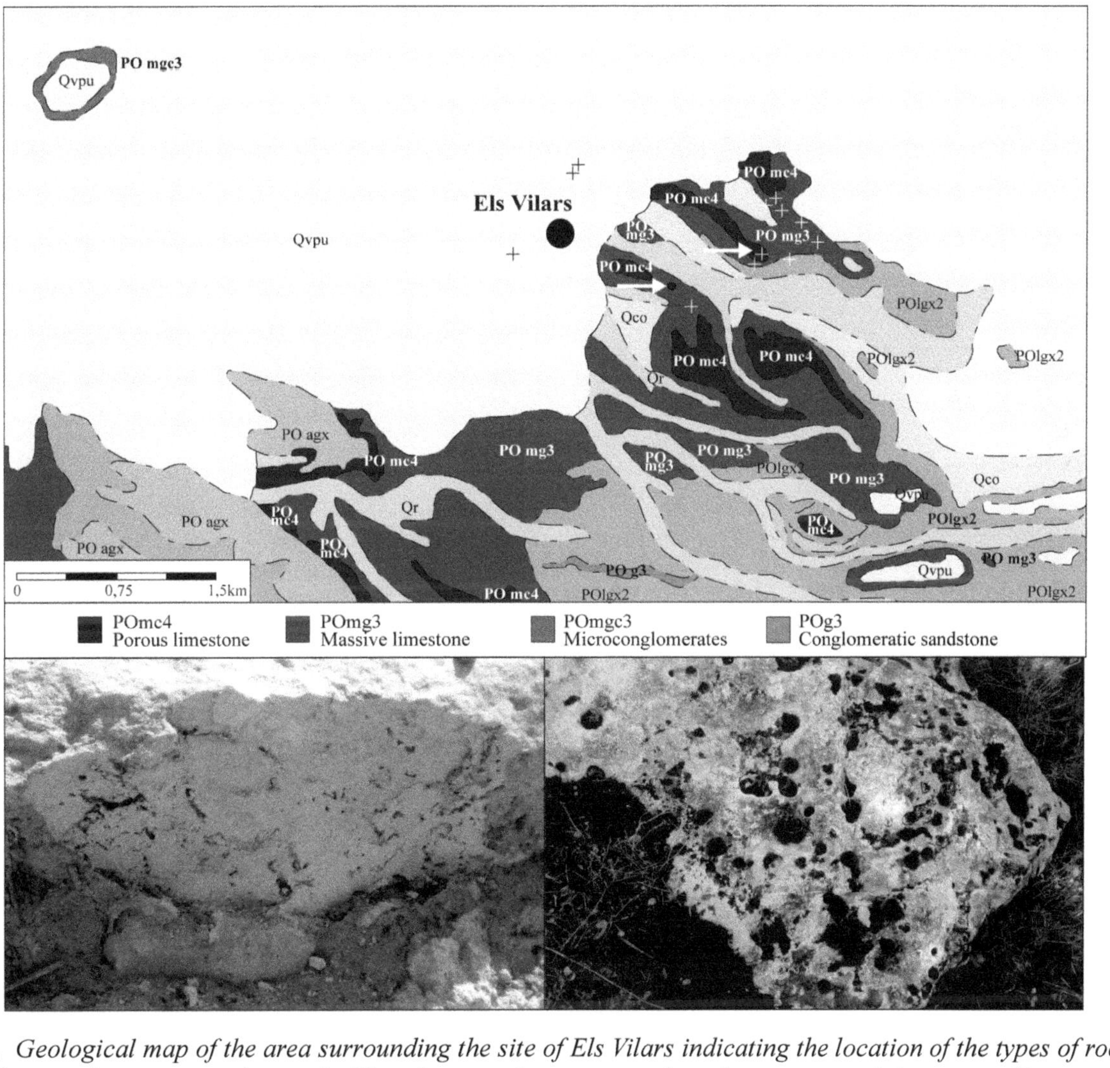

Figure 8. Geological map of the area surrounding the site of Els Vilars indicating the location of the types of rock used to manufacture the querns under study. The photographs correspond to the outcrops of the porous limestone facies (POmc4), the type rock most often used in the manufacture of rotary querns. The location of the photographed outcrops is indicated by the white arrows. The relief of the terrain is now possibly covered by quaternary deposits and/or farmland topsoil, sediments that would not have existed in the Iberian Culture period.

of walls at Els Vilars. The source of this limestone type has been identified in neighbouring areas to the south and south-east of the settlement. This massive limestone is generally interspersed with either sandy or micro-conglomeratic layers (Fig. 8). The small number of querns of this rock type recovered at Els Vilars suggests that the limestone of massive type from the surrounding outcrops was rarely exploited for rotary querns. This is probably due to its tendency to suffer from weathering.

(3) *Microconglomerates* (POmgc3)
Only one quern of microconglorametaic composition has been identified. The inherent anisotropy of this rock is probably the reason why it was not used on a regular basis by the mill makers of Els Vilars. A very specific outcrop is known to the northeast of Els Vilars (Fig. 8).

(4) *Conglomeratic sandstone* (POg3)
This small group consists of coarse textured sandstones interbedded with microconglomerate layers of the preceding facies (Pomgc3). There are very modest outcrops of this rock-type to the south of Els Vilars (Fig. 8).

As noted above, the millstone makers showed a clear preference for porous limestones, far beyond that of the other lithologies. They probably gave priority to this porous stone because it generated a higher degree of friction during the grinding process. In addition, it was also probably the only adequate stone locally available (Fig.8). It is not, however, a high quality material due to the presence of many micro-fissures (Fig.10) that provoked numerous unexpected breaks during production, as evidenced by the high quantity of broken and stray quern fragments, some of which were re-used as construction material.
Since we ignore the precise location of the quarry, we cannot define exactly the different steps of the extraction process. It would seem logical that the stones were extracted in the form of angular blocks, easily detached from the horizontal strata. Subsequently, rough 'drums' were cut from these angular blocks, like the cylinders recovered on the site (Fig. 10). This first stage of cutting might have taken place either at the quarry or at the settlement. In passing, let us state that the outcrops were not sufficiently homogenous and compact to permit cylinders to be scored directly from the bedrock, leaving circular imprints, as is the case of

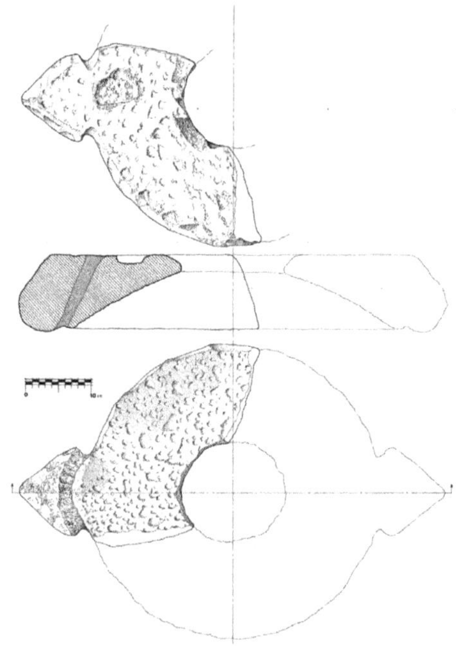

Figure 9. Drawing of quern V-11 evoking the porosity of the rock. Drawing: Antoni Llussà i Guasch.

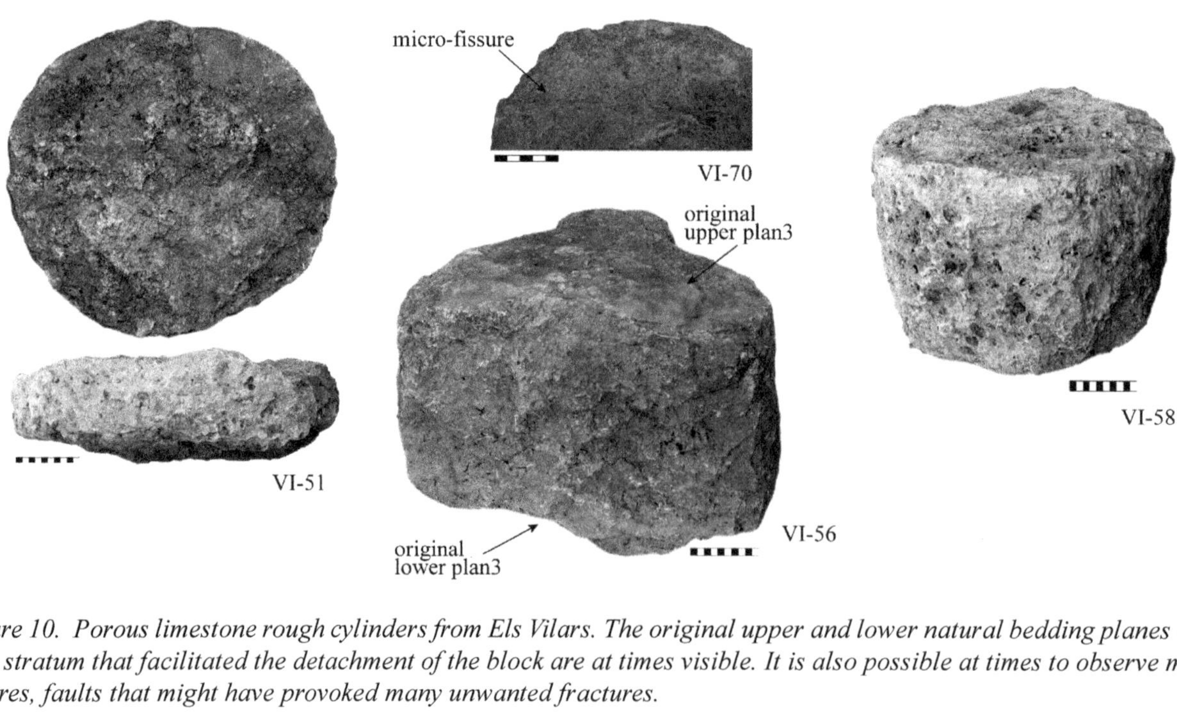

Figure 10. Porous limestone rough cylinders from Els Vilars. The original upper and lower natural bedding planes of the rock stratum that facilitated the detachment of the block are at times visible. It is also possible at times to observe micro-fissures, faults that might have provoked many unwanted fractures.

ROUGH CYLINDERS

No.	Ø	H	H/Ø x 100	H type	Without lugs	With lugs	Direction of fracture
VI-47	39	28	72	4	•		No fracture
VI-49	/	35	/	/		•	Diagonal from top
VI-50	43	29	67	3	•		No fracture
VI-51	35	24	69	3	•		No fracture
VI-56	45	33	73	4		•	Vertical
VI-58	38	23	61	3	•		No fracture
VI-67	33	33	100	4	•		No fracture
VI-70	33	30	91	4		•	Vertical
VI-71	32	44	138	4	•		No fracture

LOWER STONES

No.	Ø	H	H/Ø x 100	H type	Unfinished eye	"Blind" eye	Hollowed grinding surface	Direction of fracture
VI-30	35	45	129	P-4	•			Vertical
VI-46	38	35	92	P-4		•	?	Vertical
VI-48	36	54	150	P-4	•			Diagonal from top
VI-52	30	38	127	P-4	•		?	Diagonal from top
VI-53	35	44	126	P-4		•	•	Diagonal from top
VI-54	33	30	91	P-4		•	•	Diagonal from top
VI-65	/	/	/	/	?	?	•	Vertical
VI-73	42	17	40	P-3	•		•	Vertical

UPPER STONES

No.	Ø	H	H/Ø x 100	H type	Unfinished eye	Completed eye	Hollowed grinding surface	Lugs	Vertical cuttings	Lug with vertical cuttings	Direction of fracture
VI-26	35	15	43	A-3							Vertical
VI-30	35	12	34	A-2	•			•			Vertical
VI-32	51	25	49	A-3		•	•	•			Diagonal from top
VI-34	37	30	81	A-4		•		•			Diagonal from top
VI-35	39	30	77	A-4		•		•			Diagonal from top
VI-45	37	28	76	A-4	•			•			Vertical
VI-55	42	34	81	A-4	•		•	•		•	Diagonal from top
VI-57	34	32	94	A-4		•			•		Vertical

Figure 11. Table compiling data of rejected rough cylinders and querns (fractured during manufacture) at Els Vilars.

certain 'true' extractive quarries in Antiquity (Anderson *et al.* 2003; see other related articles by different authors in these proceedings).

The thickness of the limestone stratum, therefore, not only facilitated the extraction process, but determined the thickness of the roughout. Any attempt to cut the rock along a horizontal plane surely resulted in an unwanted fracture. In the case of several roughouts, and even several finished querns, it is possible to recognise both the original upper (exposed to air) and lower plane of the natural stratum (Figs. 10, 11 and 12).

3.2. The production sequence based on the discarded rotary querns

As we have just noted, it is difficult to determine whether the first stage of fashioning the angular block into a cylinder took place at the fortification or at the quarry. There are arguments that bolster both options. The fact that a carved cylinder is lighter than an angular block is a solid argument in favour that the roughout was shaped at the outcrop. The weight difference is, nonetheless, not significant when compared to the risk of damage during transport, especially for the more fragile cases of cylinders bearing rough handle protuberances. A second argument in favour of the option of carving the 'drum' at the quarry site is based on the find at Els Vilars of six whole rough cylinders bearing no trace of further carving. If the process at the site were to have begun with an angular block - and continued without interruption through to the final finishing touches of the quern – this high number of intact cylinders would never have been unearthed. These 'drums,' presumably already shaped at the quarry site, could thus be explained as stored roughouts 'waiting' to be finished.

In six of the nine rough cylinders recovered, it is not possible to determine whether they were destined as upper or lower stones (table of Fig. 11). For example, cylinders of simple morphology could ultimately have been carved into lower stones or upper stones equipped with cuttings, such as VI-51 or VI-58 (Fig. 10). Cylinders bearing rough lateral lugs, on the other hand, were logically destined as upper stones (Fig. 10, for example VI-56). During the process of carving, the damaged roughouts were either discarded or reused as construction material. The total of 32 examples found at Els Vilars can be classified as follows (Fig. 11): 9 rough cylinders (6 whole and 3 fragmented), 8 lower stones, 8 upper stones, and 7 unidentified fragments split in half vertically and reused in the defensive walls of the moat. This last lot is not included in the table.

The definition of the subsequent steps of the manufacturing sequence, after the shaping of the roughout, is based mainly on discarded querns (Fig. 12). These steps are completed with observations of finished querns that saw use (not included in the table of Fig. 11). As we have noted, in the quarry, the block was selected for the desired thickness of

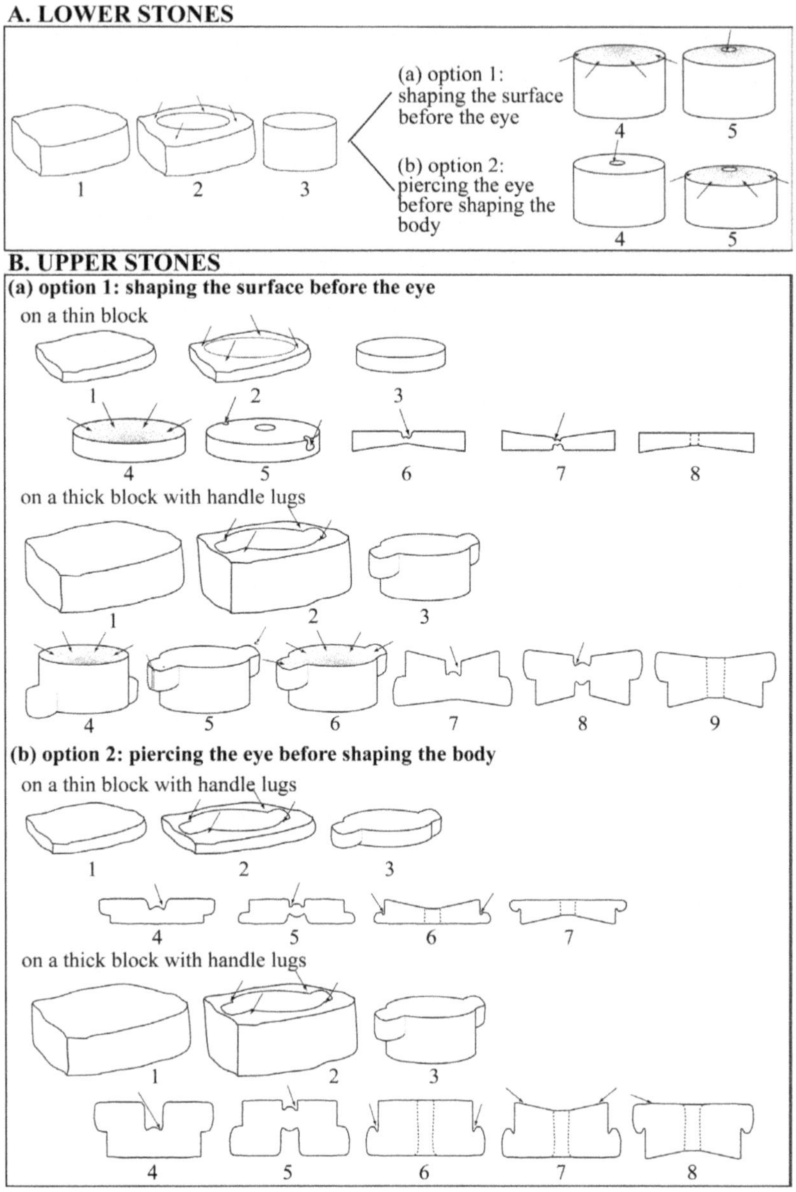

Figure 12. Production sequence of rotary querns in the 4th c. BC based on the abandoned quern fragments at Els Vilars.

the quern. The key step at this stage - the most delicate and the cause of most of the accidents - was the piercing of the eye of both upper and lower stones.

• *Lower stones*

The process of manufacture of the lower stone is quite simple (Fig.12 A). After shaping the rectangular block into a cylinder (steps 1 to 3), there are two options: (a) the inclined working surface is carved before the eye is pierced as in the case of VI-52, where it is clearly visible that the grinding surface was already hollowed when it was damaged during the perforation (Fig. 13). The second option (b) is that the piercing precedes the carving of the grinding surface. This is the case of VI-48 (Fig. 13), a large broken rough cylinder showing no work except for that of the eye. Of the seven cases, five were fashioned following the first sequence and two following the second. In addition, the perforation of the eye sometimes provoked either vertical or diagonal fractures, resulting in either 'half drums' or defective blocks with intact bases.

• *Upper stones*

Since the upper stones feature many types of driving fittings, their steps of manufacture are more complex to define than that of the lower stones. The first step, as expected, consists of carving the cylinder from an angular block with or without lateral handle lugs (B.1 to B.3). Their height, as we have noted, is that of the stratum from which the original angular block was detached. As in the case of lower stones, the next step varies (Fig. 12 B) according to when the eye is pierced.

(a) The first option is to hollow the lower grinding surface (B.a.4), independently of whether the original block is thin or thick or equipped with handles. This is followed by the carving of the other fittings, for the handles as in the case of VI-69 (Figs. 14 and 15), or the carving for the handle slots in the lugs, as for VI-55 (B.a.5), and, if it were necessary, the carving of a 'hopper' on the upper surface, such as VI-55, to facilitate the feed of the grains (Figs. 14 and 15). The final step of this option is to pierce the eye starting

Figure 13. Lower stones broken during manufacture. In some cases it is possible to distinguish the original upper and lower natural bedding plane of the limestone outcrop. Drawing: Antoni Llussà i Guasch.

from one side (B.a.7) and, once turned around, complete it from the opposite side (B.a.8). This is apparently the most critical step with a higher number of unwanted breaks in the thicker cases. VI-55, for example, broke after it was turned around to complete the eye. VI-69, a finished quern damaged during use, was presumably fashioned in this way. This is based on tool marks from the piercing of the eye that would not be preserved if the grinding surface were not previously hollowed (Fig. 14).

b) The second option, that of initially piercing the eye before hollowing the grinding surfaces and cutting the other fittings, is observed on both thin and thick stones, all bearing handle lugs. This begins, after the general cylindrical shape is defined (B.1 to B.3), first from thethe upper face (B.b.4) and then the lower face (B.b.5). The shaping of the lugs and the hollowing of the surfaces (B.b.6 to B.b.8) took place only after the eye was completed. This second option, certified in the case of 6 of 8 upper stones, seems to have been more common, or a least the option that provoked a larger number of damaged stones such as VI-45 (Figs. 14 and 15), VI-34 (Fig. 14), VI-26 (Fig. 15), and VI-57 (Fig. 15).

Although the different steps in the production sequence have been established, there is apparently no evidence, in either the manufacture sequence or the resulting final product, of any intention on behalf of the Els Vilars mill makers to produce a standardised product. This could be related to the fact that the porous limestone is not a hard, compact rock facilitating a uniform output following a consistent morphology. The factor of experience and skills of the quern makers - certainly not specialised craftsmen – would also play an important role.

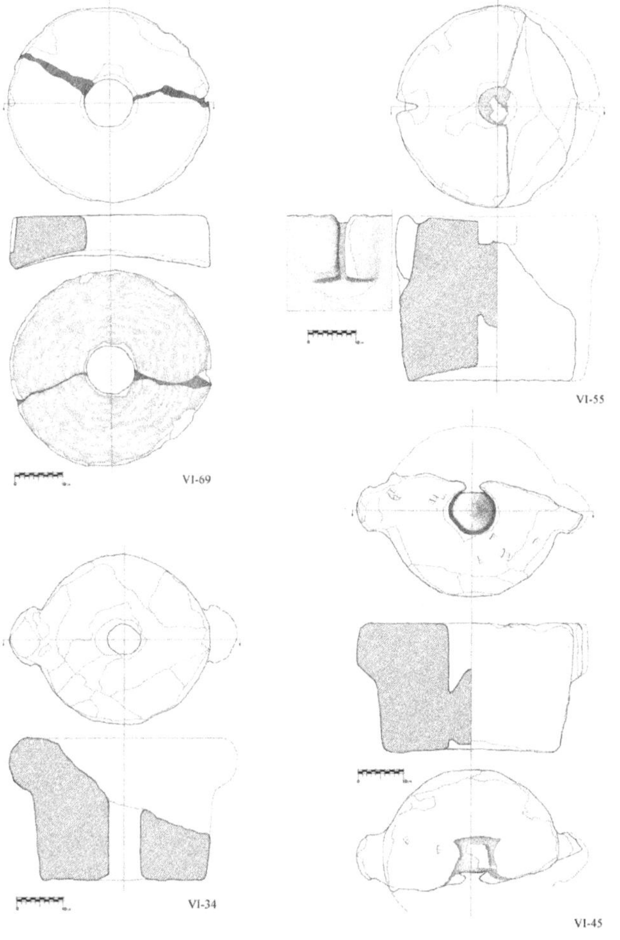

Figure 14. Upper stones broken during manufacture. Drawings: Antoni Llussà i Guasch.

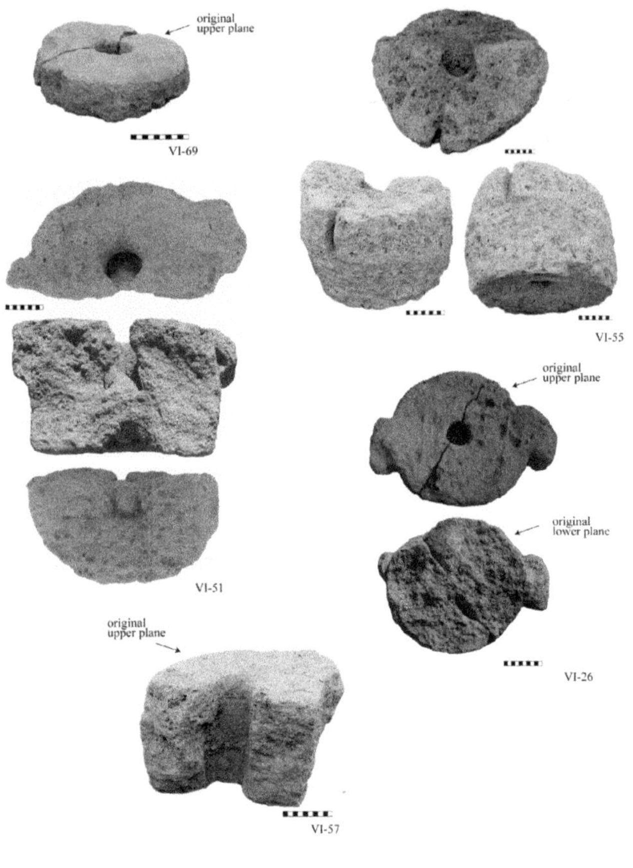

Figure 15. Upper stones broken during manufacture. In certain cases it is possible to observed the upper and lower bedding plane of the original limestone outcrop.

4. Conclusion

The manufacture of rotary querns on the site of Els Vilars in the 4th century BC is just one more of the exceptional features of this fortification. We do not know of any published or unpublished Protohistoric sites with evidence of the production of rotary querns. The only examples of comparison are found in other geographical and chronological contexts, such as that of the Roman quern quarry of Châbles-Les Saux (Switzerland), which specialised in a commercial production (Anderson et al. 2003).

The Els Vilars inhabitants were aware of the rotary mill technology already in the middle of the 5th century BC. We are inclined, however, based on the typology and exogenous petrography of one single model recovered in an early phase of the site (Fig. 7), to interpret it as an import from the Catalonian coast (Alonso 1999; 2002). But subsequently, in the 4th century BC, Els Vilars was producing its own rotary mills from a nearby stratum of porous limestone. The proximity of this outcrop in all likelihood compensated for its poor grinding properties. It is interesting to note that at the same outcrop the dwellers also exploited a massive limestone for the construction of houses, walls and other features.

It is difficult to assess whether the production of the fortress was for its own consumption or for trade. The number of finished rotary querns with worn grinding surfaces recovered in the excavations is approximately the same as the number damaged during the manufacture process. This high number of discards seems therefore disproportional to the quantity that might have been necessary for the internal use on the site. However, we must be prudent with these interpretations since we only have a partial view of this specific level of occupation conserved only on a strip of the settlement, in the well/cistern, and in the moat. A better conservation of this level might yield a very different view.

Furthermore, the study of a possible regional commerce of querns originating at Els Vilars entails a programme of research in the inventories of other archaeological sites that has not been attempted. It must also be noted that in western Catalonia, there are not many Iron Age sites that have benefited from extensive excavations yielding securely dated querns. Moreover, their petrography is usually limited to conglomerates and sandstones and the few cases of limestones are not as porous as those of Els Vilars. The absence of porous limestone querns on the other regional sites suggests that the Vilars querns were most likely not destined for commercial distribution.

The large dimensions of many of these stones, both uppers and lowers, is intriguing. This might indicate that part of the production, besides the smaller hand-operated querns, was for larger grinding mechanisms meant to be placed on elevated cylindrical supports, a type of feature that is identified not only at Els Vilars in phase IV, but on other contemporary settlements in the area as well. These larger models might therefore suggest more sophisticated mills operated by either two individuals or even a small animal.

The daily chores of the Els Vilars inhabitants were related to agriculture and livestock. They also took part in other basic activities such as iron work, pottery, and making grinding stones. Yet probably only a few possessed the skills for these activities. Those who did this work, however, cannot be considered highly specialised craftsmen. Whether it be for their own consumption or not, it is certain that this middle Iberian Culture fortified site was self-sufficient in the production of these items in both the Vilars III phase, a time when the site experienced a moment of brilliance with the urban remodelling and construction of the great moat, and the Vilars IV phase, which lasted 20 to 30 years until the site was abandoned.

Acknowledgements

We would like to thank Timothy Anderson for the translation and Jeff Hodges for the proofreading of this text.

References

Alonso, N., 1997. Origen y expansión del molino rotativo bajo en el Mediterráneo Occidental. In D. Garcia and D. Meeks (eds.), *Techniques et économie antiques et médievales. Le temps de l'innovation.* Aix-en-Provence: Ed. Errance. 15-9.

Alonso, N., 1999. *De la llavor a la farina. Els processos agrícoles protohistòrics a la Catalunya Occidental.* Monographies d'Archéologie Méditerranéenne, 4. Lattes: Publications du CNRS.

Alonso, N., 2002. Le moulin rotatif manuel au nord-est de la Péninsule Ibérique: une innovation technique dans le contexte domestique de la mouture de céréales. In H. Procopiou and R. Treuil (eds.) *Moudre et Broyer. L'interprétation fonctionnelle de l'outillage de mouture et de broyage dans la Préhistoire et l'Antiquité.* Ministère de la Recherche, Paris, 105–20.

Alonso, N., Junyent, E., Lafuente, A. and López J.B., 1998. Poder, símbolo y territorio: el caso de la fortaleza de Arbeca. *Congrés Internacional: Els Ibers, princeps d'Occident,* Barcelona, 355-372.

Alonso, N., Junyent, E., Lafuente, A., and López, J.B., 1999. Chronométrie de l'Age des Métaux dans la basse vallée du Segre (Catalogne, Espagne) à partir des datations C14. *Actes du 3ème Colloque International 14C et Archéologie (Lyon, 1998), Mémoires de la Société Préhistorique Française*, 26, París, 287-92.

Alonso, N., Junyent, E., Lafuente, A., López J.B., Lorés, J., Muñoz, D., Pérez, M. and Tartera, E., 2000. Virtual reality as an extension of the archaeological record: reconstruction of the Iron Age fortress els Vilars (Arbeca, Catalonia, Spain). In J.A. Barceló, M. Forte and D.H. Sanders (eds.), *Virtual Reality in Archaeology,* BAR International Series 843, Oxford, 225-231.

Alonso, N., Junyent, E., Lafuente, A., and López J.B., 2008. Plant remains, storage and crop processing inside the Iron Age fort of Els Vilars d'Arbeca (Catalonia, Spain). *Vegetation History and Archaeobotany*, 17 (Suppl. 1), 149–58.

Alonso, N., Junyent, E. and López, J.B., 2010. *Arbeca. La Fortalesa dels Vilars. Guies del Museu d'Arqueologia de Catalunya,* 2a edició. Museu d'Arqueologia de Catalunya, Barcelona.

Anderson, T., Agustoni, C., Duvauchelle, A., Serneels V. and Castella D., 2003. *Des artisans à la campagne: carrière des meules, forge et voie gallo-romaines à Châbles (FR).* Archéologie Fribourgeoise 19, Academic Press Fribourg, Fribourg.

Dunham, R.J., 1962. Classification of carbonate rocks according to depositional texture. In W.E. Ham (ed.), *Classification of carbonate rocks: American Association of Petroleum Geologists Memoir*, 108-121.

Equip Alorda Park 2002. Les meules rotatives du site ibérique d'Alorda Park (Calafell, Baix Penedès, Tarragone). In H. Procopiou and R. Treuil (eds.), *Moudre et broyer. L'interprétation fonctionnelle de l'outillage de mouture et de broyage dans la Préhistoire et l'Antiquité*, II, Paris: Éd. du CTHS, 155-75.

Ferrio, J.P., Alonso, N., López J.B., Araus, J.L., and Voltas, J., 2006. Carbon isotope composition of fossil charcoal reveals aridity changes in the NW Mediterranean Basin. *Global Change Biology,* 12, 1–14.

GIP 2003. Caballos y hierro. El campo frisio y la fortaleza de Els Vilars d'Arbeca (Lleida, España), siglos VII-IV ane. In N. Alonso, E. Junyent, A. Lafuente and B. López (eds.), *'Chevaux-de-frise' i fortificacio' en la primera edat del ferro europea,* Universitat de Lleida, 233–74.

Junyent, E., López, J.B., Moya, A,. and Tartera, E., 2009. L'accés fortificat i les portes en el sistema defensiu de la fortalesa dels Vilars (Arbeca, les Garrigues), *Revista d'Arqueologia de Ponent,* 19, 307-333.

Nieto, A., Gardeisen, A., Junyent, E. and López, J.E., 2010. Inhumations de foetus d'équidés dans la forteresse du premier âge du Fer de Els Vilars (Arbeca, Catalogne). In A. Gardeisen, E. Furet and N. Boulbes (eds.), *Histoire d'équidés, des textes, des images et des os*, Monographies d'Archéologie Méditérranéenne, hors série, 4, 125-48.

Portillo, M., 2006. *La mòlta i triturat d'aliments vegetals durant la Protohistòria a la Catalunya Oriental.* Doctoral dissertation. Departament de Prehistòria, Història Antiga i Arqueologia, Universitat de Barcelona, Barcelona (http://www.tdx.cat/TDX-0811106-121331).

Py, M., 1992. Meules d'époque protohistorique et romaine provenant de Lattes. *Lattara*, 5, 183-232.

Vidal, A., Tartera, E., Alonso, N., Aguiló, C., Junyent, E. and Lorés, J., 2007. Vilars AR Project. Testing the Augmented Reality paradigm in an archaeological site. *Computer Applications in Archaeology (CAA) 2005, The World in your eyes*, 21-24 de març de 2005, Tomar 2007

Weffers, S., In press. Mühlen der Latènezeit zwischen den Steinbruchrevieren Mayen und Lovosice. *Vulkanpark-Forschungen,* 9 (Mainz).

Still using your saddle quern? A compilation of the oldest known rotary querns in western Europe

Stefanie Wefers

This paper discusses the oldest known rotary querns in western Europe, based on evidence acquired from the literature or on first hand evidence given by colleagues[1]. I will begin by evaluating the earliest known rotary querns in the Mediterranean, which I will try to link to examples from the area north of the Alps. However, I will commence with a brief explanation of the construction and mode of operation of a rotary quern (Fig. 1).

Rotary querns consist of a fixed lower stone and a rotatable upper stone. Both parts are discoidal and their grinding faces are accurately coordinated. The grinding face of the lower stone is convex, that of the upper stone is concave. The upper stone has a centred perforation which is called an 'eye', which serves to fix the upper stone on the perforated lower stone which incorporates a spindle. Furthermore, the eye serves for feeding the mill. In order for the mill charge to attain between both grinding faces and to be centrifugally carried to the fringe where the grinding takes place, both faces must touch each other only at the outermost edge. At the centre and the eye respectively a space, the so called opening, must exist (Wefers, in press).

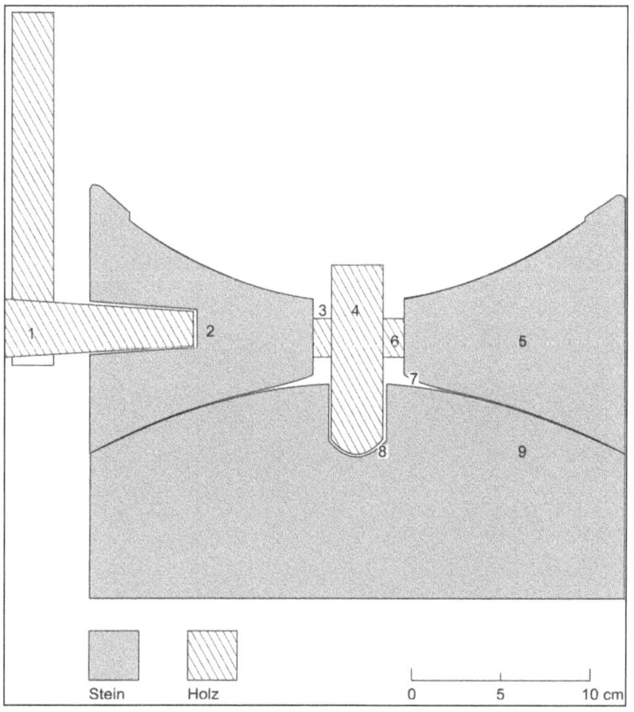

Figure 1. Schematic reconstruction of a late Iron Age rotary quern. 1 Handle. 2 Hole for the handle. 3 Eye. 4 Spindle. 5 Upper stone. 6 Spindle support. 7 Opening. 8 Centre bore. 9 Lower stone (modified after Staubitz 2007, 17 Abb. 10).

An experiment carried out by P. Holodňák using saddle querns, Olynthan mills, and rotary querns highlights the particular importance of rotary querns. The experiment enabled him to roughly estimate the effectiveness of a saddle quern compared to an Olynthan mill as well as the effectiveness of a saddle quern compared to a rotary quern. According to P. Holodňák, Olynthan mills are six-times more efficient than saddle querns and rotary querns are six- to twelve-times more efficient than saddle querns (Holodňák 2001,44). As a result it should be expected that rotary querns, if they were available, would be the preferred device for processing grain.

The oldest known rotary querns of the ancient world

Since the beginning of the 20th century assumptions have been made about the origin of rotary querns and where the inventors of the rotary quern came from, researchers postulating their invention in Greece, Italy, Spain, or Tunisia[2]. On present showing, the oldest rotary querns were found on the Iberian Peninsula[3].

In 1943 V. G. Childe first drew attention to their early appearance in the Iberian Peninsula. Because vertical handles on Iberian pieces contrasted with horizontal handles on Italian donkey mills and Greek rotary querns, he presumed two independent points of origin in the Mediterranean West and East. In 1958 L. A. Moritz reiterated this. From his point of view the invention of a large Pompeian mill, driven by animals, took place in Italy, contrasting with rotary querns, which could be powered by one person, originating in Spain. He based this on the term *mola hispaniensis* mentioned by Cato the Elder in the 2nd century BC (Quo modo oletum agri iugera CCXL instruere oporteat. [...] molas asinarias unas et trusatilis unas, Hispaniensis unas [...] Cato *agr.* X, 1-4). Moritz was convinced that Cato had described the rotary querns characteristic of Spain. However, he was not convinced that this trail-blazing invention can be attributed to two

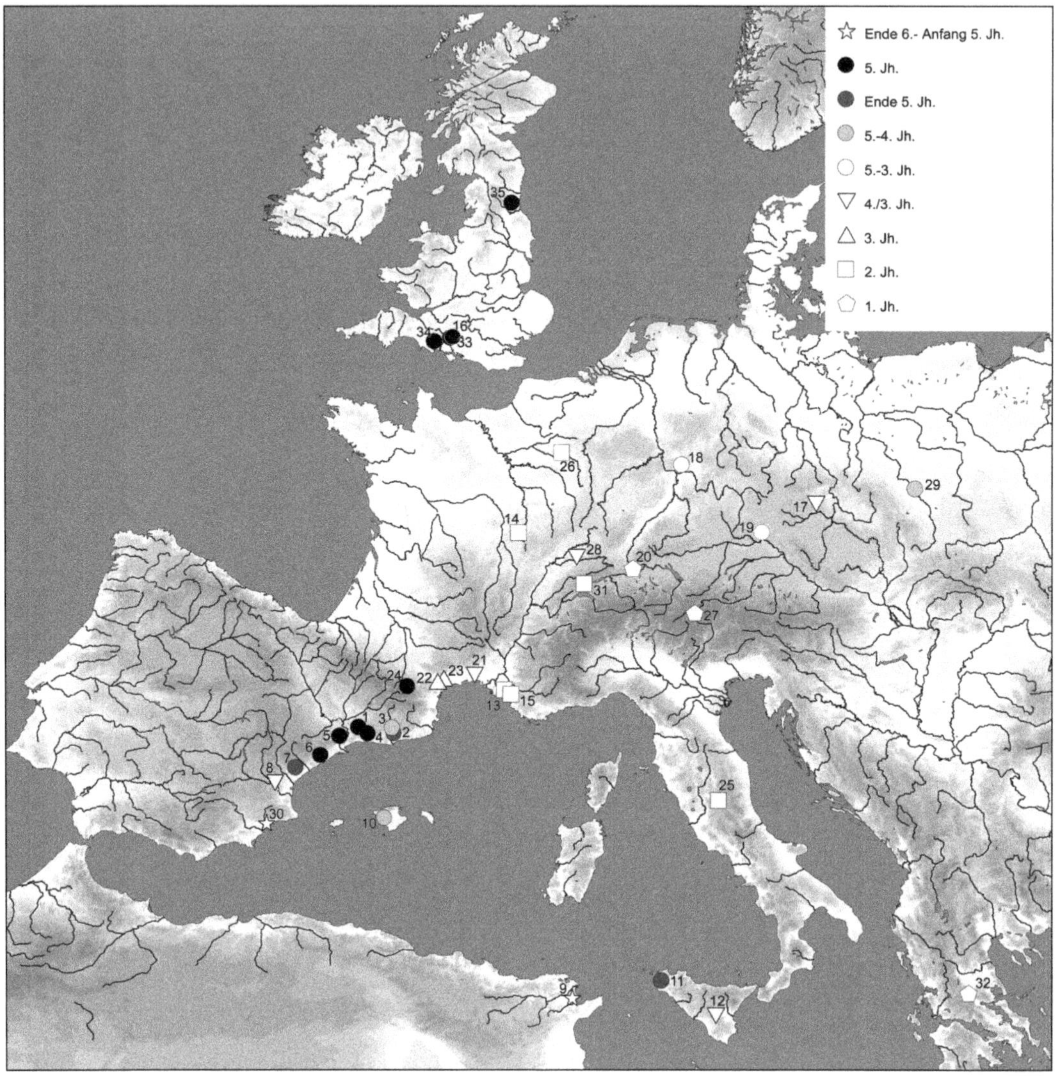

Figure 2. 1. Els Vilars d'Arbeca. 2. Turó de Ca n'Olivé (Cerdanyola del Vallès, prov. Barcelona/E). 3. Penya del Moro (Sant Just Desvern, prov. Barcelona/E). 4. Alorda Park (Calafell, prov. Tarragona/E). 5. Tossal del Moro de Pinyeres (Batea, prov. Tarragona/E). 6. Puig de la Nau (Benicarló, prov. Castelló/E). 7. La Escudilla (Zucaina, prov. Castelló/E). 8. Los Villares (Caudete de las Fuentes, prov. Valencia/E). 9. Karthago/TN. 10. El Sec (Mallorca/E). 11. Motyé (Sicilia/I). 12. Morgantina (Sicilia/I). 13. Martigues (dép. Bouches-du-Rhône/F). 14. Bibracte (dép. Saône-et-Loire/F). 15. Massalia/Marseille (dép. Bouches-du-Rhône/F). 16. Danebury (Hampshire/GB). 17. Soběsuky (okr. Kroměříž/CZ). 18. Münzenberg-Ober-Hörgern (Wetteraukr./D). 19. Regensburg-Harting (Stadt Regensburg/D). 20. Bern-Engehalbinsel (Kt. Bern/CH). 21. Lattes (dép. Hérault/F). 22. Peyriac-de-Mer (dép. Aude/F). 23. Ensérune bei Béziers (dép. Hérault/F). 24. Pech-Maho bei Sigean (dép. Aude/F). 25. Italienische Halbinsel. 26. Vallée de l'Aisne (Nordostfrankreich). 27. Ganglegg (Prov. Bozen (Südtirol)/I). 28. Orchamps (dép. Doubs/F). 29. Wrocław-Partynice, -Oporów (Stadt Wrocław/PL). 30. La Fonteta (Guardamar del Segura, prov. Alicante/E). 31. Yverdon-les-Bains (Kt. Waadt/CH). 32. Griechenland. 33. Rowbury Farm (Hampshire/GB). 34. Gussage-All-Saints (Dorset/GB). 35. Thorpe Thewles (Cleveland/GB).

areas so far apart and he suggested the concept of the rotary mill was invented in an as yet unknown area of the Mediterranean. Furthermore, due to the extraordinary productivity of the mill, it was diffused rapidly from this hypothetical place (Moritz 1958, 115). In the beginning of the 1990s, based on newly discovered and dated rotary querns, M. Py resumed the discussion concerning the origin of rotary querns. Generally he divided them into two groups. The first group encompassed small devices, which can be powered by one person and are comparable with those from late Iron Age findspots in Germany (Wefers in press). The second group is composed of large devices, which can be compared to the later and much bigger Pompeian mills. Due to their size they had to be operated by two persons. At that time the oldest rotary querns were known from the Phoenician settlements of Motyé and Morgantina on Sicily as well as from the shipwreck El Sec located offshore of the island of Mallorca (Fig. 2) which contained a partially Phoenician cargo. The devices found inside the shipwreck are dated to the second quarter of the 4th century BC, the ones from Morgantina are dated to the 4th and 3rd century BC, and the pieces from Motyé are dated at the earliest to the end of the 5th century BC (Arribas *et al.* 1987, 7; Py 1992, 195; Baatz 1994, 98; Baatz 1995, 8-9; Junkelmann 1997, 114-115; Curtis 2001, 335-336).

Hence, Py suggested that large rotary mills were a Phoenician invention (Py 1992, 195-7). In 2001 this assumption (Baatz 1994, 98; Baatz 1995, 8; Beranová and Břicháček 1993, 264; Buchsenschutz 2002, 65; Curtis 2001, 335-336; Dal Ri 1994, 65; Marzoli and Donner 1994, 84; Junkelmann 1997, 114; Watts 2002, 28) was confirmed by a fragment of a lower stone of a large device found in the Phoenician settlement of Carthage (Fig. 2). It was used as construction material for a stone chamber grave which can be dated to the last quarter of the 6th century BC. The fragment consists of an imported volcanic rock (Morel 2001). According to Py while the large rotary mills were of Phoenician origin, the so-called small devices, also referred to as rotary querns, are present in the north-east of the Iberian Peninsula, Roussillon, and the Languedoc. He suggested that at the end of the 5th century BC rotary querns were derived from large devices through contacts between the Iberian population and Phoenician seamen. Developed on the north-eastern coast of Spain, they were accepted rapidly by the population along the coastline to the east (Py 1992,192-7). N. Alonso-

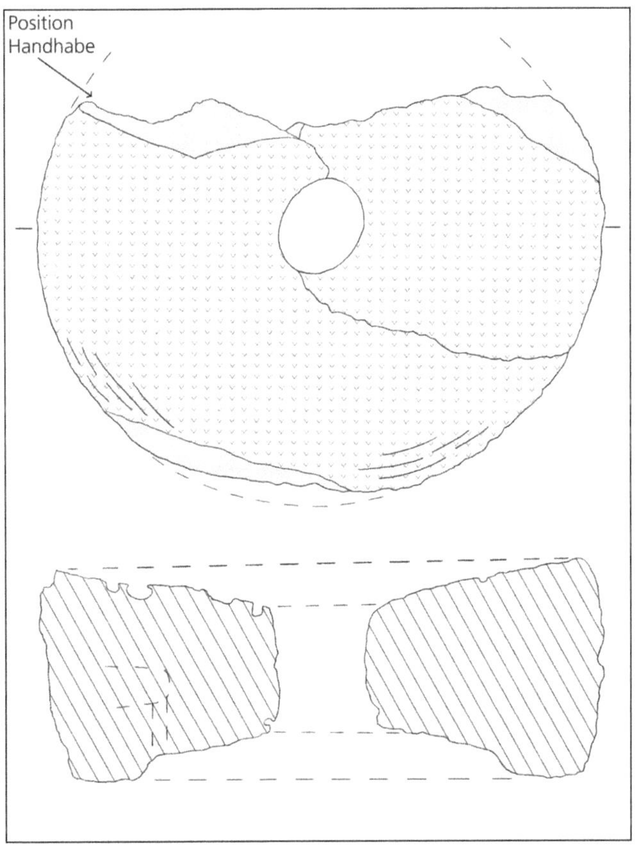

Figure 3. Fragment of a rotary quern found in Münzenberg–Ober-Hörgern (Hesse, Germany). Grey: breakage. Tickmarks: picking marks. Diagonal hatching: cross-section. White: hole. – circular lines: polished marks. Scale 1:5.

Martínez claimed an even earlier advent of rotary querns on the Spanish coast. According to her (kind personal information), objects, deriving from Spanish find spots: Els Vilars, Turó de Ca n'Olivé, Penya del Moro, Alorda Park, Tossal del Moro de Pinyeres, Puig de la Nau, La Escudilla, Los Villares are predominantly dated to the 5th century BC (Alonso-Martínez 1995, 16-18). Moreover, one rotary quern discovered at a find spot close to Murcia, 'La Fonteta, Guardamar del Segura, Alacant', can be dated to the beginning of the 6th century BC (Alonso-Martinez 1995, 16) (Fig. 2). In the Italian Peninsula, there is no evidence for rotary querns or mills before the 2nd century BC. However Varro, quoted by Pliny the Elder (*Nat Hist* xxxvi, 135), who states that millstones were invented in Volsinii, Italy, so the late occurrence of rotary querns on the Italian Peninsula surely must be ascribed to the research to date. Examination of this locally might yield interesting results, but this was not possible within the current project. However, there is one item published as a rotary quern, found in Bienno, Campolungo, attributed to an ore-mine dated to the 6th to 5th century BC, but not found *in situ* (Morin and Tizzoni 2009, 135-9, Abb. 34). It is in a poor state of preservation with worn surfaces, and apart from the perforation there are no marks of usage or production confirming its interpretation as a rotary quern for which reason I disregard it. Interestingly, rotary querns discovered in Greece are all dated no earlier than the first century BC (Fig. 2). This relatively late advent of rotary querns in Greece can be explained by the commonly used Olynthan mills, which were recently compiled by Rafael Frankel (2003, 18-9).

Currently, it can only be assumed that the two types of mills were invented in two areas at almost the same time. The other possibility would be either that the rotary querns were invented by the Celtiberians due to Phoenician influence or that they were copied by the Celtiberians from Phoenician prototypes. In form, the large devices resemble the later so-called Pompeian or hourglass mills. However, in diameter and height they are considerably smaller than the later Pompeian mills. Therefore, in contrast to later millstones, which must be powered by a capstan, they can be operated by two persons. At present, it cannot be ascertained when and where the further development to an animal driven mill occurred. However, a further development resulted in water powered, immobile mills or 'mobile' boat mills. Mobile rotary querns were used in parallel with such devices. In general it is assumed that the concept of rotary mills, whether a rotary quern or a rotary mill, spread rapidly from west to east and northeast [4].

Advent of rotary querns in Europe north of the Alps

There are two rotary querns found in Germany which suggest that rotary querns were already in use in areas north of the Alps as early as the early late Iron Age.

One of these rotary querns was found in *Münzenberg-Ober-Hörgern* in Hesse, the other one was found in *Regensburg-Harting* in Bavaria (Fig. 2). The settlement of Münzenberg-Ober-Hörgern is situated at the foot of a hillside, oriented to the southeast and in the vicinity of the River Wetter. The site was exclusively late early Iron Age to early late Iron Age in date. In 1986 it was excavated by the Study Group of Prehistory of the Historical Society of Butzbach in advance of the construction of a wastewater treatment plant. Of all features of this settlement, the content of the pit comprising the two matching fragments of the upper stone (Fig. 3) was looted. Possibly because of its weight, size, and worthlessness for treasure hunters, all that now remains is the fragment of the upper stone and there were no directly dateable associated finds. However, the remaining 18 features, documented by the excavators are all dated by ceramics[5] to the late early Iron Age and early late Iron Age and it is reasonable to assume that the feature comprising the rotary quern belongs to the same period (Heun 1992-3,71.77 Abb. 7; 82). Yet, according to C. Bergmann (cultural heritage management Wiesbaden) early late Iron Age settlements located in the Rhine-Main area were commonly settled continuously until the end of the late Iron Age or resettled at the end of the late Iron Age, putting into question the definite dating of the quern. However, it can be concluded that it was locally manufactured in Hesse as it consists of typical olivine-basalt.

The fragmentary item of *Regensburg-Harting* was found in backfill within a settlement dating to the early late Iron Age (Fig. 4). At present, it is the only known lower stone of a rotary quern in Germany dating to such an early period. According to the excavator H. Koch this lower stone should definitely be dated to the early late Iron Age since the whole excavation site yielded nothing which could be dated to the middle late Iron Age or the end of the late Iron Age. Only a few finds could be assigned to an early Medieval and modern settlement (Koch 1991, 218-9, Taf. 84).

Apart from the two afore mentioned querns, no further objects are known in Germany which can definitely be interpreted as belonging to the early late Iron Age[6,7]. Both are probably of this date, although unfortunately their circumstances of discovery are not very clear. However, they imply the possibility of an early advent of rotary querns in areas north of the Alps, particularly in the light of the possible 6th century and definitely 5th century BC dated examples found at the coast of the Iberian Peninsula. This encourages a discussion on the routes of distribution or transmission of this important invention. The map (Fig. 2) shows the oldest known rotary querns in different regions of Europe[8]. The concentration of very early dated rotary querns in the northeast of the Iberian Peninsula implies that they were firstly used in this area. The remaining European find-spots are hard to interpret the data on diffusionist models as the path of transfer of

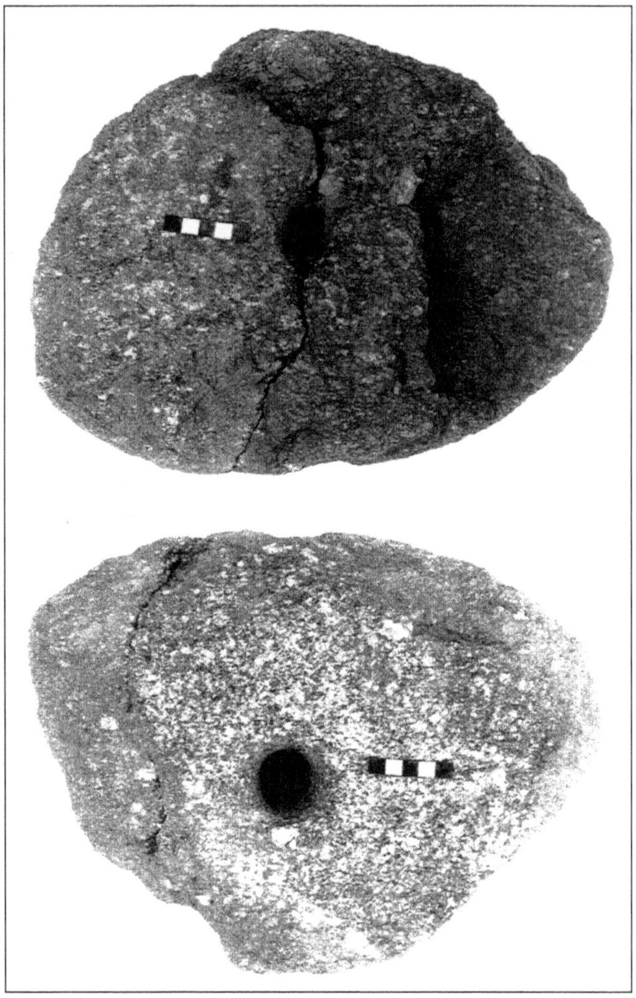

Figure 4. Lower stone of a rotary quern found in Regensburg-Harting (Bavaria, Germany). Scale 5 cm (after Koch 1991, Taf. 84).

this innovation northward and in a north-eastern direction would have big chronological gaps or leapfrogs (Lang 2002, 63-4). Already being used at the end of the 6th century BC in southern Spain, at the latest from the 5th century BC onwards, rotary querns appear on the north eastern coast of Spain and would be likely to emerge on find spots of Roussillon, the Languedoc, and Provence in the 5th/4th century BC at the earliest. Rotary querns are a common device in that area at the latest in the 3rd and 2nd century BC.

In contrast to this, very early dated rotary querns are documented in the north: they were found at Danebury (Hampshire) where they are dated by associated finds to the 'the middle of the fifth century' (Laws, Brown and Roe 1991, 396; Brown 1984, 418). Further rotary querns are known from Gussage-All-Saints (Dorset), Thorpe Thewles (Cleveland) and Rowbury Farm in Wherwell (Hampshire) dated to the 5th century BC (Fig. 2) (Wainwright 1979; Watts 2002, 28; pers comm. D. Peacock). The most obvious explanation for this early advent of rotary querns in south-eastern England is the tin-trade. At the end of the 6th century BC, the tin-trade was in the hands of the

Phoenicians. According to Strabo's *Geographica*, Pliny the Elder's *Naturalis historia* [9], the journey of Pytheas of Massilia, and Herodotus' works [10] as well as the barbary ape cranium and mandible found in Navan Fort (County Armagh, Ireland)[11], it can be assumed that the British Isles were accessible by ship[12]. Due to the long sea route, which possibly followed the shoreline of Spain and France, the equipment had to include rotary querns to allow fresh food-preparation[13]. In this way it is very likely that rotary querns reached England and were copied by the indigenous population. However, another route is possible: there could also have been a trade route via Rhône, Saône, and Seine or Marne and Seine by which means a rotary quern or the concept of a rotary quern could have been transmitted. From 535 BC onwards (Battle of Alalia), the Greeks inhabiting the colony of Massalia focussed on the trade route following the French rivers. This enabled a northward transfer of Mediterranean concepts. For instance, the scattered finds of Greek transport amphorae deriving from southern France, of attic black and red figure pottery, found in the area of the river Seine suggest a connection via Rhône, Saône, and Seine or Marne and Seine is more likely than a route along the shoreline of Spain and France (Kimmig 1983, 33-41 with fig. 27-29; Pauli 1993, 162-167 with fig. 41-43).

Furthermore, a dated rotary quern was recently found in Orchamps (dép. Jura) (Fig. 2). This implies an early advent in Gaul also and possibly a staging post on the way to southern England. This rotary quern was found inside a dwelling which measured 6.5 m to 5.2 m. The only associated find was a small piece of an unsignificant rim sherd. However, a radiocarbon date puts the dwelling as well as the rotary quern into the 4th to 3rd century BC[14].

It is not possible to decide how the early rotary querns in England are to be explained: it is very likely that rotary querns were transmitted overland as well as by rivers and the sea. All these contacts are likely to be connected to the tin-trade. However, despite this a local invention of the rotary quern in south-eastern England cannot be completely excluded.

The chronological gap between the oldest rotary querns in Switzerland, dating to the 2nd century BC, those in Germany, dating to the 5th to 3rd century BC, and the Czech Republic, dating to the 4th-3rd century BC, cannot be explained in the same way as those from the British Isles (Fig. 2). Rotary querns in regions north of the Alps appear to progress in leaps and bounds and a chronologically continuous step by step distribution emanating from the find spot with the oldest known rotary quern is untenable. The heterogeneous diffusion of the oldest known rotary querns may depict the current state of research, and the routes of distribution along Rhône and Doubs can merely be presumed, not proven. Only further dateable rotary querns, like the one discovered in Orchamps, might clarify the issue. The current distribution and the date of three rotary querns found in regions which are far away from north-eastern Spain, implies that rotary querns were already known and used in regions north of the Alps as early as the 4th century BC. This would suggest that there are huge deficits in research, possibly explicable by the focus on settlements dating to the end of the late Iron Age – namely the *oppida*. As a result it is hard to find rotary querns belonging to the early late Iron Age. This seems to be true for Switzerland, for instance; here the oldest known rotary quern is dated to the 2nd century BC. It should be noted that the circumstances of discovery of the early late Iron Age rotary querns found in Germany were quite fortuitous, and that the early late Iron Age has a long duration (5th to 3rd century BC). Firstly, it is possible that the dating of the objects to the early late Iron Age should be revised to the end of the late Iron Age. Secondly it is possible that the early late Iron Age dating is correct, but this may mean that the rotary quern actually dates to the 3rd century BC. This would lead to the fact that there is no huge discrepancy between the oldest known French or Swiss rotary querns and the one in Germany. Objects dated to the 3rd century BC and found in regions north of the Alps may be traced back to the Celtic migrations or the backflow of those who have been in the Mediterranean to regions north of the Alps. The persons who came back could have brought the concept of a rotary quern with them.

Interpreting the present distribution, not as a result of the current status of research, but as the real condition of the former situation, does not fit the diffusionist model, rather it suggests that the concept of a rotary quern was developed independently in different European areas contemporaneously, or within short chronological intervals. Contemporaneous invention at different spots is very difficult to prove as the basis of the long-distance connections in Europe from the early Iron Age onwards, are indicated by imports which comprise drinking dishes and ceramics[15]. Since rotary querns were in use in Spain from the end of the 6th century BC onwards and in France from the 4th century BC onwards, at the early Iron Age and early late Iron Age the assumed routes of connections to Central Europe alongside Rhône, Saône, and Doubs, along Rhône and across Lake Geneva, and across the Alps or around the Alps on an eastern route[16] make it possible that rotary querns were used in regions north of the Alps (Kimmig 1983, 31-2). There they could have come up either as an object of import or as a reaction on an innovative technical concept. The item found in Orchamps which is dated to the 4th to 3rd century BC supports this assumption. For the Hessian region there are known imports dating to the late Iron Age and even earlier. For instance, in Hanau Klein-Auheim (Main-Kinzig-Kreis) an attic sherd was found within an early late Iron Age settlement, reflecting long-distance contacts to the Mediterranean south (Mansel 1996, 153-162; Spindler 1973, 30-31 with pl. 15. 16. 64). In the late early Iron Age in Southern Germany and France, there are even hints of direct contacts with north-eastern Spain, the region with the oldest rotary querns currently known. In this respect there should be mentioned the 'Acebuchal' type belt hook from grave sixty-five of the Magdalenenberg near Villingen-Schwenningen (Schwarzwald-Baar-Kreis), ringshaped pendants with globular appendage made of Bronze found in Southern Germany, a flange hilted sword from Apremont (dép. Haute-Saône), and diverse finds of gold (Neumaier 1995, 495-503 with fig. 5; Pare 1989, 460-46). Similarities of objects made of gold belonging to the Hallstatt culture to those known from the Iberian Peninsula were firstly pointed out by W. Schüle (1969, 62). According to analyses by B. Armbruster, due to manufacturing technique and decorations at least a few of those objects seem to be influenced by objects made up of gold found on the Iberian Peninsula (Armbruster 2000). Among these are for example the gold torc discovered in the Vix Grave, the golden earrings found in the tumulus of La Butte (Sainte-Colombe), and the golden bangles (Armstulpe) found in the chieftain's grave of Eberdingen-Hochdorf (Kr. Ludwigsburg). Technical and stylistical similarities to those gold objects found on the Iberian Peninsula however do not suggest a production on the Iberian Peninsula (Pare 1989, 460-464; Armbruster 2003, 213-214; Haffner 2003, 183-186). Due to complex brazing techniques on an Iberian object of comparison (find spot El Carambolo (Prov. Sevilla)) to the golden bangles found in Hochdorf, the last mentioned objects can only be interpreted as imitations (Pare 1989, 460-464). And the gold torc found in Vix does not only highlight technical parallels to Iberian but also to Italian objects (Joffroy 1962, 105-106; Haffner 2003, 183-186.).

Thus the concept of rotary querns could have been inherited by the peoples north of the Alps as early as in the late early Iron Age, even though maybe only selectively. Referring to this, a proper examination of the finds made up of stone known from the Heuneburg or rural settlements around the Hohenasperg might be worthwhile. The openness to Mediterranean imports and Mediterranean technology to the inhabitants of the Heuneburg, as for example the brick wall made up of adobe, or the application of the Hippodamian plan, makes it possible that rotary querns might be as early as the late early Iron Age in this settlement.

Conclusion

It should be noted that the most likely inventors seem to be the Phoenicians and by dint of their commercial activity, rotary querns were diffused quite rapidly. Possibly they were brought to the British Isles via the tin-trade. From the 6th century BC onwards rotary querns can be anticipated in regions north of the Alps, but only as the exception, as the spectrum of all known rotary querns in the German hills suggests (Wefers in press). As recently as the end of the late Iron Age rotary querns are commonly used by the better part of the population. I would expect them primarily at find spots such as the Heuneburg, since this hill-top site played a decisive role in the trade with the Mediterranean. Later rotary querns could have reached Central Europe by the Celtic migrations or their backflow respectively (Baatz 1995, 8-9). To date all rotary querns known, are of indigenous production, clearly demonstrated as they are made up of autochthonous rock[17]. Therefore it should be assumed that only the concept was transmitted and not an actual rotary quern. In the end, rotary querns were transmitted by the Roman legions who reached regions north of the Alps in the late Iron Age and who were in contact with the indigenous population. Every *contubernium* had its own rotary quern and processed its cereals day-to-day (Junkelmann 1997, 117 with fig. 58; Junkelmann 1997, 114-6. 118; Mangartz 2008, 101). Hence, eventually direct contact with Roman legions in effect enforced the distribution of the rotary quern in regions north of the Alps and made it common kitchen equipment.

Endnotes

[1]I am much obliged to Natàlia Alonso-Martínez (Universidad de Leida), Lyn Cutler (University of Southampton), Luc Jaccottey (Inrap), David Peacock (University of Southampton) and Marco Tizzoni (Universität Bergamo) for their willingness to provide information.

[2] Jacobi 1912, 81-83; Hörter 1914, 292; Curwen 1941, 15-16; Childe 1943, 19-25; Gleisberg 1956, 14-15; Moritz 1958, 105; Beranová 1963, 217; Beranová 1980, 374; Waldhauser 1981, 217-218; Joachim 1985, 362-364; Gleirscher 1987, 260; Koch 1991, 217; Penack 1991, 58; Beranová and Břicháček 1993, 264; Baatz 1994, 98; Dal Ri 1994, 65; Marzoli and Donner 1994, 84-85; Baatz 1995, 8-9; Junkelmann 1997, 115; Curtis 2001, 335-337; Buchsenschutz 2002, 65; Frankel 2003, 18-19.

[3] Alonso-Martínez 1995; Alonso-Martínez 1997, 15-17 fig. 1, 2. A first usage or the invention of rotary querns in Asia should be ruled out. According to J. Needham (1965 115, 189-190) in China the first rotary querns emerge at the beginning of the 2nd century BC. Actually, A. Lucas does not completely exclude a transfer of technology emanating from Europe to China even if there are differences in design and raw material (Lucas 2006, 15-17).

[4] At this point I would like to refer to a rotary quern found in the vicinity of Lake Van (Turkey), which is deposited in the museum of Tbilisi. According to G. Tseretheli (1939, 21 pl. 32) it is dated to the 8th century BC. However, it was found in the 19th or 20th century. Moreover, the circumstances of its discovery cannot be reconstructed. Due to this improperly known context the dating proposed

by G. Tseretheli is not tenable. Also Ö. Wikander (1985 annotation 83) proposes a later dating for this rotary quern.

[5] Only four finds of metal were discovered. These objects are chronologically insignificant: a fragment of a knife, a fragment of a fitting comprising a rivet hole, and two nails.

[6] According to descriptions by A. Müller-Depreux (2005, 103-104) as well as studies by the author herself, three out of four fragments of lower stones discovered in the early late Iron Age earthwork I in Niedererlbach (district Landshut) can no longer be interpreted as rotary querns.

[7] Even in the Eifel region the first known rotary querns were found in features near Kottenheim belonging to period Latène C2 and they were associated with Napoleonshüten (von Uslar 1938/39; Haberey / Rest 1941, 395-398; Oesterwind / Schäfer 1989, 43; Mangartz 2008, 13). The earliest known objects found in settlements in this region are dated to the periods La Tène C and D, too (Joachim 1980, 369. 371. 375-376. 385 Abb. 16, 4; 387 Abb. 17, 4; 401 Abb. 25, 2; 405 Abb. 27, 4).

[8] Müller-Beck 1959/60, 402; Wielowiejski 1960, 123. 392; Brown 1984, 418; Blanck 1987, 112; Peacock 1987, 67; Buxo i Capdevila 1991, 22; Koch 1991, 218-219; Py 1992, 195; Heun 1992/93, 71. 82; Alonso-Martínez 1995, 16-18; Alonso-Martínez 1997, 15-18, especially fig. 2; Chausserie-Laprée 1998, 230-231; Buchsenschutz and Boyer 1999, 216; Holodňák and Mag 1999, 439; Pommepuy 1999, 131-132; Reille 2000, 261. 264; Morel 2001; Anderson, Villet and Doswald 2002, 82-83; Buchsenschutz 2002, 65; Brun 2006, 95; Gamper 2006, 112-114; kind written notification by L. Jaccottey: radiocarbon dating of a dwelling measuring 6.5 x 5.2 m in Orchamps 389-204 BC; kind written notification by N. Alonso-Martínez.

[9] Roseman 1994, 1-6. 24-25. 30-31. 41-42. 48-49. 87-88. 91-92. 124-125. 131-132. 148-158.

[10] Hdt. III, 115. – According to Herodotus tin derives from the Κασσιτερίδας. This region is to be found in the furthermost west of Europe. Today it is connected to the British Isles.

[11] Due to this find in Navan Fort, B. Raftery would like to affirm his assumption of trade connections between the British Isles and the Mediterranean. Based on feature context as well as radiocarbon analyses the cranium is dated to the 3rd and 2nd century BC (Lynn 1997; Raftery 1997).

[12] Peiffer 1972, 188-189; Kimmig 1983, 16; A. González Ruibal (2006-2007, 262-269. 650) compiles finds which are related to the Phoenicians and *Emporien* at the northwestern coast of Spain. They are dated to the early 5th century BC, whereas there are also earlier dated finds, even if they are scarce, which point to contacts to the Mediterranean. This compilation has to be understood as further evidence for long-distance marine contacts to northern Europe

[13] Besides the shipwrecks of Mahdia (Baatz 1994, 98) and De Meern 1 (Mangartz 2007, 246) there are other examples verifying rotary querns as common equipment of a ship.

[14] Kind notification by L. Jaccottey (Institut National de Recherches Archéologique Préventives).

[15] For instance Fischer 1973, 436-442. 455-459; Peschel 1979, 221. 224; Kimmig 1983, 5-75; Spindler 1983, 203-204. 323-324; Neumaier 1995, 498; Mansel 1996, 161-162; Böhr and Shefton 2000, 22; van den Boom and Pape 2000, 47; Kohler 2000, 197-208; Pape 2000; Shefton 2000, 36-39.

[16] Kind notification C. Bergmann (cultural heritage management Wiesbaden). – However, P. Jüngling (1987) who published a discovery report does not mention this attic sherd. – Concerning more goods of import see: Peschel 1979, 231 fig. 4; 236 fig. 6; Kimmig 1983, 38 with pl. 8, 2; 39 with fig. 32; 41 with fig. 35; 65-67 with fig. 57; Herrmann and Jockenhövel 1990, 247. 256. 262-264.

[17] A discovery of a rotary quern north of the Alps consisting of a typically Mediterranean rock would give an unequivocal evidence for the process of transfer. At least at this find spot the technique of rotary querns would have been conveyed directly from the Mediterranean.

References

Alonso-Martínez, N., 1995. Les premières meules rotatives manuelles dans le nord-est de la péninsule Ibérique. In M.-C. Amouretti and G. Comet (eds.), La Transmission des connaissances techniques. Tables rondes Aix-en-Provence April 1993 - Mai 1994. *Cahier Hist. Technique 3* Provence, 15-23.

Alonso-Martínez, N., 1997. Origen y expansión del molino rotativo bajo en el mediterráneo occidental. In D. Meeks and D. Garcia (eds.), Techniques et Économie Antiques et Médiévales: le temps de l'innovation. *Colloque international (C.N.R.S.). Aix-en-Provence 21-23 Mai 1996.* Paris, 15-9.

Anderson, T.J., Villet, D. and Doswald, C.,2002. Production and distribution of Iron Age and Roman handmills in Switzerland. In L. Lazzarini (ed.), Interdisciplinary Studies of Ancient Stone. *Sixth International Conference of the Association of the Study of Marble and Other Stones in Antiquity Venice 15 -18 June 2000.*

Padova, 79-84.
Armbruster, B.R., 2000. Goldschmiedekunst und Bronzetechnik. Studien zum Metallhandwerk der Atlantischen Bronzezeit auf der Iberischen Halbinsel. *Monogr. Instrumentum 15*. Montagnac.
Armbruster, B.R., 2003. Remarques sur la technique de fabrication. Relations entre l'orfèvrerie du domaine hallstattien occidental et l'orfèvrerie de la péninsule Ibérique au Bronze final et au premier Âge du Fer. In C. Rolley (ed.), *La tombe princière de Vix*. Paris, 200-15.
Arribas, A., Trias, M.G., Cerdá, D. and de la Hoz, J., 1987. Presentacion. In A. Arribas, M.G. Trias, D. Cerdá and J. de la Hoz (eds.), *El barco de El Sec (Calvià, Mallorca). Estudio de los materiales*. Mallorca, 7-11.
Baatz, D., 1994. Die Handmühlen. In G. Hellenkemper Salies, H.-H. von Prittwitz, G. and G.Bauchhenß (eds.), *Das Wrack: der antike Schiffsfund von Mahdia*. Ausst. Rheinisches Landesmus. Bonn 8. September 1994 – 29. Januar 1995. Kat. Rheinisches Landesmus. Bonn 1. Köln, 97-103.
Baatz, D., 1995. Die Wassermühle bei Vitruv X 5,2, *Saalburg-Jahrb*., 48, 5-18.
Beranová, M., 1963: Pravěké žernovy v Československu (Die vorzeitlichen Mühsteine in der Tschechoslowakei), *Vznik a počatky Slovanů*, 4, 181-219.
Beranová, M., 1980. *Zemědělství starých Slovanů*. Prag.
Beranová, M. and Břicháček, P., 1993. Beitrag zur Erkundung der landwirtschaftlichen Produktion in der späthallstattzeitlichen und latène zeitlichen Periode in Böhmen (Příspěvek k poznání zemědělské výroby v pozdně halštatském a časně laténském období v Čechách), *Arch. Rozhledy*, 45, 251-67.
Blanck, H., 1987. Utensili della cucina etrusca. In C. Vizzini (ed.), *L'alimentazione nel mondo antico*. Rom, 107-14.
Böhr, E. and Shefton, B.B. ,2000. Die griechische Keramik der Heuneburg. Kommentierter Katalog.In W. Kimmig (ed.), Importe und mediterrane Einflüsse auf der Heuneburg. *Heuneburgstud. 11 = Röm.-Germ. Forsch*. 59. Mainz, 1-26.
van den Boom, H. and Pape, J., 2000. Die massaliotischen Amphoren. In W. Kimmig (ed.), Importe und mediterrane Einflüsse auf der Heuneburg. *Heuneburgstud. 11 = Röm.-Germ. Forsch*. 59. Mainz, 43-70.
Brown. L. 1984. Objects of stone. In B.W. Cunliffe, *Danebury: an Iron Age Hillfort in Hampshire. II: The Excavations 1969-1978: The Finds*. Council British Arch. Research Report 52. London, 407-26.
Brun, J.-P., 2006. La diffusion de technologies méditerranéennes de transformation des produits agricoles dans le monde celtique durant l'Empire romain. In D. Paunier (ed.), *Celtes et Gaulois. L'archéologie face à l'histoire. La romanisation et la question de l'héritage celtique*. Actes de la table ronde de Lausanne 17. – 18. Juni 2005. Collect. Bibracte 12,5. Glux-en-Glenne, 93-108.
Buchsenschutz, O. 2002., Die Entstehung von Wirtschaftszentren in Gallien. In C. Dobiat, S. Sievers and Th. Stöllner (eds.), *Dürrnberg und Manching 2. Wirtschaftsarchäologie im ostkeltischen Raum*. Koll. Vor- u. Frühgesch. 7. Bonn, 63-76.
Buchsenschutz, O. and Boyer, F., 1999. Les meules. In O. Buchsenschutz (ed.), *Les remparts de Bibracte. Recherches récentes sur la Porte du Rebout et le tracé des fortifications*. Collect. Bibracte 3. Glux-en-Glenne, 213-16.
Buxo i Capdevila, Y., 1991. Aperçu sur les artefacts associés à la transformation des céréales en préhistoire. État des études en Espagne méditerranéenne. In F. Sigaut and D. Fournier (eds.), *La préparation alimentaire des céréales*. PACT 26. Strasbourg, 17-24.
Chausserie-Laprée, J., 1998. Les meules des habitats protohistoriques de Martigues. *Doc. Arch. Méridionale*, 21, 211-35.
Childe, V.G., 1943. Rotary Querns on the Continent and in the Mediterranean Basin. *Antiquity*, 17, 19-26.
Curtis, R.I., 2001. *Ancient food technology*. Leiden, Boston, Köln.
Curwen, E. C., 1941. More about querns. *Antiquity*, 15, 15-32.
Dal Ri, L., 1994. Die Mahlsteine, ein archäologisches Problem. In S. W. de Rachewiltz (ed.), *Korn und Mahlsteine. Das Mahlen von Getreide in Südtirol von der Urgeschichte bis ins Mittelalter*. Ausst. Südtiroler Landesmus. Schloß Tirol 27. April – 24. Juli 1994. Bozen, 51-72.
Fischer, F., 1973. KEIMHΛIA. Bemerkungen zur kulturgeschichtlichen Interpretation des sogenannten Südimports in der späten Hallstatt- und frühen Latène-Kultur des westlichen Mitteleuropa. *Germania*, 51, 436-59.
Frankel, R., 2003. The Olynthus Mill, Its Origin, and Diffusion: Typology and Distribution, *Am. Journal Arch*., 107, 1-21.
Gamper, P., 2006. Die latène zeitliche Besiedlung am Ganglegg in Südtirol. Neue Forschungen zur Fritzens-Sanzeno-Kultur. *Internat. Arch*. 91, Rahden/Westf.
Gleirscher, P., 1987. Die Kleinfunde von der Hohen Birga bei Birgitz, *Ber. RGK* 68, 181-351.
Gleisberg, H., 1956. Technikgeschichte der Getreidemühle. *Dt. Mus. Abhandl. u. Ber. 24,3*. München.
González Ruibal, A., 2006-2007. Galaicos. Poder y comunidad en el Noroeste de la Península Ibérica (1200 a.C-50 d.C.). *Brigantium 18*. A

Coruña.

Haberey, W. and Rest, W., 1941. Vorgeschichtliche und frühmittelalterliche Siedlungsreste in Kottenheim. Kreis Mayen, *Bonner Jahrb.* 146, 395-403.

Haffner, A., 2003. Type et fonction. In C. Rolley (ed.), *La tombe princière de Vix*. Paris, 176-88.

Herrmann, F.-R. and Jockenhövel, A., 1990. *Die Vorgeschichte Hessens*. Stuttgart.

Heun, S., 1992/93. Eine früh La Tène zeitliche Siedlungsstelle bei Ober-Hörgern, Stadt Münzenberg, Wetteraukreis, *Fundber. Hessen*, 32/33, 69-104.

Holodňák, P., 2001. Experiment s mletím obilnin na žernovech tzv. řeckého typu. Experiment mit dem Mahlen des Getreides an den Mahlsteinen des sog. griechischen Typs (Balkenhandmühlen), *Arch. Rozhledy*, 53, 31-44.

Holodňák, P. and Mag, M., 1999. Vývoj mlecích zařízení a provenience surovin drtidel a žernovů v Soběsukách (Okr. Chomutov, SZ Čechy). Mikrosonda do ekonomiky jednoho sídliště (Die Entwicklung der Mahleinrichtung und die Herkunft des Rohstoffes der Reib- und Mühlsteine von Soběsuky (Bez. Chomutov, Nordwestböhmen). Eine Mikrosonde in die Wirtschaftsstruktur einer Siedlung), *Pam. Arch.*, 90, 398-441.

Hörter, P., 1914. Die Basaltlava-Industrie bei Mayen (Rheinland) in vorrömischer und römischer Zeit. *Mannus*, 6, 283-294 with Taf. XII-XV.

Jacobi, J., 1912 Römische Getreidemühlen. *Saalburg-Jahrb.*, 3, 75-95.

Joachim, H.-E., 1980. Jüngerlatènezeitliche Siedlungen bei Eschweiler, Kr. Aachen. Mit Beiträgen von K.-H. Knörzer und J. Schalich. *Bonner Jahrb.*, 180, 355-459.

Joachim, H.-E., 1985. Zu eisenzeitlichen Reibsteinen aus Basaltlava, den sog. Napoleonshüten. *Arch. Korrbl.*, 15, 359-69.

Joffroy, R., 1962. *Le trésor de Vix. Histoire et portée d'une grande découverte*. Fayard.

Jüngling, P., 1987. Vorbericht über die Ausgrabungen früh latène zeitlicher Siedlungsfunde in Hanau Klein-Auheim. *Neues Magazin Hanau Gesch.*, 9, 203-05.

Junkelmann, M.,1997. *Panis militaris*. Mainz.

Kimmig, W., 1983. Die griechische Kolonisation im westlichen Mittelmeergebiet und ihre Wirkung auf die Landschaften des westlichen Mitteleuropa. *Jahrb.* RGZM, 30, 5-78 with Taf. 1-18.

Koch, H., 1991. Die keltischen Siedlungen vom Frauenberg über Kloster Weltenburg, Stadt Kelheim, und von Harting (Neubaugebiet Süd), Stadt Regensburg. *Internat. Arch.* 3. Buch am Erlbach.

Kohler, Ch., 2000. Die Obeloi der Heuneburg. In W. Kimmig (ed.), *Importe und mediterrane Einflüsse auf der Heuneburg*. Heuneburgstud. 11 = Röm.-Germ. Forsch. 59. Mainz, 197-213.

Lang, F., 2002. *Klassische Archäologie*. Eine Einführung in Methode, Theorie und Praxis. Tübingen, Basel.

Laws, K., Brown, L. and Roe, F., 1991. Objects of stone. In B. Cunliffe and C. Poole (eds.), *Danebury an Iron Age Hillfort in Hampshire 5. The Excavations, 1979-88: the finds*. Council British Arch. Research Report 73. London, 382-404.

Lucas, A., 2006: Wind, Water, Work. Ancient and Medieval Milling Technology. *Technology and Change in History 8*. Leiden, Boston.

Lynn, C.J., 1997. The date of deposition of the ape cranium. In D.M. Waterman and C.J. Lynn (eds.), *Excavations at Navan Fort 1961-71. County Armagh*. Northern Ireland Archaeological Monographs 3. Belfast, 125.

Mangartz, F., 2007. Een 'logboek' voor De Meern 1: beschrijving en determinatie van de herkomst van het natuursteen. In T. de Groot and J.-M. A. W. Morel (eds.), Een Romeinse Rijnaak, gevonden in Utrecht-De Meern. Resultaten van het onderzoek naar de platbodem 'De Meern 1'. *Rapportage Arch. Monumentenzorg 144*. Amersfoort, 246-55.

Mangartz, F., 2008. Römischer Basaltlava-Abbau zwischen Eifel und Rhein. *Monogr. RGZM 75. Vulkanpark-Forsch. 7*. Mainz.

Mansel, K., 1996. Zu einer Gürtelschließe aus dem Süden der Iberischen Halbinsel vom Magdalenenberg bei Villingen (Baden-Württemberg). In Th. Stöllner (ed.), Europa celtica. Untersuchungen zur Hallstatt- und Latènekultur. *Veröff. Vorgesch. Seminar Marburg Sonderbd. 12*. Espelkamp, 153-65.

Marzoli, C. and Donner, M., 1994. Das Mahlen: Entwicklung der Techniken und Geräte. In S. W. De Rachewiltz (ed.), *Korn und Mahlsteine. Das Mahlen von Getreide in Südtirol von der Urgeschichte bis ins Mittelalter.* Ausst. Südtiroler Landesmus. Schloß Tirol 27. April – 24. Juli 1994. Bozen, 73-98.

Morel, J. P., 2001. Aux origines du moulin rotatif? Une meule circulaire de la fin du VIe siècle avant notre ère à Carthage. In J.-P. Brun and Ph. Jockey (eds.), Techniques et sociétés en Méditerranée. *Collect. Atelier Méditerranéen.* Paris, 241-49.

Morin, D. and Tizzoni, M., 2009. Aux origines des techniques minières. L'exploitation d'un gisement filonien au Premier Âge du fer. Les mines de Silter di Campolungo et de Baita Cludona di Fondo (Val Camonica, Alpes lombardes, Italie). *Bull. Soc. Préhist. Française*, 106,109-41.

Moritz, L.A., 1958. *Grain-Mills and Flour in Classical Antiquity*. Oxford.

Müller-Beck, H., 1959/60. Grabungen auf der Engehalbinsel bei Bern 1957 und 1959. *Jahrb. Bern Hist. Mus.*, 39/40, 383-410.

Müller-Depreux, A., 2005. Die hallstatt- und früh latène zeitliche Siedlung »Erdwerk I« von Niedererlbach, Landkreis Landshut. *Materialh. Bayer. Vorgesch. A 87*. Kallmünz/Opf.

Needham, J., 1965. *Science and Civilisation in China 4*. Physics and Physical Technology. 2: Mechanical Engineering. Cambridge.

Neumaier, J., 1995. Eine iberische Fundgruppe Südwestdeutschlands in der Späthallstattzeit. *Germania*, 73, 495-503.

Oesterwind, B.C. and Schäfer, K., 1989. Die Mittellatènzeit im Neuwieder Becken. *Pellenz Mus. 4*. Nickenich.

Pape, J., 2000. Die attische Keramik der Heuneburg und der keramische Südimport in der Zone nördlich der Alpen während der Hallstattzeit. In W. Kimmig (ed.), *Importe und mediterrane Einflüsse auf der Heuneburg. Heuneburgstud. 11 = Röm.-Germ. Forsch. 59*. Mainz, 71-175.

Pare, Ch.F.E., 1989. Ein zweites Fürstengrab von Apremont – »La Motte aux Fées« (Arr. Vesoul, Dép. Haute-Saône). Untersuchungen zur Späthallstattkultur im ostfranzösischen Raum. *Jahrb. RGZM 36*, 411-72 with Taf. 34-36.

Pauli, L., 1993. Hallstatt- und Latènezeit. In H. Bender, L. Pauli and I. Stork (eds.), Der Münsterberg in Breisach II. Hallstatt- und Latènezeit, *Münchner Beitr. Vor- u. Frühgesch. 40*. München, 19-172.

Peacock, D.P.S., 1987. Iron Age and Roman quern production at Lodsworth, West Sussex. *Ant. Journal*, 67, 61-85.

Peiffer, K., 1972. Antike Quellen zur Geschichte der Atlantikfahrten. *Almogaren*, 3, 175-95.

Penack, J.-J., 1991. Eine frühgeschichtliche Handmühle aus Fritzlar-Geismar, Schwalm-Eder-Kreis. In Th. H.-D. Scholz (ed.), *Wasser- und Windmühlen in Kurhessen und Waldeck-Pyrmont 1*. Kaufungen, 55-63.

Peschel, K., 1979. Mittelmeerwelt und keltischer Norden nach archäologischen Quellen des 7. – 1. Jh. v. u. Z. In J. Preuss (ed.), *Von der archäologischen Quelle zur historischen Aussage*. Halle, 221-48.

Pommepuy, C. 1999. Le matériel de mouture de la vallée de l'Aisne de l'Age du Bronze à La Tène Finale: formes et materiaux. *Rev. Arch. Picardie*, 3/4, 115-41.

Py, M. 1992. Meules d'époque protohistorique et romaine provenant de Lattes. *Lattara*, 5, 183-232.

Raftery, B. 1997. Implications of the monkey bones from site B. In D.M. Waterman and C.J. Lynn (eds.), *Excavations at Navan Fort 1961-71. County Armagh*. Northern Ireland Archaeological Monographs 3. Belfast, 121-24.

Reille, J.-L., 2000. L'apparition des meules rotatives en Languedoc Oriental (IV s. avant J.-C.) d'après l'étude du site de Lattes. *Gallia*, 57, 261-72.

Roseman, Ch.H., 1994. Pytheas of Massalia: *On the Ocean. Text, Translation and Commentary*. Chicago.

Schüle, W., 1969. Die Meseta-Kulturen der Iberischen Halbinsel. Mediterrane und eurasische Elemente in früheisenzeitlichen Kulturen Südwesteuropas. *Madrider Forsch. 3*. Berlin.

Shefton, B.B., 2000. On the Material in its northern Setting. In W. Kimmig (ed.), *Importe und mediterrane Einflüsse auf der Heuneburg. Heuneburgstud 11 = Röm.-Germ. Forsch. 59*. Mainz, 27-41.

Spindler, K., 1973. *Magdalenenberg 3. Der hallstattzeitliche Fürstengrabhügel bei Villingen im Schwarzwald*. Villingen.

Spindler, K., 1983. *Die frühen Kelten*. Stuttgart.

Staubitz, H.-J., 2007. Die Mühlsteine des spätkeltischen Heidetränk-Oppidums im Taunus. *Kleine Schr. Vorgesch. Seminar Marburg 56*. Marburg.

Tseretheli, G., 1939. *Sakartvelos museumis urartuli Zeglebi. The Urartean Monuments in the Georgian Museum Tbilissi*. 1939.

von Uslar, R., 1938/39. Kottenheim (Kreis Mayen) *Bonner Jahrb.* 143/144, 374-75.

Wainwright, G.J., 1979. *Gussage All Saints. An Iron Age settlement in Dorset*. Department of the Environment Archaeological Report 10. London.

Waldhauser, J., 1981. Keltské rotační mlýny v Čechách (Keltische Drehmühlen in Böhmen). *Pam. Arch.* 72, 153-221.

Watts, M., 2002. *The Archaeology of Mills and Milling*. Charleston.

Wefers, S., in press. *Mühlen der Latènezeit zwischen den Steinbruchrevieren Mayen und Lovosice*. Monogr. RGZM. = Vulkanpark-Forsch. (Mainz in press).

Wielowiejski, J., 1960. Przemiany gospodarczo-społeczne u ludności południowej Polski w okresiepóźnolateńskim i rzymskim (The social and economic development of southern Poland in the late La Tène and the Roman Period). *Mat. Starożytne*, 6, 7-426.

Wikander, Ö.J., 1985. Archaeological evidence for early water-mills – an interim report. *History of Technology*, 10, 151-79.

The earliest rotary querns in southern England

David Peacock & Lynn Cutler

Introduction

In the previous paper, Steffi Wefers reviews the earliest rotary querns throughout Europe. Here we take a small part of the evidence and examine it critically at a scale that would be impossible in Wefers' broad survey. We do not differ in our outcomes, but this paper is intended to complement hers by looking in detail at southern England. Our object is to evaluate the strength or weakness of the case by analysing the evidence.

It is now more than seventy years since Curwen (1937; 1941) initiated the debate about the origin of the rotary quern. He considered it to be a domestic derivative of the Pompeian style animal mill which he thought, was based on a Greek prototype appearing in the eastern Mediterranean perhaps in second century BC. His work stimulated Gordon Childe (1943) to bring his encyclopaedic knowledge of European prehistory to bear on the problem and he, with characteristic perspicacity, pointed to north-eastern Iberia as a probable point of origin for the rotary quern. The east, he claimed, was dominated by the oscillatory Olynthus style mill and the north by saddle querns. Current research has proved his ideas fundamentally correct for Alonso (1997; 1995; 2002) and her colleagues have shown that the rotary quern in Catalonia goes back to the fifth century BC, while Runnels (1990) has demonstrated it did not appear in Greece until the first century BC. In Egypt the Olynthus mill may have co-existed with the rotary variety until the second century AD (Peacock 2007). Meanwhile, Morel (2001) has published a quern fragment from a Carthage tomb which he claims to be sixth century in date, making it the earliest rotary quern yet found. Unfortunately, Morel's illustration gives little away as it comprises a photograph of what appears to be an amorphous block of stone. However, the author is a respected and knowledgeable archaeologist, who would certainly know a rotary quern and the find cannot be easily dismissed. Equally, it could be argued that as it was found as a fragment, it is more likely to be intrusive than an integral grave good. This must be balanced with Hill's (1995) work in which he suggested that quern breakage might be a ritual act in the British Iron Age, so it is not impossible that the mill was part of the burial assemblage.

In southern England it has generally been thought that rotary querns were an innovation associated with the middle Iron Age and a conventional date of around 350 BC was often quoted. This date may owe something to Curwen's (1937, 138) view that the invention could be ascribed to a forgotten forerunner of Archimedes (c. 287-212 BC) or to Hawkes' (1959) ABC of the British Iron Age as he dated the middle Iron Age from 350 BC. However, in recent years the southern British, or more specifically Wessex Iron Age, has been the subject of extensive research, principally by Wainwright who organised a series of co-ordinated rescue excavations, and Cunliffe, who undertook major research excavations at a number of Wessex hillforts and settlements, particularly Danebury. As a result we are now in a position to re-evaluate the date of the arrival of rotary querns. Three sites seem of particular importance, Rowbury Farm in Wherwell, Danebury near Stockbridge and Gussage All Saints in Dorset. The only other potentially early site, although well outside the area, is Thorpe Thewles in Cleveland. The object of this paper is to critically evaluate the evidence and to discuss its wider implications.

1.1 Rowbury

The site at Rowbury Farm comprised an enclosure occupied between the early Iron Age to the second century AD and has produced what is potentially the earliest rotary quern yet excavated from British soil. The quern in question comes from Pit 402 (1) which is a shallow excavation about 90 cm across, but only 30 cm deep (Cunliffe and Poole 2008, fig. 5.29). It is associated with pottery, which on the nearby Danebury sequence, would be dated 550-450 BC. It comprises a rather shapeless lump of puddingstone about 115 mm thick and 225 mm across. There are traces of a spindle hole demonstrating that it was the lower stone of a rotary quern. Study of puddingstones, British and French, shows that this rock is similar to the rock from the quarries in the Bois des Hogues, St-Léonard, near Fécamp on the Normandy coast. If correctly dated it would be the earliest French puddingstone yet discovered anywhere, as well as perhaps the earliest rotary quern known from Britain.

Hertfordshire puddingstone appears to be essentially a Roman phenomenon, although Major (2004) has drawn attention to a few examples from Essex which could be pre-Conquest. French puddingstone is difficult to date but there are indications that the industry may have started considerably earlier. Thus the Avrilly production site is dated to the Augustan period (Guillier *et al.* 2005). However the Normandy Museum in Caen, has on display, a fine quern in French puddingstone which was found as part of a La Tène III grave assemblage at Mondeville in Calvados. This would seem to indicate an Iron Age origin for the industry, but there is no evidence of anything as early as Rowbury.

However, the Rowbury evidence must be treated with caution. The find came from the uppermost layer (1) of a very shallow pit and while there is no mention of contamination, this must remain a possibility. Secondly, many of the pits considered to be of middle Iron Age date contained earlier pottery (Cunliffe and Poole 2008, 131) and so it could be that the quern dates Pit 402 rather than vice-versa. On the other hand, the paucity of middle Iron Age pottery from the site is curious and the excavators postulate a period of abandonment in the second and first centuries BC (Cunliffe and Poole 2008, 132). If this is the case, the quern is most probably early or late Iron Age rather than middle Iron Age. Either way, it is important evidence for the pre-Conquest importation of French puddingstone.

Rowbury begs an important question, but does not provide a firm answer.

1.2 Gussage-All-Saints

Our second Wessex site was published almost a decade earlier (Wainwright 1979). The site is a similar enclosed settlement occupied between the earliest Iron Age and first century AD. The excavators recognised three phases of which the earliest is relevant to this discussion. According to Buckley (1979) Phase 1 produced a total of 127 fragments of worked stone of which 54 were querns. Of these nine or 17% were of the rotary type. This phase antedates phase 2 which seems to date from the third century BC. It is supported by three radiocarbon dates which suggest a corrected date of c. 750-430 BC. In other words it appears that rotary querns were introduced at some stage between the eight and fifth centuries BC.

In this case, caution is needed as three firm radiocarbon dates are available and the pits were phased by their pottery content. Once again it is possible that a middle Iron Age pit would contain nothing but early pottery and that the querns date the pits rather than vice-versa.

1.3 Danebury

Danebury is the most extensively excavated hillfort in Britain, thanks to a prolonged series of excavations by Cunliffe (1984). It was here that Cunliffe defined the ceramic phases which are used to date the Wessex Iron Age. These were backed by an extensive programme of radiocarbon dates. In all 77 dates were measured, but of these 13 were disregarded because they gave anomalous results clearly indicating something amiss. However, this leaves 64 dates which is an excellent sample on which to found a chronology.

It appears that rotary querns were introduced in ceramic phase 5. There were nine radiocarbon dates for cp4-5 which translate as broadly 450-400BC (Cunliffe 1984, 194). This suggests firmly that rotary querns were used in Wessex in the fifth century BC. Laws, Brown and Roe (1991, 396) sum up the situation admirably:
'Clearly the majority of the rotaries occur in Middle-Late Iron Age contexts (Cp 6–8). The six examples from phase 5 contexts are sufficient, however, to show that rotaries were already in use at Danebury by the middle of the fifth century. The four examples from Cp 3 or 4 are more difficult to assess. These pits contain very little datable material and what there is may be residual. On this evidence, therefore, it cannot be assumed that rotary querns began as early as the sixth century.' This modifies Brown's (1984) original assessment:
'The earliest, reliably dated, rotary quern (a Type R1) appears in a cp 5 context. There are also three examples of the type from cp 6 pits. These form only a small proportion of the Type R1 querns, however, and the general date of that type seems compatible with that of Types R2–R5. In terms of real chronology this would suggest that the rotary querns were beginning to be introduced no earlier than the 4th century BC. The argument based on the evidence from Gussage All Saints (Wainwright 1979, 89) that rotary querns were introduced much earlier in the Iron Age is, therefore, contradicted by the Danebury chronology.'

If we allow for the small radiocarbon sample, the error inherent in the method, and the broad phasing of Gussage there need be no conflict between the two.

1.4 Thorpe Thewles

This site is not in southern England but in Cleveland, Northumberland (Heslop 1987, 111). However, it must be mentioned as it has been cited in evidence for the fifth century BC introduction of the rotary quern (e.g. Watts 2002, 28). In fact only two stones are potentially of this date and the case is based on pottery thermoluminescence dates, which for the earliest settlement (phase II), have a mean of 485 ±190 BC. However, there are problems with this. Firstly, Swain's (1987), pottery report refers to the assemblage as the *'largest well stratified group of **Late** Iron Age pottery from the northeast'* (our emphasis). Equally, Bailiff's report on the dating categorically states: *'This mean date for Phase II raises the interesting question of whether the occupation of the site was longer than had been previously thought (a proposed terminus ante quem of 300 BC), although this cannot be clearly demonstrated at the 95% level of confidence using the overall error'*. At the 68% confidence level there would be a one in three chance of the date lying outside the range 675 – 295 BC and it appears, sadly, that the case is far from proven.

1.5 Discussion

It is now becoming clear that rotary querns have been in use for a much longer period than previously suspected and there is a strong case for suggesting that the idea was adopted at least as early as fifth century BC, rather than the mid fourth to early third. It remains an open possibility that future research will promote the case for a sixth century date. This means that the Wessex querns are as early as any in Europe and could even be as old as the Carthage example, although until that is fully published it is hard to place too much reliance on the date. In Wessex the earliest querns appear as the fully developed Sussex type of Curwen (1937, 1941) or Danebury type R1 (Brown 1984), which is perhaps surprising as we might anticipate a preceding phase of experimentation. However, this flat topped hopperless type is without parallel on the Continent and seems to have been a local innovation.

Having disposed of the old '*ex oriente lux*' argument it now seems that even a source in the Mediterranean is in doubt. It seems inconceivable that the idea was invented and spread southwards from Britain, leapfrogging large tracts of Europe. Perhaps it is not the evidence but the model supposing diffusion from a single point of origin that is at fault.

To make a rotary quern is not rocket science and it could easily have sprung up where the social and economic conditions dictated, without the intervention of blue prints or inspiration from another place. Here Holodňák's (2001) experiments are of greatest significance. He was able to show that the Olynthus mill was six times more efficient than a saddle quern and a rotary quern would be 6-12 times more efficient. This explains why the rotary quern eventually achieved a pan-European distribution, but it does not explain why the saddle quern continued to be used alongside it, nor does it explain the longevity of the Olynthus mill in the East. In theory an innovation with this advantage should eclipse older methods but in reality this does not seem to be the case. Clearly there are other parameters that we need to isolate before the situation will be clarified.

The choice of mill might relate to grain type, but there is very little data on this. There are two ways in which the *lacuna* might be filled. The first is by studying assemblages of carbonised seeds from sites with querns, although it is usually hard to correlate between the two types of find as they often come from different loci. The second would be to study phytoliths trapped in the cavities of the grinding surface. While this approach has been used fairly frequently on American grindstones (Piperno 2006), European attempts, although promising, have been sporadic (e.g. Ross 2004; Juan Tresseras 2002; Aranguren *et al.* 2007). It seems that identification to genus level will be feasible, but not species. This could help differentiate acorns, oats, rye, barley, and wheat for example, but not spelt from emmer as they are both species of *Triticum.*

It has often been said that necessity is the mother of invention and it is almost a commonplace that innovations relate to need and if there is no need there will be no innovation. For example, the Romans knew of wind power, but they failed to invent the windmill presumably because there was no need. Watermills had all the capacity that was required in a relatively stable economy. If society or its economic milieu changes, this might act as a catalyst for introducing new concepts and techniques. In this case increased population might lead for a need to process more grain quickly or perhaps there was a social change in which a process, once sited in every household, became the preserve of specialists producing flour for the whole community. Such factors might promote innovation without outside intervention.

References

Alonso Martínez, N., 2002. Le Moulin rotatif manuel au nord-est de la Péninsule ibérique: une innovation technique dans le contexte domestique de a mouture des cereals. In H. Procopiou and R. Treuil (eds) *Moudre et Broyer II.* Paris, 111- 27.

Alonso-Martínez, N., 1995. Les premières meules rotatives manuelles dans le nord-est de la péninsule Ibérique. In M.-C. Amouretti, and G. Comet, (eds). *La Transmission des connaissances techniques. Tables rondes Aix-en-Provence April 1993 - Mai 1994. Cahier Hist. Technique,* 3. 15-23.

Alonso-Martínez, N., 1997. Origen y expansión del molino rotativo bajo en el mediterráneo occidental. In D. Meeks, and D. Garcia, (eds). *Techniques et Économie Antiques et Médiévales: le temps de l'innovation. Colloque international Aix-en-Provence 21-23 Mai 1996.* Paris, 15-9.

Aranguren, B., Becattini, R., Mariotti Lippi, M., and Revedin, A., 2007. Grinding flour in Upper Palaeolithic Europe (25000 years bp). *Antiquity,* 81, 845-55.

Bailiff, I. K., 1987. The thermoluminescent dating of the Iron Age pottery. In Heslop 1987, 71-72.

Brown, L., 1984. Objects of stone. In Cunliffe 1984, 407-25

Buckley, D., 1979. The stone. In Wainwright, 1979, 89-97.

Childe, V.G., 1943. Rotary Querns on the Continent and in the Mediterranean Basin. *Antiquity*, 17, 19-26.

Cunliffe, B. 1984. *Danebury: An Iron Age hillfort in Hampshire Vol. 2 The excavations 1969-1978: the finds*. CBA Research Report 52.

Cunliffe, B. and Poole, C. 1991. *Danebury: An Iron Age hillfort in Hampshire Vol. 5 The excavations 1979-1988: the finds*. CBA Research Report 73.

Cunliffe, B and Poole, C., 2008. *Rowbury Farm, Wherwell, Hants, 2003.* Oxford Archaeology Monograph 71.

Curwen, E.C., 1937. Querns. *Antiquity,* 11, 133-51.

Curwen, E.C., 1941. More about querns. *Antiquity,* 15, 15-32.

Guillier, G., Biard, M., and Cherel, A-F., 2005. Un atelier augustéen de taille de meules en poudingue au 'Clos des Forges' à Avrilly (Eure). *Revue archéologique de l'Ouest*, 22, 199-220.

Hawkes, C., 1959. The ABC of the British Iron Age. *Antiquity*, 33, 170-82.

Heslop, D.H., 1987. *The Excavation of an Iron Age Settlement at Thorpe Thewles, Cleveland, 1980-1982.* CBA Research Report 65.

Hill, J.D., 1995. *Ritual and rubbish in the Iron Age of Wessex. A study of the formation of a specific archaeological record.* BAR 242.

Holodňák, P., 2001. Experiment s mletím obilnin na žernovech tzv. Řeckého typu. *Archeologické rozhledy*, 53, 31-44.

Juan Tresseras, J., 2002. Resultados de los análisis de residuos. In H. Procopiou and R. Treuil (eds) *Moudre et Broyer II.* Paris, 145-54.

Laws, K., Brown, L., and Roe, F., 1991. Objects of stone. In Cunliffe and Poole 1991, 382-404.

Major, H., 2004. The dating of puddingstone querns. *Lucerna,* 27, 2-4.

Morel, J-P., 2001. Aux origins du moulin rotatif? Une meule circulaire de la fin du VIe siècle avant notre ère à Carthage. In J-P Brun, and P. Jockey (eds). *Techniques et sociétés en Méditerranée.* Collection Atelier Méditerranéen. Paris, 241-9.

Peacock, D., 2007. Stone artefacts. In D.Peacock and V. Maxfield *The Roman Imperial Quarries. Survey and excavation at Mons Porphyrites 1994-1998, Volume 2.* London, 267-81.

Piperno, D.R., 2006. *Phytoliths: a comprehensive guide for archaeologists and paleoecologists.* Lanham and Oxford.

Ross, J.M., 2004. Phytoliths from a Norse Greenlandic quern stone: a preliminary investigation. *Environmental Archaeology*, 9, 99-106.

Runnels, C., 1990. Rotary querns in Greece. *Journal of Roman Archaeology*, 3, 147-54.

Swain, H. P. 1987. The Iron Age pottery. In Heslop 1987, 57-71.

Wainwright, G. J., 1979. *Gussage All Saints : an Iron Age settlement in Dorset*. London.

Watts, M., 2002. *The Archaeology of Mills & Milling.* Stroud.

Late La Tène Rotary querns found in the Oppidum of Heidengraben: a new type of volcanic rock and its origin

Achim Lehmkuhl

The Oppidum of Heidengraben ('Heathen Ditch') is situated in south-west Germany on the north side of the Swabian Alb near the village of Grabenstetten in the districts of Reutlingen and Esslingen (Baden-Württemberg).

The Oppidum, which is of the late La Tène period, is located on a high plateau forming a peninsula, naturally protected by steep slopes. Some paths ascend from the valleys to the plateau, and there the plain was protected by ramparts with ditches in front of them. The whole paved area measures about 1.700 ha, so it is one of the largest oppida in Europe of this period. In the south there is an inner area 153 ha in size, which was protected by another system of ramparts and ditches. Four in-turned gateways ('Zangentore') lead into the probable centre of the settlement, which was called 'Elsachstadt' by the first excavators at the beginning of the 20th Century (Fischer 1982).

Only a few small areas of the plain have been archaeologically excavated so far, mainly between 1994 and 1999 by the University of Tübingen in connection with farmland consolidation. Most of the archaeological artefacts have been collected due to systematic field surveys in the last two decades. This is necessary, as the whole Elsachstadt is intensively used for agriculture. The artefacts are unearthed by the plough and are in danger of being destroyed during the winter. A lot of fragments of glass rings of the late La Tène D1 have been found as well as some coins of Leuker and Sequaner Potin type, and numerous amphora sherds of the type Dressel 1A (Knopf 2006, 68-97). These, as well as the finds from the excavations, demonstrate far-reaching trade connections, for example to the Mediterranean.

Figure 1. The complete quernstone with the inner and outer hole, scale 10 cm (Photograph: University Tübingen, H. Jensen).

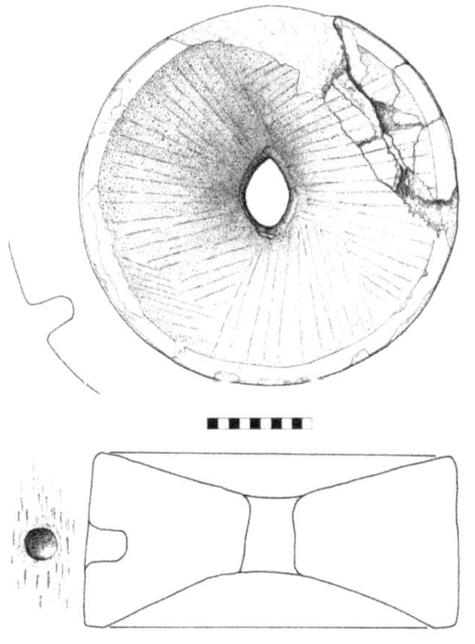

Figure 2. Drawing of the probably unused side with the well preserved marks of production (Drawing: University Tübingen, M. Möck-Aksoy).

In addition, about 300 red stone fragments were found, representing a type of rock not naturally occurring at the site. After initial inspection their function remained unclear. In the spring of 1998 a complete rotary quern of the same material was found. It was unearthed by a plough, which damaged the surface. The freshly broken red fragments highlighted the original find-site in the field. This place was then excavated and more objects, a scale and a weight stone, were found (Knopf 2000, 79-80).

The quern has a diameter of 36 cm, a height of 17 - 19 cm and a weight of 30 kg, which is relatively light for its size. The grinding faces of the upper and the lower side of the stone are inclined inwards and show only a slight abrasion. The central hole is pointed oval and measures 9 to 4.5 cm (Fig. 1). On the outer radius there is another round hole

Figure 3. Irregularly shaped inclusions in the used grinding face, scale 10 cm (Photograph: A. Lehmkuhl).

with a depth of 3 cm and a diameter of 3 cm; it is not going all the way through and may have been used for a crank, in which case this quern could be interpreted as a runner stone. The face of the outside radius is finely chiselled. On one grinding face the chamfered outer edge and the well preserved marks of production are visible, which is probably an indication that this was the unused side of the quern (Fig. 2). The surface of the used side is worn-down and of deep red colour.

Initially, the red colour of this stone suggested that it could have been made from the Permian Breccia of Schweigmatt, a sedimentary rock of the southern Black Forest, which previously was documented as a material for other quernstones from the late La Tène and early Roman period (Joos 1975, 197-9). Both stones have a similar reddish colour with bright inclusions. But the Schweigmatt material is more compact and the components consist of glittering gneiss and gleaming pieces of quartz. In contrast, the material from the Heidengraben, the stone as well as the irregularly shaped inclusions, is porous and thus relatively light and rough, which is typical for a volcanic rock (Fig. 3).

After this discovery, the author had the opportunity to compare this material with similar looking fragments which were excavated in a square ditch enclosure ('Viereckschanze') at Ladenburg near Heidelberg. The inner side of these stones was blackened, but the surface was of the same red colour as the Heidengraben quernstone. It was a volcanic material, but different from a basalt or basalt tuff. It is a so-called porphyry tuff, which was silicified to a solid breccia. Internationally the term 'Ryolithtuff' is commonly used. Because of the finding place at Ladenburg, the research for the origin of this material was now focussed on the surroundings of Heidelberg. It could be located in a small forest in the southern Odenwald mountain range. There it was formed in the Permian, at the margin of a volcanic formation, the Dossenheimer Quarzporphyr, named after the nearest little town. Some parts of this porphyry tuff, an effusive or eruptive material of an old volcanic massif, were afterwards silicified, i.e., impregnated by quartz (SiO_2). Characteristic of the silicification are needles of quartz crystals in holes, which can be seen under the microscope. As the quartz porphyry quarries at Dossenheim are still in use, the chance of finding traces of antique quarries is low. There are only a few remaining deposits in the forest, near the ruin of a castle, which is a protected monument .

In the following years, several additional quernstones made from the same material were found during excavations in two square ditch enclosures ('Viereckschanzen') of the late La Tène at Nordheim near Heilbronn. They have been burnt, broken, and thrown into a pit. Another complete quernstone was found at Nussloch near Wiesloch, also not far from the Neckar. So one can now follow the way like a string of beads from the origin of the rock near Dossenheim along the Neckar River from Ladenburg, via Wiesloch and Nordheim to the Heidengraben (Fig. 4).

There, apart from the complete quernstone, more than 250 fragments of the same silicified porphyry tuff were found. They were spread over a wider area within the Elsachstadt, and they show slight differences in colour and also minor variations in the material. Some fragments show traces of

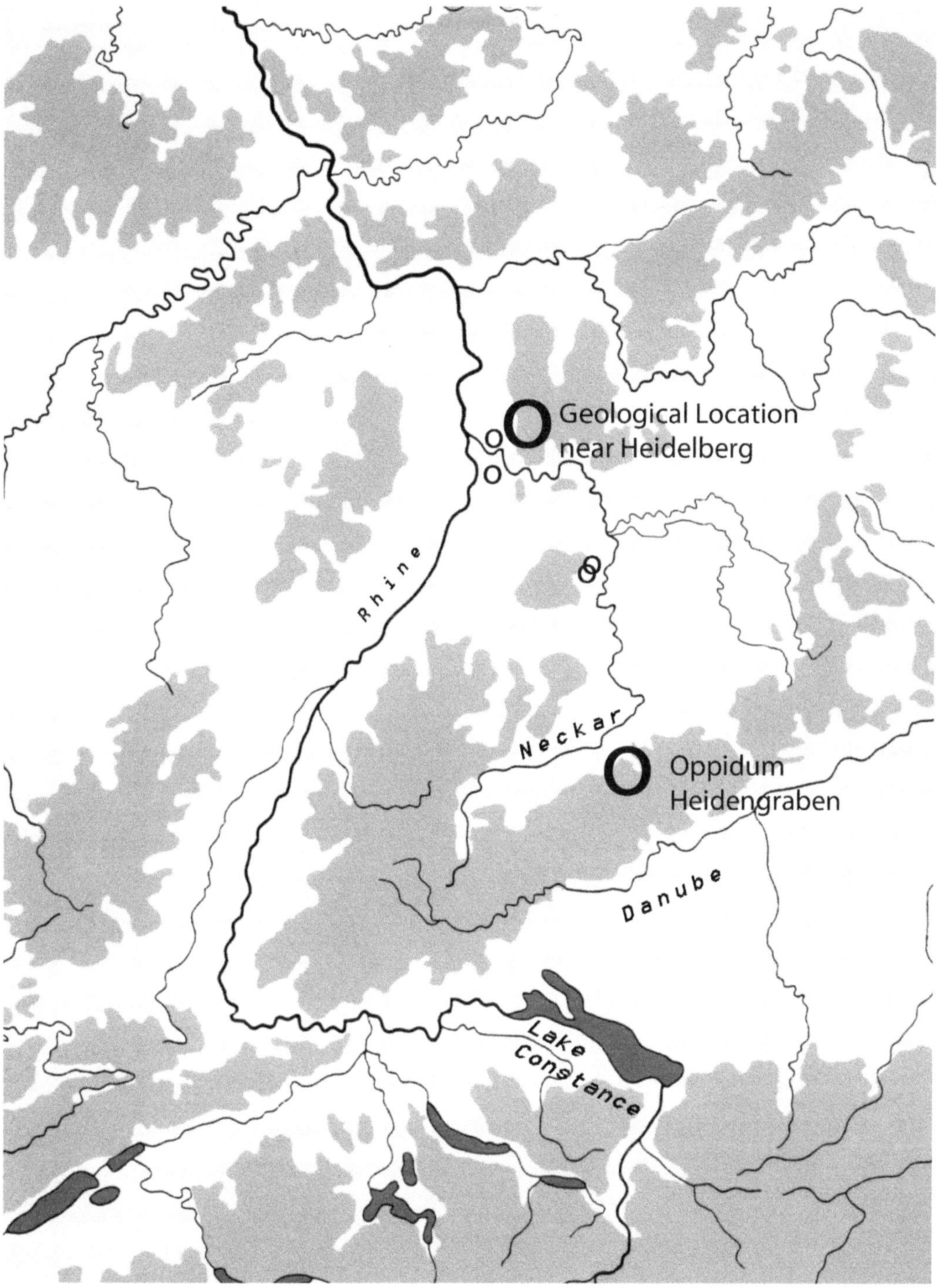

Figure 4. Origin of the geological formation of porphyry tuff and the distribution of millstones made of this material along the Neckar River up to the Oppidum Heidengraben.

production or pieces of a crafted surface. According to the shape of their surfaces with the characteristic inclination, all these fragments most likely belonged to runner stones. All this is evidence that there were at least 10 to 12 different quernstones in the Heidengraben area.

The quernstones found in the square ditch enclosures are all from the late La Tène period. But how can the quernstone and the fragments of the Heidengraben be classified chronologically, as they are survey finds from the surface? The other survey finds from the Elsachstadt, the amphorae and coins, the rings and beads of glass, are all from the late La Tène D1. Traces were neither found of the former phase, the middle La Tène, nor of the following period, the late La Tène D2. When the rampart was investigated in the beginning of the 20th Century, it showed no signs of repair. That can be seen as evidence that it was used only for a short time, probably for one generation. Even though the Eifel is not far away, there is no evidence of quernstones of basalt lava in the Heidengraben. This could also indicate for a small time frame or difficulties in trading, due to different dominions.

The number and the similarity of the quernstones and their distribution along the Neckar River, suggests a kind of industrial production. Could millstones made of this red silicified porphyry tuff be regarded as a new 'index fossil' for the late La Tène D1 in South Germany?

References

Andreae, A., and Osann, A., 1918. *Geologische Spezialkarte des Grossherzogtums Baden. Erläuterungen zu Blatt Heidelberg*. Heidelberg, 29-34.

Bock, Ch., and Lehmkuhl, A., 2006. Die eisenzeitlichen Gesteinsfunde vom Heidengraben. In Th. Knopf, *Der Heidengraben bei Grabenstetten. Archäologische Untersuchungen zur Besiedlungsgeschichte, Universitätsforschungen zur Prähistorischen Archäologie*. Bonn, 141, 175-195.

Fischer, F., 1982. *Der Heidengraben bei Grabenstetten. Ein keltisches Oppidum auf der Schwäbischen Alb bei Urach. Führer zu archäologischen Denkmälern in Baden-Württemberg,* 2. Stuttgart.

Joos, M., 1975. Eine permische Brekzie aus dem Südschwarzwald und ihre Verbreitung als Mühlstein im Spätlatène und in frührömischer Zeit. *Archäologisches Korrespondenzblatt,* 5, 197-199.

Knopf, Th., 2000. Ausgrabungen im Oppidum Heidengraben, Gde. Grabenstetten, Kreis Reutlingen. In Landesdenkmalamt Baden-Württemberg (ed.), *Archäologische Ausgrabungen in Baden-Württemberg 1999,* 79-80.

Knopf, Th., 2006. *Der Heidengraben bei Grabenstetten. Archäologische Untersuchungen zur Besiedlungsgeschichte, Universitätsforschungen zur Prähistorischen Archäologie, 141,*Bonn.

Lehmkuhl, A., and Stegmaier G., 2007. Viel Steine gab's und damit Brot. *Archäologie in Deutschland,* 5, 42.

Stegmaier, G., 2009. Stadt – Land – Fluss: Überlegungen zum Wirtschafts- und Besiedlungsgefüge des spätkeltischen Oppidums Heidengraben und seines weiteren Umlands. In R. Karl and J. Leskovar (ed.), *Interpretierte Eisenzeiten. Fallstudien, Methoden, Theorie. Tagungsbericht der 3. Linzer Gespräche zur interpretativen Eisenzeitarchäologie. Studien zur Kulturgeschichte von Oberösterreich. Linz, 22,* 253-264.

Stegmaier, G., 2009. Bemerkungen zu Lage und Bedeutung des spätkeltischen Oppidums Heidengraben. Naturraum, Topographie und Ökonomie. In *Kelten am Rhein. Akten des dreizehnten Internationalen Keltologiekongresses. Beihefte Bonner Jahrbücher*. Bonn 58.1, 161-170.

Wefers, S., 2009. Studien zu vorgeschichtlichen Drehmühlen in dem Gebiet zwischen den Produktionsstätten Mayen und Lovosice. *Archäologisches Nachrichtenblatt, 14.3,* 232-240.

Wieland, G., 1999. *Keltische Viereckschanzen: einem Rätsel auf der Spur*. Stuttgart.

First century querns of the Roman army in the light of modern texts

Florent Jodry

Text translated from the French by R.P. Symonds (INRAP)

'*... how much Antony's legions were made to suffer the loss of their querns when, during the expedition against the Parthians, they were forced to leave them behind, having insufficient beasts of burden to carry them.*'

Lieutenant-Général Joseph Rogniat *Considérations sur l'art de la guerre*, 1820

Nearly 2000 years after Cato's *De Agri Cultura* and Vitruvius' *De Architectura*, essential works mentioning different types of mills, a manuscript of 1813 (Béguin 2003, 392) presents a classification of contemporary mills. This '*Note sur les moulins d'armée, de siège et d'hôpital*' also includes hydraulic mills, windmills and mills powered by animals, as well as a manual mill called a '*Durand Père*'. By also proposing to supply 'a mill that could be carried by a single man and that would relieve marching troops of all encumbrance and equipment' this study provided the founding principles of the portable manual mill for military use.

This present paper is concerned with the portable manual mills of the Roman Army in the 1st century AD, through literary sources and finds derived from archaeological excavations. It will begin with an analysis of the civil and military literature of the 18th and 19th centuries in order to better appreciate the armies' motivations for using these devices, the circumstances of their use and their output. This first section will therefore be devoted to the numerous scientific works developed to meet the needs of marching armies.

With the help of a mill that is easily transported and adapted to all types of terrain, moving troops must be furnished with flour in all situations and in a practical way that optimizes the cereal rations and thus makes savings that are the concern of the commanders. A manual mill must at the same time be constructed with sufficient simplicity and solidity (Béguin 2003, 395) to cope with the temperaments and unwillingness of soldiers!

1. The mills of the modern army

Different types of mills continued in use into the modern period to meet the demand for flour for military units: these are mainly mills that use mechanical energy, whether wind-, water- or animal-powered. These very costly mills, an example of which was a water-powered mill introduced by Vauban and his engineers in 1688 (the Mont-Royal mill, downriver from Trier), were constructed inside fortifications in a building that would be well away from the powder magazines and resistant to cannon fire (Béguin 2003, 387). These mills were an indispensable element of a stronghold, since they permitted the continued transformation of (easily conserved) grain into flour and thus the production of bread for the inhabitants on a daily basis in the case of a siege. These machines were thus powerful and productive: we should not forget that a garrison of 6000 men consumes almost 3000 kg of flour per day. The report of Captain Belmas, *aide-de-camp* of Lt. Gen. Joseph Rogniart , is eloquent when he writes of the surrender of Mainz or the Fort Vauban on the Rhine in 1793 because of the lack of the means to mill when the wheat reserves were full! It seems that even though human-powered mills were used, these were insufficient for making flour (Belmas 1829, 288).

In a period of conflict when numerous regiments were on the move, the main concerns were, however, not only to deal with the regular problems of grain supply, but also to maintain the best possible support for the troops in flour. In order to resolve this matter, the Royal Army of France and the Napoleonic *Grande Armée* equipped the soldiers with portable manual mills that enabled processing of the ration of grain to provide good flour for the making of daily bread (Regnier 1814, 589).

1.1 A brief review of rations, value for money and milling

At the outset of a military campaign every soldier needs to be able to obtain flour very quickly, and this is the main concern of Sieur Delavault, a 'quartermaster' who in 1741 had 100 portable mills, capable of supplying a good quality flour, 'weight for weight', constructed for the

Author and date of publication	Quantity of flour obtained in 24 hours (Kg)	Quantity of flour obtained in 1 hour (Kg)	Ration-Equivalent (ration=0,734 kg)
Aubert de LCdB, mill of the XVIIth century	250	10,42	14,2
Le Courrier 1741	216	9	9,81
Marmont 1857, mill of 1811-1812	288	12	16,35
Regnier 1814	240	10	10,9
Belmas 1829, mill Durand of 1823	303	12,63	17,21

Table 1. The yield of modern military mills. [The publications of Le Courrier (1741) and Régnier (1814) mention the treated quantities of wheat rather than flour: the data has thus been adjusted, using a rate of flour extraction at 80%, corresponding to an average milling, or a flour that is brown or semi-wholemeal (Journal of Adrianor 2001)].

French army at war (in the War of the Austrian Succession) (Béguin 2003, 390). Economic efficiency and rationing are intended to meet the needs of the troops but above all to be within the means of governments. Here we focus on production, introducing the concept of a 'ration-equivalent' that represents the number of soldiers supplied with a ration of flour produced by a mill in one hour (see Table 1). The bread is the basic foodstuff of mobile troops, and its transformation from grain becomes therefore a matter of interest to the engineers whose reports will find their way to the influential people of the realm.

1.2 The portable mill of the Royal Army of France

The military dictionary (1758) of François-Alexandre Aubert de La Chesnaye-Desbois (1699-1784), presented to His Highness the Prince of Turenne, deals with tactics, discipline, engineering and artillery, but also, of particular interest here, with subsistence. In the chapter 'The Mill', we learn that a 'mill of particularly fine invention' was presented under the ministry of Louvois (second half of the 17th century), to meet the difficulties of grain supply in the event of a siege. Even if this apparatus would not be enough to sustain the resistance of a stronghold, it remains the first indispensable element of a mobile army. As we have seen, the portable mill needed to respond to several essential criteria for the good progress of an army, and to possess a number of qualities including a good shape, handling and efficiency. The apparatus in this case is made of iron, which means that it does not wear easily and does not heat up. Different sorts of flour can be produced, either coarse or fine according to wishes. It is circular, 16 inches (43 cm) in diameter, weighs 100 pounds (40 kg) and can be operated by a single man: it is thus easily moved about and adaptable to circumstances. Its efficiency is significant since the author tells us that this mill, using a crank, could be used to mill enough to feed 500 men in 24 hours, which is thus 14 ration-equivalents.

1.3 The portable mill of the Napoleonic Grande Armée

For the Napoleonic campaigns, the Maréchal d'Empire Marmont ordered an engineer of his regiment to design a rotary mill for which the conditions were fairly simple: 'Make a manual mill light enough that if necessary a soldier can carry one'; that can be 'turned by a single man' and that 'gives the finest flour'. The Maréchal Marmont, obviously concerned with the yield, goes on to insist that the mill must be able to cope with the needs of a company in four hours. Respecting these conditions, the mill produced is able, following his orders, to be turned by one man from the company, weighs 30 pounds (12 kg) and furnishes 30 pounds (12 kg) of flour per hour, or a little more than 16 ration-equivalents.

The mills adopted by Le Génie in 1823 are composed firstly of a conical piece (the core) that turns in a receptacle (the bushel), in the manner of a coffee mill. According to the figures in Table 1 these mills were powerful, since they could produce a ration-equivalent of more than 17. Yet Belmas points out that these metal devices were much less effective than the mills built of two stone discs (the mill known as a *Durand Père*) that had been abandoned (Béguin 2003, 392) for lack of skilled mechanics to maintain it! In fact, whereas the metal mill could produce 166 rations for 200 pounds of refined flour, a stone mill could produce 180 rations, or 14 more (Belmas 1829, 287). But in 1814 Régnier makes it clear that the metal mills could bear prolonged use with no alteration, which is not the case with the stone mills (Régnier 1814, 589). The critical aspects are the yield and the lack of need for maintenance.

2. The *impedimenta* of the Roman Army

Whether mills were used by the Napoleonic *Grande Armée* or by the Roman army in the 1st century AD, they constituted the basic equipment for feeding the troops and made up the *impedimenta*. For the Roman army, this meant manual mills in stone. In terms of equipment carried by each soldier, however, the Roman Army can be compared to modern period armies with one exception: the former did not use non-combatants for transporting goods, also known as *the baggage train* (Goldworthy 1998, 295; Roth 1999, 68).

2.1 The Roman legions are their movements

In general, in the 1st and 2nd centuries AD, a Roman legion was composed of ten cohorts, the first of which would be made up of 800 men and the nine others of 480 men each, adding up to a total of 5120 men. To move about, a legion could use two systems; while the first allowed a unit to be independent and to move freely despite the *impedimenta* that slowed the speed of movement and reduced the opportunities of appropriate operations, the second made use of convoys. These supplementary detachments would make return trips between a mobile unit and a fortress or an *oppidum* occupied by a Roman garrison in order to supply the troops (Goldworthy 1998, 287-8).

Even if the calculations of the quantities implied by movements and troop numbers are conjectural, we can establish that a legion on the march would have had a baggage train composed of almost a thousand pack animals either saddled or used to pull chariots or carts of two or four wheels (Goldworthy 1998, 290).The following extract mentions the needs of marching troops in a passage from Chapter 45 of Book 28 of Livy's History of Rome (trans.A. de Sélincourt) in which Scipio is engaged in Spain against Carthaginian forces:

> *The peoples of Etruria were the first to promise aid to the consul in proportion to their respective means; [...] Arretium (offered) 3000 shields, 3000 helmets, and a total of 50,000 pike, javelins, and spears, an equal number of each, together with enough axes, shovels, sickles, basins, and hand-mills to equip forty warships; also 120,000 measures of wheat [...].*

The means of land transport for portable mills is provided by Plutarch in his Parallel Lives:

> *This method of attack and defence being repeated every day, they made but little progress in their march, and were, besides, distressed for want of provisions; they could not forage without fighting; the corn they could get was but little, and even that they had not instruments to grind. The greatest part of them had been left behind; for many of the beasts of burden were dead, and many were employed in carrying the sick and wounded* (trans.J. and W. Langhorne).

The mills were thus, we have seen, an essential part of the impedimenta of mobile troops. They were distributed within a legion according to the smallest legionary division, the *contubernium*, a unit of eight men in the same tent or shelter (Daremberg and Saglio 1877, 1489). If a legion was at its full strength of 5120 men, then it would have been composed of 640 *contubernium*.

2.2 The *contubernium*

Each legionary carried, along with his weapons, twenty kilograms of items (*sarcina*) comprising clothing, kitchen utensils, tools and rations. In this sense Livy notes three major points that each soldier must observe:

> *...That it was a soldier's business to attend to these three things, – his body, that he may keep it in perfect strength and agility; his armour, that it may be always in good order; with his stores of all kinds, so as to be ready in case of a sudden order; and to rest assured that all other matters relating to him will be directed by the immortal gods and his captain* (trans.by G. Baker).

Yet the soldiers of the *contubernium* shared two types of equipment, the tent and the mill, with the weight of these two elements representing a total of about 70 kg (the tent about 40 kg and the mill 30 kg) carried by a mule belonging to the barracks. A further 75 kg of utensils were included, with the baggage train thus representing 145 kg. A mule could carry no more than 135 kg (Roth 1999, 77 and 207) and so a second animal was needed (Southern 2007, 223). The *impedimenta* of a legion at the strength of 640 *contubernium* would have thus reached around 93 tonnes of kit including more than four tonnes of wheat rations and more than nineteen tonnes of querns!

Although in his first chapter of *Supply needs and rations*, Roth (1999, 24) has fully developed the theme of the food needs of the Roman Army, we can note that the daily measure of rationed wheat (*frumentum*) represents 80 g and that, as reported in Caesar's Gallic Wars, this seems to have been distributed regularly but not every day:

> *When departing, he promises that he will return before the end of the seventh day, on which he was aware corn was due to that legion which was being left in garrison* (Book VI, Chapter 33, trans. A. Hirtius, W. A. M'Devitte and W. S. Bohn); or in Livy, *Here he received two thousand five hundred talents of silver [484,275l], which he sent to Apamea; the corn he distributed to the army* (Book 38, Chapter 37, trans.G. Baker).

2.3 The milling of the contubernium

Modern experiments have shown that a manual rotary mill (*mola manuaria*), based on the design of that of the Roman legions, could process 4 kg of wheat in an hour (4.71 ration-equivalents); in other words, it takes almost an hour and forty minutes to process the daily ration of the *contubernium*, or 6.80 kg. However, we cannot be sure whether milling took place every day or each week, and we know little of the quality of the resulting flour. Nevertheless, in Book 18, Chapter 20 of his *Natural History*, Pliny does give us some useful indications on that question:

> [...] *The proper proportion for the yield of Campanian wheat to the modius of grain is four sextarii of what is known as bolted flour; but when it is used in the rough and has not been bolted, then the yield should be five sextarii of flour;* [...] *It is evident, too, that the Romans subsisted for a long time upon pottage, and not bread* (trans. J. Bostsock and H.T. Riley).

Indeed, Junkelmann tells us that this menu is certainly valid for the army (2006, 128). Flours of different grain sizes (pastry flour, coarse flour, etc.) are obtained by adjusting the space (called the *light*) between the two querns of the mill. This is the essential element that keeps the apparatus from clogging and which can be modified with a mechanical system. A part of the following section, concerning the morphology of querns, is devoted to the use of this modulator.

3 The mills of the Roman Army: examples from archaeological excavations

In his *Panis Militaris*, Marcus Junkelmann mentions a mill dating to 80 BC found at the legionary camp of Càceres el Viejo (Spain). The military use of this mill is supported by the discovery of a *pilum* spike used as the rotational axle. Similarly, a number of presumed military mills have been found in the province of *Germania* in the camps at Strasbourg, Saalburg and Xanten and it is these implements that we will look at in this last section.

3.1 The mills of Strasbourg, Saalburg and Xanten (Fig. 1)

The 1st century AD basalt turning mill (*catillus*) found at Strasbourg at the beginning of the 20th century (Forrer 1919) boasts the singularity of bearing an inscription on its side that is the only known so far in Alsace. Although broken, there is enough information to decipher and understand its operation. It consists of a half disc of 44 cm in diameter, 10 cm thick, with the centre pierced with a circular cavity flanked by two mortises for a balance rynd. A curved lateral perforation opening towards the top allows for the insertion of a handle for turning; the handle may not follow the same curvature. This handle type is fairly rare since only eleven examples have been found so far in Alsace. By contrast the morphology and the stone used are quite common, as these are the characteristic elements of querns of the early Roman period in the region. It is also important to note that all the surfaces are grooved, which may occur in two manners: there is decorative treatment (on the sides and on the upper receptacle), and there is functional grooving, on the active surface (Lepareux-Couturier *et al.* 2010). The decorative vertical grooves that encircle the side of the quern are covered with an inscription 20 cm long by 8 cm high that reads **C-VIREI** (Fig. 2). These same grooves often appear

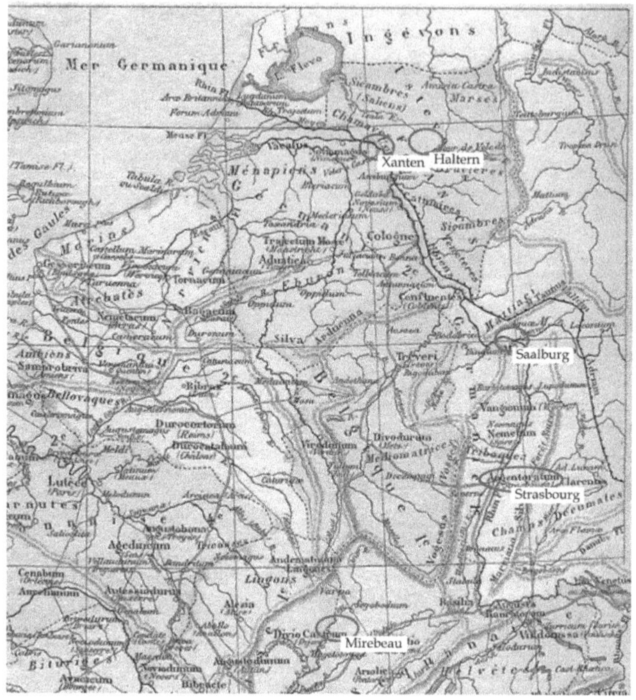

Figure 1. Locations of the military camps in question (Jodry 2010, 37.1)

on basalt querns found in Alsace but also occur on a quern without inscription dating to 30 BC found at Mirebeau-sur-Bèze (Venault 2003), not far from the Roman camp where the eighth Legion was stationed before its transfer to Strasbourg (Bénard 1994, 148).

Two basalt querns found during excavation campaigns at the Roman camps of Saalburg and Xanten are similar to our Strasbourg examples. In fact, the quern found at Saalburg has the same decorative and functional grooves on the surfaces, as well as identical functionality to the revolving quern of Strasbourg. Its central eye is also flanked by two small lobes that suggest the hypothesis that there was a balance rynd, a metal plate dependant on the axle (Jodry 2000, 70). The rynd allows one to vary the 'light' between the two querns through the use of a lever fitted under the mill when the mill is fixed on a structure, and at the same time to facilitate the transmission of rotary movement (Curwen 1937).

Another contemporary mill found at Haltern also has the same type of cut depression around the eye and with it has been preserved its metal hook equipped with a ring at the end where perhaps was set a vertical wooden handle to turn the *catillus*. This same system has been found on a recently-excavated quern discovered in the interior of the enclosure of the Roman camp of *Argentoratum*(Strasbourg) (Kuhnle and Jodry 2010) and in the Roman camp of *Trimontium* (Newstead) in Scotland (Curle 1911; Breeze 2002, 49, 54), in both cases identical querns and *a priori* in the same basalt.

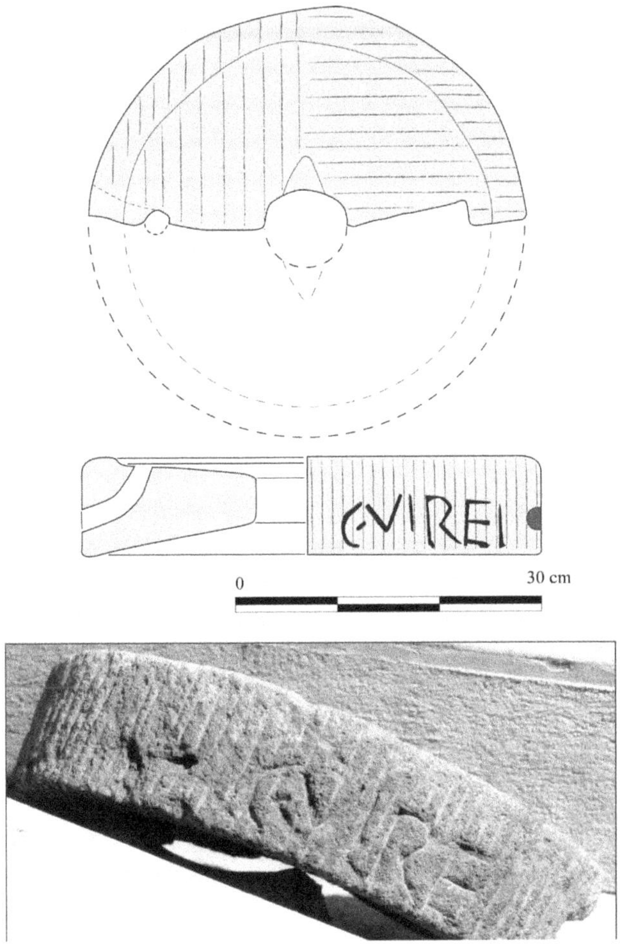

Figure 2. Quern found at Strasbourg bearing the inscription C-VIREI (Jodry 2010, 37.2)

3.2 The inscriptions

Following the Strasbourg example, the Saalburg *catillus* has the same dimensions and also has a curved lateral perforation but with the inscription **CON-BRITTONIS** (Junkelmann 2006, 117). The two inscriptions from Strasbourg and Saalburg could be interpreted as being proprietorial marks of *contubernium*, with **C[ontubernium]-VIREI** indicating 'the *contubernium* of Vireus' and **CON[tubernium]-BRITTONIS** indicating 'the *contubernium* of Britto'. By contrast, the lower quern (*meta*) at Xanten, of similar date and dimensions, carries a different inscription on its side: **TVR-ENNI**, interpreted by M. Junkelmann as: **TVR[ma]-ENNI** or 'the cavalry squadron of Ennius'. Some querns bearing a probable proprietorial inscription have also been found at the *Vindolanda*-Chesterholm camp, as has been reported by Susan Watts at the 2009 Conference: *Bread for the People* (oral presentation).

4. Standardisation of querns of the Roman Army

The most relevant criteria to account for the morphological evolution of the mill are mainly the diameter and the height. In fact, whereas during the later Iron Age the mill is thicker and more 'stocky', during the early Roman period the diameter of mills tends to grow and the thickness to diminish: the Roman-period quern produced in the Eifel is thus normally a disc with an average diameter of 40 cm with a thickness of 10 cm (Hörter 1994, 26).

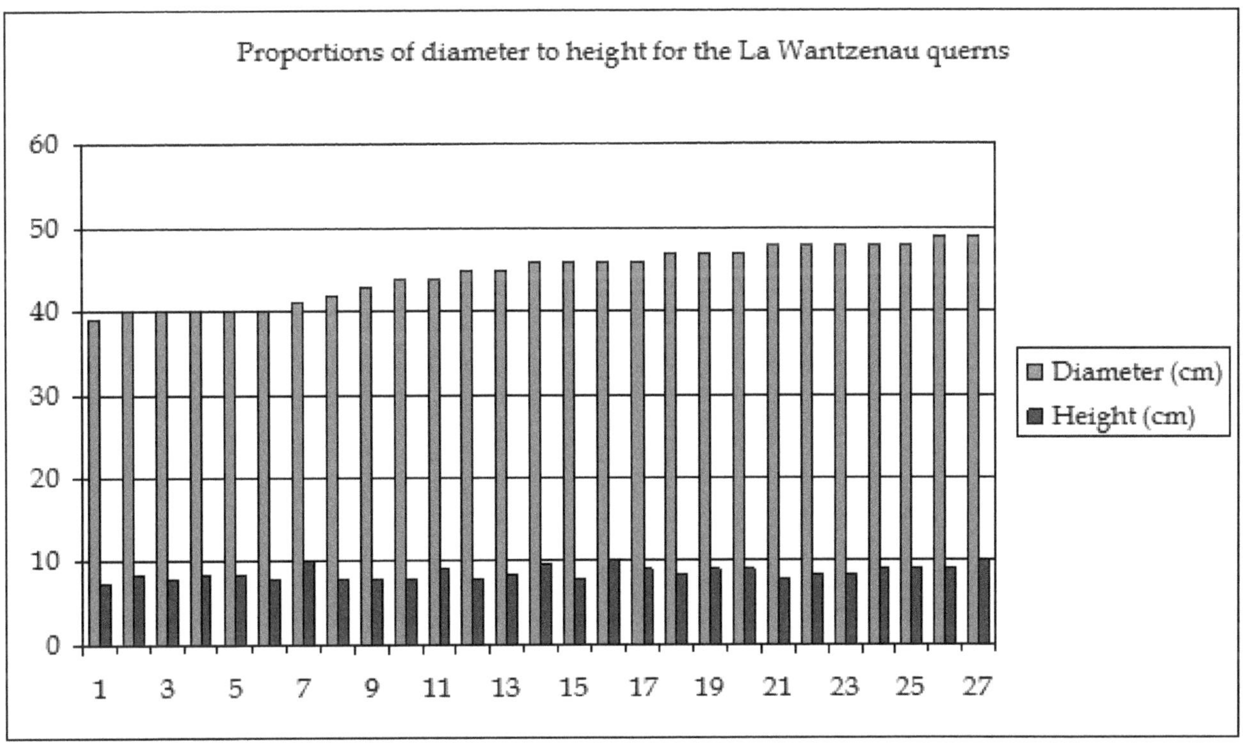

Table 2. The morphology of the La Wantzenau querns

4.1 In search of an efficient material

We can further observe that all these querns are fashioned out of the same basalt, extracted from the magnificent quarries of the Eifel region (Rhineland-Palatinate, Germany). These are located 200 km to the north of Strasbourg, about 10 km west of Koblenz, in the area of the confluence of the Rhine and the Moselle. The main deposits are in the volcanic massif of the Osteifel, framed by present day Mayen, Kottenheim, Ettringen and Niedermendig. Indeed between the first millennium BC through the 19th century, these lava flows have presented the opportunity of extraction of a very compact and very good quality stone from the basalt columns. During the Roman period these open-face quarries were exploited in sections and the extraction multiplied to represent a market of more than 17 millions querns (Mangartz 2006, 31).

4.2 The standardized querns from a boat found at La Wantzenau

In the light of this understanding of the standardization of mills, we can examine the group of basalt querns found in 1911 in a gravel quarry at La Wantzenau, a few kilometres to the north of Strasbourg, at the confluence of the Rhine and the Ill. These querns constitute the cargo of a small boat dated to AD 275 (Forrer 1911) probably from Andernach, the port of embarkation from the Eifel on the Rhine. The 27 querns preserved from the boat are all in the form of a flat disc carved from grey basalt. They average 44 cm in diameter and nine cm thick, and they were smooth on the surfaces and the sides. The only working that has been recorded is the central cavity made at the centre of the object, whose profile does not seem to be very accurate: these perforations are not suitable for immediate use (Jodry 2000, 631-632).

The La Wantzenau querns do not show any particular indication that would suggest how the object would be positioned in the mill. Indeed, the unrefined treatment of the surfaces and the approximation of the central hole do not allow us to distinguish the meta from the catillus, which leads us to see this as a consignment of rough cuts at an advanced stage of finishing, on which the trimming and central perforation were carried out at the end of the operative chain before shipment (Jodry 2000, 631-632). This hypothesis of importation confirms the importance of the *castrum* and the centre of activity that it generated, since in the second half of the 3rd century *Argentoratum* remained 'a key point essential to the new defensive system' (Baudoux and Flotté 2002, 187).

Conclusion

Did the Army have concessions in the quarries? Was there a military market? Was the Army a creator of a supply network? It is in response to this last question that Junkelmann (2006, 127) suggests that the manual mill was undoubtedly developed in order to meet military needs. It is true that it is tempting to associate the cargo of querns from the boat at La Wantzenau with a military order, but we are not in a position to confirm or refute this proposition. Nevertheless, it is clear that in 1st century AD Alsace basalt was a favoured stone, since 58% of the querns are carved from this imported material (versus 41% in local sandstone); this must have been due to the growth of commercial dealings and the militarisation of the area of the *limes* (Jodry 2010).

At the present time we cannot answer these questions with certainty, yet with the examples cited in this study we can observe that the querns in question possess common aspects that can be used to create a depiction of the typical mill of the Roman Army whose key elements are the following: the stone, the decorative and functional surface treatments, the type of handle and the rotational transmission system. But for the inscriptions that certify a level of ownership of the querns but which remain a secondary aspect, whether they come from Strasbourg, Saalburg, Xanten, Haltern, Newstead or Chesterholm, these querns answer the needs of efficiency, of simplicity of design, use and transport. For the provision of mills to armies, whether Roman or modern, these are universal criteria.

References

Aubert de La Chesnaye-Desbois, F.A., 1758. *Dictionnaire militaire, portatif, contenant tous les termes propres à la guerre*. Paris.

Baudoux J., Flotté, P., 2002. *Strasbourg, Carte Archéologique de la Gaule 67/2*, 2002, 178-9.

Béguin, F., 2003. Les 'objets d'art' du génie. Moulins de siège, d'hôpital ou de campagne. In M.Barboff, and C. Griffin-Kremer(eds.), *Meules à grains*. Paris, 387-99.

Belmas, M., 1829. *Ministère de la guerre. Comité Technique du Génie, Mémorial de l'Officier du Génie*. Paris.

Bénard, J., 1994. *Les agglomérations antiques de Côte-d'Or*. Presses Universitaires de Franche-Comté.

Breeze, D.J., 2002. *Roman forts in Britain*. Aylesbury.

Curle, J., 1911. *A Roman frontier post and its people*. Glasgow.

Curwen, E.C., 1937. Querns. *Antiquity*, 11, 133–151.

Forrer, R., 1911. Ein versunkener spätantiker Mühlsteintransport in Wanzenau bei Strasburg. *Anzeiger für Elsässische Altertumskunde* 7-8, 131-143.

Forrer, R., 1919. Römische Mühlen- Töpferei- und Handelsbetriebe, Metallwerkstätten und Waffenfunde in Strasburg. *Anzeiger für Elsässische Altertumskunde*, 988-990.

Goldworthy, A.K., 1998. *The Roman army at war*. Oxford.

Hörter, F., 1994. *Getreidereiben und Mühlstein aus der Eifel, Ein Beitrag zur Steinbruch- und Mühlengeschichte*. Mayen.

Jodry, F., 2007. A propos d'une meule militaire. *Instrumentum,* 26.

Jodry, F., 2000a. La meule de la rue de la Mésange, Strasbourg. *Fouilles archéologiques de la ligne B du Tram*, fouilles récentes 5.

Jodry, F., 2000b. Les meules de La Wantzenau, *Strasbourg, Carte Archéologique de la Gaule* 67/1, 631-632.

Jodry, F., forthcoming. Grès VS Basalte, L'exploitation du grès et l'importation du basalte en Alsace durant l'Antiquité (1er-3e siècle), in *Évolution typologique et technique des meules du Néolithique à l'an mille sur le territoire français. Chronique sur la Table rondede Saint-Julien-sur-Garonne* (F) du 2 au 4 octobre 2009, supplément Aquitania.

Junkelmann, M., 2006. *Panis Militaris, Die Ernährung des römischen Soldaten oder der Grundstoff der Macht, Kulturgeschichter der antiken welt*, band 75, Mainz, 110-127.

Kuhnle, G., and Jodry, F., 2010. *La rue Brûlée à Strasbourg, Rapport Final d'Opération, INRAP.*

Lepareux-Couturier S., Boyer F., and Jodry F., forthcoming. Le traitement de la surface active desmeules rotatives antiques. Bilan documentaire, proposition d'un vocabulaire descriptif et d'une grille d'analyse, in Évolution typologique et technique des meules du Néolithique à l'an mille sur le territoire français. Chronique sur la Table ronde de Saint-Julien-sur- Garonne (F) du 2 au 4 octobre 2009, supplément *Aquitania.* Le courrier, 26/12/1741.

Mangartz, F., 2006. Vorgeschichtliche bis mittelalterliche Mühlsteinproduktion in der Osteifel. In A Belmont and F. Mangartz 2006 *Les Meulières. Recherche, protection et valorisation d'un patrimoine industriel européen (antiquité-XXIe siècle)*. Mainz.

Mission Innovation technologique et multimedia et Hélène Jouguet Florent Lartet, 'Dictionnaire des Antiquités Grecques et Romaines de Daremberg et Saglio,' http.//dagr.univ-tlse2.fr/sdx/dagr/index.xsp.

Regnier, 1814. *Archives des découvertes et des inventions nouvelles*.

Rogniat, J., 1820. *Considérations sur l'art de la guerre*. Paris.

Roth, J., 1999. *The logistics of the Roman army at war (264 BC-AD 235).* Leiden.

Southern, P., 2007. *The Roman Army*. Oxford.

Venault, S., 2003 . Mirebeau-sur-Bèze (21), La Fenotte, Rapport Final d'Opération, INRAP.

Viesse de Marmont (duc de Raguse), A.F.L. 1857. *Mémoires du maréchal Marmont, duc de Raguse de 1792 à 1841*. Paris.

Ancient sources

Caius Plinius Secundus. *Natural History*. translation by J. Bostsock and H.T. Riley 1856, Vol. IV, 32-3.

Julius Caesar. *Gallic Wars*. Book VI, Chapter 33, translation by A. Hirtius, W A M'Devitte and W S Bohn, 1853, 157 and Book 38, Chapter 37, translation by G. Baker 1847, New York.

Marcus Porcius Cato (Mayor), *Traduction d'anciens ouvrages latins relatifs a l'agriculture et a la médecine vétérinaire. L'économie rurale de Caton* (chez P. Fr. Didot le Jeune, libraire, 1771).

Marcus Vitruvius Pollio. *de Architectura* (chez Panckoucke, 1847).

Plutarque. *Les vies des hommes illustres*. Firmin-Didot Frères, 1854. translation by J. and W.Langhorne 1832, London, 44-5.

Titus Livius. *History of Rome*, Books 41, 44, translation by A. de Sélincourt, edited by B. Radice 1965, Penguin Classics and translation by G. Baker 1836, New York

A newly identified milling artefact from Roman Britain

Martin Watts

In his review of *The Archaeology of Mills and Milling* (Watts 2002), Dr Michael Lewis drew attention to a Roman milling artefact which I had omitted from my discussion of Roman milling technology (Lewis 2003, 63). The omission was simply due to my ignorance of its existence and Dr Lewis kindly sent me a copy of an article by Professor Dietwulf Baatz, which describes and discusses in some detail the '*eiserne dosierkegel*', simply translated as an iron feed-cone (Baatz 1994, 19-35). The identification of this iron artefact as being connected with grain milling had been made by Baatz, on examining several examples which had been found during excavations on the Roman military frontier fortifications at Zugmantel and Saalburg in the Taunus hills of Germany. A search of other collections of Roman ironwork brought further examples to light and Baatz's published catalogue contains 15 from Germany, three from Belgium and one from France (*ibid.*, 30-5). None of these artefacts had previously been securely interpreted, with possible functions such as flagpole or tent-pole tips, tools and weapon points having been proposed. The connection with milling technology was made by Professor Baatz, based primarily on the context of an example recovered by H. Jacobi in his 1912 excavation of well 300 in the vicus at Zugmantel, where the mystery iron object was found with a pair of matching millstones and the spindle and pinion from a geared mill (Jacobi 1914).

In more recent milling terminology, the function of the *dosierkegel* is similar to, but not the same as, the damsel. It is an iron fitting which fits onto the head of the millstone spindle, above the rynd on which the upper millstone is hung, and it rotates with the upper stone. It both encourages and controls the feed of grain into the millstones. However, its function is more than simply that of ensuring an even flow of grain into the eye of the runner stone; by altering the elevation of the hopper, the larger coned section of the *dosierkegel* regulates the flow of grain from the hopper outlet into the stones, acting as a feed valve rather than simply as an agitator (Fig. 1). Vitruvius's account of a watermill states that a hopper was suspended above the stones (Morgan 1960, 295) and this suspension arrangement, however it was contrived, presumably allowed adjustment of the height of the hopper and thus the grain feed. A further consideration is that the action of raising and lowering the runner stone – tentering or lightering – whilst milling, to alter the distance between the working faces and thus the texture of the material being ground, is likely to have had an effect, albeit a lesser one, on the amount of feed from the hopper into the eye of the runner stone.

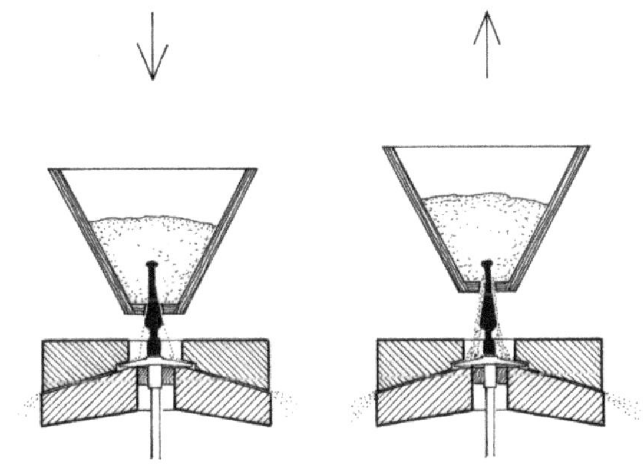

Figure 1. The working position of the feed cone with the grain feed from the hopper when lowered (left) and raised (right). (After Baatz 1994, Abb.14.)

As the available evidence indicates that power driven millstones were introduced into Britain by the Romans in the 1st century AD, I was determined to find a British example of a *dosierkegel*. Over a period of several months I searched through published catalogues of ironwork from a wide variety of Roman sites and excavations, looking particularly at illustrations, as it seemed most likely that if an example had been found it would have been identified as having a different function than milling, as were the examples listed by Baatz. My sporadic research produced no comparable artefacts and it seemed that my search was probably in vain. However, in *Furniture in Roman Britain* Joan Liversidge (1955) illustrates a '*Fragment of Iron Stool Leg (?) found at Mansfield Woodhouse*' (*ibid.*, 37a, b). The published photograph shows an iron object with a socket at one end, a projecting coned section in the centre and a point at the other end, viewed from the front and the back. Although shown upside-down, the artefact clearly resembles some of the iron feed-cones illustrated by Baatz. A subsequent examination of the artefact itself convinced me that it is not a stool leg, an identification that Liversidge

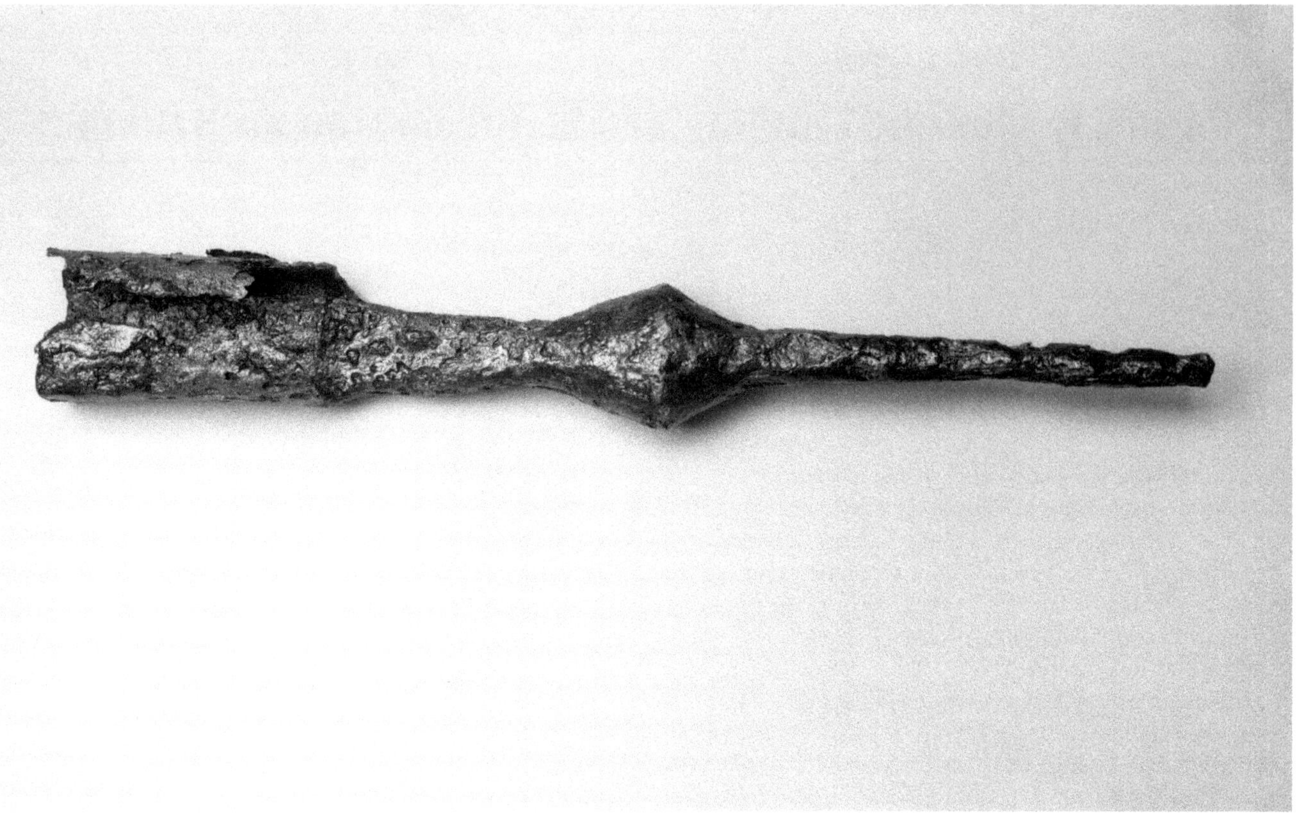

Figure 2. The iron object from Mansfield Woodhouse, Nottinghamshire. Birmingham City Museum and Art Gallery (1955 A 350)

was also rather dubious about (*ibid.*, 28), but an iron feed-cone (Fig. 2). The object in question must therefore have been associated with a Roman milling establishment.

The site at Mansfield Woodhouse in Nottinghamshire from which the artefact came is described as a Roman villa. It was originally excavated in the late 18th century by Hayman Rooke, who described it as the site of 'an elegant *villa urbana*' (Rooke 1787, 363-76). Among the finds he noted and illustrated were two quern fragments (*ibid.*, Plate XXIV, 6 and 7), neither of which is large enough to have come from a power-driven mill; the larger diameter flatter stone '*must have been thirteen inches diameter when perfect, and two inches and a half thick*' (*ibid.*, 373). No further information is given and the fate of the stones is unknown (Baddeley, pers comm). Rooke did note the presence of building evidence close to the Pleasley Water, a watercourse that forms the county boundary between Nottinghamshire and Derbyshire to the north of the main building complex (*ibid.*, 371), but this does not appear to have been explored further. The villa site was partly re-excavated in 1936-9 (Oswald 1949, 1-14) and it was during this second phase of excavation that the object in question was found. According to Adrian Oswald 'the leg of a chair in iron from the lower filling of the ditch' deserved note (*ibid.*, 13-14), although it was neither described nor illustrated in his report. No millstone fragments or other possible milling connections were recorded by Oswald, although a number of ovens forming a sequence from the late 1st to the 4th centuries AD were identified and evidence of iron smelting was also found (*ibid.*, 1, 4).

The object, which is now in Birmingham Museum and Art Gallery (accession number 1955 A 350), is visually comparable in form to one of the examples from Saalburg (Fig. 3). It is an iron forging 29.2 cm long weighing 0.52 kg and has a slightly tapered rectangular socket at its lower end, 3 cm by 2 cm internally, which would have slotted over the head of the millstone spindle. It is interesting to note that of the two Roman millstone spindles that have been identified in Britain, that found in a hoard of ironwork from Great Chesterford, Essex (Neville 1854, 9-10) has a long, slightly tapered top section, a maximum of 2.2 cm square, projecting 5.2 cm through the rynd, which is still in place. A further interesting feature is that the rynd has three arms, rather than the more usual two. Although three-armed rynds were common in Scotland in the post-Medieval and modern periods, no other Roman examples are known and no Roman millstones with three rynd chases have been positively identified. The Silchester spindle is both lighter and longer than the Great Chesterford example and has a flatter rectangular section top which tapers from about 2.5 by 1.5 cm to 2.2 by 1.1 cm over a length of 5.5 cm. Both spindle heads would therefore be suitable to locate the socket forged as part of the foot of the feed cone, although the projecting end of the Great Chesterford spindle is larger and squarer than any of the *dosierkegel* sockets recorded by Baatz. With regard to dating, a mid 4th century AD

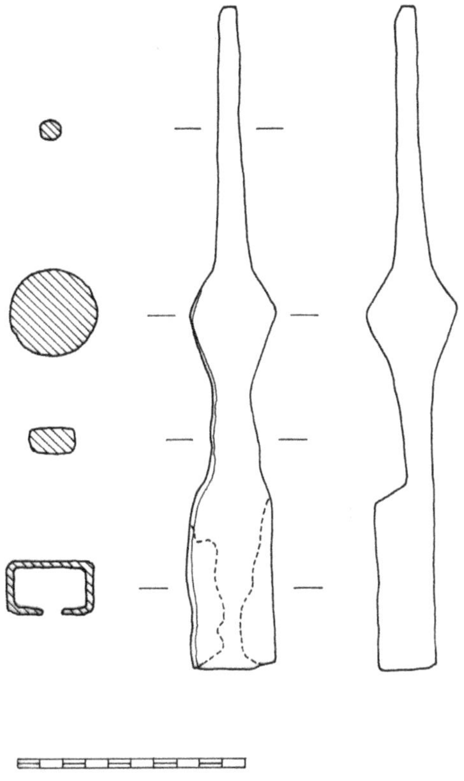

Figure 3. Iron feed cone from Mansfield Woodhouse, Nottinghamshire. The broken line shows the open centre of the socket. Scale: 5 cm

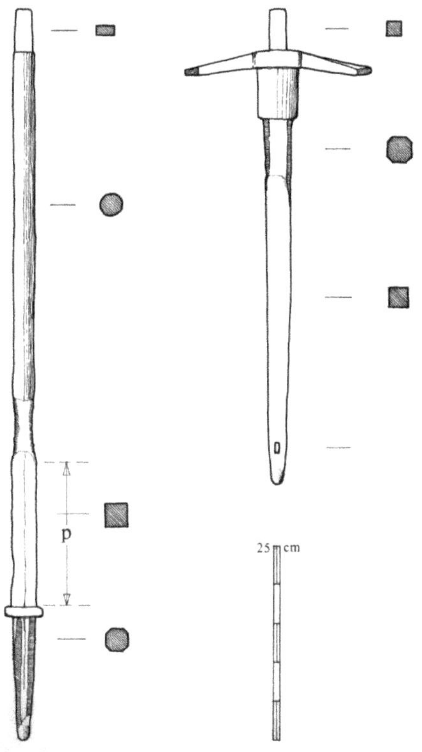

Figure 4. Roman millstone spindles from Silchester (left) and Great Chesterford (right). P indicates the working position of the drive pinion

date has been suggested for the deposition of the hoard of which the Silchester spindle formed part (Manning 1964, 38-40), while the Great Chesterford hoard was considered to be of early-5th century AD date (Neville 1854, 9-10).

To summarize Baatz's catalogue, the examples he identified vary between 20 and 40.4 cm in length and weigh between 0.83 and 1.38 kg, with the exception of two of the Belgian finds, which are lighter, weighing 0.60 and 0.42 kg respectively, although the latter example is incomplete. It appears that some of the other examples, including the Mansfield Woodhouse find, may also be incomplete, as the majority of those catalogued appear to have had a rectangular or flattened section at the tip, which it is suggested was to keep the grain moving, to prevent it bridging or blocking the outlet in the bottom of the hopper (Baatz 1994, 22). It is difficult to say whether the Mansfield Woodhouse example is incomplete; its tip now appears to be neatly rounded, rather than broken. With regard to dating evidence, the Zugmantel examples were deposited before about AD 170-185 and Saalburg was abandoned in about AD 260. Where information exists, the other examples generally date from the second and third centuries AD. A suggested date for the Mansfield Woodhouse example is second century AD, which places it comfortably with the examples identified in mainland Europe.

A preliminary survey carried out by David King in the 1990s presented evidence for 170 power-driven mills, based largely on finds of millstones that were considered too large to be classified as querns, simply worked by hand. These finds were associated with a variety of Roman occupation sites – rural, urban, military and ceremonial – in England and Wales (King 1998). To this list further sites can certainly be added and the current number of possible mill sites from Roman Britain is now in excess of 200. Apart from the use of water power to drive millstones (for a recent summary see Spain 2008), the use of animal power has also been suggested (for example, Neal 1996), while geared mills worked by human muscle power are also a probability. Much work still needs to be undertaken to identify Roman milling installations in Britain and to unravel details of the technology that was involved. It is hoped, therefore, that the identification of an iron feed cone from Mansfield Woodhouse, which is one of very few artefacts other than millstones to have been connected with milling in Roman Britain, may encourage others to look closely at any insecurely identified objects which may be locked away in collections of Roman ironwork, or may yet turn up on sites where millstones or other evidence of milling and baking are found.

Acknowledgements

I am extremely grateful to Michael Lewis, for drawing my attention to the existence of dosierkegels and to Ann-Marie Dick for translating Professor Baatz's article for me. Dr David Symons, Curator of Antiquities and Numismatics at Birmingham Museum and Art Gallery, kindly allowed me to examine the artefact and Tom Heaven supplied the photograph. I am grateful to Virginia Baddeley, Sites and Monuments Record Officer, Nottinghamshire County Council, for checking the SMR entry for Mansfield Woodhouse, and to the staff at Cambridge University Museum of Archaeology and Anthropology and Reading Museum, for allowing me to examine and measure the Roman millstone spindles from Great Chesterford and Silchester in the 1990s.

References

Baatz, D.,1994. Eiserne dosierkegel. *Saalburg Jahrbuch*, 47, 19-35.

Jacobi, H., 1914. Römische Getreidemühlen. *Saalburg Jahrbuch,* 3 (1912), 75-95.

King, D., 1998. *Mechanised Corn Milling in Roman Britain.* Unpublished presentation to the Quern Study Group, Reading, February 1998.

Lewis, M., 2003. Book Reviews. The Archaeology of Mills and Milling by M. Watts. *Industrial Archaeology Review*, XXV.1, 63.

Liversidge, J., 1955. *Furniture in Roman Britain.* London.

Manning, W.H., 1964. A Mill Pivot from Silchester, *The Antiquaries Journal*, 44, 38-40.

Morgan, M.H., 1960. *Vitruvius. The Ten Books on Architecture.* New York.

Neal, D., 1996. Upper storeys in Romano-British villas. In P. Johnson (ed.), *Architecture in Roman Britain,* CBA Research Report 94, 33-43.

Neville, R.C., 1856. Roman Antiquities at Great Chesterford, Essex. *The Archaeological Journal*, 13, 9-10.

Rooke, H., 1787. An Account of the Remains of two Roman Villae discovered near Mansfield Woodhouse. *Archaeologia,* 8, 363-76.

Spain, R., 2008. *The Power and Performance of Roman Water-mills.* BAR International Series 1786.

Walters, H.B., 1910. Romano-British Nottinghamshire. *Victoria County History of Nottinghamshire*, 2, 28-32.

Watts, M., 2002. *The Archaeology of Mills and Milling.* Stroud.

Pompeian millstones in France

Luc Jaccottey and Samuel Longepierre

In collaboration with Florent Jodry, Stéphanie Lepareux-Couturier, Paul Picavet, Bruno Robert and Boris Robin

1. Introduction

Given their very specific and particular hourglass shape, Pompeian mills are the most emblematic millstones of the Roman world. The image of this type of mill is widely disseminated throughout archaeological literature due to the photographs of the bakeries of the Roman cities of Pompeii and Ostia. These mills must have been impressive even in Antiquity, since we find their representations on bas-reliefs such as the tomb of the baker Eurysaces near the *Porta Maggiore* in Rome. This bas-relief depicts the whole sequence of operations of a Roman bakery.

But behind these scenes exists a complex reality that reflects a wide variety of types of Roman mills: rotary hand querns, animal- or man-driven mills and hydraulic mills. When consulting Roman mill publications, it would seem that the Pompeian mill is the prominent type (Poulin 2000, 17; Bayard and Massy 1983, 163). Furthermore, it is not uncommon that these mills, due to their specific shape, are given priority in museum displays, as is the case of Amiens, Reims and Lyon, to the detriment of more common domestic rotary querns that are often stored away in depositories out of sight of the visitor.

Paradoxically, no work of synthesis on the distribution of Pompeian mills in Roman Gaul has to date been undertaken. Contrary to generally accepted notions, a recent study (Longepierre 2007) has proven that this mill type was widely distributed in Gallia Nartbonenis. The first compilation of these mills reveal examples in Amiens, Lyon, Orleans, Paris, Soissons, Tours and Troyes (Beal 1996, 93). Nevertheless, the few bibliographical references available are for the most part incomplete or unverifiable and devoid of information relating to a stratigraphical context, typology, dating and petrography. Recently, in the framework of the Collective Research Project of *Group Meule* (Millstone Research Group), an inventory of Pompeian millstones was initiated focused on mainland France. We will not be citing millstones of this type beyond the mainland, for example in Corsica at the sites of Aléria (Jehasse and Jehasse 1987, 31) and Lucciana (Nucci 2006).

One of the aims of this broad study is to determine, based on a petrographical analysis, the principal production centres that supplied these hourglass mills to Gaul, in order to compare the role played by regional millstone quarries in Gaul with the major production centres, notably in the Italian peninsula. In addition, we hope to shed light on the use of Pompeian mills in relation with other types of mills from the standpoint of region, chronology and type of settlement (towns, villas or farms).

This preliminary study of Pompeian millstones is based on data stemming from still partial knowledge of local publications and finds stored in local archaeological museums. We have not had access to every millstone stored in every depository. Since the survey areas of the different contributors to this paper are spread out evenly across the French territory, we believe that our conclusions are representative of the general situation of Pompeian millstones in France during Antiquity. When possible, millstones were drawn following established standards (Jaccotey and Farget, forthcoming). Our first petrographical determinations are macroscopic, pending further, more elaborate analyses. These visual analyses allow us to establish a broad classification of each millstone into the two basic categories that are as follows: volcanic rock such as basalt or andesite and non-volcanic rock such as sandstone or conglomerate.

2. Pompeian mills

2. 1. General description and typology (Fig. 1)

The inclination of the grinding surface of the two millstones is between 50° and 65°. This strong inclination, common to both the *catillus* (upper stone) and the *meta* (lower stone) explains in part their considerable height that is always greater (or at least equal) to their width. The grinding surface of hand driven rotary querns in Gaul, in comparison, rarely surpasses 20° to 24°. The diameter of Pompeian grinding surfaces is between 50 and 90 cm. The *catilli* range in height between 45 to 100 cm and are characterized by their biconical (hourglass) shape that renders them reversible (when the grinding surface became too worn and fragile, the millstone could be turned over and used as the hopper). The *catillus* bears two large more or less prominent lugs or mounts located on opposite

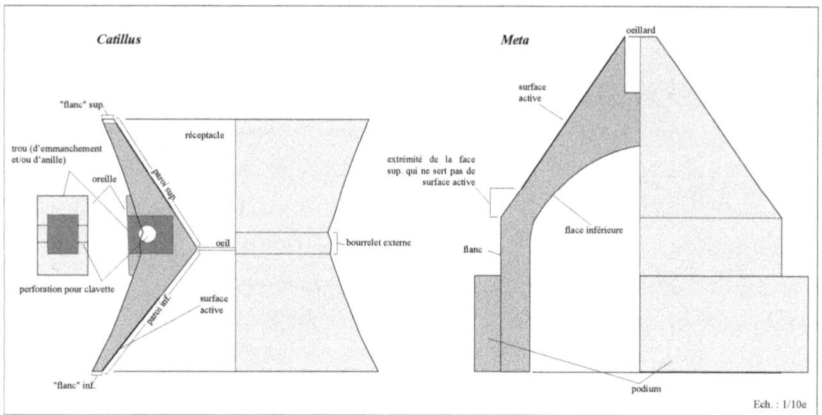

Figure 1. The two main parts of the Pompeian mill and terminology.
catillus - catillus
"flanc" sup. - upper rim
oreille - mount
trou (d'emmanchement et/ou d'anille) – mount hole and/or rynd hole
"flanc" inf. - lower rim
paroi sup. - hopper slope
paroi inf. - lower internal slope
bourrelet externe - external band
perforation pour clavette - cotter hole
réceptacle - hopper
œil - eye or hopper orifice
surface active - grinding surface
Meta - meta
oeillard - spindle socket
surface active - grinding surface
extrémité de la face sup. qui ne sert pas de surface active - extremity of the upper surface where no grinding takes place
flanc - lower rim
face inférieure - internal face
podium - podium
Ech. : 1/10e - Scale 1:10e

Rynd system that regulates the centering of the catillus as well as transmitting the driving force. This system also facilitates lifting the catillus.

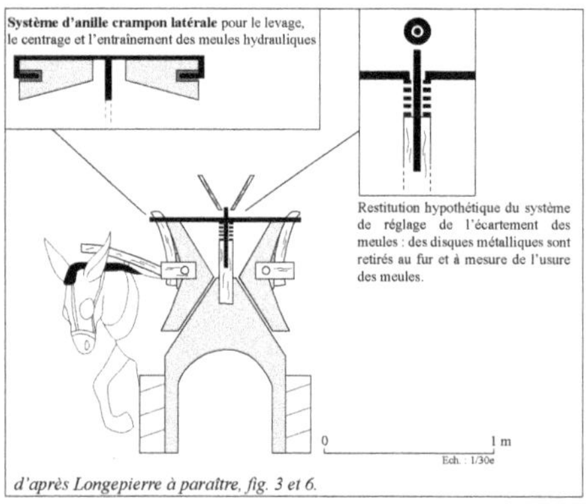

Restitution d'un moulin de type Pompéi avec système d'anille crampon latérale et réglage de l'écartement

Reconstruction of the system of regulating the grinding gap. Superimposed metal disks on the top of the spindle are removed one by one to compensate the wear of the grinding surfaces (after Longepierre forthcoming figs. 3 and 6).

Left: Reconstruction of a Pompeian mill with a lateral rynd system that regulates the grinding gap between the two stones.

sides of the midsection. Each of these bears a slot with a function that we will explain later. The vertical spindle socket of the *meta* always has a square section (measuring 7 to 12 cm). Their massive shape is determined by their considerable height, ranging from 40 to 70 cm. Due to the height their bases are often hollowed to reduce weight to facilitate transport.

Although there are many variations of this type, suggesting different dating and different production centres, the typology of Pompeian mills is in itself specific permitting it to be distinguished without difficulty from other types of mills known in ancient Gaul. However, reconstruction of the fittings and mechanisms used to operate it is still not certain. Based on various graphic representations, especially ancient bas-reliefs, we know that this type of mill was often driven by equines (Monteix 2006, fig. 2-61 to 2-69), hence the name donkey mills.

However, in these representations, only one of the two mounts in the *catillus* is fitted with means to harness the animal. The function of the other is not clear. The iconography reveals a complex mechanism, probably of wood (although iron fittings cannot be excluded), that connects the mounts with the vertical spindle of the *meta*. It seems that this hypothetical implement, labelled a "spiked" rynd (*anille crampon*), was intended not only to center the upper stone, but to finely adjust the gap between the grinding surfaces of the stones to control the grade of the flour. It is our belief that without this technical feature the mill would not function correctly. Other similar research of the Pistrina bakeries (under the direction of Nicolas Monteix, EFR), currently under progress by one of the authors of this paper, has assembled data that supports this hypothesis. However, to avoid deviating from our original subject, we will not delve further into the mechanics of this mill.

2. 2. Geographical origin and petrography

The origin of the Pompeian mill is likely found in the earlier Morgantina mill that dates to at least the 6th century BC. Morgantina type millstone finds are known in Punic sites at Carthage, Sardinia and Sicily (Alonso 1997, 17, fig. 2). Both mill types display a very similar morphology and share highly inclined grinding surfaces (more than 50 degrees). In addition, the Morgantina type shows no similarities with rotary hand querns (with grinding surfaces that never exceed 20 degrees) found in very early contexts (5th century BC) in the northeastern part of the Iberian Peninsula (Alonso 1999, 241; 2002, 115-118). The Morgantina mill takes its name from a find in this Sicilian city clearly dated to the 3rd century BC (White 1963, 204-205). Other finds of this type are a *meta* at Motya, a

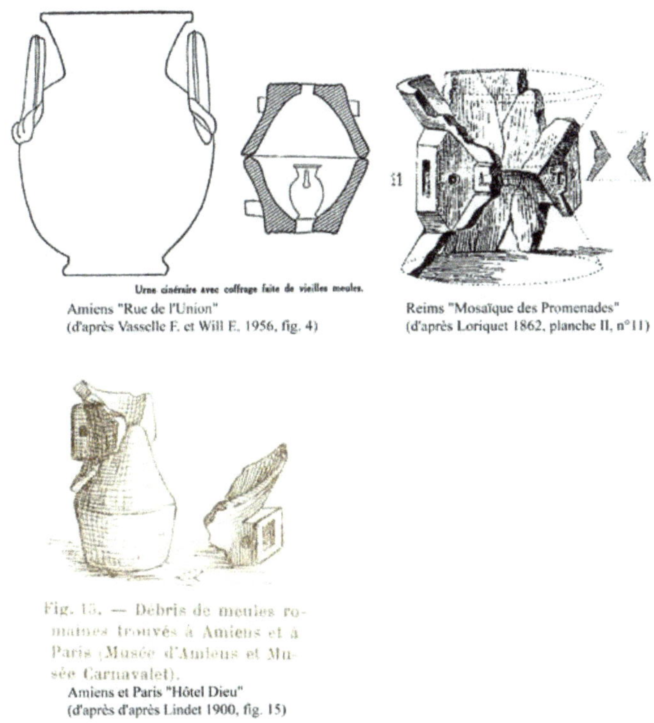

Figure 2. Examples of representations of Pompeian millstones from old texts.

Phoenician colony in Sicily destroyed in 397 BC (Moritz 1958, 55; Alonso 1995, 19-20) and the two complete examples recovered from the shipwreck of El Sec off the shores of Mallorca that date from the second quarter of the 4th century BC (Arribas 1987, 573-576). A geo-chemical analysis has determined that one of the two was fashioned from volcanic rock from Mulargia in Sardinia (Williams-Thorpe and Thorpe 1990, 133). The Morgantina mill from Carthage, dated to the first half of the 2nd century BC, was brought to light during excavations of the Punic quarters of Byrsa (Lancel 1982, 93-95). Fortunately, it was discovered in its original location, revealing that it could only have been driven by a person in a sitting or squatting position.

The earliest written reference to a Pompeian mill is found in Cato's *De Agricultura* dating to *c.* 160 BC. Cato alludes to a donkey-driven *mola asinaria* in the Italian peninsula (*De Agr.* 10.4 and 11.4) (Amouretti 1986, 144). This reference cannot be to a Morgantina type, a mill too small to be driven by an animal. It must therefore be a larger Pompeian mill. Archaeological studies, however, are lacking and so determining the precise nature of Cato's mill is not possible.

The most celebrated groups of Pompeian millstones are those from the bakeries of Pompeii dating to the 1st century AD (Peacock 1989) as well as those from of the port city of Ostia from the 2nd century AD (Bakker 1999). The Pompeian mills from the 1st century BC in the context of metallurgical centres such as Sortijón de Cuzner (Córdoba, Spain) and Domaine des Forges (Les Martys, Aude, France), reveal a morphology that is closely related to the types from the early Empire (Domergue *et al.* 1997, 49, 52-53, fig. 1 and 2). However, there are most likely older Pompeian models such as the two *catilli* from the Lascours mine (Hérault, France) bearing a morphology that clearly recalls the Morgantina *catilli* (with smaller hoppers and, in one case, handle cuttings on the upper rim) (Domergue *et al.* 1997, 53-54, 58, fig. 4). These stones, apparently hewn from volcanic outcrops of Orvieto (Latium, Italy), are unfortunately not precisely dated. Given their early typology, nonetheless, they are probably related to the initial phase of Roman exploitation of the mine between the late 2nd century and 1st century BC.

In any case, identifying the origin of the Pompeian mill is easier than determining its end. Based on the current state of research, we can only state that it was in still in use beyond Antiquity. It is known in the 4th century AD villa of Casale in Sicily and in the 5th century in Tunisia, in the town of Musti (Tunisia) a *catillus* engraved with a Christian cross (Williams-Thorpe 1988, 261) is known.

An interesting aspect of the pioneering work conducted by D.P.S. Peacock on the Pompeii group consists of petrographical analyses (Peacock 1989, 205-214). Two main sources of volcanic rock were identified. The first is local from a source from the surrounding flows of Mount Vesuvius. A second more distant source is the region of Orvieto, a production centre characterized by leucite crystal inclusions.

2. 3. The state of research prior to this study

There are very few bibliographical references to Pompeian millstones in Languedoc and Provence prior to our research. For the period between the end of the 2nd century BC and the Augustan period, nine mills are mentioned on the site of "Domaine des Forges" (Les Martys, Aude) and four from Lascours (Hérault). This high number of millstone finds could be explained by the fact that they were brought to light in the specific context of metallurgical centres run by Republican companies closely connected to the Italian Peninsula (Domergue *et al.* 1997, 52-53, 58). Elsewhere, Pompeian millstones are reported at the important Roman town of Aix-en-Provence as well as in two secondary Roman towns of Lattes (Hérault) and Lunel-Viel (Hérault). While the find from Aix-en-Provence (Bouches-du-Rhone) is not dated (Nin 2002, 242), the *catillus* fragments of Lattes and Lunel-Viel, respectively, are dated from 50 to 25 BC (Py 1992, 213-214, fig. 21) and AD 70 to 100 (Raynaud 1990, 112, fig. 56).

In the northern part of Gaul, beyond Vienne, there are few written references to Pompeian millstones. On the whole, most are reported in an incomplete manner in local or regional studies, such as those of Amiens (Bayard and Massy 1983, 163), Tours (Audin 2002, 75), Limoges

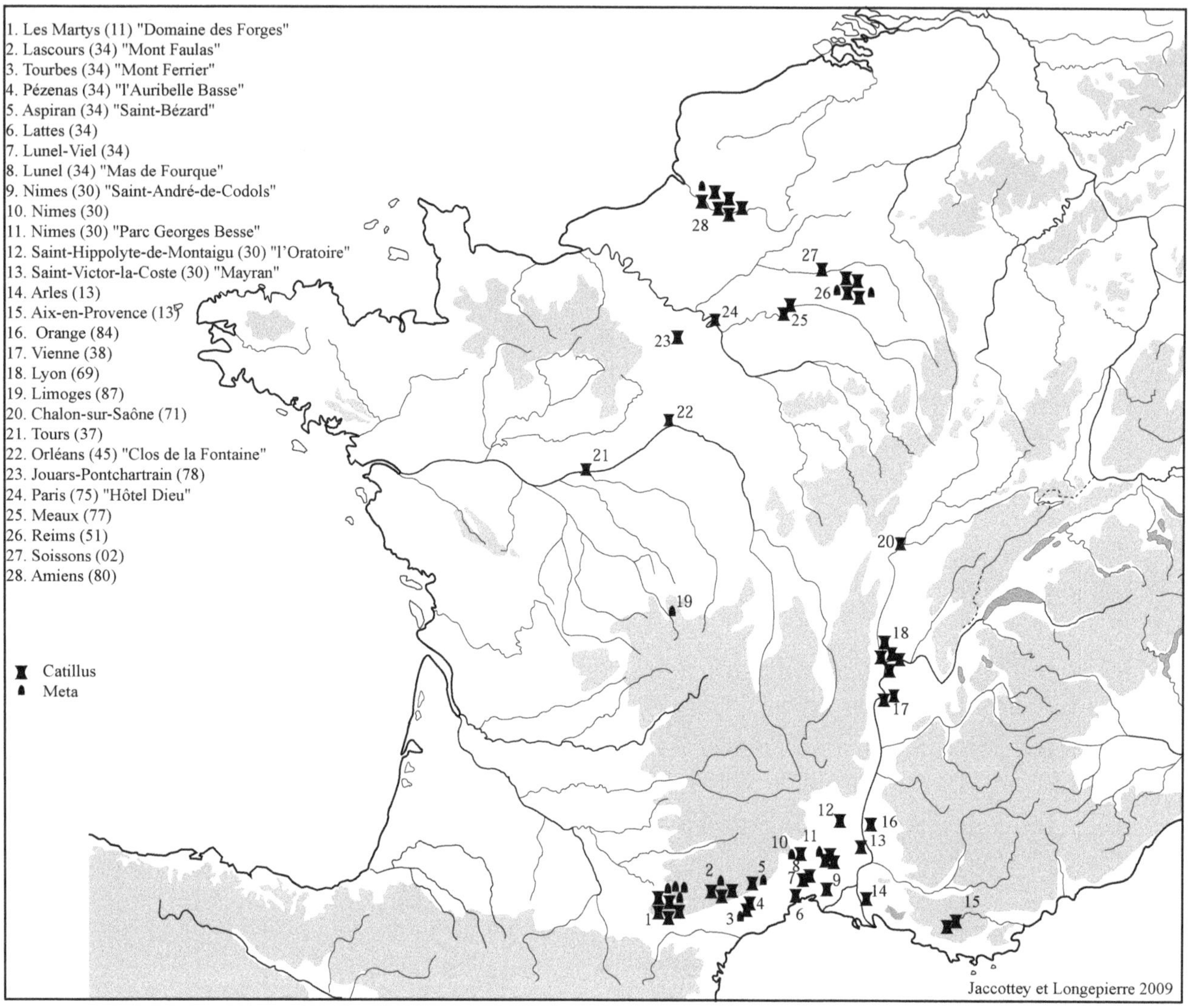

Figure 3. Distribution map of Pompeian millstones in France.

(Loustaud 1972, 35) and Reims (Poulin 2000, 15-18). Although Béal recompiled a series in an article, he did not apparently re-examine the original data (Béal 1996, 93). Particularly conspicuous in this area are two finds, the chance find of a Pompeian millstone used as funerary urn in Amiens (Vassel and Will 1956, 328) and another find brought to light during the excavation of a mosaic in Reims (Lindet 1900; Lorikeet 1862, 10) (Fig. 2). Unfortunately, the drawings of these pieces are either imprecise or non-existent, so their typological study cannot be undertaken.

3. Distribution of Pompeian millstones

After the first phase of inventory of Pompeian millstones, we quickly came to the conclusion that their existence in France is much more widespread than has been commonly thought (Py 1982). The total number identified to date is 63, of which 35 are spread out over the 16 municipalities in Narbonese Gaul alone (Figs. 3 and appendix). This remains very modest, however, compared to the total number of rotary hand querns in France. Indeed, in the database of the *Groupe Meule*, the proportion of rotary querns compared to Pompeian mills was 2044 to 7, thus less than 1% of the total of the corpus.

The existence of Pompeian mills in Narbonese Gaul is no longer to be doubted. Our research in what is modern Languedoc and Provence has identified a total of 673 grinding stones (both whole and fragmented) driven by rotary motion from 112 different sites dating for the most part from either Protohistory, Roman times and, to a lesser extent, Medieval times. This lot can be broken down into the two following major categories: 568 smaller rotary hand querns and 105 larger stones driven either by man, animal or water. Of the 105 larger millstones, almost all (102) date from the 1th century BC to the 4th century AD. About one fifth (22) are of Pompeian millstones. This proportion, however, tends to minimise the real extension of the spread of the Pompeian type. These are, in fact, always hewn from volcanic rock from Italian Orvieto (see petrographical description below), a type of stone that has the disadvantage of becoming brittle with age. Therefore, only large fragments can be unequivocally identified.

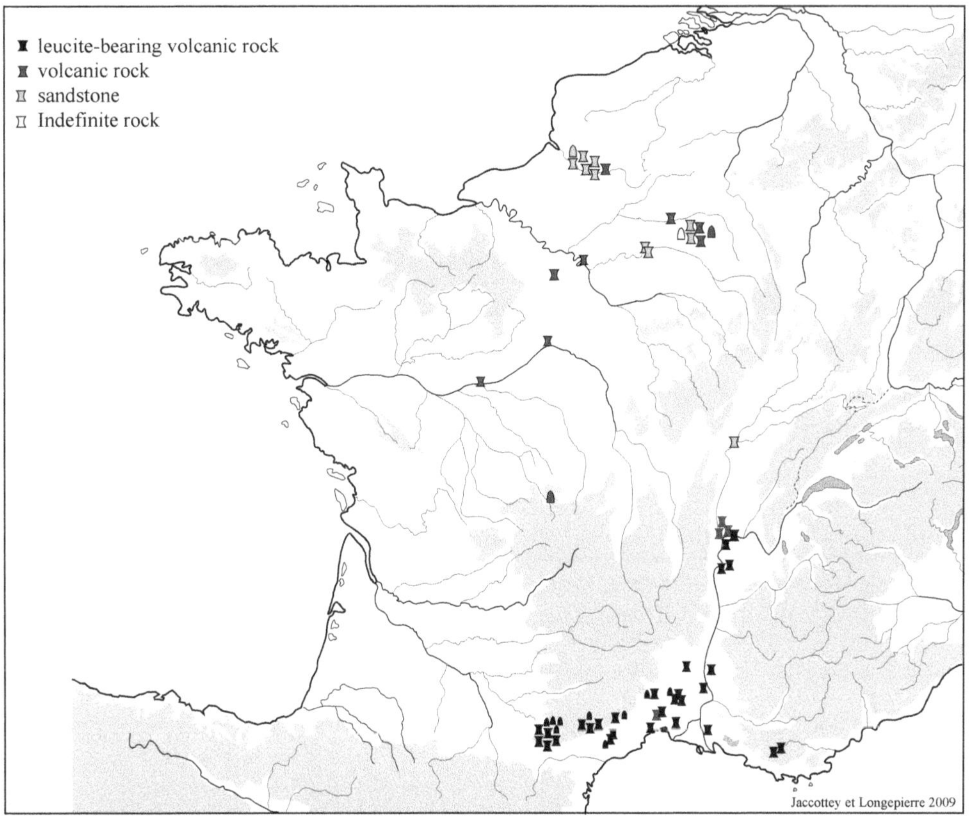

Figure 4. The distribution of Pompeian millstones in France according to petrography.

There are in this region, nevertheless, 17 smaller Orvieto stone fragments on various sites that are likely to be of the Pompeian type. It cannot be excluded that some of these smaller unconfirmed fragments could be parts of hydraulic millstones, or possibly even rotary querns, products that were also, but to a much lesser extent, exported to southern Gaul. We therefore definitively consider that the proportion of 39 Pompeian millstones (i.e., 22 certain fragments and 17 possible fragments) to 80 (other types of large millstones) to be more plausible.

It is true, on the other hand, that the number of confirmed Pompeian millstones (22) is only a small portion of the total mills that operate with a rotary motion (568). This proportion can be explained on chronological grounds and does not signify that Pompeian mills occupied a subordinate role in the diffusion of grinding stones. Indeed, our research clearly reveals that the larger millstones (including the Pompeian type) are more frequent than rotary hand querns from the 1st to the 4th century AD, while the smaller version predominates during the late Republican period and then subsequently in late Antiquity.

The number of Pompeian millstones identified in France beyond the Languedoc and Provence regions is very modest and is stretched out over a wide geographical area to the north up to the Belgian border. This border does not, however, constitute the perimeter to their distribution. Two cases are known in Germany in the Eifel region (Hörter 1994, 33; Baatz 1995, 10) and two others in the United Kingdom (oral communication by D.P.S. Peacock, 2009). The unequal spread of these finds in northern France can be explained by a lack of research in certain areas. This is due to the fact that no member of our research group resides in these areas, restricting access to museum collections and local bibliography. Similarly, no Pompeian mill has been identified to date in the southwest of France. In the Rhone Valley, however, this mill type is quite frequent, particularly in the Roman cities of Lyon and Vienne. If this observation is not only the result of the state of research in the two areas, one might speculate that the stones were transported by means of the Rhone River Valley, a natural axis that connected the Mediterranean to the inland and favored the transport of goods from Narbonese Gaul where Pompeian mills are common finds.

The proportion of finds of *meta* to *catilli* is respectively 16 to 47. The higher proportion of *catilli* is particularly evident outside Narbonese Gaul (24 to 5). A possible explanation is that these *catilli,* with their biconical shape and large quadrangular mounts, are much easier to identify (especially in a fragmented state) by the untrained eye than the *metae*. This probably explains why, for example, three of four *metae* inventoried in the north of France are whole.

4. Petrography (Fig. 4)

All of the Pompeian mills fragments identified to date in Languedoc and Provence were hewn from a vesicular volcanic rock containing white leucite crystals. In spite of the lack of systematic formal analyses, an Italian origin seems very likely. In fact, this rock is highly characteristic and can be identified by the naked eye and it is difficult

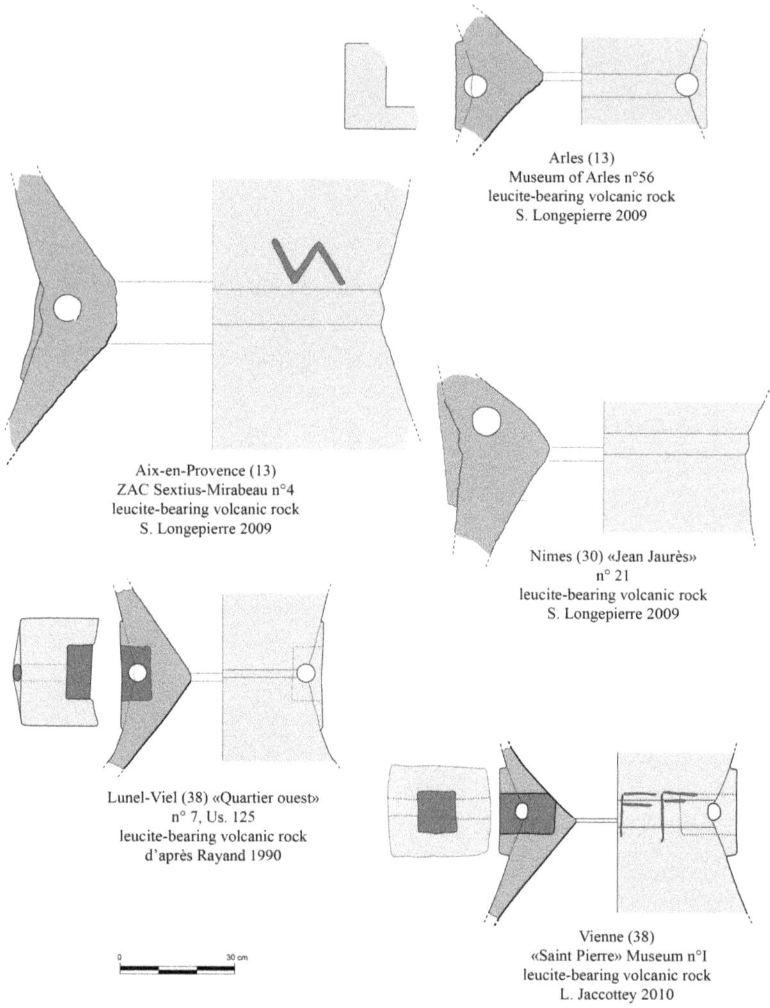

Figure 5. Catilli of leucite-bearing volcanic rock with a band at the midsection.

to confuse with the volcanic rocks of Gaul. Although there are numerous volcanic regions throughout the Italian Peninsula, it is near Orvieto that the production of Pompeian millstones is certified (Peacock 1989, 205-213). According to analyses, Orvieto appears to be the source of the Pompeian mills found in the metallurgical sites of Languedoc at Forges (Les Martys, Aude) and Lascours (Hérault), (Domergue *et al.* 1997, 51-54). Furthermore, J.-L. Reille, in his recent analyses of Protohistorical millstones in southern Gaul, also proposes Orvieto as the source of a Pompeian *catillus* that we have identified typologically from the Antique city of Aix-en-Provence (Nin 2002, 242).

Other than in the southeast of France, the situation is more complex and no petrographical analysis has been performed on the Pompeian millstones. To distinguish between volcanic and non-volcanic rock, we have had to settle for a visual identification. The origin of the volcanic mills could be either Orvieto or the volcanic district of the *Massif Central* in France, an area that had yielded rotary querns since the end of the Iron Age (Jaccottey *et al.* 2008, 377-381). Another possible origin is that of the Eifel in Germany, where different types of millstones have been manufactured in large numbers since Antiquity (Mangartz 2006).

In the Rhone valley, the Pompeian millstones from the cities of Lyon and Vienne are all volcanic. It is conceivable that some come from Orvieto and that they were transited through Narbonese Gaul, where all the identified Pompeian mills are exclusively from Orvieto. It is also conceivable that anothers could originate in the volcanic region of the Massif Central, located a short distance from the Rhone Valley. To date, however, there is no archaeological evidence to back up the hypothesis of Pompeian millstones manufacture in this region.

Beyond the Rhone Valley, the material employed to produce these millstones is either sandstone (Chalon-sur-Saone, Meaux, Reims and Amiens) or volcanic rock devoid of leucite crystals (Limoges, Orleans, Tours, Jouars-Pontchartrain, Reims, Soissons, Amiens and Troyes). If the volcanic millstones originate in the *Massif Central*, they would have been produced at a distance of over 150 km from Limoges and close to 300 km from Tours and Orleans. However, if their origin is in the Eifel of Germany, the distance would surpass 600 km. Moreover, the cities of Paris, Jouars-Pontchartrain, Amiens and Reims are 300 to 500 km from the Eifel, and the *Massif Central* in central France is even more remote. Therefore, since we are unable for the moment to determine precisely the origin of production, they must be considered as long-range exports. Toward the end of Protohistory and during Roman domination, the commerce of this mill type was surely far more long-range than that of rotary querns. The majority of the smaller rotary handmills, although known in some cases to have travelled long distances, originated in local and regional quarries (Boyer *et al.* 2006).

Sandstone Pompeian millstones identified on some sites are proof, however, of the existence of regional French productions. This is the case of the millstone conserved in the museum of Denon in the town of Chalon-sur-Saone. It was hewn from Triassic sandstone similar to outcrops known on the eastern flank of the Morvan Mountains located a few dozen kilometers away. Fragments of sandstone mills were also unearthed in the towns of Meaux (Lepareux-Couturier, forthcoming) and Reims. These are a gray coarse-grained variety (with a quartzite aspect) containing feldspar and black flint inclusions that are easily perceptible to the naked eye due to their "salt and pepper" appearance. Outcrops are known only in the communes of Fosses and Belleu, hence the name "Fosses-Belleu". The distance between the two outcrops and the millstone finds is between 40 and 130 km (Amiens-Fosses: 100 km; Amien-Belleu: 100 km; Meaux-Fosses: 40 km; Meaux-Belleu: 70 km; Reims-Fosses: 130 km; Reims-Belleu: 60 km).

5. The context and types of sites with Pompeian millstone finds

None of the Pompeian millstones in our study was discovered in its original operational position. Most were unearthed in backfill levels. Some come from the following varied contexts: recycled in walls (Meaux), in modern demolition levels of an abbey from the Chalon-sur-Saône River bottom in Lyon, or from the bottom of a well in Reims. In Orleans at "Clos de la Fontaine", a *catillus* mount was fixed as a pivot in the frame of the door of an ancient sanctuary, and at the Rue de l'Union, Amiens, the broken-off cones of a *catillus* contained a Roman burial urn (Vassel and Will 1956, 328).

The use of Pompeian mills outside the region of Narbonnese Gaul is limited, for the moment, exclusively to settlements dating from Antiquity. They are found in either major cities (eg, Amiens, Chalon-sur-Saone, Limoges, Lyon, Paris, Reims, Saint-Romain-en-Gall and Troyes) or smaller settlements (eg, Jouars-Pontchartrain, Meaux and Soissons). Inside Narbonese Gaul, they are in the main centres of Arles, Aix-en-Provence and Orange (Provence) and Nîmes (Languedoc) as well as in the more modest settlement of Lunel-Viel (Hérault). Seven rural sites, including at least five *villae,* and one smaller rural

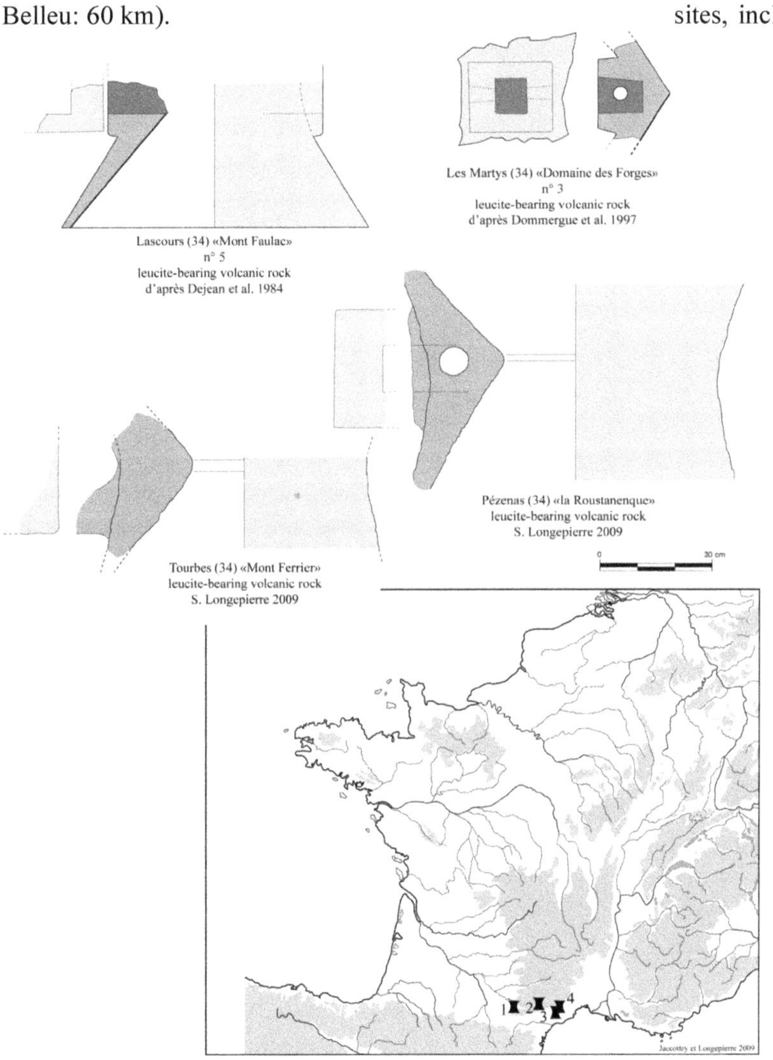

Figure 6. Catilli of leucite-bearing volcanic rock with a band at the midsection and their distribution in France.

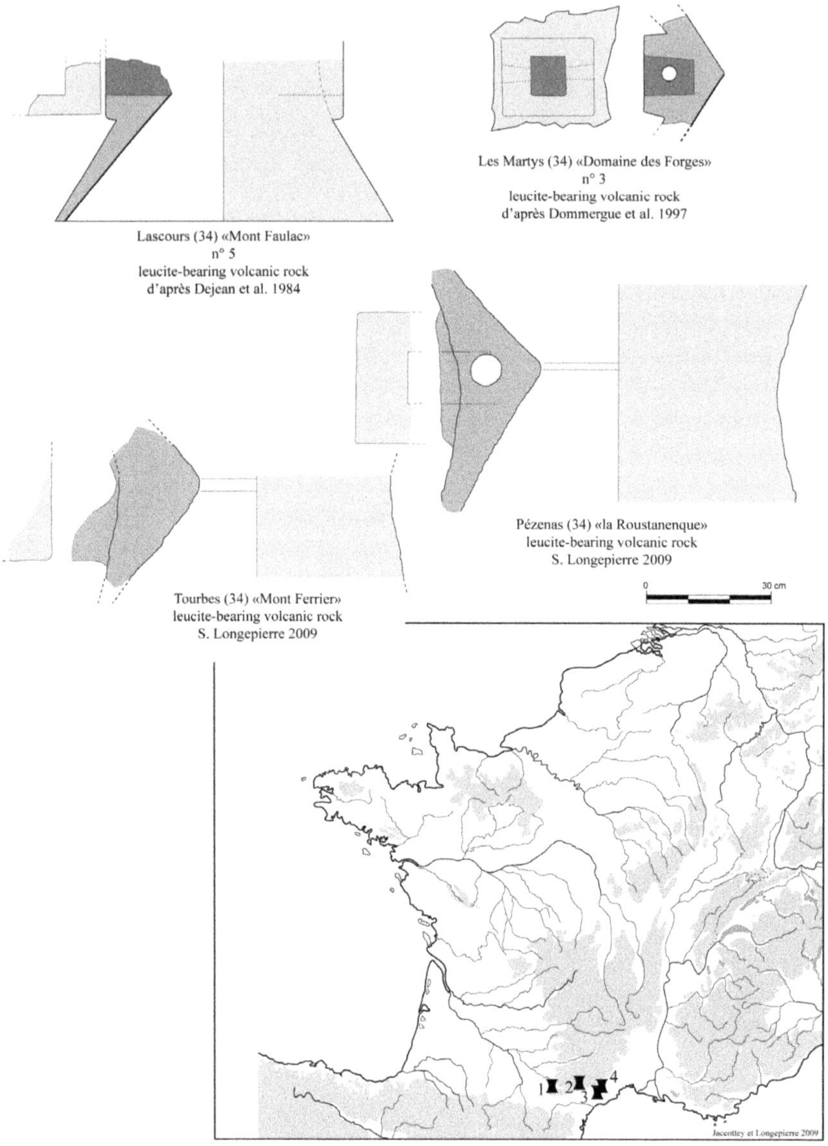

Figure 7. Catilli of leucite-bearing volcanic rock without an external band and their distribution in France.

site (that cannot possibly be classified as a *villa)* also yield this millstone type.

6. Typological aspects

The following section is an attempt, without delving into detail, to establish a preliminary typology on certain morphological criteria in combination with lithology (volcanic leucite-bearing rocks, volcanic rocks, Fosses-Beleu sandstone, etc).

6.1. *Catilli*

Finds of the upper stone (*catilli*) of Pompeian mills are much more numerous than those of lower stones (*meta*) and constitute the base for this typology. Both volcanic rocks and sandstones are represented. In the cases of volcanic origin, the typology begins with finds from Narbonese Gaul with an established Italian origin.

6.1.1. Catilli of leucite-bearing rock with a band at the midsection (Figs. 5-6)

A large proportion of the group are of volcanic rock bearing leucite crystals, and these originate in the region of Orvieto (see above). They feature a slightly rounded horizontal band at the narrowest part of the girth, which measures between 6 and 10 cm wide (only the Lunel-Viel case, at 2 cm, is thiner) and generally bears no decoration (an exception with incisions is the case of Vienne II). The section of these *catilli* is narrow and constricted (Aix-en-Provence, Lunel-Viel, and Lyon 3 and 5) or, in some cases, more open (Arles, Nîmes and Vienne I and II). Several bear inscriptions (Aix-en-Province, Vienne I and II). The mounts are rectangular, slightly higher than wide, with straight upper edges (except the cases of Lunel-Viel and Lyon 3) and slightly slanting lateral edges. These finds are limited to Narbonese Gaul and to the Rhône Valley.

6.1.2. *Catilli* of leucite-bearing rock without a band at the midsection (Fig. 7)

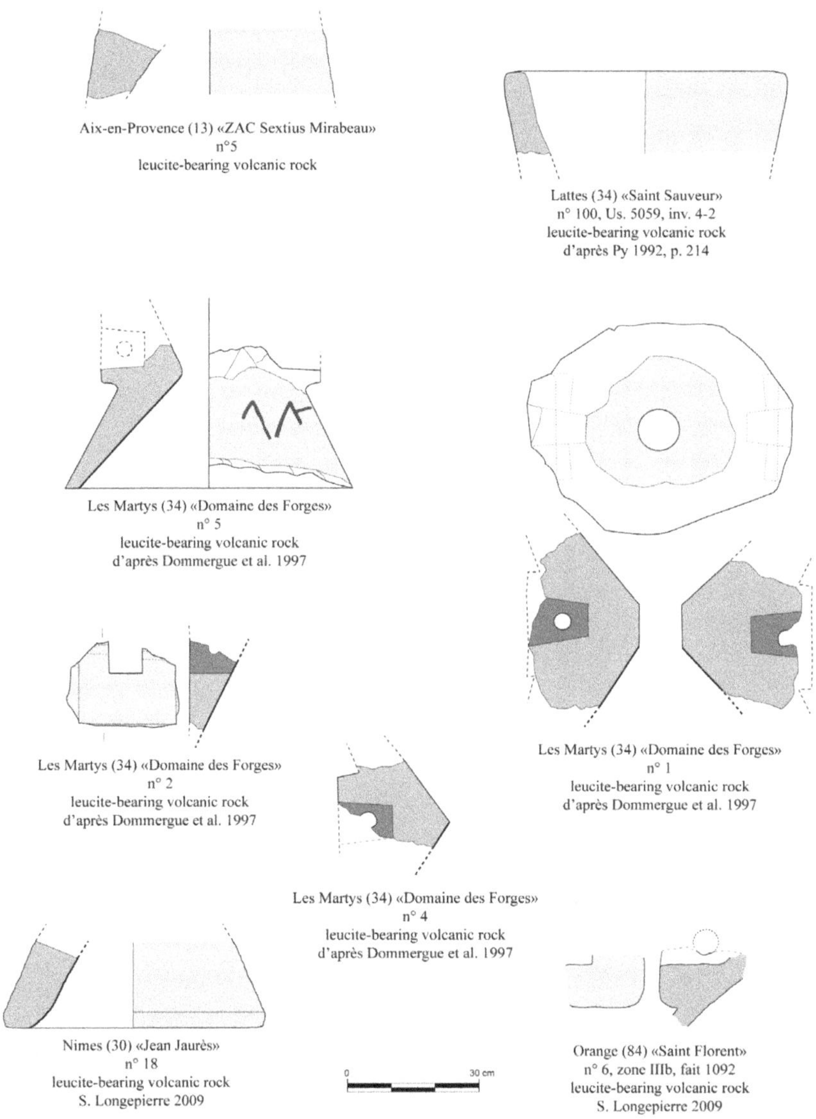

Figure 8. Catillus fragments of leucite-bearing volcanic rock.

The lithology and the other typological criteria of this group, located in Narbonese Gaul, remain the same as the preceding group. The difference is the absence of a band at the midsection. Other similar fragments, included here, are, in fact, too fragmented to determine truly whether they feature a band at the midsection (Fig. 8). Two asymmetric cases from "Mont Faulac" at Lascours have very short hoppers (Fig. 9).

6.1.3. *Catilli* of volcanic rock without a band at the midsection (Figs. 10-11)

To the north of Lyon are several *catilli* of volcanic rock devoid of leucite crystals. They also show no band at the midsection. The mounts are either square or rectangular. Contrary to the cases of *catilli* bearing leucite crystals, when the mount is of rectangular shape, it is oriented vertically (Lyon, n°1 and 4, Orléans, Tours). These few finds come from Lyon (n°1, 2 and 4), Tours and Orléans. It is probable the "Hôtel Dieu" (Paris) fragment with a square mount and no band (Fig. 2) falls into this category. This millstone type is distributed throughout the centre of France.

6.1.4. *Catilli* of volcanic rock with trapezoidal-sectioned mounts (Figs. 12-13)

The five examples of this type all are found in the north of France and they all feature mounts with trapezoidal sections. This trait is rarely seen outside of this area (the only exception being the feldspar-rich sandstone of Chalon-sur-Saône). These mounts are usually proportionally higher than wide (except for an example in the museum of Saint Rémi at Reims), and their top and bottom edges are slanting. The lateral edges of the mounts are either straight (Reims "musée Saint Rémi" and Soissons) or slanting (Amiens and Reims "rue de Cernay"). Three of these millstones (Soissons "rue des Chaperons Rouges", Reims "mosaïque des Promenades" and "rue de Cernay") have a band at the midsection decorated with geometric motifs.This motif is generally made up of three horizontal lines intersected at intervals with vertical lines.

6.1.5. *Catilli* of Fosses-Belleu sandstone (Figs. 14-15)

The eight *catilli* of this rock type are concentrated in the north of France and the eastern region of the Parisian Basin.

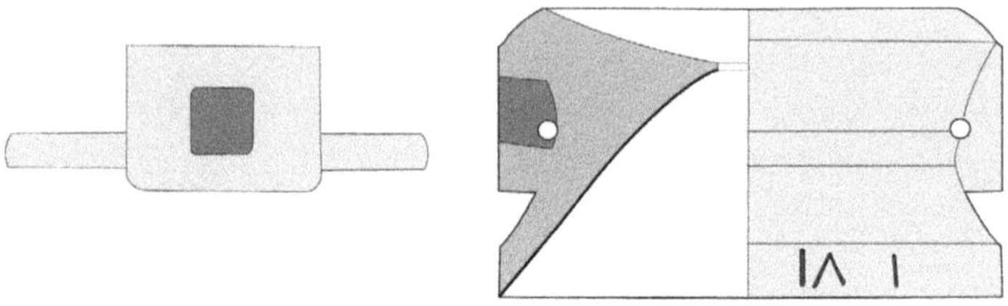

Lascours (34) «Mont Faulac»
n° 3
leucite-bearing volcanic rock
d'après Dejean et al. 1984

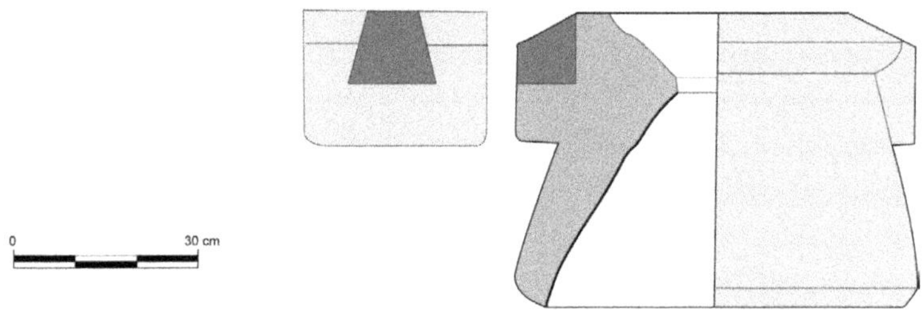

Lascours (34) «Mont Faulac»
n° 2
leucite-bearing volcanic rock
d'après Dejean et al. 1984

Figure 9. Catillus fragments of leucite-bearing volcanic rock from the site of Lascours.

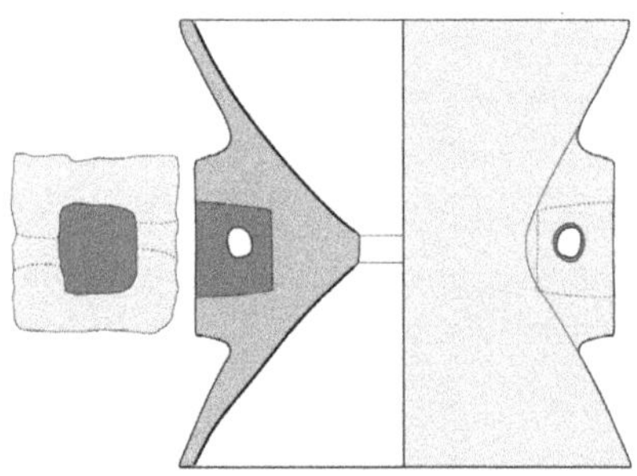

Lyon (69)
Museum of the Gallo-Roman civilization, n°1
volcanic rock
L. Jaccottey 2010

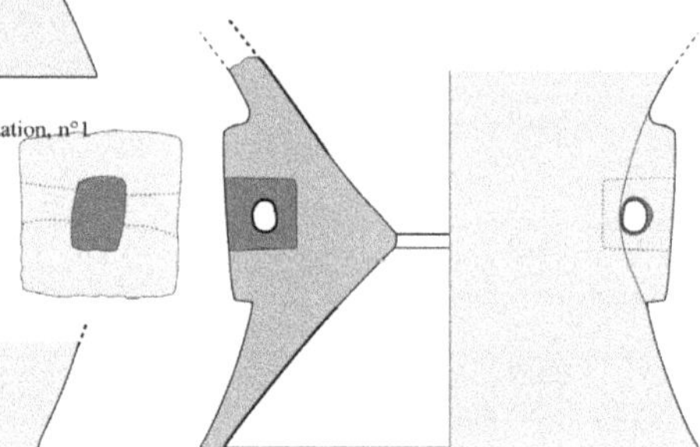

Lyon (69)
Museum of the Gallo-Roman civilization, n°2
volcanic rock
L. Jaccottey 2010

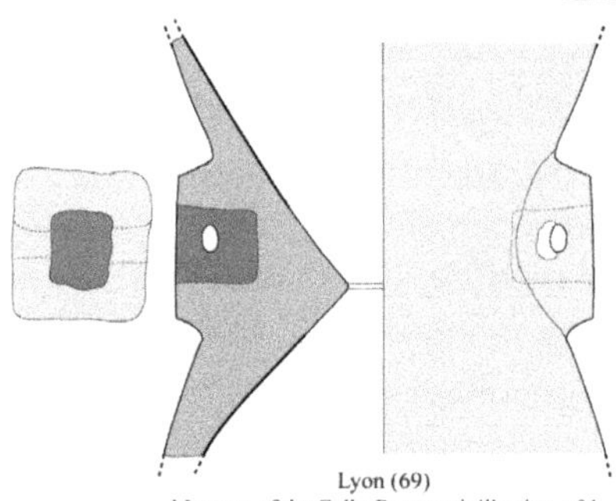

Lyon (69)
Museum of the Gallo-Roman civilization, n°4
volcanic rock
L. Jaccottey 2010

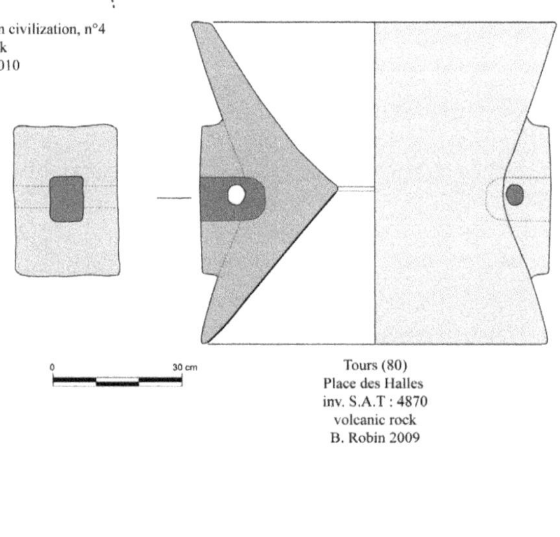

Tours (80)
Place des Halles
inv. S.A.T : 4870
volcanic rock
B. Robin 2009

Figure 10. Catilli of volcanic rock (devoid of leucite crystals) without a band at the midsection.

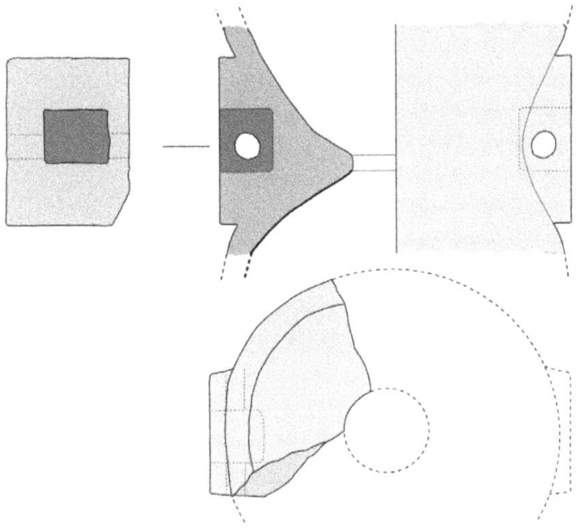

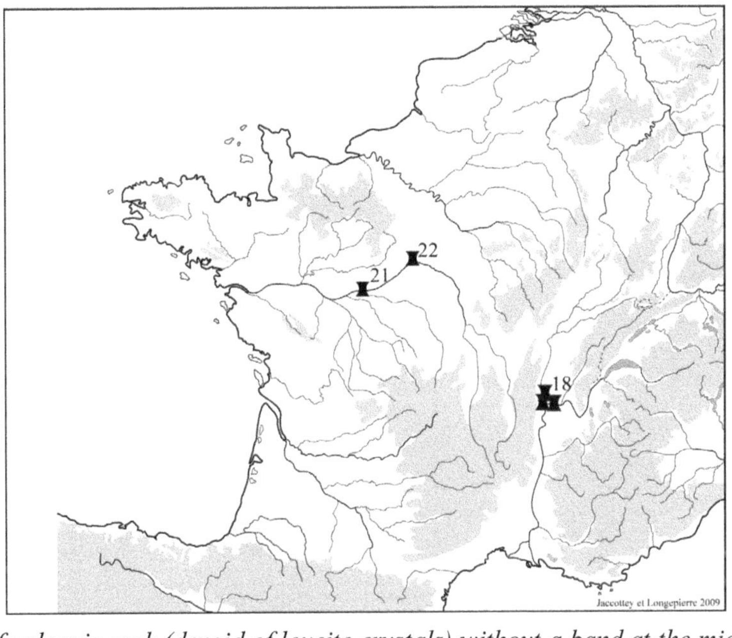

Figure 11. Catilli of volcanic rock (devoid of leucite crystals) without a band at the midsection and their distribution in France.

At Meaux and Reims their shape is more squat, while at Amiens they are higher. Their mounts are square-shaped and square-sectioned and, except for one case at Amiens (n°4), do not have a band at the midsection.

6.1.6. *Catilli* of feldspar-rich sandstone (Fig. 16)

These two upper stones hewn from local sandstone are proof of regional Pompeian type millstone production in France. For the moment, the number of finds of this type is far too modest to formulate a typology. In any case, the *catillus* of Chalon-sur-Saône does not bear an external band like the leucite volcanic *catilli* of the nearby cities of Lyon, Orléans and Tours. On the other hand, the feldspar sandstone *catillus* of Amiens, as well as the Fosses-Belleu sandstone, also of Amiens (n°4) and most of the volcanic *catilli* of the north of France (Reims "mosaïque des promenades" and "rue Cernay" and Soissons "rue des Chaperons Rouges") have a band at the midsection. It would thus appear that we find certain "regional" typological aspects present on upper stones of different rock types.

6.2. *Metae* (Figs. 17-18)

Pompeian lower stone finds are far fewer that the *catilli* as well as of quite varied typology. All of the volcanic

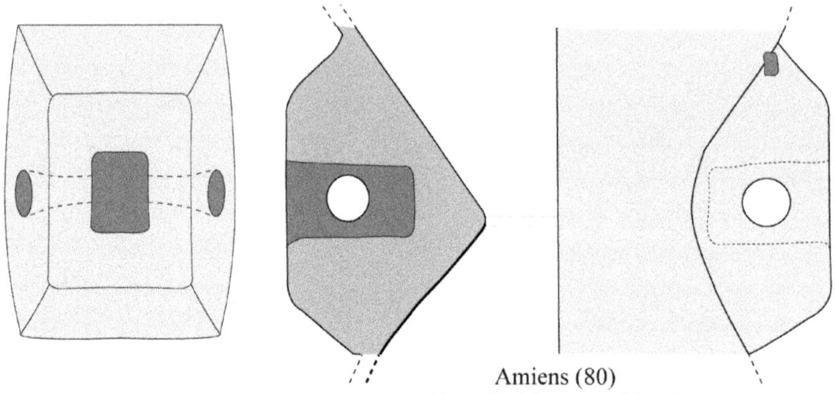

Amiens (80)
Picardie Museum n°Am 3
volcanic rock
P. Picavet 2009

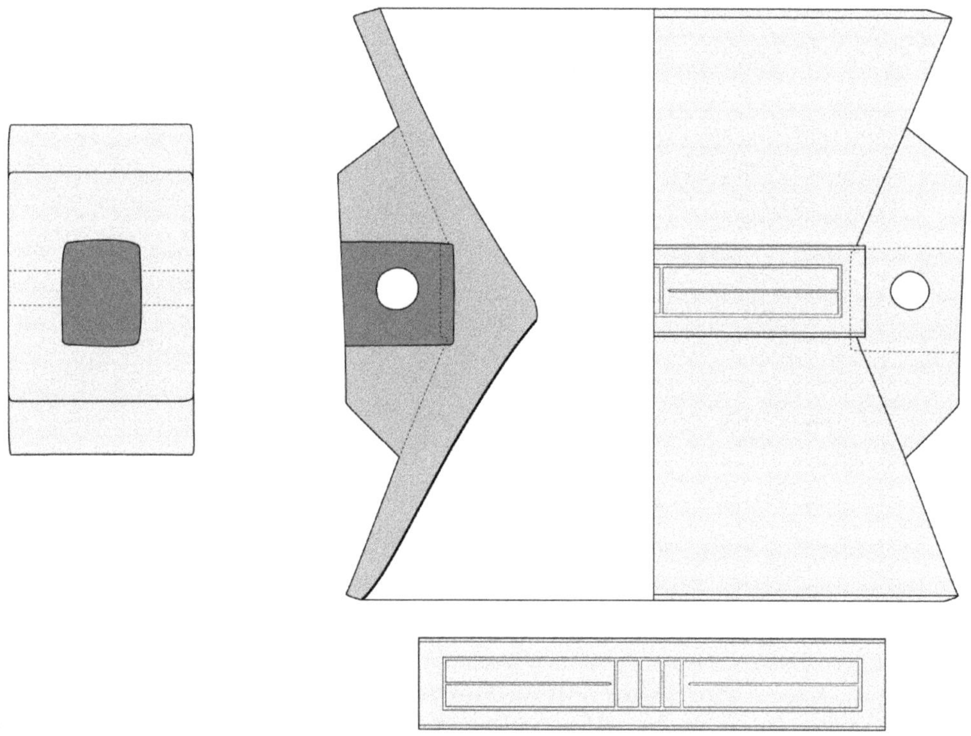

Soissons (02) "rue des chaperons rouges"
Soissons Museum 93.7.1287
volcanic rock
B. Robert et J. Lamant 2009

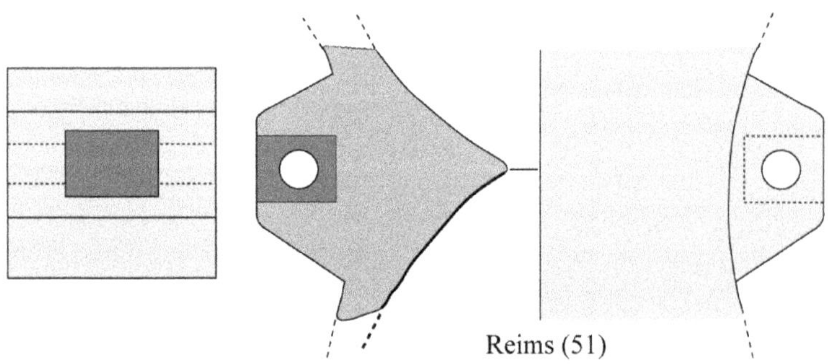

Reims (51)
Saint Rémi Museum
volcanic rock
F. Jodry et S. Lepareux-Couturier 2009

Figure 12. Catilli of volcanic rock (devoid of leucite crystals) with trapezoidal mounts.

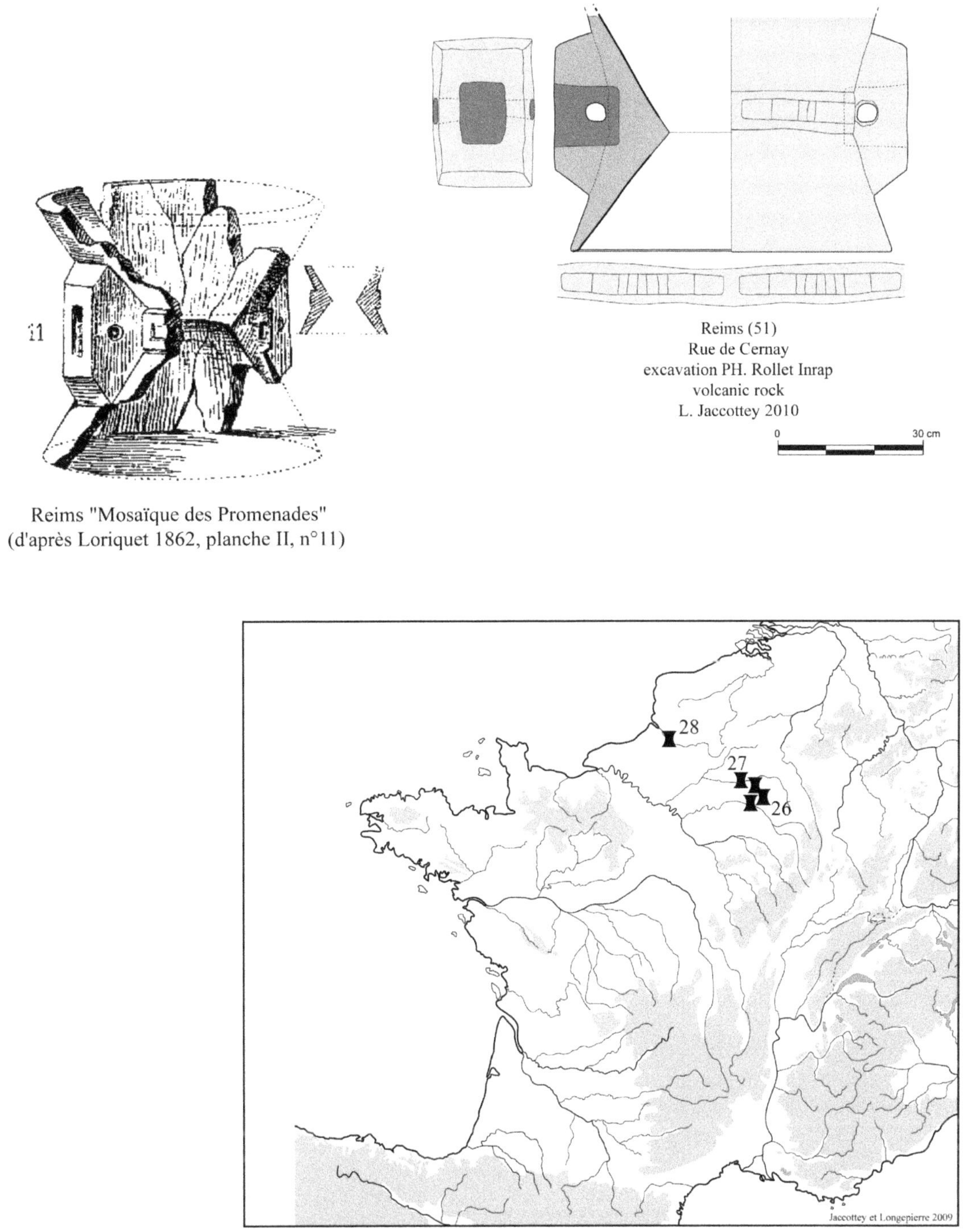

Figure 13. Catilli of volcanic rock (devoid of leucite crystals) with trapezoidal mounts and their distribution in France.

metae bearing leucite crystals have a cylindrical base. This base is sometimes high (Les Martys n°1, 2 and 4, Nimes, Saint-Victor-la-Coste). The only exception is the volcanic *meta* devoid of leucite crystals from the "boulevard du Dr Henron" at Reims. Apart from the aspect of the base, the typological features are varied, for example grinding faces are both straight (Les Martys 4, Lascours, Nimes and Saint-Victor-la-Coste) and convex (Les Martys 1 à 3) and the base is often hollowed.

7. Conclusion

This first study of Pompeian millstones in France reveals that this mill type is not limited exclusively to Narbonese Gaul. This suggests that the number of discoveries of this type of millstone will surely increase in the coming years as further research in this domain is undertaken.

Concerning the type and the spread of the raw material, the rock used in the southeast of France is evidently exclusively Italian, originating in the region of Orvieto (Latium). The situation in other areas of France is quite different, with the presence of exogenous volcanic rocks

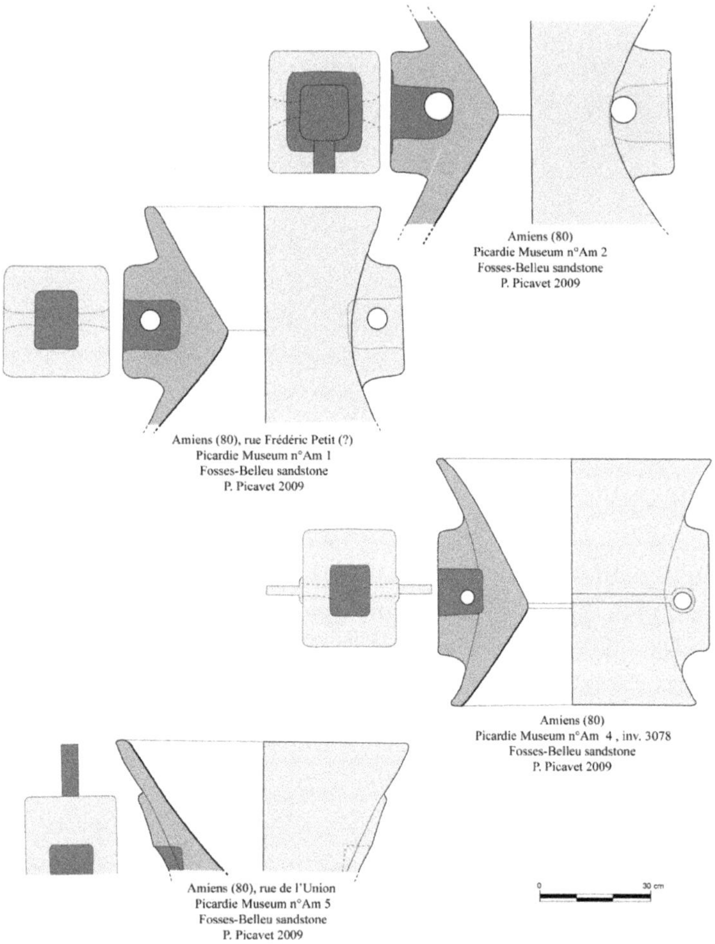

Figure 14. Catilli of Fosses-Belleu sandstone.

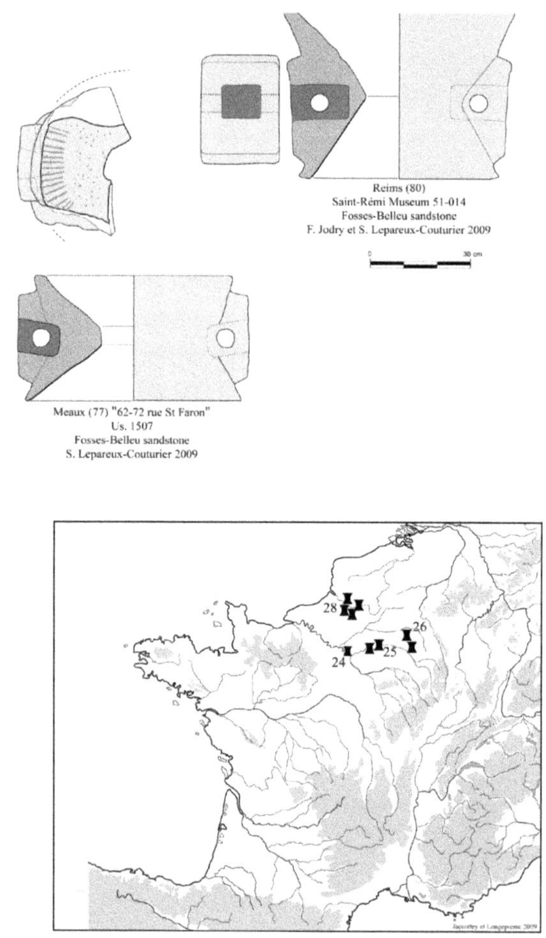

Figure 15. Catilli of Fosses-Belleu sandstone and their distribution in France.

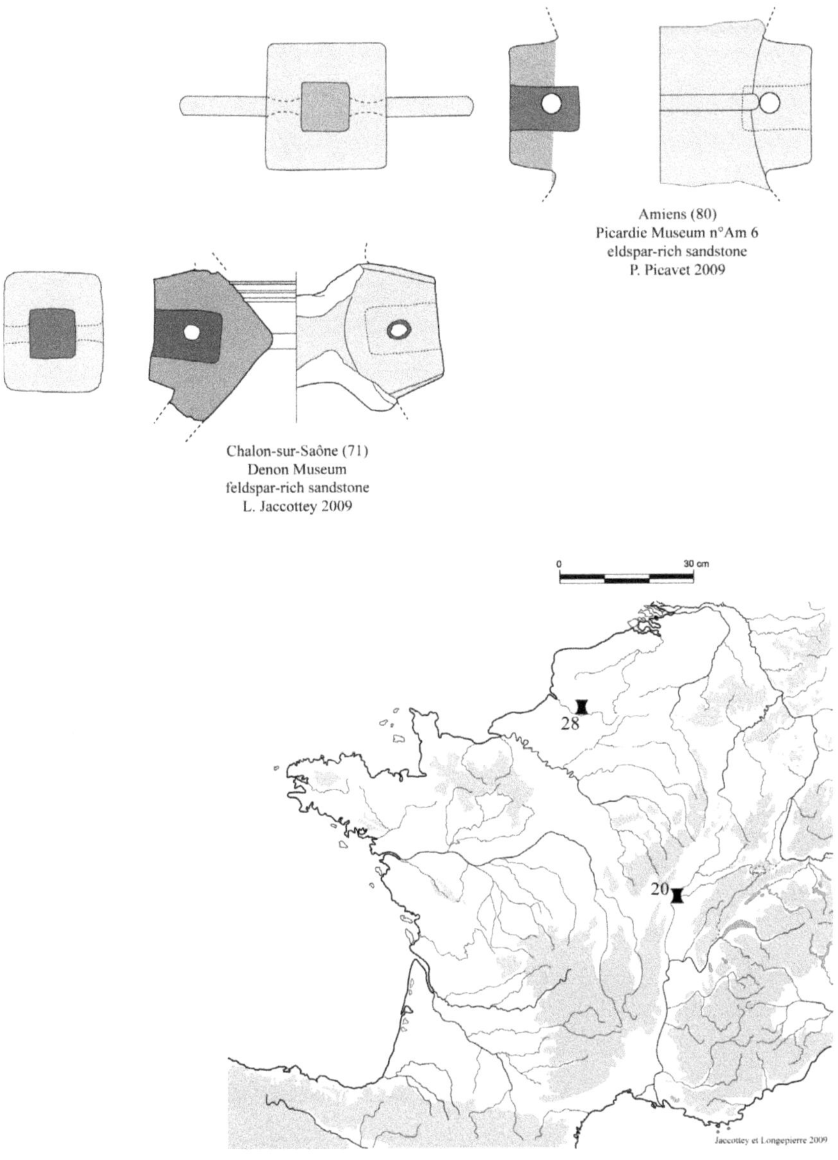

Figure 16 (left). Catilli of feldspar-rich sandstone and their distribution in France.

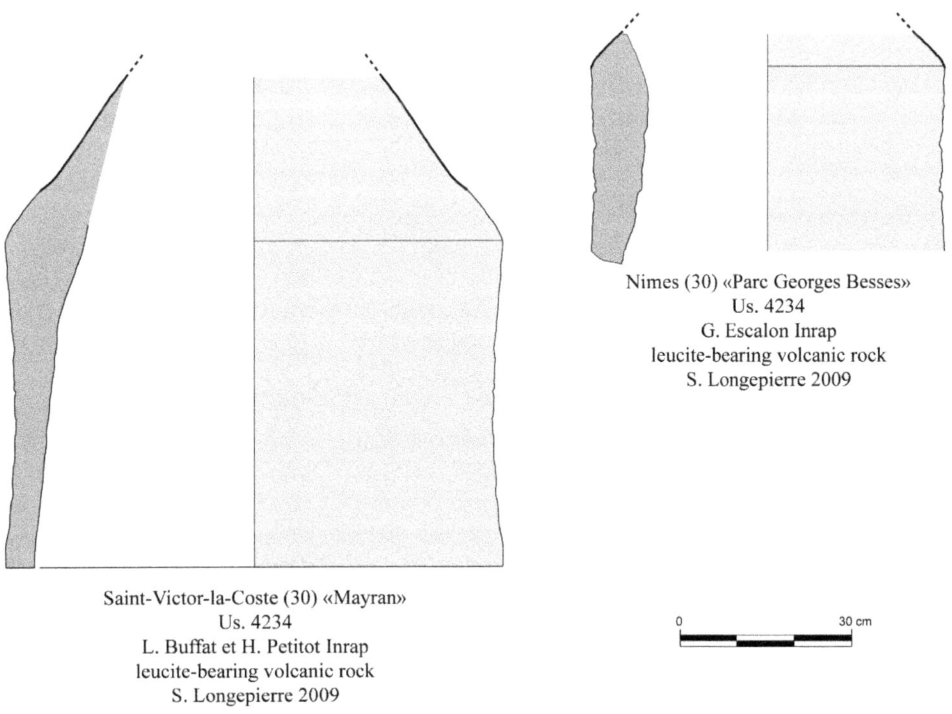

Figure 17 (right). Metae of volcanic rock.

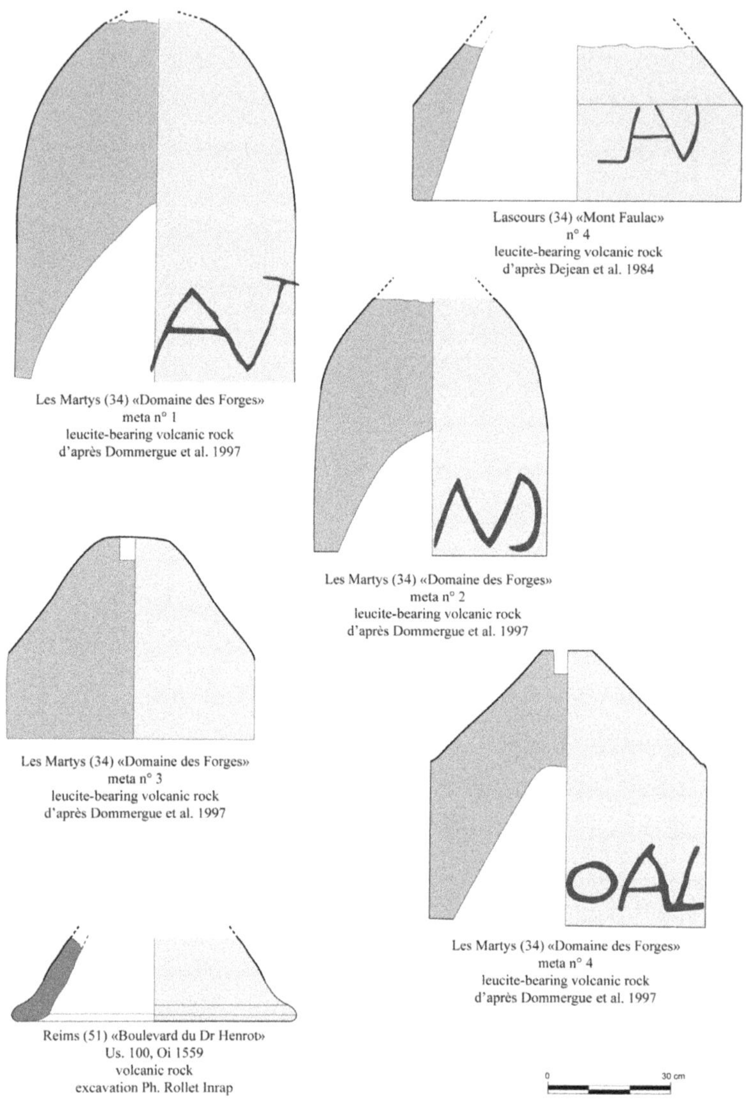

Figure 18. Metae of volcanic rock.

as well as local sandstones. This makes for an interesting comparison to Pompeii, where the stones are either long-distance volcanic imports from Orvieto or local volcanic rock exploited near the ancient city in the vicinity of Mount Vesuvius (Peacock 1989, 205-214).

Several typological assemblages begin to appear in France. In Narbonese Gaul and in the Rhone Valley appears a group of volcanic *catilli* with leucite crystals bearing a simple band at the midsection, while a second group of volcanic *catilli* devoid of both leucite crystals and band are concentrated around central France. A third assemblage of Fosses-Belleu sandstone *catilli* is found in the Parisian Basin as well as in an area slightly to the northeast. Finally, in the north is a fourth group of volcanic *catilli* devoid of leucite crystals bearing trapezoidal shaped mounts and, sometimes, decorated bands. These first typological assemblages must, however, be confirmed and validated by petrographical and chemical analyses so as to clearly distinguish the Orvieto production from other potential volcanic millstone production centres.

Acknowledgements

The study and the drawings of the different millstones were undertaken by the following researchers: Luc Jaccottey (Chalon-sur-Saône, Lyon, Reims and Vienne), Florent Jodry (Reims and Troyes), Stéphanie Lepareux-Couturier (Meaux and Reims), Samuel Longepierre (Aix-en-Provence, Arles, Aspiran, Lunel, Nîmes, Orange, Pézenas, Saint-Hyppolite-de-Montaigu and Saint-Victor-la-Coste), Paul Picavet (Amiens), Bruno Robert (Soissons) and Boris Robin (Orléans and Tours).

We would like to thank the following people for their welcome assistance: Isabelle Bollard (Musée de Bavay), Louis Bonnamour (Conservateur Musée Denon à Chalon-sur-Saône), Jean-Yves Breuil (Inrap), Loïc Buffat (Mosaïque), Michel Compan (Inrap), Frédéric Conche (Inrap), David Couturier (Inrap, Centre Ile-de-France), David Djaoui (Musée de l'Arles antiques), Virginie Durand (Musée de Vienne), G. Escalon (Inrap), Alain Genot (Musée de l'Arles antique), Noël Maheo (Conservateur Musée d'Amiens), Stéphane Mauné (CNRS, Lattes),

Jean-Marc Mignon (Service Archéologique du Vaucluse), Catherine N'Guyen (Musée de la civilisation gallo-romaine de Lyon), Nuria Nin (Service Archéologique de la Ville d'Aix-en-Provence), Mathieu Ott (Inrap), Hervé Pomarèdes (Inrap), Claude Raynaud (CNRS, Lattes), Sylvie Robin (Conservatrice Musée Carnavalet à Paris), Philippe Rollet (Inrap, Reims), Dominique Roussel (Conservateur Musée de Soissons), Hugues Savay-Guerraz (Conservateur au Musée de la civilisation gallo-romaine de Lyon) and Franck Verneau (Inrap Centre Ile-de-France). The translation is the work of Timothy Anderson.

References

Alonso, N., 1995. Les premières meules rotatives manuelles dans le nord-est de la péninsule ibérique. In M.C. Amouretti and G. Comet, *La transmission des connaissances techniques: Table Ronde d'Aix-en-Provence, Cahiers d'histoire des techniques,* 3, 15-23.

Alonso, N., 1997. Origen y expansión del molino rotativo bajo en el Mediterráneo occidental. In D. Meeks and D. Garcia (eds.), *Techniques et économies antiques et médiévales: le temps de l'innovation:* Actes du international d'Aix-en-Provence, Travaux du Centre Camille Jullian; 21, 15-9.

Alonso, N., 1999. *De la llavor a la farina, Els processos agrícoles protohistòrics a la Catalunya Occidental, Monographies d'Archéologie Méditerranéenne*, 4, Lattes, 1999.

Alonso, N., 2002. Le moulin rotatif manuel au nord-est de la Péninsule ibérique: une innovation technique dans le contexte domestique de la mouture des céréales. In H. Procopiou and R. Treuil (eds.), *Moudre et broyer, L'interprétation fonctionnelle de l'outillage de mouture et du broyage dans la Préhistoire et l'Antiquité, II, Archéologie et Histoire: du Paléolithique au Moyen Âge,* CTHS, 111-27.

Amouretti, M.-C., 1986. *Le pain et l'huile dans la Grèce ancienne, de l'araire au moulin.* Paris, Les Belles Lettres.

Amouric, H., 1997. L'anille et les meules. In D. Meek and D. Garcia (eds.)., *Techniques et économies antiques et médiévales: le temps de l'innovation:* Actes du international d'Aix-en-Provence, Travaux du Centre Camille Jullian; 21, 39-47.

Arribas, A., 1987. *El barco del Sec (Calvia, Mallorca), estudio de los materiales.* Majorque.

Audin, P., 2002. *Tours à l'époque gallo-romaine*, éd. Sutton, 75.

Baatz, D., 1995. Die Wassermühle bei Vitruv X 5,2, Ein archäologischer Kommentar. *Saalburg-Jahrbuch*, 48, 5-18.

Bakker, J.-Th., Van Dalen, J.-H., Heres, Th.-L., Meijlink, B. and Sirks, A.-J.-B, 1999. *The mills-bakeries of Ostia, Description and interpretation.* Amsterdam.

Bayard, D. and Massy, J.L., 1983. Amiens romain, *Revue Archéologique de Picardie*, 1983, n° spécial 2, 163.

Béal, J. C., 1996. Pistor et Matériarius, à propos d'une stèle funéraire de Metz antique, *Revue Archéologique de l'Est*, 47, 93.

Boyer F., Buchsenschutz, O., Hamon, C., Jaccottey, L., Lagadec, J.-P., Milleville, A., Thomas, E. and Triboulot, B., 2006. Production et diffusion des meules du Néolithique à l'Antiquité: quelques exemples français. In A. Belmont and F. Mangartz (eds.), *Les meulières. Recherche, protection et valorisation d'un patrimoine industriel européen,* colloque international. Grenoble, RGZM Tagungen 2, 5-13.

Buchsenschutz, O., autres directeurs., forthcoming. *Évolution typologique et technique des meules du Néolithique à l'an mille sur le territoire français, Table ronde de Saint-Julien-sur-Garonne*(F) du 2 au 4 octobre 2009, supplément Aquitania.

Domergue, C., 1993. Un centre sidérurgique romain de la Montagne Noire, Le Domaine des Forges (Les Martys, Aude), *Revue Archéologique de Narbonnaise*, supplément 27.

Domergue, C., Béziat, D., Cauuet, B., Jarrier, C., Landes, C., Morasz, J.-G., Olivia, P., Pulon R. and Tollon, F., 1997. Les moulins rotatifs dans les mines et les centres métallurgiques antiques, In D. Meeks and D. Garcia (eds.), *Techniques et économies antiques et médiévales: le temps de l'innovation:* Actes du international d'Aix-en-Provence, Travaux du Centre Camille Jullian, 21, 48-61.

Hörter, F., 1994.*Getreidereiben und müesli aus der Eifel, ein Beitrag zur steinbock und mühlengeschichte.* Mayen, 192.

Jaccottey, L. and Farget, V., forthcoming. Les normes de dessin des meules rotatives. In O. Buchsenschutz *et al.*, *Évolution typologique et technique des meules du Néolithique à l'an mille sur le territoire français, Table ronde de Saint-Julien-sur-Garonne*(F) du 2 au 4 octobre 2009, supplément Aquitania.

Jaccottey, L. and Gluhak, T., 2008. Etude des meules de Bibracte, *Bibracte rapport 2008*, 377-388.

Jehasse, J. and Jehasse, L., 1987. *Aléria antique.* Lyon, 31 and fig. 18.

Lancel, S., 1982. L'îlot D. In S. Lancel, J-P. Morel and J-P. Thuillier, *Byrsa II, Rapports préliminaires sur les fouilles 1977-1978: niveaux et vestiges puniques.* Rome, Collection de l'Ecole française de Rome, 41, 85-103.

Lepareux-Couturier S., forthcoming. Les meules gallo-romaines du chef lieu de la cité des Meldes (Meaux, Seine-et-Marne), première approche, In Buchsenschutz *et al.*, *Évolution typologique et technique des meules du Néolithique à l'an mille sur le territoire français, Table ronde de Saint-*

Julien-sur-Garonne(F) du 2 au 4 octobre 2009, supplément Aquitania.

Lindet, 1900. Les origines du moulin à grain. *Revue Archéologique*, 1900, 127-29.

Longepierre, S., 2007. L'apport des meules dans la connaissance des moulins hydrauliques romains en Languedoc. In J-P. Brun and J-L. Fiches (eds.), *Energie hydraulique et machine élévatrice d'eau durant l'Antiquité*, 177.

Longepierre S., forthcoming. manuelles, à sang et hydrauliques, quels critères distinctifs? In O. Buchsenschutz *et al.*, *Évolution typologique et technique des meules du Néolithique à l'an mille sur le territoire français, Table ronde de Saint-Julien-sur-Garonne*(F) du 2 au 4 octobre 2009, supplément Aquitania.

Loriquet, Ch., 1862. *La mosaïque des promenades et autres trouvées à Reims*, imprimerie P. Dubois. Reims, p. 10 and pl. II, fig. 11.

Loustaud, J.-P., 1972. Vestiges gallo-romains sur les rives de la Vienne à Limoges de la Roche-au-Go au Pont-Neuf (1964-1970), *Bulletin de la Société Archéologique et Historique du Limousin*, XCIV, 1972, 35, fig. 5.

Mangartz, F., 2006. Vorgeschichtlche bis Mittelalterliche Mühlsteinproduktion in der Osteifel. In A. Belmont and F. Mangartz (eds.), *Les meulières. Recherche, protection et valorisation d'un patrimoine industriel européen*, colloque international, Grenoble, 2005, RGZM Tagungen 2, 25-34.

Monteix, N., 2006. *Les lieux de métier: boutiques et ateliers d'Herculanum*. Thèse de doctorat soutenue en décembre 2006 sous la direction de C. Virlouvet. Aix-en-Provence, Université de Provence (Aix-Marseille I), 2 vol.

Moritz, L.A., 1958. *Grain-mills and flour in classical Antiquity.* New York.

Nin, N., 2002. L'artisanat à Aix-en-Provence durant l'antiquité. In J-C. Beal and J-C. Goyon., *Les artisans dans la ville antique.* Paris: Université Lumière-Lyon 2, Collection Archéologie et Histoire de l'Antiquité, 6, 235-250.

Nucci, F., 2006. Fragment d'un moulin antique à Lucciana, *Découvertes fortuites en Corse VI, Cahier Corsica 223.*

Peacock, D.P.S., 1989. The mills of Pompeii. *Antiquity*, 63, 205-214.

Poulin, C., 2000. Moudre chez les gallo-romain rémois, *Bulletin de liaison de la Société Archéologique Champenoise*, 93e année, n°1, janvier-mars 2000, 15-8.

Py, M., 1992. Meules d'époque protohistorique et romaine provenant de Lattes, *Lattara 5*, 1992, 213, fig 21, n°100.

Raynaud, C., 1990. *Le village gallo-romain et médiéval de Lunel-Viel (Hérault). La fouille du quartier ouest (1981-1983)*. Paris, Les Belles Lettres, Centre de Recherches d'Histoire Ancienne, 97.

Vasselle, F. and Will, E., 1956. Les cimetières gallo-romains d'Amiens, *Revue du Nord*, 38, 152, octobre-décembre 1956, 328.

White D., 1963. A survey of millstones from Morgantina. *American Journal of Archaeology*, 67, 199-206.

Williams-Thorpe, O., 1988. Provenancing and archaeology of Roman millstones from the Mediterranean area. *Journal of Archaeological Science*, 15, 253-305Williams-Thorpe, O. and Thorpe, R., 1990. Millstone provenancing used in tracing the route of a fourth-century BC Greek merchant ship. *Archaeometry*, 32, 1990, 115-37.

Appendix

Commune	Place-name	Structure, Strati. Unit, Number	Dating	Type of mill	Petrography	Height	Diameter
Aix-en-Provence	Zac Sextius Mirabeau	ZAC Sextius-Mirabeau n°4	-25 / 300	*catillus*	Leucite-bearing volcanic rock	> 72 cm	92 cm
Aix-en-Provence	Zac Sextius Mirabeau	ZAC Sextius-Mirabeau n°5 (US 005)	-25 / 300	*catillus*	Leucite-bearing volcanic rock		
Amiens	rue de l'Union	Am 5		*catillus*	Fosses-Belleu sandstone	> 70 cm	79 cm
Amiens	rue Frédéric Petit ?	Am 1		*catillus*	Fosses-Belleu sandstone	> 59 cm	64 cm
Amiens	rue Frédéric Petit ?			*meta*	Sandstone		
Amiens		Am 2		*catillus*	Fosses-Belleu sandstone	> 55 cm	69 cm
Amiens		Am 3		*catillus*	Volcanic rock	> 47 cm	58 cm
Amiens		Am 4, 3078		*catillus*	Fosses-Belleu sandstone	68 cm	76 cm

Amiens		Am 6		*catillus*	Sandstone with feldspar	> 40 cm	sup. 55 cm
Arles		musée d'Arles n°56		*catillus*	Leucite-bearing volcanic rock		sup. 49 cm
Aspiran	Saint-Bézard	Saint-Bézard US 7400	10 / 450	*catillus*	Leucite-bearing volcanic rock		
Aspiran	Saint-Bézard	Saint-Bézard US 4034	1 / 20 ap.	*meta*	Leucite-bearing volcanic rock		
Chalon-sur-Saône	Lit de la Saône			*catillus*	Triassic sandstone		
Jouars-Pont Chartrain				*catillus*	Volcanic rock		
Lascours	Mont Faulac	n°5	-100 / 1	*catillus*	Leucite basalt		80 cm
Lascours	Mont Faulac	n°4	-100 / 1	*meta*	Leucite-bearing volcanic rock	> 38 cm	83 cm
Lascours	Mont Faulac	n°2	-100 / 1	*catillus*	Leucite basalt	50 cm	65 cm
Lascours	Mont Faulac	n°3	-100 / 1	*catillus*	Leucite basalt	43 cm	70 cm
Lattes	Saint-Sauveur	Saint-Sauveur n°100, US 5059, inv. 4-2	-50 / -25	*catillus*	Leucite-bearing volcanic rock	?	64 cm
Les Martys	Domaine des Forges	catillus 1	-100 / 1	*catillus*	Leucite-bearing volcanic rock	> 40 cm	sup. 65 cm
Les Martys	Domaine des Forges	catillus 2	-50 / 1	*catillus*	Leucite-bearing volcanic rock	?	?
Les Martys	Domaine des Forges	meta 1	-100 / 1	*meta*	Leucite basalt	50 cm	62 cm
Les Martys	Domaine des Forges	meta 2	-100 / 1	*meta*	Leucite basalt	66 cm	70 cm
Les Martys	Domaine des Forges	meta 3	-100 / 1	*meta*	Leucite basalt	58 cm	70 cm
Les Martys	Domaine des Forges	meta 4	-100 / 1	*meta*	Leucite basalt	66 cm	70 cm
Les Martys	Domaine des Forges	catillus 3	-100 / 1	*catillus*	Leucite-bearing volcanic rock	?	?
Les Martys	Domaine des Forges	catillus 4	-100 / 1	*catillus*	Leucite-bearing volcanic rock	?	?
Les Martys	Domaine des Forges	catillus 5	-100 / 1	*catillus*	Leucite-bearing volcanic rock	> 33 cm	66 cm
Limoges	13 rue du Pont Saint-Martial	Arc. L179		*meta*	Basalt	90 cm	70 cm
Lunel	Mas de Fourque II	Mas de Fourque II n°4, fosse 2047	125 / 175	*catillus*	Leucite-bearing volcanic rock	?	?
Lunel-Viel	Quartier ouest	Lunel Viel n°7, Us. 125	70 / 100	*catillus*	Volcanic rock	sup. 45 cm	52 cm
Lyon		n°1		*catillus*	Volcanic rock	66 cm	66 cm
Lyon		n°2		*catillus*	Volcanic rock	60 cm	74 cm
Lyon		n°3		*catillus*	Basalt		
Lyon		n°4		*catillus*	Volcanic rock	63 cm	65 cm
Lyon				*catillus*	Basalt		
Meaux	Rue Saint Faron	Us. 1507, meule 77.049		*catillus*	Fosses-Belleu sandstone	40 cm	60 cm
Meaux	25 Rue Saint Fiacre		50 / 100	*catillus ?*	Fosses-Belleu sandstone	?	?
Nîmes	Jean Jaurès	Jean Jaurès n°16 US 4348	140 / 175	*meta*	Leucite-bearing volcanic rock		75 cm
Nîmes	Jean Jaurès	Jean Jaurès n°18 US 4411	20 / 200	*catillus*	Leucite-bearing volcanic rock		> 60 cm
Nîmes	Jean Jaurès	Jean Jaurès n°19 et 20 US 4465	75 / 200	*catillus*	Leucite-bearing volcanic rock		> 60 cm

Nîmes	Jean Jaurès	Jean Jaurès n°21 US 6369		*catillus*	Leucite-bearing volcanic rock		> 55 cm
Nîmes	Saint André de Codols	Saint André de Codols n°8, US 5112		*catillus*	Leucite-bearing volcanic rock		
Nîmes	Parc Georges Besse II-5	Parc Georges Besse II-5, n°4, US. 4415	1/100	*catillus*	Leucite-bearing volcanic rock		
Nîmes	Parc Georges Besse II-5	Parc Georges Besse II-5, n°7, US. 4234	150 / 300	*meta*	Leucite-bearing volcanic rock	> 60 cm	> 33 cm
Orange	Saint Florent	Saint Florent n°6, zone IIIB, fait 1092	-10 / 30	*catillus*	Leucite-bearing volcanic rock		> 70 cm
Orléans	Clos de la Fontaine			*catillus*	Volcanic rock	> 44 cm	64 cm
Paris	Hôtel Dieu			*catillus*	Volcanic rock		
Pézenas	La Roustanenque	La Roustanenque n°2	30 / 100	*catillus*	Leucite-bearing volcanic rock		
Reims	Rue des promenades			*catillus*	Volcanic rock	93 cm	95 cm
Reims	Rue de Cernay	Musée Saint-Remi		*catillus*	Volcanic rock	44,5 cm	73 cm
Reims		Musée Saint-Remi, meule 51.014		*catillus*	Fosses-Belleu sandstone	53 cm	70 cm
Reims		Musée Saint-Remi		*catillus*	Volcanic rock	> 46 cm	57 cm
Reims		Musée Saint-Remi		*meta*	Unknown rock (reconstrution?)		
Reims	Boulevard du Docteur Henrot	Inrap Reims		*meta*	Volcanic rock	?	70 cm
Saint-Hyppolite-de-Montaigu	L'Oratoire	L'Oratoire n°3		*catillus*	Leucite-bearing volcanic rock		
Saint-Victor-la-Coste	Mayran	Mayran n°3	200 / 750	*meta*	Leucite basalt	> 90 cm	86 cm
Soissons	Rue Saint Martin	inv. 93.7.1287		*catillus*	Volcanic rock	120 cm	92 cm
Tourbes	Mont Ferrier	Mont Ferrier n°1	50 / 200	*catillus*	Leucite-bearing volcanic rock		> 51 cm
Tourbes	Mont Ferrier	Mont Ferrier US 4005	50 / 200	*meta*	Leucite-bearing volcanic rock		> 51 cm
Tours	Ilot rue Marceau, commerce			*catillus*	Volcanic rock	75 cm	85 cm
Vienne		inv. 1058 (Musée lapidaire)		*catillus*	Leucite-bearing volcanic rock	40 cm	63 cm
Vienne		Musée lapidaire		*catillus*	Leucite-bearing volcanic rock	56 cm	75 cm

Pompeian style mills in Britain

David Williams and David Peacock

The purpose of this short paper is to draw attention to the small but nevertheless increasing number of Pompeian style mills found in Britain. These large mills consist of two mill-stones, one cone-shaped, positioned upright and stationary (*meta*) and the other hollow, hourglass-shaped and positioned on top (*catillus*). The *catillus* had two handle slots cut into the outer sides opposite to each other, into which wooden levers were inserted so that the stone could be rotated in a circular motion around the *meta* (Fig. 1). Grain would be poured from above into the top of the *catillus*, probably through a wooden hopper attachment (cf. Fig. 2) and ground between the near touching sides of the lower catillus in a circular motion around the *meta.* The resulting flour would be caught in a specially constructed trough set around the base of the *meta.* Each mill would have been operated either by manpower turning the heavy wooden beams of the *catillus* or with a donkey or horse performing the same function (Moritz 1958, 99-102). In the largest bakery at Herculaneum, the skeletons of two donkeys were found harnessed to the mills (Deiss 1966, 204). Roman households may have had one or more small rotary querns for domestic use, but the Pompeian style mill was essentially a piece of specialized equipment used to produce commercial-scale quantities of flour. They tend to be sited where large populations were present and at Pompeii, Herculaneum and Ostia, where considerable numbers of these mills have been found, they are commonly associated with bakeries (e.g. Peacock 1989). However, it is clear that they were also used for the crushing of various ores, as they are also found associated with ancient mines (Oliva *et al* 1999).

Pompeiian style mills were generally constructed from grey vesicular lava, whose sharp-edged vesicles or gas-holes were ideally suited for slicing and cutting the grain. The Pompeian style mills found in Roman Britain were last studied in 1988 by Olwen Williams-Thorpe and Richard Thorpe, who called them 'donkey mills' (they are also known as 'hourglass mills' due to the shape of the *catillus*). We prefer the less perjorative term 'Pompeian style', as these large millstones may or may not have been exclusively animal powered. In 1988 only three were known in Britain, one from London, and two from Dorset, one from Corfe Mullen and the other from Hamworthy, Poole (*ibid.*). The Hamworthy example was rightly dismissed at the time as a recent Sardinian mill, a point now reinforced by ethnographic work in Sardinia (Williams-Thorpe and Thorpe, 1989). Williams-Thorpe and Thorpe claimed that the London and Corfe Mullen examples came from the volcanic outcrop of the Chaine des Puys in central France, though from different lavas implying separate locations of production (1988, 287). The London mill was confirmed as being chemically close to the rock from Volvic, Auvergne, but the Corfe Mullen mill could not be determined with the same precision. At the time of writing, they could cite no evidence of importation of Pompeiian style mills from the important quarries at Mayen in Germany.

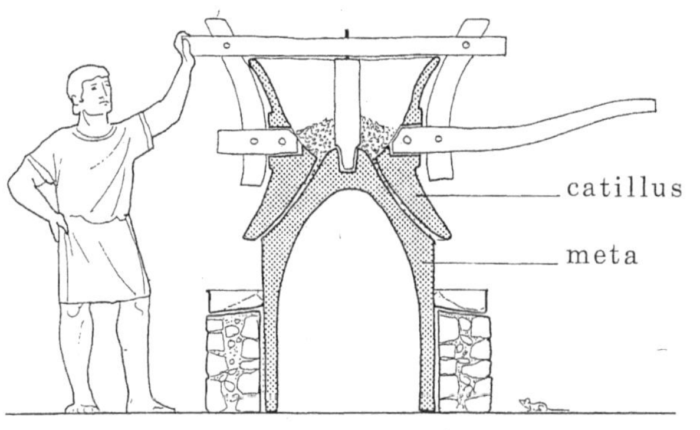

Figure 1. Cross-section of a Pompeian style mill (after Adam 1994, fig. 735).

In the intervening years more examples have come to light and in particular a rich collection of lava mills, including fragments of the Pompeian style, has recently been excavated at Poultry, London (Williams and Peacock forthcoming). In addition, a near complete *catillus* has been recovered from the early Roman fort at Clyro, near Hay-on-Wye, Wales (on display in the National Museum and Gallery, Cardiff). These further finds tell a rather different story of the sources of rock used for British Pompeian mills.

Catalogue of British Pompeian style mills

1. *London, Princes Street* (Fig. 2; on display in the Museum of London).

An almost complete *catillus* was discovered in Princes Street in 1929 (Merrifield 1965; Hall and Merrifield 1986).

Figure 2. Pompeian style mill from Princes Street, London (on display in the Museum of London).

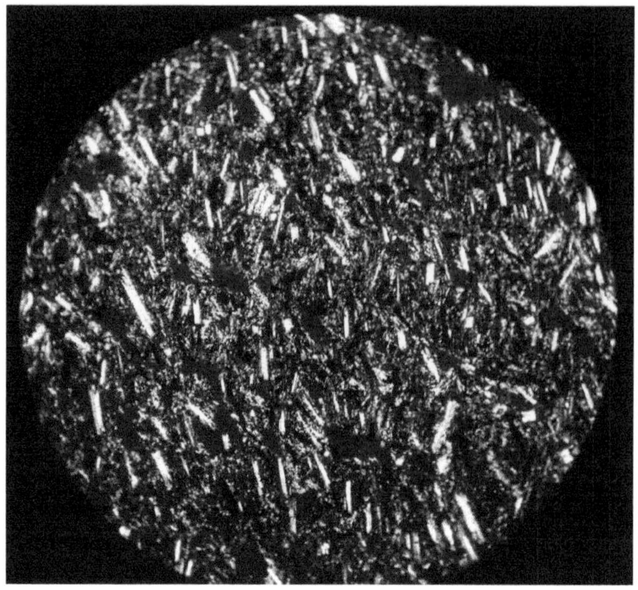

Figure 3. Photomicrograph of the Princes Street, London, Pompeian style mill. Crossed polars x 40.

A small sample was detached from the inner surface and examined in thin section under a petrological microscope. The rock proved to be a trachyandesite, dominated by alkali feldspar and sodic plagioclase together with some dark coloured mafic minerals (Fig. 3). The composition clearly differentiates it from Mayen lava (see below), which was commonly associated with the many imported small rotary querns used in Roman Britain. It matches our specimens of Volvic lava from the Massif Central region of France, and an origin in that region seems probable (cf. Williams-Thorpe and Thorpe, 1988).

2. *London, Poultry* (Fig. 4 a. b. c.)

Three large pieces of Pompeian style mills were found, two fragments come from a *catillus* and one from a *meta* (Poultry nos. 137, 144 and 168). Both of the upper stones also have squarish handle slots, which would have been part of the yoke pole attachment for turning (cf. Curle 1937, fig. 2; Šebesta 1976, 85-86). All of the fragments are in dark grey, fine-grained, rough vesicular lava. Even in the hand-specimen dark phenocrysts of pyroxene are conspicuous. Thin sectioning shows that the most prominent minerals are frequent grains of green and colourless clinopyroxene, mainly augite, together with a little olivine, set in a groundmass of small lath-shaped crystals of andesine/labradorite felspar, opacite, leucite and some xenomorphic nepheline (Fig. 5). As nepheline is not ubiquitous it serves to finger-print these mills. There are some slight differences in the texture of the groundmass which equate with the textures noted in the hand-specimen, suggesting that we are dealing with three separate mills rather than one, but overall the fabric of the rock appears fairly homogeneous for the group as a whole. The composition of this vesicular lava is particularly distinctive and it can be classed a nepheline-tephrite. This type of rock is characteristic of the lavas of the Bellerberg volcano in the Mayen area of the German Eifel Hills. This region is well-known in both Roman and later times for supplying querns and millstones (Parkhouse 1976; Kars 1980; Peacock 1980). Recent geochemical and petrographic work has shown that this lava was exported as far as Bohemia as early as the late Iron Age (Wefers and Gluhak 2010) and to Magdalensberg in Carinthia and Britain in the Roman period (Gluhak and Hofmeister 2008). Gluhak and Hofmeister have also shown that it is possible to chemically characterise every Eifel quarry and with further work it may be possible to tie down the source of the British lava even more precisely.

However, the situation is not simple for Zirkl (1955) long ago showed that the younger volcanic rocks of Austria were being used for La Tène mills from Vienna. More recently Draganits (2002) has claimed the same for a Roman mill from Zwingendorf in Lower Austria. Thus there is a strong possibility that other lava sources may have been involved.

3. *Corfe Mullen, Dorset* (Fig. 6; on display in Poole Museum).

A thin section was made from the large fragment of *catillus* found unstratified at Corfe Mullen during the 1940's (Fig. 6; Smith 1943; RCHM 1970; Williams-Thorpe and Thorpe 1988). This confirmed that the volcanic rock used was an alkali-rich basalt (Fig. 7). Chemical analysis suggested an origin in the Massif Central region of France, quite possibly originating from the Chaine des Puys (Williams-Thorpe and Thorpe, 1988). It is interesting to note that the Corfe Mullen *catillus* has a decorated horizontal band at the midsection, where it has broken off (Fig. 6). This band can be paralleled in an example of a Pompeian style mill found in the rue de Cernay in Reims (Poulain 2000) as well

Figure 4. a (top left), b (above) and c (top right), showing fragments of Pompeian style mills from Poultry, London (Mark Burch, MoL).

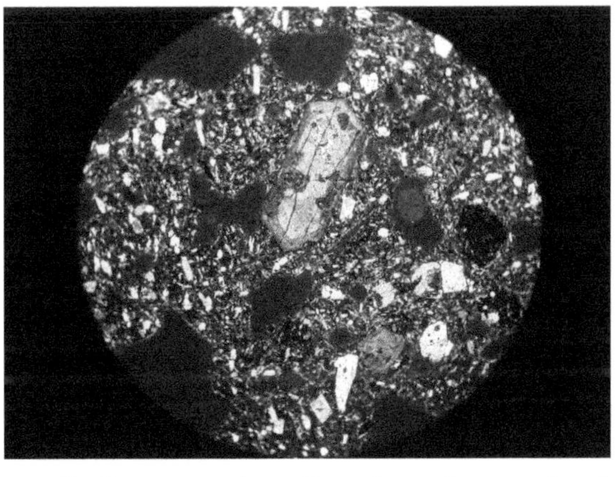

Figure 5. Representative of a photomicrograph of the Poultry, London, Pompeian style mills. Crossed polars x 40.

as a number of other sites in France, which has pointed to the likelihood of a regional Pompeian style millstone production in that country (Jaccotey and Longepierre this volume). It is interesting to note that the latter are found in a range of rock types, including non-leucite volcanic material (*ibid.*).

4. *Clyro, Powys* (Fig. 8; on display in the National Museum & Gallery, Cardiff)

An almost complete *catillus* was recovered from the early Roman fort at Clyro, although the exact find spot is not recorded. It appears to be made of a hard quartzitic sandstone, resembling the Ordovician of the Welsh borderland. If this visual identification is correct, it would be the only 'locally made' Pompeian style mill from Britain. However, this has yet to be confirmed petrographically. Sandstone, some of it quartzitic, was also used in France for Pompeian style mills (cf. Jaccotey and Longepierre this volume). As it is the only sandstone example from Britain the possibility that it is a modern reproduction must also be considered. However, the design is perfect, it reputedly comes from a Roman site and sandstone was used in France. There is thus little reason to doubt its authenticity.

5. *Canterbury, Kent* (Frere 1967, 259).

Frere makes a passing remark that a donkey mill has been recovered from Canterbury but unfortunately no details are given and a search of the Museum storerooms has proved unsuccessful in locating this piece.

Figure 6. Pompeian style mill from Corfe Mullen, Dorset (on display in Poole Museum).

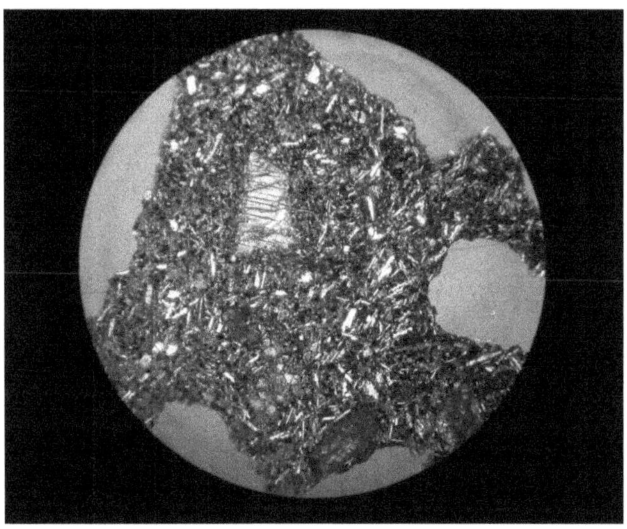

Figure 7. Photomicrograph of the Pompeian style mill from Corfe Mullen, Dorset. Crossed polars x 40

6. *Silchester*

Fragment from a *catillus* excavated in 2009. Analysis by John Allen (2010) suggests that the stone is from Mayen.

The find sites listed above are interesting. Three of them, London, Canterbury and Silchester are Roman towns. Clyro is a Roman vexillation fort comprising a large well-defended site of 26 acres with evidence of two periods, the earliest of which dates from at least *c.* AD 60. The find from Corfe Mullen may also relate to a military site established by the Second Legion under Vespasian in the early years following the conquest in AD 43, although it was not found in the fort itself (Smith 1943; RCHM 1970; Williams-Thorpe and Thorpe 1988, 276). These large mills might be expected in populated places where the demand for flour would be high, such as a town, but a Legionary fortress would also fit the bill.

London Poultry adds a new dimension to the story, as all the Pompeian style mill material is clearly from Mayen (no information is available yet on the possible Canterbury piece but a Mayen origin is also suggested for the Silchester fragment). In addition to these finds, an estimated total of 1,000 broken and discarded fragments from rotary quernstones, very largely Mayen, were also recovered from a small number of closely related individual contexts within Area 12 of the excavation sequence at Poultry, which is located near to the west bank of the Walbrook stream. These quern fragments, of variable size, had been used as an external cobbled surface and were adjacent to and thought to be contemporary with, a large timber-lined water reservoir dated by dendrochronology to *c.* AD 70-90. There is further evidence that this was a flour production area because the almost complete *catillus* from London previously mentioned above was found in Princes Street, which abuts the Poultry site. Moreover, smaller fragments of the same type of mill have been recovered from other parts of the Poultry site in previous years (Merrifield 1965) and two other Pompeian style mill fragments have been found in residual and later contexts from the recent excavations (Site 1 94 Area 12 and Site 1 94 3067). Furthermore, large flat millstones have been recovered from the bed of the Walbrook, suggesting that they may have been used to power water mills (Hall and Merrifield 1986, 37). This concentration of finds leaves little doubt that this part of the town was primarily concerned with grain processing.

The main problem with this hypothesis as an explanation for the exceptionally large quern assemblage at Poultry is that the fragments are almost exclusively hand mills, which have their place in a domestic rather than a commercial context. However, hand mills are also present in the bakeries of towns such as Ostia, Pompeii and Herculaneum and it seems that they were an adjunct of commercial as well as domestic grain processing (Peacock 1989, 205; Moritz 1958). It is true that their numbers are limited, but being smaller than power mills, which comprise large heavy pieces of stone, it is probable that they would break more frequently and so need replacement. Thus, while Pompeian style mills or water mills might have a long life, the hand mills would not and a relatively short life might explain the fresh appearance of milling on many of the Poultry querns (Williams and Peacock forthcoming).

It is particularly important to note that five potential Pompeian style mills have been recovered from the Poultry site. Previously only two donkey mills were known from Roman Britain as a whole, the one from London mentioned above and the one from Corfe Mullen, both of lava from central France

Figure 8. Pompeian style mill from Clyro, Powys (on display in the National Museum & Gallery, Cardiff).

Figure 9. Pompeian style mill from Haltern, Germany (from Schnurbein 1979).

(Williams-Thorpe and Thorpe 1988). The Poultry evidence clearly shows that the Mayen region was after all an important source for donkey mills (cf. *ibid.*, 275). This seems to be reinforced by the example from Silchester (see above).

In this volume Jaccotey and Longepierre (Chapter 10) discuss new Pompeian style mill finds from France, which shows a much broader distribution than was envisaged in the past. To this we can add an example from Haltern (Fig. 9; Schnurbein 1979). So the pattern that emerges is of a marked concentration in Mediterranean lands where they are prominent in the bakeries of towns such as Ostia, Herculaneum and Pompeii (Peacock 1989) and are also found associated with mines and iron-making centres (Oliva *et al* 1999), but with a rapid fall off northwards where they are commonly found as single pieces. If we confine our attention here to the milling of grain, clearly ground flour would have been needed everywhere, which suggests that northern millers did not rely solely on the Pompeian style mill for producing substantial quantities. However, water is more abundant in the north and it seems likely that there water mills were the norm. Thus, large millstones, which could not be hand powered, have been noted on Roman sites in the south of England. This is not the place to discuss the water mill in detail, but it may be significant that the Roman palace at Fishbourne produced no Pompeian style mills but instead several large mill stones were found which were potentially water driven (Cunliffe 1971, 153). With the exception of London with its large population and heavy demand for bread, perhaps single Pompeian mills were kept in certain bakeries to produce Mediterranean delicacies.

Acknowledgements

We are extremely grateful to Francis Grew, senior curator, Museum of London, and to Catherine Gardiner, Interpretation Officer (Collections), Poole Museum, for permission to detach small samples for thin sectioning from the Princes Street and Corfe Mullen mills respectively. Also to Craig Bowen, Collections and Research Manager, Canterbury City Council Museums, for generously trying to locate the Canterbury fragment of Pompeian style mill mentioned by Frere. Mark Burch kindly took the photographs of the Poultry Pompeian style mills.

References

Adam, J-P., 1994. *Roman Building: materials and techniques*. London.

Allan, J., 2010. Analysis of a Pompeian style mill at Silchester. *Science @Silchester* Reading.

Cunliffe, B., 1971. *Excavations at Fishbourne. Volume II: The Finds*. London.

Curle, J., 1937. Querns. *Antiquity,* 11, 133-151.

Deiss, J.J., 1966. *Herculaneum*. London.

Draganits, E., 2002. Petrologische Untersuchung eines römischen Mahlstein-fragmentes aus Zwingendorf, Niederösterreich. In N. Doneus *Die Ur- und Frühgeschichtliche Fundstelle von Zwingendorf, Neiderösterreich*. Vienna, 273-8.

Frere, S.S., 1967. *Britannia: a history of Roman Britain*, London.

Gluhak, T., and Hofmeister, W., 2008. Provenance analysis of Roman millstones: mapping of trade areas in Roman Europe. In R.I. Kostov, B. Gaydarska and M. Gurova (eds) *Geoarchaeology and Archaeomineralogy*. Sofia, 111-5.

Hall, J. and Merrifield, R., 1986. *Roman London*. London.

Horter, F., 1994. *Getreidereiben und Muhlsteine aus der Eifel*. Mayen.

Jaccotey, L., and Longepierre, S., 2011. Pompeian Millstones in France. This volume.

Kars, H., 1980. Early-Medieval Dorestad, an archaeo-petrological study. *Berichten van de Rijksdienst voor het Oudheidkundig Bodemonderzoek*, 30, 393-422.

Merrifield, R., 1965. *The Roman City of London*. London.

Mortiz, L.A., 1958. *Grain-Mills and Flour in Classical Antiquity*. Oxford.

Oliva, P., Beziat, D., Domergue, C., Jarrier, C., Martin, F., Pieraggi, B., and Tollon, F., 1999. Geological sources and use of rotary millstones from the Roman iron-making site of Les Martys (Montagne Noire, France). *European Journal of Mineralogy*, 11, 757-762.

Parkhouse, J., 1976. The Dorestad quernstones. *Berichten van de Rijksdienst voor het Oudheidkundig Bodemonderzoek*, 26, 181-188.

Peacock, D.P.S., 1980. The Roman millstone trade: a petrological sketch. *World Archaeology*, 12, 43-53.

Peacock, D.P.S., 1989. The Mills of Pompeii. *Antiquity* 63,205-14.

Poulin, C., 2000. Moudre chez les gallo-romain rémois. *Bulletin de liaison de la Société Archéologique Champenoise*, 93e année, n°1, janvier-mars 2000, 15-18.

Royal Commission on Historical Monuments, 1970. *Vol. II, Dorset part 3 (south-east)*. London.

Šebesta, G., 1976. *La via dei mulini*. Trento.

Smith, H.P., 1943. The course of the Hamworthy-Badbury Road. *Proc. Dorset Nat. Hist Archaeol. Soc*., 65, 57-58

Wefers, S., and Gluhak, T., 2010. Eifel lava – the provenance of two late Iron Age rotary querns discovered in Bohemia. *Archeologické rozhledy*, 62, 3-16.

Williams, D.F., and Peacock, D.P.S., forthcoming. A group of early Roman quernstones from Number 1 Poultry, London, EC4.

Williams-Thorpe, O., and Thorpe, R.S., 1988. The provenance and of donkey mills from Roman Britain. *Archaeometry*, 30.2, 253-305.

Williams-Thorpe, O. and Thorpe, R.S., 1989. Provenancing and archaeology of Roman millstones from Sardinia (Italy). *Oxford Journal of Archaeology*, 8.1, 89-113.

von Schnurbein, S., 1979. *Die Römer in Haltern*. Munster.

Zirkl, E.J., 1955. Zur Herkunft einiger latènezeitlicher Handmühlen. *Arch. Austriaca*, 18, 90-2.

Hertfordshire Puddingstone querns - working with a difficult rock

Chris Green

During a short period early in the Roman occupation of Britain, Hertfordshire Puddingstone, a very hard conglomerate sedimentary rock, was used to make small rotary querns rarely more than 35 cm in diameter. They are widely distributed over part of eastern England which has no other supply of suitable stone (Fig. 1). These querns have not received much study and have even been thought of as rough domestic products, though nothing could be further from the truth. In fact Hertfordshire Puddingstone quern production must have been a substantial industry, for over 700 examples have been recorded. The querns were probably expensive because they were difficult to make, and desirable because they wore slowly, and the rock was attractive.

This paper concentrates on basics: how Puddingstone querns were made and where - necessary starting points in understanding these neglected objects.

Hertfordshire Puddingstone is so-named because it largely coincides with an administrative county immediately to the north of London - today a heavily populated area of low hills, towns and motorways. In the Late pre-Roman Iron Age (c. 25 BC - AD 50) it contained important tribal settlements at Braughing, Baldock, Welwyn and Verlamion (later the Roman town of Verulamium), and its archaeology indicates strong pre-Conquest contacts with the Roman world.

The few Puddingstone querns which have been reliably excavated from dated deposits suggest that the period of production was remarkably short, mainly in the period AD 50-100; evidence for production as early as AD 25 and as late as AD 150 is still under investigation (Major 2004). After AD 100 Mayen lava querns and millstones flooded the market in Roman Britain, and particularly the east of the Province where Puddingstone querns had been so popular. The great majority of Hertfordshire Puddingstone querns, made after AD 50, form a simple typological group with only minor variations (Fig. 5). The upper stone (of unworn examples) is close to a hemisphere, though a more conical shape might be cut if a finer-grained stone allowed, perhaps to reduce weight. Querns of this general form are termed 'beehive' or 'domed' querns in England. Perforation was eggcup-shaped, with the upper part forming a large hopper, and the constriction only about 25 mm in diameter - an iron ring for the spindle to run against is sometimes found wedged below it and was probably a general feature. The lower stone is a flattened hemisphere, and never perforated; there is instead a blind hole for an iron spindle (which very occasionally survives), so the cut of the stones could never be adjusted to produce fine or coarse flour. Perhaps the chief typological variation is in provision for a handle: in many there is no trace and the use of a leather strap around the girth has been suggested; in some there is a handle hole in the side (never penetrating to the hopper); while others exhibit a groove cut around the girth for an iron 'driving-band', and these are discussed later on. The upper stone is often regularly and smoothly finished to a surprising degree considering the difficulty of working Puddingstone, while the lower stone is usually rougher.

There is no question that the Romano-British Hertfordshire Puddingstone industry was based directly on the example of Gallo-Roman material, forms, and techniques. Comparable rock, of similar geological origin (and indeed termed *poudingue*) occurs in the Haute-Normandie region of northern France, and David Peacock's and Lyn Cutler's present researches are establishing its long prehistoric use as saddle- and then rotary querns in southern Britain (Peacock and Cutler 2010). In contrast, the present author's work has confirmed that Hertfordshire Puddingstone was only very rarely used in prehistory. (However, the megalithic use of Puddingstone boulders should not be discounted: for pebbly sarsen megalithic monuments near Dorchester in Dorset see Piggott and Piggott 1939.) But the precise means by which French conglomerate quern technology came to Roman Britain, and its date of introduction, are not yet fully understood. The study is difficult because the archaeological literature is practically non-existent and because the English and French puddingstones are often hard to distinguish. It can be said that a group of roughly-finished querns with flattened profiles and small hourglass perforations, quite unlike the usual Romano-British product, seems to characterise Hertfordshire and Essex (the neighbouring county to the east) in the period c. AD 25-50 (Fig. 4). But some are English and some French, by no means all have yet been located, and it is usually difficult to establish precise dates. To complicate matters, French querns of both beehive and flatter forms may then have

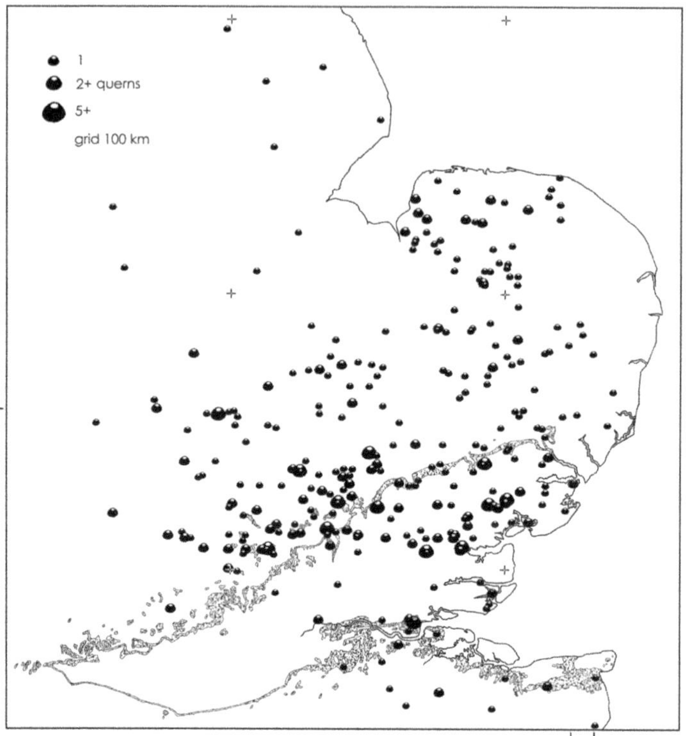

Figure 1 (top left). Distribution of Hertfordshire Puddingstone querns in eastern England (provisional November 2009).

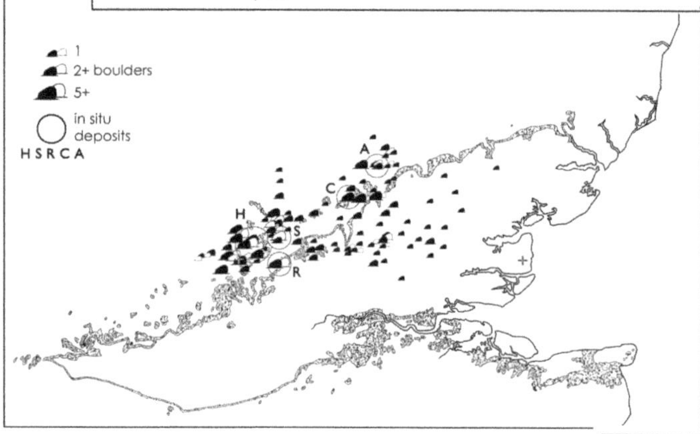

Figure 2 (left). Distribution of Hertfordshire Puddingstone boulders and in situ rock, with the 'solid' Palaeocene outcrop of the London Basin. In situ deposits are circled at Hemel Hempstead, St Albans, Radlett and the Collier's End outlier (all Hertfordshire), and Arkesden, Essex. Isolated boulders south and east of the outcrop are mainly periglacial erratics.

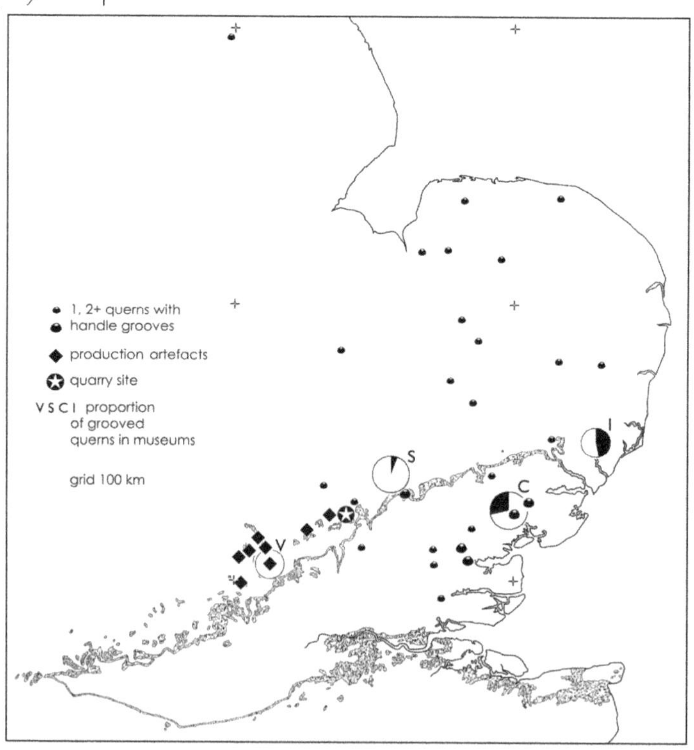

Figure 3 (right). Production evidence for Hertfordshire Puddingstone querns, with the distribution of querns with driving band grooves from provenanced finds and unprovenanced museum material at Verulamium (St Albans), Saffron Walden, Colchester, and Ipswich.

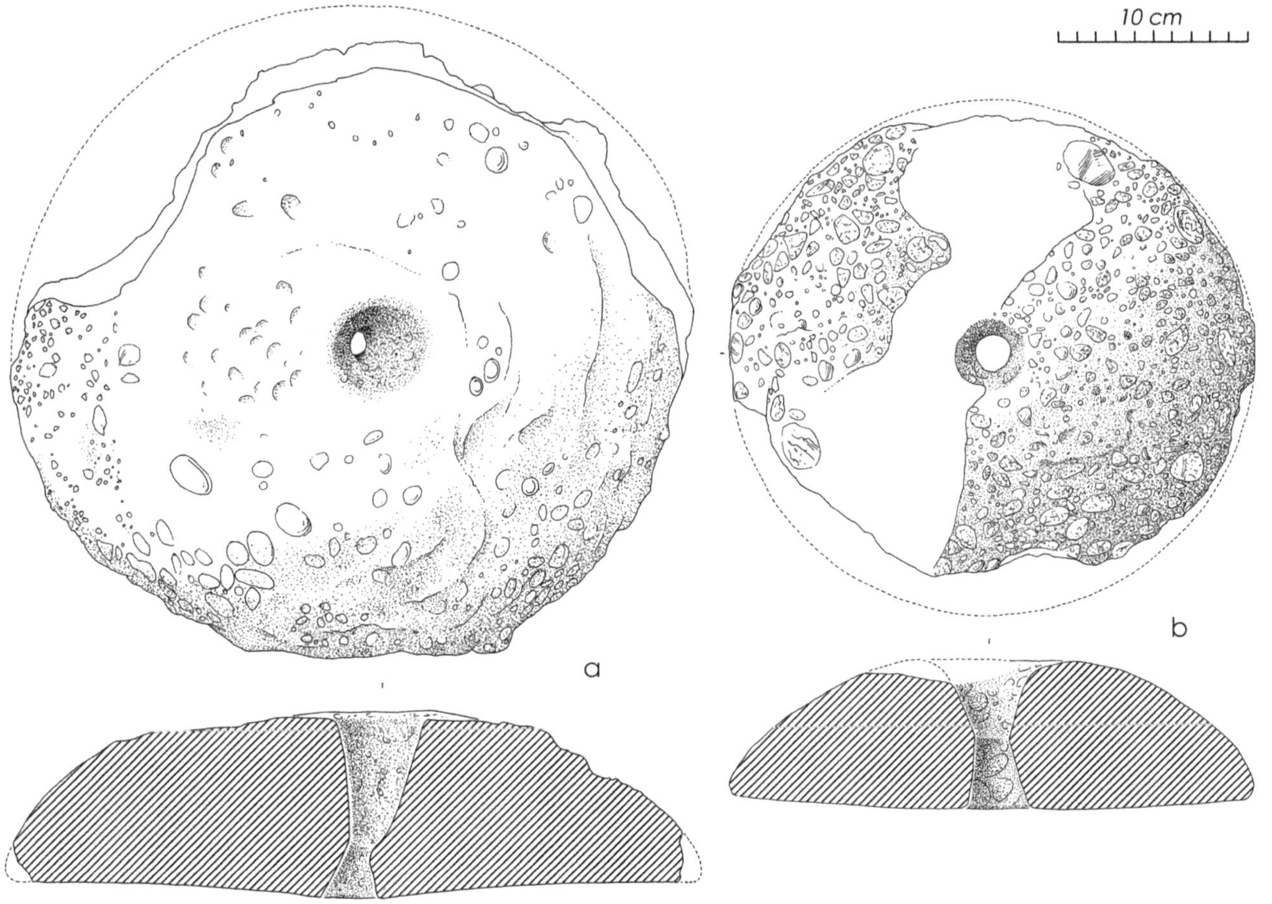

Figure 4. Early Hertfordshire Puddingstone querns (?pre-AD 50) in imitation of French querns: a) Frogmore Hall, Benington, Hertfordshire (Hertford Museum 2974.1); b) Furneux Pelham, Hertfordshire (Mr R Gibson). Scale: ¼

been imported in small quantities throughout the Romano-British period (C Green, D P S Peacock and L Cutler, work in progress).

To return to the rock: Hertfordshire Puddingstone is a flint-pebble conglomerate formed in the Palaeocene Period, 55-56 Ma, during a period of marked global warming. Palaeocene rocks in Britain appear at the surface fringing the London and Hampshire basins, and at various points their erosion has left boulders from a discontinuous layer of grey silicified sandstone, known as *sarsen*, at the surface. The trilithons of Stonehenge are the most famous instance of this rock. Locally, over a distance of about 70 km in Hertfordshire and its bordering counties, the well-rounded pebbles of a beach marking the south-eastern shore of Palaeocene Britain are incorporated in the sarsen. (The pebbles vary, but are typically around 20 mm in diameter, grey white, yellow, orange, red or black in colour, and often have a black 'rind'.) The resulting conglomerate is Hertfordshire Puddingstone (Fig. 2). Both sarsen and puddingstone are silcretes, that is, loose sediments later indurated by silica in solution to form a hard rock. In Hertfordshire Puddingstone the interstices of the flints and fine sand of the original beach deposit have been entirely filled with silica, and no voids remain. Typically the rock is 97% SiO_2. The formation of silcretes is a controversial problem in geology, but in this case we should imagine silica-rich and probably acid groundwater acting on the sand and pebbles of an upraised beach, in marshy conditions, an extremely hot climate, and within a million years of the initial beach deposit (for a recent view of the formation of Hertfordshire Puddingstone see Ellison 2004). Puddingstone is far from being a continuous layer. In the few instances where it has been seen *in situ* it has been a lens amongst pebble and sand deposits, no more than a few metres in horizontal extent, and between 30 and 90 cm in thickness. Since it is extremely hard and impervious Puddingstone resists the normal processes of erosion, and it is most commonly seen as surface boulders picked clean by glaciation, often some kilometres away from the parent Palaeocene outcrop.

The attractions of a flint conglomerate for quern-making are that the stones wear slowly, and that flints exposed in the grinding surface have many sharp facets (Fig. 7a). The advantage must have been strong, in view of the difficulty of manufacture. The sarsen matrix of the stone did little more than resist wear - sarsen itself probably makes a poor quern.

Museum visitors - and staff - often ask how the Romans made querns from such a very hard rock, and it is a

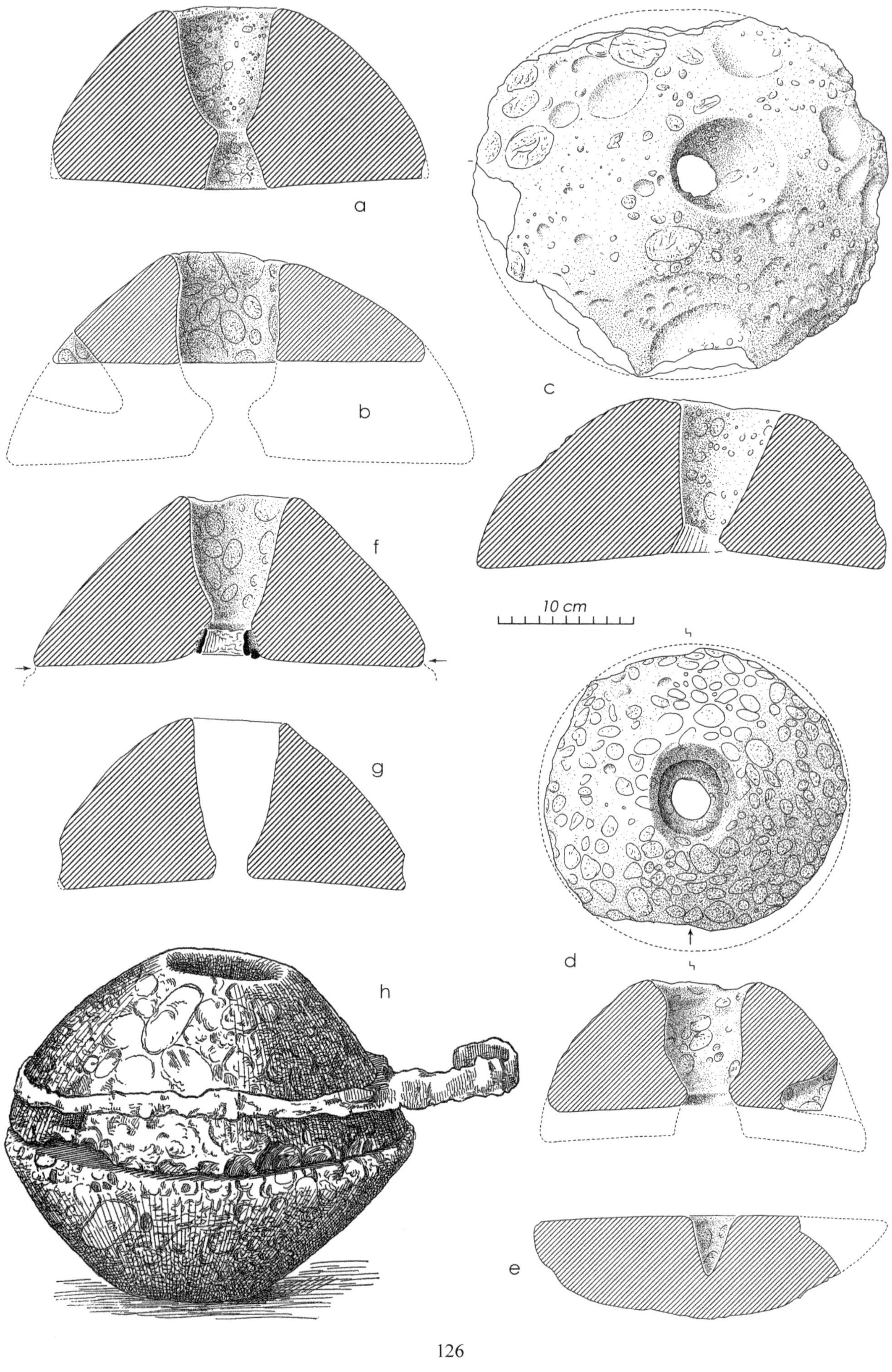
a
c
b
f
10 cm
g
d
h
e

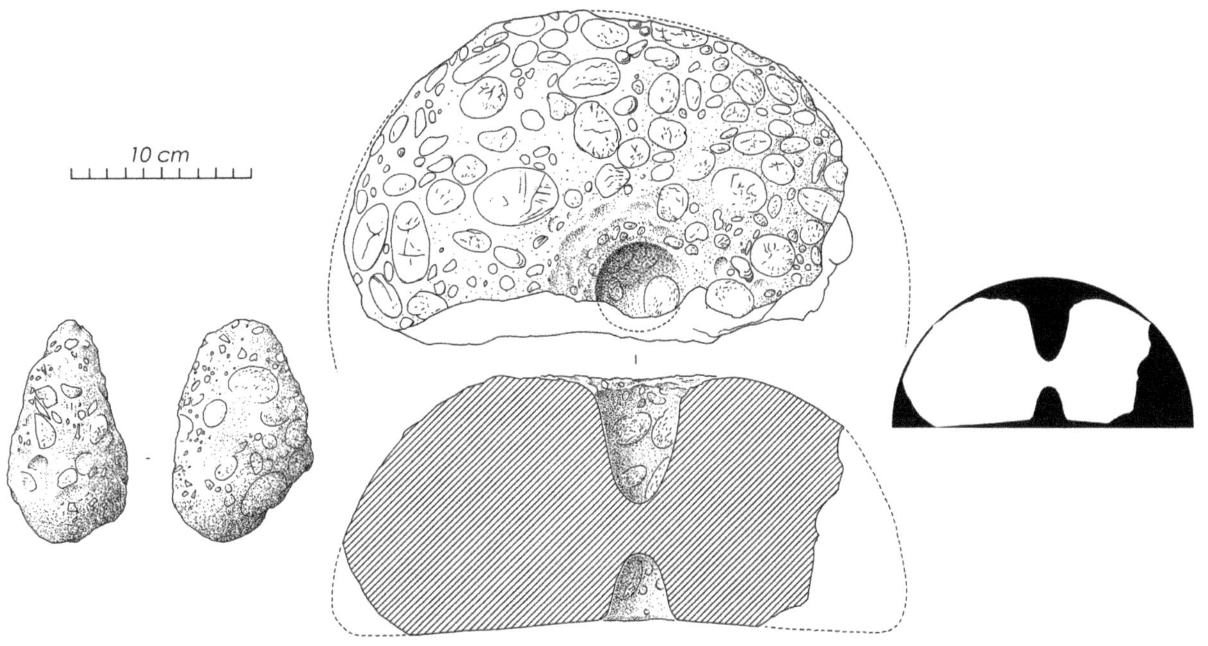

Figure 6. An unfinished quern found at Gaddesden Row, Hertfordshire (St Albans Museums 2007.166); and a Puddingstone hammerstone found near Redbourn, Hertfordshire. Sc: ¼

good question. Among the quern-making rocks from all over Europe featured at the Rome Colloquium (2009), Hertfordshire Puddingstone was probably the only one which could not be worked from start to finish with the stonemason's standard tool kit of pick, hammer and wedges, and point or chisel. With Puddingstone, the pick was probably used to start the process of splitting a quern-sized blank from a mined slab or a surface boulder (with much waste), and light blows may have accomplished the rough external shaping. Because both pebbles and matrix are solid silica, Puddingstone can be flaked - very roughly - like flint or chert, and an over-deep detachment scar is shown in Fig. 7b. Puddingstone débitage has recently been excavated near Redbourn in Hertfordshire, and is thought to derive from Roman quern-making (M1 Motorway widening, Junction 9, pers. comm. Ruth Shaffrey; report in preparation). But the final external finishing of the querns is believed to have been accomplished by crushing with hand-sized Puddingstone hammer-stones (Figs 6; 7e-f), three of which have been found in Hertfordshire in recent years; formerly all these objects would have been ascribed to the Neolithic or beyond.

The real difficulties lay in making the large central perforation of the upper stone. Many other quern types are observed from tool marks to have been gently perforated with the pick, and a Roman gravestone depicts a 'Pompeian' *catillus* being opened in this way (Porteous 1973, fig. 51). Although quern-making rocks were generally harder than tool steel, microfracture allowed slow shaping with light blows. But Puddingstone is brittle as well as hard and would undoubtedly shatter under repeated blows directed to the centre of the workpiece, rather than the tangential blows needed to shape the exterior. In fact, all the perforations (hopper, handle hole, and spindle socket), are seen to have been drilled, in both English and French Puddingstone querns. The difficulty of this procedure was demonstrated (on screen) for the Colloquium by an attempt to perforate Hertfordshire Puddingstone using a powerful modern hammer drill and a new tungsten carbide bit: ten minutes of drilling resulted in less than a millimetre of progress (and a blunt drill). It is concluded that the process was beyond the capacity of an unaided workman - he needed a drilling machine and a drill bit as hard as the stone itself.

Figure 5 (opposite). Typical Romano-British Hertfordshire Puddingstone querns, c. AD 50-100+, all upper stones except e); a)-e) are from Hertfordshire. Scale ¼, including the Leziate drawing.

a) almost unworn, Hertford Museum

b) very heavily worn, Verulamium, St Albans Museums 1980.809

c) drilled and worn at an angle, Gorhambury, St Albans Museums 1991.184

d) small quern worn to the end of its intended life, Braughing private collection

e) lower stone, Verulamium St Albans Museums 1980.970

f) retaining the iron ring which ran on the iron spindle, and remains of the driving band groove, Norfolk Natural History Museum Palaeontology Z.176

g) with groove for the driving band, Chelmsford, Essex, Chelmsford Museum 1982.029

h) the driving band and its 'eye' surviving on the quern from Leziate, Norfolk (the lower stone is probably a reconstruction, and would actually have been a flattened hemisphere); after Beloe 1893.

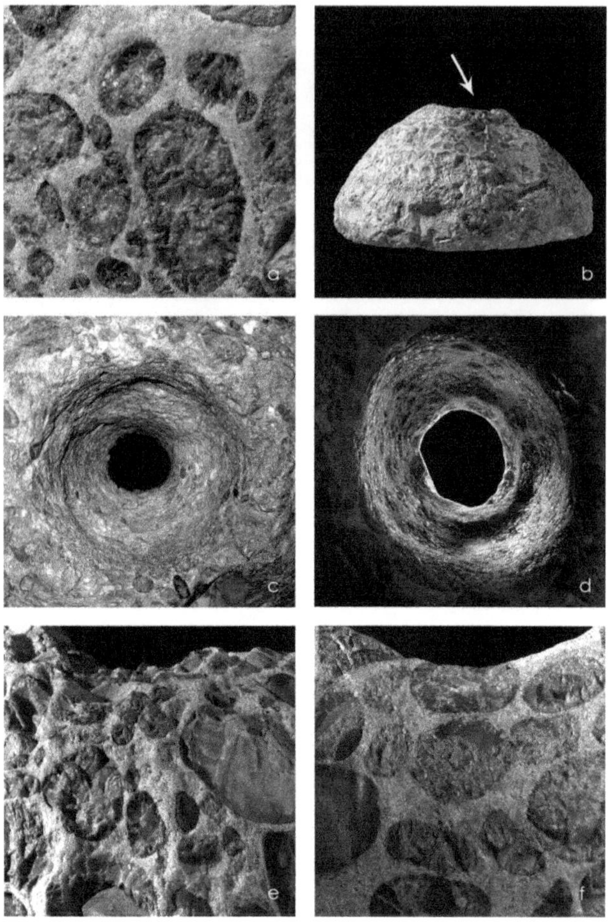

Figure 7. Surface details of Hertfordshire Puddingstone querns: a) grinding surface, showing angular fracture of flints; b) detachment scar from flaking; c, d) hoppers of two querns showing undercut and eccentric drilling; e) top surface of the unfinished quern from Gaddesden Row (broken during drilling); f) surface of a finished quern, same scale as e). (d, f, from Braughing, Hertfordshire, private collection, rest St Albans Museums).

It is no surprise that the drill bits used for Puddingstone querns have never been found, but their spear-shaped form, and the use of progressively larger sizes to open the hopper, can be established from the querns themselves, the smaller holes made for the handle and spindle, and also through the recovery from a Hertfordshire garden of an unfinished quern, evidently broken during the process of drilling (Gaddesden Row, Hertfordshire, given by Mrs D Mills: St Albans Museums 2007.166; Fig. 6). The drills worked not by direct contact with the stone, but by rotating a charge of broken Puddingstone which would cut its own kind so long as the fragments remained sharp (Fig. 8). This is established without doubt, for the hoppers of many querns are elliptical and/or undercut in ways that direct cutting could not achieve (Fig. 7c-d). When the charge had been reduced to powder and ceased to make a satisfactory cutting noise, the drill would be withdrawn, the waste scooped out, and the process begun again. It is of course an old technology for cutting hard rocks, widely used with copper tools in Pharaonic Egypt (Stocks 2003). In Roman times, wrought iron would have been preferred to steel for the drills as it was far cheaper, less likely to break, and because rock fragments could become temporarily embedded in the softer metal, helping the cutting process. Although the drills would not make direct contact with the workpiece, they must have worn very quickly.

It is contended that such a drill could not have been turned by hand for the economical production of mass-produced domestic querns. The tapering form of the drill bits created a long cutting section which in itself demands substantial downward pressure. This could have been achieved with a device such as that shown in Fig. 8, which would allow intense effort to be applied by two (or more) men at the expense of a slow speed of rotation. There is no direct evidence for the form of this hypothetical drill, but the weighted beam press is a well-established Greek and Roman mechanism, as is the capstan, which was of course employed on 'Pompeian' donkey mills (White 1984, 50-51, 67-68). Variations can be imagined in which the walls of a building take the place of a split tree to hold the beam, or a heavier weight is applied directly to the rotating spindle rather than through a lever. But certain constraints apply: the need to keep the quern fixed in place (here, imagined in a framework packed with clay); the need to hold the upper end of the drill spindle steady; and a convenient means of raising the drill frequently to clear the waste, change the bit, and ultimately reverse the stone.

Mechanical aid appears vital for one last process after the quern stones had been perforated and smoothed: grinding together or 'marrying' the upper and lower stones (often an important symbolic act involving ceremony, see Heslop 2008, 60). For Puddingstone querns the actual grinding process seems almost impossibly laborious without a machine to maintain pressure on the two stones as they were worked together.

Where did all these activities take place, or in other words, how was the industry organised? One quarry site is known and it is practically certain to be Roman, lying close to a Roman main road south of Braughing in east Hertfordshire (Lovell and Tubb 2006). The stone was no doubt also obtained from surface boulders and from larger deposits further west, at Hemel Hempstead, St Albans and Radlett, but these places are now heavily built-up and the chance of identifying a Roman quarry is remote. In each instance it is thought that *in situ* rock would have been prominently exposed on gentle hillsides in the first century AD, and a scatter of weathered-out boulders would be available down-slope. Quarrying for fresh stone might follow the outcrop, and of course cease when no further lens of rock was encountered, or the overburden became uneconomic. There are further substantial sources of Puddingstone and sarsen amongst periglacially altered clays and sands north and west of the 'solid' Palaeogene outcrop, but in antiquity they were probably much less important than *in situ* rock or surface boulders.

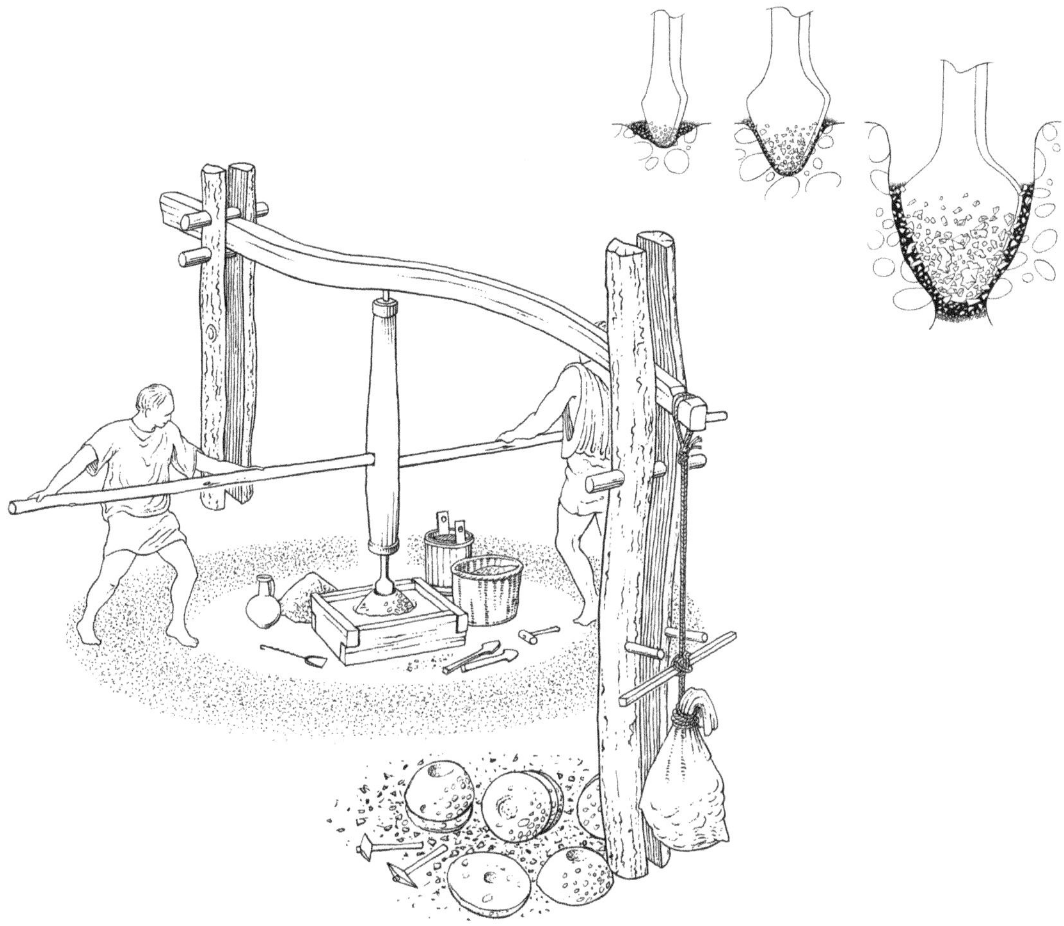

Figure 8. Hypothetical reconstruction of the drills and drilling machine used to perforate and 'marry' Hertfordshire Puddingstone querns (author)

Artefactual evidence of quern making comprises, at the time of writing, débitage from one site, four unfinished querns, and three hammer-stones which may have been used to finish the surface of upper stones. Almost all have been found in a limited area to the north and west of St Albans/Verulamium, near the western extremity of Hertfordshire Puddingstone quern distribution. But it does not follow that most of the querns were made near a quarry site. A widespread pattern of quern-making is emerging, in which stone blanks are made at the quarry and finished elsewhere. In Puddingstone, as with Millstone Grit beehive querns, it is likely that the blank left the quarry worked close to finished size, un-drilled, but with a depression worked in the top of the upper stone to locate the drill. Perforation, smoothing, fitting the stones with ironwork, and marrying them, took place elsewhere for both symbolic and practical reasons (Heslop 2008, n 10; for quarry blanks see also Peacock 1987; Keller 1989).

There is some evidence from the means used to rotate the querns that 'elsewhere' may lie many kilometres from the quarry. A substantial minority of Hertfordshire Puddingstone upper stones exhibit a groove around their girth to retain an iron band, termed the 'driving band' here. A quern dug up in 1892 at Leziate in Norfolk retained this band and its 'handle' appendage (Beloe 1893; Fig. 5h). At the Colloquium it was pointed out that the 'handle' was actually the iron ring used to drive the quern with a pole pivoted from above, in conical motion (as in some Roman military querns: Jaccottey *et al.* 2009; thanks to Luc Jaccottey for this important observation). When the occurrence of grooved Hertfordshire querns is plotted, very few prove to be close to the rock source: they are found considerably further to the east. This surprising conclusion is confirmed by a count of the unprovenanced Hertfordshire Puddingstone querns which have found their way into the collections of the larger local museums: 7 out of 15 are grooved at Ipswich Museum and 4 out of 14 at Colchester Museum in the east, 1 out of 16 at Saffron Walden, and none at all at Hertford Museum and Verulamium Museum (St Albans) in the west (Hurcombe 1981; King 1986; observations by Hilary Major, and the present author). But it is hard to imagine that quern makers in west Hertfordshire were making grooved querns specially for markets over 100 km away. The obvious alternatives are that the finished stones were grooved and fitted with a band at points of consumption in the east; or that the production of the complete quern from a transported blank took place in an eastern centre or centres, far away from the 'quarry' sites. The latter is perhaps the more likely solution: Puddingstone quern production was extraordinarily labour-intensive, so while the blank was not much heavier than a finished quern, it had only a fraction of its value. Quern blanks were therefore eminently transportable in economic terms and we should look for part-finished Hertfordshire Puddingstone querns outside Hertfordshire. More local finishing of these querns would make sense in many ways, not least for the user who occasionally needed her glazed quern re-dressed, or its spindle socket re-cut as the lower stone wore down. It is far from certain that even the re-dressing of these stones could be accomplished in the home, and if primitive machinery was required for the operation, we cannot suppose that it was only available in the south-western corner of a 200 km distribution range.

References

Beloe, E.M., 1893. Exhibit at ballot. *Proceedings of the Society of Antiquaries of London*, 14, 183

Ellison, R.A., 2004. *Geology of London*. British Geological Survey Memoir, Keyworth, 6-7, 41

Heslop, D.H., 2008. *Patterns of Quern Deposition, Acquisition and Deposition. A corpus of beehive querns from northern Yorkshire and southern Durham*. Yorkshire Archaeological Society Occasional Paper 5, Leeds.

Hurcombe, L., 1981. *Iron Age and Roman Rotary Querns in Norfolk, Suffolk, Essexand Hertfordshire*, BSc diss., University of Southampton

Jaccottey, L., avec la collaboration de F. Boyer and A. Milleville 2009. La presence de traces d'oxyde métallique sur les meules de Bibracte. *Instrumentum*, 30, 12-14

Keller, P.T., 1989. Quern production at Folkestone, south-east Kent: an interim note. *Britannia*, 20, 193-200

King, D., 1986. Petrology, dating and distribution of querns and millstones in Bedfordshire, Buckinghamshire, Hertfordshire and Middlesex. *Bulletin of the Institute of Archaeology, University of London*, 23, 65-126

Lovell, B., and Tubb, J., 2006. Ancient quarrying of rare in situ Palaeogene Hertfordshire Puddingstone. *Mercian Geologist*, 16, 185-89

Major, H., 2004. The dating of Puddingstone querns. *Lucerna*, 27, 2-4

Peacock, D.P.S., 1987. Iron Age and Roman quern production at Lodsworth, West Sussex. *Antiquaries Journal*, 67, 61-85

Peacock, D.P.S., and Cutler, L., 2010. A Neolithic voyage. *International Journal of Nautical Archaeology*, 39, 116-24.

Piggott, S., and Piggott, C.M., 1939. Stone and earth circles in Dorset. *Antiquity*, 13, 138-58

Porteous, J., 1973. *Coins*. London

Stocks, D.A., 2003. *Experiments in Egyptian Archaeology: stoneworking technologyin ancient Egypt*. London

White, K.D., 1984. *Greek and Roman Technology*. London

Hand and 'donkey' mills in North African farms

Mariette de Vos, Redha Attoui and Martina Andreoli

Ubi Lapis Lapidem Terit (Plaut. Asin. 32)

The present paper concerns the mills in rural settlements of areas in Tunisia and Algeria. The distribution patterns and statistical proportions of imported and local hand and 'donkey' mills in North Africa are discussed. Mills are ubiquitous in ancient rural sites, but a systematic study for North Africa is still lacking after the fundamental research of David Peacock, Olwen Williams Thorpe and Richard S. Thorpe; statistics are available also from the Segermes and Djerba survey, the excavation of the Nador villa, and some maps of the Carte Nationale des Sites Archéologiques et Monuments Historiques de la Tunisie, now online (www.inp.rnrt.tn/Carte_archeo/html/index_fr.htm; Ben Baaziz 2000). North Africa is always cited as the granary of Rome, and studied for that reason, but analyzing African mills means that we are dealing with that part of grain production reserved for local consumption and not to seaborne trade: supposedly the exported grain was not ground. In this sense, the topic becomes post-colonial. Finally, North Africa is studied not from a European point of view for what it contributes to Europe, but for itself.

Context

The data used for this study were collected during survey activities conducted in 1994-2008 by Trento University in north-western Tunisia around the ancient cities of Thugga and Thubursicu Bure in the Bagradas (now Medjerda) valley, 100 km from the north and east coast. In 2003-09 an area in north-eastern Algeria was studied. It was located in the coastal area of the El Kala National Park and the area of Ouled Driss, 20 km to the north of Souk Ahras, ancient Thagaste, 60 km to the south of the Mediterranean coast (Fig. 1). The Tunisian survey covered a surface of 371 km^2 (69 km^2 intensively) revealing 635 sites; the Algerian survey of 1338 km^2 revealing 341 sites.

The mills are found on the surface of ancient farms which produced olive oil and eventually wine, characterised by monolithic crushing and pressing tools made of local limestone in Tunisia and of local sandstone in Algeria. Both areas present a high settlement density and show an increasing importance of agricultural activities in late Antiquity. The area studied in Tunisia is located on the southern slope of the mid-Medjerda valley, whose bedrock consists of limestone; the area investigated in Algeria, located to the north of the Medjerda, has a bedrock of Numidian sandstone.

Figure 1. Map showing surveyed areas.

Terminology

The definition 'donkey' mill is used in this paper with quotation marks, because it is not sure that hourglass shaped mills were always animal driven: in some cases *meta* and *catillus* are so small that they would have been humanly powered (cf. Williams-Thorpe 1988, 255 and Williams-Thorpe and Thorpe 1989, 90. See also Moritz 1958, 97-102 and Leduc 2008, 482-85, figs 4-5). However, bigger hourglass shaped mills could be moved by humans as has been proposed recently for Morgantina (White 1963, 205; recently verified on an ancient example Type 3.1: pl. XLIII from Morgantina: Crisafulli 1996 in Sposito 2008, 210-11, pl. XLIV.A). A similar use has been proposed also for a *meta* which was found sunk 30 cm in the pavement of a Punic house at Byrsa (see Lancel 1982, 96-100, fig. 119). The general lack of masonry bases for these mills and their secondary position make it impossible to understand their real functioning. The alternative definition 'Pompeian style mill' is avoided because the name might imply a Pompeian provenance.

Every individual millstone was counted as one mill, because of the impossibility of reassembling the original parts. Also 'flour' catchers are mentioned with quotes: in some cases they may have been used to collect oil instead

	Surface	Total of Sites	Productive sites	Sites with olive grinders	Sites with 'Donkey' Mills and Flour Catchers	Sites with Hand Mills
NE Algeria Survey	1338 km2	341	266	111	13	24
% of total			78%	42%	5%	9%
Thugga Survey	371 km2	635	250	0	33	18
% of total			39%	0%	13%	7%

Table 1. Mill presence in surveyed areas.

of flour. Where possible the ancient Latin terminology is proposed, e.g. *gremium* for flour or oil catcher and *machina* for dough mixer (see below).

Statistics

The survey results show that rotary 'donkey' mills are preserved only in 5% of discovered agricultural sites in north-eastern Algeria (11 *metae*, 3 *catilli*, 1 flour catcher) and in 13% of the sites in Tunisia (19 *metae*, 23 *catilli*, 4 flour catcher), while at least one hand-mill was found in 9% and 7% respectively (26 and 27 items) of all sites. In Tunisia three dough-mixers or kneading machines (one of them uncertain) were also found.

The higher percentage of 'donkey' mills in the Tunisian area (371 km^2) is probably due to the dramatic absence of olive grinders. In north-eastern Algeria (1338 km^2) 42% of all productive sites possess a monolithic circular basin in which a smooth or cogged stone wheel or cylinder compacted the olives (Table 1, Fig. 2). It is, therefore, possible that hourglass-shaped mills discovered in rural sites around Thugga and Thubursicu Bure were also olive crushing mills, as verified at Volubilis in Morocco (Akerraz and Lenoir 1981-1982, 71-4). As noted by these authors at Volubilis wheat mills and olive mills differ only in the rock types used to build them: lava for wheat mills, local fossil limestone for olive mills, even if the *meta* in the press room of the 'House of the Two Oil Mills' seems to be of black lava (Riße 2001, 101, fig. 151, 'shell sandstone'). Also at Madauros in Algeria, hourglass-shaped mills of local conglomerate and fossil limestone are *in situ* in olive press rooms (see below). In the rural settlements surveyed in Tunisia, around Thugga and in north-eastern Algeria no *trapetum* was found. Brun argues that the *trapetum 'est attesté surtout sur la côte, dans le Cap Bon et dans la vallée du Bagradas (région de Dougga)'* but without citing references (Brun 2004, 209; cf. Mattingly 1996, 578-79, fig. 1 distribution map).

Provenance of mills

61% of all mills in rural sites around Thugga (Tunisia) are imported. Of 68 examples, 42 are made of black or red-brownish lava and 27 of local stone. In north-eastern Algeria the proportions are totally different: only one out of 40 discovered mills was imported. It is made of black lava and was found in the monumental villa of Ksar Laatach, some 60 km from the Mediterranean coast. All the other more modest farms between Ksar Laatach and the coast used quartz conglomerate or sandstone to make their mills (Figs. 3 and 4).

Figure 2. North-eastern Algeria, Oued Jenane-Ksar Fatma, OJ001. Olive grinder with its wheel.

The Bouchegouf survey, 40 km south of Hippo Regius (now Annaba) also produced only conglomerate and sand-limestone mills (Houamria 2009), just like the villa of Nador, 4 km south-west of Tipaza on the Mediterranean coast, where of 56 mills only one fragmentary hourglass-shaped *catillus* of lava was found (Anselmino *et al.* 1989, 192). At Madauros (M'daourouch) and Theveste (Tebessa) no imported mills have been found. The prevalence of nummulitic mills at Tebessa can be easily explained by the nummulitic deposits at Gastel on Djebel Dyr (13 km north of Tebessa; see http://www.mem-algeria.org/english/index.php?page=potentialites), where a high number of broken hand-mills still lie on the surface (Guichard and Djerrab 2009). It is also interesting that hand-mills of the same stone, similar in form to ancient ones, are still for sale today in the city souk (Fig. 5).

In contrast, Hippo Regius on the eastern Algerian coast, has more imported mills: 8 of lava (four of red-brownish lava and four of black lava) and 11 of local rocks. It is not surprising to find imported mills in this important wheat exporting port, but their number is inferior to that noticed at Carthage. In the Byrsa Museum garden 60 hand and hourglass shaped mills are on view: 37 of them are made of rhyolitic ignimbrite (red-brownish lava), seven of black lava and 16 of local rocks. So 73% of the visible material was imported and only 27% made from

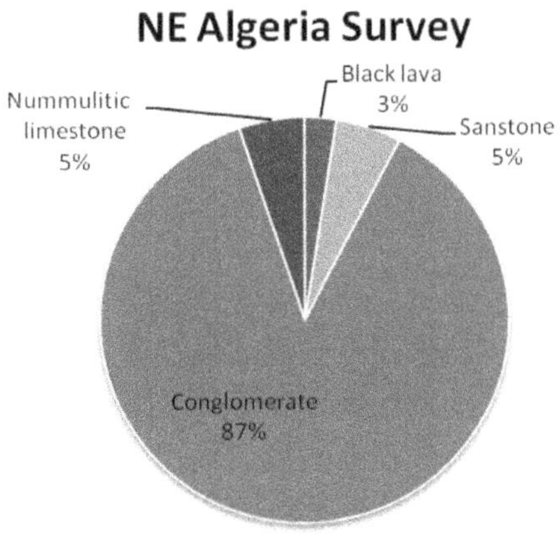

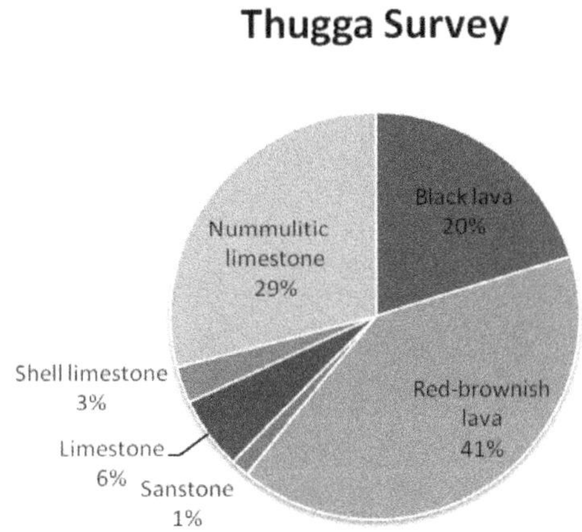

Figures 3 (far left) and 4. Comparison between north-eastern Algeria and Thugga survey: rock provenance of hand mills.

Figure 5. Algeria, Tebessa, souk. Modern hand mills.

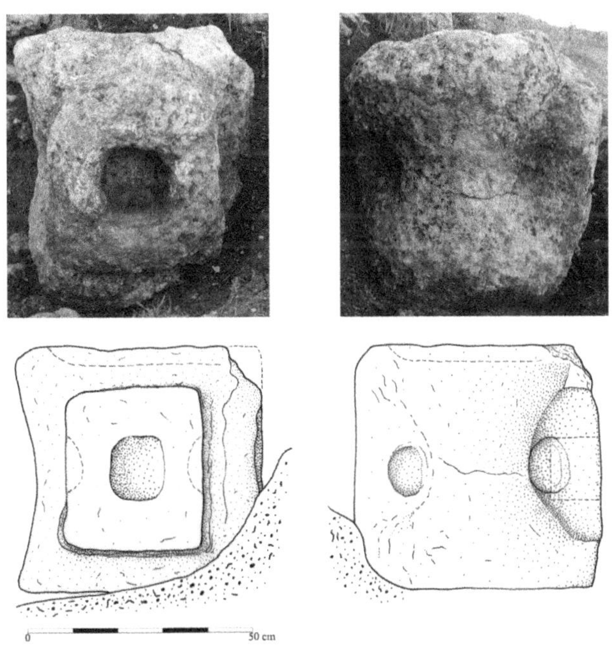

Figure 6. Tunisia, Djebel Gorra, Du131. Unfinished catillus of nummulitic limestone. Scale 50 cm.

local material. This percentage shows that between the metropolis and the hinterland, 100 km distant, there is a quite similar situation, confirmed also in the Segermes Valley, 53 km south of Tunis, where black and red-brownish lava dominate with 92% of all discovered mills (Dietz 1995, 796, fig. 13; Gerner Hansen 1995, *passim*). Carthage had an important role as importer and distributor of Italic millstones (Peacock 1980; Williams-Thorpe and Thorpe 1989, 109-110). Thugga lays in the *pertica Carthaginiensium*, a zone of tax-free cities whose territory belonged to Carthage until the reign of Septimius Severus, when the city became a municipium (AE 1963, 94. Tax free condition was limited to Roman citizens. Maurin 2000, 145-47). Imported red-brownish lava was also found in Tunisia outside of the *pertica*, as documented through at least three mills at Mustis (One red lava *catillus* with cross is already mentioned by Peacock 1989, 211, 212 fig. 4f).

The relationship of Hippo Regius and its hinterland, the Subus plain, is not as intense as in the *pertica Carthaginiensium*. Moreover, the local quartz sandstone conglomerate in the area north of Medjerda is an excellent substitute for lava.

Near Thugga we discovered a nummulitic limestone quarry area on the plateau of Djebel Gorra. This plateau, a *cuesta* or homoclinal ridge, has many nummulitic outcrops (Perthuisot 1979, 14: inferior and middle Eocene), but so far only a building and pressing stones quarry has been found. Anyway, an unfinished nummulitic *catillus* on site Du131 on Djebel Gorra suggests that its find place is not far from the quarry (The *catillus* of site Du131 is indicated on map de Vos 2007, 46, fig. 2). The final cutting of this *catillus* failed because of a horizontal crack at the level of the handle lugs (Fig. 6).

The diffusion of mills imported from the Italic peninsula and islands in Tunisia has been explained by the movement of the many grain ships from Carthage to Italy, which on their return journey north-south carried millstones as a combination of ballast and trade objects (Williams-Thorpe 1988, 286). The red-brown lava has been identified as a rhyolitic ignimbrite of Sardinian provenance, confirming the general preference for this rock which excelled in quality, availability, wear resistance and price advantages (Williams-Thorpe 1988, 276, fig. 5, 281 Table 7, 285, and Appendix 3 with results of analyses; Williams-Thorpe

O. WILLIAMS-THORPE AND R.S. THORPE

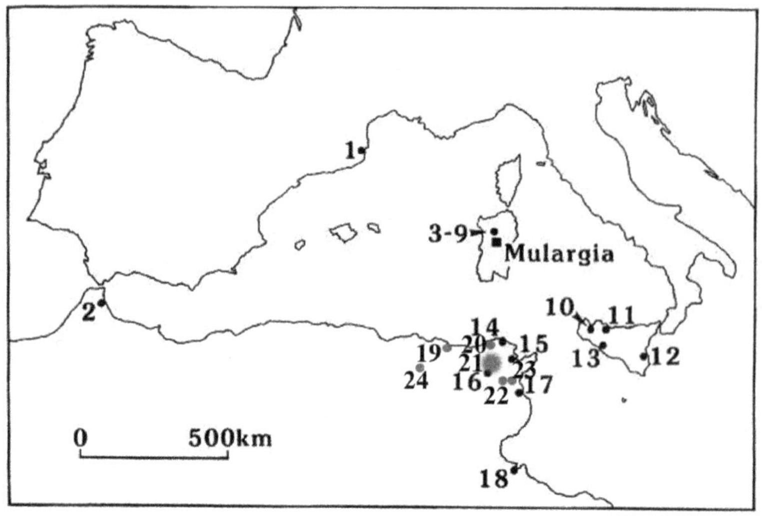

Figure 8a
Distribution of Mulargia millstones in the western Mediterranean area showing positions of the Mulargia source and sites with millstones from there. Sites are numbered: 1 — Ampurias; 2 — Tetouan museum; 3 — Sassari museum; 4 — San Pietro di Sorres; 5 — Sant'Elena; 6 — Monte Zuighe; 7 — Sa Pattada; 8 — Mulargia; 9 — Tharros; 10 — Segesta; 11 — Solunto; 12 — Megara Hyblaea; 13 — Selinunte; 14 — Utica; 15 — Carthage; 16 — Musti; 17 — Sousse museum; 18 — Gightis. Sites 3–9 are also marked on Figure 7.

19 - Hippo Regius (4), 20 - Thabraca (1), 21 - Thugga (27), 22 - Thuburbo Maius (3), 23 - Segermes (17), 24 - Tiddis (1)

Figure 7 (right). Distribution of Mulargia millstones. Updated map of Williams-Thorpe and Thorpe 1989, Fig. 8a.

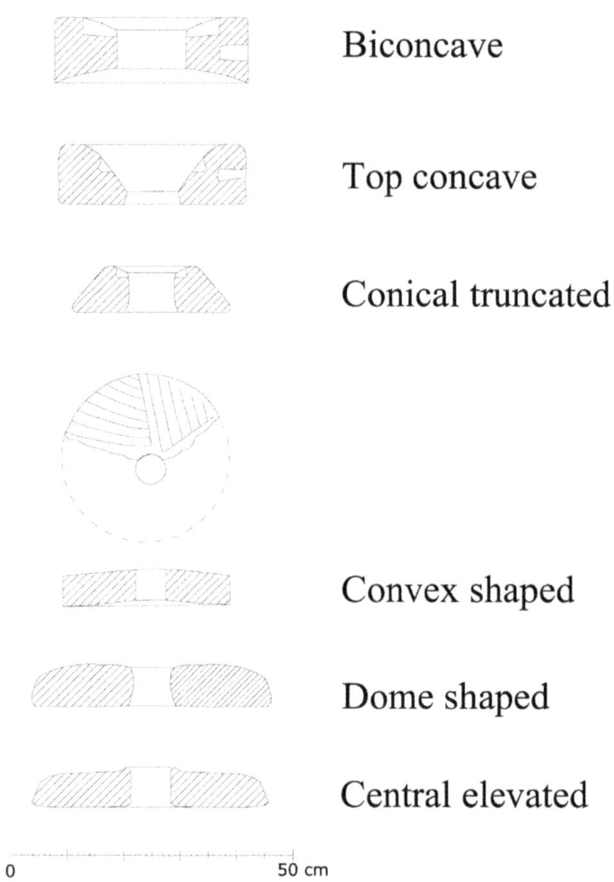

Figure 8. Typological scheme of recorded hand mills. Scale 50 cm.

and Thorpe 1989, 105). It has been attributed to the area of Mulargia, which suggestive toponym derives from Molinaria, near Macomer (the Carthaginian toponym of Macomer itself underlines the intensive interaction between Sardi and Carthaginians). Outcrops of red-brownish lava lined with distinctive greenish *mineral* and ancient quarries were noted 30 years ago in the village of Mulargia (Peacock 1980, 47; Williams-Thorpe and Thorpe 1989, 94-5, fig. 3). Red-brownish lava mills with green mineral collected in the Thugga survey have therefore been catalogued as probably quarried at Mulargia, on the basis of visual criteria mentioned by Williams-Thorpe and Thorpe (Williams-Thorpe and Thorpe 1989, 103-5. Chemical analyses of some red lava samples are in preparation at University IUAV of Venice. For attribution problems see also Antonelli-Lazzarini 2010, 2089.). Therefore, we propose to update the distribution map of red lava mills published in 1989 adding 6 new sites (Fig. 7).

After the Severi, the trade pattern may not have changed very much in the area around Thugga, it became much more intensified, as a consequence of the *lex hadriana de rudibus agris*, found in three surveyed sites (Du025, Du539, Ain Djemala: de Vos 2000, 35-6, fig. 57-8; 2004, 38, 42-5). The rural settlements in the Thugga region started in the early Empire, their development steadily increased through late Antiquity, but exploded and persisted through the 6th-7th century AD, until the Arab invasions (de Vos 2007, 43).

Hand-mills

The total number of hand-mills examined was 53. The 27 hand-mills from the Thugga survey have been classed as 7 lower and 17 upper stones (three unidentified), while the 26 hand-mills of the north-eastern Algeria survey produced 9 lower and 17 upper stones. The recorded hand-mills for both areas can be classified as (Fig. 8):

1. Biconcave upper stone consisting of concave bottom and top with straight vertical external profile;
2. Top concave upper stone with upper deep hopper, flat bottom and vertical external profile;
3. Conical truncated upper stone with sloping outer profile;
4. Convex shaped perforated lower and upper stones with grooves on top or bottom surface;
5. Dome shaped perforated lower and upper stones;
6. Central elevated perforated lower and upper stones consisting of raised collars around the hole.

The comparison between Thugga and the north-eastern Algeria survey hand-mills reveals some micro-regional preferences (Figs. 9 and 10). The survey in north-eastern Algeria has demonstrated that the most common upper stone for hand-mills (13 examples) is the biconcave type, with a height varying between 9 and 16 cm and a base width between 28 and 36 cm. Two upper stones can be attributed to the conical truncated type and one to the convex shaped type with parallel grooves in herringbone pattern on the bottom and one to the top concave type. The lower stones are divided in 6 convex shaped (one with parallel grooves in herringbone pattern and two with radial grooves on top), one dome shaped, one conical truncated, one central elevated and one unidentified. Around Thugga other types of lower and upper stones seem to be preferred: upper stones are mainly of dome shaped (7) and conical truncated type (6). Only one example of top concave type was discovered in the excavated Byzantine farm of Ain Wassel and 3 other fragments remain unidentified. Lower stones are of dome shaped (4) and conical truncated type (1). Only 3 upper stones (two of them are broken) and 2 lower stones recorded in north-eastern Algeria are complete.

Features and dimensions of the hand-mills recorded in north-eastern Algeria show more typological consistency than in the Thugga region. In north-eastern Algeria hand-mills vary only slightly in height and base width and the preferred upper and lower stone types are respectively the biconcave and the convex shaped one. Dimensions of 6 complete lower stones of the villa of Nador, with base widths between 32 and 39 cm, approach our Algerian mill measurements (Anselmino *et al.* 1989, 190). The base width of hand-mills around Thugga varies more; their upper stones comprise at least two types and the lower stone is of the dome shaped type. Dimensions of all recorded examples in both areas are anyway sensibly smaller than Djerbian hand-mills (1 lower stone and 11 upper stones) (base width 50-110 cm) (Drine 2009, 330-33) (Fig. 11).

Modern hand-mills in Africa and the Levant have the same dome-like profile or a raised collar (Williams Thorpe and Williams 1993, 270: 5th-6th cent. examples near the Dead Sea) as some ancient examples (Fig. 12). Frequently, the modern hoppers have a perforated vertical hole near the rim from the upper surface in which a rope is tied. The miller uses a piece of wood in order to turn the hopper.

Grooves on the grinding surface are present only on 6 hand-mills (3 upper stones/3 lower stones) of the north-eastern Algeria survey. Grooves are designed to
1. prevent overheating of meal and milling stones,
2. improve grain flow from centre to periphery and output of the meal in the receptacle,
3. avoid the filling of the pores of the milling stones,
4. enlarge the grinding surface (Šebesta 1977, 78; cf. Moritz 1958, 79 for grooves on *metae*).

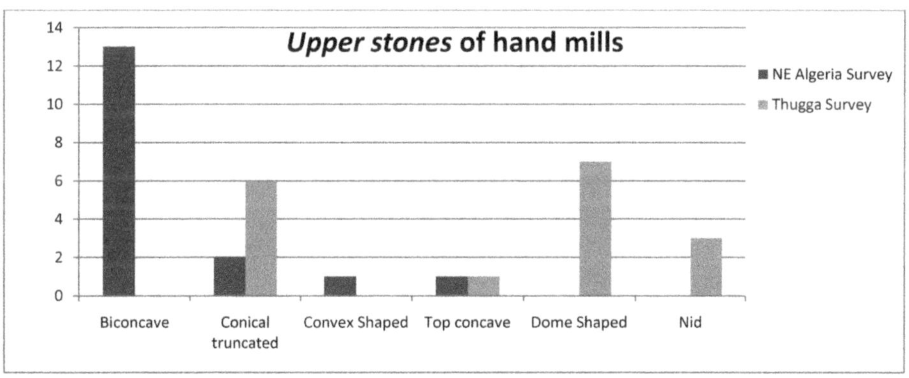

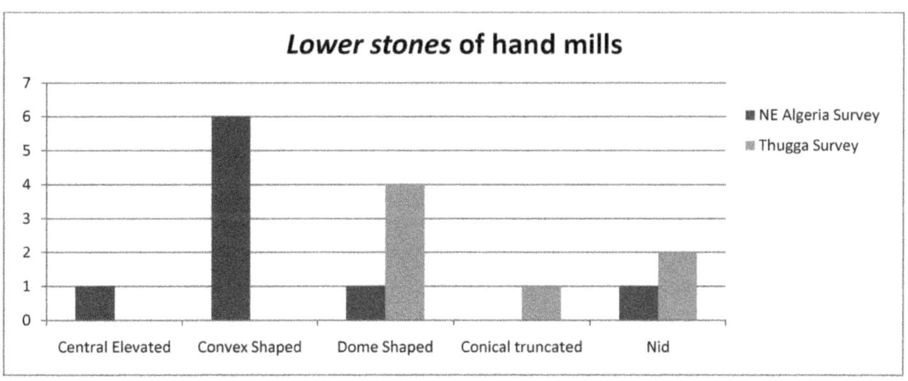

Figures 9 (top) and 10. Histogram showing preferences in hand mill shapes. Comparison between upper and lower stones in north-eastern Algeria and Thugga survey.

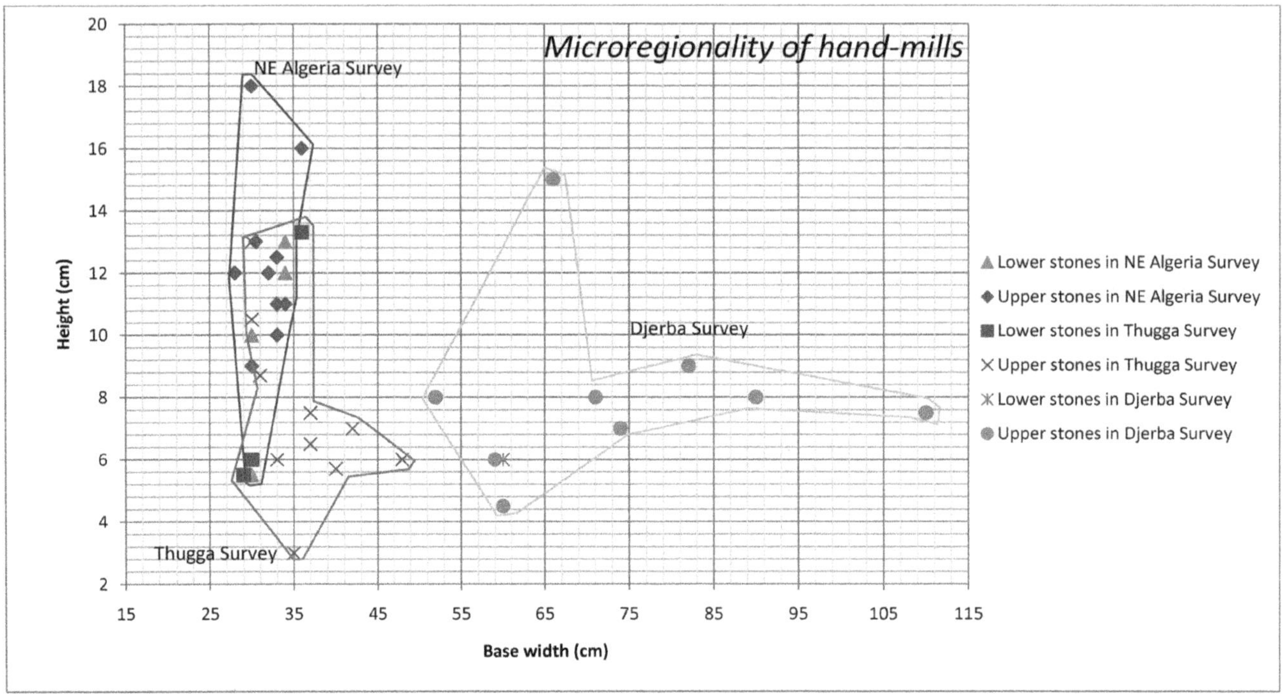

Figure 11. Scatter plot showing microregionality of hand mill features. Comparison between north-eastern Algeria, Thugga and Djerba surveys.

Figure 12. Tunisia, Maatria-Birhayl. Modern hand mill in farm.

Either the grooves were never cut on the other hoppers, or they disappeared as a result of wear - see for example, the lopsided red lava hopper of Du474 (Fig. 13). Indeed, some hand-mills are made of very porous stone, unfit to cut grooves in, and the edges of the pores carry out the function of the grooves. Maybe the porosity can prevent overheating, but it does not stimulate the grain flow. Evidently, the conglomerate and shell limestone hoppers are less efficient than the lava examples.

The smooth bottom of the upper has not been re-grooved, as nowadays practiced by the millers or farmers themselves (Šebesta 1977, 139-42) and presumably because of the lack of specialised stone cutters in the isolated rural settlements, even if the hand-mills are portable. The rural settlements in Tunisia are all a short distance of one or more cities; in Algeria they are not. Most settlements had a very long life, from the 2nd through to the 7th century AD, and some even after the Muslim invasions. So the rubbing and disappearance of the grooves may be also the result of a very long use. Šebesta reproduces a symmetrical hammer with flat cutting edges on both sides in the Museo Archeologico Nazionale of Naples, supposing it was used for cutting grooves; this hammer is identical to the examples used nowadays (Šebesta 1977, 80, fig. 224).

Provenance of hand-mills

Around Thugga 46% of hand-mills are made of imported lava: 25% (6) of black vesicular lava, possibly of Pantellerian origin. The analysis of hopper rubbers from the Sec wreck have demonstrated massive basalt quarry exploitation on Pantelleria since the 4th century BC. The proximity of this island to the investigated area near Thugga and the presence of many unworked Pantellerian lava fragments at Carthage and other Tunisian sites, could point to Pantelleria as a possible source for the black lava mills (Williams-Thorpe 1988, 285 table 8 and appendix 3; Williams-Thorpe and Thorpe 1990, 123, 127, 129-31). However, 21% (5) of Thugga hand-mills are of the rhyolitic ignimbrite from Mulargia. The remaining 54% of mills are of local material: 9 of white nummulitic limestone, 2 of local coarse shelly limestone and 2 of compact limestone. Hand-mills discovered in the north-eastern Algeria survey are 96% local rocks: 18 (68%) are made of local quartz conglomerate of small rounded white and grey pebbles cemented together in a pink or grey matrix of sandstone, 3 (11%) of sandstone, 2 (8%) of white nummulitic limestone and only 1 of imported black vesicular lava, which has

Figure 13. Tunisia, Du474. Lopsided red lava runner stone.

been found in a villa to the south east of the surveyed area, in the most distant site from the sea. Two mills are of unidentified sedimentary stone.

Also, the Nador villa near Tipaza on the west Algerian coast, contained 54 hand-mills (9 identifiable as upper and 8 as lower stones). Thirty-five of them are made of sandstone, 17 of shell limestone, 2 of conglomerate (Anselmino *et al.* 1989, 190-92 'puddingstone'). Note the high number of hand-mills in one single excavated monumental farm through the different phases (IIA, IIC, III) of three and a half centuries. All but one are fragmentary. Even if the single elements formed pairs, the number remains high and indicates an average life span of 6 or 13 years: 8-16 examples per century. The Nador hand-mills contain an example with 2 opposite side entrances, similar to one in the Museum of Hippo Regius (Annaba). This type is not present among the hand-mills of north-eastern Algeria, north-western Tunisia and Djerba. Further documentation is needed in order to pronounce any conclusion on distribution patterns and typology.

In the Djerba survey, 9 hand-mills of 'volcanic stone' and 2 mills of limestone were discovered (provenance not specified: Drine 2009, 330-33).

'Donkey' or hourglass shaped mills

The climate of the surveyed areas in North Africa is good for olive growing in Tunisia (rainfall 400-500 mm per year) and for vines in Algeria (up to 1000 mm). The stony slopes of the hilly landscape are particularly suitable for olive culture, less for grain growth, but polyculture is possible and probable, as many olive groves with barley and other crops occur between the trees nowadays. In the case of polyculture, Columella (5.9.7) and Palladius (18) recommend a greater distance between olive trees 40-60 feet instead of 25 feet (de Vos 2000, 22, figs. 66-7; 2004, 13).

In our areas we do not know if there was a surplus of cash crop. In the Thugga region it might have been barley which is more resistant to the semi-arid climate. So far, no granaries have been identified in either the countryside, or in the cities. However, epigraphic evidence can fill the gap in the archaeological record:

1. C. Artorius Bassus duovir at Thugga in AD 48/49 was also *curator lucustae* (CIL 08, 26517, AE 2006, 107, Desanges 1975; Ben Abdallah and Maurin 2000, 138-42. *Locustae* is a collective singular. A law in Cyrenaica that obliged people to kill locusts 3 times a year: as eggs, foetuses and adults, see Pliny *Nat. Hist.*11.35.105).
2. In an inscription of AD 170 a crop sale is mentioned by L. Memmius Pecuarianus Marcellinus sponsor of the *Capitolium* construction of Numluli, 10 km N of Thugga, who boasts the sale of wheat to the people at a very low price, in a period of modest market offer (CIL 8.26121; Aounallah and Ben Abdallah 1997, 81, note 26).
3. A dedication dating to the beginnings of the 2nd century AD by the *civitas* Thugga praises the generosity of Calpurnius Faustinus, imperial *flamen perpetuus*, for selling wheat to the people at a much lower price than the one in force (Aounallah and Ben Abdallah 1997, 80-1, n. 2; 245 pl. 5, 3-4). A farm of this family can be identified from the funerary inscription of Calpurnius Faustinus at site Du151, some 2,5 km to the south west of Thugga (CIL 8.27369, Aounallah and Ben Abdallah 1997, 78. The site is not well preserved, the mausoleum and marabut of Sidi Bou Mous mentioned by Carton 1895, 200 have disappeared completely).

The distribution of ancient farms and agglomerates with 1-12 presses of the Thugga survey shows that small-scale farming prevailed (de Vos 2004, 27; 2007, 44, fig. 1: distribution map of farms and agglomerates with 1-12 presses). Almost half of the farm-buildings had only one press, almost one third had two presses, one tenth had three presses (de Vos 2000, 26, map 1). The clustered settlements had 6-12 presses, but in the clusters the presses were arranged individually, without over-all organization. In Algeria we have no knowledge of ancient roads as the cork oak forests cover the evidence. Twenty seven settlements around Thugga with 'donkey' mills and/or a flour/oil catcher are provided with multiple oil/wine presses: 7 have one press, 5 have 2 presses, 6 have 3 presses, 2 have 4 presses, 4 have 5 presses, 2 have 7 presses, 1 has 10 presses. Five damaged settlements have a 'donkey' mill and no press(es), probably because of recent anthropic erosion. No farm has an olive grinder. Also, the 11 settlements with 'donkey' mills in north-east Algeria are all provided with multiple oil/wine presses and: 1 has 1 press, 3 have 2 presses, 2 have 3 presses, 1 has 7 presses, 1 has 8 presses, 1 has 10 presses, 1 has 11 presses, 1 has 13 presses. These settlements contain 25 olive grinders.

Metae

Total number of *metae* examined was 30. Around Thugga there are 19 *metae* (12 in good condition and 7 fragmentary), in the north-eastern Algerian area 11 *metae*

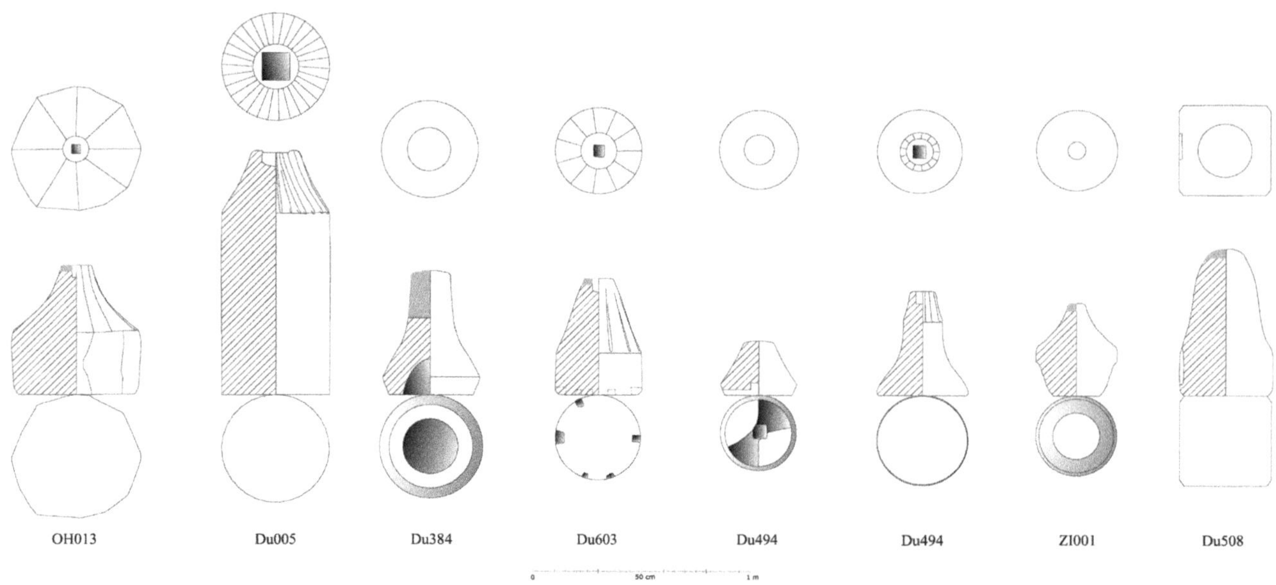

Figure 14. Typological scheme of recorded metae. Scale 1 m.

(8 complete, 3 fragmentary) were documented. Shapes and dimensions of these elements vary considerably in both areas, so that it is impossible to subdivide *metae* according to their morphological features (Fig. 14). In north-eastern Algeria some *metae* have high, only roughly prepared, conical inverse bases (4 examples) and 2 *metae* have a polygonal base (Fig. 15). Circular and flat bases are common, only one base is square. Tops are flat or rounded, the latter can be simply worn. Four examples are provided with square holes on the top (6 other *metae* are slightly broken at this point) and none is perforated. This means that the wooden vertical beam did not turn with the frame. Grooves are present in 4 examples (1 windblown, 3 radial vertical). One *meta* (ZI001) has a slightly protruding step above the conical inverse base, as to put the element in a receptacle (Fig. 14). The conical inverse base is represented in a stylized form on a gem (Blümner 1912, fig. 21; Wilson and Schörle 2009, 111, fig. 10) and a circular baker's shop sign from Pompeii (Mazois 1824, pl. XIX fig. 2; Overbeck and Mau 1884, 379, fig. 186; Moritz 1958, 80-1; Wilson and Schörle 2009, 111, fig. 11). Surprisingly, this type does not correspond with Pompeian *metae*, but it may represent an italic prototype of *meta* ZI001.

Metae recorded around Thugga have more technical features as those in north-eastern Algeria. If conserved and visible, bases are flat and circular (8), inverse conical (4), polygonal (1), square (1), hollowed (1), with a square socket in the centre of the bottom (2), with central socket and two opposite dove tails in the bottom (2) to anchor the *meta* base, or small mortises on the edge (1) (Du603). One *meta* has a shallow socket on one side of its square base (Du508). Tops (11 unidentifiable) are flat (6) or rounded (2), with square top sockets (5) and radial vertical grooves (4). No *meta* is perforated.

Three *metae* at Rusicade, Setif (Delamare 1850, pl. 31 and 75) and Morgantina (Crisafulli 1996 in Sposito 2008, 219, pl. XLIII, Type 3.3b) have a similar square socket in the bottom centre. *Meta* Du603 (Fig. 14) can be compared to the one with small holes on the circular Pompeian shop sign (see above) which could have the purpose to enhance the attachment of the *meta* to its base or to the flour/oil catcher. The *meta* of site Du005 with a very high cylindrical base (85 cm) is located near a press room (de Vos 2000, map of the site: fig. 30; fig. 31. 2: considered a grain mill), so it could have been used for olive compacting as for example at Volubilis. It has deep vertical, radial grooves

Figure 15. North-eastern Algeria, Oued el Hout, OH013. Meta with polygonal base and radial vertical grooves.

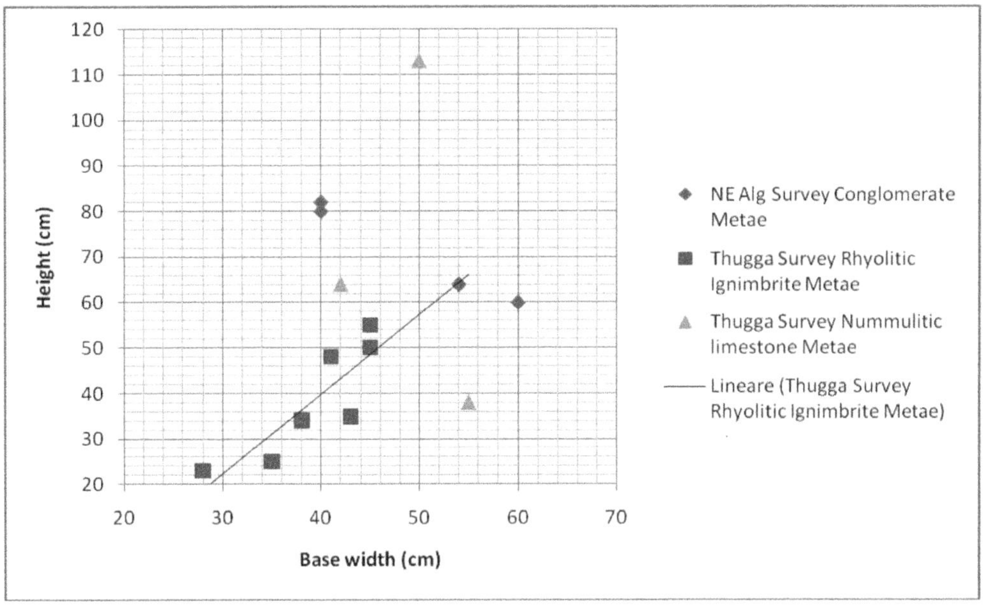

Figure 16. Scatter plot of complete meta examples. Rhyolitic ignimbrite metae are rather standard in their proportions.

on its cone, considered characteristic for oil processing (Akerraz and Lenoir 1982, 71-2). However, some *metae* in representations of bakeries on friezes of Rome and Ostia are grooved in the same way (Relief of the Vigna delle Tre Madonne, Museo Chiaramonti: Wilson and Schörle 2009, 11, fig. 17; Ostia, relief of P. Nonius Zethus: Junkelmann 1997, 121 fig. 62. For Moritz 1958, 79 the grooves on the *metae* are applied by the sculptors of the reliefs with the intention to 'convey an impression of turning and perhaps to suggest the meal flowing from the mill'). The *meta* of site Du005 is of nummulitic limestone; transport was simply by rolling the cylinder down 3 km from one of the outcrops of Djebel Gorra.

Dimensions of complete *meta* examples of both surveyed areas show different features (Fig. 16). Around Thugga heights of *metae* range from 23-113 cm, in north-eastern Algeria from 45 cm *ca.* to 82 cm. Base widths of Tunisian examples range from 28-55 cm, in north-eastern Algeria they are larger from 40-60 cm. Most of the elements in both areas have base widths of 40-55 cm, as generally marked by Williams-Thorpe (Williams-Thorpe 1988, 255) (Fig. 17). There is a higher percentage of small *metae* in Tunisia, supporting the theory that they were hand driven.

Provenance of *metae*

Metae discovered during the Thugga survey are 63% of imported red-brownish lava (rhyolitic ignimbrite). No other imported rock type was used. The remaining examples are made of local material: 32% of nummulitic limestone, 5% of sandstone. In north-eastern Algeria all *metae* are of local material: 91% of quartz-conglomerate, 9% of sandstone. One of the most interesting features concerns the relationship between used rock types and dimensions of the *metae* (Fig. 16). On imported ignimbrite *metae* of the Thugga survey, nearly standard proportions (height: base

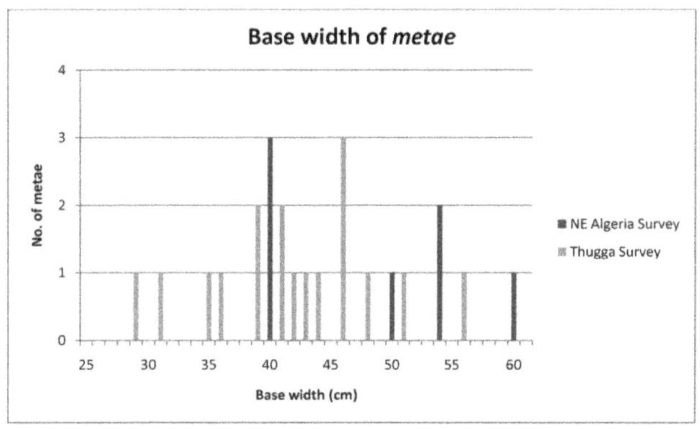

Figure 17. Histogram of base widths of recorded metae. Comparison between north-eastern Algeria and Thugga survey.

width = 1:1) can be noted. The dimensions are rather small in height and width (not more than 55 cm and 45 cm). These measurements probably depend on the producing centres, which realized elements in series. *Metae* of local origin were possibly adapted to the demands of purchasers, with very different dimensions and shapes. In north-eastern Algeria the proportions of locally produced *metae* are quite regular, with a width and height ratio ranging from 1:1 to 1:2.

Rhyolitic ignimbrite *metae* are very common also elsewhere in Tunisia and Algeria and on coastal sites the majority of *metae* are made of this volcanic rock. At least 20 examples are present at Byrsa, 2 examples at Thabraca, 3 examples at Hippo Regius. Hinterland sites in Tunisia (e.g. Mustis) show examples made of ignimbrite and local stones. In north-east Algeria, except at Hippo Regius, no ignimbrite *metae* were noted in inland sites (e.g. Thibilis, Madauros, Theveste, Thubursicu Numidarum, Calama), where local rock types were used massively. In North Africa ignimbrite seems to be preferred for *metae* and only 2 from the Byrsa were of black basalt or grey vesicular lava (Peacock 1980, 47; Williams Thorpe 1988, 285).

Figure 18. Algeria, Oued el Hout, OH010. Fragmentary catillus of local conglomerate.

Figure 19. Algeria, Oued el Hout, OH010. Inside upper slope of catillus (Fig. 18) decorated with herringbone pattern grooves.

Catilli

The total number of *catilli* examined was 26. *Catilli* are generally less well preserved than *metae*, because of their fragile structure. In north-eastern Algeria only 3 fragments were recorded while around Thugga 23 examples were found, some of them very small and hardly distinguishable. Original shape, dimensions and technical features are therefore difficult to understand. One of the 3 Algerian examples, (OH010) is nearly complete (Fig. 18). The inside upper slope of this quartz conglomerate *catillus* has herringbone grooves. (These are frequently seen on other examples, from the much earlier Olynthus mills (Frankel 2003, 11, fig. 7 b) (Fig. 19), or the ring *catilli* of Volubilis, used for olive crushing before putting them in the baskets (Akerraz and Lenoir 1982, 73-4, pl.VI.2; Brun 1997, 71-2, fig. 2; according to Riße 2001, 102)). The presence of 167 olive grinders with wheels or cylinders (see above) in the ancient settlements of north-east Algeria does not preclude the use of 'donkey' mills in the same settlements for crushing the olives; the cylinder grinders would be used probably for mixing the paste between the different pressings. The 2 other Algerian *catilli* fragments are part of the mount in the form of a closed rectangle. All these examples are probably to be considered symmetric hourglass shaped *catilli*, with two equal cones, and therefore reversible types.

The Tunisian examples are, as already noted for other mill elements, more varied in shape, material and technical features. On 14 rim or mount fragments it is impossible to specify more than their rock type. The other 9 *catilli* are better conserved and show some peculiarities. Two or maybe 3 rhyolitic ignimbrite *catilli* (Du395, Du508 and scatter find) have their handle lug not in the centre of the external profile, but on the higher lower cone; the lugs have only 3 protruding sides, the upper side being open. The 2 superposed cones are not of the same size, as in the Sec wreck (Beltrame and Boetto 1997, 183, 185: n. 115.3, dated to 375-50 BC by amphorae, pottery and bronze finds on the wreck), at Morgantina (White 1963; Sposito 2008, *passim*, pl. XLIII-XLIV, dated to the 3rd century BC) and at Byrsa (Lancel 1982, *ca.* 150 BC) examples which could not be reversed. At Hippo Regius a similar fragment of ryholitic ignimbrite is exposed in the museum garden. A complete *catillus* at Rusicade (Skikda) (Delamare 1850, pl. 160.11 and 12) is provided of similar slots (Fig. 20). This kind of lever slot allowed the horizontal traction beams to be lifted during standstill, creating circulation space in the mill room (Alonso 1996a, 235; 1996b, 184, fig. 1; Sposito 2008, 212). The upper protruding side of the mounts of at least 3 black lava *catilli* at Castello Ursino in Catania have an interruption in their central part to block the wooden framework. It is similar to the vertical groove in the upper outside wall of the *catilli* in the Casa di Laocoonte at Pompeii VII 14, 30 (fittings in upper rim of other examples are described by Moritz 1958, 80) (see below) and of insula 13 at Volubilis (Leduc 2008,

Figure 20. Rusicade (Skikda). Catillus with lugs consisting of 3 protruding sides, the upper side is open (Delamare 1850, pl. 160.11 and 12).

501, fig. 8; 502: description). At Catania the interruption was planned from the beginning, as can be seen in an unfinished item, while at Volubilis it results from a repair.

Two nummulitic *catilli* present a straight external cylindrical profile with dovetail mortises near the upper rim, instead of mounts in the centre of the outer hour glass profile (Du002 and Du038: de Vos 2007, 52, fig. 8a-b; 46 fig. 2 distribution map of mills is updated in the map published in this article, fig. 30). Two small similar examples are on show in the Byrsa Museum Garden, but they are not provided with mortises. Also, the ring *catilli* of Volubilis are provided with dovetails (Luquet 1966, fig. 3b; Akerraz, and Lenoir 1982, 71, pl. VI 2; Riße 2001, 104, fig. 157b. The ring *catillus* (height 22 cm) at Volubilis was used both for grain grinding and for olive compacting. Only 3 classical Pompeian style *catilli* (height 50 cm) have been found at Volubilis: Leduc 2008, 482). The already mentioned *catillus* (Du131) was found in unfinished form on the Djebel Gorra (Fig. 6). Its visible initial squared form (55 x 55 cm) indicates that cubic stones were roughly prepared in the quarry to be finished after in detail as symmetric hourglass shaped *catilli*. Two other *catilli* of nummulitic limestone were recorded during the Thugga survey (Du575, Du577).

Dimensions (height and base width) can be recorded in only a few cases, where a reconstruction of the piece was possible (Fig. 21): the previously mentioned nummulitic *catillus* Du575 was c. 43 cm high with an inner base width of 41 cm and outer of 47 cm; the non reversible *catillus* of ignimbrite was only 32 cm high and had a upper inner base width of 31 cm; the cylindrical nummulitic *catillus* Du002 is 40 cm high with an inner base width of 28 cm and outer width of 39 cm. All the examples have a height-width ratio of nearly 1:1. Only the fragmentary mount of ignimbrite *catillus* Du508 could be reconstructed on the

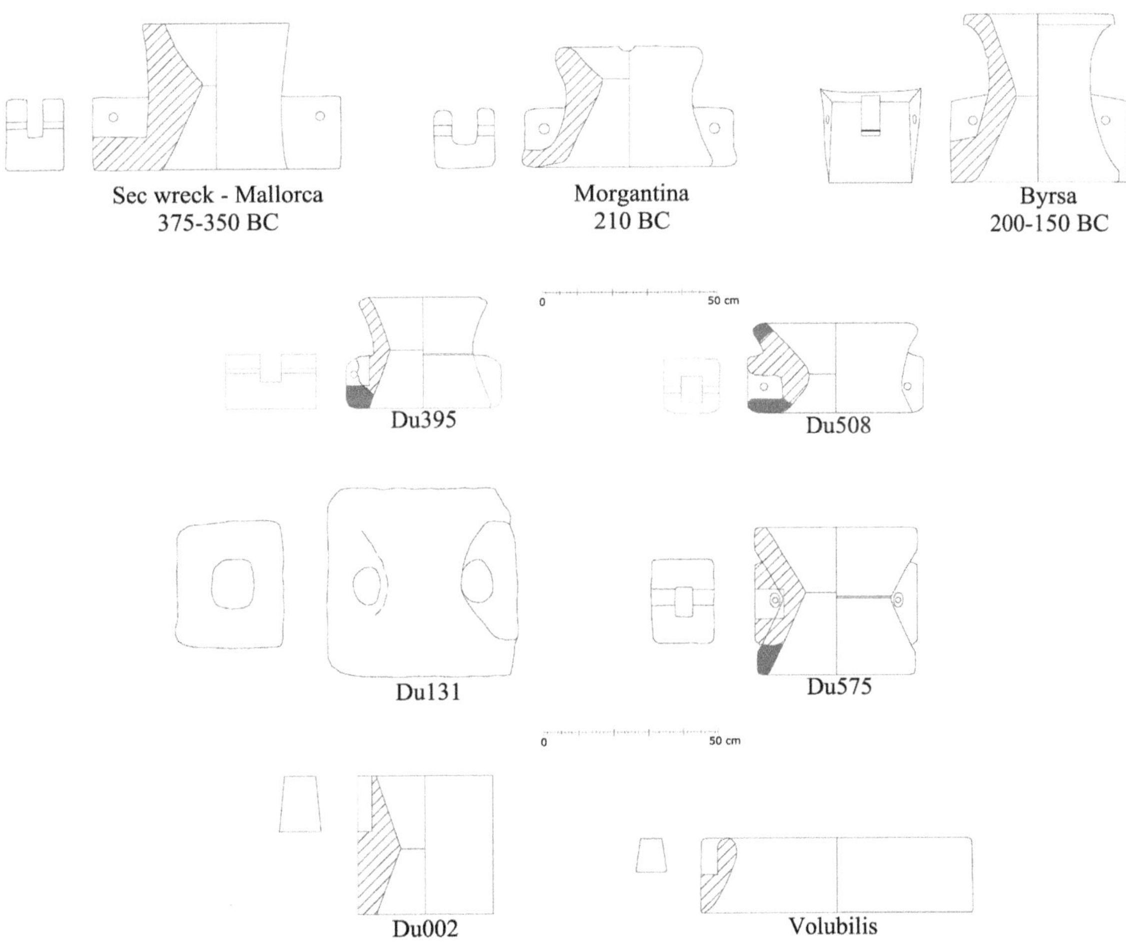

Figure 21. Typological scheme of recorded catilli and some published examples. Scale 50 cm.

basis of similar examples at Morgantina. It was a short non-reversible example similar in shape to a ring *catillus*. One broken lava *catillus* found at Djerba has a reconstructed height of 54 cm (Drine 2009, 332-33).

Provenance of *catilli*

The 3 north-eastern Algerian fragmentary *catilli* are all made of local quartz-conglomerate with small protruding pebbles. In the villa of Nador 2 fragmentary *catilli* were found: one of imported lava (origin not mentioned) and one of sandstone (Anselmino *et al.* 1989, 192). Examples found during the Thugga survey are 52% of ryholitic ignimbrite (12 examples), 26% of black vesicular lava (6 examples) and 22% of local nummulitic limestone (5 examples). Once more we can note the preference for imported ignimbrite. Maybe this evidence can be put in relationship with the very small *catillus* fragments of black lava found in the surveyed area, as to demonstrate that rhyolitic ignimbrite *catilli* were more resistant than examples of other imported volcanic rock types.

Flour catchers or *gremia*

Maybe we can propose *gremium* as the latin equivalent of flour catcher, used in the description of Simylus cleans the hand-mill and its support with the tail of his goatskin before starting to prepare flour for his Moretum: *perverrit coda silices gremiumque molarum* (Moretum 23) (TLL VI 2324, 33-35, Perutelli 1983, 89; Laudani 2004, 68; Moritz 1958, 28, note 1: compares *gremium molarum* 'lap' of the mill with *cava machina* Ov. fast. 6.381, considering both mills). *Gremium* occurs in this sense only in the poetic context of Moretum. But the poem uses a concrete language and contains some other technical terms; this seems to exclude a metaphoric use of *gremium*. *Gremium* is used in another technical sense for a seat, especially that of the judge in the court. The round form of the flour catcher with a depression in the centre reminds us of the aspect of a belly with navel. The TLL suggests a stone *gremium* (*lapis*), but it could be made also of terracotta. A terracotta channel or catcher around a granite *meta* was found in Vetulonia (Falchi 1894, 357-58, fig. 27. We thank M.G. Celuzza, director of the Museo Archeologico di Grosseto for the information that the flour catcher is no longer present in the Museum). Alternatively it could have been of wood as suggested by the illustrations of Junkelmann (1997, pls VI-VII; Moritz 1958, 76), or leather (Ben Baaziz 2000, 312: as in modern times). In Pompeian bakeries, the catchers consisted of a lead sheath which covered the top of the circular round masonry bases in which the *metae* were sunk.

Most of these troughs have disappeared and only a few are published, e.g. a drawing of one of the three lead troughs in the bakery of Casa del Laocoonte (VII 14, 30) shown by Mau (1878, 197; drawing Mau 1899, 389, fig. 220 ; cf.

Figure 22 (top). Pompeii, Casa del Laocoonte, bakery. Mill with flour catcher covered by a lead sheath (drawing by Mau 1899, fig. 220).
Figure 23. Boscoreale (Pisanella), Silver Treasury villa. Mill conserving the lead sheath of its flour catcher (De Cou 1912, Pl. CLXII).

'Laminato di piombo destinato a raccogliere la farina': Fiorelli 1873, 29) (Fig. 22). Another example, in the Chicago Field Museum since 1903, was found in the Silver Treasury of the villa at Boscoreale (De Cou 1912, 209-10, pl. CLXII, inv. 31699; Oettel 1996, 24, 184, 189) (Fig. 23). The only lead sheath *in situ* can be seen in the bakery of Sotericus (Pompeii I 12, 1-2) (Mayeske 1972, 87; de Vos and de Vos 1982, 130): its inward side is folded upwards against the *meta*, covering very well the joint between the horizontal upper surface of the circular masonry base and the vertical *meta*.

Four flour catchers have been documented in the Thugga survey (one was found in Teboursouk, ancient Thubursicu Bure, Du494). The two halves of a broken limestone flour catcher from two different spaces near the press room of the excavated Byzantine farm at Ain Wassel (site Du025, 12 km to the west of Thugga) have circular traces of the rotating *catillus* (de Vos 2000, 36, fig. 58.10-11). The stratigraphic excavation of the farm confirms the use of the flour catcher in Byzantine times. Also, the flour catchers of sites Du355 (Fig. 24) and Du494 show the same circular traces. Only the flour catcher of site Du494 is not perforated. Its central hole is 11 cm deep and 42-45 cm wide, corresponding with the average base width of local *metae*. A similar not perforated receptacle is at Mustis and

Figure 24. Tunisia, Du355. Flour/oil catcher of local limestone.

in a press room of site OH005 (Safhe 1), the only example discovered during the north-east Algerian survey (Fig. 25). Just one perforated fragmentary example was identified at Thugga city.

The 4 mentioned flour catchers (all perforated) of the Carte Nationale des Sites Archéologiques de la Tunisie, are defined '*plateau de moulin à céréales*': 050.003, 067.105, 068.088, 068.311; the latter (Ksar Tlili) in Middle Tunisia was used for olives, as the climate does not allow grain growth (Ben Baaziz in http://www.inp.rnrt.tn/Carte_archeo/html/index_fr.htm). Another 3 catchers are recorded in rural sites to the south west of Mactar (Ben Baaziz 2000, 43, photo 49; 107, photo 233; 176, photo 409). Two flour catchers are located in the baths of Ain Tounga (Thignica, near Thugga) (Ben Hassen 2006, 124, fig. 74), 5 at Mustis, 1 at Thuburbo Maius, 2 others in the House of Amphitrite at Bulla Regia (Bonini and Rinaldi 2003, 202-03, fig. 3; Ghedini 2003, 332, fig. 10), 2 at Carthage, 6 at Madauros (Brun 2004, 220), 8 at Theveste, 1 at Thamugadi, 1 at Hippo Regius, 1 at St. Charles (Ramdan Jamal) (Reinach 1893) and 1 at Tiddis (Fig. 26). The examples in the Museo Archeologico Nazionale at Naples and in the Museo Archeologico Etneo of Adrano in Sicily mentioned by Williams-Thorpe (1988, 260) do not concern a catcher *stricto sensu*: catcher and *meta* in the example of Naples are made of one piece of lava (Moritz 1958, 76-7, 94) while at Adrano a lava trapetum is exposed next to a stone oil press bed and a very small *meta*. For this reason nos. 1 and 2 are in brackets (Fig. 26). Dimensions of these receptacles seem to be very standard: data taken from published examples show outer base widths between 105-120 cm, inner hole widths between 45-63 cm and heights between 20-31 cm (only the example of north-eastern Algerian survey OH005 is very high with 47 cm).

In the typology proposed by Mattingly and Hitchner the two types of flour/oil catcher discussed in our paper are presented as olive grinders or crushers (Mattingly and Hitchner 1993, 443-4, Fig. 3, type 2 with central hole and type 3 with central depression) together with the African variant of the *trapetum* with central cylindrical pier (Fig. 3, type 1). The authors suppose a separate stone column or wooden beam in the centre pivoted with the roof structure and did not understand that a *meta* was placed in the central hole, probably because in the Kasserine survey they did not find examples on which they could have seen circular traces of the rotating *catillus*. The typology Mattingly- Hitchner is accepted by Frankel (1993, 480, note 19) and Biagini, Coletti, Vismara (Vismara 2008, 200, 272 F13, 274 F24, F26, 277 F54, 284-5 fig. 10.16-9, 452-4), for the 4 'flour/oil' catchers of Uchi Maius, defined 'sottomola' and 'macina a corona circolare'.

Provenance of flour catchers

All the discovered receptacles are made of local stones: limestone in north-west Tunisia, sandstone in north-eastern

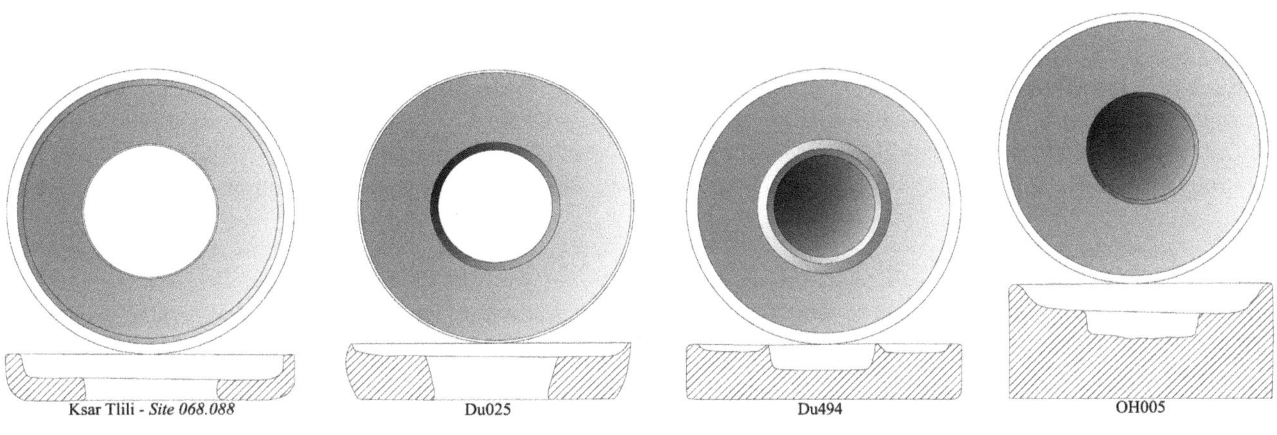

Figure 25. Typological scheme of recorded flour catchers and published examples. Scale 1 m.

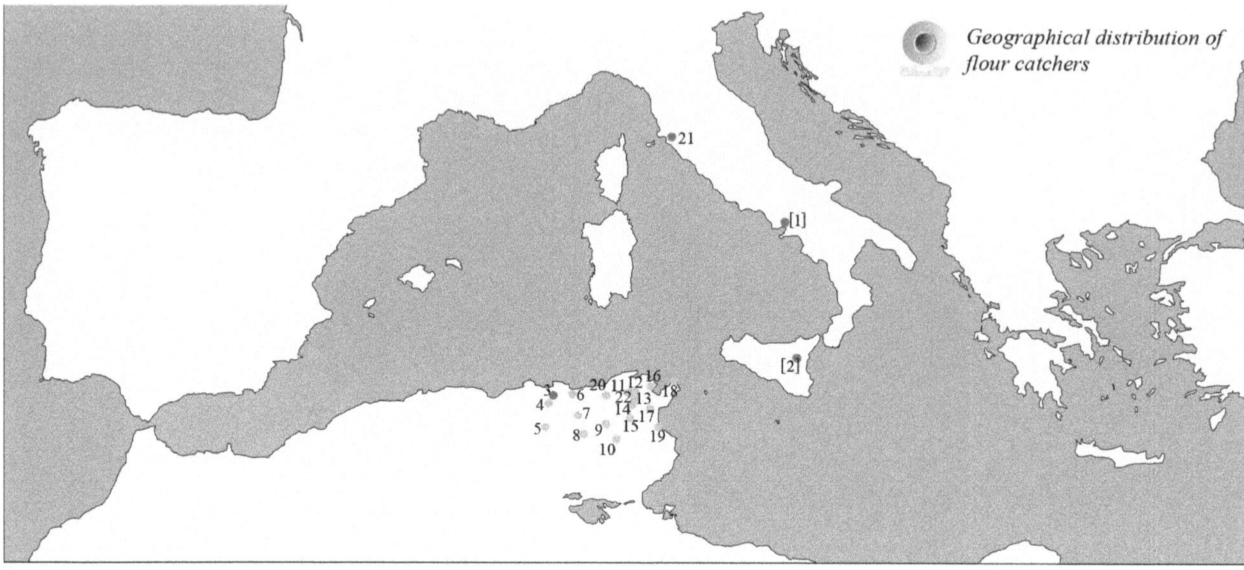

[1. Naples, 2. Adrano (CT)], 3. St. Charles (Ramdan Jamal), 21. Vetulonia (1)

4. Tiddis (3), 5. Thamugadi (1), 6. Hippo Regius (1), 7. Madauros (6), 8. Theveste (8), 9. Thala (1), 10. Ksar Tlili (2), 11.Thubursicu Bure (1), 12. Tignica (2), 13. Thugga (4), 14. Mustis (5), 15. Rohia (3), 16. Carthage (2), 17. Thuburbo Maius (1), 18. Sidi Daoud (1), 19. Sidi Ali Bou Kachabia (1), 20. Bulla Regia (2), 22. Uchi Maius (4)

Figure 26. Distribution of flour catchers. Updated map of Williams-Thorpe 1988, Fig. 3b.

Algeria. Also the other mentioned examples in Tunisia are of limestone.

Kneading machines or *machinae, quibus farinae subiguntur* (Paul. *sent.* 3.6.64)

A kneading machine or dough mixer was found in the dismembered farm of site Du369 (de Vos 2007, 48-50, fig. 5) (Figs. 27 and 28). The position of the farm in the area where the important Carthage-Theveste road enters the Khalled valley explains the presence of a bakery in this point of intensive circulation, in or near *Civitas Mizigitanorum* (The inscription AE 1921.42 walled into a French farm in site Du385, 900 m distant from Du369 mentions the toponym of this local Berber settlement containing the ethnic name Amazigh, used by the Berbers as their self-definition par excellence: Ghazi-Ben Maïssa 2006).

Cereal culture dominates the valley nowadays, the slopes are occupied by olive groves. A very damaged and doubtful fragment has been found in site Du031, nearly 1 km to the north-east of Thugga. Normally farms or villas are not provided with mechanical dough-kneaders: they are required on sites with a relatively high concentration of population, e.g. in Thugga there is a kneading machine near the Byzantine fortress, in Mustis there are 6, in Hippo Regius 3 (of white limestone), and in the Solb region, at Gigthi (Drine 2001, Solb: 254, figs 5-6; Gigthi: 255 figs 7-8. Wheat had and has to be imported in this region whose yearly rainfall of 200 mm is too low for corn growing). Examples are also known from Sabratha (Wilson and Schörle 2009, 121, note 16), Tomis (Romania) on the Black Sea, Theveste, Thagaste, Thamugadi, Thibilis have each one, while at Volubilis there are 17 in bakeries and 19 in private houses (Leduc 2008, 479, 481, 490-92, out of 36 mixers 3 are not mechanical; Riße 2001, 103, fig. 156 (one square, one round); 104, fig. 158). Volubilis is quite exceptional for the high number of both oil press rooms and bakeries in the houses. Fourteen bakeries in Volubilis (22 have been excavated) are equipped with 14 'donkey' mills and 17 mixers, 15 houses with one mixer, two mixers are scattered *ex situ*. Twenty two bakeries in Pompeii (44 have been excavated) are equipped with 79 'donkey' mills and 14 mixers (Mayeske 1989, 169; Mayeske's list of 20 bakeries has to be updated with VII,15,1-2/15, discussed by Franklin 1990, 35-42; Bakker et al.1999, 11 note 58; and with the bakery recently excavated in the Casa dei Casti Amanti IX 12 containing 4 'donkey' mills and one mixer). Mayeske's list of 8 mixers has to be updated with this mixer and the items in bakeries VI 3,3.27 (Mayeske 97, but Mazois who assisted at the excavation and described it in detail, does not mention stone vats) and VI 11, 9-10 published by Mau 1886, 46, pl. III.1; Strocka 1992; in bakeries VII 2,20-22, Mau 1886, 47, pl. III.4, and V 4, 1-2. Bakker et al. 1999, 11; Leduc 2008, 490 list only respectively 6 and 3 mixers). One damaged mixer was been reused as stepping stone on the street in front of the House of the Bear VII 2, 46 and one possible mixer is reused in the counter desk of shop VII 3,3. The relatively low number of mixers discovered at Pompeii may be due to three factors:

1. the holes in the vertical wall make them vulnerable,
2. they are identified as mixers only in 1886, so 130 years after the start of the excavation of Pompeii;
3. during post-eruption research of precious and recyclable material they have been removed through tunnels.

In the seven or eight bakeries that have been identified at Ostia there may have been 60 to 65 millstones, but many were removed in late Antiquity, when the city was abandoned and transformed into a luxurious residential

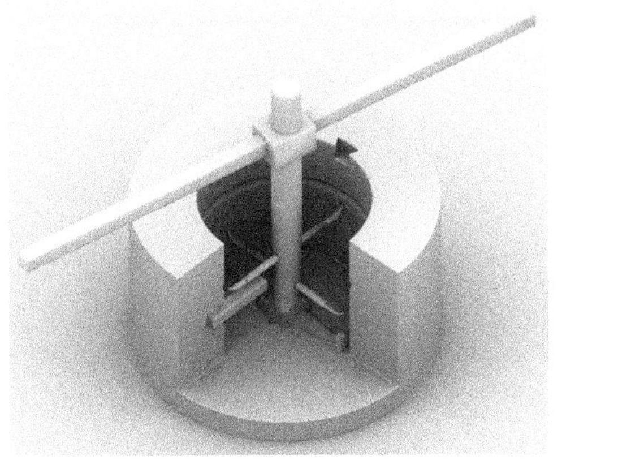

Figures 27 (above) and 28 (right). Tunisia, Khalled valley, Du369. Kneading machine and its reconstruction.

town (Bakker *et al.* 1999, 113). The 13 mixers preserved in the 7 or 8 extant bakeries of Ostia are made of black lava and one of travertine: in North Africa they are of limestone. The two big bakeries at Ostia, that each contains 5 and 6 kneading machines and respectively 10-12 and 8 mills, baked bread for the fisc (Bakker *et al.* 1999, 1-3, 59, 78, 89, 98, 99; Zevi 2008, 497-501). Two mills could deliver flour for one kneading machine. The smaller, but more numerous Pompeian bakeries are provided with three - four 'donkey' mills, had only one mixer. Eight villas out of 25 with published inventories explored in the territories of Pompeii and Stabiae were equipped with a 'donkey' mill and an oven, but none was provided with a kneading machine. The late-Republican or early Augustan Tomb of Eurysaces near Porta Maggiore in Rome was composed of at least 12 x 3 vertical and 30 horizontal monumental kneading machines (on three sides; the destroyed fourth side is reconstructed by Canina 1850 (Ciancio Rosetto 1973, pl. 4.1) with the portraits of the baker and his wife and 6 other horizontal mixers. Canina interpreted the travertine cylinders of the tomb as hand *mortaria*, mentioned by Cato *agr.* 74 and in Moretum 92, but misunderstood the function of the square mortise in the bottom (Ciancio Rosetto 1973, 34, note 27; Petersen 2006, 84-122). The tomb is the oldest evidence of these tools, so it looks almost like a celebration of the invention of the mechanical mixer. So far no kneading machine older than the ones of Eurysaces' tomb has been excavated. Vitruvius' reference 10.5.2 (Wilson 2008 *vs.* Curtis 2001, 363-4, note 75) to water powered dough kneading is contemporary to the tomb. Cato *agr.* 74 recommends to knead flour in a mortarium adding gradually water (*farinam in mortarium indito, aquae paulatim addito subigitoque*). Apart from Vitruvius and Paulus' *sententia* 3.6.64 *machinae, quibus farinae subiguntur* the mixer is not mentioned in ancient literature.

Archaeology owes the identification of the kneading machine to August Mau. In his description of the discovery in 1878 of bakery Pompeii VI 14, 34, he did not yet understand the functioning of the square mortise in the inner bottom of the lava cylinder, but he noticed that it was humanly driven, because the floor was not paved with basalt flags (Mau 1878, 196: '*pare che fossero una specie di molini, il cui meccanismo però non mi è chiaro, girati probabilmente da un uomo, giacché non vi è il selciato*'). Updating Overbeck's monograph on Pompeii in 1884, Mau proposes the hypothesis that the vessel was a mixer (Overbeck and Mau 1884, 390). The hypothesis was corroborated a year later when two examples in the Casa del Labirinto (VI 11, 9) and the Casa di Popidius Priscus (VII 2, 22) were cleared from volcanic debris fill, on Mau's demand. Here he discovered the iron bar and socket fixed in the bottom of the container and wooden remains of the vertical beam (Mau 1886 contains plan and section of 4 mixer vessels and iron elements). The reconstruction drawing of the wooden framework appears only some 15 years later. He was also the first who proposed in 1886 the correct interpretation of the scene in the frieze of Eurysaces' tomb, where a donkey driven mixer is represented.

Conclusions. Typology, chronology and distribution

Distribution pattern

The north-eastern Algeria survey results show that 65% of all millstones are spread in a small northern part of the surveyed area, to the north-east of El Tarf, while the remaining 35% of examples have been found in a vast area between El Tarf and Souk Ahras (Fig. 29). Surprisingly, 70% of all recorded hand-mills belong to the northern part of the survey. It is also remarkable that these elements are very similar in their features, demonstrating a sort of microregionality of millstone production (see below). *Metae* of 'donkey' mills were found sporadically in all areas, but more than the half of elements were recorded in the mentioned northern part. *Catilli* are very rare, due to their major fragility; a feature noticed in many surveys.

In the southern area of the survey, between El Tarf and

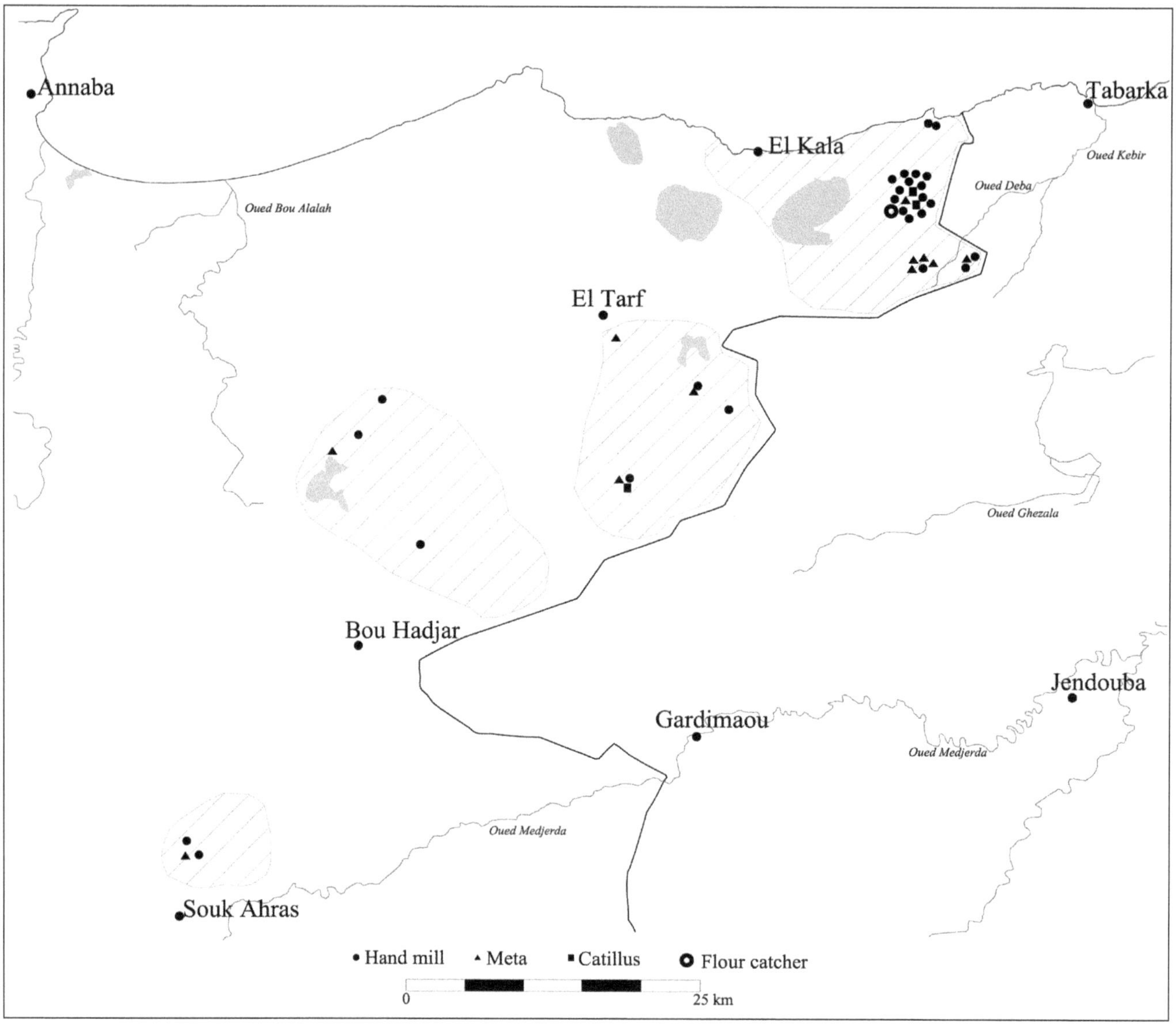

Figure 29. Distribution map of north-eastern Algeria survey millstones. Surveyed areas are hatched.

Souk Ahras, only 8 hand-mills, 5 *metae* and 1 *catillus* were seen, even if visibility is better than in the northern woodland. In the southern area, especially the territories of Ain Karma (in 32 sites only one fragmentary hand-mill was found), Bogous (in 46 sites 2 hand-mills and 1 *meta* were discovered) and Zitouna (out of 30 sites only 1 site has a *meta*, a small part of a *catillus* and a fragmentary hand-mill) are distinguished by the massive presence of big bedrock cut wine or oil presses. The intensive Thugga survey (69 km²) produced 36% (26 elements out of totally 73), so three times as many as the extensive one (Fig. 30). The whole area is densely urbanized: nine cities, two *pagi*, two *civitates*, one *res publica* with epigraphically attested toponyms. Of other villages such as sites Du005 and Du282 provided with a church we do not know the toponym nor their status. The rural settlement pattern is very dense, the sites are regularly distributed. And so are the finds, a trend of mill concentration, however, can be noted in greater settlements, that are also situated near good soils, springs and/or torrents, e.g. Maatria valley (sites Du508, Du509, Du512: 7 mill elements), site Du005 (5 items) in the fertile, but small Arkou valley, Du049 (5/6 elements) at 2 km from Thugga.

Chronology

Unfortunately, the mills, which are found generally on the surface, are not datable. In one case an upper stone of a hand-mill was reused as building material in a wall of the excavated area of the Byzantine farm at Ain Wassel. A 'flour' catcher broken in two pieces leaned against the wall in two different rooms of the farm, testifying to the use and the end of this particular common tool in Northern Africa in the 7th century AD (de Vos 2000, fig. 58.10-11; 2004, 37).

Scholarship is still discussing the origin of rotary mills, which seem particularly present in Punic or Punic influenced areas: the discovery of a fragmentary rotary hand-mill of grey lava in a closed tomb at Byrsa

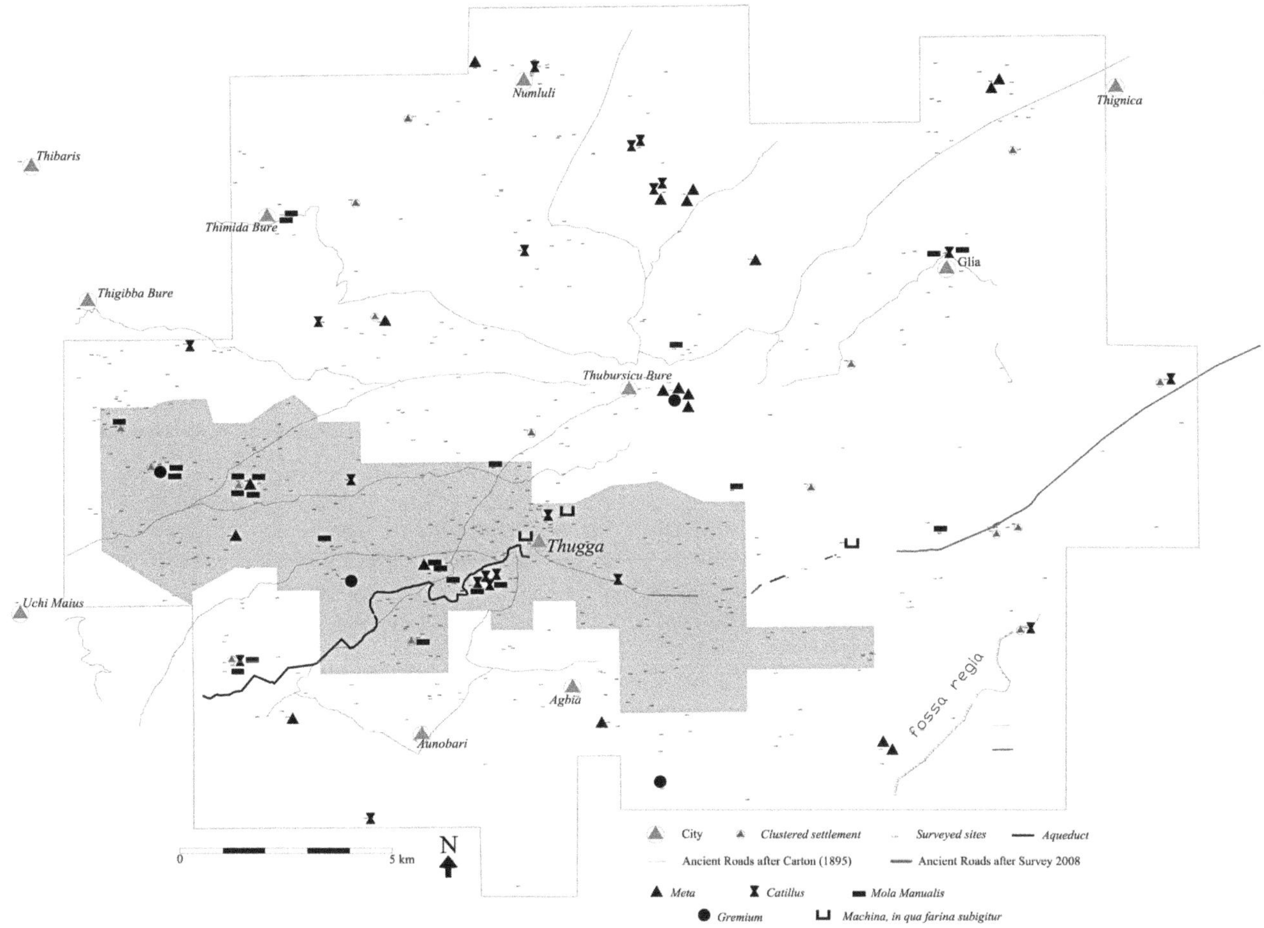

Figure 30. Distribution map of Thugga survey millstones. The grey area concerns intensive survey.

stratigraphically dated by pottery and a scaraboid seal to the end of the 6th century BC (Morel 2001) could confirm an eventual Punic origin (As supposed earlier by Py 1992; Brun 1997, 73-4; and now by Curtis 2001, 343; 2008, 375). The examples dating to the 5th century BC found in Spain induced Alonso Martínez to propose an Iberic origin of rotary hand-mills (Alonso Martínez 1996a; 1996b), but the chronology is not very sure and the introduction of the rotary hand-mill in the Iberic peninsula can be attributed to the Phoenician-Punic dominating influence in the western Mediterranean. Other examples of rotary 'donkey' mills have been recorded in Phoenician and Punic influenced areas at Motya (destroyed in 397 BC), at Megara Hyblea (destroyed in 214 BC) at Morgantina (destroyed in 211 BC) and Byrsa (1st half of 2nd century BC) (White 1963, 205; Lancel 1982, 95; Peacock 1989, 213).

The most ancient datable Morgantina Type *catillus* 3.2 (375-350 BC) provided with 3-sided lever slots instead of 4 protruding sides (Crisafulli Typ 3.2) (Crisafulli 1996 in Sposito 2008, 219 pl. XLIII). Only one example of this type of lever slot was found at Morgantina vs. 11 4-sided lever slots *catilli*), found in the Sec wreck near Mallorca is made of red Sardinian ignimbrite. This demonstrates again a potential Punic origin of rotary mills. The mentioned examples suggest a pre-Roman date for at least 3 *catilli* of the Thugga survey (Du395, Du508 and one scatter find) and for the ones in the two punic ports of Rusicade and Hippo Regius (Rusicade (now Skikda) port on the ancient *Sinus Numidicus*, in punic means Fire cape; the surname Regius 'of the King' was bestowed). Indeed at least one of the three find places near Thugga is surely of pre-Roman origin: the Numidian hill top settlement Glia (Du395) which dominates the pass and entrance to the Khalled valley, where two rims of black glazed ware have been found (Probably of local production similar to class Byrsa 401: Morel 1982, 52-7; Ferchiou 1990, 240-3).

Typology

The flour catcher was used also as oil catcher, it collected flour or compacted olives milled by a *catillus* rotating around a *meta*; consequently it cannot be considered as grinder or crusher and has to be cancelled from the typology of *molae oleariae* (Fig. 31).

The detailed type-analysis of discovered hand and 'donkey' mills in the surveyed areas in Tunisia and Algeria shows some interesting differences:

1. Uniformity of measure, form and material of hand-mills in the north-eastern Algeria survey (26 examples);

Figure 31. Complete mill discovered at Ramdan Jamal (St. Charles) in Algeria (Reinach 1893, Pl.XV).

2. Heterogeneous measures and forms of 'donkey' mills in the north-eastern Algeria survey (11 examples);
3. Relatively high number of 'donkey' mills in the small surveyed Tunisian area (371 km2 = 42) in comparison to the north-eastern Algeria surveyed area (1338 km2 = 14);
4. Heterogeneous measures, forms and material of hand (27 examples) and 'donkey' mills in the Thugga survey.

Possible explanations:
Ad 1. The grinding capacity of the local conglomerate with protruding small quartz pebbles is an excellent substitute for vesicular lava; cultural and social reasons, i.e. the Berber tribal population dispersed in a vast countryside without cities was less permeated by external influences and has conserved its identity in a higher degree, as demonstrated by the plethora of funerary Libyc inscriptions found in the same surveyed area.
*Ad*3. The 167 olive grinders in the north-eastern Algerian survey and their total absence around Thugga can explain the presence in Tunisia of more 'donkey' mills, eventually used also for olive compacting.
*Ad*4. Thugga region as part of the *pertica Carthaginiensium* had a mixed population of indigenous *peregrini* and Roman Carthaginian citizens and in this way it was integrated in the long distance trade circuit of the metropolis. The close relationship between Carthage and Sardinia started in the 6th century BC and continued through the Byzantine Exarchate (7th century AD) (Peacock 1980, 50).

Acknowledgements

The research has been conducted collectively by the three authors in the Laboratory of Archaeology of Trento University and on the field in Tunisia in collaboration with the Institut National du Patrimoine Tunis (co-director Mustapha Khanoussi), in Algeria with the Centre National de Recherche en Archéologie, Ministère de la Culture, Algiers. The competences of the individual authors of the paper regard statistical evaluation of data, typology and drawings of the mills (Martina Andreoli), survey methodologies, cartography and ethnoarchaeology (Redha Attoui), interregional framework, historical contextualization and photography (Mariette de Vos).

References

Akerraz, A. and Lenoir, M., 1981-82. Les huileries de Volubilis. *Bulletin d'Archéologie Marocaine*, 14, 69-101.

Alonso Martínez, N., 1996a. Origen y expansión del molino rotativo bajo en el mediterráneo occidental. In D. Garcia and D. Meeks (eds.), *Techniques et économie antiques et* médiévales. Le temps de l'innovation, Colloque d'Aix-en-Provence. Paris, 15-9, 235.

Alonso Martínez, N., 1996b. Els molins rotatius. origen i expansió en la Mediterrania occidental. *Rivista d'Arqueologia de Ponent*, 6, 183-98.

Antonelli, F. and Lazzarini, L., 2010. Mediterranean trade of the (most widespread) roman volcanic millstones from Italy and petrochemical markers of their raw materials. *Journal of Archaeological Science,* 37.9, 2081-92.

Anselmino, L., Bouchenaki, M. and Carandini, A., (eds.) 1989. *Il Castellum del Nador, Storia di. una fattoria tra Tipasa e Caesarea (I-VI sec. d.C.).* Roma.

Aounallah, S. and Ben Abdallah, Z., 1997. Les Calpurnii de Thugga. In M. Khanoussi, and L. Maurin, (eds.), *Dougga (Thugga) Études Épigraphiques.* Paris, 77-96.

Bakker, J.Th., Van Dalen, J.H., Heres, Th.L., Meijlink, B. and Sirks, A.J.B., 1999. In J.Th. Bakker (ed.), *The Mills-Bakeries of Ostia. Description and Interpretation.* Amsterdam.

Beltrame, C. and Boetto, G., 1997. Macine da relitti. In P.A. Gianfrotta and P. Pelagatti (eds.), *Archeologia subacquea. Studi, ricerche e documenti* II. Roma, 167-96.

Ben Abdallah, Z. and Maurin, L., 2000. Dédicace à Auguste divinisé et à l'empereur Claude par Iulius Venustus. In M. Khanoussi, and L. Maurin, (eds.), *Dougga, Fragments d'histoire.* Bordeaux Tunis, 137-42.

Ben Baaziz, S., 2000. *Rohia et le Sraa Ouertane dans l'Antiquité.* Tunis.

Ben Hassen, H., 2006. Thignica *(Aïn Tounga) son histoire et ses monuments.* Ortacesus.

Blümner, H., 1912. *Technologie und Terminologie der Gewerbe und Künste bei Griechen und Römern* I (2nd edition). Leipzig-Berlin.

Bonini, P. and Rinaldi, F., 2003. Gli ambienti di servizio. In S. Bullo, and F. Ghedini (eds.), *Amplissimae atque ornatissimae domus. saggi.* Roma, 189-220.

Brun, J.-P., 1997. L'introduction des moulins dans les

huileries antiques. In D. Garcia and D. Meeks (eds.), *Techniques et économie antiques et médiévales. Le temps de l'innovation*, Colloque d'Aix-en-Provence. Paris, 69-78.

Brun, J.-P., 2004. *Archéologie du vin et de l'huile dans l'Empire romain.* Paris.

Canina, L., 1850. *Indicazione topografica di Roma antica in corrispondenza dell'epoca imperiale.* Roma.

Carton, L., 1895. *Découvertes épigraphiques et archéologiques faites en Tunisie (région de Dougga)*. Paris.

Ciancio Rosetto, P., 1973. *Il sepolcro del fornaio Marco Virgilio Eurisace presso Porta Maggiore.* Roma.

Crisafulli, M., 1996. Le macine a tramoggia biconica nel sito di Morgantina. In Sposito 2008, pls XLIII-XLIV.

Curtis, R.I., 2001. *Ancient food technology*. Leiden, Boston, Köln.

Curtis, R.I., 2008. Food processing and preparation. In J. P. Oleson (ed.), *Engineering and technology in the Classical World.* Oxford, 369-92.

De Cou, H. F., 1912. Antiquities from Boscoreale in Field Museum of Natural History. *Field Museum of Natural History, Publication 152 Anthropological Series*, 7.4, 149-212.

Delamare, H., 1850. *Exploration Scientifique de l'Algérie. Archéologie*. Paris.

Desanges, J., 1974-75. Un curateur de la sauterelle sur la *pertica* de Carthage en 48/49 de notre ère. *Bulletin archéologique du C.T.H.S.,* nouv. Sér., fasc. 10-11b, 136-40.

Dietz, S., 1995. A Summary of the Field project. In S. Dietz, L. Ladjimi Sebaï and H. Ben Hassen (eds.), *Africa Proconsularis* II. *Regional Studies in the Segermes Valley of Northern Tunisia.* Copenhagen, 771-99.

Drine, A., 2001. Meules à grain et pétrins autour du lac El Bebèn et à Gigthi. In J.-P. Brun and P. Jockey (eds.), Technia. *techniques et sociétés en Méditerranée.* Paris, 251-57.

Drine, A., 2009. Stone objects. In A. Drine, E. Fentress and R. Holod (eds.), *An Island through time. Jerba studies* 1, *the Punic and Roman periods*, JRS, Suppl. 71, 330-34.

Falchi, I, 1894. Vetulonia. Scavi della necropoli vetuloniese durante l'anno 1893. *Notizie degli Scavi*, 335-60.

Ferchiou, N., 1990. L'habitat fortifié pré-impérial en Tunisie antique. Carthage et son territoire dans l'antiquité. In *Histoire et archéologie de l'Afrique du Nord*, Actes du IVe colloque international réuni dans le cadre du 113e Congrès national des Sociétés savantes, Strasbourg 1988, I, Paris, 229-52.

Fiorelli, G., 1873. *Gli scavi di Pompei dal 1861 al 1872.* Napoli.

Frankel, R., 2003. The Olynthus Mill, its Origin, and Diffusion. Typology and Distribution. *American Journal of Archaeology*, 107.1, 1-22.

Franklin, J.L., 1990. *Pompeii. The 'Casa del Marinaio' and its History*. Roma.

Ghedini, F., 2003. La casa romana in Tunisia fra tradizione e innovazione. In S. Bullo, and F. Ghedini (eds.), *Amplissimae atque ornatissimae domus. saggi.* Roma, 315-56.

Gerner Hansen, C., 1995. Architectural Studies. In S. Dietz, L. Ladjimi Sebaï and H. Ben Hassen (eds.), *Africa Proconsularis* I. *Regional Studies in the Segermes Valley of Northern Tunisia.* Copenhagen, 177-379.

Ghazi-Ben Maïssa, H., 2006. Et si on désignait ces fameux *Berbères* par leur vrai nom? In *L'Africa Romana*, Atti del XVI convegno di studio, Rabat 2004. Roma, 16.3, 2089-2108.

Guichard, P. and Djerrab, A., 2009. La ville disparue de Majjâna et la 'Montagne des Meules'. In *Colloque International sur l'Archéologie*, 25-9 avril 2009. Tebessa.

Houamria, S., 2009. UBUS I, Survey dans la région de Bouchegouf. Prospection extensive et documentation systématique du paysage antique de la Numidie orientale. In *Colloque International sur l'Archéologie*, 25-9 avril 2009. Tebessa.

Junkelmann, M., 1997. Panis militaris. *Die Ernährung des römischen Soldaten oder der Grundstoff der Macht.* Mainz am Rhein.

Lancel, S., 1982. 1.4 L'Ilot D. In S. Lancel (ed.), *Byrsa* II. *Rapport préliminaire des fouilles 1977-78. niveaux et vestiges puniques*. Rome-Tunis, 85-105.

Laudani, C., 2004. Moretum, *Introduzione, testo, traduzione e commento.* Napoli.

Leduc, M., 2008. Les *pistrina* volubilitains, témoins majeurs du dynamisme économique municipal. In *L'Africa Romana. Le ricchezze dell'Africa. Risorse, produzioni, scambi*, Atti del XVII convegno di studio, Sevilla 14-17 dicembre 2006. Roma, 475-505.

Luquet, A., 1966. Blé et meunerie à Volubilis. *Bulletin Archéologique Marocaine*, 4, 302-15.

Mattingly, D., 1996. Olive Presses in Roman Africa. Technical Evolution or Stagnation? In *L'Africa romana*, Atti del XI convegno di studio, Cartagine 1994. Ozieri, 577-95.

Mattingly, D. and Hitchner, B., 1993. Technical Specifications for some North African olive presses of Roman date. In M.-C. Amouretti and J.-P. Brun (eds.), *La production du vin et de l'huile en Méditerranée*, BCH Suppl. 26, Paris, 439-62.

Mau, A., 1878. Scavi di Pompei reg. VI is. 14. *Bullettino dell'Instituto di Corrispondenza Archeologica*, 186-204.

Mau, A., 1886. Su certi apparecchi nei pistrini di Pompei. *Mitteilungen des Deutschen Archaeologischen Instituts. Römische Abteilung*, 1, 45-8, pl. III.

Mau, A., 1899. *Pompeii. its life and art.* New York.

Maurin, L., 2000. Immunitas perticae Karthaginiensium.

In M. Khanoussi, and L. Maurin, (eds.), *Dougga, Fragments d'histoire*. Bordeaux Tunis, 145-47.

Mayeske, B.J., 1972. *Bakeries, bakers and bread at Pompeii. a study in social and economic history.* Ann Arbor.

Mazois, F., 1824. *Les ruines de Pompéi*, II partie. Paris.

Morel, J.-P., 1982. La céramique à vernis noir de Carthage-Byrsa. nouvelles données et éléments de comparaison. In *Actes, Colloques sur la céramique antique.* Carthage 23-24 juin 1980. Tunis, 43-76.

Morel, J.-P., 2001. Aux origines du moulin rotatif? Une meule circulaire de la fin du VIe siècle avant notre ère à Carthage. In J.-P. Brun and P. Jockey (eds.), Technia. *techniques et sociétés en Méditerranée.* Paris, 241-50.

Moritz, L.A., 1958. *Grain Mills and Flour in Classical Antiquity*. Oxford.

Oettel, A., 1996. *Fundkontexte römischer Vesuvvillen im Gebiet um Pompeji.* Mainz.

Overbeck, J. and Mau, A., 1884. *Pompeji in seinen Gebäuden, Alterthümern und Kunstwerken.* Leipzig.

Peacock, D.P.S., 1980. The Roman Millstone trade. A petrological sketch. *World Archaeology*, 12, 43-53.

Peacock, D.P.S., 1989. The mills of Pompeii. *Antiquity*, 63, 205-14.

Perthuisot, V., 1979. *Carte Géologique de la Tunisie, feuille n° 33, Teboursouk.* Tunis.

Perutelli, A., 1983. *Moretum*. Pisa.

Petersen, L.H., 2006. *The Freedman in Roman Art.* Cambridge.

Py, M., 1992. Meules d'époque protohistorique et romaine provenant de Lattes. *Lattara*, 5, 186-236.

Reinach, S., 1893. Moulin à grain découvert à Saint-Charles. *Bulletin archéologique du Comité des travaux historiques et scientifiques*, 149-50, pl. XV.

Riße, M., (ed.) 2001. *Volubilis. Eine römische Stadt in Morokko von der Frühzeit bis in die Islamische Periode.* Mainz am Rhein.

Šebesta, G., 1977. *La via dei mulini dall esperienza della mietitura all'arte di macinare (molinologia).* Trento.

Sposito, A., 2008. Le macine nell'Antichità. In A. Sposito, *Tecnologia Antica. Storie di procedimenti, tecniche e artefatti.* Palermo, 193-220.

Strocka, V.M., 1992. *Casa del Labirinto (6.11,8-10).* München.

Vismara, C., (ed.) 2007. *Uchi Maius 3. Frantoi.* Miscellanea, Sassari.

de Vos, M. and de Vos, A. 1982. *Pompei, Ercolano, Stabia.* Roma–Bari.

de Vos, M., 2000. Rus Africum. *Terra acqua olio nell'Africa settentrionale. Scavo e ricognizione nei dintorni di Dougga (Alto Tell tunisino)*, Labirinti 50. Trento.

de Vos M., (ed.) 2004. *Archeologia del territorio. Metodi materiali prospettive: Medjerda e Adige. Due territori a confronto*, Labirinti 73. Trento.

de Vos, M., 2007. Olio per Roma e per il mercato intraregionale. In E. Papi (ed.), *Proceedings Supplying Rome and the Roman Empire,* Certosa di Pontignano, 2nd-4th May 2004, JRS Suppl. 69, 43-58.

White, D., 1963. A survey of millstones from Morgantina. *American Journal of Archaeology*, 67, 199-206.

Williams-Thorpe, O., 1988. Provenancing and Archaeology of Roman Millstones from the Mediterranean Area. *Journal of Archaeological Science*, 15, 253-305.

Williams-Thorpe, O. and Thorpe, R.S., 1989. Provenancing and Archaeology of Roman Millstones from Sardinia (Italy). *Oxford Journal of Archaeology*, 8, 89-117.

Williams-Thorpe, O., and Thorpe, R.S., 1990. Millstone Provenancing used in tracing the route of a fourth-century BC Greek merchant ship. *Archaeometry*, 32.2, 115-37.

Williams-Thorpe, O. and Thorpe, R.S., 1993. Geochemistry and Trade of Eastern Mediterranean Millstones from the Neolithic to Roman Periods. *Journal of Archaeological Science,* 20, 263-320.

Wilson, A. and Schörle, K., 2009. A baker's funerary relief from Rome. *Papers of the British School at Rome*, 77, 101-23.

Wilson, A., 2008. Dough-Kneaders. In J.P. Oleson (ed.), *Engineering and technology in the Classical World.* Oxford, 358.

Zevi, F., 2008. I collegi di Ostia e le loro sedi associative tra Antonini e Severi. In C. Berrendoner, M. Cébeillac-Gervasoni, L. Lamoine (eds.), *Le quotidien municipal dans l'Occident romain*, Clermont-Ferrand, 477-505.

Volcanic quern and millstone quarries in Cabo de Gata (Almería) and Campo de Calatrava (Ciudad Real), Spain

Timothy J. Anderson, Tor Grenne and Juan Manuel Fernández Soler

1. Introduction

Research on the provenance of querns and millstones in antiquity in the framework of studies defining ancient trade routes have multiplied in Europe over the last 30 years. A specific emphasis has been placed on the study of vesicular basaltic rocks, a type of rock particularly sought after in antiquity due it its grinding properties. The fact that outcrops of this rock are restricted to certain regions, combined with specific petrographic and chemical properties or "signatures", makes it possible to determine their origin. Peacock's seminal paper of 1980 defined the volcanic sources recognised in Europe at that time. He identified various petrographic groups spread across Europe such as Orvieto on mainland Italy, Mount Etna in Sicily, Mulargia in central Sardinia, Volvic in central France, and the Eifel in western Germany, as well as other less well defined sources in the western Mediterranean. However, due to the absence of research at that time, Peacock could not cite the volcanic sources in the Iberian Peninsula (Peacock 1980, 50).

The first indication of an Iberian volcanic source, based on petrographic and chemical analyses of rotary querns from regional Catalan museum collections, dates to 1987. These analyses point to the Olot-Garrotxa volcanic province (Fig. 1) (also referred to as the north east volcanic province) in north east Catalonia (Williams-Thorpe and Thorpe 1987). Volcanologically this area is characterised by Strombolian cones and lava flows, in addition to phreatomagmatic deposits, formed 11,500-700,000 years ago (in Quaternary times) and ranging in composition from leucite basanite to alkali olivine basalt (e.g. Palli and Pujadas 1999; Cebriá *et al.* 2000). This source designation has been widely embraced in later literature (e.g. Alonso 1997; Alonso 1999, 261-265; Portillo 2000, 25-25, 86), although no specific extraction site has been identified in the field. Williams-Thorpe and Thorpe (1987) and Williams-Thorpe (1988) suggested that possible quarries were most likely shallow workings in vesicular flow-tops and large volcanic boulders and hence leave few traces in the terrain. However, we speculate that columnar jointed deeper parts of lava flows may have been equally favourable, particularly in view of later studies in the volcanic Eifel region (Germany) where millstone exploitation has taken place since Neolithic

Figure 1. Map of the Iberian Peninsula with the volcanic regions mentioned in the text (Cabo de Gata, Campo de Calatrava, Garrotxa-Olot, and Lisbon) and the principal sites mentioned in the text.

times (e.g. Harms and Mangartz 2002). At the medieval town of Castellfollit de la Roca, for example, steep cliffs expose lava with vertical columnar jointing similar to that at Eifel. If ancient millstone quarries did exist in the area, it is possible that they were destroyed by more recent quarry work, such as the exploitation still active today around Castellfollit. In any case, extraction features and marks left in these types of ancient exploitations are not simple to recognise.

Williams-Thorpe and Thorpe (1987, 52, 56, fig. 4) and Williams-Thorpe (1988, 268, fig. 4) also refer fleetingly to other potential sources in the Iberian Peninsula such as the area around Lisbon in Portugal (cf. Fig. 1), the southeast corner of Andalusia and, although indirectly and not by name, the vast area of Central Spain known as the Campo de Calatrava. The Lisbon volcanic province remains unconfirmed with respect to millstone exploitation. Lavas of alkali basaltic to basanitic composition are known (Rock 1982), but we ignore whether or not they were suitable for quern production. Their much higher age (ca. 73 million years, i.e., Late Cretaceous) suggests that their vesicles may be filled with secondary minerals, which would greatly reduce their grinding properties, compared to young porous lavas. Also, we are not aware if any

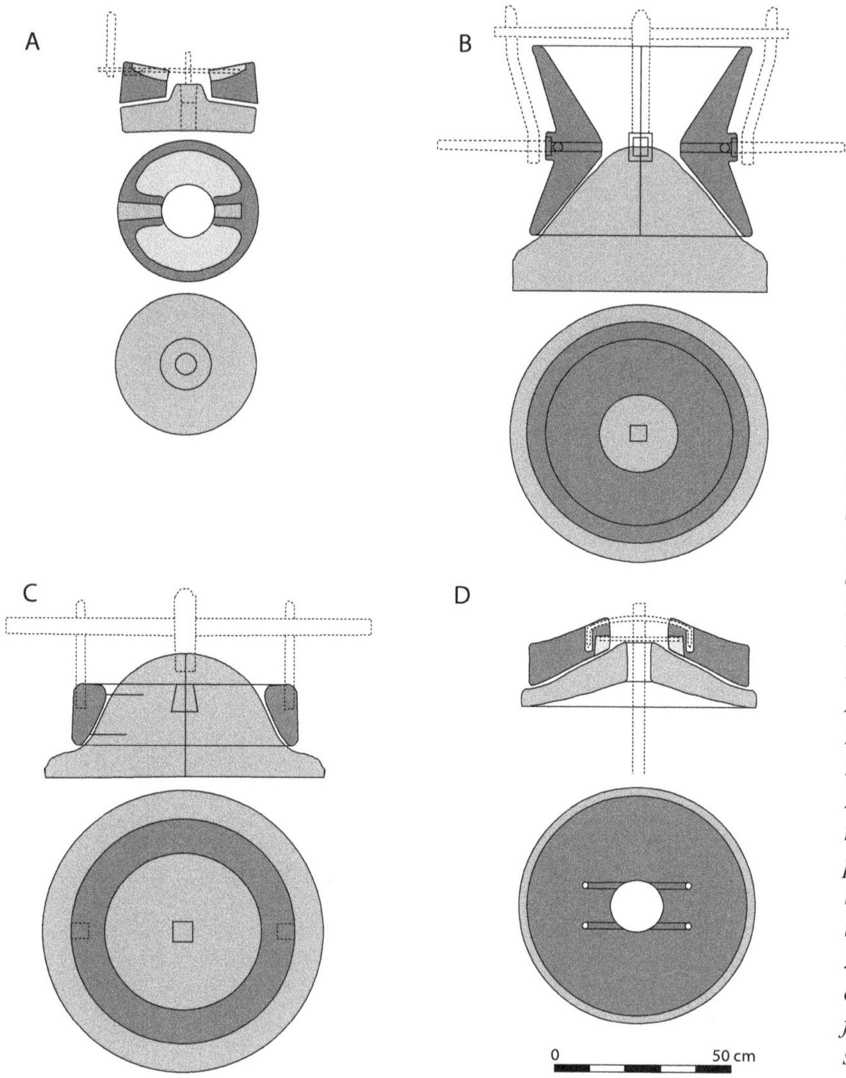

Figure 2. Schematic view of the different categories of millstones mentioned in the text. A) hand-operated rotary quern based on the model from the Cerro de Limones quarry; B) "Pompeian" hourglass-shaped upper stone based on the sketch of an example from Córdoba (Berrocal 2006, 281, fig. 4.4), the lower stone is theoretical; C) the "Volubilis" mill type with a "ring-shaped" upper stone and bell-shaped lower stone is based on the sketch by Luquet (1966, 306, fig. 3b); D) hydraulic millstone based on the models of Barbegal, Avenches and Mesclans (Leveau 2006, 13-14; Castella 1994, 47, fig. 31; Brun and Borréani 1998, 297, fig. 25). Both systems for transmitting the motive force are illustrated; 1) vertical perforations on the upper surface interpreted as lodgings for a complex iron device, plugged with molten lead; 2) rynd cuttings on the lower surface of the catillus. Both systems are rarely featured on the same millstone. The first system is presumably older.

specific field surveys have been undertaken in that area with regard to ancient quarries. Furthermore, if this were an important centre of quern and millstone production, one would expect to find basaltic querns at the Roman centre of *Conimbriga* (Coimbra) about 200 km to the north. The excellent catalogue of querns and millstones undertaken by Borges (1978), despite offering more than 50 entries with rock descriptions, does not identify any such rocks. This absence, in our opinion, reduces the probability that the Lisbon volcanic province was a major quern and millstone production area.

The last two potential source areas alluded to by Williams-Thorpe and Thorpe (1987) is the focus of the present contribution. Here we will substantiate, through evidence based on the identification of quarries in the field, millstone production in both the Cabo de Gata area (Province of Almería) in the south eastern corner of Spain and Campo de Calatrava (Province of Ciudad Real) in central Spain (cf. Fig. 1). Furthermore, we will advance preliminary petrologic descriptions based on microscopic and chemical analyses of geological samples. All the same, research on this subject is still in an embryonic state and much remains to be done.

2. Mill types in the southern half of the Iberian Peninsula

2.1. Rotary querns

The first and the most elementary type of ancient mill, driven by a single person, is the rotary quern (Fig. 2a). The mill is composed of two cylindrical stones, the *catillus* (upper stone) and the *meta* (lower stone), and measures roughly 40 cm in diameter. When assembled, the mill has a proportion of diameter roughly twice that of its height. In the last few years, much has been written about its introduction during the Iron Age Iberian Culture - a major step forward in the evolution of mechanics. Several significant assemblages have been published, for the most part from settlements in north eastern Spain (Alonso 1999; Alorda Park Team 2002; Portillo 2006). For the subsequent periods, before and after the Roman conquest, with the exception of Numantia (Soria) in north-central Spain (Checa *et al.* 1999) and *Conimbriga* in Portugal (Borges 1978), little research has been undertaken on this subject and the typology of these small mills is still not fully established. There are examples of rounded "doughnut" shaped *catilli*, poorly dated, such as the fragment on display in the courtyard

Figure 3. Volcanic millstones. A) querns from Los Cazadores (Barranquete, Almería); Archaeological Museum of Almería (except, no. 3 unknown origin); average diameter 40 cm. No. 1 fractured during the piercing of the eye; no. 6 is a discarded cylinder; B) upper stone from the Historical Museum of Priego de Córdoba (Córdoba); C) lower stone from the Archaelogical Museum of Úbeda (Jaen); D) examples of bell-shaped lower stones from Oreto y Zuqueca (Granátula de Calatrava, Ciudad Real).

of the San Miguel Castle in Almuñecar, a shape resulting from being hewn out of rounded surface blocks probably collected in riverbeds. However, based on preliminary data of querns observed in several museums spread out over the eastern half of Andalusia, a more sophisticated, standardised model, most often of volcanic rock, appears to stand out for the Roman period (Fig. 3a). The *catillus* has a biconical section resulting from hollowed-out upper and lower faces. In the centre it has a large, circular eye. Some cases present well-defined rims along their circumference that surround hopper-like hollows resembling butterfly wings. Two radial rectangular cuttings placed opposite each other were presumably fitted with a slat or crosspiece (never conserved) that stretched across the diameter (Fig. 3b). The crosspiece thus took on the role of a bridge-rynd that, assembled with the spindle of the lower stone or *meta*, securely "centred" the stone during rotation. In some cases, one of these cuttings did not attain the outer rim. This clearly indicates that the crosspiece projected beyond the rim only to one side. This projection served as a base for a single vertical handle, the means of rotating the mill. Molten lead was at times poured into the cuttings to fasten the crosspiece indicating that it was made of iron. The lower stone, with a conical grinding surface, shows less typological diversity. The central hole for the spindle usually does not traverse the stone. The principal typological variable is the presence or not of a prominent collar around the spindle hole (Fig. 3c). This feature will be examined in further detail later in this paper.

2.2 Animal- and man-powered mills (Pompeian and "*Volubilis*" types)

The Pompeian mill is the best known of the different mills of the antique world (Fig. 2b). With its typical hourglass-shaped upper stone (*catillus*) and conical or bell-shaped lower stone (*meta*), this mill was either driven by a donkey, as seen in antique iconography, or by man. Although the most celebrated examples are found in the ancient bakeries of Pompeii and Ostia (Peacock 1989), isolated models are known as far as the Eifel in Germany (Hörter 1994, 32-33) and London (Williams-Thorpe and Thorpe 1988). Recent research undertaken in France reveals that this mill, contrary to accepted notions, was widely distributed in the Narbonese province of present southern France. Most are imports of volcanic rock originating in quarries near Orvieto (Italy), but there are also models hewn from local rock (Jaccottey and Longpierre, this volume).

The level of distribution of this mill type in the Iberian Peninsula is still not established. At the time of the publication of Moritz (1979, 95-96), no cases had been reported in *Hispania*. More recently, hourglass-shaped *catilli* have surfaced in the museum collections of Málaga, Córdoba, Empúries, Mérida and Zaragoza (Berrocal 2007, 294; González 2010); yet, this model does not appear to have been widespread. In some cases, such as the two *catilli* from a Roman villa in Huesca (Zaragoza) (Mínguez and Ferreruela 1992, 134-135, 158, fig. 12: 49), they are volcanic imports, like the majority of cases observed in Narbonese Gaul.

Nonetheless, as we can see from the assemblage of the Roman city of *Volubilis* in northern Morocco, the situation of these mills is extremely complex, and the identification of a Pompeian mill based solely on a bell-shaped lower stone can be highly misleading. In *Volubilis,* most lower stones are not associated with high hourglass *catilli*, but with low, ring-shaped stones with lateral dovetail cuttings for horizontal levers (Fig. 2c) (Luquet 1966,

306, fig. 3B; Akerraz and Lenoir 2002, 198, 201). The function of the *Volubilis* mills is still a question of debate. One interpretation is that the *grès coquillier* (shell-rich sandstone) models with elongated S-shaped furrows were destined for grinding olives while the basalt millstones, rarely presenting furrows, were meant for cereal grinding (Akerraz and Lenoir 2002, 203; Brun 1997b, 71-72). On the Iberian Peninsula, an example of a "*Volubilis*" ring-shaped upper stone of granite is known in the Roman city of *Conimbriga* in Portugal (Borges 1978, 125, no. 31). There are at least a dozen of this type also reported in the Guadalquivir River valley in the province of Seville (Akerraz and Lenoir 2002, 206). An example of basaltic lava is on display in the museum of the Roman city of *Cástulo* in Linares, Jaén (inv. CEO1429). There is therefore a high probability that a large number of the bell-shaped *metae* reported throughout Andalusia (for example: Cerro Martos in Seville, Morín *et al.* 2003; Úbeda Museum (CEO372); Granada Museum (CEO14119)), including the black basalt examples from Oreto y Zuqueca (Granátula de Calatrava) (Fig. 3d) and Almedinilla (Córdoba), were coupled with "*Volubilis*" ring-shaped *catilli*, and not "Pompeian" hourglass-shaped *catilli*. As we will see later in this paper, this is an important detail related to the type of mills produced in the Campo de Calatrava region.

2.3. The problem of watermills

Until recently, it was commonly accepted that although water-powered mills were a Roman invention, as is corroborated by the writings of Vitruvius (*De Architectura* X, 5), they were not widely distributed until Medieval times. Research in the last years in France, Switzerland, England, Germany, Italy, Greece, North Africa and the Middle East suggests the contrary (Brun and Borréani 1998, 308, fig. 35). For example, in southern France, besides the celebrated mill complex of Barbegal near Arles (Leveau 2006), water mills have been identified in the *villae* of Mesclans and Saint-Pierre (Brun and Borréani 1998). A mill at Longvic near Dijon in eastern France has also been documented recently (Jaccottey and Labaune 2010). In Switzerland, four watermills have been excavated at the sites of Hagendorn, Zug (Gähwiler and Speck 1991), Rodersdorf-Klein Bühl, Solothurn (Harb 2002), Avenches/En Chaplix (Castella 1994) and Avenches/Les Tourbières (oral information from D. Castella). In addition to this Swiss fieldwork, a methodological study of 168 ancient millstones stored in Swiss collections clearly substantiates that these hydraulic installations were widely distributed throughout the countryside of Roman Switzerland (Anderson *et al.* 2004).

The typology of Roman hydraulic millstones (Fig. 2d) differs substantially from the other contemporary millstone types. These stones, quite often long distance imports of basaltic lava, measure between 55 and 90 cm in diameter and present, compared to the *Volubilis* lower stone, a more flattened bell-shaped section. Two systems of transmitting motive power from the vertical spindle are known from the archaeological record. The first is a complex H-shaped apparatus lodged on the upper surface in a series of perforations linked by grooves (in French "*anille-crampons*"; Castella 1994, 47, fig. 31). The second, the system that has endured until recent times, is found on the lower surface and consists of cuttings of various types (rectangular, dovetail, cross-shaped) for metal rynds.

The situation of watermills in the Iberian Peninsula, however, is still not settled. With the exception of *Conimbriga* in Portugal (Brun 1997a), no watermill installation or millstone has been unequivocally identified and published (Philippe Leveau, according to recent information from Jean-Pierre Brun, has observed features of a hydraulic mill at the Roman villa of El Munts, Altafulla, in Tarragona). With the growing number of discoveries elsewhere in Europe, it seems unlikely that these mills were not present in *Hispania*, a land well-known for its hydraulic works. Texts as far back as the 8th century, in the early phases of the Islamic domination, indirectly point to the presence of watermills in the southern Guadalquivir Valley in Andalusia (Palomo and Fernández 2006-2007, 499-500). It is also well established that during the Medieval and Modern periods, hydraulic works, including watermills, were widespread throughout the Iberian landscape. The Census of the Marques de la Ensenada in the 18th century (1750-1754) and the geographical compilations of Pascual Madoz (1845-1850) cite literally thousands of these installations in almost every Spanish town, and they outnumber by far their wind-powered counterparts. It therefore seems unlikely that hydraulic milling technology was not adopted in Roman *Hispania*. This absence is therefore probably related to the present state of research.

3. The Cabo de Gata quern production

3.1. Cabo de Gata: historical, geographical and geological context

At the time of the Roman conquest toward the beginning of the 2nd century BC, the Cabo de Gata at the south eastern tip of the Iberian peninsula had already known a long tradition of maritime commerce by Phoenician, Greek and Carthaginian navigators and merchants. Its strategic position made its ports natural halting points in the passage from the eastern Mediterranean toward the Straits of Gibraltar and the north of Africa. The great number of amphorae recovered in the region (presumably from shipwrecks) imply that navigation through this cape could at times be treacherous (Martínez Maganto 1994, 199, fig. 1, 204). The Roman conquest, besides consolidating the established maritime trade routes, opened a network of inland thoroughfares in the region (Martínez Maganto 1994, 201, fig. 2) that connected with the *Via Augusta*

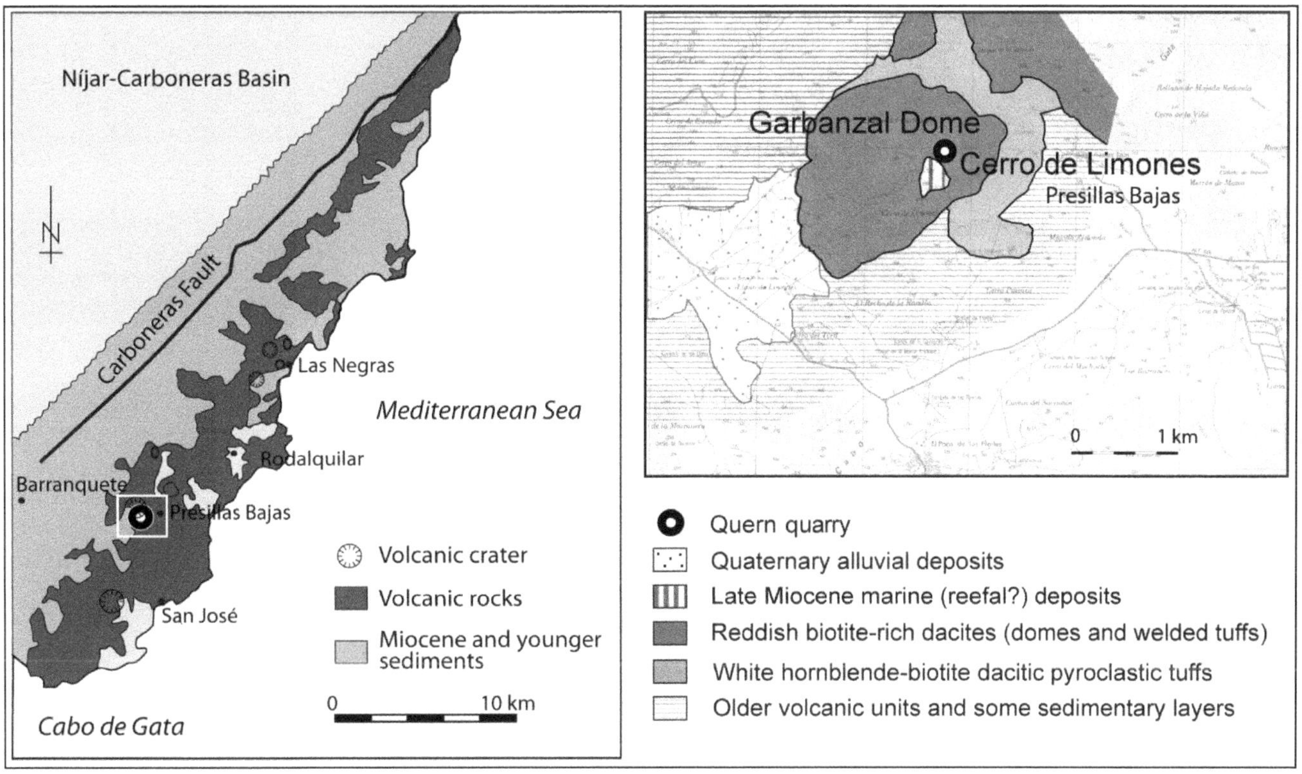

Figure 4. Simplified geological map of the Cabo de Gata region (left, modified from Bourillot et al. 2009) and geological map of the area surrounding the Garbanzal dome with the location of the Cerro de Limones quern quarry.

which traversed *Hispania* from the Straits of Gibraltar to the north eastern Pyrenees. Along the Almería coastline the Romans profited in particular from salt mines and fisheries for the production of the *garum* condiment (Cara Barrionuevo 1988, 64, fig. 7) and purple imperial dye extracted from snails (Martínez Maganto 1994, 211-213). The Roman conquest also brought about the exploitation of inland mineral resources such as lead and silver ores from the mountain ranges of Gador, Alhamilla, and Almagreda (Cara Barrionuevo and Rodríguez López 1986, 11) and white marble from the quarries of Macael, some 40 km from the coast (Padilla Monge 1999, 272). Locally in the area of Rodalquilar, abundant deposits of manganese were exploited in the 19th century. Gold was also intermittently exploited in the 19th and 20th centuries, while alunite was probably mined since Roman times (Cunningham *et al.* 1990, 22; Rytuba *et al.* 1990, 33).

Figure 5. Prominent columnar jointing on the south west slope of the Cerro de Limones that forms part of the Garbanzal volcanic dome. These were the types of formations that were exploited for the manufacture of querns.

In a geological context, the area belongs to the Cabo de Gata volcanic province (Fig. 4), a part of the Betic Cordillera that forms the westernmost segment of the Alpine orogenic belt in Europe. The volcanic rocks were emplaced during Miocene times, about 15-7 million years ago, as a consequence of extensional tectonic processes following the Alpine collision (e.g. Di Battistini *et al.* 1987; Scotney *et al.* 2000; Arribas *et al.* 1989). They constitute a typical calc-alkaline series, ranging in composition from basaltic andesite to dacite and rhyolite. Volcanism was characterised by the formation of large explosive craters and pyroclastic deposits, as well as by dacitic domes (Fernández Soler 1992). The volcanic rocks are intercalated with, and overlain by, marine sedimentary rocks. The Garbanzal dome represents the youngest phase of volcanism in the area. In geological terms it is a "low-lava dome" or "Torta" (*cf.* Blake 1989), composed largely of biotite-hornblende dacite and rhyodacite with a prominent columnar jointing (Fig. 5). Penetrative hydrothermal alteration is responsible for a reddish appearance in most of the rocks in the region. Similar rocks are exposed further east, at Rellana and Caldera of Majada Redonda, and extend to the Rodalquilar area where hydrothemal alteration is extreme ("red-ochre volcanism" in local geological terminology).

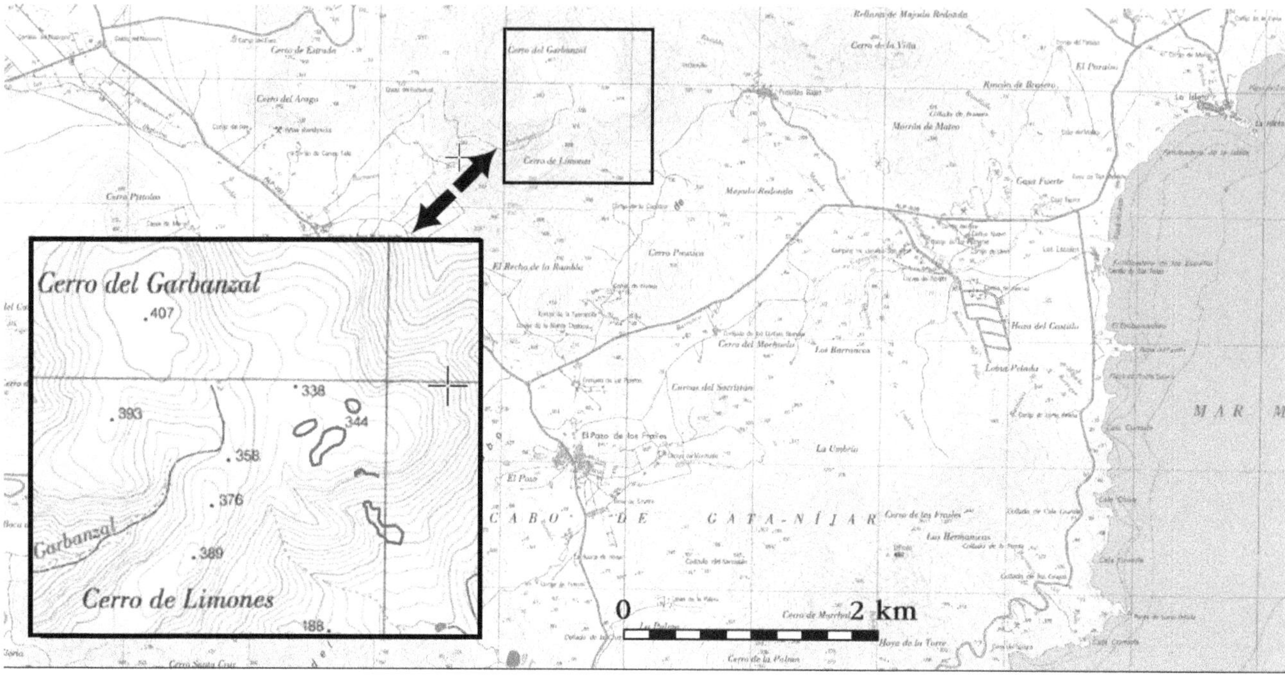

Figure 6. Extract of the topographical map with the outlines (in the insert) of the working areas of the Cerro de Limones quern quarry (extract from SIGPAC, Sistema de Información Geográfica de Parcelas Agrícolas).

3.2. The Cerro de Limones quern quarries

These quern quarries are located in the municipality of Níjar in the Cabo de Gata Geopark, perched on the top of the Garbanzal volcanic dome (Bordet 1985) that comprises the mountains Cerro de Limones and the Cerro del Garbanzal (Fig. 6). The material evidence of quern manufacture consists of scores of cylindrical blanks and roughouts scattered among heaps of stone working debris (Fig. 7a) spread over a surface of several thousand square metres.

The working areas consist of large, poorly defined shallow pits where the workers pried out angular shaped blocks with levers. The largest working area, about 140 m long and up to 45 m wide, is toward the southern edge of the plateau. A second, slightly to the north west, measures about 70 m long and 30 m wide. A third working area, further to the north, is about 30 x 40 m. A fourth extraction area is on the south slope of the mountain. This area differs from the others in that workers exploited a vertical escarpment about 100 m long toward the top of the mountainside. The blocks detached from this cliff apparently tumbled down the steep mountainside and were processed over an area of about 200 by 50 metres on top of the natural scree.

No tools have been found along on the surface at the site, and there is no evidence that the querns were finished at the quarry (Fig. 7b). First observations indicate that after initial extraction, the blocks were roughly knapped by means of direct percussion (probably with an iron hammer) into cylindrical form (Fig. 7c) with an average diameter of 39.7 cm, based on the measurement of 80 examples. This step resulted in the accumulation of a massive quantity of large flakes that litter the site. Initial steps of the fashioning process, for example, the rough carving of the eye and radial cuttings of the *catillus*, and the flange of the *meta*, were carried out with hammer and different types of chisels. But the carving of the rim, along with the hollowing of the surfaces and the hoppers, appears to have taken place elsewhere. Two examples, one of them, notably, a *meta* that broke during the piercing of the eye (*cf.* Figs. 3a1 and 3a6), were found at Los Cazadores, El Barranquete, about 7 km to the west of Cerro de Limones (outside of the volcanic area), and thus indicate that at least part of the final cuttings for the different fittings took place elsewhere (the Cazadores querns, recovered during rescue work in 1983, are stored in the depository of the Archaeological Museum of Almeria (no. 318-319). There is, unfortunately, no data about the character or chronology of the site).

Based on several models at the Archaeological Museum of Almería, we have been able to reconstruct a typological model of the Cerro de Limones product. The upper stone seems to adhere to a typology (rim, symmetrical hoppers and radial cuttings, cf. Fig. 2a) that is widespread in southern *Hispania* and not restricted to one specific rock type. The lower stone, however, appears to be original to the Cabo de Gata. Numerous abandoned *meta* roughouts bear a central protuberance or knob (Fig. 7d). This feature, along with the rynd, was probably meant to help maintain the mill centred during rotation. At present, we have identified only two parallels of this type outside the Almería area, in *Conimbriga*, Portugal (Borges 1978, 127, no. 52, Pl. VII) and in the Archaeological Museum of Úbeda in the province of Jaén (CEO0370). The first is of rose granite and the second of dark basaltic lava (this

Figure 7. Cerro de Limones (Cabo de Gata). A) view of one of the working areas on the plateau; B) defective rotary querns abandoned during the process of manufacture; C) a roughly knapped cylindrical blank with surrounding working debris; D) example of an unfinished flanged lower stone abandoned. Diameters approx. 40 cm.

type of lower stone should not be mistaken for flanged collars of upper stones of hand querns known elsewhere in Europe in Roman or later contexts, e.g.Welfare 1985, 159; Curwen 1941, 23; Désirat 1996, 139, fig. 59). There is no evidence that other types of mills, from Roman or later times, such as the larger animal-powered millstones or watermills, were manufactured at the site. In any case, the significant number of abandoned blanks and roughouts on the top of the mountain and decorating the neighbouring towns of Presillas Bajas and Rodalquilar suggest a massive production that should be counted in the thousands.

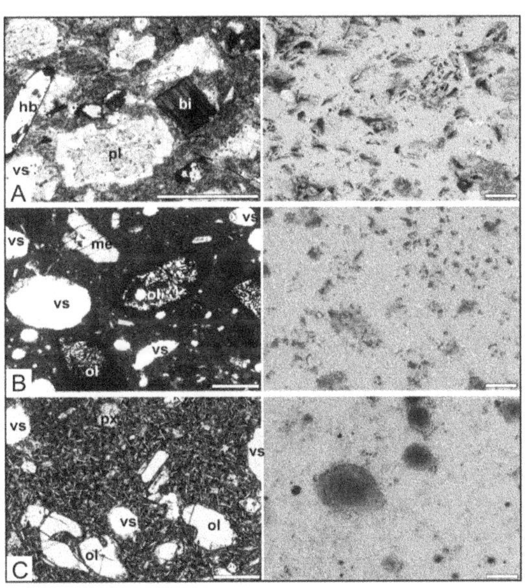

Figure 8. Photomicrographs (left) and back-scattered electron (BE) images (right) of volcanic quern sources. A) dacite from Cerro de Limones (Cabo de Gata); B) melilitite from La Bienvenida (Campo de Calatrava); olivine basalt from Oreto y Zuqueca (Campo de Calatrava). Photomicrographs by plane polarised light. Abbreviations: hb=hornblende; bi=biotite; pl=plagioclase; ol=olivine; me=melilite; px=pyroxene; vs=vesicle. All scale bars are 1 millimetre. The BE images show the porous nature of the surface due to open vesicles.

The exploited rock is a hydrothermally altered, reddish dacite with ca.50% phenocrysts of euhedral plagioclase, biotite and hornblende in addition to partially resorbed quartz (Fig. 8a). Magnetite, ilmenite and pyroxene are minor phases. The phenocrysts are mostly 1-4 mm and are set in an extremely fine grained (originally glassy) groundmass showing fluidal distribution of scarce plagioclase microlites. A porosity on the order of 10% is due to highly irregular vesicles (voids) 0.1-1 mm across (cf. Fig. 8a). Due to the hydrothermal alteration, plagioclase is partly transformed to an aggregate of clay minerals. Hornblende and pyroxene are largely replaced by fine grained iron oxides, while biotite is better preserved and shows only thin oxidised rims. The groundmass is altered to an extremely fine grained, hard aggregate of silica, K-feldspar and minor hematite, the latter giving the reddish coloration to the rock. Quartz is not affected by the alteration and contributes significantly to the hardness. Geochemically, the rock is characterised by a very high content of K^2O (more than 10%) and low CaO and Na^2O (ca. 1.2 and 0.5%, respectively), all of which may be attributed to the hydrothermal alteration of a normal dacite as seen elsewhere in the Cabo de Gata volcanic province (Table 1).

4. The Campo de Calatrava quern and millstone quarries

4.1. Campo de Calatrava: historical, geographical and geological context

The region of Campo de Calatrava covers a vast area of the Autonomy of Castilla-La Mancha. Its traditional borders are based on geographical factors related to ancient

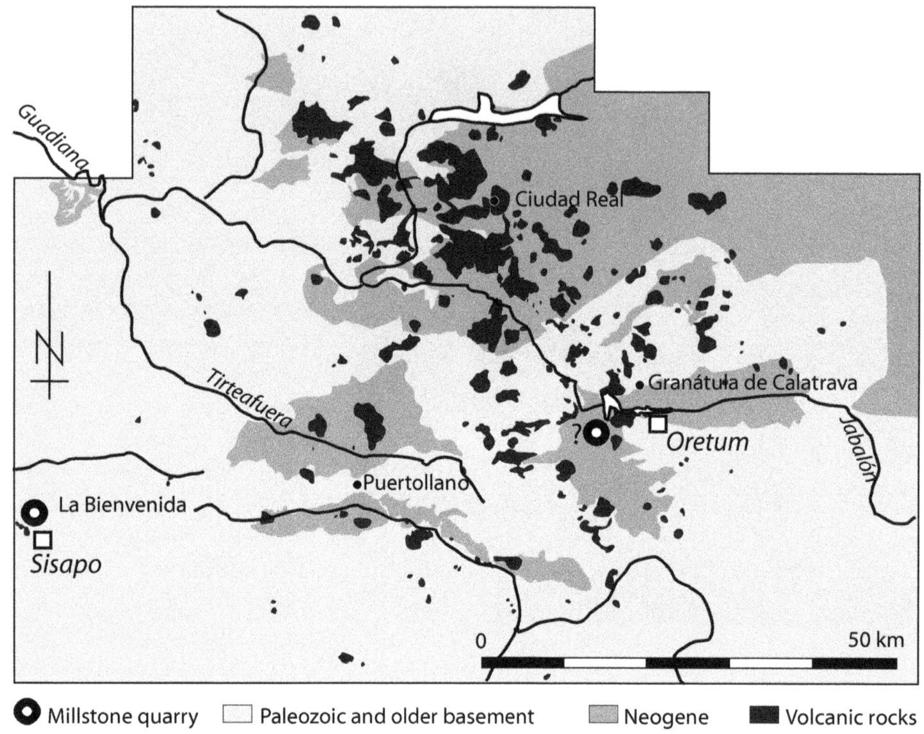

Figure 9. Simplified geological map of the Calatrava Volcanic Province (modified from Martín-Serrano et al. 2009).

volcanic activity (Fig. 9) as well as frontiers dating to the struggle between the Christian and Islamic kingdoms in medieval times. The name derives from the religious and military Calatrava Order dating to the 12th century. In our brief visit to this region, we were able to identify two quern and millstone extraction sites.

The first quarry is located in the south western corner of the Campo de Calatrava region on the outskirts of the hamlet La Bienvenida in the municipality of Almodóvar del Campo. La Bienvenida, with less than one hundred people now, had an illustrious past. An epigraphic inscription brought to light during excavations in 1982 is one of several arguments that point to it as being the Roman city of *Sisapo*, built on the foundations of a pre-existent Iberian *oppidum.* This main administrative settlement that once dominated the centre of the Alcudia Valley is cited by the ancient authors Pliny the Elder, Strabo, Cicero, and Ptolemy (Fernández *et al.* 1982-1983). The ruins cover a surface of about 10 acres with its surrounding walls and towers (and possible amphitheatre). Through written sources, *Sisapo* is known to have played an extremely important administrative role in the exploitation of the mining resources in the northern Sierra Morena region. These exploitations included numerous lead and silver ores as well as cinnabar ore that transited through *Sisapo* in an unrefined state on its way to workshops at Rome for the manufacture of red pigment and mercury (Arboledas, 266-267, note 22).

The second site is inferred from several unfinished querns exhibited at the archaeological site of Oreto y Zuqueca (Granátula de Calatrava). This excavation, which has brought to light ruins dating from pre-Roman times to the Islamic domination, is directly related to the neighbouring Roman city of *Oretum,* located about two kilometres to the west on the Cerro de los Domínguez (López Domech 1996, 29). *Oretum* lent its name to the ancient *Oretanian* region and probably drew its wealth from raising livestock and farming the surrounding fertile lands. With its 1st century AD bridge spanning the Jabalón River, *Oretum* was at the crossroads of several major thoroughfares linking Andalusia in the south, the Castilian *Meseta* to the north, and the Iberian *Levante* to the east opening its way to the Mediterranean. It is believed, based on citations in ancient texts, that cargoes of different mineral products transited through *Oretum* from, for example, *Sisapo,* about 70 km to the west, to *Cástulo* (Linares, Jaén), about 75 km to the south. In any case, *Oretum* was known to have been linked by road to the other economic centres of *Hispania* such as *Cástulo* and *Corduba* (Carrasco 2007, 372).

The Calatrava Volcanic Province (CVP), of Late Miocene to Quaternary age (8.7-0.7 million years), is a conspicuous feature of the region. Like volcanic millstone quarry areas such as Garroxta-Olot in Catalonia, the Massif Central in France and the Eifel in Germany, the CVP represents a phase of intraplate, dominantly alkaline, volcanism related to the development of an extensive lithospheric rift system in the northern foreland of the Alps (Wilson and Downes 2006; Martín-Serrano *et al.* 2009). The province was characterised by Strombolian and phreatomagmatic eruptions, now represented by more than 200 volcanic centres spread over an area of about 5000 km2. Late Miocene-Quaternary fluvial and lacustrine sediments deposited in a series of local fault-controlled basins are intercalated with the volcanic deposits. The Strombolian volcanic activity produced small, mostly monogenetic, cinder cones and lava flows, while the phreatomagmatic

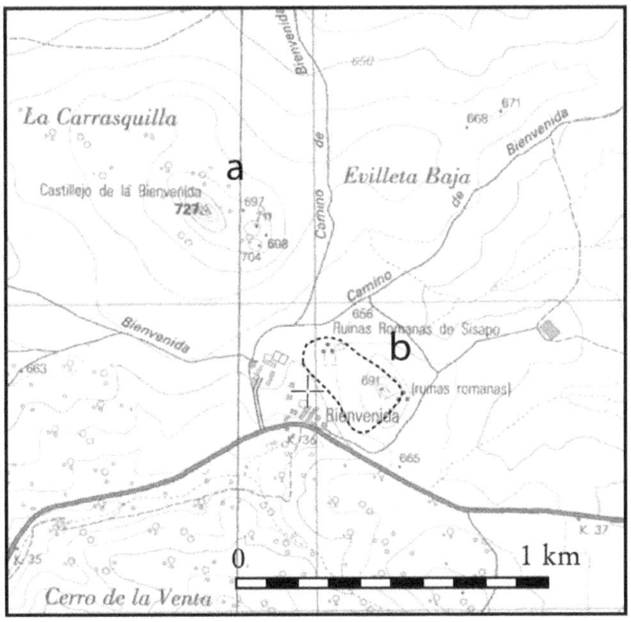

Figure 10. Topographic map of the area of La Bienvenida (SIGPAC). A) indicates the location of the millstone quarries and B) the location of the ruins of the Roman city of Sisapo.

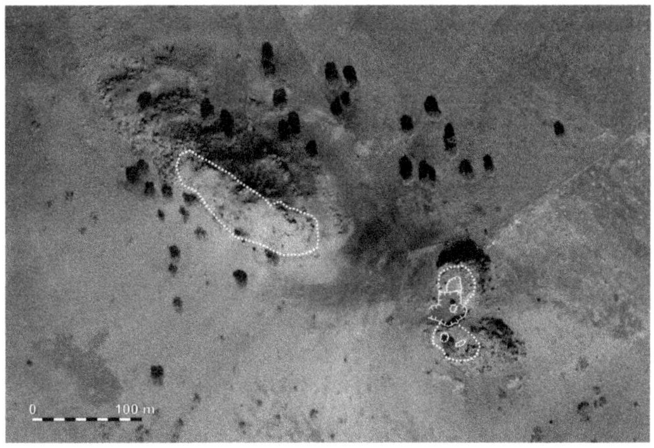

Figure 11. Aerial photograph (orthophoto-SIGPAC) of the Bienvenida volcanic domes with the location of the quarries. The dashed outline indicates scattered extractions.

eruptions are represented by maar craters and pyroclastic deposits. Three main petrologic groups have been distinguished: olivine melilitites, olivine nephelinites, and alkali olivine basalts; in addition olivine leucitites are found locally (Cebria and López-Ruiz 1995; López-Ruiz *et al.* 1993; Carracedo-Sánchez *et al.* 2009).

4.2. The Bienvenida quern and millstone quarry

The quern and millstone quarries are evidenced by multiple circular extraction hollows still well visible today resulting from scoring directly from the lava flow (Fernández Ochoa *et al.* 2002, 121). The hollows are found on three small volcanic hills (spatter cones) about half a kilometre northwest of the centre of the Roman city (Fig. 10). A lava flow stretches about one kilometre to the north west but shows no evidence of millstone quarrying. The largest cone is 250 metres long and 80 metres wide and is spotted with shallow quern and millstone extractions (Fig. 11). About 100 metres to the south east are two very small cones together covering a surface of 100 by 50 metres. This second area has both scattered superficial extractions and a large central, semicircular pit 2-4 metres deep covering an area of about 650 m2 (Fig. 12a). On the northern edge of this large pit, there is evidence of both tiered quern extraction and ashlar extractions (Fig. 12b). Shallower pits of the order of 50-150 m2, probably for querns and millstones, were also worked further north and south.

Figure 12. La Bienvenida quarry. A) The largest extraction area viewed from the north; B) tiered extractions of small hand-operated querns; C) detail of a hand quern extraction 40 cm in diameter surrounded by a channel cut by pick; D) abandoned cylinder measuring 80-90 cm in diameter.

	Campo de Calatrava		Cabo de Gata	
	La Bienvenida	*Oretum*	LMN-1	B-127
SiO_2	37.40	43.70	67.05	65.3
Al_2O_3	9.35	12.30	14.81	15.19
Fe_2O_3 (total)	12.30	12.20	3.35	3.75
TiO_2	3.13	2.79	0.31	0.35
MgO	12.80	10.50	0.79	1.80
CaO	12.80	11.70	1.21	3.00
Na_2O	2.11	2.07	0.53	2.10
K_2O	1.23	1.14	10.39	3.05
MnO	0.20	0.17	0.08	0.05
P_2O_5	1.58	0.63	0.09	0.05
L.O.I	5.76	2.58	0.87	4.52

Table 1. Chemical data (X-ray fluorescence analysis of fused glass beads) of quern quarry samples and related rocks. L.O.I.: Loss on ignition. LMN-1: quern quarry sample, Cerro de Limones (hydrothermally altered dacite). B-127: dacite, Garbanzal Dome (Bordet, 1985). La Bienvenida: quern quarry sample (olivine melilitite). Oretum: waste from quern production (alkali olivine basalt).

Based on preliminary observations, the sizes of the extractions of the Bienvenida quarry seem to fall into two large categories. The smaller extractions correspond to hand-operated cylindrical rotary querns about 40 cm in diameter and about 20 cm high (Fig. 12c), similar to those of Cerro de Limones. The size of the larger extractions appears to vary between 80 and 90 cm and corresponds to an unidentified type of mill that is too large to be operated by hand (Fig. 12d). These querns and millstones were scored directly from the volcanic deposits following horizontal planes, an extraction technique that resulted in the numerous circular negatives. However, due to weathering, the tool marks are poorly visible so it is not possible to specify precisely all of the different phases of the extraction sequence.

From the study of similar types of Roman extraction sites, we infer that the centre was pecked to serve as a reference to sketch the circumference by means of a compass (or simply a nail at the end of a string) of the future cylinder. Then circular channels were cut along the circumference with a long-hafted iron pick following a well-defined systematic extraction method common to quarries in antiquity (e.g. Bessac 1996, chap. 6, 205-236; Anderson *et al.* 2003, 48-51). A pictogram of a quarryman wielding a variation of this type of tool was discovered in the ancient volcanic quarry of Kruft of the Eifel in Germany (Röder 1957, Taf. 21, Abb.1). The width of the trench depended on the size of the cylinder. For the smaller hand querns a trench about 10 cm wide was suitable. For the larger cylinders (around 70 cm in diameter) a trench about double that size was necessary. In the case of one of the larger extractions we have observed triangular-shaped cavities along its base. This suggests the use of either iron or wooden wedges to split the cylinder from the bedrock.

Based solely upon the surface features, this quarry cannot be dated precisely. However, according to the dimensions of the extractions and the proximity to ancient *Sisapo*, the site can be assigned to Roman times.

The Bienvenida quarries exploited a dark grey lava in the form of variably welded scoria, some places grading into agglutinated fragmental deposits. Petrologically the rock is an olivine melilitite, comprising abundant, up to 2 mm phenocrysts of olivine (variably weathered) in a vitreous matrix with scattered olivine and melilite microcrysts; melilite is also a subordinate phenocrysts phase (cf. Fig. 8b). Vesicles are mostly less than 2 mm across and are in

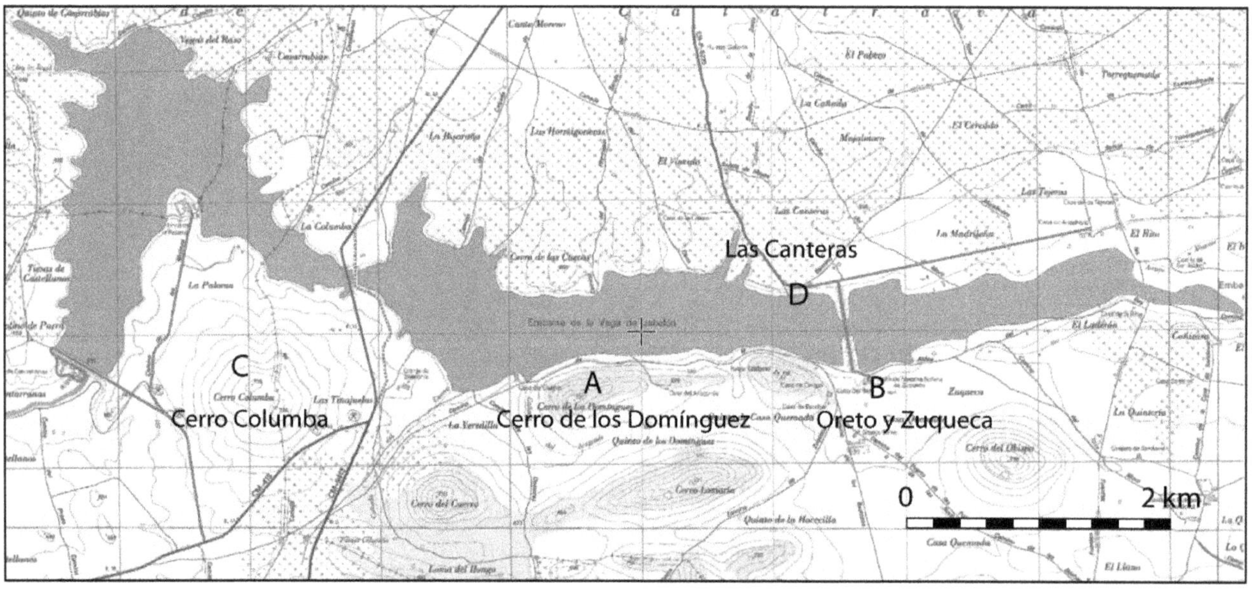

Figure 13. Topographical map (SIGPAC) of the area of the ancient city of Oretum. A) Cerro de los Domínguez is the location of the city; B) the archaeological site of Oreto y Zuqueca; C) the volcano Cerro Columba; D) the approximate location of the modern millstone quarry of Las Canteras.

Figure 14. Examples of dark, highly vesicular, basaltic lava quern cylinders at the archaeological site of Oreto y Zuqueca (Granátula de Calatrava) that suggest the presence of a quern quarry in the vicinity. The example on the right has large vesicles lined with zeolites and/or calcite while the fragment in the foreground and the roughout on the left have vesicles with little or no linings.

some cases lined with white carbonates and/or zeolites. Vesicularity of the individual scoria clasts rarely exceeds 20%, but the weakly welded character in some places seems to leave the rock with a higher bulk porosity. Chemical composition (cf. Table 1) is similar to olivine melilitites elsewhere in the Calatrava Volcanic Province, characterised by low SiO^2 and high P^2O^5 contents (*cf.* Cebria and López-Ruiz 1995).

4.3. The unidentified source of millstones related to *Oretum*

There are about a dozen basaltic lava querns and mills exhibited at the archaeological site of Oreto and Zuqueca (Fig. 13). A part of these might be the "*molinos*" referred to in a brief notice of archaeological finds (Alañon 1982, 230). According to Helena Romero Salas, an archaeologist that has excavated part of the site, some were recovered during the excavation (for the most part in secondary context, such as in walls of later occupations), while others were surface finds recovered in the area.

The rotary querns, all between 38 and 40 cm in diameter, include two blanks (Fig. 14a), two aborted querns (a *catillus* and a *meta*), a finished *meta* with signs of wear, and two fragments of finished *catilli*. Besides the querns, there are two bell-shaped "*Volubilis*" type lower stones (*cf.* Fig. 3d), for either cereal or oil production. The fact that four rotary querns, almost half of the lot, are fabrication rejects clearly indicates that a workshop is to be found in the surrounding area. The exact location of the workshop and the quarry remains unknown. According to the initial petrographic analyses, these stones do not hail from La Bienvenida by *Sisapo*. The nearest volcano is that of Cerro Columba, about 2 km to the west of the ruins of *Oretum*. It is generally thought that natural escarpments through the volcanic rocks were created when the Jabalón River cut through the northern flank of this volcano and that these escarpments were exploited in Roman times for construction material (oral information from Juan Manuel Donoso Gómez). These potential quarries, which would have benefited of the river for transport, are now unfortunately under the waters of a dam.

Querns at *Oretum* were made using dark grey, highly porous basalt. Individual vesicles are up to 2 cm in size, in some cases with thin, white linings of carbonates and/or zeolites. A sample of waste material collected on the surface of a neighbouring field, identical macroscopically to the millstones seen at Oreto and Zuqueca, shows the petrologic and geochemical characteristics of alkali olivine basalt; phenocrysts of olivine and minor clinopyroxene (mostly less than 1 mm) are set in an intersertal matrix consisting of lath-shaped plagioclase and interstitial glass and cryptocrystalline material (cf. Fig. 8c). The high vesicularity and the large size of the vesicles may suggest that the *Oretum* querns and millstones were hewn from volcanic bombs, not from volcanic bedrock such as at *Sisapo*; however, columnar jointed lava flows comparable to those of the Eifel millstone quarries in Germany are known in the area and may have been exploited, supposing that they are equally vesicular. Further studies are required to identify possible quarries. In any case, it is noteworthy that the geochemical composition of our sample is comparable to that of olivine basalt from the Cerro Columba volcano (cf. Pellicer Bautista, unpublished report) .

Figure 15. Views of a lower stone bearing measuring 65 cm in diameter on display at the entrance of Oreto y Zuqueca (Granátula de Calatrava, Ciudad Real). The morphology of this stone is similar to that of hydraulic millstones known elsewhere in Europe. The radial furrows might be indicative of ore grinding.

5. Conclusions and perspectives

Although we have alluded briefly in recent conferences to the quern and millstone production in the regions of the Cabo de Gata and Campo de Calatrava (Anderson in press; Anderson and Scarrow, this volume), this is the first paper devoted specifically to these Iberian volcanic millstone production centres.

Based on the proportions of the extraction hollows, the Bienvenida quarry produced small hand-driven rotary querns, similar in size (and probably in typology) to those of the Cabo de Gata sites. La Bienvenida also produced larger millstones, about twice the diameter. However, due to the absence of roughouts, the exact nature of these larger extractions remains uncertain. Theoretically, small Pompeian mills could have been hewn at La Bienvenida. But this is unlikely due to the scarcity of this type in the Iberian Peninsula. The larger and deeper cylindrical hollows are also perfectly compatible with the dimensions of hydraulic millstones, like the basalt examples known in France and Switzerland. As we have noted earlier, with the exception of *Conimbriga*, Portugal, and possibly El Munts, Altafulla, in Tarragona, there is still no Roman watermill installation documented on Iberian soil. We have, however, observed one case of dark basaltic lava having a form homologous to that of Roman hydraulic millstones published elsewhere in Europe. It is a lower stone with a circular eye, 65 cm in diameter (Fig. 15), exposed at the entrance of the archaeological site of Oreto y Zuqueca in the heart of the Calatrava region. It would not be incongruous to imagine that this surface find, produced in a regional Calatrava quarry, might have been a component of a Roman watermill near the Jabalón River related to the city of *Oretum*. The third option, and at the moment the most likely, is that the larger extractions at La Bienvenida correspond to the more widespread "*Volubilis*" mills, both the ring-shaped *catilli*, like the case in the Museum of Linares, and the bell-shaped *metae*, like the examples at Oreto and Zuqueca (Ciudad Real) and Almedinilla (Córdoba).

Extraction techniques that can be observed on the surface differ significantly from the Cabo de Gata sites and La Bienvenida. While the first quern makers profited from surface blocks and columnar jointing, techniques that leave little or no visible traces, quern makers of La Bienvenida cut out their cylinders with picks directly from the bedrock, leaving circular imprints or hollows still perfectly visible today. It is nonetheless difficult to establish the "*chaîne opératoire*" of the extraction process without benefiting from a proper archaeological excavation.

The chrono-typology of these productions is still not established with certainty. Most volcanic querns stored in museum collections are not correctly published and do not derive from secure, stratified contexts, and therefore, unfortunately, provide no chronological data. Nevertheless, the rock type, the general typological aspects and the massive volume of the production, especially in the case of the Cerro Limones, point broadly to Roman times. Future research, especially through archaeological interventions, should focus on these essential facets.

For the moment, the Cerro de Limones is the only quern quarry identified in the Cabo de Gata (we have recently learned of the existence of a second quern quarry in the Cabo de Gata, a few kilometres from the Cerro de Limones. Unfortunately, we have not yet been able to corroborate it). The absence of dark grey or black basalt querns in the Almería museum gives the impression that only one type of rock, the reddish dacite, was exploited in this district, suggesting that exploitation was limited to this specific area. This hypothesis has yet to be confirmed since the number of querns (7) observed in the museum collection, for the most part from the site of Los Cazadores (El Barranquete) is too modest for a statistical base. More field and museum surveys are necessary to corroborate the possibility of other ancient quern quarry operations in the Cabo de Gata. This is obviously applicable also to the vast area of the Campo de Calatrava, with a potential of numerous ancient millstone exploitations.

The sphere of distribution of these rocks is also still far from established. The survey of millstones in several museum collections, still in an initial stage, seems to indicate that the majority of the hand-operated querns in the region surrounding the Calatrava and Cabo de Gata areas were hewn from volcanic rock, suggesting that these two areas dominated the local and regional markets. As we mentioned above, the Roman conquest brought a vast network of inland roads linking the different commercial centres. The Campo de Calatrava quern and millstone quarries could have benefited from these inland trade routes originally designed for military logistics and the transport of other products. Parts of the Calatrava production could also have been conveyed by fluvial transport. The Guadiana River, the fourth largest in the Peninsula and with its source near the Calatrava region, was navigable from the Atlantic coastline, present day Huelva, to at least *Colonia Iulia Augusta Emerita* (Mérida), about 200 km inland. It is possible that the Calatrava quarries profited from the Guadiana tributaries such as the Jabalón, which flows beside the city of *Oretum*, to transport their millstones to the east and the south. In the case of the Cabo de Gata, there also existed the option of exporting querns by sea, profiting from long established maritime commercial routes. Cerro de Limones is just four kilometres uphill from the coastline, near several deep water coves where querns could have been loaded onto ships.

Although volcanic rocks make up a great percentage of the material exploited for querns and millstones, they did not in antiquity dominate the whole of the market of the south of the Peninsula. A number of other rocks were also exploited. In the region of Granada, we have seen querns and millstones hewn from local sandstones and conglomerates. A highly porous, cream-coloured limestone tufa (similar to travertine), a type of material with outcrops throughout Andalusia, was also carved for mills dating from both the end of the Iberian Culture and Roman times, such as the examples in the museums of Alcalá la Real (Jaén), Priego de Córdoba, and Almedinilla (Córdoba) (Quesada *et al.* 2010, 89). A reddish-tinted granite, probably from Pedroches, near Villanueva de Córdoba), was exploited for Roman bell-shaped lower stones such as those in the museums of Úbeda and Linares. A third, probably more important source of querns and millstones, is the brown-yellow, shell-rich sandstone called "*piedra ostionera.*" This stone seems to have played an essential role in the southern coastal market. An *ostionera* quern quarry, dating presumably from both Roman and later Medieval times, is known on the coast of Trafalgar Bay (Anderson, this volume). A large assemblage of *ostionera* querns are found, for example, in the Roman coastal city of *Baelo Claudia,* Cádiz (oral information provided from Ángel Muñoz Vicente). A second group of larger *ostionera* unfinished cylinders (65 cm in diameter) are exhibited in the Castle of San Miguel, Almuñecar (Granada). The quantification of the production of these other sources of querns and millstones, as well as their economic interaction with the volcanic productions, is a subject that is also still far from established.

In south eastern France, Amouric has noted that at a certain moment between the Roman period and the Middle Ages, a major change took place regarding the material exploited for millstones. Volcanic rocks from sites such as Agde were abandoned in favour of other materials, notably conglomerates (Amouric 1997, 45-47). This change is also discernible in the south of the Iberian Peninsula. At the Cerro de Limones quarry site, for example, there is no indication of exploitation beyond Roman times. The more recent mills, in particular windmills, of the Cabo de Gata were not equipped with volcanic rocks. Moreover, all of the documented post-Roman regional millstone quarries, such as the Rambla Honda near Albox (Martínez *et al.* this volume), Caniles (Madoz 1846, 5, 461), Adra, Guardias Viejas, and Ugíjar (Cara Barrionuevo *et al.* 1999, 153), exploited conglomerate or similar sedimentary outcrops. The supplanting of basaltic volcanic rocks after the Roman period also appears to have occurred in the Campo de Calatrava region. A presumably recent millstone quarry exploiting a sedimentary rock (*Las Canteras*), for example, is located in the heart of the volcanic district in the municipality of Granátula de Calatrava on the northern bank of the Jabalón River, less than one kilometre from *Oretum*. In this same area, the Medieval (12th - 14th centuries) animal-driven mill in the courtyard of the Castle of Calatrava la Nueva was equipped with conglomerate millstones (Heras 2008, 148).

The reasons behind this "mysterious substitution" are still unknown. Amouric suggests that the change might be related to the arrival of new cereals and to questions of efficiency connected to shape, size, rate of rotation and grinding properties of the new stones (Amouric 1997, 46-47). The widespread medieval adoption of hydraulic, and later aeolic, milling installations, suggests a more accelerated rotation (between 60 and 90 RPMs). This higher speed would apparently be more "efficient", yielding more flour in a shorter time, especially with larger, flatter, pebble-incrusted, tight-fitting millstones that permitted grain to transit once through the mill contrary to the earlier slow rotating installations where grains had to be ground in repeated passages. Other researchers second these notions and advance the opinion that volcanic stones rotating at high velocity might "burn" the flour (Brun and Borréani 1998, 301). But the fact is that the debate of the substitution of the grinding stone material remains unresolved.

There are less tangible aspects related to the ownership and the personnel of the quarries that also remain hazy. Are the exploitations of Cabo de Gata the property of the master of a local (unidentified) villa? Or are these exploitations, like certain construction quarries, the property of the authorities established in a city, such as *Sisapo*? Were these quarries exploited directly by their owners or leased to specialists as concessions? In any case, these exploitations, and others that remain to be identified, certainly brought prosperity to their owners. Were the quern and millstone makers freemen or slaves, where did they live and did they work at the sites seasonally or permanently? We also know nothing about the other logistical aspects related to quarry work, such as the location of the smithies for the maintenance of the iron tools, and how the querns, the larger millstones in particular, were transported and where they were finished and commercialised.

In any case, the typology of the rotary querns, in particular the protruding flange of the *metae* and the opposite cuttings for handle and rynd fittings of the *catilli,* do not seem to be imported features and seem to follow the tradition of querns and millstones known in the indigenous Iberian Culture.

The sites that we have presented are the first representatives of quern and millstone quarries in the Cabo de Gata and Campo de Calatrava volcanic districts. These Iberian volcanic districts now join the ranks of other production areas, such as the Eifel in Germany, the Massif Central and Agde in France, Olot-Garrotxa in Catalonia, Murlargia in Sardinia, Orvieto in central Italy, Etna in Sicily, and the Middle Atlas in Morocco, and form part of the complex chapter of quern and millstone production and trade throughout the Mediterranean Basin and beyond.

Acknowledgments

This study would not have been possible without assistance from the Norwegian Millstone Quarry Landscape research project directed by the Dimension Stone team of the Norwegian Geological Survey (NGU), Trondheim. Special recognition therefore goes to Gurli Meyer and Tom Heldal of the Norwegian Geological Survey for their assistance and support. We owe the identification of the Cabo de Gata quarries to the geographer and guide Loïc Martínez and to Antonio Hernández of the Isleta del Moro. Daniel Castella, Luc Jaccottey, Stephanie Lepareux-Couturier, Nicolas Milleville, Olivier Buchsenschutz and Jean-Pierre Brun assisted with their comments on the different mill categories. We would also like to thank the following museum directors and technicians for opening up the doors to their collections: Ana Navarro Ortega, director of the Archaeological Museum of Almería; Marcelo Castro, director of the Linares Archaeological Museum, Jaén; María del Mar Capel and José Luis Latorre Bonachera (director) of the Úbeda Archaeological Museum, Jaén; Ignacio Muñiz Jaén, director of the Ecomuseum of the Caicena River, Almedinilla, Córdoba; Rafael Carmona Ávila, director of the Muncipal Historical Museum of Priego de Córdoba, Córdoba; Ángel Muñoz Vicente, director of the Centre of Interpetation of the Roman city of Baelo Claudia, Cádiz; and Domingo Ortiz, director of the Municipal Museum of Vera, Almería. We also thank Jane Scarrow and Fernando Bae Barredo, of the Department of Petrology and Mineralogy of the University of Granada, for their data about possible granite sources, and Bjørn Willemoes-Wissing (NGU) for help with the back-scatter electron images. We are also in debt to the historian Juan Palomo of Villanueva de Córdoba for his pertinent information concerning watermills and to Helena Romero Salas and Juan Manuel Donoso Gómez, archaeologists, for their knowledge of *Oretum* and pointing out the location of the *Canteras* quarry. Finally, we want to recognise Jeff Hodges for proofreading the text and Aurora Pulido, Mara Pulido, and Jaime Medina for their patience, support and encouragement.

References

Akerraz, A. and Lenoir, M., 2002. Instruments de broyage en Maurétonie Tingtane à l'époque romaine, le cas de Volubilis. In Procopiou, H. and Treuil, R., (dir.), *Moudre et broyer II: Archéologie et histoire : du Paléolithique au Moyen Âge,* Paris, 197-207.

Alañon Flox, L., 1982. Arqueología del bajo Jabalón, Mapa arqueológico de la Zona de Aldea del Rey (Ciudad Real). *Cuaderno de estudios Manchegos,* 13, 1982.

Alonso, N., 1997. Origen y expansión del molino rotativo bajo en el Mediterráneo Occidental. In Garcia D. and Meeks. D. (coord.), *Techniques et économie antiques et médievales. Le temps de l'innovation, Colloque international CNRS, Aix-en-Provence 21-23 mai 1996*, 15-19.

Alonso, N., 1999. *De la llavor a la farina, Els processos agricoles protohistorics a la Catalunya occidental* (*Monographies d'archéologie méditerranéenne 4*), Lattes.

Alorda Park Team, 2002. Les meules rotatives du site ibérique d'Alorda Park. In Procopiou, H. and Treuil, R., (dir.), *Moudre et broyer II: Archéologie et histoire : du Paléolithique au Moyen Âge,* Paris, 155-174.

Amouric, H., 1997. L'anille et les meules. In Meeks, D. and Garcia, D. (éds.),*Techniques et économies antiques et médiévales, le temps de l'innovation, Colloque d'Aix-en-Provence (CNRS) 21-23 mai 1996*, Paris, 39-47.

Anderson, T., In press. Un premier bilan sur la production de meules dans la Péninsule Ibérique depuis la Protohistoire à la période médiévale. In Buchsenschutz, O. and Jaccottey, L. (eds.), *Évolution typologique et technique des meules du Néolithique à l'an mille sur le territoire français, Actes de la Table Ronde du Groupe Meules à St. Julien-sur-Garonne du 2 au 4 octobre 2009, Aquitania.*

Anderson, T., Agustoni, C., Duvauchelle, A., Serneels, V. and Castella, D., 2003: *Des artisans à la campagne. Carrière de meules, forge et voie gallo-romaines à Châbles (FR), (Archéologie fribourgeoise 19), Academic Press Fribourg,* Fribourg.

Anderson, T., Castella, D., Doswald, C. and Villet D., 2004. Meules à bras et meules hydrauliques en Suisse romaine: répartition et pétrographie. *Minaria Helvetica,* 24a, 3-16.

Anderson, T. and Scarrow, J. H., Millstone quarries in Southern Spain: Preliminary Pinpointing of Provenance and Production - Exploiting the Internet. This volume.

Arboledas Martínez, L., 2007. *Minería y Metalurgía Romana en el Alto Guadalquivir: Aproximación desde las Fuentes y el Registro Arqueológico,*

Doctoral Thesis, University of Granada.

Arribas, A. Jr., Rytuba, J. J., Rye, R. O., Cunningham, C. G., Podwysocki, M. H., Kelly, W. C., Arribas, A., Mckee, E. H. and Smith, J.G., 1989. Preliminary study of the ore deposits and hydrothermal alteration in the Rodalquilar Caldera Complex, southeastern Spain. *USGS open-file Report*, 89, 327.

Berrocal, L., 2007. De la Mecánica a la molienda: Un ensayo sobre los molinos giratorios de la España Antigua. In Blánquez Pérez, J., Roldán Gómez, L., and Jiménez Vialás, H. (coord.), *Augusto Fernández de Avilés, en homenaje,* Madrid, 275-297.

Bessac, J.-C., 1996. *La pierre en Gaule Narbonnaise et les carrières du Bois des Lens (Nîmes): histoire, archéologie, ethnographie et techniques,* Journal of Roman Archaeology, Suppl. series, 16, Ann Arbor.

Blake, S., 1989. Viscoplastic models of lava domes. *IAVCEI Proceddings in Volcanology*, 2. *Lava flows and domes*, Heidelberg, 88-126.

Bordet, P., 1985. Le volcanisme miocène des Sierras de Gata et de Carboneras (Espagne du Sud-Est). *Documents et Travaux IGAL*, Paris, 8.

Borges, N. C., 1978. Mós Manuais de Conimbriga. *Conimbriga (Coimbra),* XVII, 113-32.

Bourillot, R., Vennin, E., Kolodka, C., Rouchy, J.-M., Caruso, A., Durlet, C., Chaix, C. and Rommevaux, V., 2009. The role of topography and erosion in the development and architecture of shallow-water coral bioherms (Tortonian–Messinian, Cabo de Gata, SE Spain). *Palaeogeography, Palaeoclimatology, Palaeoecology*, 281, 92-114.

Brun, J.-P., 1997a. Um primeiro moinho hidraulico romano na Peninsula Ibèrica em Conimbriga. In Moutinho Alarcão, A., *Portugal romano, exploracao dos Recursos Natura*, 30-1.

Brun, J.-P., 1997b. L'introduction des moulins dans les huileries antiques. In Garcia D. and Meeks. D. (coord.), *Techniques et économie antiques et médievales. Le temps de l'innovation, Colloque international CNRS, Aix-en-Provence 21-23 mai 1996*, 69-78.

Brun, J.-P. and Borréani, M., 1998. Deux moulins hydrauliques du Haut-Empire romain en Narbonnaise: Villae des Mesclans à la Crau et de Saint-Pierre/Les Laurons aux Ares (Var). *Gallia: Archéologie de la France antique,* 55, 279-326.

Cara Barrionuevo, L. and Rodríguez López, M.J., 1986. Notas para el estudio de la minería almeriense anterior al siglo XIX. *Boletín del Instituto de Estudios Almerienses*, 6, 11-24.

Cara Barrionuevo, L., Cara Rodríguez, J. and Rodríguez López, M.J., 1988. Las Cuevas de la reserva (Roquets) y otras factorías pesqueras de época romana en la Provincia de Almería. *Boletín del Instituto de Estudios Almerienses,* 8, 53-72.

Cara Barrionuevo, L., García López, J. L., Lentisco Puche, J. D. and Ortiz Soler, D., 1999. Los molinos hidráulicos tradicionales de la Apujarra (Almería). *Instituto de Estudios Almerienses,* 32. Almería.

Carracedo Sánchez, M., Sarrionandia, F., Arostegui, J., Larrondo,E. and Gil Ibarguchi, J.I., 2009. Development of spheroidal composite bombs by welding of juvenile spinning and isotropic droplets inside a mafic eruption column. *Journal of Volcanology and Geothermal Research*, 186, 265-79.

Carrasco Serrano, G.., 2007. Vías de comunicación y moneda en torno a *Sisapo* en época romana. *Gerión,* Vol. Extra, 363-73.

Castella, D., 1994. *Le moulin hydraulique gallo-romain d'Avenches 'En Chaplix'. Fouilles 1990-1991, Cahiers d'archéologie romande*, Lausanne.

Cebriá, J.-M. and López-Ruiz, J., 1995. Alkali basalts and leucitites in an extensional intracontinental plate setting: The late Cenozoic Calatrava Volcanic Province (central Spain). *Lithos,* 35, 27-46.

Cebriá, J. M., López-Ruiza, J., Doblasa, M., Oyarzunb, R., Hertogenc J. and Benito, R., 2000. Geochemistry of the Quaternary alkali basalts of Garrotxa (NE Volcanic Province, Spain): a case of double enrichment of the mantle lithosphere. *Journal of Volcanology and Geothermal Research,* 102, 3-4, November 2000, 217-35.

Checa, A., Jimeno, A., Tresserras, J.J., Benito, J.P. and Sanz, A., 1999. Molienda y economía doméstica en Numancia. In Burilllo, J. (coord.), *IV Simposio sobre los Celtíberos, Economia, Homenaje a José Luis Argente Oliver*, Daroca 1997, Zaragoza, 63-8.

Cunningham, C.G., Arribas, A. Jr., Rytuba, J. J. and Arribas, A., 1990. Mineralized and unmineralized calderas in Spain; Part I, evolution of the Los Frailes Caldera. *Mineralium Deposita, 25 (Suppl.)*, 21-8.

Curwen, C., 1941. More about Querns. *Antiquity,* 15, 15-32.

Desirat, G.,1996. *Bagnols en Fôret, Var, Contribution à l'étude de la Provence orientale*, 1, Millau.

Di Battistini, G., Toscani, L., Iaccarino, S. and Villa, I. M., 1987. K/Ar ages and the geological setting of calc-alkaline volcanic rocks from Sierra de Gata, SE Spain. *Neues Jahrbuch für Mineralogie Monatshefte*, H8, 369-83.

Fernández Ochoa, C., Caballero A. and Morano, C., 1982-1983. Nuevo documento epigráfico para la localización de Sisapo. *Cuadernos de prehistoria y arqueología,* 9-10, 211-22.

Fernández Ochoa, C., Zarzalejos Prieto, M., Burkhalter Thiébalt, C., Hevia Gómez , P. and Esteban Borrajo, G.., 2002. *Arqueominería del sector central de Sierra Morena. Introducción al estudio del Área Sisaponense. Anejos de AespA, XXVI*, Madrid.

Fernández Soler, J. M., 1992. *El Volcanismo Calco-Alcalino de Cabo de Gata (Almería): Estudio volcanológico y petrológico*, Tesis Doctoral, Universidad de Granada.

Gähwiler A. and Speck, J., 1991. Die römische Wassermühle von Hagendorn bei Cham ZG: Versuch einer Rekonstruktion. *Helvetia Archaeologica,* 86, 34-75.

Gonzalez Tascón, I. 2010: http://club.telepolis.com/nachoben/TrydacnTelepolis/arquitecturarural/FTP/los_molinos_y_las_acenas.htm#BIBLIOGRAF%CDA. Accessed: 2/11/2010.

Harb, P., 2002. Roder/Klein Büel. *Archäologie und Denkmalpflege im Kanton Solothurn,* 7, Solothurn, 68-72.

Harms, E. and Mangartz, F., 2002. *Von magma zum mühlstein. Eine zeitreise durch die lavaströme des Bellerberg-Vulkans, Mainz, Erschienen zum hundertfünfzigjährigen jubiläum des Romisch-Germanischen Zentralmuseums*. Mainz.

Heras, de las, J., 2008. *La orden de Calatrava, religión, guerra y negocio*. Madrid.

Hörter, F., 1994. *Getreidereiben und Mühlsteine aus der Eifel. Ein Beitrag zur Steinbruch- und Mühlengeschichte*, Mayen.

Jaccottey, L. and Labeaune, R., 2010. Découvert d'un moulin hydraulique Antique à Longvic, (Côte-d'Or). *Revue Archéologique de l'Est*, 59, 665-68.

Jaccottey, L. and Longepierre, S., Pompeian Millstones in France. *This volume.*

Leveau, Ph., 2006. Les moulins de Barbegal (1986-2006). *TRAIANVS.* <http://www.traianvs.net/index.php>.. Accessed: 18/11/2010

López Domech, R., 1996. *La Región Oretana, Anejos de Antigüedad y Cristianismo III, Universidad de Murcia*, Murcia.

López-Ruiz, J., Cebriá, J.M., Doblas, M., Oyarzun, R., Hoyos, M. and Martín, C., 1993. Cenozoic intra-plate volcanism related to extensional tectonics at Calatrava, central Iberia. *Journal of the Geological Society*, 150, 915-22.

Luquet, A., 1966. Blé et meunerie à Volubilis. *Bulletin d'archéologie marocaine,* 6, 301-16.

Madoz, P., 1846. *Diccionario geográfico-estadístico-histórico de España y sus posesiones de Ultramar,* 5, Madrid.

Martin, J.M., Braga, J.C. and Betzler, C., 2003. Late Neogene uplift of the Cabo de Gata volcanic province, Almeria, Spain. *Geomorphology*, 50, 27-42.

Martínez López, F.J., Anderson, T. and Granero Gallego, A., The Quern and Millstone Quarry of the Rambla Honda, Almería, Spain. This volume Chapter 00 ??

Martínez Maganto, J., 1994. El litoral del LE. peninsular en época romana. Algunas cuestiones en torno a su explotación económica y comercial, *Gerion, 12*, 197-215.

Mínguez Morales, J.A. and Ferreruela Gonzalvo, A. 1992. La villa romana de las Coronas (Pallaruelo de Monegros, Huesca. *Bolskan,* 9, 49-50, 133-58.

Morin, J., Sánchez, M., Barroso, R., Escolà, M., González, L., López M. and Sánchez F., 2003. El Cerro Martos (Herrera, Sevilla): una almazara de época altoimperial. *Bolskan,* 20, 151-62.

Moritz, L. A., 1979. *Grain-Mills and Flour in Classical Antiquity*, New York, reprint, originally published in 1958.

Padilla Monge, A., 1999. Consideraciones en torno a la explotación del Mármol en la Bética durante los sigles I-II. *Habis,* 30, 271-81.

Pallí, L. and Pujadas, A., 1999. El Campo Volcánico Catalán. *Enseñanza de las Ciencias de la Tierra*, 7.3, 229-36.

Palomo, J. and Fernández, M.P., 2006-2007. Los molinos hidráulicos en la Antigüedad. *Espacio, Tiempo y Forma*, Serie II, *Historia Antigua,* 19-20, 499-524.

Peacock, D.P.S., 1980. The Roman millstone trade: a petrological sketch. *World Archaeology,* 12,1, 43-53.

Peacock, D.P.S., 1989. The mills of Pompeii. *Antiquity,* 63, 205-14.

Pellicer Bautista, M. J., Unpublished report. Memoria explicativea sobre la ubicación geológica y materiales geológicos empleados en las construcciones del yacimiento arqueológico de la Ermita de Oreto-AZuqueca, Granátula de Calatrava (Cuidad Real). In Romero Salas, H. and Garcés Tarragona, A., *El Complejo Arqueológico de Oreto-Zuqueca. Campañas 1996-1997. Granátula de Calatrava. Ciudad Real, II. Diputación Provincial de Ciudad Real.*

Portillo M., 2006. *La mòlta i triturat d'aliments vegetals durant la protohistòria a la Catalunya Oriental,* Doctoral thesis, University of Barcelona. http://www.tesisenred.net/es.

Quesada Sanz, F., Kavanagh de Prado, E. and Moralejo Ordax, J., 2010. El asentamiento de época Ibérica en el Cerro de la Cruz. In Muñiz Jaén, I. and Quesada Sanz, F. (eds), *Oikos, 2, Cuadernos Monográficos del Ecomuseo del Río Caicena*, 75-95.

Rock, N.M.S., 1982. The Late Cretaceous Alkaline Igneous Province in the Iberian Peninsula and its tectonic significance. *Lithos*, 15, 111-31.

Röder, J., 1957. Die antiken Tuffsteinbrüche des Pellenz, Zur Steinbruchgeschichte des Pellenz- und Brohltaltuffs. *Bonner Jahrbücher*, 157, 213-71.

Rytuba, J.J., Arribas, A. Jr., Cunningham, C.G., McKee, E.H., Podwysocki, M.H., Smith, J.G.,

Kelly, W.C. and Arribas, A., 1990. Mineralized and unmineralized calderas in Spain; Part II, evolution of the Rodalquilar caldera complex and associated gold-alunite deposits. *Mineralium Deposita, 25 (Suppl.)*, 29-35.

Scotney, P., Burgess, R. and Rutter, E. H., 2000. 40Ar/39Ar age of the Cabo de Gata volcanic series and displacements on the Carboneras fault zone, SE Spain. *Journal of the Geological Society,* 157, 1003-8.

Welfare, A.T., 1985. The milling-stones. In Bidwell, P., *The Roman Fort of Vindolanda at Chesterholm, Northumberland, Historic Buildings and Monuments Comission for England, Archaelogical Report 1*, London, 154-64.

Williams-Thorpe, O. and Thorpe R., 1987: Els origens del molins romans de pedra del NE de Catalunya. *Vitrina, 2,* Museu Comarcal de la Garrotxa, 50-8.

Williams-Thorpe, O., 1988. Provenancing and Archaeology of Roman Millstones from the Mediterranean Area. *Journal of Archaeological Science*, 15.3, 253-305.

Williams-Thorpe, O. and Thorpe R., 1988. The Provenance of Donkey Mills from Roman Britain. *Archaeometry*, 2, 275-88.

Wilson, M. and Downes, H., 2006. Tertiary-Quaternary intra-plate magmatism in Europe and its relationship to mantle dynamics. In Gee, D. G.. and Stephenson, R. (eds.), *European lithosphere dynamics, Geological Society of London Memoire* 32, London: Geological Society of London, 147-66.

Querns as markers for the determination of Medieval northern European trade spheres

Meinrad Pohl

1. Trade spheres and goods

Looking at distribution maps for non-luxury items in the North Sea and Baltic Sea region, northern Europe seems to be separated into a western and an eastern trade sphere, roughly overlapping with the influence sphere of the Holy Roman Empire and England in the West and the Scandinavian influence sphere in the East. In contrast to luxury items which can be found all over northern Europe, there seems to be a boundary between the distribution areas of utilitarian items and basic commodities roughly running through Jutland from the western present day mouth of the Limfjord to the Veilefjord and from there southwards along the Baltic Sea coast prolonged by the River Elbe (Carelli and Kresten 1997; Sindbæk 2005; Steuer 1987). This is, of course, rather an area of intersection of those two trade spheres than a strict borderline.

Even though the trade spheres are distinguishable, the goods are often incomparable due to differences in category and time. Those of the western trade sphere are, for example, querns of Mayen basalt, Rhenish pottery and Rhenish tuff as building material, and sceattas as coins. While the querns were found from the Roman period throughout the time analysed, there are different types and time horizons for pottery and coins, and the building material is restricted between the mid 11th century AD to the end of the 12th century (Haiduck 1992). In the East, the goods were querns of Hyllestad garnet-muscovite-schist, Baltic pottery, soapstone vessels, other soapstone items and Arabic silver coins. The querns date from the 10th century onwards, the soapstone items at least from the 8th century until the 11th century (Baug 2006, 10), the Arabic silver coins date from the 10th century.

2. Querns as markers

For a meaningful comparison comparable goods are needed. Besides pottery, the only comparable goods in time and category are the querns. Pottery could be packaging for other goods or could have been traded on again after it already had reached its destination. Some pottery could even have been that prestigious to be subject to gift exchange, as for example the Tating vessels.

While these factors rule out pottery as a marker, querns seem to be suitable. A quern was an essential item since the European diet consisted mainly of cereals. Combined with the fact that a quern was worn out after five to seven years (Hörmann and Richter 1983, 93) this must have created a continuous, inelastic basic demand, dependent only on population but independent of economic cycles. So, it can be assumed that they really mark the desired goods and are not just a by-product of transport and trade. In addition, it can be assumed that they would remain at their destination until they were worn out. Querns could possibly be a marker for other goods which shared similar properties, had a similar origin or belonged to the same consignment of a merchant, but did not leave visible traces in the archaeological record. Examples could be wine, cereals, or walnuts, etc. (McCormick 2001, 610).

The result of accepting querns as markers for Medieval trade spheres, the Mayen lava querns representing the western sphere and the Hyllestad querns the eastern sphere, should show the following picture: the area of distribution of Mayen querns extends from the quarries down the river Rhine spreading from the Netherlands along the British Channel and North Sea coast with the area around York as the northern-most find spot in England. Then eastwards along the North Sea coast north to the present day mouth of the Limfjord in northern Jutland, and upstream with the rivers flowing into the North Sea (Fig. 1). The area of distribution of the Hyllestad querns extends from the quarries south along the Norwegian coast spreading from the Skagerrak westwards along the coast of northern Jutland to the mouth of the Limfjord and eastwards along the Kattegat, the coasts of the Danish Isles, the Baltic Sea coast of Jutland and eastwards to the Baltic Sea to Bornholm and up the rivers flowing into the Baltic Sea at those coasts (Fig. 2). The boundary between these spheres marked by querns is as mentioned above (Fig. 3).

3. Determination

The next question is how these trade regions were constituted. What was the reason for their development? Were they politically, ethnically, technically or economically determined?

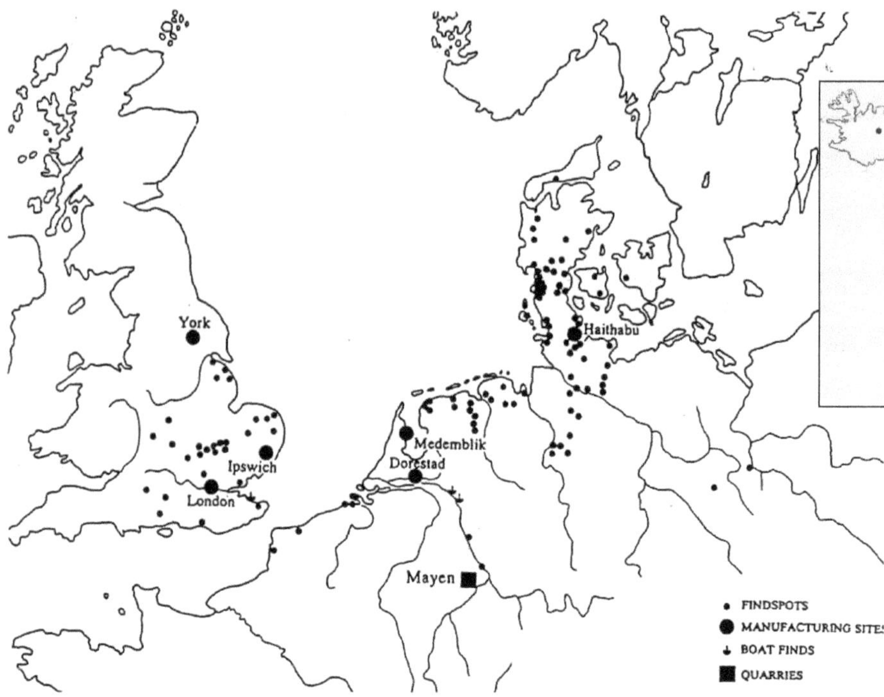

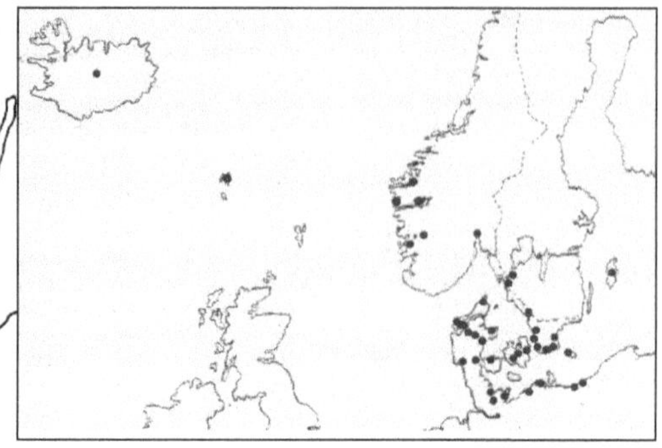

Figure 2. Distribution of Hyllestad querns (Baug 2002).

Figure 1. Distribution of Mayen querns (Parkhouse 1997).

Carelli and Kresten (1997) discuss political and cultural reasons as possible explanations for the distribution of querns. They do mention also 'technical transport considerations' but remain somewhat unspecific about what that means. Likewise, they dismiss them to be constitutive since there had been no technical obstacles for the distribution of the querns and come to the conclusion that these trade spheres had been - at least partly - culturally determined (Carelli and Kresten 1997, 125ff.). In the following paragraphs different reasons for the determination of these trade spheres will be discussed.

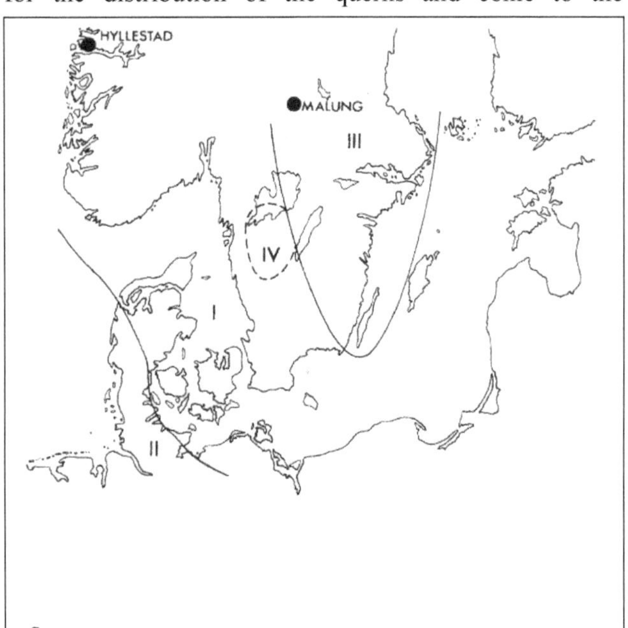

Figure 3. Northern European quernstone regions (Carelli and Kresten 1997), showing the eastern trade sphere represented by the Hyllestad querns (I) and the western trade sphere represented by the Mayen querns (II) and the boundary in between.

3.1. Political

As Carelli and Cresten point out, the Hyllestad querns spread along the area of Viking Age and Medieval Denmark limited by the Medieval border between Sweden and Denmark (Carelli and Kresten 1997, 126) (Fig. 3). So it can reasonably be assumed that the distribution stops at political borders. But the contrary is the case. The Mayen stones in the west are not only found in the former territory of the Holy Roman Empire, but also in England, and along the North Sea coast of Viking Age and Medieval Denmark in the important towns of Ribe and Hedeby and its successor Schleswig at the Baltic Sea coast (Fig. 1). The River Elbe as the eastern border of the Holy Roman Empire applies just until the 11th century. Thus it is doubtful that those trade regions were only politically determined.

3.2. Ethnical

Another explanation for those trade spheres could be of ethnic nature, counting the North Sea as a Frisian domain and the Baltic Sea as a Scandinavian domain. But this assumption is difficult to verify. Even if ship types were accepted as indicators for the ethnicities of the merchants, assuming that ships of an Anglo-Scandinavian type have been used by Scandinavians and cogs by Frisians, we get anything but a clear picture. Both ship types have been identified, for example, in the archaeological record

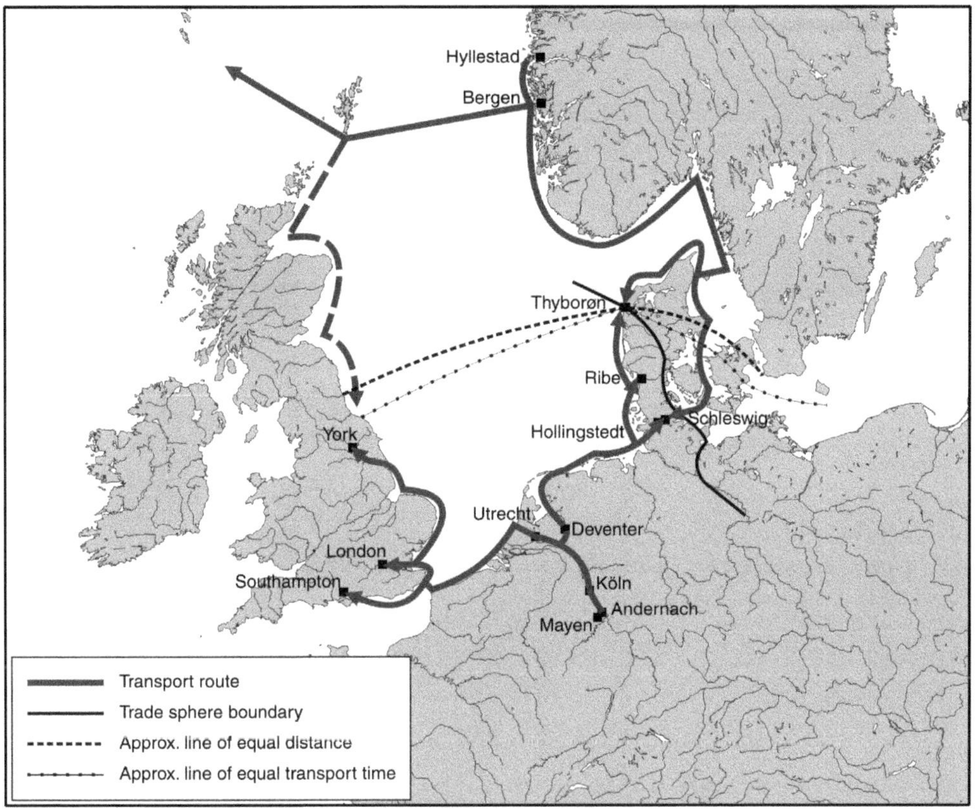

Figure 4. Trade routes, approximate lines of equal distances and equal transport times and boundary separating eastern and western trade sphere.

of Hollingstedt, which served as a North Sea harbour for Hedeby and Schleswig (Brandt 2005, 274). And the possibility of a Frisian merchant owning a Scandinavian ship and vice versa cannot be dismissed. An ethnical determination can also be ruled out.

3.3. Technical

3.3.1. Distances

For a technical determination of those trade areas it is worthwhile to take a look at distances and thus at the boundaries of the distribution areas. Places on those boundaries should share the same distance from the places of origin of the goods traded within the spheres. Before distances can be measured, the trade routes have to be taken into account. Researchers agree that coastal sailing was predominant in the Middle Ages thus avoiding the open sea and dangerous waters whenever possible (Ellmers 1972, 227).

With this prerequisite, the trade route for the Mayen querns was as follows: the querns were transported overland from Mayen to Andernach and then shipped down the Rhine to the trade emporia in the Netherlands, first Dorestad, later Tiel which was succeeded by Deventer and Utrecht. From Utrecht the ships to England followed the coast westward to cross the channel at the sites of present day Calais and Dover, following the coast and the rivers to end in Southampton in the south and in York in the north. From Deventer the ships went north through the Zuiderzee and eastward along the Frisian coasts up the Cimbrian Peninsula with the area around the present day mouth of the Limfjord near present day Thyborøn as an end to the North. Also, here trade went on up the rivers which were flowing into the sea at those coasts. If going to Hedeby or Schleswig, the Mayen querns were shipped via the rivers Eider and Treene to Hollingstedt, from where they were transported further overland.

The Hyllestad querns would be taken south along the Norwegian coast, crossing the Kattegat between the sites of present day Gothenburg and Skagen. There the route would bifurcate into a western route along the northern coast of Jutland to Thyborøn and a southern route following the coasts of the Kattegat along the coast of Scania, the Danish Isles and the coast of the Cimbrian Peninsula and the Baltic Sea coast to Bornholm in the East (Fig. 4).

Measuring the distances, the results seem to confirm the assumption that the trade spheres were dependent on distances, at least in most cases. In the case of York the explanation might be that Mayen querns had to be transported over a distance of c.1200 km while the Hyllestad querns would have to cover a distance of c. 1600 km (Fig. 5). Mayen querns are dominant around York, where no Hyllestad querns are known. The point of equal distances would be around Newcastle upon Tyne, so we could expect Hyllestad querns north and Mayen querns south of there. But the distribution of the Mayen querns reaches its largest extent in the area around York. Northern England and Scotland seem to have been self-sufficient and produced their own querns, which had been competing successfully against imported querns (Campbell 1987;

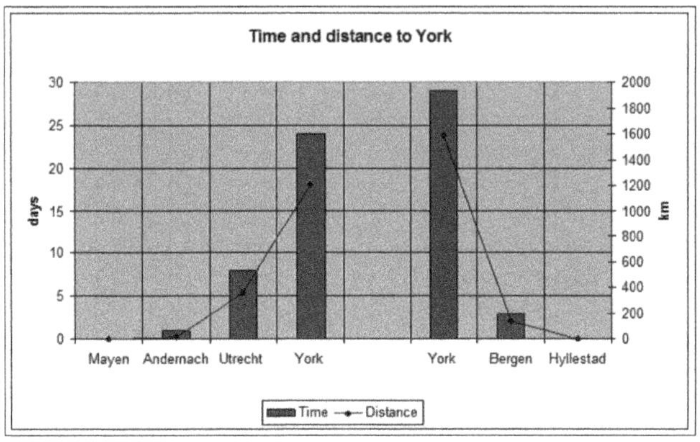

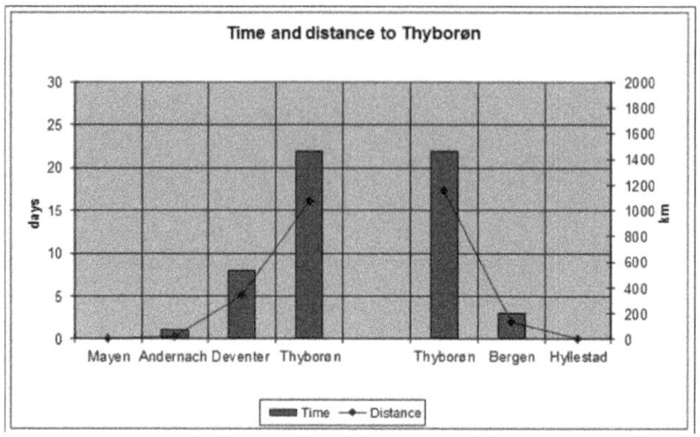

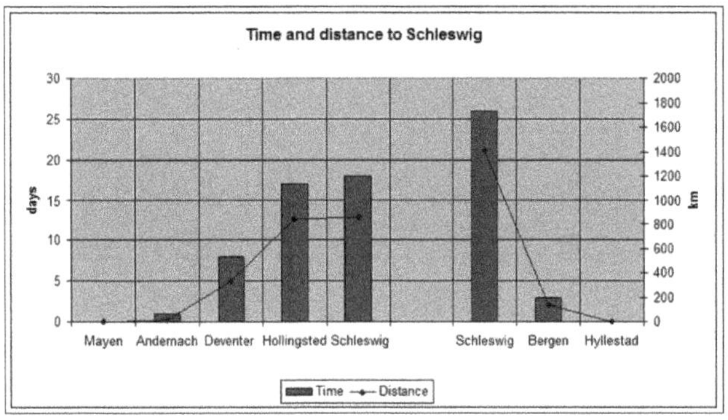

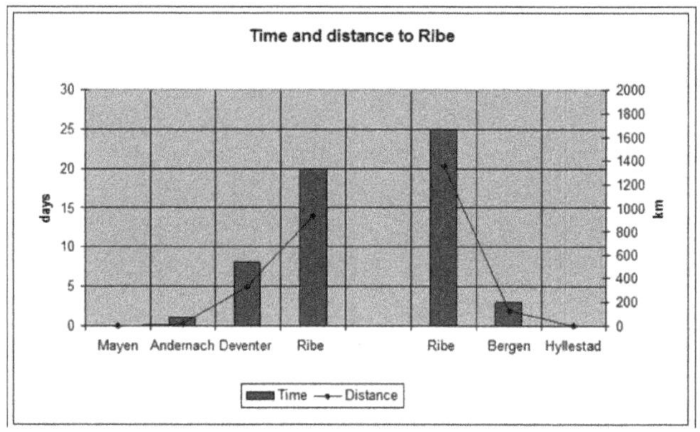

Figures 5-8. (clockwise from top left). Distances and transport time from Mayen and Hyllestad to different markets. Showing increasing distances and transport time from Mayen from left to right and from Hyllestad from right to left.

Tucker 1984; Watts 2006). We find Hyllestad querns firstly in the Faeroes and again in Iceland (Baug 2002) (Fig. 2). For northern Jutland the results more or less fit. To Thyborøn the Mayen querns had to cover 1080 km against 1150 km for the Hyllestad querns (Fig. 6). Distances can explain the meeting of the distribution spheres in northern Jutland in the area of present day Thyborøn. Neither is it surprising finding Mayen querns dominating the material of Ribe and the rest of the North Sea area, since they had to cover just 925 km to Ribe, while the Hyllestad querns which dominate in the Baltic Sea area had to cover 1350 km (Fig. 7).

The case of Hedeby and its successor Schleswig is very interesting, since it does not fit into this scheme. Hyllestad querns are present in the material, but Mayen querns dominate, although Hedeby and Schleswig are situated closer to the Baltic Sea coast and therefore should also belong to the sphere of the Hyllestad querns. The Mayen querns had to cover around 860 km to Schleswig while the Hyllestad querns had to cover around 1400 km (Fig. 8). Looking only at distances, the Mayen querns should dominate not only in Schleswig, but also in the Baltic Sea area up north to a line Århus - Copenhagen - Malmö, which roughly represents a line of equal distances. Thus distances can work as a part explanation, but this is insufficient for the case of Hedeby and Schleswig.

3.3.2. Duration of journeys

Another important factor in transportation, and closely connected with distances, is time. The distances divided by the pace of appropriate cargo carriers should result in the time needed for travelling from the quarries to the final destinations. The speed of cargo carriers can be found in different sources. As quoted by Ellmers, this would be 30 NM per day for sea ships, 50 km per day for riverboats downstream (Ellmers 1972, 248-254). For overland transport with wagons or carts 20 km per day can be assumed (Bartels 1994, 106; Dennecke 1987, 217; Straube 1997, 218). Considering the routes and conditions as above the result would be as follows. The cases of York, Thyborøn and Ribe would not be altered significantly (Figs 5-7). On the other hand neither distances nor transport time can explain the case of Hedeby and Schleswig. While it took the Hyllestad querns 25 days to reach Schleswig it took the Mayen querns just 18 days (Fig. 8). Regarding time, the Mayen querns should dominate Schleswig and the Baltic Sea area north to a line from Århus, separating Sjælland from the northwest to the southeast and prolonged through the Baltic Sea leaving Bornholm to the sphere of Hyllestad (Fig. 4). But with the exception of Schleswig, the Baltic Sea area is dominated by the Hyllestad querns. So neither political borders, ethnicity of the merchants nor distance and transport time can wholly explain the distribution of the querns.

3.4. Economical - Transport expenditure

As seen before, distances and time work as an explanation for certain points but fail in the case of Schleswig. So maybe that an economic explanation, transport costs as a factor connected to distances and time, could be an explanation. Every transport involves costs. Ships need to be built, sailors paid, etc. The problem in calculating transport costs is that there is no information about rates for transportation costs or other services available. Neither do we know in which way these 'costs' were compensated. In an attempt to calculate the possible transport expenditures for the querns from Mayen and Hyllestad, I set up the following simplified model based on certain assumptions. In this model I shall try to measure transport expenditures in terms of the required workforce, using the unit 'man-day', which means the workforce of one man per day.

The trade spheres seem to have been stable over a longer period but many conditions changed in the meantime, so that it was necessary to define a point in time to get stable conditions. As a point in time the year 1100 is arbitrarily chosen. Within this time frame the geographical framework is set up. At that time, Andernach served as a port of transhipment for the Mayen querns, Cologne was not yet a staple port, Utrecht and Deventer were the important hubs between the river Rhine and the North Sea, Southampton, London and York did exist, Schleswig had succeeded Hedeby and Bergen was established as a town and presumably already served as a staple port for western Norway where the Hyllestad querns were brought to (Hansen 2008, 27) (Fig. 4). For this point in time sizes of ships are known, it is possible to estimate the number of their crew (Crumlin-Pedersen 1989, 75) and there are sources which tell how much time a certain journey took (Ellmers 1972, 250). So it should be possible to calculate how many people were needed over a certain time to transport goods. Routes and distances used above will be applied. Furthermore it will be assumed that transhipments were avoided wherever possible. Eventual tolls will be neglected.

3.4.1. Routes

Even though the routes mentioned above will be applied, some specifications need to be made. From Mayen the querns would be transported overland by ox cart, possibly also on a flat bottomed boat on the River Nette to Andernach. To keep the model simple, equal transport expenditures will be assumed for the transport over the River Nette and overland. The latter will be used in the calculations. To Britain, the querns would be shipped via Utrecht, to Denmark via Deventer. The crossing of the North Sea from the Rhine estuary to Lowestoft Ness and the crossing of the Skagerrak as shown in a map by Ellmers (Ellmers 1972, Karte 1) will be omitted due to incompliance with the assumptions in 3.4.4. Mayen querns destined for Schleswig would be shipped to Hollingstedt and transported over the remaining 17 km by ox cart to Schleswig (Brandt 2005, 276 f.). Bergen is assumed as a port of transhipment for Hyllestad querns (Fig. 4).

3.4.2. The cargo

In the wreck finds from Norway, querns were the only part of the cargo that has been preserved. Even though it is difficult to imagine that a ship was loaded only with querns and a merchant would have rather had a bundle of different goods on board his ship; a cargo of querns only will be assumed. The quernstones found on the way to the markets were blanks weighing between 20 and 25 kg (Parkhouse 1997, 102; Mangartz 2008, 125; Baug 2002, 75; Feveile 2010). For a quernstone 25 kg will be assumed, 50 kg for a complete quern.

3.4.3. Cargo carriers

3.4.3.1. Land carriage

For overland transport, wagons or carts drawn by animals with a cargo capacity of one ton will be assumed. Since reliable data for the actual point in time are lacking, data from later wagons are taken into account and their cargo capacity reduced slightly (Bartels *et al.* 2007, 234, 273; Fessner 1998, 23-28, 181; Straube 1997, 214).

3.4.3.2. River boats

In the 11th century the sea ships in the North Sea and Baltic did not exceed a cargo capacity of 50 t. For river boats the capacity seems to be somewhat lower. Even though flat bottomed ships with a cargo capacity up to 60 t had been used by the Romans on the Rhine (Bockius 2000, 479 ff.), there is no evidence that these cargo capacities had been reached again in the 11th century. According to research to date all the finds until that period suggest that the maximum cargo capacity did not exceed 20 t (Ellmers 2002; 1978; 1980; Hoffmann 2002; Obladen-Kauder and Peiss 2000). A cargo capacity of 20 t for a river boat will be assumed.

3.4.3.3. Sea ships

Until the end of the 11th century a cargo capacity of 50 t was possible for a sea ship (Crumlin- Pedersen 1989, 79; Ellmers 1972, 257) but a couple of facts do speak against assuming ships with the maximum cargo capacity of 50 t in this model. Presumably no other goods than quernstones and some other stone products were exported from Hyllestad. Ships were apparently travelling to Hyllestad with little or no cargo, since they had to travel with ballast (Baug 2002, 74). In the Alverstraumen a cargo of 505 Hyllestad quernstones weighing 13 t was found. Both these facts would suggest a ship with a cargo capacity slightly more that 15 t. In addition to that, a sea ship exceeding the cargo capacity of 20 tonnes would be no longer able to pass through the Treene to Hollingstedt (Rohde 1986, 321) or the Riber Å to Ribe, the same can be assumed for the Ouse to York. Feveile (1995, 42) assumes that ships with a cargo capacity between 15 and 20 t were common at that time. For the applied model these restrictions suggest assuming a sea ship with a cargo capacity of maximum 20 tonnes that was able to sail the Rhine to Cologne, the North Sea Coast, the Thames to London, the Humber and Ouse to York, Eider and Treene up to Hollingstedt and the Riber Å to Ribe.

3.4.4. Speed

For water transport it will be assumed that coastal sailing was used in favour of open water sailing whenever possible, sailing in daytime only and anchoring/harbouring during the night. As used above, a coverable distance per day of 50 km will be assumed for river downstream and 30 NM on sea (Ellmers 1972, 250). For overland transport a coverable distance per day of 20 km will be assumed (Straube 1997, 218; Dennecke 1987, 217).

3.4.5. Crew

For the ships Skuldelev 1 and 3, Crumlin-Pedersen assumes a crew of five to eight people (Crumlin-Pedersen 1989, 75). For riverboat transport of stones for the building of Utrecht Cathedral, a crew of six seemed to be usual (Jappe Alberts 1954, 24). Thus an equal crew of six people will be assumed for riverboats and sea ships. For a wagon or cart a crew of one man will be assumed.

3.4.6. Cost of manpower

For Mayen and Hyllestad querns equal production costs, thus an equal price, will be assumed. For the whole area of interest also equal costs of manpower will be assumed. For a change of the cargo carrier expenditures of half a man-day per metrical ton will be assumed. Prior investments in ships or oxcarts are neglected as sunk costs. The costs of draught animals will also be neglected since they are not transferable to man-days.

3.4.7. Transport expenditure

Taking all these assumptions in to account, the formula for calculating the transport expenditures is shown in figure 9.

4. Evaluation of the results

Calculating the transport expenditures according to the formula above leads to the following results. For York the picture remains the same. The transport of 20 t or 800 querns from Mayen would take 138 man-days of shipping, plus 20 man-days for overland transportation to Andernach, plus 20 man-days for transhipments there and in the Netherlands, resulting in 178 man-days in total and 8.9 man-days average per ton or 0.445 man-days per quern. The transport from Hyllestad would take 174 man-days for shipping, adding 10 man-days for transhipment in Bergen, resulting in 184 man-days in total or 9.2 man-days average per ton or 0.46 man-days per quern, leaving the Mayen querns cheaper at York (Fig. 10).

For Thyborøn the picture tilts over in favour of the Hyllestad querns. The transport of the Mayen querns would take 126 man-days for shipping, plus 20 man-days for overland transportation, plus 20 man-days for transhipment in Andernach and the Netherlands resulting in a total of 166 man-days or an average of 8.3 man-days per ton or 0.415 per quern. The transport of the Hyllestad querns would take 132 man-days in shipping, plus 10 man-days for transhipment in Bergen resulting in 142 man-days in total, leaving the Hyllestad querns with an average of 7.1 man-days per ton or 0.355 man-days per quern, with lower transport expenditures than the Mayen stones (Fig. 11).

In Ribe the picture stays in favour of the Mayen querns. It would take them 114 man-days for shipping, plus 20 man-days for overland transport, plus 20 man-days for transhipments in Andernach and the Netherlands, resulting in a total of 154 man-days or an average of 7.7 man-days per ton or 0.385 man-days per quern. The cargo of Hyllestad querns would take 150 man-days in shipping, plus 10 man-days for transhipment in Bergen, resulting in 160 man-days in total or an average of 8 man-days per ton or 0.4 man-days per quern (Fig. 12).

For the case of Schleswig the transport expenditures of Mayen querns and Hyllestad querns would be the same with 166 man-days in total or an average of 8.3 man-days per ton or 0.415 per quern.

But the interesting thing is the composition of the expenditures. For the Mayen querns they would consist of just 96 man-days in shipping, but 40 man-days in overland transport and a further 20 man-days in transhipment. The transport of querns from Hyllestad to Schleswig would take 156 man-days in shipping but just 10 man-days in transhipment (Fig. 13). This congruence can be seen as an arithmetical coincidence caused by the general conditions of this model. But it shows a possible explanation for the overlap of the trade spheres at Schleswig.

$$\text{overland transport} + \text{river transport} + \text{sea transport} + \text{transhipments} =$$

$$\text{man-days} \times \frac{\text{quern stones}}{\text{cargo capacity cart}} + \text{man-days} \times \frac{\text{quern stones}}{\text{cargo capacity riverboat}} + \text{man-days} \times \frac{\text{quern stones}}{\text{cargo capacity sea ship}} + \text{transhipments} \times \frac{\text{quern stones}}{80}$$

Figure 9. Transport expenditures.

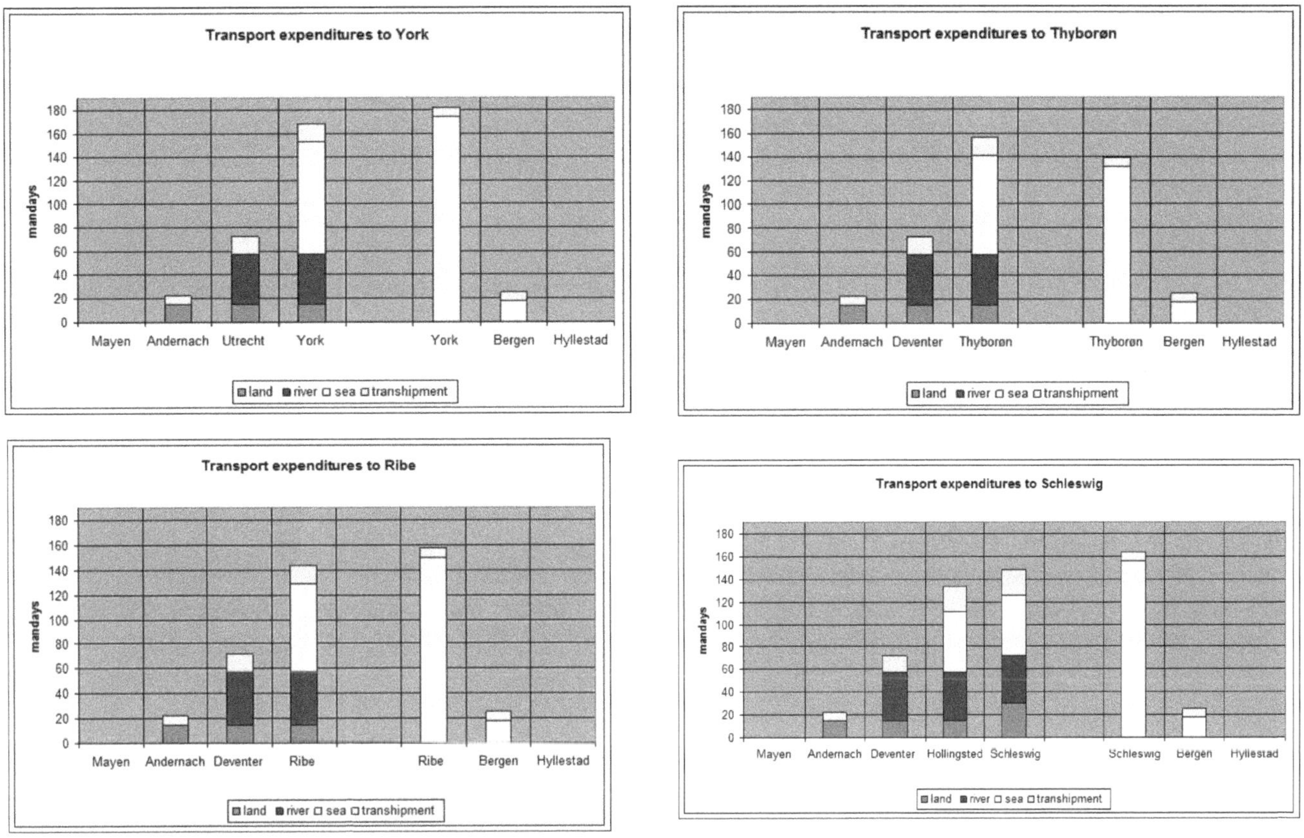

Figures 10-13 (clockwise from top left). Transport expenditures from Mayen and Hyllestad to different markets and their composition. Showing increasing transport expenditures from Mayen from left to right and from Hyllestad from right to left.

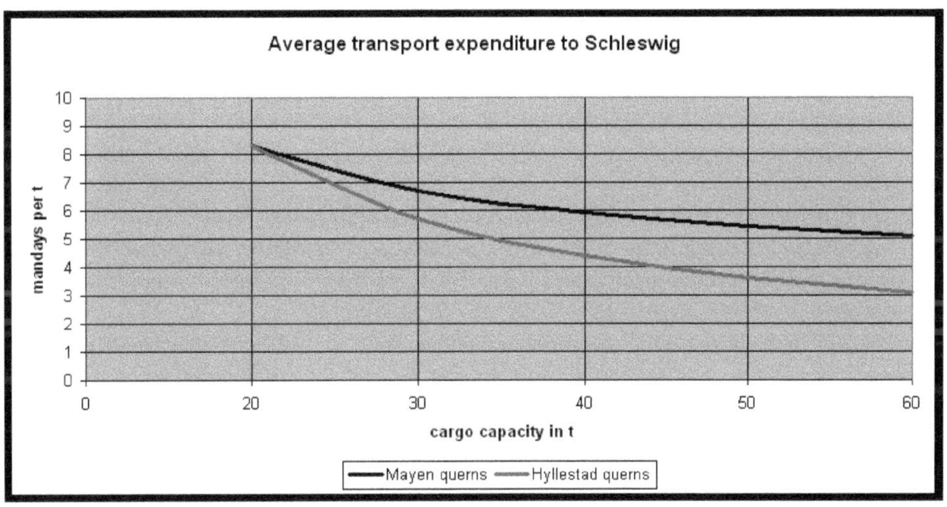

Figure 14. With increasing cargo capacities average transport expenditures to Schleswig from Mayen and Hyllestad are decreasing.

5. Shortcomings

The model as developed above provides a plausible explanation for the determination of Medieval northern European trade spheres on transport expenditure. While it remains a simplified model constructed on assumptions, in turn based on research results as much as possible, it also has some shortcomings. The trade routes are not known for certain. The same applies for the ports of transhipment. But in the context of this model it does not matter whether the querns were transhipped onto sea ships in Utrecht or Deventer or even in Cologne, which might have been possible until the 13th century. Distances, transport times and thus transport expenditures would have varied but a little if at all. Another problem concerning the number of transhipments is that it is impossible to tell where and how often the querns changed their owner on their way to the markets and if this would have had repercussions to their price.

Some data were either not available and could not have been taken into account, would have been insignificant for the results or would have complicated the model. Among the unconsidered circumstances were currents and winds, longer days in summer in Scandinavia allowing longer coverable distances, costs for building ships and carts, cost for draught animals, real production costs and prices of querns, toll tariffs, local proportions of power, personal relations of merchants and nobles, routines and possible product affinities.

Another significant shortcoming is cargo capacities. Even though there are enough reasons to assume a cargo capacity of circa 20 t, a carrier with a greater cargo capacity would have been possible. Applying bigger carriers to this model, the balance would tilt in favour of the Hyllestad querns (Fig. 14). They could use economies of scale because they would not have been subject of numerous transhipments or longer overland transport. A calculation involving bigger carriers becomes problematic in this model, because they would not be able to pass the waterways to Hollingstedt, Ribe and York. Additional transhipments would become necessary without knowing where. But even with additional transhipments it can be expected that the advantage would stay on the side of the Hyllestad querns.

Another problem is the understanding of time. The concept of time underlying this model is modern. It is unknown and will possibly remain unknown how time was valued in the period of observation. Thus we know neither how manpower nor how working time was paid. Also, the assumption that a merchant would sail with a cargo of only quernstones seems improbable. He would rather have sailed with a diversified bundle of goods of more or less different origins. This would have made his considerations of transport expenditure a mixed calculation which is almost impossible to reconstruct. But these considerations had repercussions on the economic reach of merchants and their trade goods. Part of a Rhenish bundle could also be wine, pottery, cereals and cloth, while a Norwegian bundle could also contain whetstones, soapstone, timber and stockfish. A Medieval merchant would most probably not calculate in the same way as a merchant today, but he would certainly know whether it paid off to make a journey with a certain bundle.

6. Conclusion

Neither political nor ethnical reasons seemed to be plausible for the determination of Medieval northern European trade spheres. Distances and transport times provide at least a partial explanation. Despite all uncertainties and shortcomings, this model shows that transport expenditures are one possible explanation. Whether the figures inserted or the results are correct or not, this model shows that transport expenditure must have been an important factor in the determination of these trade spheres. Transhipments and overland transports increased the transport expenditures of Mayen querns, thus limiting their economical reach as they possibly did for other Rhenish trade goods. On the other hand Eider, Treene and Schlei, enabled a more economical crossing of the Cimbrian Peninsula, just making it possible for Mayen querns to reach the shores of the Baltic. So it is true that there were no technical obstacles to continued trade (Carelli and Kresten 1997, 126) and it would have been possible to transport the Mayen quernstones further east. But it would not have been economically reasonable.

Given that these assumptions are correct, the Cimbrian Peninsula would have been the barrier for the distribution of the Mayen querns, preventing them from penetrating into the markets of the Baltic, while the Skagerrak would have been the barrier for the distribution of the Hyllestad querns to the North Sea. The long distance from Hyllestad to the markets in the Baltic did not pose a problem, since costly overland transports and transhipments could be avoided and with rising cargo capacities the advantage of a long distance transport over sea without numerous transhipments came into effect. These economies of scale could not be utilised by the Mayen querns.

This model could possibly also be applied to the distribution of other goods with similar properties and competing goods in the other trade sphere. But as Sindbæk points out, these trade spheres seemed to be stable for several hundred years and were apparently also independent of competing goods (Sindbæk 2005). The distribution of the Mayen querns started about 200 years earlier in the western sphere than the distribution of the Hyllestad querns in the eastern sphere, where they began to spread in the same area where soapstone imports were common but Mayen querns were absent. He sees these distribution patterns caused by deep-rooted contact patterns and routine trade of spheres connected to the Rhineland trade or turning their back towards it (*ibid.*, 149f.). This estimate is not necessarily a contradiction to the model developed above but shows the possible interdependence of numerous factors.

There may not of course, be a single explanation for the constitution of those trade spheres. Also transport expenditure could be just one of numerous interdependent factors of which some were not considered, some could not be considered in this model and some may remain unknown.

References

Bartels, C., Fessner, M., Klappauf, L., and Linke, F.-A., 2007. *Blei und Silber aus dem Goslarer Rammelsberg*. Bochum.

Bartels, C., 1994. Zur Geschichte des Dachschieferbergbaues im Mittelhunsrück um Bundenbach. *Schriftenreihe des Schiefer-Fachverbandes in Deutschland e.V.*, 3, 91-132.

Baug, I., 2006. Stein som handelsvare. In F.B. Førsund (ed.), *Stein som Handelsvare.* Hyllestad, 6-15.

Baug, I., 2002. *Kvernsteinsbrota I Hyllestad.* Kongsberg.

Bockius, R., 2000. Antike Prahme. Monumentale Zeugnisse keltisch-römischer Binnenschiffahrt aus der Zeit vom 2. Jh. v. Chr. bis ins 3. Jh. n. Chr. *Römisch-Germanisches Zentralmuseum Jahrbuch*, 47, 439-493.

Brandt, K., 2005. Die Schleswiger Landenge in der Wikingerzeit und im Mittelalter. In C. von Carnap-Bornheim and H. Friesinger (ed.), *Wasserwege. Lebensadern – Trennungslinien.* Neumünster, 269-288.

Bremer, E., 2001. *Die Nutzung des Wasserweges zur Versorgung der römischen Militärlager an der Lippe.* Münster.

Campbell, E., 1987. A cross-marked quern from Dunadd. *Proc Soc Antiq Scot,* 117, 105-117.

Carelli, P. and Kresten, P., 1997. Give us this Day our daily Bread. *Acta Archaeologica*, 68, 109-137.

Crumlin-Pedersen, O., 1989. Ship Types and Sizes AD 800-1400. In Crumlin-Pedersen, O. (ed.) *Aspects of Maritime Scandinavia AD 200-1200.* Roskilde, 69-82.

Dennecke, D., 1987. Straße und Weg im Mittelalter als Lebensraum und Vermittler zwischen entfernten Orten. In B. Herrmann (ed.), *Mensch und Umwelt im Mittelalter.* Stuttgart, 207-223.

Ellmers, D., 2002. Baumschiff und Oberländer. In K. Elmshäuser (ed.), *Häfen, Schiffe, Wasserwege: zur Schiffahrt des Mittelalters.* Hamburg, 97-106.

Ellmers, D., 1980. Mittelalterliche Schiffe am Rhein. *Beiträge zur Rheinkunde,* 32, 27-41.

Ellmers, D., 1978. Die Bagger brachten Geschichte an den Tag. In H. Körschen (ed.), *Panta Rhei(n) alles fließt.* Krefeld, 22-27.

Ellmers, D., 1972. *Frühmittelalterliche Handelsschiffahrt in Mittel- und Nordeuropa.* Neumünster.

Fessner, M., 1998. *Steinkohle und Salz.* Bochum.

Feveile, C., 1995. Tufstenskirkerne i Sydvestjylland - set i handelshistorisk arkæologisk Belysning. *By, marsk og geest*, 8, 31-51.

Feveile, C., 2010. Basalt fra udgravningerne 1970-76. *Ribe Excavations 1970-76, Volume 6*, Ribe (forthcoming).

Jappe Alberts, W., 1954. Leveranties van steen uit het Rijnland voor de Dombouw te Utrecht en tollheffing op de Rijn. In W. Jappe Alberts and Ketner. *Nederrijnse Studien XIIIe –XVe eeuw.* Groningen, 1-48.

Haiduck, H., 1992. *Beginn und Entwicklung des Kirchenbaues im Küstengebiet zwischen Ems- und Wesermündung bis zum Anfang des 13. Jahrhunderts.* Aurich.

Hansen, G., 2008. Konger og byfolk i det eldste Bergen. In H. Andersson *et al.* (eds), *De første 200 årene.* Bergen, 15-40.

Hoffmann, P., 2002. Konservierung und Präsentation des Flussschiffes Karl im Deutschen Schiffahrtsmuseum. In K. Elmshäuser (ed.), *Häfen, Schiffe, Wasserwege: zur Schiffahrt des Mittelalters.* Hamburg, 86-96.

Hörmann, P. and Richter, A., 1983. Vergleichende mineralogisch-petrographische Untersuchungen an Mühlsteinresten in Haithabu und Bruchsteinproben aus der Eifel. In K. Schietzel (ed.), *Archäometrische Untersuchungen.* Neumünster, 93-108.

Mangartz, F., 2008 . *Römischer Basaltlava-Abbau zwischen Eifel und Rhein.* Mainz.

McCormick, M., 2001. *Origins of the European Economy.* Cambridge.

Obladen-Kauder, J., and Peiss, A., 2000. Ein Flußkahn aus der Zeit Karls des Großen. In H.G. Horn (ed.), *Millionen Jahre Geschichte. Fundort Nordrhein Westfalen.* Mainz, 378-380.

Parkhouse, J., 1997. The Distribution and Exchange of Mayen Lava Quernstones in Early Medieval Northwestern Europe. In G. DeBoe (ed.), *Exchange and Trade in Medieval Europe.* Bruges, 97-106.

Rohde, H., 1986. Überlegungen zur mittelalterlichen Wasserstraße Eider/ Treene/ Schlei. *Festschrift für Albert Bantelmann zum 75. Geburtstag.* Offa, 43, 311-333.

Sindbæk, S. M., 2005. *Ruter og rutinisering.* Copenhagen.

Steuer, H., 1987. Der Handel der Wikingerzeit zwischen Nord- und Westeuropa aufgrund archäologischer Zeugnisse. In K. Düwel (ed.), *Untersuchungen zu Handel und Verkehr der vor- und frühgeschichtlichen Zeit in Mittel- und Nordeuropa.* Göttingen, 113-197.

Straube, M., 1997. Notwendigkeiten, Umfang und Herkunft von Nahrungsmittellieferungen in das sächsische Erzgebirge zu Beginn des 16. Jahrhunderts. In E. Westermann (ed.) *Bergbaureviere als Verbrauchszentren im vorindustriellen Europa.* Stuttgart, 203-220.

Tucker, D.G., 1984. Millstone making in Scotland. *Proc. Soc. Antiq. Scot.,* 114, 539-556.

Watts, S., 2006. *Rotary querns c. 700-1700.* The Finds research group AD 700-1700, Datasheet 38. www.findsresearchgroup700-1700.org.uk

Of cakes and kings: bread-making in early Medieval England

Carolyn Coulter

Bread was the staple food of Anglo-Saxon society, and control of production was a key element in the elite struggle for power. This study is part of ongoing research into the varieties of stone imported for grinding cereals in early Medieval England. A survey of stone distribution reveals that the types of quern and millstone used were determined not simply by locally available sources, but also by cultural affiliations and foreign contacts. By examining chronological and regional variation, as well as technological change, this paper suggests that different bread traditions existed throughout Anglo-Saxon England; choice of stone and of baking method were both part of establishing regional identity and reinforcing social status.

Introduction

England's most famous early Medieval king is perhaps best remembered for his incompetence in the kitchen. The apocryphal tale of the cakes first appears a hundred years after King Alfred's death in the late 10th century life of the Cornish saint, Neot, and recounts how England was terrorised in AD 878 by the Danish king Guthrum and his Viking army. Alfred retreats into the salt marshes of Athelney in western Britain to await the time when God would see fit to grant him his kingdom again. While staying *in cognito* with a swineherd, Alfred is left in charge of the daily bread baking in the fire, but occupied by his own troubles and suffering, he allows the loaves to burn. The swineherd's wife, understandably furious, scolds him. She does not know that this is the king, and Alfred, instead of enlightening her, humbly accepts her admonishment and attends to the baking of the bread. The story ends with Alfred rewarded for his patience and humility: St Neot appears in a vision, promising to lead him to victory against the Viking army. The story of Alfred and the Cakes exists in several versions: the earliest, contained in the anonymous late 10th century *Vita S. Neoti*, was copied and embellished in the early 12th century for a homily collection, preserved in the manuscript B.L. Cotton Vespasian D.xiv (Dumville and Lapidge 1985, lxxv, cxvi-ii). The latter rendition depicts Alfred as a deserter and weak king in need of God's guidance, rather than as the humble sinner patiently enduring trials and chastisement. Another 12th century adaptation, in the *Annals of St Neots*, Trinity College MS R.7.28, describes Alfred neither as sinner or deserter but as 'the unconquerable king'. English translations of all three texts with a commentary on the story's development can be found in Keynes and Lapidge (1983, 197-202).

The association between the elite and bread production in early Medieval England was of long-standing. The English title 'lord' derives from the Anglo-Saxon for 'loaf-guardian' (*hlafweard*) and the word 'lady' from 'loaf-kneader' (*hlæfdige*). In the early 7th century laws of Æthelberht of Kent, the dependents of a *ceorl* or commoner are described as 'loaf-eaters' (*hlafæta*) (Whitelock 1955, 358). The same law code reveals that female grinding slaves were part of the king's personal retinue (*ibid.*, 357).

References to bread-making were not confined to etymology and legal documents in this period. The 10th century *Exeter Book* contains a riddle that plays on the idea of leavened bread dough rising. The riddle (no. 45) suggests that this was the work of noblewomen (Tupper 1910, 36):

... hrægle þeahte/ Þrindende þing þeodnes dohtor.
[... the daughter of a king covered that swelling thing with a cloth]

Riddle 49 from the same collection may also refer to bread-making (*ibid.*, 37-8):

Ic wat eardfæstne anne standan,/ deafne, dumban, se oft dæges swilgeð/
þurh gopes hond gifrum lacum... þa æþelingas oft wilniað,/cyningas ond cwene...
[I know something that stands earthfast;/ deaf and dumb, who often by day swallows/
from a servant's hand useful gifts... which princes, kings and queens often desire...]

Both 'bread oven' and 'mill' have been put forward as solutions (Trautmann 1905; Doane 1987, 249-53); but whether this riddle describes the activity of milling or baking, the implication remains the same: desired by princes, kings and queens, this was something essential for the survival of the elite, the production of which upheld the very structure of Anglo-Saxon society. The 10th century scholar Ælfric explained the elite's association with bread-making in terms of Christian metaphor (Thorpe 1844, 258-9):

Figure 1. Grinding grain in a handmill similar to the type used in early Medieval England (Gunn 1909, 270).

Se rica him sylð þone hlaf bið to meoxe awend,
and se ðearfa sylð þam rican þæt ece lif...
[The rich give to the poor bread...
and the poor give to the rich everlasting life...]

The elite were thus perceived in Old English as the persons responsible for providing, or at least ensuring the provision of bread - the staple of society - if they wanted to maintain their power and authority here on earth, and secure their places for the afterlife in the kingdom of heaven.

Control of bread production clearly lay in the interests of the king, and it is in considering this premise that this paper briefly sets forward the preliminary results of a survey into the archaeological evidence for Anglo-Saxon bread-making, from the decline of the Roman Empire until just after the Norman Conquest (AD 450-1100). The sites in this survey are listed in an appendix and are divided into three main periods: Early Anglo-Saxon with the arrival of the Anglo-Saxons (AD 450-700), mid Anglo-Saxon with the renewal of towns and long-distance trade (AD 700-900), and finally late Anglo-Saxon with the consolidation of royal power (AD 900-1100). While rotary querns similar to the type in Fig. 1 were used across England during the early Middle Ages, there are differences from site to site in the archaeological record. Different bread traditions existed in England throughout the period with quernstone choice and baking method both employed as a means of establishing regional identity and reinforcing social status.

Stones for Milling

The oldest form of handmill, the saddle quern, persisted in use in the western parts of Britain into the early Medieval era (Fig. 2). They may also have been used in southern and eastern England, at Southampton, Spong Hill and West Stow (Appendix). Unfortunately, the saddle quern fragments from these three sites are difficult to date, and

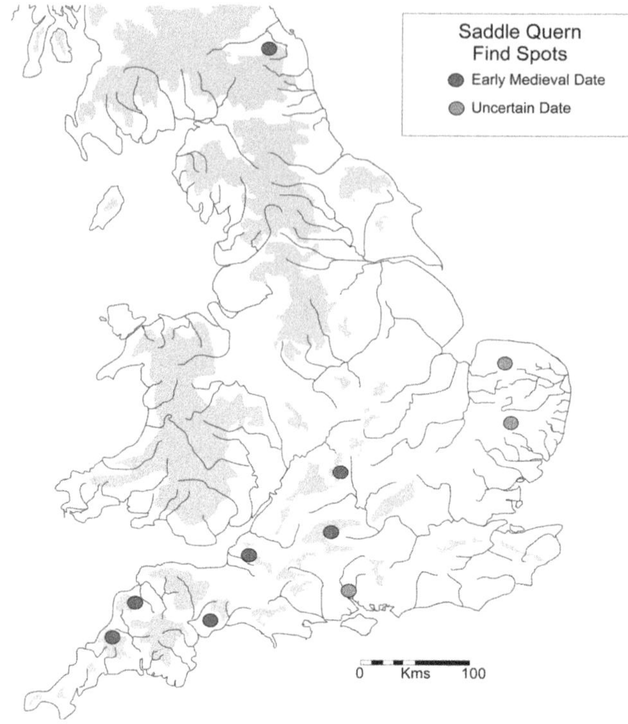

Figure 2. Early Medieval saddle quern find spots.

the excavators suggest that they are residual from an earlier period. The Old English word for quern, *cwern* or *cwyrnstan*, specifically describes a revolving handmill (Curwen 1937, 134), and no Anglo-Saxon text - to the author's knowledge - mentions the saddle quern. Yet, the practice of using saddle querns, or at least reusing Iron Age and Roman examples, continued until the mid Anglo-Saxon period, as the find at Trowbridge demonstrates (see Appendix). At the hill-fort of Cadbury-Congresbury in Wiltshire and at the small enclosed settlement of Trethurgy in Cornwall, saddle querns were found alongside rotary querns, as well as at early Medieval sites excavated in North Wales and Anglesey, and in Ireland (Quinnell 2004, 151; Rahtz 1992, 113). Regina Sexton (1998, 81) posits reuse of prehistoric Irish saddle querns for kneading troughs; this could also have been the case in England. Or perhaps these stones were reused for crushing and rolling substances other than bread cereals. There is, however, another possibility; that is, that the English saddle querns from early Medieval sites were used for grinding grain, and that their presence alongside rotary forms indicates the existence of social distinctions within settlements during this period, with cultural divergence across Britain.

All the saddle querns in this sample were made from local stone (Fig. 3). In contrast, the rotary querns are frequently stones from further afield. The main types quarried or surface-collected during this period were Millstone Grit from Derbyshire, various types of sandstones, and limestone. Other stones include greensand, granite and puddingstone, all available from native sources within Britain. These were quarried and traded within a delimited regional area, or as in the case of the boulder used at

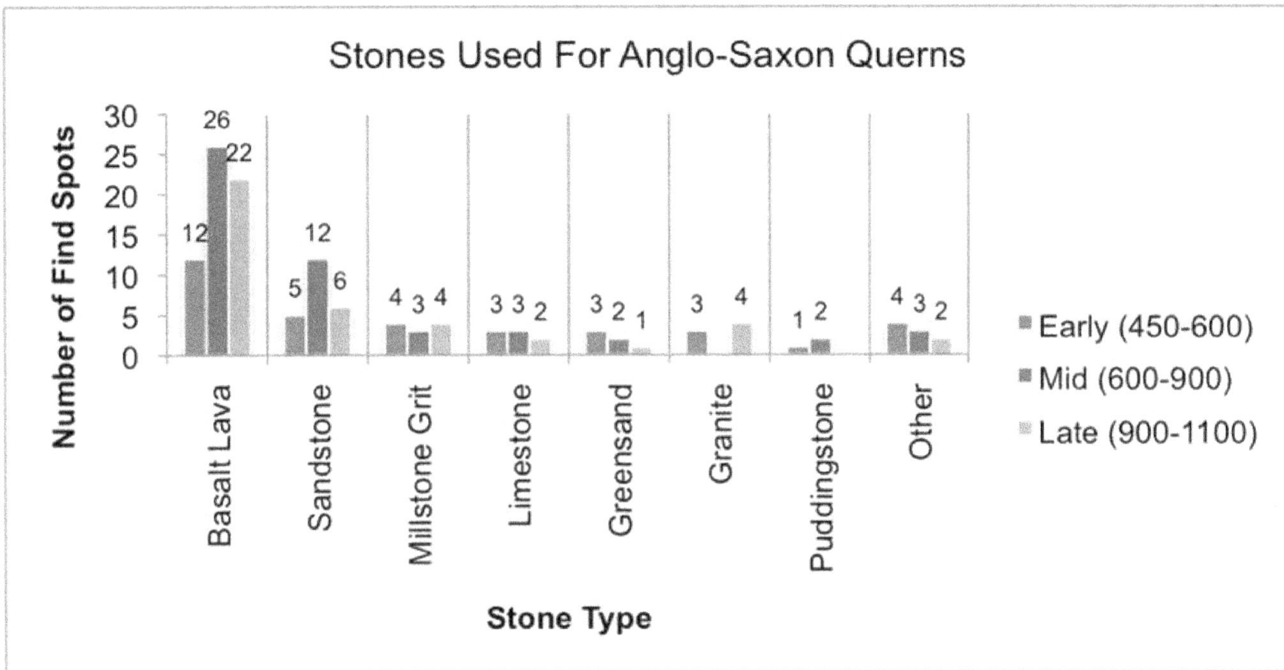

Figure. 3. Stones used for Anglo-Saxon querns.

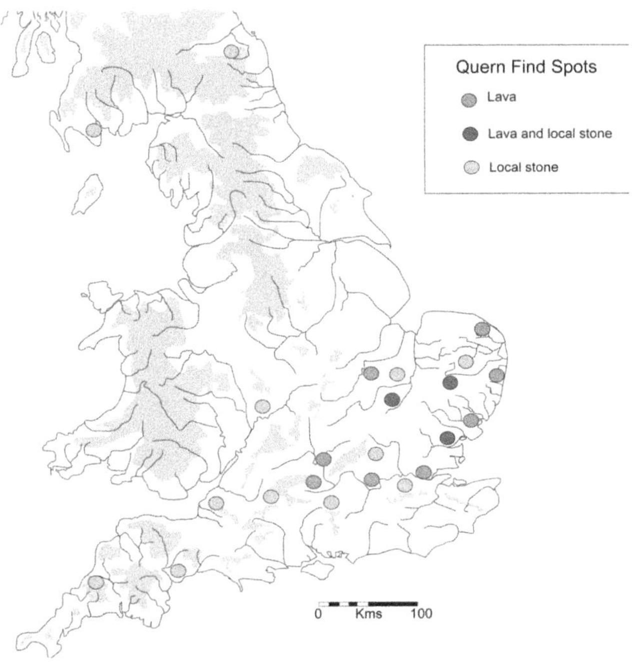

Figure 4. Early Anglo-Saxon quern find spots (AD 450-600)

Yeavering in Northumberland, picked up from local outcrops. The most ubiquitous stone for milling at this time was imported from the Continent. Basalt lava from the Rhineland was first introduced into England with the arrival of the Romans in the 1st century AD. By the 3rd and 4th centuries lava querns are no longer found with such frequency, and it has previously been suggested that lava quernstones were not reintroduced into England until the seventh and eighth centuries with the revival of foreign trade (Hodges 1982; Parkhouse 1997).

The initial results from this survey reveal the contrary; that lava querns occur at sites dating from the earliest period of Anglo-Saxon settlement (Fig. 4). Finds of these stones come from settlements in eastern England and along the major artery of the River Thames. This distribution corresponds closely with what are believed to be the areas of the earliest Anglo-Saxon settlement (Fig. 5 and Appendix). At Heybridge in Essex, it is even suggested that the Anglo-Saxon settlers who used lava querns were living on the outskirts of the contemporary late Roman town (Drury and Wickenden 1982, 34). The siting of many early Anglo-Saxon settlements close to former Roman occupation areas has resulted in excavators imagining that lava fragments may be residual. Accurate dating is certainly difficult given the broken and abraded nature of lava finds. Similar form and design of Roman and early Medieval querns mean that typological comparisons are, in any case, of little value.

Where lava querns have been dated to 5th and 6th century deposits (Fig. 5), excavators have all too often dismissed them as reused Roman material. This assumption should be resisted, for the earliest Anglo-Saxon settlers to England may in fact have been familiar with Rhenish milling stones. In northern Germany and Jutland, the homelands of the Anglo-Saxons as reported by Bede, excavations have revealed the use of lava querns at settlement sites in the 4th century and before (Sherley-Price and Farmer 1990, 63; Christensen and Hardt 1996). It may well be that when the Anglo-Saxons were first invited to England in the 5th century to defend the British shore from northern invaders, they reopened trade routes that had previously been closed by pirate raids in the Channel in the 3rd and 4th centuries (Sherley-Price and Farmer 1990, 62; Wood 1990, 94; This argument is developed more fully in my forthcoming doctoral thesis).

SITE	COUNTY	DATING BY POTTERY FINDS
Bloodmoor Hill, Carlton Colville	Suffolk	5-7th
Chieveley	Berks	Early?
Dorney-on-Thames	Berks	5-6th
Gamlingay	Cambridges	5-6th
Grays	Essex	Early
Heybridge	Essex	5th
Higham Ferrers	Northants	5-6th
Little Oakley	Essex	Early
Spong Hill	Norfolk	5-6th
Upton	Berks	5-6th
West Stow	Suffolk	5-late 6th
Witton, North Walsham	Norfolk	5th

Figure 5. Early Anglo-Saxon sites with lava fragments.

With the spread of Anglo-Saxon culture in the mid Anglo-Saxon period, a corresponding pattern of distribution occurs for lava querns, now found at more northern sites and further inland, such as at Chicheley, Eynsham and Flixborough (Fig. 6). Find spots tend to cluster along river estuaries, possibly reflecting the need for water transportation. It is during this period that the *wics*, the proto-towns of early Medieval England, emerge at London, Ipswich, York and Southampton. All four served as ports of trade with the Continent, and while lava querns predominate at these sites, other stones were also used for grinding: sandstone and limestone at York, and puddingstone at Ipswich (Appendix). Later Medieval accounts show that different stone types were sometimes paired for milling. However, in the case of the mid Saxon emporia, stone variation may indicate that social stratification already existed in what would eventually become urban areas. This pattern becomes more prevalent in the Later Anglo-Saxon period, when the greatest range of stones available for milling are found at sites such as Colchester and Winchester, which were developed in order to establish political control and consolidate trade networks.

Sites with only local stone resources in the mid Anglo-Saxon period are, as for the earlier period, mostly restricted to the western and northern parts of England. This dichotomy is even more marked in the late Anglo-Saxon period, as can be seen at the northern monastic centre of Whithorn and at the south-western settlements of Mawgan Porth and Beere (Fig. 7 and Appendix). These sites did not have to depend on foreign imported stones as sources for good millstones were locally available. Lying on the periphery of the areas settled by the Anglo-Saxons, these regions were defined by their strong affiliations with the Celtic Christian world (Dark 1996). Cornwall, in particular, maintained its social identity and traditions through the use of material culture during the early post-Roman centuries (Quinnell 1993). In contrast, the material culture of the eastern and southern parts of England reflects regional links with the Continent: lava querns are frequently found alongside Badorf and Pingsdorf wares, coins and glass produced in the Rhineland.

The Viking invasions of the 9th and 10th centuries do not

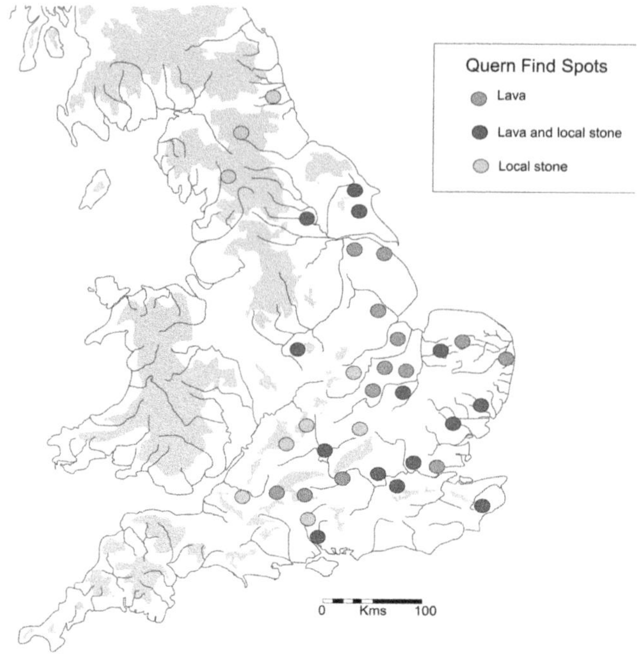

Figure 6. Mid Anglo-Saxon quern find spots (AD 600-900).

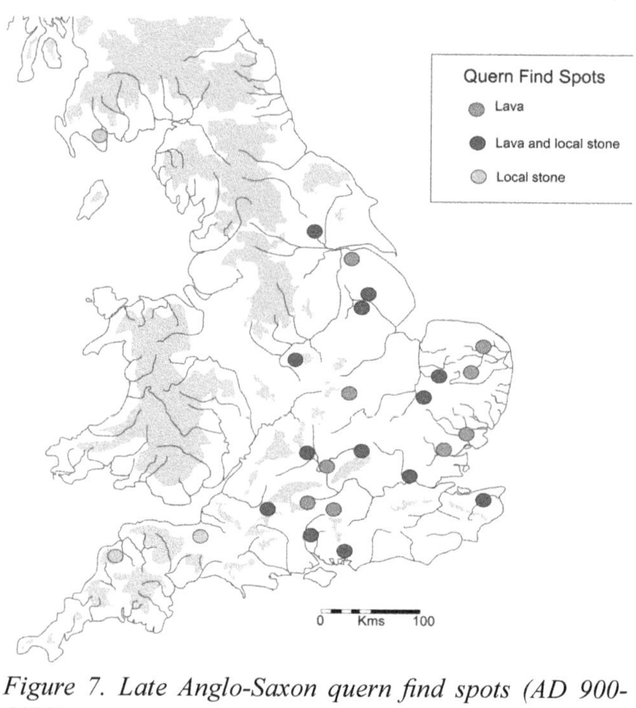

Figure 7. Late Anglo-Saxon quern find spots (AD 900-1100).

seem to have disrupted the distribution pattern, or indeed the import, of lava querns (compare Figs. 6 and 7). When England was united under a Danish king in the early 11th century, amicable relations with Continental Europe continued to play a part in English foreign policy. King Cnut made a pact with Conrad II, Emperor of Rome, and sometime before his death in 1035, married his daughter Gunhilda to Conrad's son Henry III, King of Germany (Lawson 2004, 104). Almost a century earlier, 'men of the Emperor' are named as 'worthy of good laws' among the merchants arriving in London in the Billingsgate toll list (Thorpe 1840, I, 300). Arrangements such as these can only have strengthened England's commercial contacts with the Rhineland and its millstone market.

SITE	COUNTY	DATE	HORIZONTAL	VERTICAL	MILLSTONES
Worgret	Dorset	c5/6-c7th	?		
Old Windsor	Berkshire	late c7th		x	
Dorney-on-Thames	Berkshire	c7-8th			Triassic sandstone (580mm)
Barking	Essex	c8th	x		
Corbridge	Northumberland	c8th	x		
Tamworth	Staffordshire	c9th	x		Lava sandstone (700-800mm)
Northampton	Northants	Mid			Red sandstone (650mm), brown siltstone (740mm)
West Cotton, Raunds	Northants	c10th		x	Millstone grit, sandstone, lava (900-1100mm)
Goltho	Lincolnshire	c10-11th			Oolitic limestone (860mm), millstone grit (740mm)
Castle Donnington	Leicestershire	c11th		x	
Springfield	Essex	Late			Lava (450mm)
Stroud	Gloucestershire	Late			
West Fen Road, Ely	Cambridgeshire	Late			Lava (800mm)

Figure 8. Archaeological evidence for early Medieval mills. (Data: Maynard 1988; Wilson and Hurst 1958; Foreman et al. 2002; MacGowan 1996; Nenk et al. 1996; Rahtz and Meeson 1992; Williams 1985; Windell et al. 1990; Beresford 1987; Clay 1986; Rahtz 1981; Mortimer et al. 2005).

DATE	SITE	COUNTY	MILL TYPE	OWNERSHIP
602-3	-	Kent	Slave (female)	King
762	Chart	Kent	-	St Augustine's Abbey and King
838	Holborough	Kent	Water	King Egbert of Wessex and Bishop of Rochester
949	Reculver	Kent	Tidal	Christchurch Abbey, Canterbury
c10-11th	Huntingdon and Yaxley	Cambridgeshire	Animal (oxen)	Thorney Abbey
1170	Swineshead	Lincolnshire	Wind	Knights Templar

Figure 9. Early Medieval mill types and the earliest reference to each in the historical record. (Data: Whitelock 1955, 357; Hardwick 1858, 325-6; Earle 1888, 287-8; Earle 1888, 188; Robertson 1939, 254; Lees 1935, 131).

Equally important at this period for explaining variation in bread-making across England is the topography of the landscape. In both the north and the west of Britain, barley and oats were the predominant cereals (Allen 1976, 140). Botanical evidence, where recovered during archaeological investigation, confirms this preference. These grains were ground and baked in the ashes or as flat-cakes on griddles, as was the case at Beere in Devon where a griddle pan has been discovered (Appendix). Bread may also have been cooked on bakestones: at Mawgan Porth broken querns appeared to have been reused for this purpose (Bruce-Mitford 1997, 129). While the milder, low-lying counties of eastern and southern England had fewer sources for suitable millstones, the land was more amenable for wheat-growing. Barley and rye were grown too, but from Old English literature it is clear that wheaten loaves were the favourite. Bede's 8th century *Life of St Cuthbert* describes the bread of heaven as white as the lily with the aroma of roses and the taste of honey (Webb and Farmer 1988, 52). Only the poor or penitent wanted to eat barley bread (Banham 1990, 77-8).

The chemical properties of wheat make it more suitable for leavening, and thus for oven-baking (Lyons and D'Andrea 2003, 515). Ovens in this period consisted of a single-chambered clay structure, sometimes with a cobble base, in which a fire was lit, then the ashes were raked out and the loaves placed inside to bake. Examples of such ovens have been excavated at sites across England, both in external courtyards and inside buildings that may have been kitchens or bakehouses (Appendix). The latter were a requirement for the residences of both monks and lords (Doyle 1948, 66; Swanton 1975, 27).

A Stone Fit for a King?

In addition to its oven-baking properties, wheat is a tough grain, requiring stones with a fine cutting edge to grind it (Collins 2003, 339-340). Texts from the 16th and 17th centuries suggest that before the French burrstone was introduced into England, basalt lava from the Rhineland was considered the best stone for wheat-grinding; the implication being that querns of this material were valued and prized (Steer 1950, 41; Best 1986, 209). In this context, it is perhaps possible that the much-disputed black stones of Charlemagne's letter to King Offa in AD 796 refer to lava millstones, as opposed to the more recent cases put forward for Tournai marble and Roman columns (Whitelock 1955, 779; Rahtz and Meeson 1992, 73; Peacock 1997). Later Medieval lords often went to great expense and sent their stewards great distances to procure the desired millstone (Farmer 1992; Langdon 2004, 162-172). Early Medieval kings would surely have gone to the same lengths to ensure that they too enjoyed the finest-ground bread.

Mechanical Milling

In Charlemagne's correspondence with Offa, the dispute about black stones is in regard to their length. If this reference is indeed to millstones, Offa was no doubt seeking a stone large enough for a mechanical mill. Watermills, both horizontal- and vertical-wheeled, are known from Roman Britain (Spain 1984; 2008), but in the early Anglo-Saxon period they suddenly seem to disappear from the archaeological record. This is in accord with other technological and cultural changes at the time, such as the transition from wheel-turned pottery and stone villas to grass-tempered wares and timber halls. The earliest evidence for an early Medieval watermill comes from Worgret in Dorset. The remains of the timber structure at this site have been dated by radiocarbon to the 5th/6th century, and by dendrochronology to the 7th century (Maynard 1988; Hinton 1992). There has been discussion as to the type of mill housed; the most recent and convincing hypothesis to date posits a simple horizontal wheel (Spain 2008, 85).

The horizontal wheel is the most frequently excavated mill type for Anglo-Saxon England (Fig. 8). The two early Medieval vertical wheels in this survey are both believed to have been replaced by horizontal wheels, in the 9th century at Windsor and in the 11th century at West

Cotton in Northamptonshire (Wilson and Hurst 1958, 184; Windell *et al.* 1990, 30). In England there thus appears to be no evolutionary transition from the primitive horizontal-wheel to the more complex vertical system. There were, of course, mechanical mills other than the water-powered during the early Medieval period. Excavations at Dorney, Northampton, Goltho, Springfield and Ely have not produced evidence for watermilling but have revealed fragments of large millstones that could have been turned by horses, oxen or indeed by humans. Fig. 9 lists the first mention of each mill type in the historical record. In each of these cases, the mill belonged to the king or a monastic community.

The idea that water-milling was a Christian innovation in the Dark Ages - or at least that Christianity provided the impetus for watermill construction - is not a new theory. The 6th century Rule of St Benedict and 9th century Plan of St Gall both demand the building of presumably water-powered mills for the use of the monastic house (Doyle 1948, 94; Horn 1975). In England the first documentary reference to a *molina in torrente* ('a mill on the river') comes from Kent, which with its close proximity to the continent was the first Anglo-Saxon kingdom to convert to the Christian faith in the early 7th century (Earle 1888, 287-8; Sherley-Price and Farmer 1990, 76-7). The etymology of the Old English word for mill (*mylen*), deriving from the Latin *molina*, certainly places mechanical milling firmly in the context of the Roman world and indicates the extent to which continental ideas infiltrated the fabric of Anglo-Saxon society. The foundation of monastic houses following the arrival of Christian missionaries from Rome in England in the late 6th century and the transition from feuding tribes to petty kingdoms in the mid Saxon period provided opportunities for the formation of estate centres with large mills, such as the type exemplified at Tamworth in Staffordshire (Rahtz and Meeson 1992). Possessing both the land and resources necessary to construct and oversee the operation of mechanical mills, kings and religious communities could now ensure a regular supply of ground corn without relying on slave labour or payments in kind.

By the time of the Norman invasion and Domesday survey of the 11th century, there were over 6,000 mills in England (Holt 1988, 8). So numerous were watermills in the 12th century that a law was passed to prevent their further impeding traffic on the country's major highways and waterways (Thorpe 1840, i, 447). Of these, some were perhaps communally-owned, like the Trerice 'common' mill in Cornwall mentioned in a pre-Norman boundary clause charter (Herring and Hooke 1993, 73). The design simplicity, inexpensive construction and low maintenance of the horizontal-wheel could well have encouraged peasants to pool resources for a communal mill-house, as often occurred in the north of the British Isles in Shetland (Goudie 1886, 258). After the Norman Conquest, however, mill owners were required to pay a double tax, both the tithe and the ninth (Thorpe 1840, i, 445), and the basic, but not very efficient, horizontal wheel simply became unprofitable. In the 12th and 13th centuries, at the same time as Medieval lords began to invest exclusively in vertical-wheel mills (Holt 1990, 53), the earliest references to milling rights and privileges appear. Now if a bond tenant wanted to prepare his daily bread, he had to bring his grain to the demesne mill. The domestic quern was prohibited by manorial custom, but its continued use among serfs was to provoke bitter dispute and rebellion in the later Middle Ages, culminating most famously in the incident at St Albans Abbey during the Peasant Revolt of 1381 (Bennett and Elton 1898, 210-221).

Alfred and the Cakes: Bread-making and Early Medieval Kingship

The legend of Alfred and the cakes was developed and reworked twice during the turbulence and upheavals of the 11th and 12th centuries. Alfred lost his Christian asceticism and became a maligned king at a time when the person of the king had become distanced from the production of the staple that ensured his survival. Secular and monastic lords had replaced the petty kings that had once ruled England. The earliest account of King Alfred and the burnt bread had been merely one incident in a body of literature, the aim of which was to establish the myth of a unified English nation (Simpson 1989, 400-404). This was, in the words of David Rollason (1989, 155), 'all part of [the process for]... the consolidation of royal power'. The 12th century renderings of Alfred and the cakes thus acted as a literary reminder that the king's control reached into every corner of early Medieval life: with rights over milling and later, the baking of bread. These versions were perhaps also a caution to the king himself about the limits of royal power and his responsibility as ruler. In a time of oppression from a foreign invader – this time, the Normans – the story of Alfred and the cakes harked back to a golden age when an English king could defend his people. By mastering the art of bread-making, the humblest of his subjects' tasks, the Alfred of legend reaffirms his role as an early Medieval lord, as *hlafweard* (the loaf-guardian), and as rightful king, he receives the aid of God and the saints to defeat the Vikings and save the kingdom of England.

Acknowledgements

I would like to thank the following people: Caroline MacDonald and Jonathan Parkhouse for details of the Ipswich querns; Martin Watts for the reference to the 11th century communal mill in Cornwall; and Professor David Hinton for his comments on an earlier version of this paper. This research is funded by the AHRC.

References

Addyman, P.V., 1964. A Dark Age Settlement at Maxey, Northamptonshire. *Medieval Archaeology,* 8, 20-73.

Addyman, P., 1969. Late Saxon Settlements in the St Neots [Huntingdon] area: II. The Little Paxton Settlement and Enclosures. *Proceedings of the Cambridge Antiquarian Society,* 62, 59-93.

Allen, D. E., 1976. Regional Variations in Food Habits. In D. Oddy and D. Miller (eds.), *The Making of the Modern British Diet.* London, 135-147.

Andrews, P., 1995. *Excavations at Redcastle Furze, Thetford, 1988-9.* Dereham.

Andrews, P. and Crockett, A., 1996. *Three Excavations along the Thames and its Tributaries, 1994: Neolithic to Saxon Settlement and Burial in the Thames, Colne, and Kennet Valleys.* Salisbury.

Andrews, P., Mepham, L. and Seager Smith, R., 2000. Excavations at Wilton, 1995–6: St John's Hospital and South Street. *Wiltshire Archaeological and Natural History Magazine,* 93, 181–204.

Best, M.R., (ed.) 1986. *The English Housewife by G. Markham, 1615.* Kingston and Montreal.

Banham, D.A.R.,1990. *The Knowledge and Uses of Food Plants in Anglo-Saxon England.* University of Cambridge, unpub. PhD thesis.

Barford, P.M., 2002. *Excavations at Little Oakley, Essex, 1951-78: Roman villa and Saxon settlement.* Chelmsford.

Barrowman, R.C., Batey, C. and Morris, C.D., 2007. *Excavations at Tintagel Castle, Cornwall, 1990-1999.* London.

Bennett, R. and Elton, J., 1898: *History of Corn Milling. I: Handstones, Slave and Cattle Mills.* London.

Beresford, G., 1987. *Goltho: The Development of an Early Medieval Manor c. 850-1150.* London.

Biddle, M., 1990. *Artefacts from Medieval Winchester. 2 vols.* Oxford.

Birbeck, V., 2005. *The Origins of Mid-Saxon Southampton: Excavations at the Friends Provident St. Mary's Stadium 1998-2000.* Salisbury.

Blockley, K., Blockley, M., Blockley, P., Frere, S.S. and Stow, S.,1995. *Excavations in the Marlowe Car Park and surrounding areas, The Archaeology of Canterbury. 3 vols.* Canterbury.

Bruce-Mitford, R., 1997. *Mawgan Porth. A Settlement of the Late Saxon Period on the North Cornish Coast.* London, English Heritage Archaeological Report, 17.

Butler, C., 2000. *Saxon Settlement and Earlier Remains at Friars Oak, Hassocks, West Sussex.* Oxford.

Canham, R., 1979. Excavations at Shepperton Green, 1967 and 1973. *Transactions of the London and Middlesex Archaeological Society,* 30, 97-124.

Christensen, L. and Hardt, N., 1996. Kværnsten af Basalt – En Overset Romersk Oldsag? *Arkæologi i Slesvig,* 5, 61-68.

Clay, P., 1986. Castle Donington. *Current Archaeology,* 102, 208-11.

Coggins, D., Fairless, K.J. and Batey, C.E., 1983. Simy Folds: An Early Medieval Settlement Site in Upper Teesdale. *Medieval Archaeology,* 27, 1-26.

Collins, E., 2003. Grain Quality and Milling Requirements. In A. Belmont (ed.), *Meules à grains: Actes du Colloque International de la Ferté-Sous-Jouarre 16-19 Mai 2002.* Paris, 338-343.

Cramp, R., 2006. *Wearmouth and Jarrow Monastic Sites. 2 vols.* Swindon.

Crummy, N., 1988. *The Post-Roman Small Finds from Excavations in Colchester 1971-85.* Colchester.

Cunliffe, B., 1976. *Excavations at Portchester Castle. Vol 2: Saxon.* London.

Curwen, E.C., 1937. Querns. *Antiquity,* 11, 133-51.

Dallas, C., 1993. *Excavations in Thetford by B.K. Davison between 1964 and 1970.* Dereham.

Dark, K.R., (ed.) 1996. *External Contacts and the Economy of Late Roman and Post-Roman Britain.* Woodbridge.

Darling, M. and Gurney, D., 1993. *Caister-on-Sea. Excavations by Charles Green, 1951-55.* Dereham.

Dickens, A., Mortimer, R. and Tipper, J., 2006. The Early Anglo-Saxon Settlement and Cemetery at Bloodmoor Hill, Carlton Colville, Suffolk: A Preliminary Report. *Anglo-Saxon Studies in Archaeology and History*, 13, 63-79.

Doane, A.N., 1987. Three Old English Implement Riddles: Considerations of Numbers 4, 49 and 73. *Modern Philology,* 84 (3), 243-257.

Doyle, L., (trans.) 1948. *St. Benedict's Rule for Monasteries.* Collegeville, Minnesota.

Drury, P.J. and Wickenden, N.P., 1982. Early Saxon Settlement within Romano-British Small Town at Heybridge, Essex. *Medieval Archaeology,* 26, 1-40.

Dumville, D. and Lapidge, M., (eds) 1985. *The Annals of St. Neots with Vita Prima Sancti Neoti.* Cambridge.

Dunning, G.C., 1932. Bronze Age settlements and a Saxon Hut near Bourton-on-the Water, Gloucestershire. *Antiquaries Journal,* 12, 279–293.

Earle, J., 1888. *A Handbook to Land Charters and Other Saxonic Documents.* Oxford,

Emery, P.A., 2007. *Norwich Greyfriars: Pre-conquest Town and Medieval Friary.* Dereham.

Fairbrother, J., 1990. *Faccombe Netherton: Excavations of a Saxon and Medieval Manorial Complex.* London.

Farley, M., 1976. Saxon and Medieval Walton, Aylesbury: Excavations 1973-1974. *Records of Buckinghamshire,* 20, 153-290.

Farley, M., 1980. Middle Saxon Occupation at Chicheley,

Buckinghamshire. *Records of Buckinghamshire,* 22, 92-104.

Farmer, D., 1992. Millstones for Medieval Manors. *Agricultural History Review,* 40, 97-111.

Foreman, S., Hiller, J. and Petts, D., (eds) 2002. *Gathering the People, Settling the Land. The Archaeology of a Middle Thames Landscape Anglo-Saxon to post-Medieval*. Oxford.

Freshwater, T., 1996. A Lava Quern Workshop in Late Saxon London. *London Archaeologist,* 8 (2), 39-45.

Gardiner, M., 2001. Continental Trade and Non-urban ports in Mid Anglo-Saxon England: Excavations at Sandtun, West Hythe, Kent. *Archaeological Journal,* 158, 161-290.

Goudie, G., 1886. On the Horizontal Watermills of Shetland. *Proceedings of the Society of Antiquaries of Scotland,* 20, 257-297.

Graham, A. and Davies, S., 1993. *Excavations in the Town Centre of Trowbridge, Wiltshire, 1977 and 1968-88: The Prehistoric and Saxon Settlements, the Saxo-Norman Manorial Settlement, and the Anarchy Period Castle.* Salisbury.

Gunn, J., 1909. *The Orkney Book: Readings for Young Orcadians.* London.

Haldenby, D., 1990. An Anglian site on the Yorkshire Wolds. *Yorkshire Archaeological Journal,* 62, 51-63.

Hamerow, H., 1993. *Excavations at Mucking. 2: The Anglo-Saxon Settlement*. London.

Hardwick, C., (ed.) 1858: *Historia Monasterii S. Augustini Cantuariensis. By Thomas of Elmham*. London.

Hardy, A., Dodd, A. and Keevil, G., (eds) 2003. *Aelfric's Abbey: Excavations at Eynsham Abbey, Oxfordshire 1989-1992*. Oxford.

Hardy, A., Charles, B. M. and Williams, R. J., 2007: *Death and Taxes: The Archaeology of a Middle Saxon Estate Centre at Higham Ferrars, Northamptonshire*. Oxford.

Haslam, J., 1980. A Middle Saxon Iron Smelting Site at Ramsbury, Wiltshire. *Medieval Archaeology,* 24, 1-68.

Herring, P. and Hooke, D., 1993. Interrogating Anglo-Saxons in St Dennis. *Cornish Archaeology,* 32, 67-75.

Hey, G., 2004. *Yarnton: Saxon and Medieval Settlement and Landscape: Results of Excavations 1990-96.* Oxford.

Hill, P., 1997. *Whithorn and St Ninian. The Excavation of a Monastic Town, 1984-91*. Stroud.

Hinton, D.A., 1992. Revised Dating of the Worgret Structure. *Dorset Natural History and Archaeology Society Proceedings,* 114, 258-9.

Hodges, R.A., 1982. *Dark Age Economics: The Origins of Towns and Trade, AD 600-1000.* London.

Holt, R., 1988. *The Mills of Medieval England.* Oxford.

Holt, R., 1990. Milling Technology in the Middle Ages: The Direction of Recent Research, *Industrial Archaeologist Review,* 13 (1), 50-8.

Hope-Taylor, B., 1977. *Yeavering: an Anglo-British Centre of Early Northumbria*. London.

Horn, W., 1975. Water Power and the Plan of St Gall. *Journal of Medieval History,* 1, 219-58.

Hurst, D., 1997. *A Multi-Period Salt Production Site at Droitwich: Excavations at Upwich.* York.

Jope, E.M., 1953. Late Saxon Pits under Oxford Castle Mound: Excavations in 1952, *Oxoniensia,* 16-17, 77-111.

Jope, E.M., and Threlfall, E., 1958. Excavation of Medieval Settlement at Beere, North Tawton, Devon. *Medieval Archaeology,* 2, 112-140.

Keynes, S. and Lapidge, M., (eds) 1983. *Alfred the Great: Asser's* Life of King Alfred *and Other Contemporary Sources.* Harmondsworth.

King, A., 1978. Gauber High Pasture, Ribblehead. In R. A. Hall (ed.), *Viking Age York and the North.* London, Council for British Archaeology Research Report 27, 21-5.

Langdon, J., 2004. *Mills in the Medieval Economy: England 1300-1540.* Oxford.

Lawson, A.J., 1983. *The Archaeology of Witton, near North Walsham, Norfolk.* Dereham.

Lawson, M.K., 2004. *Cnut: England's Viking King.* Stroud.

Lees, B.A., (ed.) 1935. *Records of the Templars in the 12th Century.* London.

Loveluck, C., 2007. *Rural Settlement, Lifestyles and Social Change in the Later First Millennium AD. Anglo-Saxon Flixborough in its Wider Context*. Oxford.

Lyons, D. and D'Andrea, A.C., 2003. Griddles, Ovens and Agricultural Origins: An Ethnoarchaeological Study of Bread Baking in Highland Ethiopia, *American Anthropologist,* 105 (3), 515-530.

MacGowan, K., 1996. Barking Abbey. *Current Archaeology*, 149, 172-8.

Malcom, G. and Bowsher, D., 2003. *Middle Saxon London. Excavations at the Royal Opera House 1989-99.* London.

Mainman, A. and Rogers, N.S.H., 2000. *Craft, Industry and Everyday Life: Finds from Anglo-Scandinavian York. The Archaeology of York 17/14.* York.

Mann, J., 1982. *Early Medieval Finds from Flaxengate 1: Objects of Antler, Bone, Stone, Horn, Ivory, Amber and Jet.* Archaeology of Lincoln 14/1.

Manning, W.H., 1974. Upton, Berkshire. *Berkshire Archaeological Journal*, 67, 1-62.

Maynard, D., 1988. Excavations on a Pipeline near the River Frome, Worgret, Dorset. *Dorset Natural History and Archaeology Society Proceedings,* 110, 77-98.

Millett, M. and James, S., 1983. Excavations at Cowdery's Down, Basingstoke, Hampshire. 1978-81. *Archaeological Journal,* 140, 151-279.

Morris, C.D., and Harry, R., 1997: Excavations on the Lower Terrace, Site C, Tintagel Island 1990-94, *The Antiquaries Journal*, 77, 1-143.

Mortimer, R., 2000. Village Development and Ceramic Sequence: The Middle to Late Saxon Village at Lordship Lane, Cottenham, Cambridgeshire. *Proceedings of the Cambridge Antiquarian Society*, 89, 5-33.

Mortimer, R., Regan, R. and Lucy, S., 2005. *The Saxon and Medieval Settlement at West Fen Road, Ely: The Ashwell Site.* Cambridge, Cambridge Archaeology Unit and East Anglian Archaeology 110.

Mudd, A., 2002. *Excavations at Melford Meadows, Brettenham, 1994: Romano-British and early Saxons occupations.* Oxford.

Mudd, A. 2007. *Bronze Age, Roman and Later Occupation at Chieveley, West Berkshire: The Archaeology of the A34/M4 Road Junction Improvement*. Oxford.

Murray, J. and McDonald, T., 2006. Excavations at Station Road, Gamlingay, Cambridgeshire. *Anglo-Saxon Studies in Archaeology and History*, 13, 173–330.

Nenk, B.S., Margeson, S. and Hurley, M., 1996. Medieval Britain in 1995: Corbridge. *Medieval Archaeology*, 40, 276-7.

Quinnell, H., 1993. A Sense of Identity - Distinctive Cornish Stone Artefacts in the Roman and Post-Roman periods. In M. Carver (ed.), *In Search of Cult: Archaeological Investigations in honour of Philip Rahtz.* Woodbridge, 69-78.

Quinnell, H., 2004. *Trethurgy: Excavations at Trethurgy Round, St Austell. Community and status in Roman and post-Roman Cornwall.* Truro.

Parkhouse, J., 1997. The Distribution and Exchange of Mayen Lava Quernstones in Early Medieval Europe. In G. de Boe and F. Verhaege (eds.), *Exchange and Trade in Medieval Europe. Papers of the 'Medieval Europe Brugge 1997' Conference. III*. Zellik, 97-106.

Peacock, D.P.S., 1997. Charlemagne's Black Stones: The Reuse of Roman Columns in Early Medieval Europe. *Antiquity*, 71, 709-15.

Rahtz, P., 1962-3. The Saxon and Medieval Palaces at Cheddar. *Medieval Archaeology*, 6-7, 53-66.

Rahtz, P., 1979. *The Saxon and Medieval Palaces at Cheddar: Excavations 1960-62.* British Archaeological Report.

Rahtz, P., 1981. Medieval Milling. In D. W. Crossley (ed.), *Medieval Industry*. London, Council for British Archaeology, 1-15.

Rahtz, P., 1992. *Cadbury Congresbury 1968-73: A Late/ Post-Roman Hilltop Settlement in Somerset.* Oxford.

Rahtz, P. and Meeson, R., 1992. *An Anglo-Saxon Watermill at Tamworth: Excavations in the Bolebridge Street area of Tamworth, Staffordshire in 1971 and 1978*. Oxford, Council for British Archaeology Research Report 83.

Richards, J.D., 1999. Cottam: An Anglian and Anglo-Scandinavian Settlement on the Yorkshire Wolds. *Archaeological Journal*, 156, 1-110.

Rickett, R., 1995. *The Iron Age, Roman and Early Saxon Settlement at Spong Hill.* Dereham.

Robertson, A.J., 1939. *Anglo-Saxon Charters*. Cambridge.

Rogerson, A., 1984. *Excavations in Thetford 1948-59 and 1973-80.* Dereham.

Rogerson, A., 1995. *A Late Neolithic, Saxon and Medieval Site at Middle Harling, Norfolk.* London.

Rollason, D., 1989. *Saints and Relics in Anglo-Saxon England.* Oxford.

Sexton, R., 1998. Porridges, Gruels and Breads: The Cereal Foodstuffs of Early Medieval Ireland. In M. Monk and J. Sheehan (eds.), *Early Medieval Munster: Archaeology, History and Society*. Cork, 76-86.

Sherley-Price, L. and Farmer, D.H., (eds) 1990. *Bede: A History of the English Church and People.* Harmondsworth.

Simpson, L., 1989.The King Alfred/St Cuthbert Episode in the *Historia de Sancto Cuthberto*: Its Significance for Mid-Tenth Century English History. In G. Bonner, D. Rollason and C. Stanclife (eds), *St Cuthbert, His Cult and His Community to AD 1200.* Woodbridge, 397- 412.

Spain, R., 1984. Romano-British Watermills. *Archaeologica Cantiana,* 100, 101-28.

Spain, R., 2008. *The Power and Performance of Roman Watermills. Hydro-mechanical Analysis of Vertical Water-mills*. Oxford, British Archaeological Report.

Stamper, R. A. and Croft, R. A., 2000. *The South Manor Area, Wharram*. York, York University Archaeological Publications 10.

Steedman, K., 1994. The Excavation of an Anglo-Saxon Site at Riby Cross Roads, Lincolnshire. *Archaeological Journal,* 151, 212-306.

Steer, F. W., 1950. *Farm and Cottage Inventories of Mid-Essex 1635-1749.* Essex Records Office Publication 8.

Swanton, M., (trans. and ed.) 1975. *Anglo-Saxon Prose.* London.

Taylor, G., 2003. An Early to Middle Saxon settlement at Quarrington, Lincolnshire. *Antiquaries Journal,* 83, 231–280.

Thorpe, B., (ed.) 1840. *Ancient Laws and Institutes of England. 2 vols*. London.

Thorpe, B., (ed.) 1844. *The Homilies of the Anglo-Saxon Church.* London.

Trautmann, M., 1905. Alte und neue Antworten auf altenglische Rätsel. *Bonner Beiträge zur Anglistik,*19, 183-4.

Tupper, F., (ed.) 1910. *Riddles of the Exeter Book.* London.

Tyler, S., and Major, H., 2005: *The Early Anglo-Saxon Cemetery and Later Saxon Settlement at Springfield Lyons, Essex.* Chelmsford

Wade-Martins, P., 1980. *Excavations in North Elmham Park, 1967-72.* East Anglian Archaeology Report 9.

Webb, J.F. and Farmer, D.H., (eds) 1988. *The Age of Bede.* London.

West, S. E., 1985. *West Stow: The Anglo-Saxon Village. 2 vols.* Ipswich.

Whitelock, D., (ed.) 1955. *English Historical Documents: AD c.500-1042.* London.

Wickenden, N.P., 1988. *Excavations at Great Dunmow, Essex: a Romano-British small town in the Trinovantian Civitas.* Chelmsford.

Wilkinson, T.J., 1988. *Archaeology and Environment in South Essex: Rescue Archaeology along the Grays Bypass, 1979-80.* Chelmsford.

Williams, J., 1981. Excavations in Chalk Lane, Northampton. *Northamptonshire Archaeology,* 16, 87-135.

Williams, J., 1985. *Middle Saxon Palaces at Northampton.* Northampton.

Williams, P., 2006. *Market Lavington, Wiltshire, an Anglo-Saxon cemetery and settlement: excavations at Grove Farm, 1986-90.* Salisbury.

Wilson, D., and Hurst, J.G., 1958. Medieval Britain in 1957: Old Windsor. *Medieval Archaeology,* 2, 183-5.

Windell, D., Chapman, A. and Woodiwiss, J., 1990. *From Barrow to Bypass: Excavations at West Cotton, Raunds, Northamptonshire 1985-1989.* English Heritage, Northamptonshire County Council.

Wood, I., 1990. The Channel from the 4th to the 7th centuries AD. In S. McGrail (ed.), *Maritime Celts, Frisians and Saxons. Papers Presented to a Conference at Oxford in November 1988.* Oxford, 93-7.

Wymer, J.J., 1995. *Excavations at North Shoebury: Settlement and Economy in South-East Essex 1500BC-AD1500.* Chelmsford.

Site	County	Date	Stones for milling	Baking utensils/ features	Reference
Beere, North Tawton	Devon	Late	Granite	Iron griddle pan	Jope and Threlfall 1958
Bloodmoor Hill, Carlton Colville	Suffolk	Early	Lava	Oven bases	Dickens *et al* 2006
Bourton-on-the-water	Gloucests	Mid	Oolitic limestone (saddle, partially burnt)	Hearth of ashes and burnt stones	Dunning 1932
Cadbury Congresbury	Somerset	Early	Limestone, sandstone, millstone grit (saddle and rotary)		Rahtz 1992
Caister-on-Sea	Norfolk	Mid	Lava		Darling and Gurney 1993
Canterbury	Kent	Mid-Late	Lava, millstone grit	Single-flue, clay-lined oven with reused Roman tile, within portico area	Blockley *et al* 1995
Cheddar	Somerset	Mid	Andesitic lava, grit, sandstone		Rahtz 1979
Chicheley	Buckinghams	Mid	Lava		Farley 1980
Chieveley	Berks	Early	Lava		Mudd 2007
Colchester	Essex	Late	Lava		Crummy 1988
Cottam	Yorks	Mid	Lava, grit		Richards 1999
Cottenham	Cambridges	Mid	Lava		Mortimer 2000
Cowdery's Down	Hants	Early	Sandstone		Millett and James 1983
Dorney-on-Thames	Berks	Early	Lava		Foreman *et al* 2002

Dorney-on-Thames	Berks	Mid	Lava, sandstone, Greensand		Foreman *et al* 2002
Droitwich	Worcestershire	Early	Sandstone, limestone		Hurst 1997
Eynsham	Oxfords	Mid	Old Red Sandstone		Hardy *et al* 2003
Eynsham	Oxfords	Late	Lava, white sandstone		Hardy *et al* 2003
Faccombe	Hants	Late	Lava		Fairbrother 1990
Flixborough	Lincolns	Mid	Lava		Loveluck 2007
Flixborough	Lincolns	Late	Lava	Domed fired-clay ovens north of refuse area and buildings	Loveluck 2007
Friars Oak	West Sussex	Mid	Lava, sandstone		Butler 2000
Gamlingay	Cambridges	Early	Lava, millstone grit		Murray and McDonald 2006
Gauber High Pasture, Ribblehead	Yorks	Mid	Millstone grit	Oven/kiln (north-east corner of stone structure)	King 2004
Goltho	Lincolns	Late	Lava, sandstone	Baking oven (burnt daub) in centre of kitchen	Beresford 1987
Grays	Essex	Early	Lava		Wilkinson 1988
Great Dunmow	Essex	Mid	Puddingstone, lava		Wickenden 1988
Heybridge	Essex	Early	Lava, millstone grit		Drury and Wickenden 1982
Higham Ferrers	Northants	Early	Lava		Hardy *et al* 2007
Higham Ferrers	Northants	Mid	Lava		Hardy *et al* 2007
Hurst Park	Berks	Early	Greensand, grit and quartzite		Andrews and Crockett 1996
Ipswich	Suffolk	Mid-Late	Lava, sandstone, puddingstone		
Lincoln: Flaxengate	Lincolns	Late	Lava, millstone grit, spilsby grit		Mann 1982
Little Oakley	Essex	Early	Lava		Barford 2002
Little Paxton	Cambridges	Mid	Lava, limestone	Small oven with burnt clay lining and stone base (21m from hut)	Addyman 1969
London: Royal Opera House	London	Mid	Lava, sandstone, limestone	Circular ovens (made of Roman tile/ragstone/flint cobbles and daub) in open area	Malcom *et al* 2003
London: Thames Exchange	London	Late	Lava		Freshwater 1996
Market Lavington	Wilts	Early	Greensand, cornish granite		Williams 2006

Market Lavington	Wilts	Mid	Lava		Williams 2006
Mawgan Porth	Cornwall	Late	Granite (burnt, saddle?)	Blackened hearth areas in northern part of each of 3 stone courtyard houses	Bruce-Mitford 1997
Maxey	Northants	Mid	Lava		Addyman 1964
Medmerry Farm	Sussex	Early	Limestone		Stamper and Croft 2000
Melford Meadowms, Brettenham, Thetford	Norfolk	Early	Millstone grit (burnt)	3 oval ovens with burnt clay linings in open area	Mudd 2002
Middle Harling	Norfolk	Late	Lava		Rogerson 1995
Mucking	Essex	Early	Sarsen	External bowl-shaped hearth	Hamerow 1993
North Elmham	Norfolk	Mid	Lava	Clay dome (3-phase oven) in centre of bakehouse	Wade-Martin 1980
North Shoebury	Essex	Mid	Lava		Wymer 1995
Northampton	Northants	Mid	Sandstone, quarrtzite, siltstone		Williams 1985
Northampton: Chalk Lane	Northants	Late	Lava		Williams 1981
Norwich Greyfriars	Norfolk	Late	Lava (reused as fire surround)	3-phase oven	Emery 2007
Oxford Castle Mound	Oxfords	Late	Lava		Jope 1953
Portchester	Hants	Mid-late	Lava, arkose grit	2-phase oven centrally-placed against east wall of bakehouse(?)	Cunliffe 1976
Quarrington	Lincolns	Mid	Lava	Charcoal pit with burnt stones	Taylor 2003
Ramsbury	Wilts	Mid	Lava		Haslam 1980
Riby Cross Roads	Lincolns	Mid	Lava	Dumped debris from ovens/ hearths?	Steedman 1994
Sandtun	Kent	Mid	Lava, sandstone		Gardiner 2001
Shepperton Green	London	Mid	Lava, sandstone		Canham 1979
Simy Folds	Upper Teesdale	Mid	Tufa		Coggins 1983
Southampton	Hants	Mid	Lava, shelly limestone, coarse sandstone, tufa (saddle?)		Birbeck 2005
Spong Hill	Norfolk	Early	Lava, greensand/ sarsen, sandstone	Circular oven?	Rickett 1995
Springfield Lyons	Essex	Late	Lava		Tyler and Major 2005
Tamworth	Staffs	Late	Lava, sandstone		Rahtz and Meeson 1992

Thetford	Norfolk	Early	Sandstone		Dallas 1993
Thetford	Norfolk	Late	Lava, millstone grit, limestone	Centrally-placed oven in posthole structure	Rogerson 1984/ Andrews 1995
Tintagel	Cornwall	Early	Gabbro?		Morris and Harry 1997
Trethurgy	Cornwall	Early	Elvan, granite (rotary and saddle)	Hearth pits	Quinnell 2004
Trowbridge	Wilts	Mid-late	Lava, sarsen (saddle)		Graham and Davies 1993
Upton	Berks	Early	Lava		Manning 1974
Walton, Aylesbury	Buckinghams	Early	Sandstone		Farley 1976
Walton, Aylesbury	Buckinghams	Late	Lava, limestone		Farley 1976
Walton, Aylesbury	Buckinghams	Mid	Greensand		Farley 1976
Wearmouth	Tyne and Wear	Mid	Sandstone (reused for c11-13th hearth)		Cramp 2006
West Fen Road, Ely	Cambridges	Late	Millstone grit, lava		Mortimer *et al* 2005
West Stow	Suffolk	Early	Lava, puddingstone (saddle?)	Large oval clay oven in structure	West 1985
Wharram	Yorks	Mid	Lava, grit	Oven in centre of structure	Stamper and Croft 2000
Whithorn	Dum and Gall	Early	Greywacke, granite		Hill 1997
Whithorn	Dum and Gall	Late	Granite, sandstone		Hill 1997
Wickhams Field, Reading	Berks	Mid	Lava		Andrews and Crockett 1996
Wilton	Wilts	Late	Lava	Hearths inside structures	Andrews *et al* 2000
Winchester	Hants	Mid	Sandstone		Biddle 1990
Winchester	Hants	Late	Sandstone, lava, granite, limestone, carstone		Biddle 1990
Witton, North Walsham	Norfolk	Early	Lava		Lawson 1983
Yarnton	Oxfords	Mid	Lava, millstone grit (burnt)		Hey 2004
Yeavering	Northumb	Early	Boulder (saddle)		Hope-Taylor 1977
York: Coppergate	Yorks	Mid	Lava, sandstone		Mainman and Rogers 2000
York: Coppergate	Yorks	Late	Lava, sandstone, limestone	Large clay-based hearths inside structures	Mainman and Rogers 2000

Les meulières de l'Ile de Minorque: trente-neuf sites industriels d'Époque Andalousí (Xè – XIIIè siècles)

Joaquin Sanchez Navarro

1. Introduction

Les prospections faites entre 1997 et 2003 par l'équipe d'investigateurs dirigée par Miquel Barceló et Félix Retamero, de l'Université Autonome de Barcelona, aux sites de résidence des paysans d'époque andalousí (du Xè au XIIIè siècles) de l'île de Minorque ont révélé la présence de morceaux de meules. Ces sites, qui avaient été occupés dans d'autres périodes, ne permettaient pas d'assurer l'âge de ces meules.

Par ailleurs, la connaissance de quelques carrières où on y trouvait des marques d'extraction de pierres de la même composition et grandeur que celles des sites islamiques a fait entreprendre une investigation pour déterminer si l'industrie des meules était locale.

J.Mascaró Pasarius (1967, 2283-2308) donne la première information sur 'roues a moulin ?' (sic) aux îles Baléares, avec la description de la meulière de Punta de Sa Dent (Lluchmajor, Majorque) et ce même auteur, en 1992, a fait une relation de 9 carrières de pierres rondes à Minorque, mais change son idée originale, l'extraction de meules par l'extraction de pierres rondes pour éboucher les silos souterrains.

Efectivement, ces pierres ont pu être utilisées pour éboucher les silos, mais la régularité du diamètre, les marques pour y mettre une manivelle, et aussi les volumes d'extraction observés, ne permettent pas assurer que la finalité principale de ces pierres fût pour éboucher les silos, bien qu'elle en ait eu un usage éventuel.

La datation des meulières a été, aussi, un autre problème : Jusqu'à l'an 2000, on n'avait trouvé aucun reste céramique ni métallurgique environ des premières carrières trouvées par J.Mascaró Pasarius. Cet investigateur a publié que ce sont des 'monuments anhistoriques' (*sic*). D'autres auteurs se limitent à indiquer que ce sont 'des carrières antiques'.

L'auteur commence l'investigation des sites avec une litologie similaire aux morceaux de meules trouvée par F.Retamero, et a fait un premier inventaire de 13 meulières et l'étude de 20 morceaux de meules trouvés. La présence de céramique andalousí dans trois de ces meulières pourrait indiquer, temptativement, que l'âge des meulières serait andalousí. Ainsi, la localisation d'un site islamique avec des morceaux de meules (Sant Joan Gran, Ferreries, Minorque) qui ne présentait que céramique andalousí avait l'air de confirmer ces afirmations (Sánchez Navarro 2001, 49-179). Ce même auteur, a approfondi l'étude antérieur, et a fait des considérations environ des analytiques des meulières et des meules, et annonce les premières corrélations meulière-meule. (Sánchez Navarro 2005, 235-267). Ainsi, (2006, 155-196) il développe l'investigation, donne à connaître 18 nouveaux lieux d'extraction, les compositions, les lithotypes et les niveaux d'exploitation des meulières.

2. Situation actuelle

Comme conséquence d'autres études du même auteur, on a trouvé 8 nouvelles meulières à l'île de Minorque, qui permettent présenter cette communication.

À la appendix 1 on peut voir la dénomination du site, le numéro de la fiche de l'Atlas des Meulières européennes où elle a été décrite, sur quels matériaux ont été taillées, la surface d'exploitation, et le minimum et maximum des meules obtenues. Ces quantités ont été obtenues avec la considération que l'exploitation moyenne a besoin d'1 m^3 de pierre pour obtenir 5 meules. Le minimum serait si le rendement de la meulière ne fût qu'un 50% et le maximum, d'un 100%. En tout cas, on a constaté que dans certaines meulières l'extraction des meules a profité excellemment les materiaux.

La première considération sur les meulières de Minorque qu'il faut indiquer, c'est sa localisation: elles sont toutes situées à côté de la mer, et on peut voir leur emplacement à la Fig.1. Celà nous fait penser que le transport serait par mer. Quelques meulières présentent un lieu pour faciliter l'approchement de petits bateaux et la charge des meules, et quelques témoins du lieu de tricherie.

Un autre considération de l'emplacement de la meulière à côté de la mer c'est parce que les materiaux où on exploite la meulière sont mieux classés et mieux consolidés que les coétanes de l'intérieur de l'île. Une partie très importante des meulières sont taillées sur des dépôts éoliens. Au côté

Figure 1. Localisation des meulières de l'Île de Minorque.

de la mer, les sables qui vont originer les grès actuels, sont lavées, classées et très bien cimentées, ce qui donne une roche dure, estable, et, au même temps, avec une dureté qui peut être facilement taillée.

L'aportation d'eau avec des sels marines aux sables en train de consolidation dans un moyen aérien a fait qu'il y ait un échange des ions $Ca^{++} \leftrightarrow Na^{+} \leftrightarrow Mg^{++}$, une légère solution des grains calcaires et une recristallisation aux interstices. L'étude des lames fines de roche des meulières a constaté que ces roches ont pû avoir jusqu'à 5 fases de cimentation, ce qui nous rapproche à l'histoire érosion/sédimentation des matériaux.

Il faut dire que tous les grès des meulières de Minorque ont la considération de 'grès coquilier' avec une proportion variable de grains siliceux. Cette proportion variable peut avoir corrélation avec la proximité de roches siliceuses à l'entour.

Les analytiques obtenues pour certains sites sont répértoriés à la appendix 2.

On a classé les meulières selon le materiel exploité. La description de chacune d'elles est à continuation.

3. Description résumée des meulières

3.1 Meulières en dépôts éoliens.

Les meulières suivantes ont été exploitées en dépôts éoliens. Ces dépôts sont d'âge quaternaire, produits entre les glaciations Riss – Würm. Ce sont de petits dépôts (de 5 m jusqu'à 200 m de longueur) qui fossilisent les paléoreliefs d'âge quaternaire de la côte minorquine.

Figure 2. Bloc de la meulière de Cala Mica.

Elles ont tous des caractéristiques similaires: Les roches exploités sont grès coquilliers avec une proportion variable de grains calcaires (du 60 au 92% selon le lieu) et siliceux (du 8 au 40%) de moins d'1 mm de diamètre. Les grains sont très bien classés, arrondis, partiellement récristallisés, sans matrix limeuse ou argileuse, et le ciment est toujours calcaire. On trouve, ainsi, de petits morceaux de pierre intercalés de la base des zones proximales au dépôt. Les grains calcaires sont d'origine biologique : petits morceaux de cloîtres, foraminifers, et d'autres produits par l'érosion des roches calcaires. Les grains silliceux sont toujours produits par l'érosion des sédiments du Trias inférieur (Buntsandstein), très fréquents au Nord de Minorque, mais les dunes avec une proportion plus haute de quars sont celles qu'on trouve environ de Cala Trebalúger, au Sud de Minorque, comme conséqüence du transport fluvial des interglaciers quaternaires. Le modèle stratigraphique est dunar, et la lamination visible est prédominant d'envers 20°. Cela fait que certaines exploitations suivent ces lignes de discontinuité de la stratification pour y adapter leurs travaux d'extraction des meules.

Cala Caldes

Le site se trouve situé au côté Ouest de l'entrée de Cala Caldés, à la commune de Maó.
Lat/long/alt : (N 39°59'51.70" – E 4°13'42.42" – 3 m).
Fiche de l'Atlas des Meulières européennes : 350.
Description du site: Il occupe la plupart de la dune, vers 300 m^2. La meulière a profité un niveau plus compact de 60 cm d'épaisseur. Les alvéoles ont entre 80 et 100 cm de diamètre. La taille est faie avec masse et bédane, et pour enlever les pièces, burin. Les marques des outils sont très érodés par l'action de la mer.
Production: 180 m^3 roche - 150 à 200 meules.

Cala del pilar

Le site se trouve situé au côté Est de Cala del Pilar, à la commune de Ciutadella de Menorca.
Lat/long/alt : (N 40°03'10.42" – E 3°58'54.52" – 3 m).
Fiche de l'Atlas des Meulières européennes : 404.
Description du site: Il occupe une petite partie au Nord de la dune (vers 3 m^2). La meulière a profité un niveau plus compact de 40 cm d'épaisseur. Les alvéoles ont entre 80 et 100 cm de diamètre. La taille est faite avec masse et bédane, et pour enlever les pièces, burin.
Production: 3 m^3 roche exploitée – 3 à 5 meules.

Cala mica

Le site se trouve situé au côté Est de Cala Mica, à la commune des Mercadal.
Lat/long/alt : (N 40°03'32.45" – E 4°04'08.22" – 1 a 5 m).
Fiche de l'Atlas des Meulières européennes : 344.
Description du site: Il occupe une petite partie au Nord de la dune (vers 50 m^2), mais le niveau inférieur de l'exploitation est actuellement couvert par la mer. La meulière profite un niveau plus compact de 120 cm d'épaiseur. La surface totale d'exploitation sous la mer n'a pu être constatée. La meulière, vers 50 m au Sud, présente un autre lieu d'extraction, (aussi grès coquilier, dépôt plus antique et mieux cimentée) avec un bloc qui nous fait rappeler un panneau "ici, meulière" par sa situation. (Fig. 2). Cette autre exploitation a vers 15 m^2 et on y a trouvé 30 alvéoles. La meule en procés d'extraction a 54 cm de diamètre. Les alvéoles ont entre 80 et 100 cm de diamètre. La taille est faite avec masse et bédane, et pour enlever les pièces, burin.
Composition chimique: SiO2 11,40%, CaO 87,60%.
Production: 30 à 60 visibles – centains sous la mer?

Cala morell – pas des mosquits

Le site se trouve situé à la base de la falaise à l'Ouest de Cala Morell, à la commune de Ciutadella de Menorca.
Lat/long/alt : (N 40°03'14.89" – E 3°52'41.14" – 5-10 m).
Fiche de l'Atlas des Meulières européennes : 379.
Description du site: Il occupe la totalité de la dune vers 300 m^2, mais le site est fisuré et une partie importante peut s'être enfoncé dans la mer à une grande profondeur. L'extraction présente un échelon progressif, pour faciliter la taille. Les alvéoles ont entre 80 et 100 cm de diamètre. La taille est faite avec masse et burin.
Production: 1000 meules constatables – la totalité est indeterminée.

Clot de Sa Cera

Le site se trouve situé dans une petite cave près de Clot de Sa Cera, à la commune de Ciutadella de Menorca.
Lat/long/alt : (N 39°58'43.30" – E 3°50'00.70" – 0-5 m).
Fiche de l'Atlas des Meulières européennes : 391.
Description du site: Il occupe la plupart de la dune, vers 30 m^2, mais on ne peut bien préciser sa longueur parce que l'extérieur est très érodé par la mer. Les alvéoles ont entre 80 et 100 cm de diamètre. La taille est faite avec masse et bédane, et pour enlever les pièces, burin.
Production: 25 m^3 - 100 meules.

Codolar de Biniatram

Le site se trouve situé à la zone Est du Codolar de Biniatram, à la commune de Ciutadella de Menorca.
Lat/long/alt : (N 40°03'01.02" – E 3°54'43.40" – 3-5 m).
Fiche de l'Atlas des Meulières européennes : 389.
Description du site: Il occupe la partie centrale de la dune, d'environ 150 m^2, et on y trouve, aussi, des alvéoles et des meules ébauchés quelques mètres environ. Cette dune est fisurée, ce qui réduit notamment sa production et, parce que son extérieur est très érodé par la mer. Les alvéoles ont entre 80 et 100 cm de diamètre. La taille est faite avec marteau et burin.
Production: 200 m^3 - 300 – 500 meules.

Còdols dels Dàtils

Le site se trouve situé entre Cala Mitjana i Trebalúger à la zone de côte connue comme els Còdols dels Dàtils, à la commune de Ferreries.
Lat/long/alt : (N 39°55'48.21" – E 3°59'14.69" – 1 a 5 m).
Fiche de l'Atlas des Meulières européennes : 349.
Description du site: Il occupe diverses parties de la dune distribuées irrégulairement et l'épaisseur d'extraction est, aussi, irrégulière. On a trouvé 29 alvéoles et des meules ébauchées. Les alvéoles ont entre 80 et 100 cm de diamètre et l'épaisseur pour une meule est environ de 20 cm. La taille est faite avec marteau et burin en V ouverte à l'extérieur.
Production: 20 m^3 - 30 a 50 meules.

Enfonsat de Binissaid

Le site se trouve situé à la falaise entre Cala Mitjana i Cala Galdana à la zone de la côte connue comme Enfonsat de Binissaid, à la commune de Ferreries.
Lat/long/alt : (N 39°55'53.84" – E 3°57'55.04" – 10 m).
Fiche de l'Atlas des Meulières européennes : 334.
Description du site: Il occupe une petite partie (environ 5 m^2) de la dune. On a trouvé 9 alvéoles. Les alvéoles ont entre 80 et 100 cm de diamètre et l'épaisseur pour une meule est environ de 20 cm. La taille est faite avec marteau et burin en V ouverte à l'extérieur.
Masse volumnique (t/m^3): 2.44.
Production: 3 m^3 - 10 meules?

Maresos de Cala Pudent
Le site se trouve au Sud de la zone connue comme Maresos de Cala Pudent, à la commune des Mercadal.
Lat/long/alt: (N 40°04'20.33" – E 4°05'47.08" – 0-5 m).
Fiche de l'Atlas des Meulières européennes : 403.
Description du site: Il occupe une petite partie (environ 5 m^2) de la dune. On a trouvé 5 alvéoles. Les alvéoles ont entre 80 et 100 cm de diamètre et l'épaisseur pour une meule est environ de 20 cm. La taille est faite avec marteau et burin en V ouverte à l'extérieur.
Production: 5 – 10 meules.

Maresos de Sa Marjal - Cala'n Turqueta
Le site se trouve situé à la falaise entre Cala'n Turqueta et Macarella à la zone de la côte connue comme Maresos de Sa Marjal, à la commune de Ciutadella de Menorca.
Lat/long/alt: (N 39°55'56.15" – E 3°55'15.06" – 10 m).
Fiche de l'Atlas des Meulières européennes : 336.
Description du site: Il occupe une petite partie (vers 5 m^2) de la dune. On a trouvé 5 alvéoles. Le reste de la meulière s'est déjà enfoncé dans la mer par les tempêtes, et aussi par l'extraction de blocs carrés pour des constructions plus modernes. Les alvéoles ont entre 80 et 100 cm de diamètre et l'épaisseur pour une meule est environ de 20 cm. La taille est faite avec marteau et un burin en V ouverte à l'extérieur.
Production/Production: 20 m^3 visibles - 20 à centaines de meules.

Morro de Ponent de Cales Coves
Le site se trouve situé à la falaise de la sortie de Cales Coves vers l'Ouest, à la commune de Alaior.
Lat/long/alt: (N 39°48'44.4" – E 4°08'33.00" – 9 m).
Fiche de l'Atlas des Meulières européennes : 444.
Description du site: Il occupe une petite partie (vers 5 m^2) de la dune. On a trouvé 2 meules ébauchés de 55 cm de diamètre et un alvéole de 80 cm. Le reste de la meulière a disparu par l'extraction de blocs carrés pour des constructions plus modernes. La taille est faite avec marteau et burin en V ouverte à l'extérieur.
Production: 2 meules visibles.

Penyals de Binideufà
Le site se trouve situé au côté Nord de Penyals de Binideufà, à la commune de Ferreries.
Lat/long/alt: (N 40°03'35.43" - E 3°59'43.54" - 10 m).
Fiche de l'Atlas des Meulières européennes : 325.
Description du site: Il occupe deux zones de la même dune: celle de l'Est de 15 m de longueur et 5 m de large, et celle de l'Ouest, de 15 m de longueur et entre 5 et 7 m de large. La surface d'exploitation est d'environ 100 m. Les alvéoles ont entre 80 et 100 cm de diamètre et l'épaisseur pour une meule est environ de 20 cm. La taille est faite avec marteau et burin en V ouverte à l'extérieur. Cette meulière est très érodée par l'action de la mer.
Composition chimique: SiO_2 8,10%, CaO 91,90%.
Production: 100 m^3, 300 - 500 meules.

*Plage de Binimel·*là
Le site se trouve situé au côté Est de la Plage de Binimel·là, à la commune des Mercadal.
Lat/long/alt : (N 40°03'20.97" – E 4°03'18.58" – 1 m).
Fiche de l'Atlas des Meulières européennes : 343.
Description du site: Il occupe une petite partie de la dune Est de la plage, de 10 m de longueur et 3m de large, et on a localisé 30 alvéoles d'extraction. Les alvéoles ont entre 80 et 100 cm de diamètre et l'épaisseur pour une meule est environ de 20 cm. La taille est faite avec marteau et burin en V ouverte à l'extérieur. Cette meulière est très abîmée par l'action de la mer.
Composition chimique: SiO_2 25,30% CaO 74,70%.
Masse volumnique (t/m^3): 2.36.
Production/Production : 30 à 50 meules.

Plage de Trebalúger
Le site se trouve situé au côté est de la Plage de Trebalúger, à la commune des Migjorn Gran.
Lat/long/alt : (N 39°55'52.17" – E 3°59'33.95" – 2 m).
Fiche de l'Atlas des Meulières européennes: 423.
Description du site: Ce site est composé par trois lieux d'extraction: celui du Nord, situé près des restes d'une ancienne maison, qui a trois alvéoles. Le central, de 16 m^2, profite la portion de dune plus homogène, d'1 m d'épaisseur, et au Sud, située à côté du lieu des bateaux pour touristes, une quinzaine d'alvéoles d'1 m et de forme conique. Les alvéoles ont entre 80 et 100 cm de diamètre et l'épaisseur pour une meule est environ de 20 cm. La taille a été faite avec marteau et burin et en V ouverte à l'extérieur. Cette meulière est très abîmée par l'action humaine à la zone de tricherie d'embarcations.
Production: 16 m^3 de roche – 80 à 100 meules.

Punta de Sa Miloca – es Corral Fals
Le site se trouve situé au lieu connu comme Punta de Sa Miloca entre Trebalúger et Cala Fustam, à la commune des Migjorn Gran.
Lat/long/alt: (N 39°55'38" - E 3°59'40" E - 5m).
Fiche de l'Atlas des Meulières européennes: 315.
Description du site: Ce site est l'exploitation plus importante de l'Île de Minorque. Il est composé de trois zones d'extraction: celle de l'Ouest, de 85 m de longueur et 11 à 20 de largeur, et avec un épaisseur de près de 3 m. La central, plus irrégulier, de près de 150 m de longueur et 20 de large, l'extraction n'est pas continue, et a une épaisseur maximum d'1 m. Celle qui est située a l'Est, d'une centaine de mètres de longueur, présente des groupes de 5-20 alvéoles dispersés. Les alvéoles ont entre 80 et 100 cm de diamètre et l'épaisseur pour une meule est environ de 20 cm. La taille a été faite avec marteau et burin. Les zones avec une bonne qualité de la roche présentent les marques d'extraction en "nid d'abeille", pour mieux profiter le matériel. (Fig. 3).
Au Nord, on trouve les déchets de la meulière, des meules ébauchées, et les restes du finissage des meules. C'est l'unique meulière de Minorque qui présente ce dépôt, et on a localisé de petits morceaux de céramique andalousí

Figure 3. Extraction en "nid d'abeille" de Punta de Sa Miloca – Corral Fals.

Figure 4. Vue de la partie centrale de la meulière de Punta Ferragut.

de cuisine (s. XIII). La fouille de ces restes pourra nous donner une meilleure information sur la datation de la meulière. La surface d'exploitation, les alvéoles et meules laissés où ébauchées pendant l'extraction, occupent, à l'Ouest, 1120 m^2, à la partie central, près de 3000 m^2, et à l'Est, des centaines de m^2. La plupart des meules ont un diamètre entre 55 et 58 cm, et on a trouvée une meule de 80 cm de diamètre et 40 cm d'épaisseur. L'alvéole a près de 120 cm et la taille a été faite avec marteau et burin et en V ouverte a l'extérieur.
Composition chimique: SiO2 36,60% CaO 63,40% (moyenne des analytiques des échantillons de la meulière).
Production: 8000 m^3 - entre 16000 et 25000 meules.

Punta de Son Escudero
Le site se trouve situé au côté Ouest de la Punta de Son Escudero, à la commune de Ciutadella de Menorca.
Lat/long/alt : (N 40°03'21.43" – E 3°51'13.82" – 50 m).
Fiche de l'Atlas des Meulières européennes : 421.
Description du site: Il occupe une petite partie de la dune d'environ 3 m^2, et on a localisé 5 alvéoles d'extraction. Les alvéoles ont entre 80 et 100 cm de diamètre et l'épaisseur pour une meule est environ de 20 cm. La taille est faite avec marteau et burin en V ouverte à l'extérieur.
Production: 5 - 10 meules.

Punta Ferragut
Le site se trouve situé au côté Ouest de Platja Ferragut, à la commune des Mercadal.
Lat/long/alt : (N 40°03'39" – E 04°04'24" - 1 m).
Fiche de l'Atlas des Meulières européennes: 317.
Description du site: Il occupe l'extrème Nord de Punta Ferragut, à l'Est, une exploitation trapézoidelle de 12 m de longueur et 10 m de largeur maximum et 5 m de minimum, avec une épaisseur d'1 m, (Fig. 4) et à l'Ouest de la Punta, d'une vingtaine de mètres de longeur, 1 à 6 m de largeur et 1,5 à 2 m d'épaisseur, mais le niveau inférieur de l'exploitation est actuellement couvert par la mer. La surface total de l'exploitation sous la mer n'a pu être constatée. Les meules en procés d'extraction ont environ de 54-58 cm de diamètre. Les alvéoles ont entre 80 et 100 cm de diamètre. La taille est faite avec marteau et burin et en V ouverte à l'extérieur. Au Sud, en un bloc de roche isolé, on a trouvé une meule sans finir de 120 cm de diamètre et 20 cm d'épaisseur.
Composition chimique: SiO2 8%, CaO 92%.
Production: 300 m^3 de roche - 600 à 1000 meules.

Punta de Ses Fontanelles – La Vall
Le site se trouve situé au côté Est de la Punta de Ses Fontanelles, à la commune de Ciutadella de Menorca.
Lat/long/alt: (N 40°02'57.22" – E 3°55'02.32" – 3-5 m).
Fiche de l'Atlas des Meulières européennes: 390.
Description du site: Il occupe une petite partie de la dune d'extension indéterminée, car il y a une exploitation de blocs qui afecte la meulière, et où on a localisé 3 alvéoles d'extraction. Les alvéoles ont entre 80 et 100 cm de diamètre et l'épaisseur pour une meule est environ de 20 cm. La taille est faite avec marteau et burin en V ouverte à l'extérieur.
Composition chimique: SiO2 16,80%, CaO 83,20%.
Production/Production: 10 m^3 - 5 à 20 meules.

Raconada Cova d'en Sastre – Son Parc
Le site se trouve situé à la zone connue comme Cova d'en Sastre, a l'Est de la Plage de Son Parc, à la commune des Mercadal.
Lat/long/alt: (N 40°01'04.32" – E 4°11'36.61" – 1-3 m).
Fiche de l'Atlas des Meulières européennes: 452.
Description du site: Il occupe une petite partie de la dune et on peut voir plusieurs meules en procés d'extraction de 90 cm de diamètre et d'autres de 45 cm. Les alvéoles ont de 80 jusqu'à 130 cm. L'épaisseur profité est de 40 à 80 cm. On a localisé 20 alvéoles de 130 cm et d'autres altéres par l'action de l'érosion marine. L'épaisseur pour une meule est environ de 20 cm. La taille est faite vec marteau et burin en V ouverte à l'extérieur.
Production: 80 à 100 meules.

S'Olleta
Le site se trouve situé à la zone connue comme S'Olleta, entre les plages de Cala Mitjana et Trebalúger, à la

commune de Ferreries.
Lat/long/alt: (N 39°55'49.19" – E 3°58'55.48" – 5 m).
Fiche de l'Atlas des Meulières européennes: 424.
Description du site: Il occupe une petite partie de la dune et les alvéoles sont disperses. L'épaisseur profité est de 40 cm. On a localisé 14 alvéoles d'extraction. Les alvéoles ont entre 80 et 100 cm de diamètre et l'épaisseur pour une meule est environ de 20 cm. La taille est faite avec marteau et burin en V ouverte à l'extérieur.
Production: 14 à 20 meules.

Sa Reganeta - es Canutells
Le site se trouve situé à l'Ouest de Sa Reganeta (es Canutells), à la commune de Maó.
Lat/long/alt: (N 39°50'57.71" – E 4°10'18.49" – 0 - 5 m).
Fiche de l'Atlas des Meulières européennes : 326.
Description du site: Il occupe une partie de la dune de 15 m de longueur et 5 à 7 de largeur, et l'épaisseur est de 60 cm à 3 m. C'est une exploitation importante parce qu'on a profité beaucoup ce dépôt éolien, l'unique apte pour faire des meules au Sud-Est de Minorque. (Fig. 5). Les alvéoles ont entre 80 et 100 cm de diamètre et l'épaisseur pour une meule est environ de 20 cm. La taille est faite avec marteau et burin en V ouverte à l'extérieur.
Production: 300 m³ de roche - 350 à 1050 meules.

Figure 5. Meulière de Sa Reganeta – Es Canutells.

Sa Torreta
Le site se trouve situé au côté est de la Plage de Sa Torreta, à la commune de Maó.
Lat/long/alt: (N 39°57'53.79" – E 4°15'29.90" – 0-3 m).
Fiche de l'Atlas des Meulières européennes: 402.
Description du site: Il occupe une petite partie de la dune d'environ 5 m² et les alvéoles sont disperses. L'épaisseur maximum profitée est de 40 cm. Le volume de roche traitée est d'environ 5 m³. Les alvéoles ont entre 80 et 100 cm de diamètre et l'épaisseur pour une meule est environ de 20 cm. La taille est faite avec marteau et burin en V ouverte à l'extérieur.
Production: 5 à 20 meules.

Ses Pedrissades
Le site se trouve situé à la zone connue comme Ses Pedrissades, entre les plages de Cala Mitjana et Trebalúger, à la commune de Ferreries.
Lat/long/alt: (N 39°55'49.41" – E 3°59'01.48" – 1 à 5 m).
Fiche de l'Atlas des Meulières européennes : 348.
Description du site: Il occupe une partie de la dune d'environ 40 m², et des dizaines de mètres a l'Est, on trouve des alvéoles disperses. L'épaisseur profitée est de 80 à 100 cm. On a localisé 39 alvéoles d'extraction. Les alvéoles ont entre 80 et 100 cm de diamètre et l'épaisseur pour une meule est environ de 20 cm. La taille est faite avec marteau et burin en V ouverte à l'extérieur.
Production: 50 m³ de roche - 60 meules.

3.2 Meulières en dépôts de *talus* récifals

Presque la moitié de l'Île de Minorque a été formée pendant le Messiniense (Miocène – Tertiarie) dans un complexe système de plages et de récifs coralins. Ces matériaux sont calcaires partiellement dolomitisés (substitution des ions Ca++ par les Mg++) et les modèles sédimentaires sont variés selon le lieu du dépôt. Pour y faire des meulières, les artisans ont cherché les niveaux plus homogènes, à grains réguliers bien classés, d'environ 1-2 mm de diamètre, avec peu de morceaux de fossiles, et, préféremment, sans diaclasses. Cette recherche nous porte aux dépôts de l'extérieur du récif, bâti par les vagues, les fortes courants, qui classe, plus ou moins, les morceaux du récif arrachés aux coraux. A vue d'oeil, ce sont des grès uniformes, et on peut voir, au microscope, qu'une partie importante ce sont des morceaux de coquilles, équinodermes, briozoos, foraminifers, coraux, algues calcaires et d'autres indéterminés. Les analytiques faites nous montrent un contenu de grains siliceux très bas: 1-2%, de grains dolomitiques variable du 60 au 95% et la reste grains calcaires (5 au 40%). Cette composition fait que les meules produites sont plus altérable et de moins durée que celles qui ont une proportion plus haute de silliceux. Mais on a fait de nombreuses meulières sur ces matériaux, et cela voudrait dire que, après la taille, c'était une bonne meule. Nous ne pouvons pas constater la durée de ces meules, mais on peut croire que ces exploitations ont été faites parce qu'il n'y avait pas de meilleurs matériaux pour y travailler.

Caló Blanc – zona Cap d'en Font
Le site se trouve au côté Sud du Caló Blanc, à la commune de Sant Lluís.
Lat/long/alt: (N 39°49'45.06" – E 4°12'37.88" – 2 m).
Fiche de l'Atlas des Meulières européennes: 377.
Description du site: Il occupe une surface de 2 m², avec une épaisseur profitée de 60 cm. On a localisé 1 meule et un alvéol d'extraction. L'alvéol a entre 80 et 100 cm de diamètre et l'épaisseur pour une meule est environ de 20 cm. La taille est faite avec marteau et burin en V ouverte à l'extérieur. Les marques d'outils sont très visibles.

La production limitée nous fait penser que cette meulière aurait été un essai pour constater l'idonéité des materiaux.
Production: 2 meules.

Caló d'en Fus
Le site se trouve dans un îlot au côté Ouest du Caló d'en Fus, à la commune de Sant Lluís.
Lat/long/alt: (N 39°49'20.86" – E 4°13'42.73" – 1 m).
Fiche de l'Atlas des Meulières européennes : 352.
Description du site: Il occupe une surface de 50 m^2, irrégulière par l'îlot, avec une épaisseur profitée de 60 cm. Les alvéoles ont entre 80 et 100 cm de diamètre et l'épaisseur pour une meule est environ de 20 cm. Les meules en taille ont entre 56 et 60 cm. La taille est faite avec marteau et burin en V ouverte à l'extérieur. Les marques d'outils sont très visibles. Les diaclases du matériel fait que leur production soit réduite.
Production: 30 m^3 de roche - 50 à 100 meules.

Caló Morlà
Le site se trouve au petit bout au côté Est du Caló Morlà, à la commune de Sant Lluís. C'est l'unique site de Minorque que son étymologie est relationnée avec la production de meules (Morlà per molar, site des meules).
Lat/long/alt: (N 39°49'10" – E 4°14'03" - 5 m).
Fiche de l'Atlas des Meulières européennes : 316.
Description du site: Il occupe la plupart du bout, et au centre, une excavation de 40 m de longueur et entre 5 et 10 de largeur, avec une épaisseur exploitée de 3 m de profondeur. La meulière a une surface total d'environ 500 m^2. Les alvéoles ont entre 80 et 100 cm. de diamètre et l'épaisseur pour une meule est environ de 20 cm. Les meules en taille ont entre 52 et 60 cm. La taille est faite avec marteau et burin en V ouverte à l'extérieur. Les marques d'outils sont très visibles.
Production: 900 m^3 de roche - 3000 à 4500 meules.

Caló Pla – Est
Le site se trouve au côté Est du Caló Pla, à la commune de Sant Lluís.
Lat/long/alt : (N 39°49'21.59" – E 4°14'02.55" – 5 m).
Fiche de l'Atlas des Meulières européennes : 401.
Description du site: Il occupe une surface de 5 m^2, avec une épaisseur variable. Les alvéoles ont entre 80 et 100 cm de diamètre et l'épaisseur pour une meule est environ de 20 cm. Les meules produites ont entre 56 et 60 cm. La taille est faite avec marteau et burin en V ouverte à l'extérieur.
Production: 5 à 20 meules.

Caló Pla – Ouest
Le site se trouve au côté Ouest du Caló Pla, à la commune de Sant Lluís.
Lat/long/alt : (N 39°49'23.13" – E 4°13'48.75" – 3 m).
Fiche de l'Atlas des Meulières européennes : 378.
Description du site: Il occupe une surface de 10 m^2, avec une épaisseur d'environ 60 cm. Les alvéoles ont entre 80 et 100 cm de diamètre et l'épaisseur pour une meule est environ de 20 cm. Les meules en taille ont entre 56 et 60 cm. La taille est faite avec marteau et burin en V ouverte à l'extérieur. Les marques des outils sont specialement visibles. On a trouvé 6 alvéoles et 2 meules ébauchés, et cela nous fait penser avec un essai d'idonéité des matériaux.
Production: 5 à 10 meules.

Caló Tancat
Le site se trouve au côté Nord du Caló Tancat, à la commune de Sant Lluís.
Lat/long/alt : (N 39°49'13.27" – E 4°13'58.67" – 5 m).
Fiche de l'Atlas des Meulières européennes : 353.
Description du site: Il occupe une surface de 10 m^2, avec une épaisseur excavée d'environ 3 m. Les alvéoles ont entre 80 et 100 cm de diamètre et l'épaisseur pour une meule est environ de 20 cm. Les meules en taille ont entre 56 et 60 cm. La taille est faite avec marteau et burin en V ouverte à l'extérieur. Les marques des outils sont especialment visibles. Au Sud de cette extraction on a trouvé un alvéole de 150 cm avec des marques de cales métalliques de 8 à 12 cm de largeur pour enlever la meule.
Production: 150 meules - 30 m^3 roche traitée.

Cap d'en Font – Est
Le site se trouve au côté Est du Cap d'en Font, à la commune de Sant Lluís.
Lat/long/alt : (N 39°49'39.09" – E 4°12'42.21" – 5 m).
Fiche de l'Atlas des Meulières européennes : 376.
Description du site: Il occupe une surface de 10 m^2, avec une épaisseur d'environ 60 cm. Les alvéoles ont entre 80 et 100 cm de diamètre et l'épaisseur pour une meule est environ de 20 cm. Les meules en taille ont entre 56 et 60 cm. La taille est faite avec marteau et burin en V ouverte à l'extérieur. Les marques des outils sont spécialement visibles.
Production/Production : 9 à 30 meules.

Cap d'en Font – Ouest
Le site se trouve au côté Ouest du Cap d'en Font, à la commune de Sant Lluís.
Lat/long/alt: (N 39°49'39.96" – E 4°12'25.46" – 10 m).
Fiche de l'Atlas des Meulières européennes : 375.
Description du site: Il occupe une surface de 5 m^2, avec une épaisseur d'environ 60 cm. Les alvéoles ont entre 80 et 100 cm de diamètre et l'épaisseur pour une meule est environ de 20 cm. Les meules en taille ont entre 56 et 60 cm. La taille est faite avec marteau et burin en V ouverte à l'extérieur. Les marques des outils sont spécialement visibles.
Production/Production : 5 à 15 meules.

Illot d'en Marçal
Le site se trouve dans un îlot près de la côte entre Cap d'en Font et Binisafúller, à la commune de Sant Lluís.
Lat/long/alt: (N 39°49'41.32" – E 4°12'57.03" – 0-3 m).
Fiche de l'Atlas des Meulières européennes : 328.
Description du site: Il occupe une surface de 450 m^2,

Figure 6. Vue de le meulière de Illot d'en Marçal.

distribuée en deux zones principales dans l'îlot, et des meules et alvéoles isolés. (Fig. 6). L'épaisseur traitée est variable, jusqu'à 2 m. Les alvéoles ont entre 80 et 100 cm de diamètre et l'épaisseur pour une meule est environ de 20 cm. Les meules en taille ont entre 52 et 60 cm. La taille est faite avec marteau et burin en V ouverte à l'extérieur. Les marques des outils sont spécialement visibles.
Production/Production: 900 m^3 de roche traitée - 1000 à 3000 meules.

Illot de Binisafúller
Le site se trouve dans un îlot près de la côte à l'Est de Binisafúller, à la commune de Sant Lluís.
Lat/long/alt: (N 39°49'23.60" – E 4°13'19.95" – 0-3 m).
Fiche de l'Atlas des Meulières européennes : 456.
Description du site: Il occupe trois aires dans l'îlot: celle du Nord, de 30 m^2, on a extrait des meules de 80 cm et les alvéoles ont 120 cm. La zone centrale, de 150 m^2 avec des meules de 55 cm, et les alvéoles de 80 cm, et celle qui est située à l'extrème Sud, de 15 à 20 m^2, avec les alvéoles entre 80 et 120 cm. L'exploitation totale est d'environ 200 m^2, et l'épaisseur profitée est variable, de 40 cm a 1.5 m. La taille est faite avec marteau et burin en V ouverte à l'extérieur.
Production: 300 m^3 de roche traitée - 400 à 600 meules.

Illot de sa Barca
Le site se trouve dans un îlot près de la côte à l'Est de Binisafúller, à la commune de Sant Lluís.
Lat/long/alt: (N 39°49'30.90" – E 4°13'28.87" – 0-1 m).
Fiche de l'Atlas des Meulières européennes : 455.
Description du site: Il occupe une petite partie au Nord de l'îlot, où on a localisé 5 alvéoles de près d'1 m, sans que cette extraction ait continuité. Le niveau profité a 60 cm d'épaisseur. On pense que c'est un essai d'idonéité du matériel. La taille est faite avec marteau et burin en V ouverte à l'extérieur.
Production: 5 m^3 de roche traitée - 5 à 15 meules.

Morro Llevant – ses Anglades – Cap d'en Font
Le site se trouve à l'Ouest de Cap d'en Font et jusqu'à el Morro de Llevant, à la commune de Sant Lluís.
Lat/long/alt: (N 39°49'52.24" – E 4°12'17.89" – 3-10 m).
Fiche de l'Atlas des Meulières européennes : 331.
Description du site: Il occupe une surface de 1000 m^2, et environ de cette zone on trouve des groupes d'alvéoles et des meules de 8-10 unités distribués irrégulairement. La zone principale a une épaiseur de 3 m. Les alvéoles ont entre 80 et 100 cm de diamètre et l'épaisseur pour une meule est environ de 20 cm. Les meules en taille ont entre 56 et 60 cm. La taille est faite avec marteau et burin en V ouverte à l'extérieur. Les marques des outils sont spécialement visibles dans une petite cave. La zone est très diaclasé et cela fait que la production soit incerte avec un si grand volume de roche traitée.
Production: 3000 m^3 de roche - 750 à 2250 meules.

Na Moix
Le site se trouve à l'Ouest de Na Moix, à la commune de Sant Lluís.
Lat/long/alt: (N 39°48'58.14" – E 4°14'18.35" – 3 m).
Fiche de l'Atlas des Meulières européennes : 351.
Description du site: Une seule meule en procés d'extraction. On pense que c'est un essai d'idonéité des matériaux. L'alvéole a 100 cm de diamètre et l'épaisseur pour la meule est de 15 cm. La meule en taille a 60 cm. La taille est faite avec marteau et burin en V ouverte à l'extérieur.
Production: 1 meule sans finir.

Punta d'en Quintana
Le site se trouve au Nord de Cala Blanca, à un petit bout connu comme Punta Quintana, à la commune de Ciutadella de Menorca.
Lat/long/alt: (N 39°58'16.42" – E 3°49'50.91" – 5 m).
Fiche de l'Atlas des Meulières européennes : 327.
Description du site: Il occupe une surface de 225 m^2, dans une zone de 30 m de longueur et entre 5 et 20 de largeur, et profite un niveau plus compact d'1 m d'épaisseur. Les alvéoles ont entre 80 et 100 cm de diamètre et légèrement coniques et l'épaisseur pour une meule est environ de 20 cm. Les meules en taille ont entre 52 et 54 cm. La taille est faite avec marteau et burin en V ouverte à l'extérieur.
Composition chimique: SiO2 7,30%, MgO 88,70%, CaO 4%.
Production: 225 m^3 de roche traitée - 600 à 1200 meules.

Punta d'Enmig – Biniparratx
Le site se trouve a l'Ouest de Biniparratx, à la commune de Maó.
Lat/long/alt: (N 39°49'56.21" – E 4°12'10.18" – 5 m).
Fiche de l'Atlas des Meulières européennes: 388.
Description du site: Il occupe une surface de 20 m^2 et profite un niveau plus compact d'1 m d'épaisseur. Les alvéoles ont entre 80 et 100 cm de diamètre. L'épaisseur pour une meule est environ de 20 cm. Les meules en taille ont entre 55 et 60 cm. La taille est faite avec marteau et burin en V ouverte à l'extérieur.
Production: 20 m^3 de roche traitée – 60 meules.

Figure 7. Meule de S'Arangí.

Figure 8. Premières incisions pour tailler la meule. Punta de Sa Miloca – Corral Fals.

Figure 9. Extraction columnaire de la meulière de Punta de Sa Miloca - Corral Fals.

Punta sud de Cala Blanca

Le site se trouve au sud de Cala Blanca, à la sortie de la cala, à la commune de Ciutadella de Menorca.
Lat/long/alt: (N 39°58'00.46" – E 3°50'01.22" – 5 m).
Fiche de l'Atlas des Meulières européennes : 335.
Description du site: Il occupe deux zones, celle du Nord avec une surface de 100 m^2, dans une zone de 20 m de longueur et 5 de largeur, et profite un niveau plus compact de 80 cm d'épaisseur. Celle du Sud, en demie-lune de 5 m de diamètre et 3 m de largeur qui se prolonge vers le Sud. Les alvéoles ont entre 80 et 100 cm de diamètre et légèrement coniques et l'épaisseur pour une meule est environ de 20 cm. Les meules en taille ont entre 52 et 54 cm. La taille est faite avec marteau et burin en V ouverte à l'extérieur.
Composition chimique: SiO_2 1,10%, MgO 77,90%, CaO 21%.
Production: 80 m^3 de roche traitée - 200 à 400 meules.

4. Meules obtenues

La plupart des meules obtenues sont des meules manuelles rotatives d'entre 54 – 56 cm de diamètre. (Fig. 7).

Les exceptions qu'on a constaté sont :

1.Meules d'environ de 45 cm. On a trouvé un morceau de meule à Santa Agueda, des alvéoles à la meulière de Penyals de Binideufà, et des meules en procés d'extraction à Cova d'en Sastre – Son Parc.
2. Meules d'environ de 80 cm. On a trouvé une meule en procés d'extraction, et cinq alvéoles de 100 cm à la meulière du Corral Fals – Sa Miloca; des meules et ses alvéoles à Cova d'en Satre – Son Parc, et aussi des alvéoles de 100 cm au côté Nord d'Illot de Binisafúller.
3. Meules environ de 120 cm. On a trouvé une meule de 120 cm et 20 cm d'épaisseur, en procés d'extraction, à la meulière de Punta Ferragut, et un alvéole de 140 cm a la meulière de Caló Tancat.

On a trouvé aussi des alvéoles pour meules monolithes aux meulières de Punta d'en Quintana, Punta sud de Cala Blanca et Platja Trebalúger.

5. Tecniques extractives

L'étude a permis d'observer les différents stades du procés extractif. Une première incision avec compas de pointes, après une taille en V ouverte, et l'enlevé de la pierre avec incisions chaque 8-10 cm environ de la meule (Fig. 8). Les meulières présentent une exploitation échelonnée à la plupart d'elles. D'autres, seulement quelques meules en cherchant la roche la plus idoine. La meulière des Corral Fals- Sa Miloca a une extraction columnaire tout à fait verticale et sans aucun échelon, ce qui fait penser qu'elle serait considérée finie et abandonnée, où l'extraction s'est dispersée aux environs de la meulière principale, en cherchant d'autres sites aptes pour y faire de nouvelles carrières. (Fig. 9). Au côté Nord de cette meulière, on trouve les déchets de la meulière et du finissage, et aussi des meules ébauchées. Les marques des outils utilisés sont le burin à pointe carrée jusqu'à 30 mm et poinçons et

Figure 10. Marques des outils à Morro de Llevant – Ses Anglades.

Figure 11. Marques des outils à Morro de Llevant – Ses Anglades.

bédane fines (17 mm à la meule de S'Arangí), pics, maillets, masses, marteaux à manches courtes, et pointeroles. (Figs. 10 et 11).

6. Considérations sur leur datation

On n'a trouvé aucune référence à ces meulières à la documentation antique, et cela nous fait penser qu'après la conquête de Minorque par les rois catalans aux sarrasins, l'exploitation était déjà arrêtée. L'absence de ce type de meules aux fouilles des sites prérromains et romains nous a fait rejeter leur utilisation dans ces périodes. Les sites prérromains de Minorque présentent des meules à va-et-vien en grès siliceux. Les sites romains ont donné des meules manuelles rotatives et catilli et metae en roches basaltiques d'importation. La présence des meules manuelles aux sites andalousins nous a fait penser à l'exploitation et utilisation dans cette période. La présence de céramique du XIII siècle près des meulières de Punta Ferragut, Penyals de Binideufà, Codolar de Biniatram, Corral Fals – Sa Miloca nous a permis de rapprocher leur datation à cette époque. La récente trouvaille à Torre den Gaumés d'une cuisine du XIII siècle avec tout le mobilier, inclus une meule, permet de constater l'utilisation en époque andalousí. Les rois catalans autorisent la construction des moulins à partir de la conquête (1287), et commence l'importation de meules de Catalogne, parce qu'à l'île il n'y avait pas de roches aptes pour les moulins d'eau, de vent ou d'animaux. La considération de "non aptes" a suivi jusqu'à aujourd'hui. Mais il faudra attendre les fouilles des meulières pour pouvoir tout assurer.

Bibliographie

Joncheray, J. P. and Sénac, P., 1995. Une nouvelle épave sarrasine du haut Moyen Âge. *Archeologie Islamique,* 5, 25-34.

Mascaró Pasarius, J., 1967. *Corpus de Toponimia de Mallorca.* Palma, 2283-308.

Mascaró Pasarius, J., 1982. Carta arqueológica de Menorca. *Geografia e Historia de Menorca,* III-IV,441-447.

Mascaró Pasarius, J. and Nicolás Mascaró, J., 1979. Un pequeño enigma resuelto. Los sitjots de Mussuptà y las canteras antiguas. *Diario Menorca,* 02-11, 14-15.

Nicolás Mascaró, J., 1990. *El camí de cavalls de Menorca ahir i avui.* Institut d'estudis Baleàrics. Palma.

Sanchez Navarro, J., 2001. Estudi de les pedres de molins manuals i de les seves zones d'extracció a Menorca. Ciutadella de Menorca. *Publicacions des Born,* 10, 49-179.

Sanchez Navarro, J., 2005. Estudi de les pedres de molins manuals i de les seves zones d'extracció a Menorca. In M. Barcelo y F.Retamero (eds) *Els Barrancs tancats l'ordre pagès al sud de Menorca en* època *andalusina.* Maó. Recerca, 11, 235-67.

Sanchez Navarro, J., 2006. Estat actual de les investigacions sobre les pedres de molins manuals i de les seves zones d'extracció a Menorca, Ciutadella de Menorca, *Publicacions des Born,* 15-16, 155-96.

Visquis, A., 1973. Premier inventaire du mobilier de l'épave des jarres a Agay. *Cahiers d'archeologie subaquatique,* II, 157-67.

Ximenes, S., 1976. Étude préliminaire de l'épave sarrasine du Rocher de l'Esteou. *Cahiers d'archeologie subaquatique,* V, 139-50.

Appendix 1

UBICACIO PEDRERA	NUM ATLAS	MATERIALS	SUP.M2	MOLES MIN.	MOLES MAX.
PUNTA QUINTANA	327	MIOCÈ	225	600	1200
CALA BLANCA	335	MIOCÈ	100	200	400
CLOT DE SA CERA	391	DUNA QUAT.	25	50	100
MARESOS DE SA MARJAL	336	DUNA QUAT.	20	20	200
ENFONSAT DE BINISAID	334	DUNA QUAT.	5	10	10
S'OLLETA	424	DUNA QUAT.	50	14	20
SES PEDRISSADES	348	DUNA QUAT.	50	30	60
CODOLS DELS DATILS	349	DUNA QUAT.	50	30	50
PLATJA TREBALUGER	423	DUNA QUAT.	10	80	100
SA MILOCA-ES CORRAL FALS	315	DUNA QUAT.	4500	16000	25000
MORRO DE PONENT - CALES COVES	444	DUNA QUAT.	2	2	2
SA REGANETA-ES CANUTELS	326	DUNA QUAT.	300	350	1050
PUNTA ENMIG-BINIPARRATX	388	MIOCÈ	20	30	60
MORRO LLEVANT-SES ANGLADES	331	MIOCÈ	1000	750	2250
CAP DEN FONT - EST	376	MIOCÈ	10	9	30
OEST CAP DEN FONT	375	MIOCÈ	5	5	15
CALO BLANC	377	MIOCÈ	1	1	2
ILLOT DEN MARSAL	328	MIOCÈ	450	1000	3000
ILLOT BINISAFULLER	456	MIOCÈ	300	400	600
ILLOT DE SA BARCA	455	MIOCÈ	8	5	15
ILLOT CALO DEN FUS	352	MIOCÈ	50	50	100
OEST CALO PLA	378	MIOCÈ	5	5	10
CALO PLA - EST	401	MIOCÈ	5	5	20
CALO TANCAT-C.MORLA	353	MIOCÈ	10	150	200
CALO MORLA	316	MIOCÈ	300	3000	4500
NA MOIX	351	MIOCÈ	1	1	1
SA TORRETA	402	DUNA QUAT.	20	5	20
CALA CALDES	350	DUNA QUAT.	300	150	200
SON PARC - COVA D'EN SASTRE	452	DUNA QUAT.	100	80	100
MARESOS DE CALA PUDENT	403	DUNA QUAT.	5	5	10
PUNTA FERRAGUT	317	DUNA QUAT.	500	1000	4000
BONA ESPERANÇA-CALA MICA	344	DUNA QUAT.	450	2000	4000
BINIMEL·LA	343	DUNA QUAT.	30	30	50
PENYALS BINIDEUFA	325	DUNA QUAT.	100	300	500
EL PILAR	404	DUNA QUAT.	2	3	5
SES FONTANELLES	390	DUNA QUAT.	10	5	20
CODOLAR DE BINIATRAM	389	DUNA QUAT.	150	300	500
CALA MORELL -PAS DELS MOSQUITS	379	DUNA QUAT.	300	1000	1000
PUNTA DE SON ESCUDERO	421	DUNA QUAT.	5	5	10
				27680	49410

Appendix 2

MEULIÈRE	ÉCHANTILLON	SiO2	CaCO3	MgCO3	CaMg(CO3)2	SUMA	ARAGONITO
PUNTA FONTANELLES	AFN-01	16.8	83.2	0	0	100	
PLATJA DE BINIMEL·LÀ	BNM-01	25.3	74.7	0	0	100	
PENYALS DE BINIDEUFÀ	BDF-01	8.5	91.5	0	0	100	
PENYALS DE BINIDEUFÀ	BDF-02	8.1	91.9	0	0	100	
PENYALS DE BINIDEUFÀ	BDF-02 b	8.7	91.3	0	0	100	
PUNTA SUD CALA BLANCA	CBL-01	1.1	0	77.9	21	100	
CODOLAR DE BINIATRAM	CBT-01	8.9	84.6	6.5	0	100	
CODOLAR DE BINIATRAM	CBT-02	7.2	92.9	0	0	100	
CALA MICA	CMC-02	26.2	49.5		24.3	100	
CALA MICA	CMC-01	11.4	87.6	1	0	100	
PUNTA DEN QUINTANA	PQT-01	7.3	4	88.7	0	100	
CALÓ MORLÀ	MRL-01	2.2	1.2	96.6	0	100	
CALÓ MORLÀ	MRL-02	10.6	18.4	71	0	100	
PUNTA DE SA MILOCA-CORRAL FALS	SMI-01	36.6	63.4	0	0	100	
PUNTA DE SA MILOCA-CORRAL FALS	SMI-02	47.1	52.9	0	0	100	
PUNTA DE SA MILOCA-CORRAL FALS	SMI-03	21.1	78	0.9	0	100	TR
PUNTA DE SA MILOCA-CORRAL FALS	SMI-04	28.6	67.9	3.5	0	100	TR
PUNTA DE SA MILOCA-CORRAL FALS	SMI-05	47.6	42.9	9.5	0	100	
PUNTA DE SA MILOCA-CORRAL FALS	SMI-06	37.6	60.9	1.5	0	100	

Rotary hand-querns in volcanic stone in the Medieval Mediterranean

Paul Arthur

Years of research have revealed the significance of the Mayen-Eifel millstone lava quarries in the early Medieval trade in northern Europe (Parkhouse 1997; Sindbaeck 2007, 310). Their success was perhaps largely due to three factors: their location along the Rhine, which permitted shipment of products down the river and overseas; Frisian trade during the period of economic growth under the Carolingian Empire; and, above all, their great suitability for providing rotary hand-querns for grinding wheat in areas were appropriate stones were lacking. The archaeological distribution of Mayen quernstones shows their success as trade items, whilst they may themselves have helped boost local economies by rendering flour and bread production more efficient. In lands where clays, gravels and soft stones predominated, the importation of lava quernstones was fundamental.

Many lands of the Mediterranean do not provide suitable rocks for production of wheat grinding millstones. The sandstones of North Africa, for example, whilst able to grind wheat, would have led to the production of particularly gritty bread. Unsuitable limestones dominate many other Mediterranean lands. For this reason, the more suitable and vesicular volcanic lavas were widely exported from suitable areas in classical antiquity, as David Peacock (1980) and others have demonstrated. The aim of this paper is to draw attention to the production and trade in lava rotary querns and millstones in the Mediterranean after antiquity. Although the available data is scanty, it probably represents just a small part of the evidence waiting to be tapped. Indeed, most studies of Medieval economy and rural labour in the Mediterranean have concentrated on the water-mills and the later windmills, often of community use (villages, monasteries, etc.; see, for instance, Harvey 1989, 128ff., or various papers in the *Economic History of Byzantium;* cf. the words of Bryer 2002, esp. 110), although the overall economic effects of tens of thousands of peasant households acquiring and using the small rotary querns should not be underestimated, nor should the effects of large-scale quarrying on communities and landscapes.

It is quite understandable that domestic rotary querns do not generally figure in the sources, which devote more attention to larger and more costly and productive mills. Nonetheless, archaeological finds of Medieval rotary hand-querns are by no means infrequent on Mediterranean sites though, more often than not, when published, they are simply mentioned, rather than being illustrated and identified to rock source. I suspect that the scarcity of publication or analysis of archaeological discoveries is due to a lack of awareness of the economic questions that such finds can raise, matched by the fact that broken querns are by no means the most appealing of archaeological discoveries. This, of course, leads to a significant loss of information regarding sources, exchange and peasant activity. Nonetheless, a number of discoveries and identified examples lead to some interesting considerations.

Millstones are fairly frequently found on wreck sites in the Mediterranean (cf. Beltrame and Boetto 1997, where the stones are illustrated). This indicates long-distance commerce since classical times of items that, at the same time, could serve as ideal saleable ballast. As Medieval archaeologists have rarely considered the provenances of millstones, the work on the 11th century Serçe Liman shipwreck, found off the south-west coast of Turkey (Van Doorninck 1986; Bass *et al.* 2004; Bass *et al.* 2009), stands out as being of particular note. Petrological analyses of lava rotary querns from the wreck indicated the Island of Melos in the Cyclades as their likely source (Williams-Thorpe and Thorpe 2004).

Melos, indeed, appears to have been a source of stone querns since the Neolithic, and manufacture also seems to have taken place in Roman times and much more recently, as has been shown by Malcolm Wagstaff (1982; cf. also Runnels 1981; 1990; Kardulias and Runnels 1995). Wagstaff further suggested that Melos was a source of millstones during the Middle Ages, particularly through quarrying in the locality of Rema, on the island's eastern coast. The international importance of the Rema quarry and its operation into the twentieth century, until it closed in 1956, has most recently been discussed by Vrettou-Souli (2002), who suggests that in Byzantine times the capital of the island was even transferred from the site of Klima to Zephyria, to take advantage of the quarry. Rema stone is a resistant whitish-grey rhyolitic tuff, whose vesicular structure makes it ideal for grinding. Nonetheless, the existence of rather similar rocks on nearby islands, as well as in volcanic areas of western Turkey and the Levant,

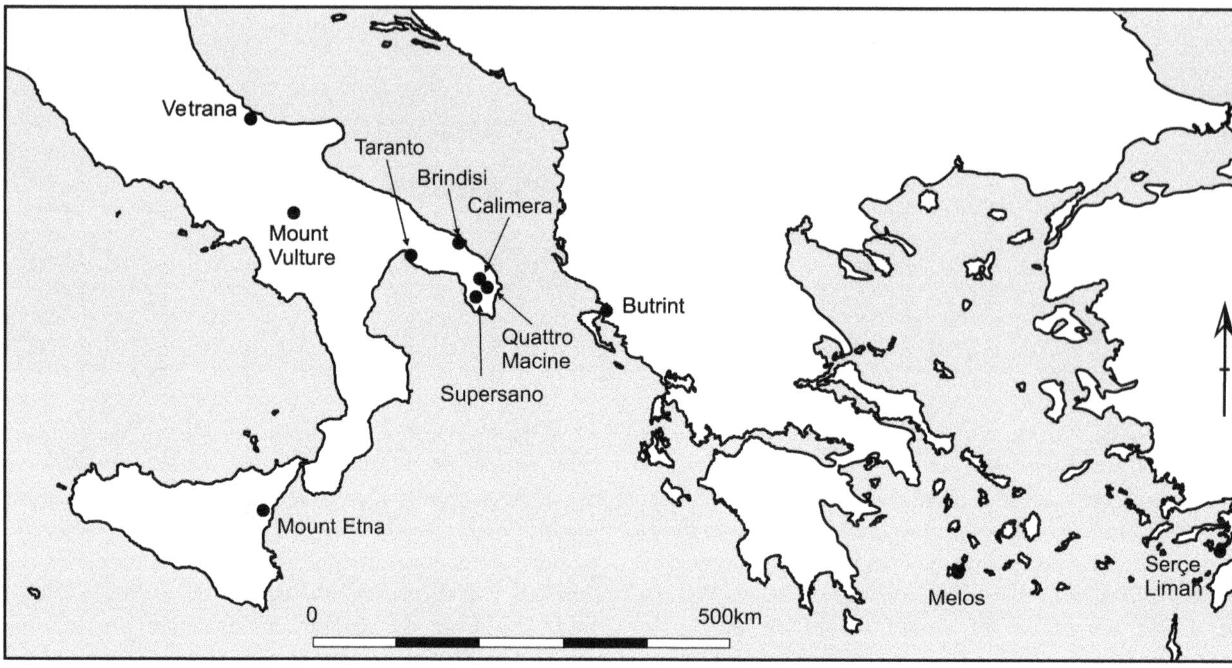

Figure 1. The Medieval Mediterranean

make macroscopic identification of finds rather uncertain, and underscore the need for petrological and geochemical analyses.

Thus, the probable attribution through analysis of the Serçe Liman millstones to Melos, and more specifically to Rema, is of great interest for understanding Medieval commerce, suggesting exportation from the island from, at least, the early 11th century (Runnels 2004). Further Rema millstones appear to have been found on sites in the Argolid dating to the 13th and 14th centuries, which suggested to Kardulias and Runnels (1995, 127) that the Franks may have held a monopoly on quarrying on the island. Whether or not this was the case, both Byzantium and Venice may have profited from the extraction and commerce of millstones from Melos prior to the 13th century, as a decree of the Emperor Alexius III Angelos dating to 1198 attests to the Venetians being permitted to trade there (Malamut 1988, 331). It is hard to believe that the island was 'nearly or totally abandoned from the early 9th until the early 13th century' as Guy Sanders suggested (1996, 159).

During almost twenty years of research on Medieval sites in Salento, the heel of Italy, I have come across abundant fragments of rotary quernstones during both excavation and field survey (Arthur 2000). Over ten years ago thirteen of these stones from the Byzantine and later Medieval village of Quattro Macine were examined by David Williams through petrological thin-sections, showing that more than half of them almost certainly came from Mt. Etna in Sicily, and four were quite plausibly products of the Rema quarry on Melos. Of the remaining two, one was possibly from Mt. Etna, whilst the other is in a distinctive, though unknown, lava stone that, however, finds a parallel in a quern uncovered during excavations at Carthage. Many of these samples were not closely datable, though Etna stone already appears in 10th-11th century Byzantine contexts. Other querns, macroscopically similar to both Melos and Etna stones, were found during excavation of the 7th-8th century Byzantine village at Supersano.

The Etna stones indicate an alternative to Melos as a source of supply of rotary querns to Salento during the Middle Ages. The Sicilian volcanic deposits were already noted for their millstones by Strabo (VI.2.3 and X.5.16), along with the island of Nisyros in the Dodecanese, some 250 kilometres to the east of Melos. Unfortunately, there is even less archaeological evidence for the commerce of Etna querns in the Middle Ages, despite a wealth of evidence for Roman times (cf. Wilson 1990, 240 and 396, note 24).

Leaving aside Roman-period imports, these and other finds indicate that imported lava querns reached the Salento from the 7th/8th through to the 15th century and beyond. An indication of their significance may be suggested by their presence at the early Medieval village of Supersano, dated to the 7th and 8th centuries, where very few other imported items were found (Arthur *et al.* 2008). Clearly, the calcareous Salento area, lacking in suitable stone for querns, was a recipient of imports, and the finds almost certainly represent just a very minor part of what must have been a quite noteworthy trade. Enticing is the discovery of millstones in the so-called Taranto A wreck which has been identified as a Byzantine ship that sunk sometime between the 5th and the mid 7th century off the northern Salento coast, though details of the stones' provenance is missing (Parker 1992, wreck 1131). On the Adriatic side of Salento and at the other end of the Middle Ages is the notice of a cargo of millstones directed to Brindisi from Crete in the 16th century (Gianfrotta 1988, 125). To complete the news of post-classical millstones in Salento, it is worth mentioning the presence of a probable example from Melos in the Museum of Calimera, which appears to be a 19th century import.

Since first publishing the paper on the Medieval quernstones from Salento, various other imported examples have come to light in the area, strengthening the case for substantial trade in these items during the Middle Ages. Unfortunately, however, the relatively abundant evidence from Salento is not matched elsewhere in the Medieval Mediterranean zone. Outside the region, the only new provenanced find that I can cite is that of a rotary quern from a Medieval context (BUT03 5337) at Butrint (Albania), indicated to have been quarried on the island of Melos, but not from the well-known Rema outcrops (Gerke *et al.* 2006, 143).
Other finds indicate different quarries as sources for Medieval rotary querns in southern Italy. An example from the hilltop settlement at Vetrana, near Guglionesi, Molise, datable between the 10th and 12th centuries, indicates the area of the Vulture volcano as a probable source (Hodges and Wickham 1981, 497). Furthermore, millstones from the Alps supplied various parts of northern Italy. Other sources are awaiting discovery or clarification.

Despite archaeological evidence for the trade in Medieval millstones in the Mediterranean first published some thirty years ago, the list of finds is still disappointingly small. The identification, provenancing and dating of quernstones in the Medieval Mediterranean is an important issue in the analysis of exchange, both from the point of view of extraction/production sites and consumer sites, and I hope that one of the effects of these proceedings will be to stimulate interest in the use of such objects during the Middle Ages and bring further examples to light.

Acknowledgements

Various people have been generous in information concerning Medieval millstones. David Williams has carried out the petrological analyses of Salento millstones, first helping to establish Melos and Etna as their probable sources. I should also particularly like to thank Sophia Germanidou, who introduced me to various studies concerning Melos and the archaeological discovery of millstones in Greece. Pamela Armstrong has kindly discussed with me the archaeological evidence for Byzantine Melos. Thanks are due to Penny Copeland for drawing figure 1. Last, but not least, my research into millstones has been supported by a grant from the Society for Medieval Archaeology.

References

Arthur, P., 2000. Macine intorno al Mille: Aspetti del commercio dalla Grecia e dalla Sicilia in età Medievale. In G.P. Broglio (ed), *II Congresso Nationale di Archeologia Medievale*. Florence, 485-9.

Arthur, P., Fiorentino, G., and Leo Imperiale, M., 2008. L'insediamento in Loc. Scorpo (Supersano, LE) nel VII-VIII secolo. La scoperta di un paesaggio di età Medievale. *Archeologia Medievale,* 35, 365-80.

Bass, G.F., Matthews, S.D., Steffy J.R., and van Doorninck, F.H., 2004. *Serçe Liman, An Eleventh-Century Shipwreck, vol. I: the ship and its anchorage, crew and passengers*. Texas.

Bass, G.F., Lledò, B., and Matthews, S.D., 2009. *Serçe Liman, vol. II. The glass of an Eleventh-Century Shipwreck*. Texas.

Beltrame, C., and Boetto, G., 1997. Macine da relitti. *Archeologia subacquea, studi, ricerche e Documenti, 2,* 167-96.

Bryer, A., 2002. The means of Agricultural Production: Muscle and Tools. In A. Laiou (ed.) *The Economic History of Byzantium from the Seventh through the Fifteenth Century* 39,101-13. Dumbarton Oaks.

Gerke, T.L., Stocker, S.R., Davis, J.L., Maynard, J.B., and Dietsch, C., 2006. Sourcing Volcanic Millstones from Greco-Roman Sites in Albania, *Journal of Field Archaeology* 31, 137-46.

Gianfrotta, P.A., 1988. Archeologia sottomarina in Puglia. In C. Marange (ed.), *La Puglia in età repubblicana, Atti del Convegno di Studi sulla Puglia Romana (Mesagne 1986.,* Galatina, 119-25.

Harvey, A., 1989. *Economic expansion in the Byzantine Empire 900-1200.* Cambridge.

Kardulias, P.N., and Runnels, C., 1995. The lithic artifacts: flaked stone and other non-flaked lithics. In C. Runnels, D.J. Pullen and S. Langdon (eds.), *Artifact and Assemblage. the Finds from a Regional Survey of the Southern Argolid, Greece, vol. I, the Prehistoric and Early Iron Age Pottery and the Lithic Artifacts.* Stanford, 74-139.

Malamut, E., 1988. *Les Iles de l'Empire byzantin: VIIIe-XIIe siécles.* Paris.

Parker, A.J., 1992. *Ancient shipwrecks of the Mediterranean and the Roman provinces*. BAR IntS, 580, Oxford.

Parkhouse, J., 1997. The distribution and exchange of Mayen Lava Quernstones in Early Medieval North-western Europe. In G. De Boe and F. Verhaeghe (eds.), *Exchange and Trade in Medieval Europe, Papers of the 'Medieval*

Europe Brugge 1997' Conference, Volume 3. Zellik, 97-106.

Peacock, D.P.S., 1980. The Roman millstone trade: a petrological sketch. *World Archaeology,* 12, 43-53.

Runnels, C., 1990. Rotary querns in Greece. *Journal of Roman Archaeology,* 3, 147-54.

Runnels, C., 2004. The Querns. In Bass *et al.* 2004, 255-9.

Sindbaeck, S.M., 2007. Part 2: Northern Europe. In P. Arthur and S.M. Sindbaeck, Trade and exchange. In J. Graham-Campbell and M. Valor (eds.), the *Archaeology of Medieval Europe, Vol. I, Eighth to Twelfth Centuries AD.* Aarhus, 289-315.

Sanders, G.D.R., 1996. Two *kastra* on Melos and their relations in the Archipelago. In P. Lock and G.D.R. Sanders (eds.), the *Archaeology of Medieval Greece.* Oxford, 147-77.

van Doorninck, F.H. Jr., 1986. The Medieval shipwreck at Serçe Liman: An early 11th-century Fatimid-Byzantine commercial voyage. *Graeco-Arabica,* 5.

Vrettou-Souli, M., 2002. *H Milopetra tis Milou. Apo tin exorixi stin emporiki a diakinisi* (*the millstone of Milos : from extraction to commerce*). Athens.

Williams, D.F., 2000. Appendice I: A petrological note on the rotary querns from Quattro Macine and Centoporte., In Arthur, 2000, 488.

Williams-Thorpe, O., and Thorpe, R.S. 2004. Comparison of millstone samples from the Serçe Limani shipwreck with possible source rocks from Rema. In Bass *et alii.* 2004, 259-62.

Wilson, R.J.A., 1990. *Sicily under the Roman Empire, the archaeology of a Roman province, 36BC-AD535.* Warminster.

Production, commercialisation et qualitè de meules à main et de meules à moulin dans l'Italie médiévale: un bilan de la recherche historique et archéologique

Paola Galetti

La recherche médiéviste sur le 'thème des moulins' s'est développée dans les dix dernières années essentiellement dans deux directions: l'une s'inscrit dans une perspective d'histoire économique et sociale et prend surtout en considération le lien moulin-societé, en suivant la démarche de Marc Bloch; l'autre se concentre sur la structure moulin et sur la problématique technologique. Ensuite, la plupart du temps, on a essayé d'éviter les généralisations et de restreindre le champ de recherche à des études 'régionales' (Mousnier 2002; Galetti and Racine 2003; Cavaciocchi 2003; Belmont and Mangartz 2006; Racine 2006; Arnoux 2008; Galetti and Andreolli 2009; Bettoni and Ciuffetti 2010. Pour une synthétique bibliographie des études régionales: Galetti 2009, 21, note 21).

C'est dans le cadre de cette première perspective que je vous propose maintenant quelques réfléxions sur deux aspects de la problématique 'des moulins'. Je voudrais surtout porter ma réfléxion sur les différents modes de mouture des céréales, leur technologie et leurs conséquences au niveau de l'histoire économique et sociale. Pour simplifier: les céréales pouvaient être moulues, soit dans un moulin, qui présupposait une plus grande complexité de construction et de technologie, soit, simplement, avec des meules à main. L'histoire du moulin remonte à l'Antiquité. Le moulin, placé dans un bâtiment, impliquait le recours à l'énergie motrice humaine et animale et, au Moyen Âge, en grande partie à l'énergie hydraulique, en plus d'un appareillage complexe afin d'activer le fonctionnement des meules. Ce qui supposait, par conséquant, quelques opérations de dérivation et canalisation des eaux (Figs.1, 2, 3). Toute aussi antique est l'usage des meules à main, qui deviennent simples outils de travail, sans avoir besoin de prévoir des structures et appareils de soutien, même si certainement nous nous trouvons, de toute façon, dans ce cas aussi, face à un outil sûrement plus évolué et moins rudimentaire que le simple mortier (Figs. 4, 5, 6) (Bradford Smith, and Wolfe, 1997; Czysz 1998; Squatriti 2000; Lohrmann 2002; Racine 2003; Comet 2003, Galetti 2010).

Un problème historiographique qui a été – et est encore – débattu parmi les chercheurs, est de savoir si l'essor des

Figure 1. Moulin à eau. (René I d'Anjou, La mortification du vaine plaisance, XV siècle, New York, Pierpont Morgan Library).

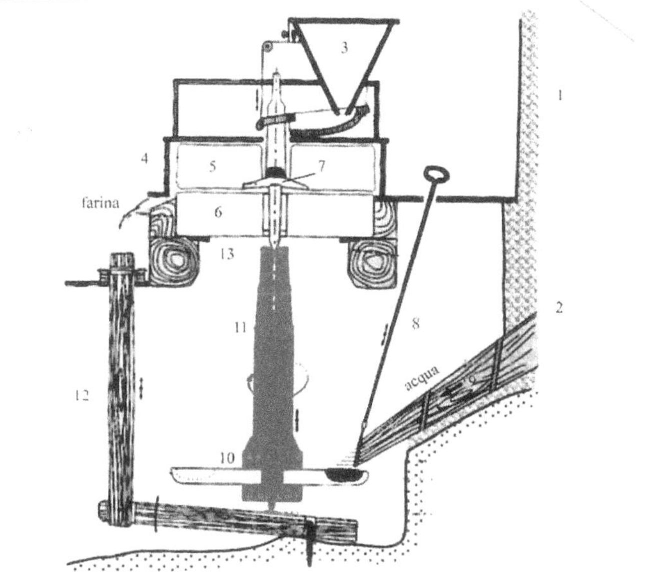

1) Camera delle mole; 2) Camera delle acque; 3) Tramoggia; 4) Cassa delle mole; 5) Macina o mola superiore, o rotante; 6) Macina o mola inferiore, o dormiente; 7) Nottola; 8) Chiusura; 9) Cannone o bocchettone; 10) Ruota orizzontale a cucchiaio, o ritrecine; 11) Palo o fuso; 12) Staffa per la regolazione dell'altezza della ruota e delle mole; 13) Cunei per porre in piano la mola dormiente.

Figure 2. Le dessin d'un moulin horizontal. (Nejrotti 2007, 63).

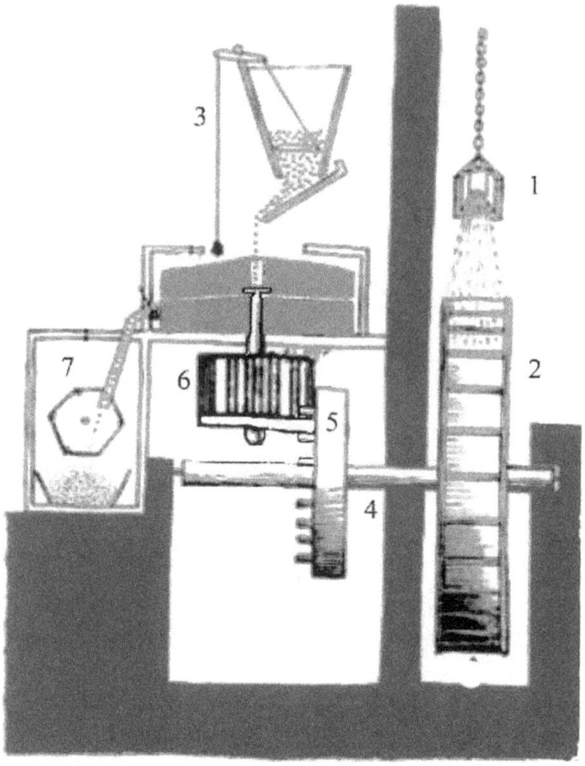

1) doccia; 2) Ruota verticale; 3) Sistema d'allarme per la tramoggia vicina allo svuotamento; 4) Albero orizzontale; 5)Lubecchio; 6) Lanterna; 7) Buratto.

Figure 3. Le dessin d'un moulin vertical. (Nejrotti 2007, 64).

Figure 5. Meules à main romaines. (Roda, Setti 2004, 35).

34

Figure 4. Meules à main préromaines. (Roda, Setti 2004, 34).

moulins seigneuriaux, en régime de monopole, ou bien des moulins publiques (je pense ici aux moulins communaux), au delà de leurs formes concrètes de gestion (de manière directe ou indirecte par des contrats de location) a constitué un obstacle à l'usage répandu des meules domestiques à main, qui pouvait permettre de se soustraire au paiement d'une taxe sur la mouture. Marc Bloch dans son célèbre essai de 1935 «Avènement et conquête du moulin à eau» parle de « la multiplicité des meules domestiques, obstinées au travail, depuis des siècles, presque dans chaque chaumière. Les seigneurs leur déclarèrent la guerre ». Mais à propos de cette longue guerre, il nous dit aussi que « la victoire, cependant, n'était pas si complète à la fin du Moyen Âge, qu'il ne substistât çà et là, employés de façon plus ou moins intermettente, beaucoup des vieux instruments manuels » (Bloch 1935; Bloch 1970 (éd.it.), 98-100). Il est indubitable, si nous suivons la trace des sources écrites, qu'à partir du VIIIe siècle et surtout de la seconde moitié du IXe siècle, les témoignages sur les moulins - qui sont toujours à eau - sont de en plus en plus nombreux et augmentent encore au siècle suivant. Cela va de pair, dans notre peninsule, avec le plein essor de la grande propriété foncière et de son organisation sous formes «curtensi», dès l'instant que nous sommes en présence, en général, de moulins situés sur des terres seigneuriales; mais va aussi de pair avec la crise du pouvoir central et la lente affirmation de la seigneurie rurale territoriale, avec l'institution de quelques monopoles seigneuriaux.

Figure 6. Montarrenti (SI), reconstruction de la 'pars dominica' avec meule à main. (Valenti 2004, 102).

Et ce n'est pas par hasard. En fait, l'établissement d'un moulin hydraulique présupposait la possession de droits publiques sur l'eau, obtenus par des autorisations royales, en plus des dépenses pour les travaux de dérivation de l'eau, pour la construction et les réparations de l'installation, et la disponibilité de main-d'œuvre pour son fonctionnement et sa gestion, impliquant aussi un usage destiné à la mouture des céréales en quantité imposante. Si le moulin n'était pas géré directement par la propriété, son affermage pouvait constituer pour elle une source de revenu et de contrôle social, tandis que pour le concessionaire, en plus d'une source de revenu, pouvait représenter un tremplin pour une promotion sociale dans la communauté (Chiappa Mauri 2003; Galetti 2003. Pour la "curtis" en Italie: Andreolli and Montanari 1983; Sergi 1993; Toubert 1995; Pasquali 2008; Mancassola 2008).

Dans cet ordre d'idées, les éléments que nous pouvons tirer de l'analyse des « polittici »/polyptiques des grands monastères des IXe et Xe siècles ne manquent pas d'interêt. En particulier, celui du monastère de femmes de Santa Giulia de Brescia (879-906) et ceux du monastère de San Colombano de Bobbio (à partir du milieu du IXe siècle jusqu'à la fin du Xe siècle et début du XIe siècle) attestent en ce qui concerne la « pars dominica » (toujours et seulement elle) de différentes « curtes » de moulins, pour lesquels était perçu une redevance et qui devaient probablement servir à faire face aux exigeances, non seulement de la 'pars dominica' et des paysants du 'massaricio' d'une seule ferme, mais de plusieurs fermes, dans le cadre d'une rationalisation de la gestion de l'ensemble de la proprieté monastique (*Inventari altomedievali* 1979: *Santa Giulia di Brescia*, 56, 62-64, 66-70, 73-75, 77-78, 80, 90; *San Colombano di Bobbio*, *I*, 136,139, 144; *II*, 158, 160, 165; *III*, 169, 173; *IV*,186. Pour les polyptiques carolingiens: Champion 1996).

Avant cette chronologie, aussi bien les sources écrites que matérielles semblent, en revanche, renvoyer à un recours concomitant tant à des structures meunières qu'à des meules à main, avec une différenciation dans l'usage liée à la quantité de céréales à moudre et, donc, une préference dans la mouture domestique, familiale, pour la meule à main, dont la possession pouvait peut-être aussi représenter un signe distinctif sur le plan social. Ainsi dans les Lois lombardes, nous trouvons trois rubriques destinées aux compensations pecunières pour les dommages causés aux moulins, auxquels on attachait beaucoup de valeur. Par exemple, si le moulin venait à brûler, celui qui avai mis le feu devait payer à la propriété le triple de la valeur estimée du moulin et de tout ce qu'il contenait (Blühme 1869; Azzara and Gasparri 1992: Rotari, rr.149, 150, 151). En outre, dans cet ordre d'idées, il faut souligner les résultats significatifs d'une fouille effectuée à Pietradurante de Bisaccia (AV) d'une nécropole du haut Moyen-Âge qui porte des traces d'une fréquentation des Goths, des

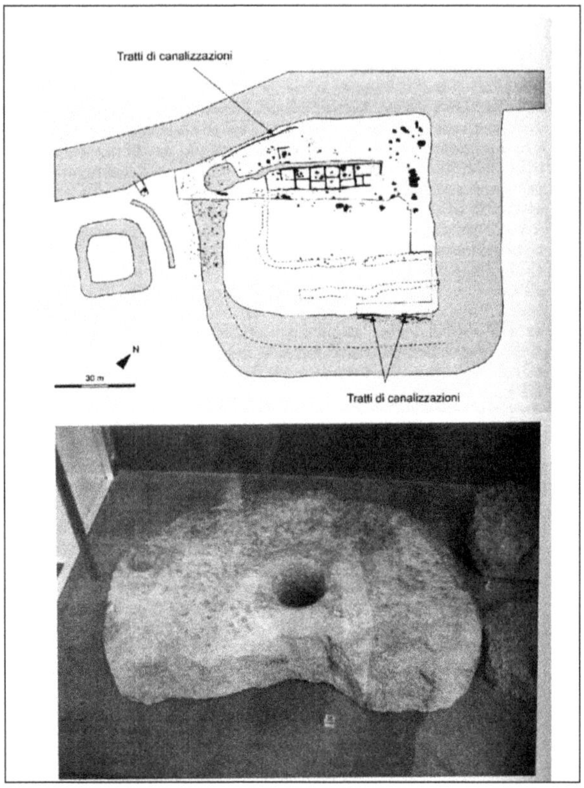

Figure 7. Planimétrie du site de Sant'Agata Bolognese (localité Crocetta)-X siècle; meule à main (Gelichi, Librenti 2009, 356).

Lombards et des Bizantins, et où 11 tombes ont révélé du mobilier funéraire, parmi lesquels on a trouvé aussi des meules à main en pierre de lave du Vulture (Peduto 1979; Peduto 1984, 58, 100). En revanche, dans les sources écrites à partir des IXe-Xe siècles, des références claires à la mouture domestique des céréales avec des meules à main semblent se raréfier. Dans les contrats de location, on précise, par exemple, que la part a payer en céréales prévue pour la redevance soit remise «battue», c'est à dire en grains, et non en farine (Mancassola 2008. Bloch 1970, 97 signale des paiements de la redevance en farine par les fermiers de Saint-Bertin, au IXe siècle, et par les serfs de Saint-Denis à Concevreux au Xe siècle = mais, s'-agit-il de farine?). Dans ce cadre, les découvertes faites dans certains sites de Toscane sont très intéressantes. Elles témoignent des procédés de transformation de quelques villages, qui depuis l'Antiquité tardive présentent une évolution continue des sites habités vers des formes «incastellate» au milieu du Moyen Âge et au bas Moyen Âge. Ainsi, à Montarrenti (SI), entre la moitié du VIIIe siècle et la moitié du IXe siècle, le village, qui à ses débuts, entre la moitié du VIIe siècle et la moitié du VIIIe siècle, était un ensemble de petites maisons rudimentaires dispersées sur toute la superficie d'un paysage de collines, subit une transformation qui concerna la partie supérieure du relief, où se concentrèrent des structures destinées non seulement à la récolte des produits agricoles, mais aussi à leur transformation. On a effectivement rétrouvé ici un petit four pour le séchage des céréales et une meule. On peut dire la même chose pour l'établissement de Miranduolo (SI). Nous sommes face au développement d'un processus de hiérarchisation des espaces présupposant la création d'un espace 'seigneurial' séparé, avec une habitation de type distinctif entourée de quelques maisons modestes et d'édifices (magasins, grenier à grain) indiquant la présence d'une figure patronale capable de rationaliser les prélevements sur la production agricole (il en va ainsi à Montarrenti, Poggibonsi, Miranduolo, peut-être Scarlino); de concentrer les structures pour la fabrication de biens (forges et fours à Poggibonsi, Scarlino, peut-être Donoratico et Rocchette Pannocchieschi) ou pour le traitement de produits alimentaires (fours pour l'assèchement des céréales, structures pour la mouture des grains, attestées par les meules découvertes, pour l'abattage et la transformation de la viande à Montarrenti, Poggibonsi, Donoratico); et aussi, capable d'éxiger certains travaux de la part de ses paysans (construction de clôtures ou de murs, creusement de fossés à Montarrenti, Miranduolo, peut-être Scarlino) ou d'emboucher des ouvriers pour des interventions spécifiques. Au delà du fait que pour les sites de Montarrenti et Miranduolo il semble difficile d'évoquer la présence d'une structure moulin, la dimension même des meules renvoie à une mouture à main. Mais ce qui est significatif c'est qu'il s'agit d'une activité destinée à la partie patronale d'un regroupement d' habitations, qui en constitue ainsi un trait distinctif. On pourrait dire la même chose des résultats des fouilles du château toscain de Rocca S. Silvestro, bâti aux Xe-XIe siècles pour l'exploitation de gisements de cuivre et de plomb. On y a découvert à l'intérieur du château la présence de meules à main et, en plus, d'un four communautaire (Valenti 1996; Valenti 2004; Francovich and Valenti 2005; Valenti 2005; Valenti 2008; Bianchi 2008, 77-87).

Une autre situation intéressante, plus complexe, est celle des fouilles menées aux environs de Sant'Agata Bolognese (localité Crocetta, Possessione Canale, 1994-97). Il s'agit d'un ensemble d'habitations (dont nous n'avons pas de traces dans les sources écrites de cette époque) s'étant formée au cours du IXe siècle: il présentait les caractéristiques d'une grande ferme remarquablement organisée, entourée de fossés concentriques, mais sans vocation defensive, à côté de laquelle s'écoulait un cours d'eau assez important. Dans la phase successive, au Xe siècle, l'agglomération se développa comme un village fortifié, caractérisé par la prévalence de l'activité artisanale et commerciale, à côté d'une « motta ». Il semble avoir été abandonné au début du XIe siècle. Il s'agissait, probablement, d'un centre de gestion des activités ou d'un site artisanal et commercial se référant à quelque important propriétaire: non loin de là se trouvait en effet le monastère de S. Silvestro de Nonantola (Fig. 7). Les fouilles ont mis à jour une série de meules (environs 12), se référant surtout à la seconde vie du site, toutes du type rotatif et qu'on peut diviser en deux groupes. Le premier (11 exemplaires) était équipé d'une traction à l'aide d'un pivot vertical avec des pierres d'un diamètre important (entre 50 et 70 centimètres), avec une ouverture centrale pour le grain et un creux sur le côté opposé pour introduire un élément à papillon, en fer, pour la rotation. Leur mouvement ne pouvait être mis en route qu'en transférant l'énergie produite par la force hydraulique d'un axe horizontal vers un axe vertical grâce un couple d'engrenages. Le second (1 exemplaire) avec un diamètre plus petit (25 centimètres) était probablement à entraînement manuel. Toutes les meules avaient une surface plane et conservaient des rayures de burin, traces des opérations qui au cours du temps avaient cherché à mantenir le frottement sur les grains. Le premier type, prédominant et caractéristique, renvoie à des installations de moulin exploitant l'énergie hydraulique, tandis que le second implique un travail domestique, à caractère non artisanal. On a découvert aussi des traces de systèmes de dérivation et du contrôle des eaux pour le fonctionnement des moulins. L'ensemble semble évoquer un site à vocation surtout artisanale, se référant à un domaine seigneurial: et c'est dans tel cadre que nous devons inscrire la présence des moulins, sans cependant passer sous silence l'usage domestique de petites meules à main pour les besoins quotidiens et de petites quantités de céréales. Nous pourrions parler dans ce cas de coexistence de deux systèmes (Gelichi and Librenti 2005; Galetti 2006; Gelichi and Librenti 2009; Gelichi and Librenti 2010, 17-22). C'est probablement la conclusion à laquelle on pourrait arriver par rapport à la première question que je me suis posée, au moins pour les IXe Xe siècles.

La dimension domestique de la mouture a dû coexister quoi qu'il en soit, au Moyen Âge, au cours des siècles suivants aussi, avec des procédés de mouture de type artisanal et à vocation collective, faisant référence aux pouvoirs seigneuriaux ou publics. Je voudrais ajouter qu'aujourd'hui la recherche historique tend à estomper encore plus le poids du moulin 'bannal/seigneurial', retrouvant des témoignages sur des manifactures réalisées à l'initiative de personnes privées ou de communautés, ou bien d'organismes privés ou autres organismes, ayant échappé aux pouvoirs 'supérieurs'. L'activité meunière développée par des structures spécifiques permanentes était de toute façon toujours à mettre en rélation avec la satisfaction des exigeances de groupes de population importants et avec la

Figure 8. Meule fouillée dans le site de Piadena. (Brogiolo, Mancassola 2005, 138).

possibilité de gérer et dévolopper un réseau hydraulique, capable de fournir la force motrice nécessaire, étant donné qu'il s'agissait de moulins hydrauliques (Rivals 1987, 59-60; Delucca 1991, 626-634; Galetti 2003; Delucca 2008, 84-90; Nico Ottaviani 2008, 81-95, 115-142). La dimension domestique est moins facilement perceptible à l'examen des sources écrites, à la différence de l'abondance des informations concernant les moulins. Mais, malgré tout, il existe des attestations significatives, par exemple les «estimi», registres fiscaux, dans lesquels nous pouvons trouver des indications sur les paysans, qui s'endettaient (les dettes étaient indiquées dans les «estimi» parce qu'ils étaient déduits de l'estimation totale de biens soumis à une taxation des « foyers »), recourant à des emprunts pour des besoins variés: par exemple, pour l'acquisition d'une meule à usage familiale (estimo de Galliera, BO, 1235) (Zanarini 2007, 67). La dimension domestique apparaît plus clairement dans les résultats materiels de recherche, là où, à côté de découvertes de meules qu'on peut sans hésiter associer (même en ce qui concerne la dimension) à des moulins, nous trouvons des meules à traction manuelle, que ce soit à l'intérieur de centres de fondation récente (XIIe-XIIIe siècles), ou que ce soit au niveau de site disséminé du bas Moyen Âge (Gelichi 1991; Comba 1993; Gelichi and Librenti 2009, 352-355; Nejrotti 2008. Cfr. aussi, en general: Watts 2002).

Un second problème historiographique est celui de rechercher, par l'analyse typologique des meules (des moulins ou à main) et des matériaux qui les constituaient, des informations utiles afin de recréer un aspect de la dynamique économique: en particulier, le réseau des trafics commerciaux à court, moyen et long terme. Ceci en raison du caractère propre des pierres à meules, qui, pour fonctionner, devaient être dures et à grains grossiers, pour éviter le glissement des graines sur la surface, et qui pouvaient être destinées à des lieux de production très diverses et même très éloignés du lieu de découverte et, probablement, de leur utilisation (Mannoni and Giannichedda 2003, 98-99, 158, 228-231, 274-277; Valenti 2008, 200-201). La découverte de meules de matériaux particuliers pouvant être associées à des zones de production bien précises peut permettre de localiser les différents sites à l'intérieur d'un circuit économique élargi ou bien, à l'inverse, d'en déduire un isolement relatif (Galetti 2006, 74-77). Considérons, par exemple, la situation de la zone régionale d'Emilie-Romagne entre les IXe et Xe siècles. Deux fragments d'une meule réalisée en « cloritoscisto a granati », du diamètre de 50 centimètres et d'une épaisseur de 6,5 centimètres, proviennent de Piadena (CR). D'autres meules fabriquées en même lithotype ont été retrouvées à Canolo (RE), dans le territoire de Medicina (BO), à Villa Clelia (*Imola*-BO) ; à Sant'Agata Bolognese on a retrouvé des meules en « talcoscisto », dont la diffusion à dû se poursuivre à l'échelon de la province, jusqu'au XIIIe siècle (Fig. 8). Il s'agit, dans le premier cas, d'un lithotype présent dans l'arc alpin occidental, là où les carrières d'extraction et de production des biens manufacturés sont situés dans la zone d'Aoste, où, ils sont encore présents, inachevés, dans des carrières, et en cours d'éxtraction. En ce qui concerne le deuxieme cas, il s'agit d'un lithotype provenant de l'arc alpin central (Val Chiavenna, Valtellina) (Tirabassi 1980, c. 55; Gelichi 1982, 57; Gelichi 1990, 189-190; Grillo 1993; Brogiolo and Mancassola 2005, 138; Malaguti 2005, 176; Galetti 2006,74-77). Des récipients en pierre ollaire pour la cuisson et la conservation de la nourriture et de forme semblable proviennent aussi de la même zone. Cette association, qui, dirais-je, est constante, nous montre un lieu de production et de commerce de produits diversifiés (meules et récipients) vers les mêmes destinations et utilisant probablement le réseau fluvial vaste et ramifié. Il s'agit d'un trafic commercial qui, en provenance de la zone alpine, devait se répandre en larges ramifications dans tout le territoire de la plaine du Pô, jusqu'au commerce de détail dans les petites fermes, et qui a dû parvenir à se diffuser à la même période aussi dans l'arrière-pays de la Romagne (par exemple, le Decimano-RA), tandis qu'auparavant il semblait avoir concerné aussi le port de Classe, là où ont été retrouvés des meules ou fragments de meules essentielment en chloritosciste (Mannoni and Messiga 1980; Gelichi 1983; *La pietra ollare* 1987; Mannoni *et al.* 1987; Gelichi 1987; Gelichi 1990; Alberti 1997; Malaguti and Zane 1999; Montevecchi and Novara 2000; Alberti 2003; Malaguti 2005; Pantò and Uggè 2007, 142, 148,152; Alberti 2009). Au delà des lithotypes, on se trouve face à des meules de dimensions diverses avec une prédominance pour celles de dimension superiéure, qui évoquent une utilisation dans les moulins, à comparer à celles plus petites et d'emploi plus souple, à usage domestique. La découverte dans le même contexte des deux typologies, provenant de la même zone, avec des diverses objets en pierre ollaire de la même provenance, peut amener à la prise en considération des secondes (les meules à main) comme élément de distiction sociale pour celui qui était en mesure de s'en procurer l'usage. De l'objet, la *mola*, nous sommes ainsi remontés à l'homme qui la produisait, la commercialisait et, enfin, l'utilisait, de façon privé ou collective. En partant de la *mola*, donc, nous sommes arrivés, comme nous le voulions, à la dynamique économique et sociale.

Bibliographie

Alberti, A., 1997. Produzione e commercializzazione della pietra ollare in Italia settentrionale tra tardoantico e Altomedioevo. In S.Gelichi (éd.), *I Congresso Nazionale di Archeologia Medievale*. Pisa, 335-339.

Alberti, A., 2003. *La pietra ollare,* in *Vivere nel Medioevo. Un villaggio fortificato del X secolo nella pianura padana,* San Giovanni in Persiceto (BO).

Alberti, A., 2009. La pietra ollare in Toscana. In G.Volpe and P.Favia (éds.), *V Congresso Nazionale di Archeologia Medievale*. Firenze, 630-633.

Andreolli, B., and Montanari, M., 1983. *L'azienda curtense in Italia. Proprietà della terra e lavoro contadino nei secoli VIII-XI,* Bologna.

Arnoux, M., 2008. *Les* moulins à eau en Europe occidentale (IXe-XIIe siècles). Aux origines d'une économie institutionelle de l'énergie hydraulique. In *L'acqua nei secoli altomedievali*. Spoleto, 693-746.

Augenti, A., De Breasi, G., Ficara, M., and Mancassola, N., 2005. L'Italia senza corti ? L'insediamento rurale in Romagna tra VI e IX secolo. In G.P. Brogiolo, A.Chavarria Arnau and M.Valenti (éds.), *Dopo la fine delle ville : le campagne dal VI al IX secolo*, Mantova, 17-52.

Azzara, C., and Gasparri, S., 1992. *Le Leggi dei Longobardi. Storia, memoria e diritto di un popolo germanico.* Milano.

Belmont, A.,and Mangartz, F., (éds.) 2006. *Les meulières. Protection et Valorisation d'un patrimoine industriel européen.* Mainz.

Bettoni, F., and Ciuffetti, A., (eds.) 2010. *Energia e macchine. L'uso delle acque nell'Appennino centrale in età moderna e contemporanea.* Narni (TR).

Bianchi, G., (ed.) 2008. *Guida all'archeologia medievale della provincia di Livorno.* Livorno.

Bloch, M., 1935. Avénement et conquête du moulin à eau. *Annales d'histoire économique et sociale*, 7, 538-563.

Bloch, M., 1970. Avvento e conquiste del mulino ad acqua. In M.Bloch, *Lavoro e tecnica nel Medioevo.* Bari, 73-110.

Blühme, F., (ed.) 1869. Edictus ceteraeque Langobardorum Leges. In M.G.H., *Fontes iuris germanici antiqui in usum scholarum*, Hannover.

Bradford Smith, E. and Wolfe, M. 1997. *Technology and resource use in medieval Europe*, Adelshot.

Brogiolo, G.P. and Mancassola, N., 2005. Il sito e gli scavi. In G.P.Brogiolo and N.Mancassola (éds.), Scavi al castello di Piadena (CR).In S.Gelichi (éd.), *Campagne medievali. Strutture materiali, economia e società nell'insediamento rurale dell'Italia settentrionale (VIII-X secolo),* Mantova, 121-41.

Cavaciocchi, S., (éd.) 2003. *Economia e energia. Secc. XIII-XVIII*, Firenze

Champion, E., 1996. *Moulins et meuniers carolingiens dans les polyptiques entre Loire et Rhin,* Paris.

Chiappa Mauri, L., 2003. Acque e mulini nella Lombardia medievale. In P.Galetti and P.Racine (éds.), *I mulini nell'Europa medievale*, Bologna, 233-68.

Comba, R., (ed.) 1993. *Mulini da grano nel Piemonte medievale. Secoli XII-XV,* Cuneo.

Comet, G., 2003. Pour une histoire des moulins entre technique et idéologie. In P.Galetti and P.Racine (éds.), *I mulini nell'Europa medievale*, Bologna, 7-36.

Czysz, W., 1998. *Die ältesten Wassermühlen*, Thierhaupten.

Delucca, O., 1991. *L'abitazione riminese nel Quattrocento. I. La casa rurale,* Rimini.
Delucca, O., 2008. *Morciano nel Medioevo. Fonti e spunti per un itinerario storico,* Rimini.
Francovich, R., and Valenti, M., 2005, Forme del popolamento altomedievale nella campagna toscana (VII-X secolo). In S.Gelichi (éd.), *Campagne medievali. Strutture materiali, economia e società nell'insediamento rurale dell'Italia settentrionale (VIII-X secolo),* Mantova, 245-58.
Galetti, P., 2003. I mulini nell'Italia centro-settentrionale dell'altomedioevo : edilizia e tecnologia. In P.Galetti and P.Racine (éds.), *I mulini nell'Europa medievale*, Bologna, 269-86.
Galetti, P., 2006. Moulins, meules, meulières et carrières de meules dans l'Italie médiévale. In A.Belmont and F.Mangartz (éds.), *Les meulières. Protection et Valorisation d'un patrimoine industriel européen*, Mainz, 71-80.
Galetti, P., 2009. Acque e mulini tra Età Medievale e Moderna. In P.Galetti and B.Andreolli (éds.), *Mulini, canali e comunità della pianura bolognese tra Medioevo e Ottocento*, Bologna, 17-26.
Galetti, P., 2010. Acque e mulini tra età medievale e moderna. In F.Bettoni and A.Ciuffetti (éds.), *Energia e macchine.L'uso delle acque nell'Appennino centrale in età moderna e contemporanea*, Narni (TR), 4-14.
Galetti, P. and Andreolli, B., (éds.) 2009. *Mulini, canali e comunità della pianura bolognese tra Medioevo e Ottocento*, Bologna.
Galetti, P. and Racine, P., (éds.) 2003. *I mulini nell'Europa medievale*, Bologna.
Gelichi, S., 1982. Materiale medievale da Bagnarola (BO). In *Il territorio di Budrio nell'antichità,* Budrio, 51-9.
Gelichi, S., 1983. La pietra ollare. In *Ravenna e il porto di Classe. Venti anni di ricerche archeologiche tra Ravenna e Classe,* Bologna, 176-77.
Gelichi, S., 1987. La pietra ollare in Emilia-Romagna. In *La pietra ollare dalla Preistoria all'Età Moderna*, Como, 201-13.
Gelichi, S., 1990. La pietra ollare, in Contesti tardo-antichi e altomedievali dal sito di Villa Clelia (Imola-Bologna). *Archeologia Medievale*, XVII, 189-90.
Gelichi, S., (ed.) 1991. *Archeologia e insediamento rurale in Emilia Romagna nel Medioevo*, Bologna.
Gelichi, S., and Librenti, M. 2005. Un villaggio fortificato dei secoli centrali del Medioevo nei pressi di S.Agata Bolognese (BO). In S.Gelichi (éd.), *Campagne medievali. Strutture materiali, economia e società nell'insediamento rurale dell'Italia settentrionale (VIII-X secolo),* Mantova, 101-17.
Gelichi, S., and Librenti, M., 2009. Tracce di attività molitoria in un villaggio del X secolo della Pianura Padana. In P.Galetti and B.Andreolli (éds.), *Mulini, canali e comunità nella pianura bolognese tra Medioevo e Ottocento*, Bologna, 347-58.
Gelichi, S, and Librenti, M., 2010. Edilizia abitativa tra IXe X secolo nell'Italia settentrionale : stato della questione. In P.Galetti (éd.), *Edilizia residenziale tra IX-X secolo. Storia e Archeologia,* Firenze, 15-30.
Grillo, P., 1979. *Inventari altomedievali di terre, coloni e redditi,* Roma.
Grillo, P., 1987. *La pietra ollare dalla Preistoria all'Età Moderna*, Como.
Grillo, P., 1993. Il commercio delle mole nel Piemonte del basso Medioevo (inizi XIV-inizi XV secolo). In R.Comba (éd.), *Mulini da grano nel Piemonte medievale. Secoli XII-XV,* Cuneo, 215-31.
Lohrmann, D., 2002. Mühle, historisch, in *Reallexicon der germanischen Altertumskunde*, 20, 281-87.
Malaguti, C., 2005. La pietra ollare. In G.P.Brogiolo and N.Mancassola (éds.), Scavi al castello di Piadena (CR). In S.Gelichi (éd.), *Campagne medievali. Strutture materiali, economia e società nell'insediamento rurale dell'Italia settentrionale (VIII-X secolo),* Mantova, 173-87.
Malaguti, C., and Zane, A., 1999. La pietra ollare nell'Italia nord-orientale. *Archeologia Medievale*, XXVI, 463-79.
Mancassola, N., 2008. *L'azienda curtense tra Langobardia e Romania. Rapporti di lavoro e patti colonici dall'età carolingia al Mille,* Bologna.
Mannoni, T. and Messiga, B., 1980. La produzione e la diffusione dei recipienti in pietra ollare nell'alto Medioevo. In *Atti del VI Congresso Internazionale di Studi sull'Alto Medioevo,* « Longobardi e Lombardia: aspetti di civiltà longobarda », Spoleto, II, 501-22.
Mannoni, T., and Giannichedda, E., 2003. *Archeologia della produzione*, Torino.
Mannoni, T., Pfeifer, H.R. and Serneels, V., 1987. Giacimenti e cave di pietra ollare nelle Alpi. In *La pietra ollare dalla preistoria all'età moderna*, Como, 7-45.
Montevecchi, G. and Novara, P., (eds.) 2000. In Agro Decimano. *Per un catalogo del patrimonio storico archeologico del territorio a sud di Ravenna,* Ravenna.
Mousnier, M., (ed.) 2002. *Moulins et meuniers dans les campagnes européennes. IXe-XVIIIe siècles*, Toulouse.
Nejrotti, L., 2007. Il mulino di Becua : i presupposti di una ricerca. In G.Chiarle (éd.), *Boschi e castelli. Itinerari medievali nelle terre dei Visconti di Baratonia*, Torino, 62-72.
Nejrotti, L., 2008. I mulini sulla Ceronda. Ricognizioni archeologiche. In G.Chiarle (éd.), *Boschi e controllo del territorio nel medioevo*, Torino, 77-92.
Nico Ottaviani, M.G., 2008. *Statuti, territorio e acque nel*

medioevo. Perugia e Marsciano, Tevere e Nestore, Spoleto.

Pantò, G., and Uggé, S., 2007. Vasellame dall'insediamento di età gota e longobarda. In E.Micheletto (éd.), *Longobardi in Monferrato. Archeologia della 'Iudiciaria Torrensis',* Chivasso, 137-57.

Pasquali, G., 2008. *Sistemi di produzione agraria e aziende curtensi nell'Italia altomedievale*, Bologna.

Peduto, P., 1979. Archeologia medievale in Campania. In *Cultura materiale. Arti e territorio in Campania*, n.15, ins. 'La voce della Campania', VII.

Peduto, P., 1984. *Villaggi fluviali nella Pianura Pestana del Secolo VII. La chiesa e la necropoli di S.Lorenzo di Altavilla Silentina,* Salerno.

Racine, P., 2003. Du moulin antique au moulin médiéval. In P.Galetti and P.Racine (éds.), *I mulini nell'Europa medievale*, Bologna, 1-15.

Racine, P., 2006. *Le paysage des moulins en Europe occidentale au Moyen Âge*, 'Nuova Rivista Storica', XV/II, 409-46.

Rivals, C., 1987. *Il mulino : l'avventura del pane quotidiano,* Firenze.

Roda, R. and Setti, G., (éds.) 2004. *Le ruote del pane. Mulini natanti e cultura molitoria nella Pianura padana. Storia, cultura materiale e immaginario*, Mantova.

Sergi, G., (éd.) 1993. *Curtis e signoria rurale : interferenze fra due strutture medievali*, Torino.

Squatriti, P., 2000. *Working with water in medieval Europe. Technology and Resource Use*, Leiden-Boston-Köln.

Tirabassi, J., 1980. Notizie geologiche e archeologiche. In *Storia popolare di Rio Saliceto,* Reggio Emilia.

Toubert, P., 1995. *Dalla terra ai castelli. Paesaggio, agricoltura e poteri nell'Italia medievale,* Torino.

Valenti, M., (éd,) 1996. *Poggio Imperiale a Poggibonsi : dal villaggio di capanne al castello di pietra. I. Diagnostica archeologica e campagne di scavo 1991-1994*, Firenze.

Valenti, M., 2004. *L'insediamento altomedievale nelle campagne toscane. Paesaggi, popolamento e villaggi tra VI e X secolo*, Firenze.

Valenti, M., 2005. La formazione dell'insediamento altomedievale in Toscana. Dallo spessore dei numeri alla costruzione di modelli. In G.P.Brogiolo, A.Chavarria Arnau and M.Valenti (éds.), *Dopo la fine delle ville : le campagne dal VI al IX secolo*, Mantova, 193-219.

Valenti, M., (éd.) 2008. *Miranduolo in Alta val di Merse (Chiusdino-SI)*, Firenze.

Watts, M., 2002. *The Archaeology of Mills and Milling*, Charleston.

Zanarini, M., 2007. *Società ed economia: estimi e altre fonti tra XIII e XIV secolo*, In P. Galetti (ed.) *Una terra di confine. Storia e Archeologi di Galliere nel Medievo*, Bologna, 63-84.

Ore grinding in the Middle Ages: the example of Brandes-en-Oisans (Isère, France)

Nicolas Minvielle Larousse and Marie-Christine Bailly-Maître

Introduction

Across the millennia, mills have played the role of 'power transformers' for multiple purposes: milling and grinding food, of course (grain, olives, etc.), and for activities unrelated to food such as tanning and metallurgy as well.

The silver coins that fed most of the economy of the west in the Middle Ages were produced by means of a long process in which mills were used to grind ore extracted from the mines. The activity was carried out on large industrial sites like the Argenteria de brandis. Brandes-en-Oisans (fig. 1) was a mining community, a part of the Dauphiné Principality, fully devoted to the extraction and treatment of several argentiferous lead veins between the 12th and 14th centuries (Bailly-Maître and Dupraz 1994.). The site, which comprised a village with some 80 dwellings, fortifications, a parish church and large industrial areas, has been excavated by the CNRS since 1977. So far, it has revealed more than 206 millstones, plus fragments and chips, the remains of the ore mills that operated on the plateau at that time as links in a large chain of mineralurgical operations. The ore leaving the galleries was trapped in gangue consisting mainly of barite and quartz. A process was required to separate the precious metal from this waste rock and several methods were implemented, such as crushing, decrepitation and grinding. While crushing was done manually and decrepitation was done by fire, grinding was mechanised by water-mills. However, instead of pouring grain onto the millstones, the miners cast ore and gangue onto them and recovered a powder that was then treated in washing and decantation basins to collect even the smallest gram of silver.

Very few documents describe this activity and this is no doubt the reason for which the mills were only acknowledged as a result of the excavation of several mining sites. Archaeologists have addressed the subject, first separately (Benoit 1997; Domergue *et al.* 1997, 48-61; Bailly-Maître 2002, 224) and then collectively to summarise acquired knowledge and set the foundations for future research in an article published in 1997 for the Aix-en-Provence symposium (Benoit *et al.* 1997, 62-69).

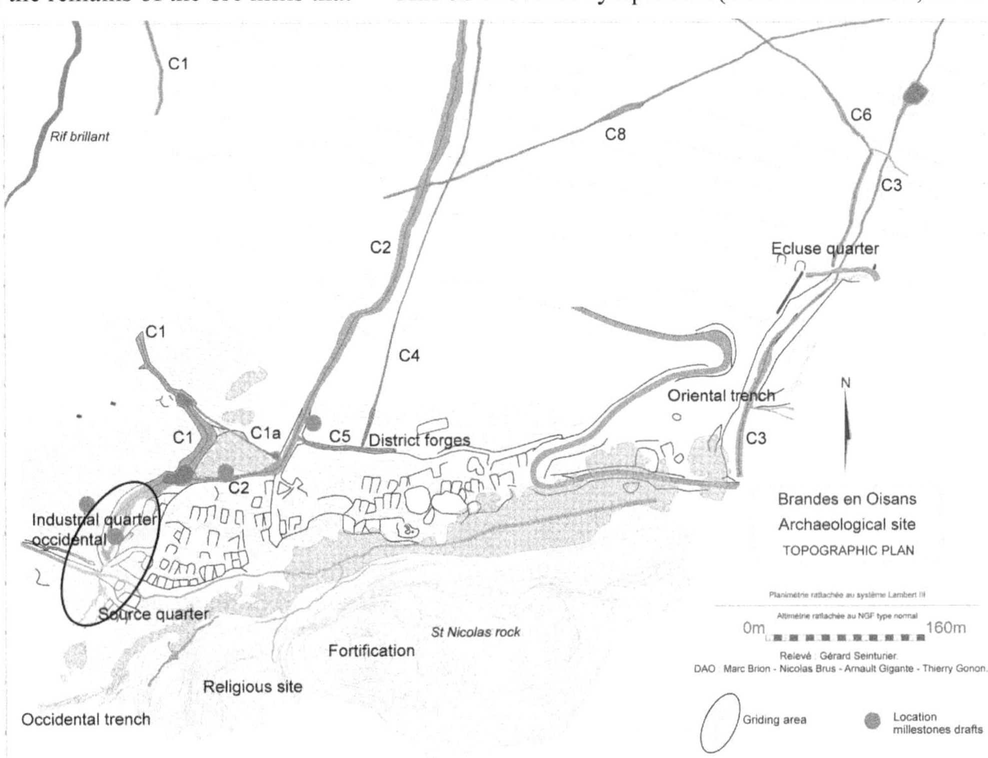

Figure 1. Brandes-en Oisans.

Figure 2. Millstone Grit B106.

Figure 3. Rough-out 2009-1-009.

This study starts with three main themes. The first one will focus on the production of millstones, their extraction and shaping, as shown by fragments that still show traces of cutting, and also by the rough-outs found on the plateau. The second theme will then evoke the 'memory' of grinding, in particular by analysing traces of wear on the grinding surfaces of millstones. The third theme consists of establishing the context in which the millstones were used in the mills, proposing interpretation schemes based on elements noted by observing the millstones and also through the contribution of other types of sources.

From quarry to millstone

1.1 Millstone grit in Brandes-en-Oisans

The Brandes plateau, composed of various outcrops and innumerable blocks, was worked to supply the millstones required for the mills on the site. The miners used the millstone grit over and over since it provided them with stones they could extract on location, without the difficulty and cost of transporting stones weighing several hundreds of kilos. Rather than working rocky massifs, however, they preferred to extract their millstones from blocks dispersed across the plateau. These blocks were the remains of glaciers or had crashed down from the slopes of the surrounding summits and many of them were indeed the right size for shaping millstones (Figs. 2 and 3).

The millstone grit quarries of Brandes are highly representative of 'village millstone grits' for local production, situated as close as possible to the mills (Belmont 2006, 77). Ore mills are no different from grain mills in this respect. Our vision is not optimal though, since few blocks of this sort have been preserved. They were probably fully whittled down by extraction and the few minute scraps created by cutting are most likely to have been dispersed or reused.

Whatever the case may be, whether the millstone grits were obtained locally or further away, this practice is not an archaism, as it might be presumed in view of the period of the operations and use (13th century AD), and given that the industry was destined to expand and become specialised. As regards the grain mills of the neighbouring community of Huez, for example, millstones continued to be extracted in this manner over the centuries that followed, up to the 19th century (AD 38, 7 S 1/1, 1809 survey on mills in France. Cited by Alain Belmont, Atlas des meulières de France et d'Europe, 2006, http://meuliere.ishlyon.cnrs.fr/index.htm). There are several reasons for this. First of all, it would appear that the financial investment was much too high for the intended use. But, considering the fact that the Huez inhabitants remained faithful to this type of quarry for wheat mills for at least 600 years, it is important to turn our attention to the quality of the stone.

1.2. Rocks

The quality of the millstone rocks cannot be emphasized too strongly. It has now been established that this point was of key importance (Belmont 2006, 89-113.). Certainly, the millstones used to grind ore did not necessarily meet the same requirements as grain millstones, but the elements that break away from them, as small as they may be, nevertheless add pollution to an ore that is supposed to be in the enrichment phase. Millstone wear is another factor to be considered. The faster the millstone wears out, the faster it needs to be changed.

The millstones from Brandes are shaped from two main types of clast (Fig. 4): 56.4% conglomerate and 33% sandstone. Added to this are a few gneiss and granite inclusions (4.3%), but they remain marginal. These are indeed the rocks that form the blocks scattered across the plateau, apart from the gneiss and granite which come from the rocky massifs (This classification was carried out by Stéphane Roumêjon (Université Joseph Fourrier, Grenoble I) to whom we are deeply grateful).

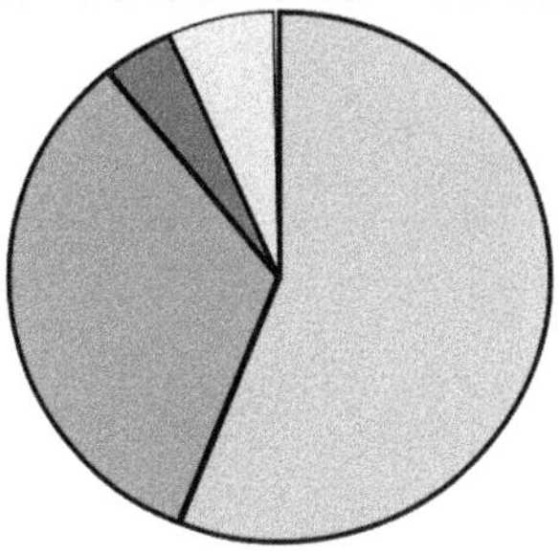

Figure 4. Distribution of types of rocks.

Is there a constant to be detected in the rocks chosen by the miners? Nothing is less certain. Resistance tests can provide part of the answer, provided that a comparative approach is used. In 2005, Yves Orengo, of the Laboratoire Interdisciplinaire de Recherche Impliquant la Géologie et la Mécanique (LIRIGM) of the Université Joseph Fourier (Grenoble I) conducted tests on three samples from the millstone grit B 106 excavation that took place during the 2004 campaign. The samples are a siliceous micro-conglomerate. The results are summarised in the table below (Fig. 5):

Sample	Density (t/m3)	Compression strength (MPa)	Static modulus (GPa)	Tensile strength (MPa)	Deere-Miller
BRANDES 1	2.693	136	44.8	13.3	BM
BRANDES 2	2.693	156	39.5	10.5	BM
BRANDES 3	2.692	111	40	9.8	CM-BM

Figure 5. Results of analyses carried out in 2005 by Yves Orengo (UJF, Grenoble I).

For each of the samples analysed, the compression strength (col. 3) is greater than 110 mega Pascal. On a scale used for the tests carried out on a number of European millstone grits (Fabre *et al.* 2006, 91-97.), the Brandes samples are situated beside the silex of La Ferté-sous-Jouarre (80 MPa) or Corfélix (240 MPa) in the Brie region (France). Hence, this is a high quality rock. Moreover, this rock was used equally for both ore and grain. This dual usage is visible on three grain millstones made from the same conglomerate and used for the Brandes miners' food. Caution is advisable however as regards the characteristics of these millstones. We do not know exactly which qualities were required for the rocks intended to grind ore. Other mining works, for their part, such as those in Pampailly (Rhône - France) (Benoit 1997, 70.), or Castel Minier (Ariège - France) have revealed granite millstones (Benoit *et al.* 1997, 62-69.).

The diversity of the cases encountered shows that the miners tended to make do with the rocks at their disposal. Whether the sandstone and conglomerate of Brandes, or the granite of Pampailly, what these rocks have in common is that the majority are present *in situ*. It was not as much about making choices as it was about adapting to the situations encountered.

1.3. Extraction techniques

The methods used to extract millstones are fully representative of the various quarries studied in Europe (Anderson 2003, 49-50; Jaccotey and Milleville 2008, 16-23; Belmont 2008). We can differentiate however between the blocks used to extract several millstones and the blocks intended to produce a single item.

First, there was the cutting out of a ring-shaped channel to outline the future millstone. This operation was carried out using a unique tool with a circular tip slightly more than 1 cm in diameter, in three cuts (Fig. 6.): the first one as close as possible to the rough-out, the second 10 cm further

First cup

Third cup

Second cup

Figure 6. Details of the ring-shaped channel of Millstone Grit B 106.

Figure 7. Overview of the rough-out 2009-1-9.

Figure 8. Details of the rough-out 2009-1-9.

along and the last one in the middle, after the second cut was made. This method, which in all likelihood involved the oblique 'thrown percussion' or 'pecking' technique, is common in the millstone grits studied. The squaring off of the edges of the blocks was done using rougher tools, which fractured the rock. The millstone makers worked step by step in order to 'test' the rock. They tackled it in different ways and in different locations to make sure it was worthwhile to continue working on it. If the rock yielded at any time, they did not pursue their art any further, so as not to waste time fully shaping a part of the millstone whereas the other part was bound to fracture.

Regarding the blocks cut for the production of a single millstone, the cutting method differs slightly. The various steps involved in the cutting of these blocks proposed by Gary Adams and Gérard Serres are fully applicable to the rough-outs of Brandes (6 steps of cutting. Cited by Hockensmith 2006, 193-194; Serres unpublished, 1995. Diagram produced from the medieval ore millstones in Peyrusse-le-Roc (Aveyron - France)). First, the gross block is roughly hewn to a shape that is rounded at best or at least plane-parallel. This step can consist initially of rough work, leaving corners, after which the operation is finalised on the edges with a sledge hammer or a pick. There is little trace of this phase still visible on the rough-

Figure 9. Rough-out of an eye on the 2009-1-10.

outs present, since the millstone makers continued their work. However, on the 2009-1-09, a clean cut can be observed on the bottom of the block, attesting to the first actions of the tradesmen of Brandes (Fig. 7). At the second step, the reduced block was cut, much more rigorously this time in order to give it its rounded shape. Again, on the 2009-1-09, traces of a pick or chipping chisel 1 cm wide resulting from this operation are visible on the major part of the edge of the block (Fig. 8). Then, the upper surface of the block was flattened. Apart from preparing a possible grinding surface, this task was also used to test the rock. If the rock proved to be inappropriate for cutting, it might break and by testing, the millstone makers could avoid wasting their time.

The methods used to extract millstones for ore grinding are therefore no different from those intended for the milling of grain. The millstones could only be adapted for specialised use after extraction.

1.4 Shaping

The coring out of the eye is a step still common to both usages. In the centre of the potential millstone, the millstone maker would first cut a circular vein that was progressively enlarged with the aim of being sunk in as work progressed. The task was carried out using a tool with a round tip 7 mm in diameter as shown by the marks left at the bottom of the incision made on rough-out 2009-1-10 (Fig. 9). Once the incision was made, the millstone cutter removed the first section in order to complete the circle, still using the same tool. The eye was only completed after the future millstone was the right thickness.

The next step consisted of making the notch in the millrynd. The widths vary from 4.5 cm to 7 cm for the 2009-1-38 (Fig. 10). As for lengths, they are seldom preserved in their entirety. For the 2009-1-38, for example, one is 29.5 cm long and another is slightly more than 28 cm. The thickness varies from 2 to 3 cm, no notches being fully regular.
Finally, the furrowing of the millstones is clearly similar

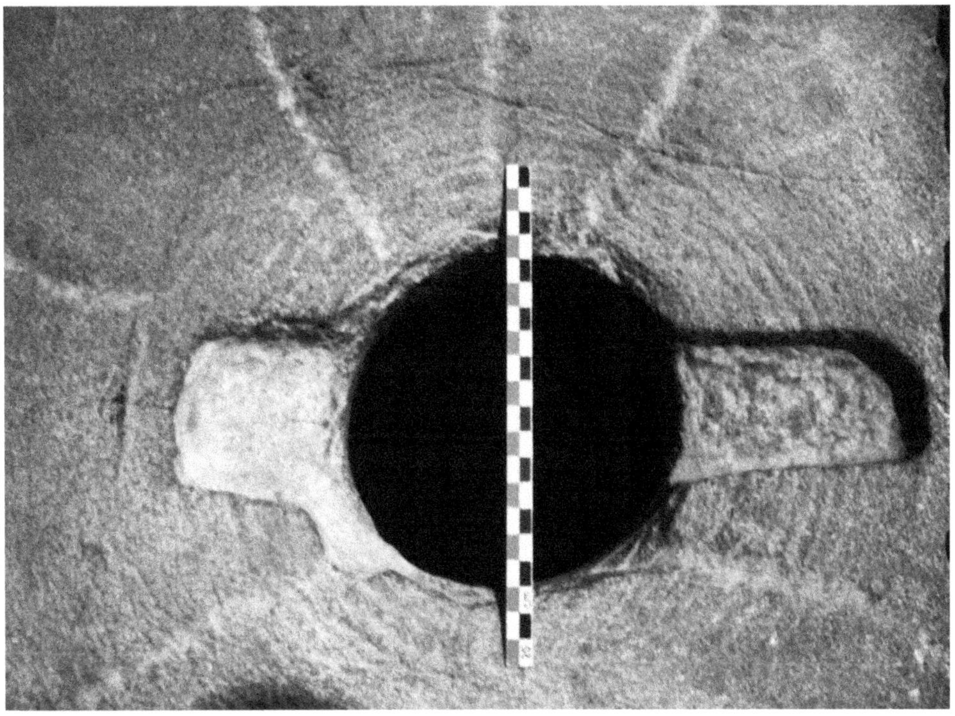

Figure 10. Eye and millrynd of the 2009-1-38.

Figure 11. View of millstone 2009-1-52. It contained 16 furrows across the surface.

here for the grinding of ore. Although during Antiquity, the grinding surfaces of grain millstones could be furrowed or at least covered by multiple grooves (Belmont 2006, 41), this process appears to have been abandoned afterwards, only reappearing at the end of the 18th century (Amouric 1984, 272; Perry 198, 155). In the meantime, the millers preferred to peck at their stones randomly. The furrowing technique, however, did not disappear forever during the Middle Ages. It was adapted for the grinding of ore not only in Brandes, but throughout the mining works equipped with such artifices. The furrowing of wheat millstones is a more complicated task than the cutting out of furrows for ore, but the skilled Medieval tradesmen were familiar with and mastered both types of know-how as they used them both at the same time.

A high degree of consistency is noted in the method used to cut out the furrows. The quarry workers appear to have preferred a particular type of tool fitted with a circular tip ranging from about 5 mm to 1 cm in diameter. In a few more isolated cases, they may also have used rougher tools such as chipping chisels, leaving marks measuring more than a centimetre. They began by cutting out a central recess which they would then enlarge by one or two shallower cuts using a pick (Fig. 11).

Several types of furrowing are observed on the ore millstones. Factors such as the number of furrows, their distribution on the millstone and their dimensions are taken into account. This typology is in the process of being drawn up using a diachronic approach (Lepareux-Couturier and Jodry 2010; Minvielle Larousse and Bailly-Maître 2010).

In making a final assessment, we can state that ore millstone quarries do not differ from those that produced wheat millstones. The methods and steps of extraction were similar as well. From squaring off to cutting out the notch for the millrynd, all of the Brandes tradesmen's actions are representative of a more general context. Only the size of the furrows and, perhaps, the thickness truly differentiate ore millstones, the variations in diameters being too fine to opt for one of the two uses.

After having compared the methods of shaping ore millstones with those used for wheat millstones, we are also in a position to link up millstone making techniques with quarrying techniques. This parallel has already been made, centred more on the way of working underground extractions (Bailly-Maître 2006, 163-170.). We propose to extend these considerations to encompass the entire chain of operations. The tool marks left by the tradesmen are clearly reminiscent of the tools used in mining work: tapers, sledge hammers, chipping chisels, picks, etc. The miners used their skills and their tools in the open air to shape the millstones which we are now going to present in detail.

2. Millstones, 'memory' of grinding

As regards traces left on the grinding surfaces, ore millstones have a sizeable advantage over wheat millstones, for two main reasons. The first concerns the material precipitated into the eye of the upper millstone. Grain is no rival to rock and always ends up being cleaned out by the asperities, without leaving any evident traces. Ore, on the other hand, mixed with gangue, is a much tougher adversary for millstones. The results are, admittedly, practically the same, with a fine powder at the outlet, but in the meantime the millstones were lacerated by the passage of ore which left thousands of indelible marks, traces that will enable us to better understand the rotation and operation of the grinding mills. The second reason is related to millstone maintenance. By regularly dressing the grain millstones, the millers also eliminated the minute traces of the previous grinding. Miners, though, did not do the same for their ore stones, thereby leaving a multitude of testimony. To propose an interpretation of the traces of grinding, we will first analyse the surfaces of fragments before more closely examining the ridges.

2.1 Intensive use

It is a useful statement of the obvious to say that the rotation of millstones progressively wears away the grinding surface. What is remarkable about this phenomenon is its intensity. The millstones that have been preserved are worn down to the chord. By extension, it is likely that ore millstones received no intermediate treatment as grain mills did.

Not content to mercilessly whittle down one side of the millstone, when the millstone became unsuitable for grinding, rather than re-employ it differently, the miners sometimes preferred to turn it over in order to extend its service life. Out of a corpus of 84 millstone fragments (excluding small fragments and rough-outs), 14 items, i.e. 16.6% were used successively on both sides. Among these items, despite a high proportion of undetermined ones (a flat surface or fragments still too small), three points stand out. First of all, no fragments are convex on both sides at the same time. The use of the millstone as a bedstone on both sides is therefore to be eliminated. However, two specimens are concave on both sides: runner stones that were always turned over to serve the same purpose. The rest of the time though (6 items), they are concave on one side and convex on the other. In this case, dual usage is indeed to be considered, but in an order that is still to be defined.

Figure 12. 2009-1-36.

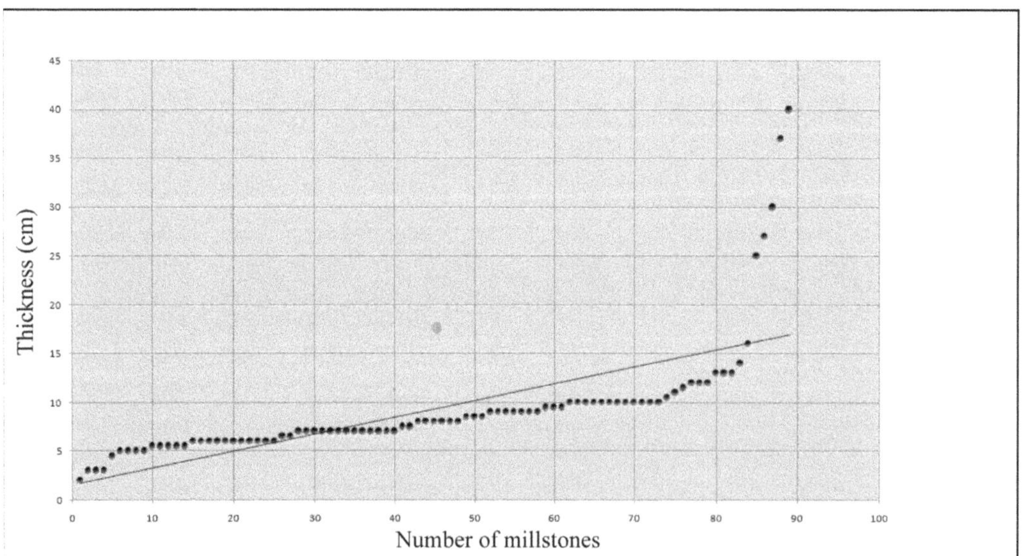

Figure 13. Millstone thickness distribution chart.

Figure 14. Negative view of the 2009-1-39.

Regarding grain mills, normally runner stones arriving at the end of their service lives are re-employed as bedstones, which have less stringent quality demands (Belmont 2006, 39.). But the grinding of ore does not have to meet the same constraints as the grinding of grain, so the question is still open. The study of the millstones concerned here has not answered it, due to similar wear on both sides. It is still likely that the miners worked not according to habit but according to the conditions encountered in particular cases.

The consequence of such intensive usage is the inexorable reduction of millstone thickness. From an initial 30 cm, if we can trust the rough-outs or the almost-new half-millstone still in context on the right bank of the source (Fig. 12), much more modest figures are found for most of the fragments. The chart below (Fig. 13) summarises the data and highlights a distribution of maximum fragment thicknesses of between 5 and 15 cm. The initial thickness of the millstone, which may appear very large at first glance, is actually in line with an average. In the 14th century, the bread baking stones purchased by the Lords of the Dauphiné were massive blocks that could be up to 2 feet thick (66 cm) for bedstones and 1.5 feet (49.5 cm) for runner stones (AD 38, 8B 83, Accounts of the Lord's domain of Claix 1382-1383.).

It would appear to have been established that the requirements for ore millstones were not the same as those for grain millstones. What remains to be found is if this type of millstone management was intentional, without any compromising of ore enrichment, or, to the contrary, if it shows that the miners were growing away from the process.

2.2 Ridges

The other visible manifestation and testimony of grinding on the millstones are ridges, discreet or obvious traces of the passage of ore and gangue (Fig. 14).

The immense majority of millstones or millstone fragments that have been preserved display these ridges. Out of a group of 84 items, 72 have ridges, i.e. 85.7%. This figure is even liable to be increased, since it only accounts for ridges that are visible to the naked eye or those appearing on tracing paper (Regarding the methodology of the acquisition of ridges: Minvielle Larousse and Bailly-Maître 2010). Generally speaking, the marks become finer from the centre of the millstone towards the edge as shown in the example of the 2009-1-39 (Fig. 14.). This is directly linked to the distance between and relative placement of the rotating millstone in relation to the bedstone. First of all, the shape of the stones, convex for the bedstone and concave for the runner stone, prepare for this flow. The runner millstones are placed in such a way that the material poured into the eye is accompanied as it is reduced. The distance between the millstones is greater at the centre of the stones, since the ore module is at its largest, and it is then smaller towards the outside so as to reduce the ore. Marks of grinding surfaces result from this arrangement. Near the eye, the ridges are wide and rough, or even absent.. They become finer towards the outside as the millstones move closer together, through the breast and finally the edge to make the ore become similar to flour. Moreover, this is exactly the same procedure as the one used for grain, minus the ridges cut by a material much harder than rye or oats.

All of these steps are found almost systematically on all of the whole millstones, half millstones and millstone segments. The fragments are classified according to their position on the millstone, by centre, breast and edge (Fig. 15). The first observation that can be made upon reading the graph is that a large majority of the fragments have been identified without difficulty. A mere 13.10% of them have remained outside classification for various reasons, related in particular to too few samples to be able to take a decision. Nevertheless, in most cases, there is hesitation between the edge and the breast. And these same edges and breast account for the lion's share. With 32.14% of edges and 17.86% of breast, the 2.38% of centres is very much a minority. This is all the more valid if double zones are added (15.48%). As a result, even if segments and whole millstones are added up, centres and eyes are underrepresented. It makes sense for the millstone to become more fragmented at the outer edge, which is more fragile than the massive central part. This does not mean however that centres would not be found as well, fewer of them certainly, but here there is a virtual absence. Nor can this difference be totally ascribed to salvaging and successive reuse. This point deserves further attention, which it will be given further on.

Once the classification is carried out, close and attentive examination of the ridges shows that under cover of surface consistency, the ridges are actually highly unequal and recut in many locations. The examples are infinite and an example is illustrated from fragment 2007-0-1515 (Fig.

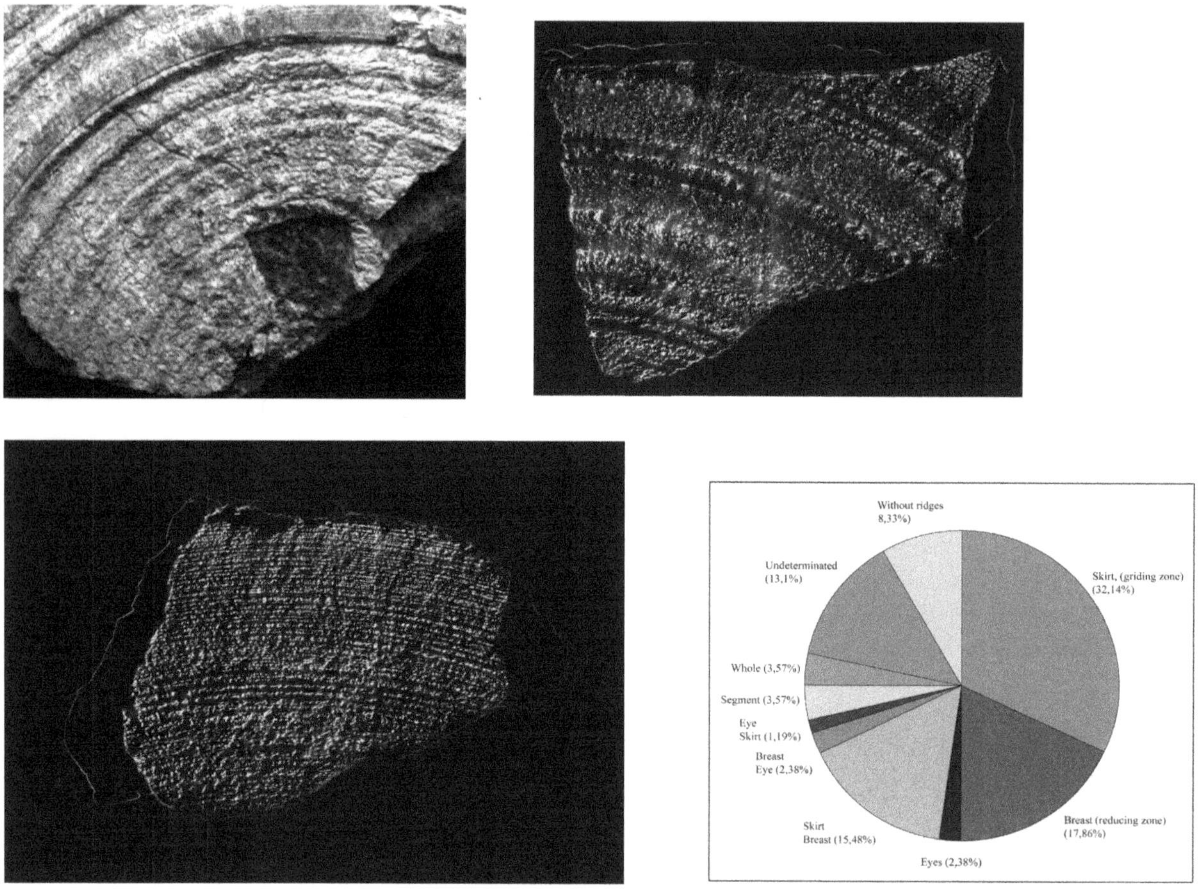

Figure 15 a, b, c and d. Views of a core (top left, 2009-1-39), a breast (top right, 2007-0-1515) and an edge (bottom left, 2007-0-1488) and distribution of fragments according to position on millstone (bottom right).

16). An initial 3 mm wide ridge is very clearly perceived and it splits a first time. The lower ridge then disappears a bit further on while the upper ridge, which had started as a fine bevel, progressively widens and ends up splitting and finally evaporating. The ridges are therefore not rigorously concentric. They evolve as the millstone rotates over time. Here, we have seen changes of direction, creation and destruction, but many fragments display ridges that have been whittled down by repeated use.

The grinding surfaces have therefore been kept 'in memory' with a quantity of information on grinding techniques, as well as on the workmen's method and actions. The ridges underwent a complex evolution, linked directly to the operation of the mills. They were furrowed consistently by the first rotations, but then continued grinding ended up by whittling them down and segmenting them. Since the axis of rotation was more or less flexible, the matching of ridges between the two millstones was not ensured for long and the ore finished cutting out the summits it had initially formed, sometimes in a direction that had more or less shifted. All of these observations have only been made macroscopically. It therefore clearly appears that a study of this type would call for a much more detailed approach than ours. Microscopic analyses would no doubt provide a number of additional elements in the quest for detail that is inherent in tribology. As an example, micro centre samples

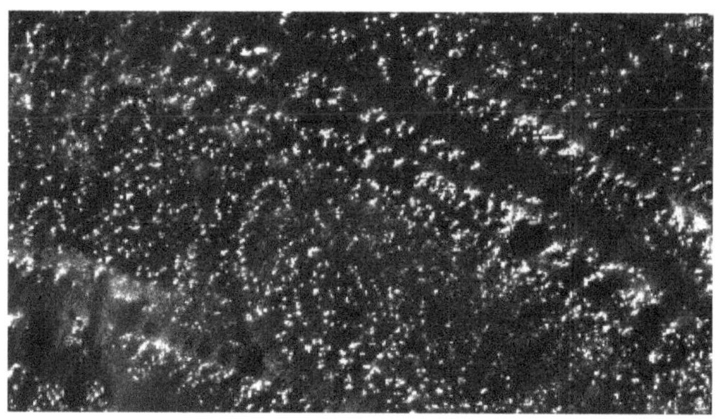

Figure 16. Details of negative of 2007-0-1515.

taken from antique ore millstones from the Les Forges domain (hand mill) have revealed grain still embedded at the end of a ridge that has been characterised chemically to define the type of material ground (Domergue *et al.* 1997, 48-61). Having said that, based on observations made, we are in a position to propose a number of interpretation schemes regarding ore grinding in Brandes.

3. Millstones in their mills

'L'impossible description technique' is the title chosen by Aline Durand to propose a typology of Carolingien mills in the Languedoc region (Durand 2002, 33.). Despite the

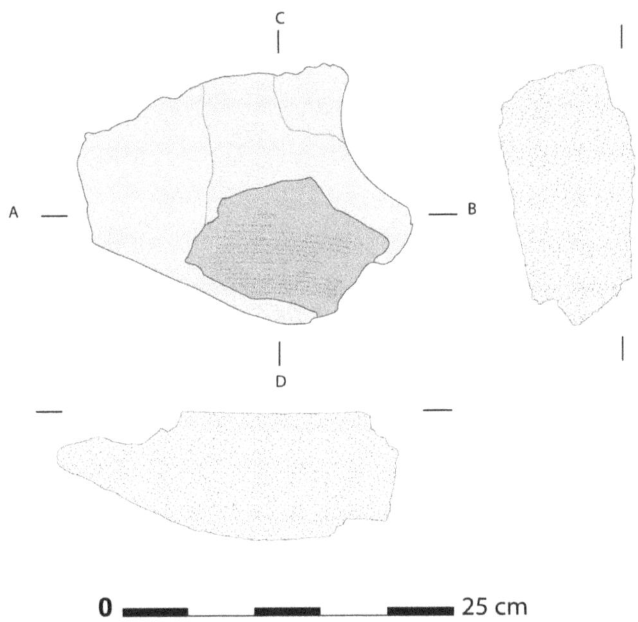

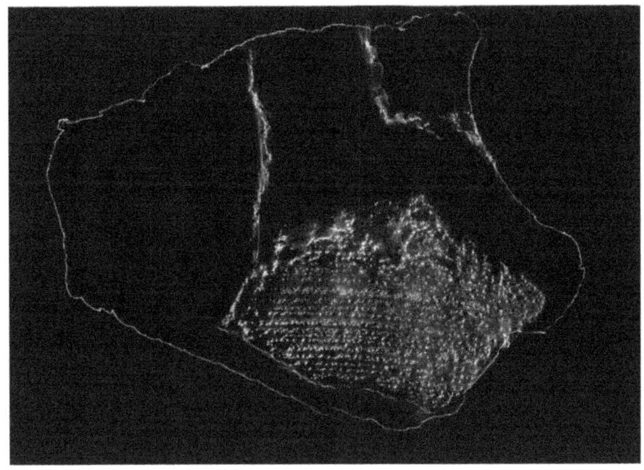

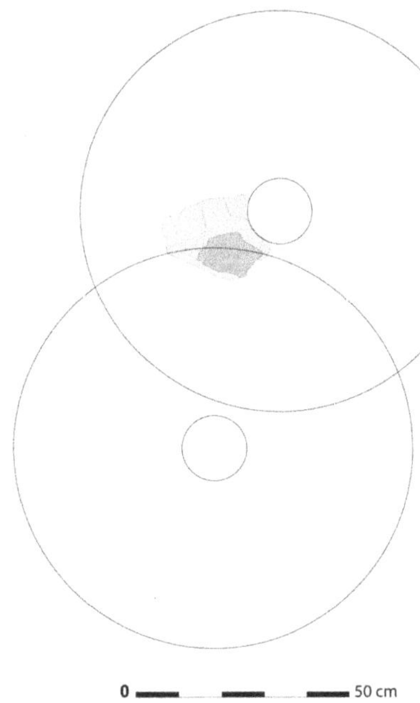

Figures 17 a (top left), b (top right) and c (right). Views of fragment 2007-0-1504 and projection of the presumed axes of rotation.

poverty of sources during the early Middle Ages, this observation has subsequently been widely contradicted in the development that has since taken place. Yet, understanding a discreet building and a complex technology is not easy, even if such artifices literally abound in western towns and rural areas (Regarding the distribution of mills in rural areas, refer to: Pichot 2002, 260.). Although, in the Oisans, nearly each parish, including Brandes, had at least one wheat mill available, there was evidently not the same demand for ore mills. *De facto*, the sources reporting on their activity are truncated and leave this type of artifice in the dark. It will be necessary to use every detail of every source, whether textual, iconographic or archaeological, in order to endeavour to shed light on certain aspects of the issue. To do so, we will first analyse the only tangible remnants of the mills, namely their millstones. Secondly, we will propose an interpretation based on iconographic sources combined with the historical context. And thirdly, we will situate ore grinding within its process, mineralurgy, in order to understand the role of the ore mill in Brandes.

3.1 Segmented millstones?

Side by side with monolithic millstones, it would appear that the miners also used segmented millstones.

The subject of segmented millstones, addressed by Henri Amouric in 1984, is a thorny issue that is continually being revisited. At the time he wrote his thesis, no examples prior to the 19th century were known for segmented millstones, regardless of the type (Amouric 1984, 266-267). Since then, the first references have been pushed back further and further as historians look into the question. Still in 1984, Dorothée Kleinmann reported on the appearance of this new type of millstone between 1760 and 1775 (Kleinmann 1984, 1107-1116). The perspective of the late 18th century would appear to receive the assent of researchers. For Jean Bruggeman, who studied the millstones of Flanders for the symposium in La Ferté-sous-Jouarre in 2003, there is no possible doubt: 'Medieval millstones are always monolithic. Black basalt stones were still used in the centuries that followed. White stones were used too, up to the end of the 18th century. But bedstones could be made of several irregular-shaped pieces. They were bonded with plaster, held together by an iron or wooden band and sometimes rested on a bed of cemented bricks' (Bruggeman 2003, 232). However, Benoît Deffontaine, at the same symposium, dated these stones back yet another 150 years or so thanks to fortunate textual references to 'milling stones' in 1647, 1652 and 1680. According to him, 'nothing prevents us from believing [that production] could even date back to before the first half of the 17th century' (Deffontaine 2003, 297-298). Alain Belmont, again at the La Ferté symposium, reported on an agreement dated 3 May 1452 whereby 'a merchant from Rouen named Robert Le Cornu agreed to convey one or more ships loaded with 35 millstones, 5 eyes, 100 tiles and a tombstone to Normany' (Belmont 2003, 282). The above-mentioned tiles obviously came from the famous quarry of La Ferté. That is the extent of research at the present time. Remember, however, that this brief review of the question only concerns wheat millstones.

In Brandes, one fragment of the corpus stands out (Fig. 17). At first glance, it is a core fragment, since a piece of the eye and millrynd notch have been saved. Below the eye, a ridged surface is present, displaying very fine ridges, of the edge type. This is atypical, but may remain possible under multiple conditions. However, the outstanding element is the direction of the ridges in relation to the eye. They are completely reversed and it is not possible for the axis of rotation to have simply been modified, since the difference is far too great. This would indeed appear to be a 'piece' of a millstone and not just a fragment.

In the light of this assumption, we must reconsider certain cases such as these half-millstones, photographed in the context in which they were abandoned at the beginning of the 20th century (Fig. 18). The break is very clear for the three visible items. It is therefore also possible that these were millstones put together from two segments.

The miners would appear to have used all of the methods at their disposal. If, and only if this assumption proves to be true, some ore millstones would have to be seen as an assembly of extremely diverse pieces of stone, difficult to connect to any typology whatsoever as such. When a piece was worn, it was replaced straight away by another one cut for the occasion, or sometimes even by a used piece that was recut. This ranges from modest-sized pieces to half-millstones. There would appear to be no general *de facto* trend in the matter. This possibility would explain the module of most of the fragments preserved and thereby the small numer of known monolithic millstones for operations that lasted nearly 150 years. It would also explain the distribution of the same fragments: many edges and breasts for very few centres. On the other hand, this amounts to saying that it is impossible to estimate a minimum number of items (MNI) for the millstones that were used in Brandes. Several points can be put forward against these elements:

1. Firstly, the existence of a single explicit fragment may appear rather insufficient. It is true that its presence is just an exception, in the end. An exception, but not one that confirms the rule; instead it raises questions.
2. Secondly, it is known that considerable salvaging of millstones was done on the plateau. As soon as the operations were abandoned in the mid-14th century, the neighbouring community of Huez was certain to have hurried over to salvage the millstones in good condition to take them down to the mills of the Sarenne, treat them and 're-dress' them to make them more suitable for grinding wheat. This type of salvaging has gone on up to the present day, with the original stones becoming objects to decorate the gardens of the Alpe d'Huez resort. There were many more monolithic millstones in situ. However, we cannot conclude that there were no segmented millstones simply by the presence of monolithic millstones. Both forms may very well have cohabited.
3. Thirdly, a question is raised by this operating method in relation to the way of holding the pieces together. Customarily, they are bonded together with plaster and made solid with an iron or wooden band, or with masonry mountings. However, so far, we have found nothing of this sort on the millstones. The ends of some of them are even so irregular that it may be wondered how strapping could have been put in place. There are no traces of iron fasteners in the fragments either.
4. Lastly, we are not familiar so far with any rough-outs cut out for this process. But we saw above that this is a more general problem. Up to now, only three millstone grit quarries have been recognized. Nonetheless, this is an element that can contribute considerably in terms of understanding the fragments. Provided, of course, that their working is not confused with that of ashlar.

Figure 18. View of half-millstones in the context in which they were abandoned. Taken by Hyppolite Müller, at the start of the 20th century. Musée Dauphinois collection.

The question therefore remains open and can only be resolved by the contribution of other comparable mining works, lacking new elements from Brandes. Regarding the chronology of the process, this would move back its date of appearance by more than 150 years. Reference is made to circular millstones, but was there not another type of mill that already used segmented millstones some 1400 years earlier? In was in 1987 on the Island of Délos that 'G. Siebert discovered the remains of a dislocated mill on the ground of one of the rooms of the main floor of the Maison des Sceaux' (Brunet 1997, 29-38). It was equipped with millstones in separate parts assembled by clamps for the rotating stone and simply held in a wooden frame for the bedstone. Of course, these are cone-shaped millstones like in the mills found in the bakeries of Pompei, and not circular, but this process had already been conceived of by man, probably back in the middle of the 3th century BC.

Figure 19. Representation of an ore mill according to Agricola.

4. Mills

The sole representation of an ore mill comes to us from Georgius Agricola for the 16th century, in his *De re metallica* (Agricola 1556). In this work, he proposes a functional though highly schematic model, without being truly attached to the proportions of the component parts (Fig. 19).

Several comments can be made about this representation. First of all, there is Agricola's desire to be didactic. All of the parts are shown clearly, sometimes to the detriment of reality. For example, the eye of the millstone is out of proportion to the millstone. It is true that on the proper scale, it would have been nearly invisible for his readers. The same is observed for the diameters of the millstones, which are much too small. This representation must therefore be seen as a model more theoretical than practical. Nevertheless, nothing is missing and no frills have been added. There is no doubt that Agricola knew his subject well. On the other hand, it is unlikely that this type of mill existed in Brandes. The wheel was the stumbling block. It is what determines the mechanism that follows and how water is to be brought within its range. In other words, it is the wheel that makes the mill. For Brandes, it is necessary to look at a much wider technological context. Claude Rivals has shown that the breakdown between horizontal wheels ('*rodets*') and vertical wheels noticeably matches the cultural areas of the '*langue d'oïl*' and the '*langue d'oc*' (Rivals 2000). A majority of horizontal wheels are used in the south whereas vertical wheels blossom in the north. There are always exceptions, but on the whole, the various regional studies carried out before and after these findings have not challenged this fact, the reasons for which are still relatively unknown, but, on the contrary, have confirmed it (Amouric 1984). This distinction was moreover reported several centuries ago with Belidor, who spoke in 1739 of the mills of Provence and the Dauphiné region customarily built with a horizontal wheel (Bélidor 1739, 301). Regarding the Dauphiné and hence the Lord›s domain of Oisan to which the village belonged, the registers of the 'Chambre des comptes et des maîtres des œuvres' [Chamber of accounts and contractors] have led to same conclusion (Minvielle Larousse 2009, 130-132). The operating method for this mill was simpler than as described by Agricola. Henri Amouric even speaks of a 'childish mechanism': a gear point with wheel and trundle, but a single engine shaft linking the horizontal wheel and the rotating millstone by means of the millrynd (Amouric 1984, 130).

This gives a process comprising a flume, a holding basin with its weir, running stone, mill and tail race to finish. The simple fact of including a running stone in water-mills can completely separate the mill from the water supply channels. It may be very short or several metres long like a mill serving in Valence in the late 15th century and including a running stone 4 'toises' long (approximately 7 m) (AD 38, B 3128, not numbered. Bail à prix fait du moulin de Valence du 5 février 1490 : 'Item due canales longitudinis qualibet quatuor tesiarulm et grossitudinis unius pedis cum dymidio'). This point is important since in the field, a hiatus may be formed in this way between the major hydraulic facilities and the place of grinding.

These ore mills are difficult to observe by excavation because of the materials used (wood), the relatively small size of the structures and the salvaging of iron parts after they were abandoned for use in other mills. In this respect, the quest for details in excavations is essential.

5. Mills and ore enrichment

Mills are only one item among others in the mining context. Based on this, their place in the system and the role they play are called into question. In fact, the industrial installations of Brandes must only have included a small number of mills. When this is put into perspective with the moderate number of fragments, the possibility of counting segmented millstones and remnants in the field, all of this combined with the period of activity of the mine, we are able to propose this conclusion. Moreover, the situation appears to be the same in Pampailly where documents mention during the middle of the 15th century, barely 'three mills to grind the ore and gangue in Cosne, Brussieu and Vernay' (Benoit 1997, 70).

To explain this, it is necessary to recall the purpose of mineralurgy: the whole process is to separate the ore from gangue. The smaller the module of the mix, the longer and more difficult it is for the miners. The 'flour' produced by the grinding of the ore and gangue is therefore often complicated to treat, even in the washing basins (Marconnet

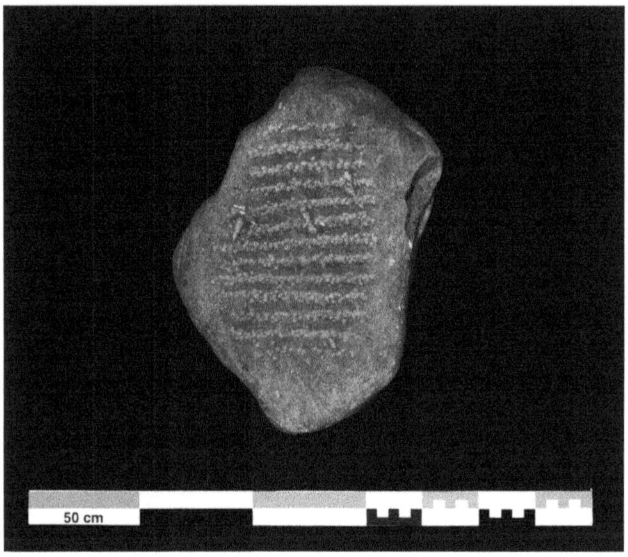

Figure 20 a (above) and b (right). Views of 2009-1-56.

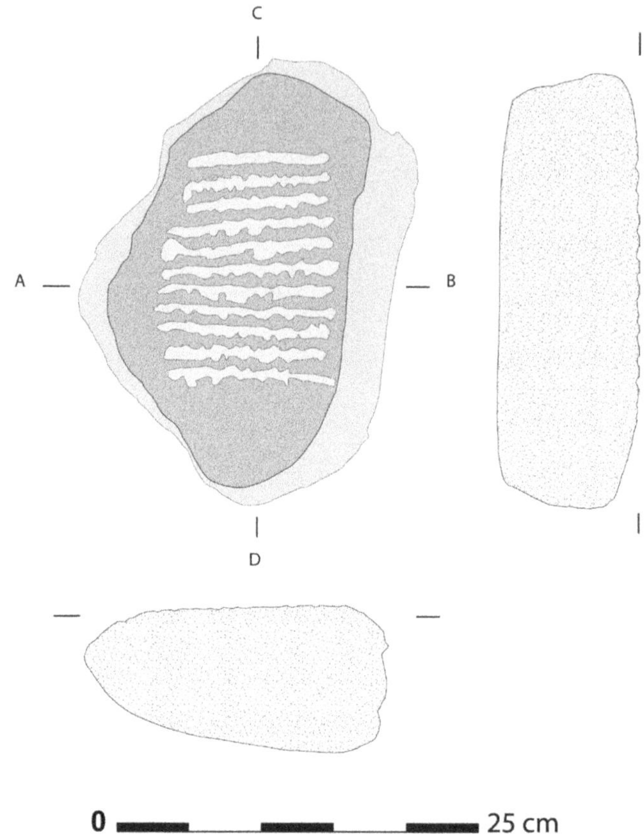

2006). In Brandes, since the gangue is essentially barite, the use of mortar and strikers is sufficient to reduce it to powder, as witnessed by the numerous experiments carried out. The very high number of experiments points in favour of the massive use of these tools.

In these conditions, ore mills played only a complementary role in the enrichment of ore. The miners used them in certain cases, which still need to be precisely defined. Was it for a particular module or for a specific composition? Additional analyses will be useful to determine this. Whatever the case may be, we have observed a wide variety of techniques used for the grinding of ore. Mortar and strikers are by far the most used, followed by water-mills and then a third, more singular but also more marginal. The corpus of millstone includes an alternative hand mill, with parallel 'furrows' made with a pick using the same method as described above (Fig. 20).

Conclusion

At the end of the study, it goes without saying that many questions have still not been answered and that further queries have arisen.

We have observed, in particular, that the extraction of ore millstones is in many ways similar to the extraction of wheat millstones, which shows that use of the methods was widespread. Generally speaking, the extraction or cutting techniques used by the millstone makers, miner and quarry workers are very similar to each other, with respect to both tools and actions. This raises the question of the specialisation of rock tradesmen in specific areas. For Brandes, we do not know, for instance, if the extraction and cutting of millstones were done by a millstone making tradesman who was called in for the occasion or by the miners themselves. Regarding the use of millstones, it appears that they were not serviced regularly by the miners, apart from the cleaning out and recutting out of the furrows as the thickness of the millstone decreased. The immense majority of the items show grinding surfaces that are extremely smooth unlike wheat millstones which are dotted, whereas the context in which they were abandoned is the same. This point relating to the properties of the rocks, the maintenance of the millstones and the miners' grinding quality requirements can only be truly clarified by experimental archaeology. By taking into consideration the diversity of the methods linked to Medieval mineralurgy (crushing, decrepitation, grinding) and knowing that ore mills were probably not numerous in Brandes as in Pampailly, mechanical grinding, by means of water-mills should not be viewed as an innovation in relation to the manual treatment undergone by the ore. They are situated in an auxiliary branch of the process and did not replace crushing, since they were not suitable for that work. Innovation would come much later with the stamp mill which transformed rotational motion into alternative motion by means of a camshaft. This device reproduced the same actions performed earlier by the miners equipped with strikers, but at a virtually industrial stage. Stamp mills appeared for the first time in the literature with the 'Graduel de Saint-Dié' produced between 1494 and 1513. But the camshaft is attested to at least in the 10th century and since the advances in archaeology regularly date technologies further back in time, there is nothing against prior use of these mills. Later excavation may clarify this point.

Acknowledgements

We are particularly grateful to: Timothy Anderson, Alain Belmont, Kathleen Lewis, David Peacock, Flavien Perraza, Francis Pierre, Stéphane Rouméjon, Gérard Serres and David Williams

References

Agricola, G., 1556. *De re metallica*. Bâle, Forben and Bischoff, re-edited in French translation by A. France-Lanord, Thionville, 1987.

Amouric, H., 1984. *Moulins et meuniers en Basse-Provence occidentale du Moyen-âge à l'ère industrielle*. Doctoral thesis, Université de Provence Aix-Marseille I.

Anderson, T. (dir.), 2003. *Des artisans à la campagne. Carrières de meules, forge et voie gallo-romaine à Chables*, Archéologie Fribourgeoise, n° 19.

Bailly-Maître, M.C. and Dupraz, J., 1994. *Brandes-en-Oisans. La mine d'argent des Dauphins (XIIIe-XIVe s.) Isère*, Document d'archéologie en Rhône-Alpes. Lyon.

Bailly-Maître, M.C., 2002. Concasser et broyer sur un site minier des XIIe-XIVe siècles. In H. Procopiou and R. Treuil, *Moudre et broyer*. Paris, 213-27.

Bailly-Maître, M.C., 2006. Techniques meulières – Techniques minières ? Exemple des carrières souterraines du sud de la France. In A. Belmont and F. Mangartz (2006), 163-70.

Bailly-Maître, M.C., Martin L. and Herrcher E., 2008. Se nourrir en haute montagne aux XIIIe- XIVe siècles le village de Brandes (Huez-Isère). Apports comparés de l'archéologie, des textes et de l'anthropologie. In 7th convention of the Association Internationale pour l'Histoire des Alpes, Vevey (Suisse), Septembre 2007, *Alimentation et santé dans les sociétés de montagne du Moyen âge au XXe siècle*. Zûrich, 83-100.

Bélidor, B., 1739. *Architecture hydraulique, ou l'Art de conduire, d'élever et de ménager les eaux pour les différens besoins de la vie*, Charles Antoine Jombert, Libraire de l'Artillerie and du Génie, à l'Image Notre-Dame, à Paris, rue Saint-Jacques.

Belmont, A., 2003. Commerce et diffusion des meules de La Ferte-sous-Jouarre aux XVe et XVIe siècles. In M. Barboff, F. Sigaut, C. Griffin and R. Kremer, *Meules à grains. Actes du colloque international de La Ferté-sous -Jouarre, 16-19 May 2002*. Paris, 282-9.

Belmont, A. and Mangartz, F., (eds) 2006. *Mühlsteinbrüche. Erforschung, Schutz und Inwertsetzung eines Kulturerbes europäischer Industrie. Les meulières. Recherche, protection et valorisation d'un patrimoine industriel européen (Antiquité- XXIe s.)*, (Proceedings of the Grenoble symposium, 22-26 September 2005). Mainz.

Belmont, A., 2006. *La Pierre à pain. Les carrières de meules de moulins en France, du Moyen Âge à la révolution industrielle*. Grenoble.

Belmont, A., 2008. *Les carrières de meules de Claix, lieu-dit Les Meulières, département de la Charente*. Report on scheduled excavation. Grenoble.

Benoit, P., 1997. *La mine de Pampailly, XVe-XVIIIe siècles, Brussieu-Rhône*. Document d'archéologie en Rhône-Alpes. Lyon.

Benoit, P., Bailly-Maître, M.C., Dubois, C. and Serres, G., 1997. Meules rotatives médiévales pour le broyage des minerais. In D. Garcia and D. Meeks, *Techniques et économie antiques et médiévales* (the Aix-en-Provence symposium, May 1996), Paris, 62-9.

Bruggeman, J., 2003. Les meules en Flandre d'après les comptes et prisées du Moyen-âge à la Révolution. In M. Barboff, F. Sigaut, C. Griffin and R. Kremer, *Meules à grains. Actes du colloque international de La Ferté-sous -Jouarre, 16-19 May 2002*. Paris, 231-9.

Brunet, M., 1997. Le moulin Délien. In D. Garcia and D. Meeks, *Techniques et économie antiques et médiévales* of the (Aix-en-Provence symposium, May 1996, Errance, 29-38.

Deffontaine, B., 2003. La production et la commercialisation de meules de moulins à Cinq-Mars-la-Pile (Indre et Loire). In M. Barboff, F. Sigaut, C. Griffin and R. Kremer, *Meules à grains. Actes du colloque international de La Ferté-sous -Jouarre, 16-19 May 2002*, Paris, 297-8.

Domergue, C. *et al.* 1997 . Les moulins rotatifs dans les mines et les centres métallurgiques antiques. In D. Garcia and D. Meeks, *Techniques et économie antiques et médiévales* (the Aix-en-Provence symposium, May 1996). Paris, 48-61.

Durand, A., 2002. Les moulins carolingiens du Languedoc (fin VIIIe siècle début XIe siècle). In M. Mousnier, (dir.), *Moulins et meuniers dans les campagnes européennes (IXe –XVIIIe siècle)*. Toulouse.

Fabre, D., Carrio, E., Orengo, Y. and Malacour, C., 2006 . Analyses pétrographiques et mécaniques d'un ensemble de roches meulières françaises. In A. Belmont and F. Mangartz (2006), 91-7.

Hockensmith, C., 2006. The preservation, ownership, and interpretation of American millstone quarries. In A. Belmont and F. Mangartz (2006), 193-4.

Jaccotey, L. and Milleville, A., 2008. Les carrières du massif de la Serre (Jura) Sept millénaires d'exploitation meulière. *Archéopage,* n° 22, 16-23.

Kleinmann, D., 1984. La fabrication de meules à Cinq-Mars la Pile (Indre-et-Loire), *Bulletin de la Société des amis du vieux Chinon*, 8 (n°8), 1107-16.

Lepareux-Couturier, S. and Jodry, F., 2010. Le traitement

des surfaces actives des meules dites 'rayonnées'. Bilan documentaire, proposition d'un vocabulaire descriptif et d'une grille d'analyse. *Table ronde de Saint-Julien sur-Garonne, Évolution technologique et typologique des meules du Néolithique à l'an mil sur le territoire français.* (Forthcoming).

Marconnet, C., 2006. Analyse sédimentologique du bassin 495. In M.C. Bailly-Maître, T. Gonon and A. Gigante, *Brandes en Oisans, Programme 25*, Rapport 2006. Grenoble.

Minvielle Larousse, N., 2009. *La meunerie delphinale en Grésivaudan à la fin du Moyen-âge*, Master 1 dissertation prepared under the direction of A. Belmont. Grenoble.

Minvielle Larousse, N. and Bailly-Maître, M.C., 2010. Elément de méthodologie pour l'étude de meules et moulins à minerai. *Table ronde de Saint-Julien sur-Garonne, Évolution technologique et typologique des meules du Néolithique à l'an mil sur le territoire français.* (Forthcoming).

Perry, L., 1986. *Le moulin et le meunier dans la société rurale auvergnate du XVIIIe siècle*, Thèse de 3e cycle, université de Clermont 2.

Pichot, D., 2002. *Le village* éclaté . *habitat et société dans les campagnes de l'Ouest au Moyen-Âge.* Rennes.

Rivals, C., 2000. *Le moulin et le meunier*. Portet-sur-Garonne.

The study of America's millstone quarries: past research and future directions

Charles D. Hockensmith

Introduction

Millstone quarries once operated across much of the United States producing millstones for grist mills and other industries. Most of these quarries were located in the eastern half of the country but a few quarries were known to exist in the western half of the country. According to surviving records, millstones were quarried on the East Coast as early as the 1650s (Hockensmith 2004d; 2009a, 10). By thc late 19th century, the American millstone industry was in decline due to the introduction of the roller mill for grinding wheat. In fact, Garnett (1883, 477) reported that between half and two-thirds of the larger flour mills had already shifted to metallic rollers in 1882. The demand for millstones continued to decline into the early 20th century. Only in New York, North Carolina, and Virginia did the millstone industry continue until the mid-20th century. This late survival date of the industry was made possible by several industries (mustard, cement, paint, feldspar, fertilizers, etc.) using millstones as well as some rural grist mills continuing to use millstones (Hockensmith 2009a, 117). Historically, France, England, and Germany dominated much of the American market for millstones used in grist mills. In some instances, American stones were tried as an alternative to foreign millstones, especially the French Burr. American millstones manufactured from local stones were often used for grinding corn while French Burr millstones were used for grinding wheat.

Several types of stone were exploited for making American millstones, including conglomerate, granite, flint or burrstone, quartzite, sandstone, and some miscellaneous rocks (Hockensmith 2009a, 19-20). In areas lacking suitable bedrock, granite boulders dropped by glaciers were sometimes fashioned into millstones (Hockensmith 2009a, 86-87). Of the stones used, conglomerate and granite were the two most widely used for American millstones. The geological literature indicated that New York and Virginia had the most important conglomerate quarries and that North Carolina had the most important granite quarries. However, the majority of the American millstone quarries are only briefly mentioned in various records. Some of the American millstone quarries (especially those in New York and Virginia) were quite extensive in size while most quarries were much smaller. The numerous millstone quarries located across the United States offer scholars tremendous opportunities for future research.

This paper has two major objectives. Firstly, to provide a brief summary of the limited research undertaken on the American millstone industry to date. A detailed overview of American millstone quarries has been previous published by the author (Hockensmith 2009a, 19-87) which contains the available information for the quarries discussed in this paper. The present discussion addresses the distribution of quarries by specific stone types and provides their approximate date ranges. The second part of the paper strives to present objectives for future research on the American millstone industry. These objectives focus on conducting additional archival research and undertaking fieldwork to document many other millstone quarries.

Past research

While many millstone quarries were mentioned in the geological and historical literature, there was no scholarly effort to examine the millstone industry as a whole until the mid-20th century. Kentucky archaeologist, William S. Webb wrote two articles about millstones in the 1930s (Webb 1933, 1935). Webb did not appear to be aware of the millstone quarries located in Kentucky. Some researchers reported on millstones in general, on individual quarries or about quarries within a certain county of a state. Paul Flory (1951a; 1951b) wrote about the millstones of Lancaster County, Pennsylvania and reported on interviewing an old millstone maker (Flory 1951a; 1951b). In their book, *The Mill at Philipsburg Manor and a Brief History of Milling*, Howell and Kellar (1977, 67-91) included a chapter that commented on foreign and American millstones and where they were obtained. During 1982, millwright Jon Sass (1982) documented a small granite millstone quarry in Dinwiddie County, Virginia. The following year, archaeologist Ronald Michael (1983) prepared a National Register of Historic Places Nomination form for Game Lands 51 Millstone Quarry, Fayette County, Pennsylvania. Michael (1983) described the remains associated with this conglomerate millstone quarry. Sass' (1984, vii-viii) book *The Versatile Millstone: Workhorse of Many Industries* briefly discussed the location of some American millstone quarries as well as discussing foreign millstones. In 1985, Lela Beaman wrote about the granite millstone quarry at Parkewood in Moore County, North Carolina (Beaman 1985). This brief article provided some very useful history about this 19th century quarry that used a blue granite containing white flint inclusions (Beaman 1985). Also during 1985, Charles Howell (a millwright) briefly discussed sources of American and foreign millstones in his paper on Colonial Water Mills (Howell 1985, 144-149). The following year, geologist Thomas N. Berg (1986) wrote about a small conglomerate millstone quarry in Tioga County, Pennsylvania. Robert Davis (1990) published a brief article about the Georgia Burr millstone quarry in Burke County, Georgia. The Georgia Burr millstones were manufactured from a porous siliceous type stone which was compared to the French Burr. Finally, a brief millstone article written by Charles Howell in 1970 was published in Old Mill News sometime after his death (Howell 1997). This article commented on both foreign and American millstones and where they were produced (Howell 1997).

Beginning in 1987, the author became involved in documenting millstone quarries in Powell County, Kentucky. Four brief field seasons were spent documenting six quarries that produced monolithic conglomerate millstones between the 1790s and the mid- to late 19th century (Hockensmith 1994; 2009a; Hockensmith and Meadows 1996; 1997; 2001). This research eventually culminated in the book *The Millstone Quarries of Powell County, Kentucky* (Hockensmith 2009b). Archival information was also compiled for millstones made in Kentucky (Hockensmith 2003b; 2008c; Hockensmith and Meadows 2006; 2007) and millstones imported into Kentucky (Hockensmith 2008a; 2008c). Over a 22 year period, a great deal of information was found on millstone quarries across the United States and in other countries. As sufficient information was assembled for certain areas, a series of articles were published. The millstone industries of several states were summarized. Studies were published for Alabama (Hockensmith 2005), Georgia (Hockensmith 2004a), Missouri (Hockensmith 2004e; 2006a), New York (Hockensmith 2010a; 2010b; 2010c; Hockensmith editor 2010; Hockensmith and Coy 2010a; 2010b), North Carolina (Hockensmith 2004c), Ohio (Hockensmith 2003c; 2007a; 2008c), Tennessee (Ball and Hockensmith 2005; 2007d; Hockensmith 2004b), and Virginia (Hockensmith 1999; Hockensmith and Coy 1999; Hockensmith and Price 1999). The Virginia research included a book entitled *Millstone Manufacture in Virginia: Interviews With the Last Two Brush Mountain Millstone Makers* (Hockensmith, editor 1999).

During the same period, a number of regional and nationwide studies were published on topics related to American millstones. In 1993, information was compiled on the American millstone industry in an article entitled the 'Study of American Millstone Quarries' (Hockensmith 1993a; 1993b). The same year, another article was published entitled 'Millstone Quarrying in the Eastern United States: A Preliminary Overview' (Hockensmith 1993c). Many early sources on millstones were complied into a publication entitled *Early American Documents and References to Millstones: 1628-1829* (Hockensmith 2004d). Ball and Hockensmith (2007a) later published their book *Millstone Studies: Papers on Their Manufacture, Evolution, and Maintenance*. The book included several studies: 'Preliminary Directory of Millstone Makers in the United States' (Ball and Hockensmith 2007b), 'Occupational Disease and Work Place Hazards Associated With Millstone Making' (Ball and Hockensmith 2007c), 'A Bibliography of the Millstone Industry in the United States' (Ball and Hockensmith 2007e) and 'The Granite and Gneiss Millstone Quarries of the United States' (Hockensmith 2007b). A manuscript of that period was entitled 'American Millstones Similar to the French Burr: 19th Century Attempts to Find Substitutes' (Hockensmith 2008d). Summary papers were also written for two previous international millstone conferences including 'The Conglomerate Millstone Industry in the Eastern United States' (Hockensmith 2003a) and 'The Preservation, Ownership, and Interpretation of American Millstone Quarries' (Hockensmith 2006b). Finally, the author produced a book bringing together over 20 years of worldwide research entitled *The Millstone Industry: A Summary of Research on Quarries and Producers in the United States, Europe, and Elsewhere* (Hockensmith 2009a). Taken together, the above literature by the author and other researchers has established a substantial foundation for future research on the American millstone industry.

American millstone quarries

Several types of stone were quarried and shaped into millstones within the United States. These include conglomerate, granite, flint/burrstone, sandstone, quartzite, gneiss, dolomite, syenite, and unspecified raw materials. The following discussion is organized by each of these material types. Information is provided on their distribution (by state) and their approximate date ranges. The date ranges do not accurately reflect the actual years that millstones were quarried in these states but include the dates mentioned in sources or the date of the publication for quarries still operating. It is often difficult to identify the geological formations used in millstone manufacture. In many instances, the geological formations are not mentioned in publications and often older works refer to geological nomenclature that is outdated and no longer in use.

Conglomerate deposits were quarried for millstones in Alabama (Blount, Jackson, Jefferson, Madison, and

Marion counties), Arkansas (Independence County), Connecticut (at Mount Tom), Kentucky (Letcher, Madison, Marshall, Powell, Rockcastle, and Whitley counties), New York (Ulster County), North Carolina (Chatham, Midland, and Moore counties), Pennsylvania (Berks, Bradford, Carbon, Fayette, Lancaster Somerset, and Tioga counties), Tennessee (Cannon, Coffee, and Trousdale counties), Vermont (Rutland County), Virginia (Montgomery and Pulaski counties), and West Virginia (Fayette County) (Hockensmith 1999; 2003a; 2006b; 2008b; 2009a 20-54; 2010a; 2010b; Hockensmith and Coy 1999; 2010a; 2010b; Hockensmith and Price 1999). Unfortunately, little is known about most of these quarries since they are only mentioned briefly in geological and historical reports. The bulk of the literature focuses on the quarries in New York and Virginia, and to a lesser extent on Pennsylvania. Of all the stones used for American millstones, the author is most familiar with conglomerate. He documented six quarries in Powell County, Kentucky (Hockensmith 2009b; Hockensmith and Meadows 1996; 1997; 2001) and has visited millstone quarries in New York (Hockensmith 2010; Hockensmith and Coy 2010a; 2010b), Pennsylvania, and Virginia (Hockensmith and Coy 1999; Hockensmith and Price 1999). The scant literature available suggest the following approximate date ranges for conglomerate quarries: Alabama (1848-1926), Arkansas (?-1858), Connecticut (unknown), Kentucky (1790s- c. 1880), New York (1732-1955), North Carolina (1850s-1890s), Pennsylvania (1790s-1918), Tennessee (1818-?), Vermont (unknown), Virginia (1790s-1943), and West Virginia (unknown).

The conglomerate millstone quarries have the best documented geological formations. In Alabama, the Pennsylvanian age Pottsville Formation was used (Dean 1996). The Powell County, Kentucky quarries exploited the Pennsylvanian age Lee Formation (McDowell 1978). The quarries in Ulster County, New York were composed of Shawangunk grit conglomerate (Newland 1907:43-44). In North Carolina, a Triassic conglomerate (Ladoo 1925, 9) or a conglomerate of the Pekin Formation was used (Reinemund 1955, 122). The Olean conglomerate was exploited in Tioga County, Pennsylvania (Berg 1896, 3). The quarries at Montgomery County, Virginia exploited the Ingles conglomerate (Campbell 1925, 26). Finally, the Fayette County, West Virginia quarries used the Pottsville conglomerate (Peters and Carden 1929, 119). From the author's personal observations, the New York and Virginia conglomerates are very similar, both being a whitish to light gray color with small quartz pebbles. The Kentucky and Pennsylvania conglomerates are also similar, both being a more tan color with larger quartz pebbles.

Limited information is available for the sizes and prices of conglomerate millstones. A June 13, 1799 ad in the *Kentucky Gazette* mentioned that Red River millstones (from Powell County, Kentucky) came in the following sizes: 4 feet [1.22 m], 3 feet 10 inches [1.17 m], 3 feet 8 inches [1.12 m], 3 feet 6 inches [1.065 m], and 3 feet [0.915 m] (Kentucky Gazette 1799). Another ad for Red River millstones appeared in the April 1, 1818 edition of the Kentucky Reporter with the following rates per pair of stones: 5 feet [1.525 m] ($200), 4 feet [1.22 m] ($150), 3 feet [0.915 m] ($80), and a general reference to other sizes available (Kentucky Reporter 1818). Prices are available for conglomerate millstones made at Charles Colyer's millstone quarry in Rockcastle County, Kentucky. The February 26, 1821 edition of *The Argus of Western America*, advertised millstones with the following values: 5 feet [1.525 m] ($100), 4 feet 1 inch [1.245 m] ($90), 4 feet [1.22 m] ($70), 3 feet 9 inches [1.145 m] ($60), 3 feet 6 inches [1.065 m] ($50), 3 feet 3 inches [0.99 m] ($40), 3 feet [0.915 m] ($35), and 2 feet 6 inches [0.765 m] ($25) (The Argus of Western America 1821a). In Ulster County, New York, Newland (1907, 43-44) reported that millstones were manufactured in sizes between 15 and 90 inches [0.385 and 2.285 m] in diameter, with the 24 [0.61 m], 30 [0.765 m], 36 [0.915 m], 42 [1.065 m], and 48 [1.22 m] inches diameter sizes in more demand. It was also noted that a pair of 30-inch [0.765 m] millstones were valued at $15, a single 60-inch [1.525 m] millstone sold for $50, and larger sizes of millstones sold for $50-100 (Newland 1907, 43-44). Abandoned millstones at the Chestnut Ridge quarry in Fayette County, Pennsylvania yielded millstones between 24 [0.61 m] and 48 [1.22 m] inches in diameter (Michael 1983). An ad appearing in the April 13, 1825 edition of The Eagle provided the following sizes and values for millstones from the Laurel Hill quarry in Fayette County, Pennsylvania: 2 feet 6 inches [0.76 m] ($20), 3 feet [0.915 m] ($30), 3 feet 6 inches [1.07 m] ($45), 4 feet [1.22 m] ($60), 4 feet 6 inches [1.37 m] ($80), and 5 feet [1.525 m] ($100) (The Eagle 1825). At the Brush Mountain quarry in Montgomery County, Virginia, millstones were made the following diameters: 12 [0.31 m], 18 [0.46 m], 24 [0.61 m], 30 [0.765 m], 36 [0.915 m], and 54 [1.37 m] inches (Hockensmith and Coy 1999, 37). Another source for Brush Mountain millstones indicates that they were made between 12 and 60 inches [0.31 and 1.525 m] in diameter with many stones made at the 3 [0.915 m] and 4 [1.22 m] feet diameters (Hockensmith and Price 1999, 80).

Millstones were manufactured from granite in Alabama, Arkansas (Pulaski and Saline counties), Connecticut (New London County), Louisiana (at Point Pleasant), Maine, Maryland, Massachusetts (at the communities of Easton, Lynn, Medfield, Quincy, Saugus, and Worcester), Minnesota, Missouri (Iron County), New Hampshire (at Enfield), North Carolina (Moore, Richmond, and Rowan counties), Rhode Island (at Westerly), South Carolina (at Abbeyville), Tennessee (Carter County), and Virginia (Dinwiddie and Louisa counties) (Hockensmith 2007b; 2009a, 54-60). Very little information is available for millstone quarries in most of these states. North Carolina was the leading producer of granite millstones and is the best represented in the literature. Very approximate date ranges are available for some of these quarries: Alabama (1923-1925), Arkansas (1860-?), Connecticut (1737-?), Louisiana (1810-?), Maine (c. 1925), Maryland (unknown), Massachusetts (1828-1880s), Minnesota (c. 1925), Missouri (1870s), New Hampshire (c. 1850-1925), North Carolina (1850-1963), Rhode Island (unknown), South Carolina (unknown), Tennessee (1870s), and Virginia (1835-?). At the Parkewood, North Carolina millstone quarry, granite millstone with white flint inclusions that were quarry dressed had the following sizes and values: 14-inch [0.36 m] ($25), 20-inch [0.51 m] ($60), 30-inch [0.765 m] ($90), 36-inch [0.915 m] ($110), 42-inch [1.07 m] ($145), and 48-inch [1.22 m] ($185) (North Carolina Millstone Co. n.d.).

Various types of flint or buhrstone (also spelled burrstone) were used for millstones in Alabama (Jackson and Winston counties), Arkansas (Izard, Lawrence, and Polk counties), California (near Laguna Seca), Georgia (Burke and Jefferson counties), Illinois (Union County), Indiana (Harrison and Jennings counties), Kentucky (Franklin and Woodford counties), Massachusetts (at Pittsfield), Missouri (Carroll, Madison, Maries, Moniteau, Montgomery, Osage, Ozark, Pulaski, Washington, Webster, and Wright counties), North Carolina (Jones and Montgomery counties), Ohio (Athens, Jackson, Licking, Muskingum, and Vinton counties), South Carolina (near Barnwell and Orangeburg), Tennessee (Clairborne, Jefferson, Knox, Sumner, and Williams counties), and Virginia (Floyd and Smyth counties) (Hockensmith 2003c; 2004a; 2007a; 2008b; 2008d; 2009a, 60-77). Many of these millstones were made from porous flints with cavities while other buhrstones were composed of silicified shells or limestone. While some of the millstones may have been monolithic, others were obviously composite millstones. Frequently, these millstones were compared to the French Burr stones in terms of their quality. While most of these millstones were obviously inferior to the French Burr, they were attempts to find a cheaper, local alternative. While we don't have detailed information on the date ranges for these quarries, some information is available: Alabama (c. 1880s), Arkansas (1850s-?), California (1850s), Georgia (1792-1850s), Illinois (1820s-?), Indiana (1830s-1850s), Kentucky (1800-1821), Massachusetts (1850s), Missouri (1830s-1870s), North Carolina (1850s-1870s), Ohio (1805-1850s), South Carolina (1840s-1860s), Tennessee (1860s-1870s), and Virginia (1830s-1860s). The flint millstones produced in Georgia and Ohio were the best known. Very limited information is available for other states.

Some information is available for the sizes and prices of flint or buhrstone millstones. The Georgia Burr was made in a variety of sizes ranging from 15 inches [0.38 m] to 7 feet [3.135 m] (Davis 1990, 6). The *Manufacturer and Builder* (1876) mentioned that a 24-inch [0.61 m] Georgia Burr millstone sold for $250. Prices are available for flint millstones at two early quarries in Franklin County, Kentucky. First, the firm of Miller, Railsback & Miller advertised their millstones in the August 9, 1821 edition of *The Argus of Western America.* The millstones had the following values: 5 foot [1.525 m] ($150), 4 foot [1.245 m] ($100), and 3 foot [0.915 m] ($50) (The Argus of Western America 1821b). In the November 8, 1821 issue of *The Argus of Western America*, Jeremiah Buckley advertised the flint millstones at his quarry. Buckley had a much greater assortment of millstones with the following values: 5 foot [1.525 m] ($180), 4 foot 6 inches [1.375 m] ($150), 4 foot [1.245 m] ($125), 3 foot 9 inches [1.145 m] ($100), 3 foot 3 inches [0.99 m] ($75), 3 foot [0.915 m] ($60), 2 foot 9 inches [0.84 m] ($50), 2 foot 6 inches [0.765 m] ($40) (The Argus of Western America 1821c). Finally, in Ohio, the Raccoon Creek Buhr millstones brought a tremendous sum for the period (Mather 1838, 33-34). Between the years 1814 and 1820, a pair of 4 ½ feet [1.375 m] millstones sold for $350, a pair of 7 feet [2.135 m] millstones sold for $500 (Mather 1838, 33-34). By 1838, prices had dropped and millstones 4 feet [1.22 m] in diameter were valued at $150 (Mather 1838, 33-34).

Sandstone was rarely used for making millstones to be used for grinding in grist mills since the sand particles could be dislodged and contaminate the meal or flour. Instead, sandstone millstones were adequate for grinding grains used in distilleries since the mash would be discarded after use (Garber 1970, 82). Several states produced millstones made from sandstone: Alabama (Jackson County), Arkansas, Massachusetts (Easton), Mississippi (Claiborne and Lauderdale counties), Missouri (Laclede and Pulaski counties), Ohio (Summit County), and Texas (Jasper County) (Hockensmith 2009a, 80-82). With the exception of a quarry in Ohio, little is known about these millstone quarries. Approximate ages for these quarries are: Alabama (1914-1917), Arkansas (1819), Massachusetts (c. 1700s), Mississippi (1850s), Missouri (1870s), Ohio (1880s), and Texas (1890s). In Jackson County, Alabama, these millstones were made from a Pennsylvanian age sandstone (Katz 1916, 554) while the Berea grit, a white sandstone, was exploited at Summit County, Ohio (Read 1883, 478-479).

Quartzite was quarried for millstones in Alabama, Connecticut (near Canton), and North Carolina (Madison County) (Hockensmith 2009a, 77-78). The age of the Alabama quarry is unknown but some information is available for Connecticut (1830s) and North Carolina (1860s-1870s). Apparently, quartzite was rarely used for millstones.

In a few states, gneiss was used for millstones: Alabama (Lee County), Massachusetts (near Washington), New York (Jefferson County), North Carolina, and Tennessee (Johnson County) (Hockensmith 2009a, 78-79). Date ranges for these quarries are as follows, Alabama (unknown), Massachusetts (1840s), New York (1805-1828), North Carolina (1870s), and Tennessee

(1860s-1870s). Gneiss, similar to granite, had limited popularity in certain areas. Haddock (1895, 434) noted that gneiss millstones sold for $100 per pair in 1805 at Jefferson County, New York.

Materials rarely used for millstones included dolomite and sienite (Hockensmith 2009a, 82-83). A dolomite quarry was reported near Knoxville, Tennessee which dated before 1900 (Hockensmith 2009a, 82-83). Likewise, one sienite quarry was reported. This quarry was located in Essex County, Massachusetts during the 1840s (Hockensmith 2009a, 83).

The geological and historical literature mentioned millstone quarries in several states where the raw material was not specified. Undoubtedly, many of these quarries utilized the stone types mentioned above but others could represent additional types of raw materials. Quarries included in this unspecified category include: Connecticut (Litchfield County), Delaware (in White Clay Creek State Park), Georgia (Bulloch, Burke, Early, Jefferson, and Screven counties), Illinois (Pope and Saline counties), Kentucky (near South Union and McCreary County), Maryland (Baltimore County), Missouri (Carroll, Montgomery, and Washington counties), New Jersey (Bergen County), New York (Allegany County), Ohio (Lawrence County), Pennsylvania (Forrest County), Tennessee (Tipton County), Virginia (Aquia Creek), and West Virginia (Randolph County) (Hockensmith 2009a, 83-87). In terms of date ranges, the quarries are: Connecticut (1830s), Delaware (unknown), Georgia (unknown), Illinois (1870s), Kentucky (1802-1819), Maryland (1753-1813), Missouri (1830s), New Jersey (unknown), New York (c. 1900), Ohio (1803), Pennsylvania (1880s), Tennessee (1870s), Virginia (unknown), and West Virginia (1880s) (Hockensmith 2009a, 83-87).

Future directions

When compared to European millstone research, the study of American millstone quarries is still in the early stages. American historical and industrial archaeologists have been content with studying more traditional site types. With the exception of the author's archaeological research in Powell County, Kentucky (Hockensmith 2009b), most of the research efforts to date have focused on compiling archival materials from many diverse sources. Most researchers have not been archaeologists but interested individuals from other backgrounds. While the past research was an important endeavor and very essential for laying an initial groundwork, much work remains to be undertaken in the future. This work can be divided into two broad components. First, additional archival research is needed in most areas of the United States. Second, millstone quarries across America need to be located and documented in detail. The following sections will deal with these future goals.

Archival research

A great deal of archival material has been discovered to date on the American millstone industry. Undoubtedly, much more information on millstone quarries remains buried in libraries and courthouses across the United States. The author's previous publications have drawn upon information available in libraries from Kentucky and elsewhere as well as websites. As most researchers are aware, to find certain types of information, you must travel to the state containing the quarry and also to the nearby communities. This level of effort will be very time consuming but most fruitful. It is obvious that the author does not have the time or resources to personally visit the 29 states and the numerous counties within the states that contain the known millstone quarries. Instead, such research will require many different scholars, working independently, to accomplish such a large task. It is the author's goal in this section to share some of what he has learned since 1987 in order to assist and guide future researchers. By making other researchers aware of useful resources available to them, it is hoped that these sources will yield valuable new information and save time for the researchers.

There are several key sources that are likely to yield information on the American millstone industry. Early geological reports are a wonderful source since the geologists frequently mentioned millstone quarries. Most states have an extensive series of geological reports, many of which date to the 19th century. A variety of these reports may contain information on millstone quarries. Early geologists often took personal interest in reporting on millstone quarries that were still in operation. Geological reports occur in several formats, including early statewide overviews, studies of certain regions, studies of selected counties, studies of specific formations, studies of particular mineral resources, volumes dealing with economic geology, etc. Some of these reports are well indexed but other reports lack indexes. For the reports that are not indexed, a few minutes of time will be required to quickly visually scan through them. Another factor to keep in mind, is that not all sources will be listed under the word millstone. They may be listed under abrasives, burr stones or buhr stones. It is useful to look under all these terms in the index before dismissing a report. The millstone quarry overviews presented in this paper can guide researchers to the types of stones to search under as well as the counties and geographical regions to search. Also, it would be advisable to check the early geological reports (1860 and earlier), especially in the eastern U.S., for those states that do not currently have recorded millstone quarries. Such research may identify millstone quarries that we are unaware of presently.

Another important source for millstones are the U.S. Census schedules. For states with well established millstone industries, the Census of Manufacturing may include millstone quarries. The Census of Manufacturing is a wonderful source for industries in general but they are usually available only on microfilm. These hand-written records can sometime be difficult to read because of poor handwriting, fading or spotting on original forms. Some published census summaries are available (i.e., Coxe 1814) which just give the number of certain industries by state. However, only the hand written census schedules will provide actual listings for individual businesses. The Census of Population is most useful after 1850 when occupations are first recorded. While some people may be listed as millstone makers, others may be listed as stone cutters. Some of the earlier Population Census records for particular counties have been transcribed and are available in book form. Also, websites operated by Ancestry and Heritage Quest, make U.S. Population Census records available online.

Newspapers are another key source for millstone researchers. Larger communities frequently had newspapers. Many of these papers are available on microfilm at local libraries and historical societies. Some libraries and historical societies have digitized versions of newspapers allowing them to be searched remotely by computer. My experience with Kentucky newspapers indicates that 19th century millstone quarries occasionally advertised in local newspapers. Also, ads were placed in newspapers at considerable distances from the quarries with the hope of attracting additional customers. Some of the millstone ads that the author has encountered contain very important details. Ads sometimes list the sizes of millstones available and the prices for each size. Also, the ads may also give the names(s) of key people and the nearest city to the quarry.

Once the exact location of a millstone quarry is known, a search of deeds can provide a lot of information. The deeds establish the ownership of the quarry through time. Also, the deeds may contain references to the quarry at different points in time, thus helping to determine the years that it was active. One obstacle to deed research is the loss of records through courthouse fires which leave gaps in coverage. Another factor to keep in mind is that county boundaries may have changed through time. Thus, deeds for a single quarry could be in different courthouses for different periods of time. Consequently, it is useful to become familiar with the development of county boundaries before searching for deeds.

In Kentucky, lawsuits have been an excellent source of information (Hockensmith 2009b; Hockensmith and Meadows 2006, 2007). When millstone makers failed to deliver completed stones on schedule, sometimes they were sued. Other reasons were also responsible for suits being filed such as the failure to pay wages to workers. Such law suits can provide new details about the quarries and the people involved. However, the names of the millstone makers are required before you can begin searching for lawsuits. Some lawsuits may be indexed on library computer files while many other indexes may only be available in handwritten form. The author has discovered that millstone makers might be involved in various types of lawsuits during their life time. Thus, a series of lawsuits may be checked that may or may not be related to the millstone industry.

If the millstone industry played an important role in the local economy or if prominent people were involved, millstone quarries have greater potential of being mentioned in various histories. Sometime the millstone industry will be mentioned in statewide histories if the state was a leading producer of millstones. Lesser known quarries may be mentioned in county or local histories. Finally, once the names of millstone makers have been identified, other types of records (family histories, biographical sketches, tax records, etc.) can be consulted for additional information.

Early directories are another potential source for the millstone industry. Some states have statewide gazetteers and business directories. These publications contain alphabetical listings of businesses which sometimes includes millstones. Also, the statewide gazetteers and business directories can include ads in addition to listings. City directories are frequently available for many American cities. They usually include an extensive business section which can be checked for millstones. When millstones were still in great demand, these directories afforded companies and individuals opportunities to promote their stones.

Fieldwork

As noted earlier, there are numerous millstone quarries across the United States that are yet to be located and documented. Currently, we are aware of reported millstone quarries in 29 states. Some of the remaining states in the eastern half of America may also contain millstones quarries that we are unaware of presently. Many of the western states may possess stones suitable for millstones but were settled after the demise of the American millstone industry. Interested researchers can simply select one or more counties containing quarries from the state they reside in or from an adjacent or nearby state. There are more than enough millstone quarries to keep researchers busy for years.

There are many conglomerate millstones quarries to be documented. Conglomerate millstone quarries are known to exist in 31 counties within eleven states. These include Alabama (Blount, Jackson, Jefferson, Madison, and Marion counties), Arkansas (Independence County), Connecticut (at Mount Tom), Kentucky (Letcher, Madison, Marshall, Powell, Rockcastle, and Whitley counties), New York (Ulster County- quarries extend for a distance of 10 miles), North Carolina (Chatham, Midland, and Moore counties), Pennsylvania (Berks, Bradford, Carbon, Fayette, Lancaster, Somerset, and Tioga counties), Tennessee (Cannon, Coffee, and Trousdale counties), Vermont (Rutland County), Virginia (Montgomery and Pulaski counties), and West Virginia (Fayette County) (Hockensmith 1999; 2003a; 2006b; 2009a, 20-54; 2010a; 2010b; Hockensmith and Coy 1999; 2010a; 2010b; Hockensmith and Price 1999).

Granite millstone quarries are known to exist in 24 counties within 15 states. These include Alabama, Arkansas (Pulaski and Saline counties), Connecticut (New London County), Louisiana (at Point Pleasant), Maine, Maryland, Massachusetts (near the communities of Easton, Lynn, Medfield, Quincy, Saugus, and Worcester), Minnesota, Missouri (Iron County), New Hampshire (at Enfield), North Carolina (Moore, Richmond, and Rowan counties), Rhode Island (Westerly), South Carolina (at Abbeyville), Tennessee (Carter County), and Virginia (Dinwiddie and Louisa counties) (Hockensmith 2007b; 2009a, 54-60).

Flint or buhrstone millstone quarries are known to exist in 41 counties contained within 14 states. Various types of flint or buhrstone quarries are located in Alabama (Jackson and Winston counties), Arkansas (Izard, Lawrence, and Polk counties), California (near Laguna Seca), Georgia (Burke and Jefferson counties), Illinois (Union County), Indiana (Harrison and Jennings counties), Kentucky (Franklin and Woodford counties), Massachusetts (at Pittsfield), Missouri (Carroll, Madison, Maries, Moniteau, Montgomery, Osage, Ozark, Pulaski, Washington, Webster, and Wright counties), North Carolina (Jones and Montgomery counties), Ohio (Athens, Jackson, Licking, Muskingum, and Vinton counties), South Carolina (near Barnwell and Orangeburg), Tennessee (Clairborne, Jefferson, Knox, Sumner, and Williams counties), and Virginia (Floyd and Smyth counties) (Hockensmith 2003c; 2004a; 2007a; 2008d; 2009a, 60-77).

Sandstone millstone quarries are known to exist in at least nine counties within seven states. Reported states producing millstones made from sandstone include Alabama (Jackson County), Arkansas, Massachusetts (Easton), Mississippi (Claiborne and Lauderdale counties), Missouri (Laclede and Pulaski counties), Ohio (Summit County), and Texas (Jasper County) (Hockensmith 2009a, 80-82).

Less common millstone materials include quartzite, gneiss, dolomite and syenite. Quartzite millstone quarries are located in Alabama, Connecticut (near Canton), and North Carolina (Madison County) (Hockensmith 2009a, 77-78). Gneiss millstone quarries are known to exist in at least five counties within five states including Alabama (Lee County), Massachusetts (near Washington), New York (Jefferson County), North Carolina, and Tennessee (Johnson County) (Hockensmith 2009a, 78-79). The only known dolomite quarry was reported near Knoxville, Tennessee (Hockensmith 2009a, 82-83). Finally, the only sienite quarry was reported in Essex County, Massachusetts (Hockensmith 2009a, 83).

Millstone quarries of unspecified raw materials are known to exist in at least 22 counties within 14 states. These quarries include Connecticut (Litchfield County), Delaware (in White Clay Creek State Park), Georgia (Bulloch, Burke, Early, Jefferson, and Screven counties), Illinois (Pope and Saline counties), Kentucky (near South Union and McCreary County), Maryland (Baltimore County), Missouri (Carroll, Montgomery, and Washington counties), New Jersey (Bergen County), New York (Allegany County), Ohio (Lawrence County), Pennsylvania (Forrest County), Tennessee (Tipton County), Virginia (Aquia Creek), and West Virginia (Randolph County) (Hockensmith 2009a, 83-87).

The millstone quarry locations listed above for each material type provides numerous research opportunities for archaeologists wanting to make a contribution to the study of America's millstone industry. Undoubtedly, some of the quarries have been destroyed by modern development while others quarries have survived intact. The surviving quarries are most likely to be located in remote or rugged areas where little modification to the landscape has occurred. In many instances, the local citizens may be totally unaware of the existence of millstone quarries near their neighborhoods. The lack of public familiarity with these quarries can make it difficult to locate them, especially for an outsider. A good place for researchers to start is with the local historical society or local historians. While they may or may not be aware of the millstone quarry, they can often point you to other knowledgeable individuals. Older citizens may have seen or heard about the quarries. Also, avid outdoorsmen who spend numerous hours walking through the woods may have seen the quarries. If you can obtain an approximate location, then local land owners can be questioned. Even after checking these resources, it can still be difficult to pin point their exact locations. The author recently tried to locate a flint millstone quarry dating to c. 1821 just a few miles from his home. Despite his efforts, the quarry has yet to be located. Similar difficulties may await other researchers for the more remote millstone quarries.

Once a quarry is located, it can be a challenge to identify the various types of quarry remains. It is easy for anyone to recognize a completed millstone. However, most of the quarries will be in forested areas with dense leaf cover partially obscuring the ground surface. Without some familiarity of the quarrying and shaping of millstones, it may be difficult to recognize a worked boulder or an early stage millstone. This will probably vary to some extent depending on the type of stone being worked and the nature of the surviving remains. Other types of remains will include stone fragments discarded from the shaping of millstones. Also depressions, benches or pits associated with the uncovering stone. Potential remains could include remnants of structures (blacksmith shop, shelters, etc.), abandoned roads and tramways, etc. Metal detecting at the quarries may locate lost or broken tools used by the millstone makers.

To facilitate the documentation of millstones and boulders that have been worked, the author has developed two recording forms. Both of these forms have been reproduced in the author's book *The Millstone Quarries of Powell County, Kentucky* (Hockensmith 2009b). The form for documenting millstones is included as Appendix A (Hockensmith 2009b, 167-169). The form for recording boulders and drill holes is included as Appendix B (Hockensmith 2009b, 170-172). These forms make it easy for recording standardized information during the fieldwork. The forms should be supplemented with sketches for each millstones and boulders as well as additional notes on unusual characteristics observed. Numerous photographs should be taken of the quarry from different directions. Also, each millstone should be photographed from different angles. Boulders, pit and other excavations, shaping debris, and other features should be photographed as well. A detailed sketch map (with distances noted) should be prepared for the quarry as a whole and with additional sketch maps documenting the details of major clusters of stones or excavation. All millstones and boulders should be numbered and these numbers placed on the forms and maps.

When armed with the completed forms, photographs, and maps, the quarry and manufacturing sequence can be reconstructed back in the office. By carefully studying each millstone and the tool marks that they contain, it is often possible to determine the steps that millstone makers followed in making a completed stone. Certain unfinished millstones can provide key information. The author's work in Kentucky (Hockensmith 2009b) has revealed that millstones were abandoned in various stages of completion. Sometimes the stone had a hidden flaw or perhaps critical mistakes were made during the shaping process. Many things could go wrong during the manufacturing process. By looking at the assemblage as a whole, various types of errors can be identified which provide considerable insight into the making of millstones. Each abandoned millstone is essentially frozen in time. Work ceased on that stone at the moment that a critical error occurred. Since errors occurred at different stages, different millstones can reveal different characteristics. As a result, a rejected millstone preform is far more important to a researcher than a perfect finished specimen.

Boulders with drill holes and tool marks require careful documentation. Stones included within this category can either be huge natural stones sitting on the surface or large slabs removed from intact bedrock. At some quarries, the millstone makers carefully shaped natural boulders into millstones where they lay. At other quarries, the overburden of soil or unusable rock was removed and rectangular slabs were split from the bedrock. The type of boulders found at a quarry indicate the type of reduction sequence being employed. Drill holes encountered at a quarry (on boulders, millstones or quarry faces) merit careful documentation. This includes measuring the hole diameters, depths, and spacing. After looking at all the drill holes, it is possible to determine the sizes of drills used, the typical hole spacing for each drill size, and the depths of holes required to split the stone. Surviving accounts indicate that wedges and feathers were placed into the drill holes and tapped with a hammer to cause the stone split apart. Various tool marks on millstones and boulders can provide clues about the type of tools used in working the stone. From the author's work in Virginia (Hockensmith ed. 1999; Hockensmith and Coy 1999; Hockensmith and Price 1999) and New York (Hockensmith ed. 2010; Hockensmith and Coy 2010a; 2010b), it was learned that the millstone makers had a very diverse set of specialized tools to accomplish the tasks required. In order to cut the hard stone, each tool had to be tempered for that particular type of stone. If the tool was too hard it would break but if it was too soft it would not cut the stone. Consequently, a blacksmith was usually at the quarry or part of the crew working on the millstones. The tools constantly dulled and had to be re-sharpened. Millstone makers were not common laborers but as Katz (1926, 328-329) has noted, they were master craftsmen. It took many years of experience and tremendous skill to create a perfectly rounded millstone with a flat grinding surface.

Conclusion

Numerous millstone quarries were once operated in at least 29 states within the United States. There is a high probability that many of these quarries have survived, especially in more remote or rugged areas. While a substantial amount of information has been extracted from various archival sources during past research, much remains to be accomplished to in order to develop a better understanding of the American millstone industry. These millstones were essential components of early flour mills and performed grinding functions for several industries. This paper has mentioned the states, and counties when known, where these quarries were located. Also, a brief summary of the literature has been provided to guide future researchers. The overview of American millstone quarries make scholars aware of the different types of stones that were utilized for making millstones and the approximate distribution of such quarries across the United States. The section on future directions points researchers to archival resources that they can consult and to fieldwork that needs to be conducted.

The longer that research is delayed on the American millstone industry, the fewer quarries will remain to be studied. It is essential that fieldwork begin while a larger group of quarries are available for study. As time passes, some quarries will be destroyed by modern developments, quarrying or perhaps mining (Hockensmith 2006b). The millstone quarries are waiting to be studied and the written records are available to aid in their interpretation. The main component lacking is interested researchers. Both historical and industrial archaeologists appear to be content in researching more traditional topics. There is a tendency in archaeology to focus on what is considered popular at the moment. As a result, people do not usually seek out new site types, like millstone quarries, but focus the more commonly studied site types. The author has devoted a substantial amount of time to studying millstone quarries during the past 22 years. He has tried to make other archaeologists aware of millstone quarries through his publications and presentations at conferences. To date, he is not aware of any other archaeologists in America that have a major research interest in millstone quarries. A few individuals have shown a passing interest in millstones but no major long term research interests. However, the author's books and articles, have led to a number of people from others states contacting him with questions about the millstones they encounter. While this is a positive development, most of these people are not archaeologists but local people with an interest in the past. It is hoped that American archaeologists will expand their research interests to include millstone quarries while we still have quarries to study. There is so much for scholars to learn from the quarries that once supplied millstones for America's grist mills and other industries.

Acknowledgements

The author expresses his gratitude to David Peacock and David Williams, University of Southampton, for permitting him to submit a contributed paper to this volume. Susie Hockensmith, the author's wife, kindly read the paper and suggested useful corrections.

References

Ball, D. B., and Hockensmith, C. D., 2005. Early Nineteenth Century Millstone Production in Tennessee. *Ohio Valley Historical Archaeology* 20, 1-15.

Ball, D. B., and Hockensmith, C. D., 2007a. *Millstone Studies: Papers on Their Manufacture, Evolution, and Maintenance.* Published by the Symposium on Ohio Valley Urban and Historic Archaeology, Murray, Kentucky and Society for the Preservation of Old Mills. East Meredith, New York.

Ball, D. B., and Hockensmith, C. D. , 2007b. Preliminary Directory of Millstone Makers in the United States. In D. B. Ball and C. D. Hockensmith (eds.), *Millstone Studies: Papers on Their Manufacture, Evolution, and Maintenance.* Published by the Symposium on Ohio Valley Urban and Historic Archaeology, Murray, Kentucky and Society for the Preservation of Old Mills. East Meredith, New York, 1-98.

Ball, D. B., and Hockensmith, C. D., 2007c. Occupational Disease and Work Place Hazards Associated With Millstone Making. In D. B. Ball and C. D. Hockensmith (eds.), *Millstone Studies: Papers on Their Manufacture, Evolution, and Maintenance.* Published by the Symposium on Ohio Valley Urban and Historic Archaeology, Murray, Kentucky and Society for the Preservation of Old Mills. East Meredith, New York, 99-105.

Ball, D. B., and Hockensmith, C. D., 2007d. Nineteenth Century Millstone Production in Tennessee. In D. B. Ball and C. D. Hockensmith (eds.), *Millstone Studies: Papers on Their Manufacture, Evolution, and Maintenance.* Published by the Symposium on Ohio Valley Urban and Historic Archaeology, Murray, Kentucky and Society for the Preservation of Old Mills. East Meredith, New York, 117-133.

Ball, D. B., and Hockensmith, C. D., 2007e. A Bibliography of the Millstone Industry in the United States. In D. B. Ball and C. D. Hockensmith, (eds.) *Millstone Studies: Papers on Their Manufacture, Evolution, and Maintenance.* Published by the Symposium on Ohio Valley Urban and Historic Archaeology, Murray, Kentucky and Society for the Preservation of Old Mills. East Meredith, New York, 208-222.

Beaman, L., 1985. Parkewood, NC, Millstone Quarry. *Old Mill News*, 13.3, 18-19.

Berg, T. N., 1986. A Sesqicententennial Story. Early Millstone Quarry in Tioga County. *Pennsylvania Geology*, 17.1, 3-6.

Campbell, M. R., *et al.* 1925. *The Valley Coal Fields of Virginia.* Virginia Geological Survey, Bulletin No. 25. Charlottesville, Virginia.

Coxe, T., 1814. *A Statement of the Arts and Manufacturers of the United States of America, for the Year 1810.* A. Corman. Philadelphia.

Davis, R. S. Jr., 1990. The Georgia Buhrs. A Forgotten Millstone. *Old Mill News*, 18.4, 6-7.

Dean, L. S., 1996. Letter to Charles D. Hockensmith concerning the millstone industry in Alabama. *Alabama Geological Survey*, Tuscaloosa, November 21.

Flory, P. B.,1951a. Old Millstones. *Papers Read Before the Lancaster County Historical Society*, 55.3, 73-86. Lancaster County, Pennsylvania.

Flory, P. B., 1951b. Millstones and Their Varied Usage, *Papers Read Before the Lancaster County Historical Society*, 55.5, 125-136. Lancaster County, Pennsylvania.

Garber, D. W., 1970. *Waterwheels and Millstones. A History of Ohio Gristsmills and Milling*, The Ohio Historical Society. Columbus.

Garnett, H., 1883. Abrasive Materials. In *Mineral Resources of the United States, Calendar Year 1882*, Part II- Nonmetals. Department of the Interior, U.S. Bureau of Mines, Government Printing Office. Washington, D.C., 476-481.

Haddock, J. A., 1895. *The Growth of a Century. As Illustrated in the History of Jefferson County, New York, From 1793 to 1894*. Weed-Parsons Printing Company. Albany, New York.

Hockensmith, C. D., 1993a. Study of American Millstone Quarries. Part I, *Old Mill News* 21.1, 5-7.

Hockensmith, C. D., 1993b. Study of American Millstone Quarries. Part 2, *Old Mill News* 21.2, 4-6.

Hockensmith, C. D., 1993c. Millstone Quarrying in the Eastern United States. A Preliminary Overview, *Ohio Valley Historical Archaeology* 7 and 8, 83-89.

Hockensmith, C. D., 1994. *The Pilot Knob Millstone Quarry. A Self Guided Trail*. Pilot Knob State Nature Preserve, Powell County, Kentucky. Ten page booklet published by the Kentucky Nature Preserves Commission and the Kentucky Heritage Council. Frankfort.

Hockensmith, C. D., 1999. The Millstone Industry in Southwest Virginia. In Charles D. Hockensmith (ed.), *Millstone Manufacture in Virginia. Interviews With the Last Two Brush Mountain Millstone Makers*, Society for the Preservation of Old Mills. Newton, North Carolina, 1-3.

Hockensmith, C. D., 2003a. The Conglomerate Millstone Industry in the Eastern United States. In M. Barboff, F. Sigaut, C. Griffin-Kremer, and R. Kremer (eds.), *Meules À Grains. Actes du Colloque International de La Ferté sous Jouarre 16-19 Mai 2002*, Ibis Press. Paris, France, 197-216.

Hockensmith, C. D., 2003b. The Millstone Industry in Kentucky. Brief Glimpses From Archival Sources. *The Millstone* 2.1, 6-17. Kentucky Old Mill Association. Clay City.

Hockensmith, C. D., 2003c. The Ohio Buhr Millstones. The Flint Ridge and Raccoon Creek Quarries. *Ohio Valley Historical Archaeology* 18, 135-142.

Hockensmith, C. D., 2004a. Additional Information on the Georgia Burr Millstone. The Lafayette Burr Mill Stone Company Years. *Old Mill News* 32.3, 22-25.

Hockensmith, C. D., 2004b. The Millstone Industry of Tennessee. *The Millstone* 3.1, 9-15, Kentucky Old Mill Association. Clay City.

Hockensmith, C. D., 2004c. The Millstone Industry of North Carolina. *Old Mill News* 32.4, 17-24.

Hockensmith, C. D., 2004d. *Early American Documents and References to Millstones. 1628-1829*. Kentucky Old Mill Association. Clay City.

Hockensmith, C. D., 2004e. The Millstone Industry of Missouri. *The Millstone* 3.2, 29-34. Kentucky Old Mill Association. Clay City.

Hockensmith, C. D., 2005. The Millstone Industry of Alabama. *The Mill Monitor*, Special Edition, No. 2, (Winter 2005). Quarterly Newsletter of The International Molinological Society of America, Sterling. Virginia, 1-9.

Hockensmith, C. D., 2006a. The Millstone Quarries of Missouri. *Old Mill News* 34.3, 20-23.

Hockensmith, C. D., 2006b. The Preservation, Ownership, and Interpretation of American Millstone Quarries. In A. Belmont and F. Mangartz (eds.), *Les Meulièrs. Recherche, Protection et Valorisation d'un Patrimoine, Industriel Européen, Antiquité-XXIe siècle*, Die Mühlensteinbrüche. Enforschung, Schutzund Inwertsetzung eines, Europäischen Kulturerbes (Antike 21. Jahuhundert). Actes du Colloque Internation de Grenoble, 22-25/9/2005. Römisch-Germanischen Zentralmuseums, Vulkanpark GmhH. Mainz, Germany, 193-204.

Hockensmith, C. D., 2007a. Ohio Buhr Millstones. The Flint Ridge and Raccoon Creek Quarries. In D. B. Ball and C. D. Hockensmith, *Millstone Studies. Papers on Their Manufacture, Evolution, and Maintenance*, Published by the Symposium on Ohio Valley Urban and Historic Archaeology, Murray, Kentucky and Society for the Preservation of Old Mills. East Meredith, New York, 134-143.

Hockensmith, C. D., 2007b. The Granite and Gneiss Millstone Quarries of the United States. In D. B. Ball and C. D. Hockensmith (eds), *Millstone Studies. Papers on Their Manufacture, Evolution, and Maintenance*. Published by the Symposium on Ohio Valley Urban and Historic Archaeology, Murray, Kentucky and Society for the Preservation of Old Mills. East Meredith, New York, 144-159.

Hockensmith, C. D., 2008a. The French Burr Millstone in Kentucky. Insight From Early Ads. 1792-1890. In C. D. Hockensmith (ed.), *Foreign and Domestic Millstones Used in Kentucky. Papers Examining Archival Records*. Kentucky Old Mill Association. Clay City, 5-38.

Hockensmith, C. D., 2008b. Millstones From Ohio and Pennsylvania Imported into Kentucky. Raccoon Buhrs and Laurel Hill Stones. In C. D. Hockensmith (ed.), *Foreign and Domestic Millstones Used in Kentucky. Papers Examining Archival Records*, Kentucky Old Mill Association. Clay City, 39-54.

Hockensmith, C. D., 2008c. Early References to Red River Millstones in Kentucky. Newspaper Ads and Other Sources. 1803-1839. In C. D. Hockensmith (ed.), *Foreign and Domestic Millstones Used in Kentucky. Papers Examining Archival Records*. Kentucky Old Mill Association. Clay City, 55-62.

Hockensmith, C. D., 2008d. American Millstones Similar to the French Burr. 19th Century Attempts to Find Substitutes. Draft Manuscript.

Hockensmith, C. D., 2009a. *The Millstone Industry. A Summary of Research on Quarries and Producers in the United States, Europe, and Elsewhere*. McFarland Publishing. Jefferson, North Carolina and London.

Hockensmith, C. D., 2009b. *The Millstone Quarries of Powell County, Kentucky*. Contributions to Southern Appalachian Studies 24. McFarland Publishing. Jefferson, North Carolina and London.

Hockensmith, C. D., 2010a. The Millstone Industry in Ulster County, New York, 1732-1955. In C. D. Hockensmith (ed.), *The Historic Millstone Industry in New York State With an Emphasis on Ulster County*, Society for the Preservation of Old Mills. Newton, North Carolina. (Draft book).

Hockensmith, C. D., 2010b. Archaeological Observations of Esopus Millstone Quarry Remains at Accord, New York. In C. D. Hockensmith (ed.), *The Historic Millstone Industry in New York State With an Emphasis on Ulster County*, Society for the Preservation of Old Mills. Newton, North Carolina. (Draft book).

Hockensmith, C. D., 2010c. The Millstone Industry in Jefferson and Sullivan Counties, New York. In C. D. Hockensmith (ed.), *The Historic Millstone Industry in New York State With an Emphasis on Ulster County*, Society for the Preservation of Old Mills. Newton, North Carolina. (Draft book).

Hockensmith, C. D., (ed.) 1999. *Millstone Manufacture in Virginia. Interviews with the last two Brush Mountain Millstone Makers*. Society for the Preservation of Old Mills, Newton, North Carolina.

Hockensmith, C. D., (ed.) 2010. *The Historic Millstone Industry in New York State With an Emphasis on Ulster County*. Society for the Preservation of Old Mills. Newton, North Carolina. (Draft book).

Hockensmith, C. D., and Coy, F. E. Jr., 1999. Early Twentieth Century Millstone Manufacture in Southwest Virginia. An Interview With Millstone Makers Robert Houston Surface and W.C. Saville. In C. D. Hockensmith (ed.), *Millstone Manufacture in Virginia. Interviews With the Last Two Brush Mountain Millstone Makers*, Society for the Preservation of Old Mills. Newton, North Carolina, 5-62.

Hockensmith, C. D., and Coy, F. E. Jr., 2010a. Twentieth Century Millstone Manufacture Near Accord, Ulster County, New York. An Interview With Vincent Lawrence And Wallace Lawrence. In C.D. Hockensmith (ed.), *The Historic Millstone Industry in New York State With an Emphasis on Ulster County*, Society for the Preservation of Old Mills. Newton, North Carolina. (Draft).

Hockensmith, C. D., and Coy, F. E., Jr., 2010b. The Esopus Millstone Industry at Accord, Ulster County, New York. An Interview with Lewis Waruch. In C. D. Hockensmith (ed.), *The Historic Millstone Industry in New York State With an Emphasis on Ulster County*, Society for the Preservation of Old Mills. Newton, North Carolina. (Draft book).

Hockensmith, C. D., and Meadows, L. G., 1996. Historic Millstone Quarrying in Powell County, Kentucky. *Ohio Valley Historical Archaeology* 11, 95-104.

Hockensmith, C. D., and Meadows, L. G., 1997. Conglomerate Millstone Quarrying in the Knobs Region of Powell County, Kentucky, *Old Mill News* 25.2,17-20; 25.3,24-26.

Hockensmith, C. D., and Meadows, L. G., 2001. Conglomerate Millstone Quarrying in the Knobs Region of Powell County, Kentucky. In *Mills & Power From the Past*, compiled by L. G. Meadows. Red River Historical Society & Museum. Clay City, Kentucky, 14-20.

Hockensmith, C. D., and. Meadows, L. G., 2006. Red River Millstone Quarries in Lawsuits. Part I. *The Millstone* 5.2, 9-17. Kentucky Old Mill Association. Clay City.

Hockensmith, C. D., and Meadows, L. G., 2007. Red River Millstone Quarries in Lawsuits. Part 2. *The Millstone* 6.1, 31-38. Kentucky Old Mill Association. Clay City.

Hockensmith, C. D., and. Price, J. L., 1999. Conglomerate Millstone Making in Southwest Virginia. An In-Depth Interview With Millstone Maker Robert Houston Surface. In C. D. Hockensmith (ed.), *Millstone Manufacture in Virginia. Interviews With the Last Two Brush Mountain Millstone Makers*. Society for the Preservation of Old Mills. Newton, North Carolina, 63-89.

Howell, C., 1985. Colonial Water Mills in the Wooden Age. In B. Hindle (ed.), *America's Wooden Age. Aspects of its Early Technology*. Sleepy Hollow Press, Terrytown, New York, 120-159.

Howell, C., 1997. Millstones: An Introduction. *Old Mill News*, 25.4, 18-22.

Howell, C., and Kellar, A., 1977. *The Mill at Philipsburg Manor and a Brief History of Milling*. Sleepy Hollow Restorations. Terrytown, New York.

Katz, F. J., 1916. Abrasive Materials. In *Mineral Resources of the United States, 1914*, Part II-Nonmetals. Department of the Interior, U.S. Bureau of Mines, Government Printing Office. Washington, D.C., 549-568.

Katz, F. J., 1926. Abrasive Materials. In *Mineral Resources of the United States, 1923*, Part II-Nonmetals. Department of the Interior, U.S. Bureau of Mines, Government Printing Office. Washington, D.C., 327-337.

Kentucky Gazette 1799. Ad for five pairs of Red River millstones at Cleveland's Landing. *Kentucky Gazette*, June 13, 1799. Lexington, Kentucky.

Kentucky Reporter 1818. Ad for Red River millstones at quarry by Spencer Adams and James Daniel. *Kentucky Reporter*, April 8, 1818. Lexington, Kentucky.

Ladoo, R. B., 1925. *Non-Metallic Minerals. Occurrences-Preparation-Utilization.* McGraw-Hill Book Company. New York.

Manufacturer and Builder 1876. Burr Stones. *Manufacturer and Builder* 8.4, 82. New York.

Mather, W. W., 1838. *First Annual Report of the Geological Survey of Ohio.* Samuel Medary. Columbus, Ohio.

McDowell, R. C. 1978. *Geologic Map of the Levee Quadrangle, East-Central Kentucky.* U.S. Geological Survey. Reston, Virginia.

Michael, R. L., 1983. National Register of Historic Places Nomination form for Game Lands 51 Millstone Quarry, Fayette County, Pennsylvania. Prepared by California University. California, Pennsylvania.

Newland, D. H., 1907. Report of Operations and Productions During 1906. In *The Mining and Quarry Industry of New York State 1906.* New York State Museum Bulletin No. 305. Albany, New York.

North Carolina Millstone Co. n.d.. North Carolina Millstone Co. brochure describing the 'Moore County Grit' stone and price lists for mills and millstones, Parkewood, North Carolina. University of North Carolina Library. Chapel Hill.

Peters, J. T., and Carden, H. B., 1926. *History of Fayette County, West Virginia.* Jarrett Printing Company. Charlestown, West Virginia.

Read, M. C., 1883. Berea Grit. In *Mineral Resources of the United States for 1882*, U.S. Bureau of Mines. Washington, D.C.

Reinemund, J. A., 1955. *Geology of the Deep River Coal Field, North Carolina.* Geological Professional Paper No. 246, U.S. Geological Survey. Washington, D.C.

Sass, J., 1982. A Virginia Millstone Quarry. *Old Mill News*, 10.4, 6-7.

Sass, J., 1984. *The Versatile Millstone. Workhorse of Many Industries.* Society for the Preservation of Old Mills. Knoxville.

The Argus of Western America 1821. Ad for Charles Colyer's millstone quarry in Rockcastle County, Kentucky. *The Argus of Western America*, February 27, 1821. Frankfort, Kentucky.

The Argus of Western America 1821a. Ad for Miller, Railsback & Miller's millstone quarry in Franklin County, Kentucky. *The Argus of Western America*, August 9, 1821. Frankfort, Kentucky.

The Argus of Western America 1821b. Ad for Jeremiah Buckley's millstone quarry in Franklin County, Kentucky. *The Argus of Western America*, November 8, 1821. Frankfort, Kentucky.

The Eagle 1825. Ad for Laurel-Hill Millstones from Pennsylvania. *The Eagle*, April 13, 1825. Maysville, Kentucky.

Webb, W. S., 1933. The Millstone as an Antique. *Kentucky School Journal*, March, 30-34.

Webb, W. S., 1935. Old Millstones of Kentucky. *The Filson Club History Quarterly*, 9.4, 209-221, Louisville.

La fabrication d'une meule en emeri et ciment magnésien

Jean Paul Duc

1. Pourquoi un mélange émeri – silex

Nous sommes en 1911, la fabrication de meules de moulin a subi depuis un certain temps des attaques de toute part avec l'évolution des moulins en minoterie, le remplacement des meules par les cylindres, la quantité à produire, le coût de production des meules relativement élevé à la vente, et surtout aller vers une facilité du travail et aussi une certaine préservation de la santé humaine. Il faut donc essayer de trouver d'autres procédés de fabrication, voire une fabrication aussi avec plus de qualité, c'est pour cette raison que ce nouveau procédé va voir le jour en cette année 1911.

L'idée d'utiliser ces deux produits, l'émeri et la pierre meulière voit le jour dans une usine de cristallerie, celle de Pantin, où l'on utilise des petites meules en émeri pour polir le verre, cette pierre l'émeri est particulièrement dure. Ainsi Charles Thill, crée sa première société, tout d'abord à Courbevoie : meules et pierres à émeri et produits à polir – Thill et Friren, 62 rue Victor Hugo. Cette société n'a pas au départ, l'ambition de rivaliser pour les meules de moulin, mais plutôt pour les meules à polir.

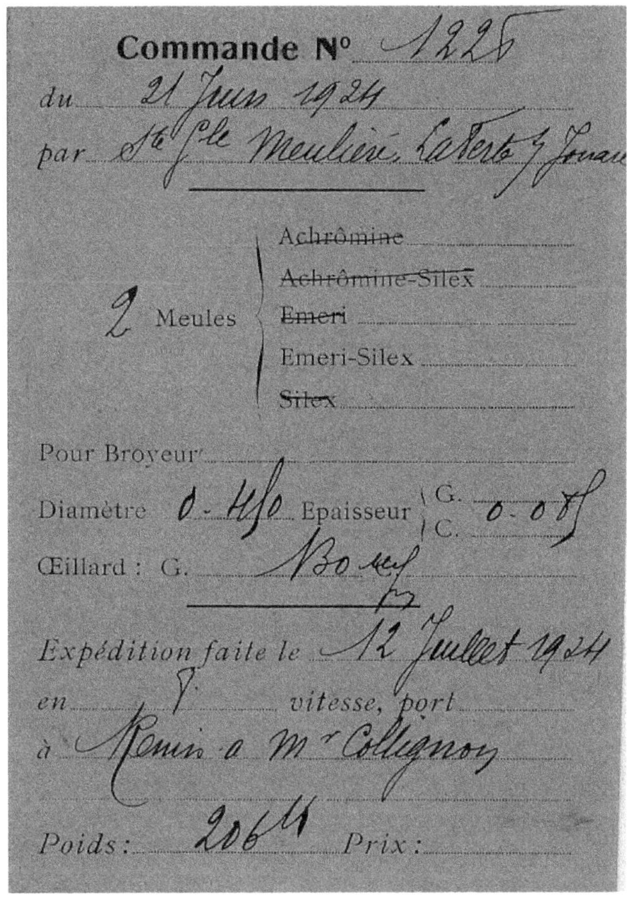
Commande N° 1225
du 21 Juin 1924
par Sté Gle Meulière, La Ferté s/ Jouarre

2 Meules
~~Achrômine~~
~~Achrômine-Silex~~
~~Emeri~~
Emeri-Silex
~~Silex~~

Pour Broyeur
Diamètre 0.40 Epaisseur G. C. 0.08
Œillard : G. Boué
Expédition faite le 12 Juillet 1924
en P vitesse, port
à Remis à Mr Collignon
Poids : 206 k Prix :

Figure 1. Fiche de commande pour la SGM (recto)

2. L'historique

L'occasion va se présenter, Charles Thill se trouve avoir sur place, à Hanches, commune voisine d'Epernon, de la famille au Château de Savonnière. Seconde circonstance, monsieur Olivier d'Argenteuil se trouve à avoir des terres à vendre sur Epernon et sa région, Monsieur Friren, co-directeur de la société citée ci-dessus est comme par hasard d'Argenteuil, hasard! De plus, les gisements d'Epernon sont de plus en plus réputés pour la qualité de leurs pierres dite d'ailleurs 'Pierres d'Epernon'. Les événements vont se précipiter, le 27 mars alors que la nouvelle société n'est pas encore créée, Messieurs Olivier, Thill et Friren s'engage à titre d'option à acheter les terres d'Epernon.

Monsieur Olivier s'engage à titre d'option avec messieurs Thill et Friren pendant une période de cinq mois à la cession d'un terrain d'une contenance d'environ un hectare (sans engagement de la part du propriétaire) situé à Epernon, lieu dit Cady, entre la route de Nogent le roi et la route de la Boissière moyennant le prix de 3000 francs, frais à la charge de l'acquéreur. En cas de réalisation, il est bien entendu que les acquéreurs devront s'entendre avec le locataire, monsieur Goumand. Lors de la réalisation, la valeur de l'acquêt en sera versé comptant entre les mains de maître Lemercier, notaire à Hermeray, rédacteur de l'acte (Actes notariés d'achat).

Reste donc à créer l'entreprise pour concrétiser l'ensemble, ce qui se fait et cette dernière se donne le nom provisoire de 'Société générale des meules artificielles en silex d'Epernon et de la Ferté sous Jouarre' et établi son siège à Paris, 40 rue des mathurins. La délibération et création de la société a lieu le 17 août 1911. On ne retrouve plus Charles Thill dès lors mais uniquement Friren, des sparnoniens

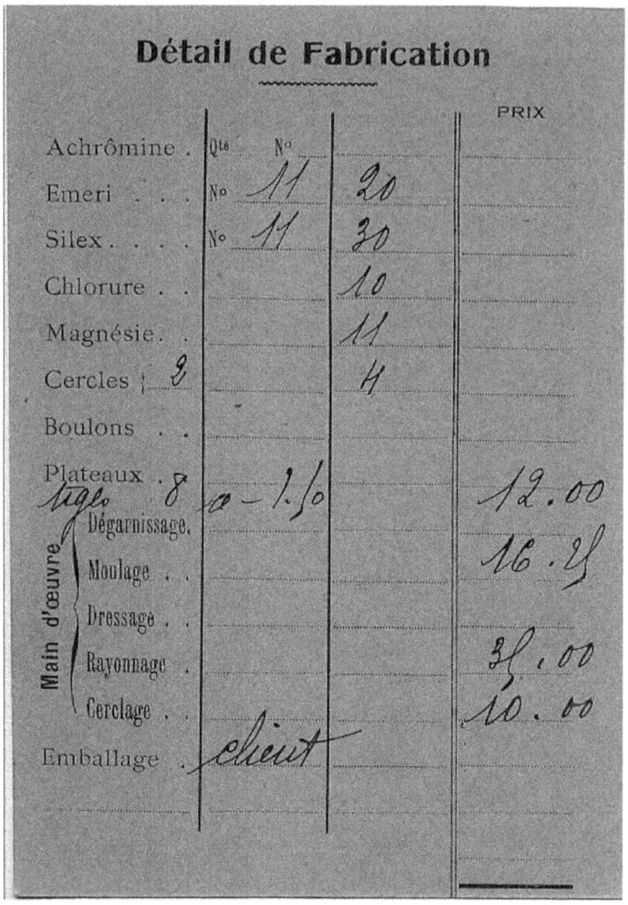

Détail de Fabrication

	Qté	N°		PRIX
Achrômine				
Emeri	N°	11	20	
Silex	N°	11	30	
Chlorure			10	
Magnésie			11	
Cercles 8			4	
Boulons				
Plateaux 8 / tiges 8	o-1/o			12.00
Main d'œuvre: Dégarnissage				16.25
Moulage				
Dressage				
Rayonnage				35.00
Cerclage				10.00
Emballage	client			

Figure 2. Fiche de commande pour 2 meules de 45 cm (verso)

(habitant d'Epernon) qui y mettent quelques finances dont un certain Ledru, conseiller municipal. Immédiatement après l'assemblée générale constitutive de la société, le conseil d'administration s'est réuni à l'effet de nommer son président et son secrétaire et de prendre les premières mesures nécessaires : monsieur de Combary est nommé le premier président du conseil et ce pour un an et monsieur Renant est nommé secrétaire. Dès le lendemain, après une nouvelle délibération, le nom de la société est modifiée en Société générale des meules artificielles et agglomérés, le siège social est transporté au 9 rue du Louvre à Paris. Immédiatement, l'achat des terres est programmé, et se concrétise le 3 octobre 1911.

Il est maintenant plus que nécessaire de bâtir une usine, celle-ci se fera dès lors, immédiatement et sera agrandie dans un second temps. L'entreprise aura l'occasion de racheter sur le lieu dit Vinerville, commune de Hanches, un ancien moulin en cessation d'activité, cela permettra avec le flux de l'eau de pouvoir broyer la pierre meulière. La superficie de l'entreprise sera au final de 2 703 m².

Lors de sa construction, en 1911, les pierres meulières étaient extraites des terres environnantes, toutes les pierres ramassées, gros ou petits calibres faisaient affaire. Nous ne sommes plus dans cette entreprise sur une fabrication de meules à carreaux ou de meules monolithe, nous sommes sur une meule reconstituée à base d'émeri, de silex et de ciment magnésien.

3. La fabrication en usine

C'est en visitant cette entreprise que nous avons la possibilité de découvrir la fabrication.

3.1 Les matières premières

Commençons tout d'abord par la visite des réserves de matières premières. Il n'y a là que d'énormes tas de cailloux de diverses couleurs; de couleur argileuse, pierre meulière caverneuse, ceux-ci à partir des années 1930 seront remplacées par des cailloux de couleur gris foncé avec des reflets bleuâtre, des galets de mer venant de Cayeux, (société Delarue) plage de la Baie de Somme – ceux-ci plus tard, après guerre ne viendront plus sous forme de galets mais arriveront broyés ; d'autres tas de cailloux plus loin, de couleur gris foncé avec des reflets couleur fer, cailloux (terme impropre) qui laisseraient indifférent le visiteur non averti. C'est pourtant de l'émeri, matière assez rare dans le monde. L'utilisation des galets a été impactée dans la société dans les années 1935/1936, pour sa consistance siliceuse – sous l'effet de la corrosion par l'air marin, les rognons de silex sont libérés de leurs matrices crayeuses et roulés par la houle devenant galets.

Figure 3. Moulin broyeur Teisset Rose et Brault utilisant des meules de l'Abrasienne

Figure 4. Usine de Vinarville dans un ancien moulin à eau, on y broyait la pierre meulière

Au fur et à mesure, il y aura des évolutions sur le vrac en ce qui concerne l'émeri. L'émeri arrivait tout d'abord sous formes de roche en vrac de Naxos (Grèce) par bateau jusqu'à Anvers et de là par le train. Arrivé à la gare, les blocs étaient chargés par des entreprises de transport locales, traversaient la ville (au grand dam des habitants qui ainsi profitaient de la poussière – puis étaient déchargés par les ouvriers de l'entreprise et entreposés dans l'entrepôt. Plus tard, l'émeri arrivait en sacs triés suivant la granulométrie. Afin de protéger les sacs de l'humidité du sol, un parquet de sapin avait été dressé (Expertise Galtier de l'entreprise de 1924).

La première opération, une fois déchargé des camions était d'amener l'émeri près des tapis mais auparavant, les plus gros blocs d'émeri devaient être cassés au couperet par un ouvrier le couperet est une masse de forme ovale et d'un poids pouvant atteindre les trente kilo.

3. 2 Le triage

Du tapis, les blocs d'émeri sont transportés dans la salle des concasseurs dans des wagonnets Decauville de type plate forme, sur une voie d'écartement 0,50 avec aiguillage. Il règne dans cet atelier un bruit assourdissant. Le minerai passe sur des tapis jusqu'à des cylindres d'acier dits 'à mâchoires' de type 'Dalbouze' qui le broient. Après avoir circulé dans des bluteries dite de forme hexagonale pour y être triées suivant la grosseur, il repasse dans d'autres concasseurs de type 'Dalbouze'. A part le chargement du premier concasseur, tout se passe automatiquement, la matière circulant sur des courroies ou des tapis roulants.

On peut comparer cette entreprise à un moulin, très spécial, qui traite une des matières les plus dures au monde.
Après concassage, broyage, le tri se fait ensuite de manière magnétique par déferrage, vient ensuite le calibrage, l'émeri termine son passage dans l'usine dans des petits sacs de toile numérotés. Ce sont ces sacs qui seront expédiés aux usines fabriquant la toile émeri ou celles ayant besoin d'assez gros grains d'émeri pour polir, usiner différents métaux.

Une partie importante de la fabrication sera transportée à la deuxième usine, celle de Vinerville pour y subir un autre traitement.

3.3 La fabrication des meules pour moulins

La seconde activité de l'usine de Cady est la fabrication des meules de moulin. Monsieur Blondel (le directeur de l'époque 1930) un des grands spécialistes français en la matière, ne tient pas et nous le comprenons à nous révéler les secrets de son métier. Nous apprenons tout de même que les meules sont fabriquées avec un mélange d'émeri, de silex et d'un ciment magnésien , le tout est mis dans un pétrin mécanique de type Jouclard avec une cuve de fonte de diamètre 1,05 mètre mis en forme dans les moules, le tout sera compressé par un système de pilon pneumatique. Pour les moules utilisés, nous pourrons utiliser des plateaux de 1,60 à 2,10 mètres de diamètre, des ceintures, oeillards et frettes. 'l'Abrasienne' fabrique des

Figure 5. Usine de la rue de Cady – Epernon, c'est là qu'était broyé l'émeri et que se fabriquait les meules

meules dont le diamètre va de trente Centimètres Cubes à près de trois mètres.

Fabrication d'une meule de 1,40 de diamètre par 0,28 de hauteur, il sera ajusté lors du moulage, un cylindre de bronze pour l'oeillard de 0,32 de diamètre, sur la meule du dessus sera collé une anille de 0,52 de longueur, pour 0,08 de large et d'une épaisseur de 0,19 d'épaisseur.
Il sera coulé 140 Kg d'émeri 12, pour obtenir cette granulométrie, il sera utilisé des tamis avec 6 fils au pouce ; du silex n° 11 – 13 utilisation de tamis pour le 11 de 8 fils et pour le 13 de 4 fils et pour un poids à 50/50 de 534 kilo. Le ciment magnésien sera quant à lui composé de 125 kilo de chlorure, 131 kilo de magnésie. Le tout fini donnera un poids à l'expédition de 930 kilo auquel sera rajouté un cerclage de 17 kilo.
Le prix : outre un emballage de 40 francs par meule, la fabrication a un coût de 51 francs et un coût de rayonnage de 82 francs, il est à rajouter un coût pour une main de levée dans ce cas précis. Nous nous sommes amusés à essayer de ramener un coût de fabrication de meules en 1921 à celui d'un coût de 2010. A ce jour, un plateau de fonte reviendrait à 82 euros, la fabrication à 10 ; le rayonnage à 4 ; un cerclage à 4 et l'emballage à 10 soit à un coût de 110 euros. Nous n'avons tenu compte que d'un coût d'indice car il faudrait voir les augmentations des coûts de la matière première, ceci n'est que par rapport à un coût d'époque.

'L'Abrasienne' répond actuellement non seulement aux besoins du marché national, mais exporte aussi beaucoup hors des frontières et la mise en route du marché commun européen ne peut que donner un nouvel essor à cette usine.

3.4 Les travaux annexes

Notre visite se termine par les ateliers d'électricité (l'usine comporte de nombreux moteurs électriques), de mécanique et de menuiserie. Un tour est en train de rectifier un cylindre d'une broyeuse. Un peu plus loin des ouvriers réparent des trémies de calibrages en toile métallique. Les réparations des appareils et leur entretien sont le principal travail des ouvriers employés dans ce moulin à émeri. On va ainsi retrouver par rapport aux fiches de commande, les divers actes de réparations. Agrandissement de l'oeillard, dégarnissage des plateaux, cerclage ou décerclage de meules usagées, création et transformation en oeillard conique pour une 'courante', rayonnage à la main telles sont les diverses opérations annexes retrouvées sur un cahier de chantier.

Mois par mois, jour par jour, le contremaître notait sur son cahier de bord, le nombre de meules terminées, le nombre de meules en chantier, le tout avec le numéro de la commande, le nombre restant à faire, les diverses opérations autres que la fabrication pure et simple. Sur un second cahier, nous avons l'emploi du temps sur deux ou trois ans de chacun des ouvriers, jour par jour et heure par heure et nous savons ainsi que la fabrication par elle-même n'est pas le plus important de l'entreprise.

On peut ainsi noter que le chargement par un de ces ouvriers de 3 wagons de meules va mettre 8 heures 30 ; et qu'il faudra 9 heures pour un chargement de 20 tonnes d'émeri en sacs par deux ouvriers ainsi les travaux de chargements et déchargements des silex, émeris, galets de mer, pierres à concasser, à amener vers les concasseurs seront une des plus importantes parties du travail humain, le reste (tamisage, broyage.. se faisant mécaniquement); le travail consistera pour les autres, à un travail de nettoyage des ateliers pour l'accumulation des poussières tout en sachant qu'il y avait un système d'aspiration, mais sûrement peu développé, nettoyage des fosses, mais également des bureaux, des sanitaires, nettoyage des bluteries, nettoyage des sacs vides et triages des sacs. Il faudra également rajouter tous les jours un minimum de deux heures d'un ouvrier pour le chargement du gazogène pour le fonctionnement des machines, l'entretien du camion. Et bien sur de la production qui peut aller pour un ou deux ouvriers de 3 à 4 heures maximum. D'après les cahiers de productions, celle-ci peut aller de 1 à 2 par jour jusqu'à 18 à 19 par jour en tant que démoulage pour des commandes pouvant être de 200 à 300 pour les sociétés les plus importantes comme Faucheux, Lafon, Teisset Rose et Brault mais plus surprenant pour la SGM.

Les clients : essentiellement des agriculteurs ayant acheté de petits moulins pour production de farine animale comme les marques ci-dessus citées. Mais également toutes les sociétés de matériels agricoles où là les commandes sont très importantes : UCAM, Sambron (plus particulièrement pour le cidre), Tripette et Renault, Bellanger de Dreux ou Tarré Dautin d'Avallon.

4. La seconde entreprise, le moulin de Vinerville

Là, changement de décor et d'atmosphère. Autant à la première usine, c'est le bruit et la poussière qui y règnent en maîtres, ici, c'est le silence et l'eau qui joue un grand rôle dans la fabrication.

Nous entrons dans un vaste hall où sont entreposés des centaines de sacs contenant dans l'émeri en grains provenant de la première usine. Sur presque toute la longueur du bâtiment, une rangées de grands tonneaux en fer, munis de plusieurs bondes, dominent sur un plancher construit dur un bras de dérivation de la rivière ' la Drouette' qui passe sous l'usine. Au fond, une espèce de four à coke. Dans une pièce adjacente fonctionne un broyeur calibreur.

L'usine de Vinarville fabrique de la poudre d'émeri par lavage, l'opération se pratique en plusieurs temps. Dans de grands tonneaux, on met une certaine quantité de grains d'émeri assez fins. On délaye ces grains dans une certaine quantité d'eau. On laisse cette eau se reposer environ deux heures. On renouvelle l'eau. On délaye à nouveau le dépôt et, après décantation, on enlève la première bonde, la plus haute du tonneau et on laisse couler le résidu dans des petits bacs.
Nouveau délayage, nouvelle décantation. On fait sauter la seconde bonde. Même opération pour la troisième et dernière bonde et on obtient de la poudre d'émeri d'une très grande finesse et de trois grains différents mais chargés d'humidité.

Les bacs sont ensuite placés dans un four pendant deux heures et demie pour le séchage et, quand cette opération est terminée, le produit est prêt pour l'expédition aux utilisateurs.

Quand vous vous servirez d'une toile émeri pour nettoyer ou dérouiller une pièce de métal ; quand vous utiliserez une pâte à nettoyer les couteaux ; quand vous verrez votre garagiste employer une potée d'émeri pour roder les soupapes de votre voiture et ceci n'est qu'un vaste aperçu de l'utilisation de l'émeri, vous pourrez penser que cela ne peut se faire que grâce à la fabrication des usines de 'l'Abrasienne' d'Epernon.

La journée des ouvriers : triage des sacs vides, broyage, approche de l'émeri aux concasseurs, transfert du charbon pour le chauffage et les génératrices, production, entretien des véhicules.

Bibliographie

Actes notariés d'achat – Archives Conservatoire des meules et pavés.

Expertise Galtier de l'entreprise de 1924 (archives Conservatoire des meules et pavés du bassin d'Epernon) – article du journal Echo républicain (AD PER 40) – film tourné lors de l'inauguration de l'allée de l'Abrasienne à Epernon avec le dernier directeur'Epernon –

La Fouille du Moulin à vent de Roissy-en-France (Val-d'Oise), France

Jean-Yves Dufour

Le moulin de Roissy-en-France était implanté dans la commune de même nom, localisée à la limite orientale du Val-d'Oise. Géographiquement, le village est au centre du Pays de France, vaste ensemble de plaines situé immédiatement au nord et au nord-est de Paris. Les données de cet article sont issues de la fouille préventive, menée 500 m à l'ouest du village de Roissy-en-France (Fig. 1). Sur trois petites parcelles était situé le moulin à vent de Roissy, dont la plus grande part a pu être fouillée.

Le moulin à vent est à la fois une machine et un lieu important sur le plan social et culturel, comme sur le plan économique et technique. Alors qu'on en dénombrait

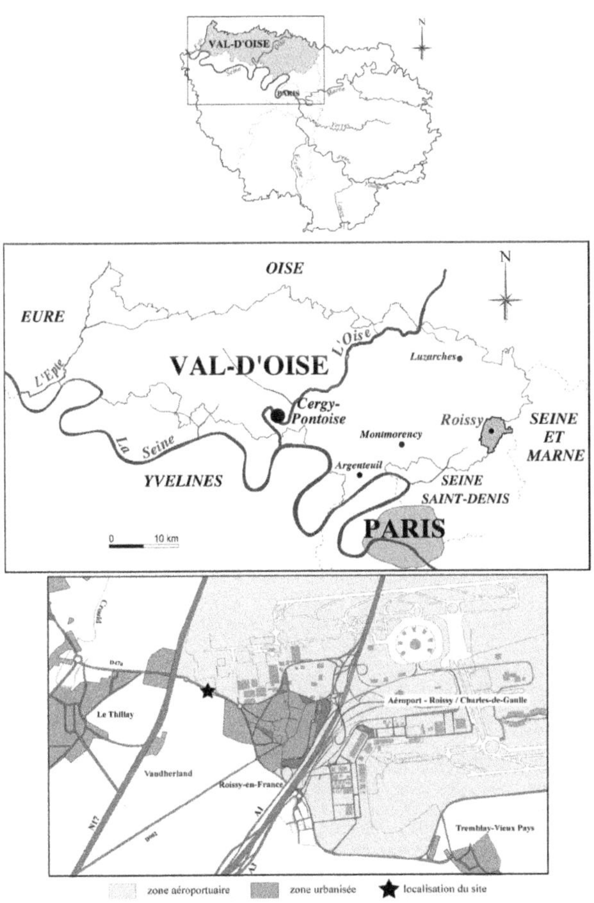

Figure 1. Localisation du site. Site location. (Jean-Yves Dufour et Mehdi Belarbi, Inrap)

15887 au recensement de 1809, seuls 36 moulins à pivot subsistent dans toute la France en 1997. Le potentiel archéologique et sociologique est donc énorme.

1 Les vestiges du milieu XVIe s

Le moulin à vent est apparu vers 1180 de part et d'autre des côtes de la Manche (Rivals 1973, 141; Holt 1988, 20), sur les côtes normandes et en Angleterre, et s'est répandu dans le bassin Parisien et les Flandres au cours du XIIIe s., c'est-à-dire au début de la lente construction du pouvoir royal. Le sol limoneux loessique de la plaine de France et la proximité du grand marché parisien sont propices au développement d'une économie céréalière poussée. Le grain dominait l'économie de l'Ancien Régime de tout le royaume. Parce que le pain est confectionné à partir de farines, le moulin est l'une des principales institutions économiques de l'Ancien régime, au même titre que l'Eglise ou la taverne. Parce que des dérèglements météorologiques (inondations) peuvent interrompre l'activité des moulins, des crises sont possibles en dépit d'une abondance de grains (Kaplan 1988, ch. 10). Le commerce de la farine supplante celui du blé à l'époque moderne. Le moulin est donc au cœur de l'activité agricole, alimentaire, commerciale et économique. Un tel contexte est éminemment favorable à un rôle fort du meunier dans la société traditionnelle.

La fouille de Roissy permet pour la première fois de confronter le moulin et la maison du meunier (Fig. 2), à un ensemble de données archivistiques, ethnographiques et agricoles. C'est à notre connaissance une opportunité unique en France.

Les vestiges les plus anciens sont ceux du premier moulin et de la butte qui le supportait. A Roissy, le moulin est bien distinct des autres habitations, positionné 'avant le village' sur les couloirs des vents dominants, à proximité d'une route passante.

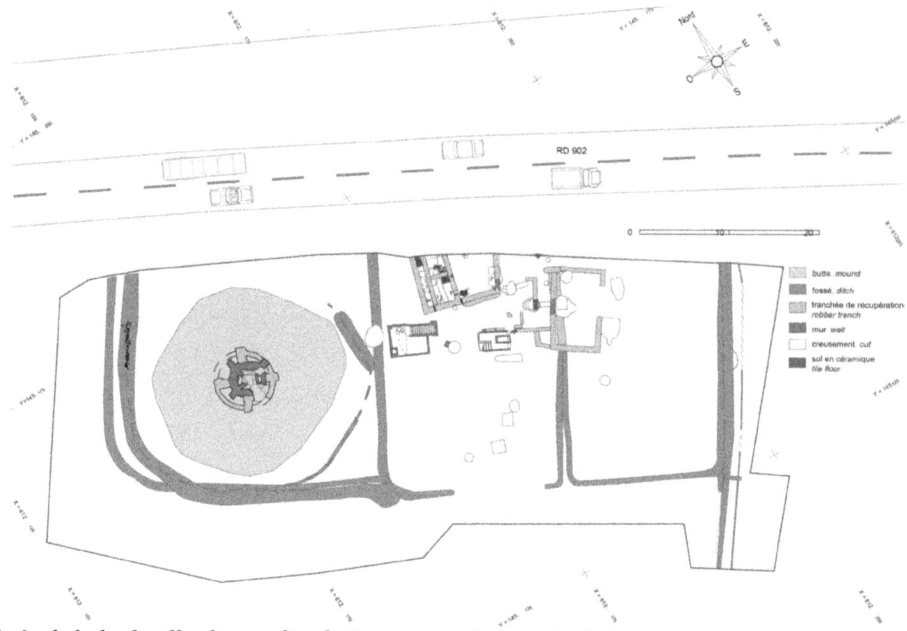

Figure 2. Plan général de la fouille du moulin de Roissy-en-France (Val-d'Oise, France). Main excavation plan. (Jean-Yves Dufour et Mehdi Belarbi, Inrap)

1.2.1 Description de la butte

Le moulin de Roissy est implanté au sommet d'un léger coteau dominant la plaine. Une petite butte bien visible au décapage rehaussait encore un peu le moulin au dessus des champs environnants. La butte (U.S. 25) est globalement circulaire, avec un diamètre entre 19 et 21 m, soit une soixantaine de pieds à la base. Après décapage, cette butte (U.S. 25) est épaisse de 50 cm au pied du moulin. L'épaisseur est naturellement moindre en périphérie de butte. La butte est composée d'un limon clair légèrement grisé, qui tranche assez nettement sur les limons bruns observés sur le reste du site. Seuls deux tessons résiduels du XIIIe s. et deux tessons du XVI e s. ont été trouvés dans la butte du moulin.

1.2.2 Description des fondations du premier moulin

Deux moulins successifs ont été érigés sur la butte (Fig. 3). Les maçonneries n'ayant pas les mêmes axes et diamètres, la stratigraphie des deux moulins était facilement lisible dès la phase terrain. Le premier moulin (Fig. 4) est daté du milieu XVIe s. par les archives. Son plan est connu par les tranchées de récupération. Quatre tranchées (F. 40, 42, 44 et 46) conservées sur une longueur de 1,6 à 1,8 m, sont disposées en étoile autour du centre du moulin. L'intense remaniement de la partie centrale ne permet pas de dire si ces tranchées sont issues de murs continus et se joignant en leur milieu, auquel cas nous aurions eu une croisée de deux murs perpendiculaires longs de 7,2 m, soit 22 pieds. Ces murs étaient larges de 97 cm, soit précisément 3 pieds. Leurs tranchées de récupération sont conservées sur 40 – 50 cm de profondeur. Elles sont comblées de gravats ou le plâtre domine, accompagné de pierres calcaires. Cette croisée de murs est reliée par une maçonnerie annulaire

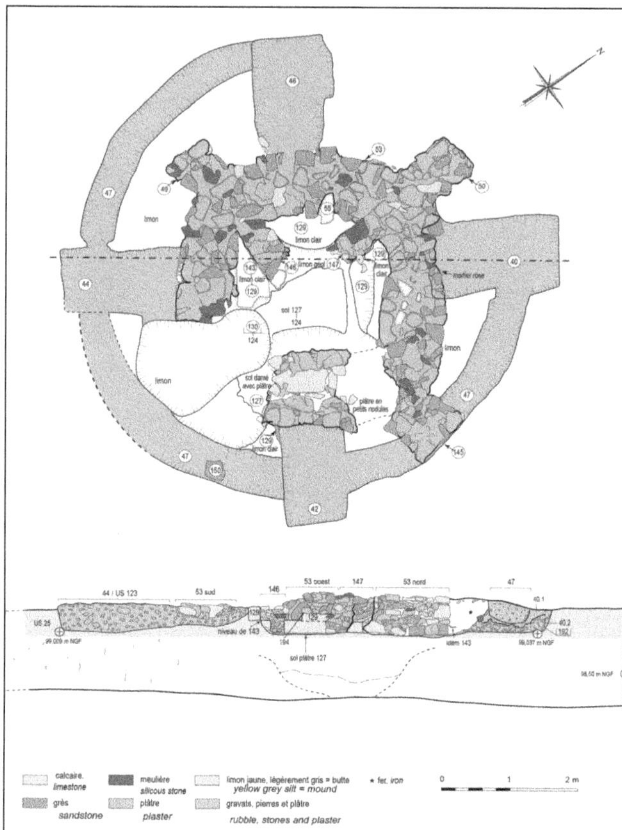

Figure 3. Plan et coupe des fondations des deux moulins successifs de Roissy-en-France. Plan and section of the foundations of the two successive Roissy mills. (Jean-Yves Dufour et Iliana Pasquier, Inrap)

bien visible sur les ¾ d'un cercle. F. 47 est la tranchée de récupération de cette maçonnerie dessinant un anneau de 6,6 m (soit 20 pieds) de diamètre hors d'œuvre. La fondation annulaire (F. 47) ne portait sans doute pas d'élévation conséquente (pas de tour en pierre), mais servait

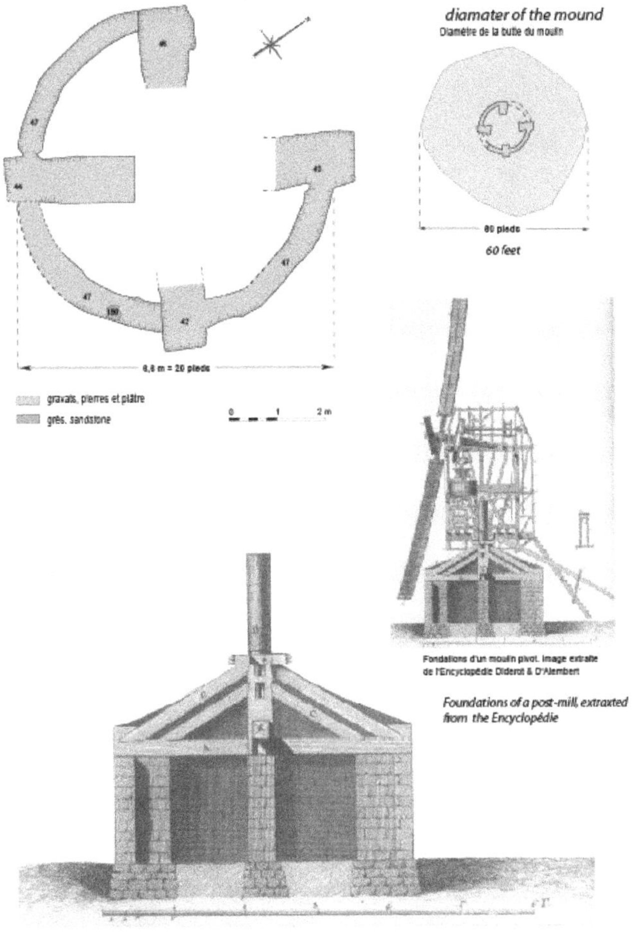

Figure 4. Les fondations du moulin à vent de Roissy-en-France (95), milieu du XVIe siècle. The windmill foundations, mid XVIth century. (Jean-Yves Dufour et Iliana Pasquier, Inrap)

avant tout à contreforter les croisées maçonnées (F. 40, 42, 44 et 46). L'histoire postérieure du site montre qu'elle n'y suffit pas. Un seul tesson du XVIe s. a été trouvé sur la fouille en dehors du secteur du moulin. C'est dire que la maison du meunier n'existe vraisemblablement pas encore au XVIe s.

Le moulin doit pourtant fonctionner aussi longtemps que possible pour être rentabilisé. Le meunier dort sur place. Dans l'ancien comté de Flandres, les comptes des XIVe et XVe siècles ne permettent pas de savoir si le meunier habite le village ou à proximité du moulin. Par ailleurs certains anciens moulins possédaient une couchette destinée au meunier (Coutant 1990, I, 1 29).

1.3 Fonction et comparaisons de la butte

Les mottes supportant les moulins à vent prennent l'apparence de monticules circulaires herbeux. Nombre d'entre eux ont été confondus avec des tertres protohistoriques et fouillés comme tels. Les archives donnent peu de détails sur les tertres des moulins des Flandres. Sur plus de 1000 textes d'archives étudiés par Yves Coutant sur les moulins des XIVe et XVe s., une seule hauteur est donnée ; le moulin de Verrebrock est construit en 1484 sur une butte de 6 pieds de haut. La surface (Coutant 1990, I. 130) aplanie au sommet de la butte aurait une circonférence de 33,45 m.

En Angleterre, les fouilles et les témoignages documentaires indiquent que les buttes élevées pour renforcer la stabilité du moulin ont pu être crées par une longue période d'accumulation, plutôt qu'en une seule fois lors de la construction du moulin (Watts 2002, 106). L'absence de tessons répartis sur une longue durée nous empêche de retenir cette hypothèse à Roissy. Les buttes observées ont un diamètre de 12 à 25 m de (40 à 80 pieds) et 1 à 3 m de hauteur.

Le sol est parfois creusé et damé avant la pose de la croisée et érection de la butte (Jarvis 1981-1982, 14). A Waltham Abbey, Essex, Grande-Bretagne (Medieval Britain in 1971, 211), la motte d'argile a un diamètre de 40 à 50 pieds, et une hauteur de 2 pieds. Un fossé circulaire (diamètre 120 pieds) entoure la motte sur les deux tiers de sa circonférence.

Rappelons les finalités d'une butte support de moulin :

- la butte rend l'accès plus facile aux ailes
- elle permet de drainer le site. Elle est généralement ceinturée d'un vaste fossé d'où est extraite la terre ayant servi à son élévation. Des fossés ont été trouvés à Roissy ; par absence de mobilier du XVIe s. dans leur comblement, il est difficile de dire s'ils sont contemporains de l'érection de la butte du moulin.
- placées en hauteur, les ailes captent des vents moins turbulents
- enfin, le poids de la terre stabilise la croisée (Jarvis 1981-1982, 41).

1. 4 Interprétation et comparaisons des fondations

Le socle, les ailes et le frein sont des éléments techniques neufs, novateurs dans le moulin à vent, inexistants dans le moulin à eau. Le moulin à vent est un parfait exemple d'innovation technique médiévale. Des progrès concernant le socle peuvent éclairer sur l'origine et la diffusion du moulin à vent. On comprend ainsi mieux l'intérêt à porter aux fondations des moulins. Les comptes liés à la construction des moulins sont généralement avares d'informations sur les fondations. Si les fouilles de moulins apportent justement des informations sur les fondations, en revanche elles ne nous apprennent rien sur les mécanismes de la superstructure. Les représentations médiévales (Holt 1988, 137) montrent des moulins à vent construits de façon similaire à ceux des XIVe et XVIe s., aussi sans aucune hésitation invitons nous le lecteur à se reporter aux moulins peints par Bruegel.

Le moulin pivot était le type le plus répandu dans les plaines du Bassin Parisien. Le pivot est le puissant axe

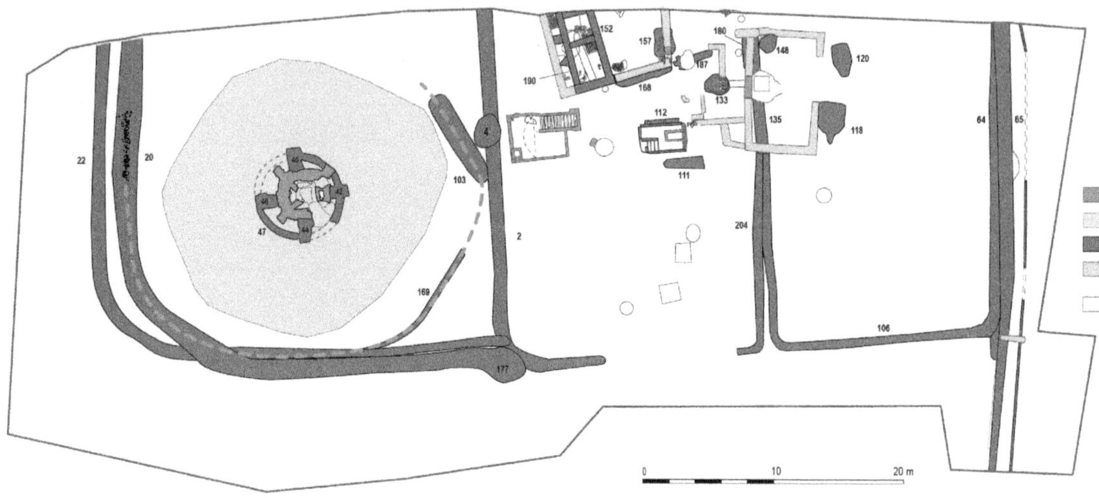

Figure 5. Plan des vestiges du milieu du XVIe siècle au premier quart du XVIIIe siècle. Plan of the structures mid XVIth - first quarter XVIIth century. (Jean-Yves Dufour et Iliana Pasquier, Inrap)

vertical autour duquel s'articule la cabine du moulin. Le socle d'un tel moulin est composé de 4 dés ou piliers maçonnés reliés par une maçonnerie circulaire. Sur ces piliers maçonnés, deux soles en bois se croisent à angle droit. Quatre à huit liens obliques emmortaisés dans ces soles soutiennent le pivot qui est encastré sur la croisure formée par les deux soles. Les maçonneries F. 40, 42, 44 et 46 observées à Roissy correspondent aux piliers ou dés qui préservent les soles de la putréfaction, soutiennent la croisure et rehaussent le moulin. Leur longueur est proche de celle du seul texte des Flandres (Coutant 1990, I, 139) nous indiquant la longueur des blocs soutenant la croisure: 6 pieds de longueur. La maçonnerie circulaire qui relie les dés participe à leur maintien.

A Roissy, si on restitue des soles aussi hautes que sont larges leurs maçonneries supports (les faits 40, 42, 44 et 46), alors la croisée faisait au moins 1 m de hauteur. Si elle était enterrée, on peut alors restituer une butte d'au moins 5 pieds de hauteur.

2 Les vestiges du XVIIe jusqu'au premier quart du XVIIIe s

2.1 Les fossés parcellaires (Fig. 5)

Le moulin, la maison du meunier et ses annexes sont insérées dans trois petites parcelles bien délimitées par des fossés. L'exploitation du moulin à vent de Roissy est composée de 3 petits enclos rectangulaires accolés, perpendiculaires au chemin qui mène de Roissy à Goussainville. Le moulin à vent occupe naturellement la parcelle à l'ouest, la plus exposée aux vents. Deux fossés (F. 22 et 20) et une barrière/palissade ceinturent le moulin. Deux autres fossés (F. 2 et 204) délimitent la zone centrale où était construite la maison du meunier. Les fossés 106 et 64 entourent la parcelle la plus à l'est. Si l'on prolonge ces fossés parcellaires jusqu'à l'ancien chemin (sous la route actuelle), on peut estimer la superficie de ces parcelles à respectivement 910 m2 (pour le moulin), 640 m2 (pour la maison) et 590 m2 (pour les dépendances). La superficie totale de l'habitat affermé au meunier occupe donc environ

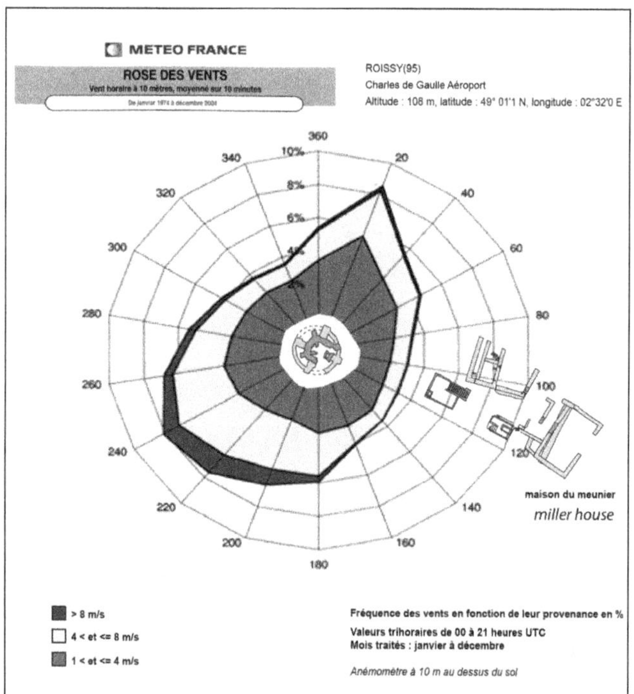

Figure 6. La maison du meunier est positionnée dans l'angle mort des vents. The miller's house is located where it least interferes with the wind circulation. (Jean-Yves Dufour et Iliana Pasquier, Inrap)

2140 m2, soit un demi-arpent (Pour mémoire, l'arpent du roi utilisé à Roissy-en-France équivaut à 4221 m2). De quoi entretenir un potager. Tous ces fossés sont largement remblayés avec des tessons datés du milieu du XVIIe s. Il est cependant probable qu'ils soient en place depuis l'origine du moulin, c'est-à-dire depuis le milieu du XVIe s. En effet, le bail de 1542 indique que le moulin est assis sur une pièce de terre d'environ un demi-arpent.

2.2 La maison du meunier

La maison du meunier est une petite ferme positionnée au bord de la route. Elle contrôle les accès de la clientèle au moulin, construit en retrait du chemin. Comme le veut la logique d'utilisation optimale des forces éoliennes, la maison du meunier est positionnée dans l'angle 'mort' des vents, là où elle gêne le moins les vents. La maison et ses annexes sont regroupées dans un gisement compris entre 80° et 120° à l'est du moulin (Fig. 6).

2.2.1 Tentative de restitution du plan de la maison du meunier au XVIIe s (Fig. 7).

A partir des murs connus, et de ce document d'archives, nous pouvons proposer de restituer le plan de l'habitation du XVIIe s. La longueur du mur pignon ouest est restituée à 6,89 m (21 pieds) hors-œuvre. La paroi sud de l'habitation mesure 10, 7 m (33 pieds) selon les données de terrain. Or, la longueur de la toiture est de 12,67 selon les archives. Si l'on suppose un débord d'un pied (32,5 cm) de la toiture de part et d'autre des murs de la maison, la longueur de la maison peut être estimée à 12 m en son milieu. Cette mesure concorde avec l'idée d'une maison légèrement trapézoïdale, information émanant de la seconde phase d'occupation de la maison du meunier. Divers bâtiments parsèment la cour de la maison du meunier. A 4 m au sud de la maison, une petite construction rectangulaire semi-enterrée est interprétée comme un cellier.

Le mobilier céramique récolté sur le site invite à rechercher un bâtiment destiné à la fabrication de produits laitiers. Sous les aspects d'une cave trop propre, le bâtiment F. 9 observé dans la ferme du meunier de Roissy présente la plupart des critères recommandés par les ouvrages d'agronomie pour le bon fonctionnement d'une laiterie. Cette interprétation répond pleinement à l'analyse du vaisselier faite par l'étude céramologique. Les formes liées aux préparations laitières (tèles à lait, pots de conservation, pots à beurre)

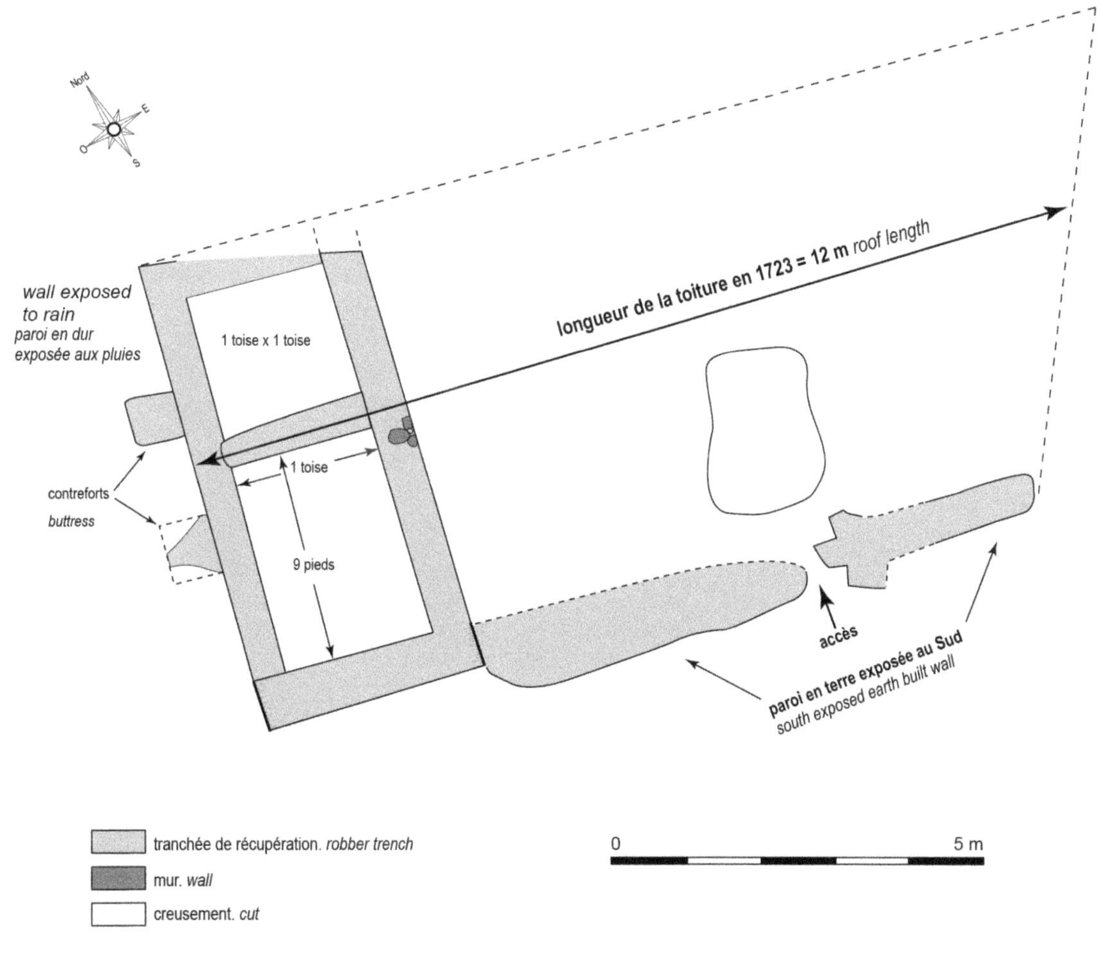

Figure 7. Interprétation des vestiges associés à la maison du meunier - phase du XVIIe siècle. Interpretation of the structures associated with the miller's house - XVIIth century. (Jean-Yves Dufour et Iliana Pasquier, Inrap)

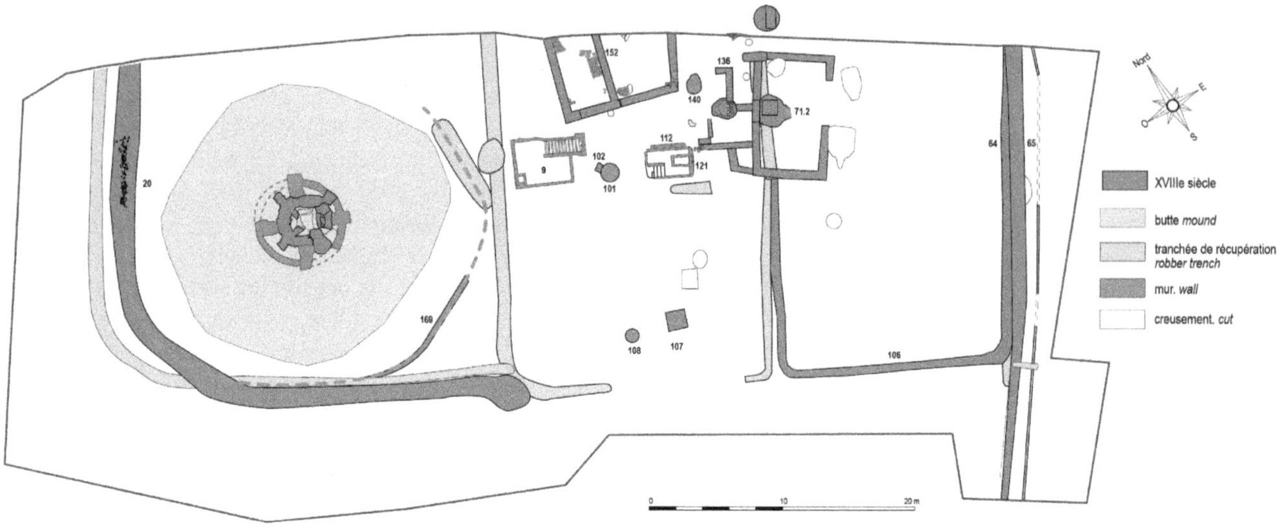

Figure 8. Plan des vestiges du XVIIIe siècle. Plan of the XVIIIth century structures. (Jean-Yves Dufour et Iliana Pasquier, Inrap)

sont présentes à hauteur de 39 % du mobilier en céramique récolté sur le site. Cette interprétation est également confirmée par un document d'archives de 1720. Un puits et un hangar (?) complètent les installations domestiques du meunier. Aux XVIIe et XVIIIe s., le meunier de Roissy trouve le temps de développer une petite activité de laiterie, sans doute menée par la meunière. Cette activité suppose l'élevage de quelques bêtes laitières et un réel travail complémentaire de laboureur.

3 Les vestiges du second quart du XVIIIe jusqu'au milieu du XIXe s (Fig. 8).

Le moulin, la maison du meunier et ses annexes connaissent des modifications bien perçues archéologiquement. Les documents d'archives nous aident à préciser la datation de ces reconstructions.

3.1 Le second moulin (Fig. 9)

3.1.1 Description des fondations du second moulin

Au cours de la période moderne, les fondations du moulin sont entièrement reconstruites au milieu des fondations plus anciennes. Une maçonnerie annulaire de forme carrée relie entre elles quatre piliers plus ou moins bien conservés. La maçonnerie annulaire (F. 53) présente un aspect massif: elle mesure deux toises (respectivement 3,94 et 4 m au milieu des côtés) d'envergure. Cette maçonnerie fourrée large de 3 pieds (0,9 à 1 m) mêle des moellons de meulière, grès et calcaire au sein d'un liant massif de plâtre blanc, solide, avec de grosses mais rares inclusions de gypse. Le parement externe des ces fondations utilise de façon plus marquée les moellons de grès. Le blocage et le parement interne sont moins soignés.

A ces angles, la maçonnerie annulaire est liée à trois murs ou piliers rayonnants depuis le centre des fondations. Ces nouveaux piliers sont axés selon les points cardinaux.

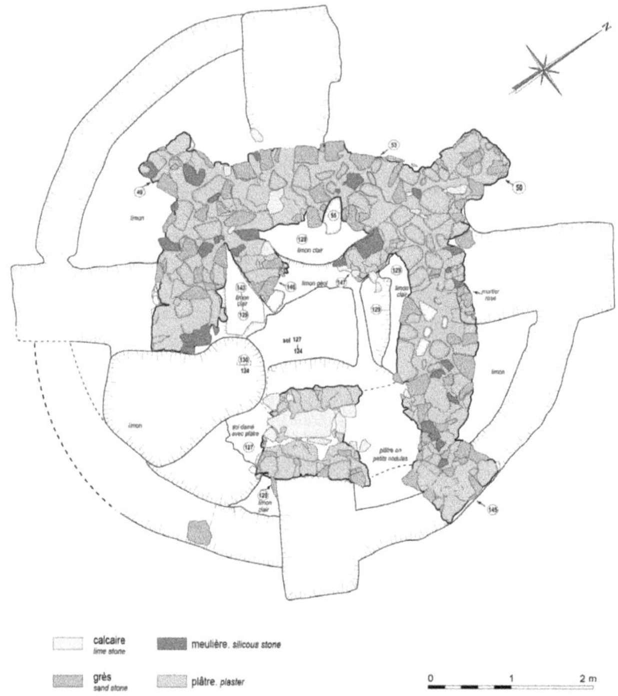

Figure 9. Plan des fondations du second moulin de Roissy-en-France (XVIIe siècle). Plan of the foundations of the second Roissy mill (XVIIth century). (Jean-Yves Dufour et Iliana Pasquier, Inrap)

Pour résumer, nous ne sommes pas en présence d'une construction quadrangulaire (type tour de moulin) contre-fortée à chaque angle, mais bien de quatre piliers rayonnants selon les points cardinaux, ces piliers étant renforcés par une maçonnerie annulaire. Ce principe de construction déjà reconnu pour la première phase est caractéristique des moulins à pivot. Les maçonneries rayonnantes (F. 147/51, F. 146/49, F. 145/ ? et F. 124 par défaut) supportaient les soles de bois entrecroisées, dans lesquelles les liens obliques emmortaisés soutenaient le pivot central.

3.1.2 Datation, comparaison et interprétation des fondations du second moulin

A Roissy, les fondations d'un second moulin sont enquillées dans la base du premier moulin daté du XVIe s. Un important problème technique nécessitant la reconstruction lourde du moulin est donc intervenu à une date que l'étude archivistique nous aide à fixer.

Un marché de charpenterie (Arch. dép. Val-d'Oise, 2 E 14/196), daté du 4 février 1737 et destiné à réparer le moulin à vent de Roissy est conclu entre Louis Fouard, charpentier demeurant à Villiers-le-Bel et le comte de Caraman, seigneur de Roissy, représenté par son concierge (Travail d'archives d'Olivier Bauchet, Inrap). Parmi les nombreux travaux prévus, on note une réfection des fondations. La reconstruction partielle du moulin de Roissy est sans doute intervenue suite à un incident d'origine météorologique. Tous les moulins sont vulnérables face aux caprices météorologiques. Les moulins à vent sont très sensibles aux tempêtes ; sous trop forte pression du vent, un moulin peut être arraché de ses fondations, ou bien riper, enfin ses liens peuvent être désarticulés (Jarvis 1981-1982, figs. 15, 16 et 17).

Un moulin à vent est un investissement si attractif, que le seigneur a intérêt à le maintenir en état de fonctionnement ; propriétaire, le seigneur de Roissy se doit de régler les grosses dépenses. Le lieu choisi étant théoriquement le plus favorable au sein d'un territoire, les moulins sont logiquement reconstruits au même endroit. Avec un recul et une vue d'ensemble des moulins sur le territoire national, l'ethnologue Claude Rivals pose la question (Rivals 1976, 127) de constructions plus carrées dans le Bassin parisien, peut être dues à une moindre richesse en bois d'œuvre que dans les Flandres ? Dans le parc de son château, le seigneur de Roissy dispose cependant régulièrement de bois d'œuvre (Un inventaire exhaustif de 1730 signale " quarente pieds d'arbres coupez propres à faire bois de charpente dans le parc, ") ; Bourdon, le concierge du château, actant le marché de charpenterie de février 1737 pour la réfection du moulin, s'engage d'ailleurs à fournir une partie du bois nécessaire. La maison du meunier connaît également des modifications au cours d'une seconde phase que nous tendons à lier au XVIIIe s.

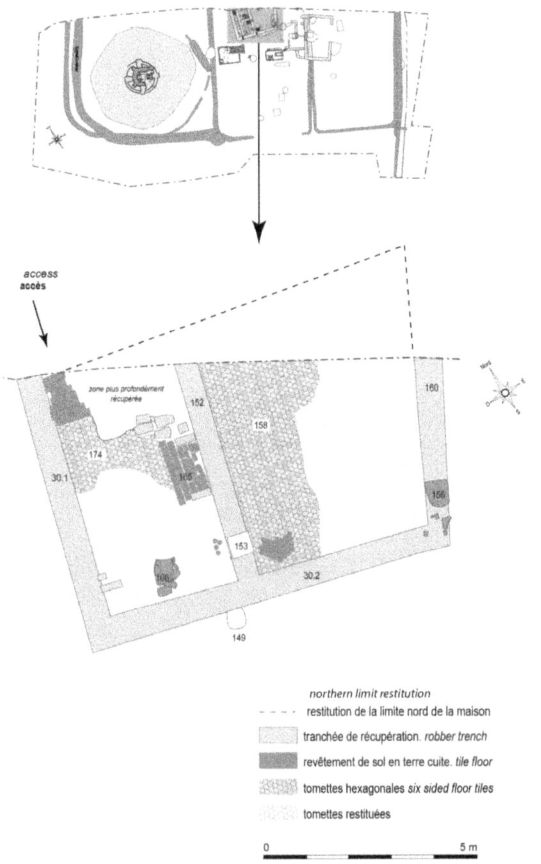

Figure 10. Description et interprétation des vestiges associés à la maison du meunier - phase du XVIIIe et XIXe siècles. Description and interpretation of the structures associated with the miller's house - XVIIIth and XIXth centuries. (Jean-Yves Dufour et Iliana Pasquier, Inrap)

3.2 La maison du meunier (Fig. 10)

Au cours d'une seconde phase de construction, le logis du meunier connaît un léger décalage vers l'ouest. La longueur du logis est donc de 7,9 m (= 4 toises = 24 pieds) dans œuvre sur son côté sud. Le mur oriental forme un angle de 100 ° avec le mur gouttereau sud, aussi la maison du meunier est elle trapézoïdale. Le mur gouttereau nord de la maison reste en dehors de notre fouille.

Le mur de refend (F. 167/152) déjà présent lors de la première phase est toujours utilisé. Dans la moitié est de la maison le sol est revêtu de petites tomettes hexagonales en terre cuite.

A l'ouest du mur de refend, l'espace semble constitué d'une pièce unique large de 3 m. Appuyée au mur de refend, au milieu de la pièce, est construite une cheminée.

Les deux tiers sud de la pièce à feu étaient revêtus (U.S. 174) de tomettes hexagonales de moyennes dimensions (tomettes de 16cm de largeur) posées sur un lit de plâtre.

Le logis du meunier occuperait 50 m2, apparemment divisé entre une pièce à feu et une plus vaste pièce de couchage

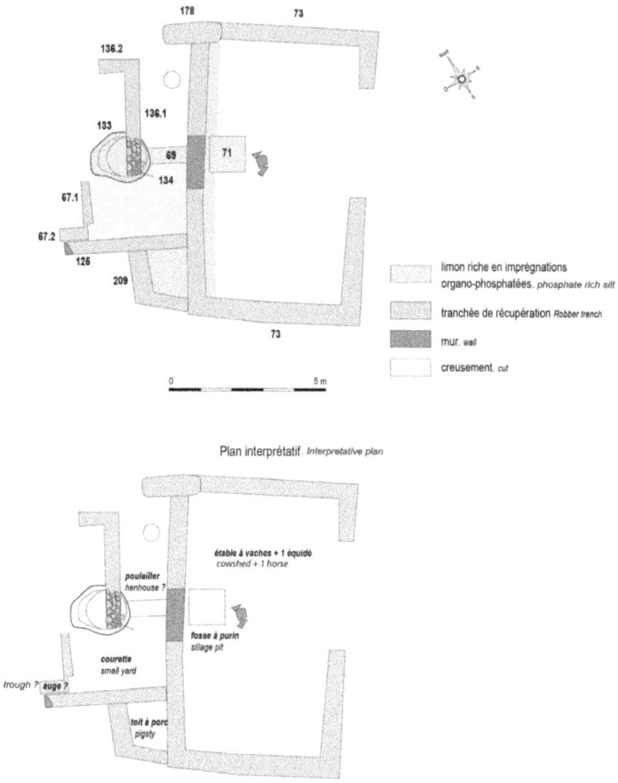

Figure 11. Plan des bâtiments annexes, F.73 et ses appentis. Plan of the ancillary buildings F. 73 and its pentices. (Jean-Yves Dufour et Iliana Pasquier, Inrap)

et sans soute de stockage. Cette surface relativement forte au sol nous rappelle que la maison du meunier est une construction entreprise par le seigneur, et surtout, qu'elle ne dispose pas d'étage. Le stockage même temporaire du grain et de la farine dans l'habitation explique sans doute cet étalement relatif de la maison.

3.3 Le bâtiment d'élevage (Fig. 11)

A l'est de la maison du meunier, un bâtiment est construit au cours d'une seconde phase de l'occupation moderne dans le troisième lopin de terre enclos. Le bâtiment 73 est composé de 5 pans de murs connus par une tranchée de récupération continue. Le bâtiment rectangulaire mesure 9,4 m x 5,5 m hors-œuvre. Il est axé nord-est/sud-ouest. Son ouverture au sud-est est marquée par une interruption du mur sur 3,4 m de largeur (soit 10 pieds). L'espace utile intérieur est de 8,2 x 4,4 m, soit 36 m2.

Dans le bâtiment, une fosse carrée est aménagée au milieu du grand côté ouest. F. 71 est un creusement originellement carré (115 x 115 cm), d'une profondeur conservée de 120 cm. Le fond plat de ce creusement, et ses parois abruptes sont fortement imprégnées de matière organique (limon vert). Les parois montrent également les traces nettes d'un habillage en bois décomposé. Une analyse micromorphologique réalisée sur le comblement inférieur confirme l'hypothèse de latrines.

Tout moulin nécessite des dépendances d'importance variable : logements, écurie, grange, terres cultivables,... La présence de bâtiments d'élevage dans la fermette du moulin de Roissy s'explique aisément ; à partir de 1783 le meunier dispose de 2,12 ha de terres labourables, soit de quoi récolter annuellement 27 à 35 hl de grain (Si nous supposons la pratique de l'assolement triennal, l'emblavement annuel de 1,41 ha et un taux de rendement entre 19 et 25 hl/ha (Moriceau 1994, 455 et suivantes) et entretenir quelque têtes de gros bétail, essentiellement à partir des résidus de la céréaliculture (chaume, paille, résidus de battage et mouture).
La présence d'une laiterie reconnue parmi les vestiges archéologiques et une mention archivistique, sous entend une étable destinée au logement de quelques vaches laitières.

On sait que les moulins disposaient ordinairement deux paires de meules, une pour la farine destinée aux hommes, l'autre pour la nourriture des animaux. Les locataires de moulins banaux sont souvent astreints par leur bail à chercher le grain chez les tenanciers du seigneur (Kaplan 1988, 220). Les archives ne permettent pas de savoir si c'était le cas à Roissy, mais une voiture et des chevaux ou mules sont indispensables au meunier qui veut commercer.

L'étude archéozoologique menée sur le site (Benoît Clavel and Alessio Bandelli, archéozoologues au CRAVO) confirme la présence (successive) d'au moins 10 « bêtes chevalines » au moulin, des petits chevaux et trois ânes. Le cheval du meunier, est comme celui du vigneron, du petit producteur de fruits ou du curé, de petite taille, cavale, canasson ou âne, loin de la force de travail des grands chevaux de labour.

Employé aux champs et surtout pour le transport des grains à moudre et de la farine, l'équidé utilisé par le meunier de Roissy nécessite une étable au moins durant la saison hivernale.

Le bâtiment 73 a logiquement servi à la fois d'étable pour un équidé et 2-3 vaches laitières, de remise à carriole et de grange pour la petite production du meunier. Les liens forts et permanents du meunier avec les boulangers, blatiers et marchands de grain, rendent logiquement inutile la présence d'un grenier à grain.

Le petit espace carré de stabulation bien marquée, défini par les maçonneries (73, 125 et 209) couvre 1,85 m2 (soit 16 pieds en carré). Cette surface correspond pleinement à celle définie par les agronomes anciens pour le logement d'un cochon à l'engrais. Accolé au nord de ce toit à porc, l'espace composé par les murs 67, 125, 73 et le fossé 69 représente une surface de 3,2 x 2,3 m, soit 7,36 m2 (= 70 pieds en carré). Cette surface dont le sol est marqué par les rejets de lisier peut tout à fait s'interpréter comme la courette nécessaire au dégourdissement du porc élevé en stabulation. La présence d'un toit à porc dans la fermette du meunier n'a rien d'exceptionnel. Un moulin produit des déchets céréaliers, le son, qui est principalement composé des débris de l'enveloppe des grains. Le son représente 20 à 25 % du poids du grain (Lachiver 1997, 1551), mais il est encore possible d'en tirer de la farine. Au final, Kaplan (1988, 220) estime les résidus de mouture à 5 %. Dans les faits, le % de déchets dépend beaucoup de la pratique du meunier. Ces déchets favorisent l'élevage 'les cochons du meunier sont vite gras' est un proverbe fréquent (Rivals 2000, 2, 66; Kaplan 1988, 234). En Seine-et-Marne, les ethnologues ont recueilli un autre proverbe plus précis:

'*Meunier larron*
Voleur de son pour son cochon;
Voleur de blé
C'est son métier'
(Rivals, 2000– Le moulin et le meunier, 2, 66).

A ces accusations traditionnelles, les meuniers répondent que la graisse de cochon est nécessaire pour graisser les mécanismes du moulin. A Roissy, le cochon entre dans la redevance du meunier due au seigneur, confirmant la facilité d'élevage des suidés au moulin.

L'entretien de quelques porcs est facilité par la présence d'une laiterie ; le petit lait, ou simplement le lait invendu, peut servir à nourrir les cochons. 55 ossements de porc ont été identifiés parmi les vestiges fauniques extraits de la fouille. Les observations relatives au porc sont très limitées, mais l'on perçoit un abattage probable d'animaux d'un âge compris entre un et deux ans, correspondant au stade de rentabilité bouchère maximum (Etude de Benoît Clavel et Alessio Bandelli).

Loin des grands élevages porcins qui ne trouvent plus leur place dans les campagnes céréalières du Pays de France à partir du XVIIe s., l'engrais d'un ou deux porcs reste une activité traditionnelle de toute ferme en milieu rural, voir des habitations villageoises ou citadines. Cet élevage domestique est facilité chez le meunier. Le meunier possédait certainement plusieurs dizaines de volailles dès le XVIe siècle, dont une dizaine puis une douzaine de chapons étant destinée au seigneur. Les rejets de mouture sont une fois de plus à la base de l'alimentation de ces animaux de basse cour.

4 L'abandon du moulin

La destruction du moulin à vent intervient vers le milieu du XIXe s., comme en témoignent les archives et le mobilier issu des tranchées de récupération. Les fondations du premier moulin (F. 40, 42, 44, 46 et 47), et partiellement celles du second moulin (surtout F. 124 et le quart sud-est de F. 53) sont récupérées simultanément. C'est logiquement la volonté de ne pas fragiliser le point d'ancrage du moulin, la motte, qui explique que les fondations du premier moulin n'aient pas été récupérées lors de la construction du second, mais seulement lors de l'abandon de ce dernier. Au centre des fondations des deux moulins, une série de couches de gravats (U.S. 58, 116, 123, 124 et 161) comble une vaste fosse centrale et vient recouvrir une large part des fondations du second moulin. De nombreux clous ont également été trouvés parmi les déchets de démolition des moulins. Les comptes flamands indiquent que des milliers de clous sont nécessaires pour couvrir les combles et la cage du moulin de petits bardeaux de chêne (Coutant 1990).

Le moulin et ses dépendances sont vendus en 1851. A cette date, il ne restait plus que 5 moulins à vent actifs sur les 52 moulins de l'arrondissement de Saint-Denis en 1810 : Le Mesnil-Aubry, Le Plessis-Gassot, Roissy, Survilliers et Tremblay (Blazy 1993, 380 et 391). Le sursis ne dura pas longtemps : le moulin à vent de Roissy fut détruit en 1862 et la maison du meunier en 1872. Après cette date, les terrains ont été convertis en terres cultivées.

Conclusion

Le moulin à vent est à la fois une machine et un lieu important sur le plan social et culturel, comme sur le plan économique et technique (Rivals 1976, 17). Alors qu'on en dénombrait 15887 au recensement de 1809, seuls 36 moulins à pivot subsistent dans toute la France en 1997 (Bruggeman 1997). Le potentiel archéologique et sociologique est donc énorme. Plusieurs moulins à vent ont été fouillés en Angleterre, mais à chaque fois, seules les fondations du moulin ont fait l'objet de recherches. Il en est de même dans les publications des sociétés molinologiques régionales : seul le moulin et ses mécanismes retiennent l'attention des observateurs contemporains. La fouille de Roissy permet pour la première fois de confronter le moulin et la maison du meunier, à un ensemble de données archivistiques, ethnographiques et agricoles. C'est à notre connaissance une opportunité unique en France.

Autorisé par lettres patentes du roi en 1541, le moulin à vent de Roissy est un moulin banal édifié dans l'une des trois petites parcelles destinées à l'exploitation du meunier. Implanté sur une butte de 60 pieds de diamètre, il nous est connu par les vestiges d'une croisée de murs reliée par une maçonnerie annulaire. Il s'agit d'un moulin sur pivot. Dés le début du XVIIe s., il est ceint d'une palissade et les fossés des enclos sont remaniés. Une maison est construite pour le meunier au XVIIe s. Outre la meunerie, de nombreux tessons indiquent une activité de laiterie peut être exercée dans un premier temps au sein d'un cellier, et plus assurément dans une laiterie véritable (le bâtiment F.9) à partir de la seconde moitié du XVIIe s.

Sans doute en 1737, un moulin plus trapu est reconstruit à l'emplacement du premier. La maison du meunier connaît également des modifications. Au XVIIIe s., elle est composée de deux pièces dont les sols sont revêtus de

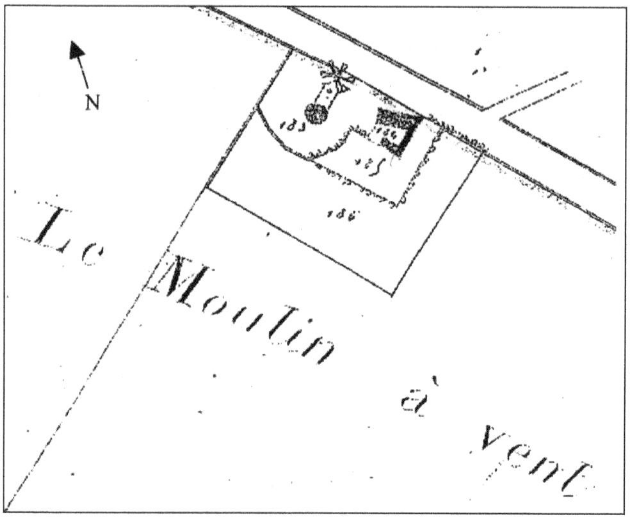

Figure 12. Le moulin sur le cadastre de Roissy en 1872

petites tomettes hexagonales. Une cheminée est adossée au mur de refend. Dans la cour, un toit à porc et un poulailler sont construits en appentis sur une étable logeant quelques vaches laitières et l'âne ou le canasson du meunier.

Le meunier de Roissy habite une véritable petite ferme à cour fermée, « pliée » par la contrainte fonctionnelle du moulin à vent qui positionne toutes les constructions de la ferme dans un angle limité et étroit, dans un petit espace trapézoïdal.

Dans la tradition rurale (française et britannique), la mentalité paysanne et le folklore, le meunier est perçu comme un individu physiquement fort, cupide et riche du droit qu'il prélève sur la mouture. Assis au cœur des terres céréalières les plus fertiles du royaume, et à proximité du plus grand centre de consommation de l'Europe médiévale et moderne, le moulin de Roissy enrichissait-il son meunier ? Le meunier de Roissy-en-France était-il riche ? La qualité de la construction, les rejets céramiques, la consommation de viande du meunier et le travail d'archives nous livrent des informations nous donnant plusieurs images successives contrastées de la qualité de vie du meunier. Des hauts et des bas dans l'activité du meunier de Roissy expliquent logiquement ces contrastes remarqués dans les vestiges faunistiques, la céramique ou les baux d'archives.

Bien qu'exerçant dans une grande plaine céréalière à proximité de Paris, le meunier de Roissy n'exploita qu'un petit moulin fonctionnant sans doute peu de temps. Le travail sur un moulin à vent est plus imprévisible, pénible et dangereux que le travail régulier sur un moulin à eau (Kealey 1987, 28). Si au XVIIIe s. les meuniers sont globalement des gens riches, il existe pourtant de nombreux meuniers en situation précaire qui survivent en pratiquant diverses activités. La subsistance du meunier de Roissy dépend d'une pluralité d'activités rurales de faible rapport (Gindin 1992) ; le meunier de Roissy fait un peu d'élevage et développe une activité de laiterie. A partir de 1783, le seigneur de Roissy doit lui concéder 5 arpents de terre; cela permet au meunier de développer une petite exploitation agricole qui lui donne quelques revenus.

Bibliographie

Blazy, J.-P., 1995. Du plat pays à la capitale, destins d'une famille de boulangers forains, les Destors de Gonesse (de la fin du règne de Louis XIV à la monarchie de juillet). *Paris et Ile-de-France, Mémoires*, 46, 195 – 225.

Bruggeman, J., 1997. *Moulins. Maîtres des eaux, maîtres des vents*. Paris.

Coutant, Y., 1990. *Lexique et technique du moulin à vent destiné à la mouture du blé d'après les comptes flamands des XIVe et XVe siècles.* Thèse de doctorat, Université de Lille III.

Gindin, Cl., 1992. Les moulins de la République. *Annales historiques de la Révolution française*, n° 290, 589 – 595.

Holt, R., 1988. *The Mills of Medieval England.* New York.

Jarvis, P.S., 1981-1982. *Stability in Windmills and the Sunk Post Mill.* Reading.

Kaplan, S.L., 1988. Le *meilleur pain du monde. Les boulangers de Paris au XVIIIe siècle.* Paris.

Kealey, E.J., 1987. *Harvesting the air – Windmill pioneers in twelfth century England.* Woodbridge.

Lachiver, M., 1997. *Dictionnaire du monde rural. Les mots du passé.* Paris.

Medieval Britain 1971. *Medieval Britain. Medieval archaeology*, 16, 1972, 211.

Moriceau, J.-M., 1994. *Les fermiers de l'Ile-de-France. L'ascension d'un patronat agricole (XVe-XVIIIe siècle).* Paris.

Rivals, Cl., 1973. Moulins à vent de France ; Les moulins à vent des plaines septentrionales: hégémonie du moulin sur pivot. *Ethnologie française*, III, n° 1-2, 141 – 166.

Rivals, Cl., 1976. *Le moulin à vent et le meunier dans la société française traditionnelle.* Paris.

Rivals, Cl., 2000. *Le moulin et le meunier.* Paris

Watts, M., 2002. *The Archaeology of Mills and Milling.* Stroud.

Millstone Quarries in Southern Spain: preliminary pinpointing of provenance and production - exploiting the internet

Timothy Anderson and Jane H. Scarrow

1. Introduction – millstone quarries across the Iberian landscape

In the 16 volumes (approximately 12000 pages) of the Spanish Dictionary of Geography and History, published in the middle of the 19th century, Pascual Madoz literally cites thousands of *molinos harineros* (flour mills) spread across the Spanish landscape, in almost every town and village, not to mention the cities. But from where did the thousands of millstones that equipped these mills originate? At the moment of the Madoz publication, the industrial revolution and the railroad network had not yet arrived on the Spanish Peninsula. Higher grade millstones from Monjuïc near Barcelona and the excellent siliceous millstones from France (Belmont 2006, 132), for example, were therefore not yet readily available, especially for the local village mills. Hence, the majority of the millstones were hewn from hundreds of local and regional quarries, the subject of this paper (Fig. 1).

Since the sheer scale of the subject is so vast and daunting, we have limited our survey area for the most part to the southern half of the peninsula, excluding Portugal. This encompasses all of the autonomous communities of Andalusia, Murcia, Valencia, Extremadura, Castilla La Mancha and Madrid. The few sites from the smaller community of Madrid, on the northern fringe of the study, are also retained due to their petrography. We also have only retained the exploitations of millstones which ground cereals for human consumption, and excluded the numerous oil roller and whetstone extraction sites. Similar research beyond our area of study, in La Rioja and Castilla León, is being conducted by P. Pascual Mayoral and P. García Ruiz, as well as J. Sánchez Navarro on the island of Menorca.

The state of research on the early quern production centers from Protohistory and Antiquity in Spain was presented recently at the Round Table Meeting of the French *Groupe Meules* at St. Julien-sur-Garonne (Anderson forthcoming). Most of this current research falls beyond our area of survey, especially in the northern half of the peninsula. Without going into details, let us cite the work of Alonso on the Iron Age sites of Catalonia, Checa on the Celtiberian city of Numacia in the central Province of Soria and Borges on Roman city of Conimbriga in Portugal (*ibid*).

Furthermore, the recent finds of this early period in our study area, in particular the volcanic rock exploitations of Cabo de Gata (Almería) and Campos de Calatrava (Ciudad

Figure 1. Moclín, Granada. View of the millstone quarry located on a promontory. Working debris is seen on the slope to the left.

Real), will only be cited briefly in this paper since the petrographical analyses have not all been completed (see Anderson, Grenne and Fernández Soler, this volume).

The sources available for the identification of millstone quarries are varied. Most derive from texts, in particular old geographic studies. The new techniques of searching and consulting these old sources through the internet facilitate this procedure. A second important source is the many local and regional molinological studies that at times point to the whereabouts of the quarries. A third source is geological studies, both old and new. A final source that cannot be neglected is hiking itineraries and descriptions of the local history posted on the internet. At this moment, based on these different sources, 88 flour millstone production sites have been identified (Fig. 2). Of the total only about 20%, for the most part in Andalusia (our area of residence), have been confirmed in the field. This footwork is often very time consuming due to the magnitude of the survey area and the fact that most of these exploitations have vanished from the memory of the local residents.

2. Rotary querns, watermills, windmills and animal mills

The latest research, on the origin of the oldest revolving mill, the rotary quern, points to the Iron Age Iberian Culture of the 5th century BC (Alonso 1999). The introduction of the more complex mills during Roman times, however, did not eclipse the rotary quern. We find it also widespread throughout Antiquity and Medieval times. This hand driven mill is even widespread in the 15th century AD associated with different types of levers and gears. It is even known to have been used, but certainly on a limited scale, as recently as in the harsh times during and after the Spanish Civil War (oral communication by P. Casado).

Concerning the more sophisticated mills, contrary to the image popularized by Don Quixote, the most common mill in the Spanish peninsula is not the windmill but the watermill. Two major types of watermills are the common *molino* with their horizontal wheel (*rodezno*), better adapted to small streams or channels, and the *aceña*, placed directly in the river flow. Wind-powered mills, equipped with the same type of millstones as watermills, are located in specific areas where water is scarce. The third major type of flour mill, the *tahona*, probably less frequent, was driven by animals (apparently of the horse family).

3. A sketch of the geology of the southern half of the Iberian Peninsula

The following description of the geology of the southern part of the Iberian Peninsula is modified and extended from the Introductory Chapter of La Geología de España (Vera 2004).

The geology of the Iberian Peninsula is dominated by the high central Iberian Massif. After considerable debate this Massif, that covers most of the western half of the Peninsula, was divided into six main zones. Broadly from north to south these zones are as follows: Cantabrian (Fig. 3, sector A); West Asturian-Leonese (Fig. 3, sector B); Galicia Tras-Os-Montes (Fig. 3, sector C); Central Iberian (Fig. 3, sector D); Ossa-Morena (Fig. 3, sector E) and South Portuguese (Fig. 3, sector F). The central region has an average elevation of 660m and is bordered to the north by the basin of the Duero river, to the northeast by the

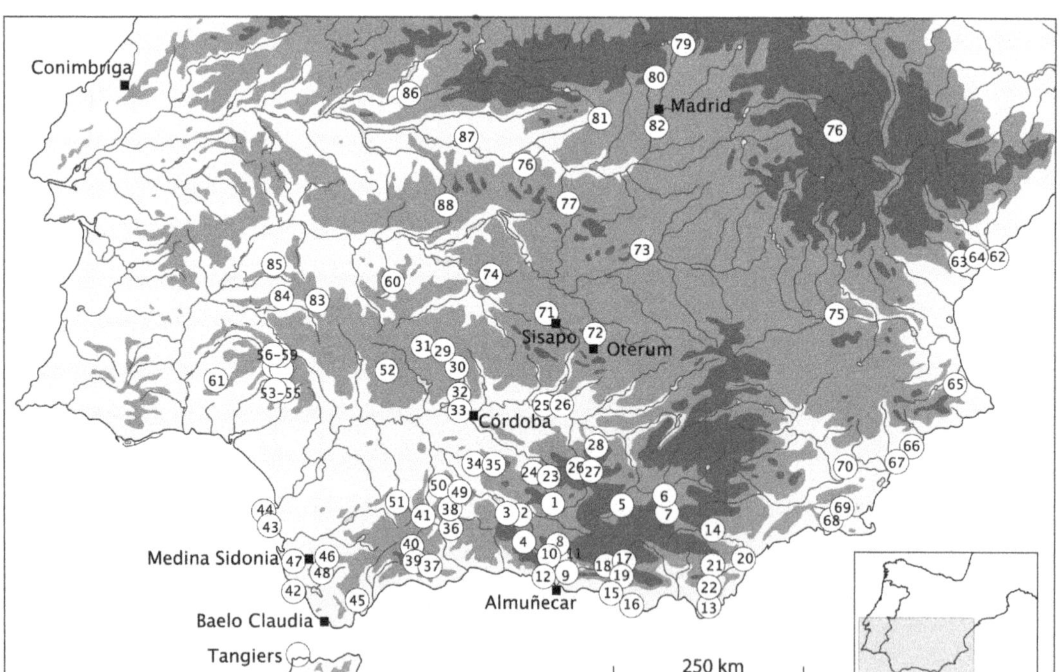

Figure 2. Distribution map of 88 millstone quarries for flour mills located in the southern half of the Iberian Peninsula. The numbers refer to the sites as they appear in the text; topography is represented by different shades of grey.

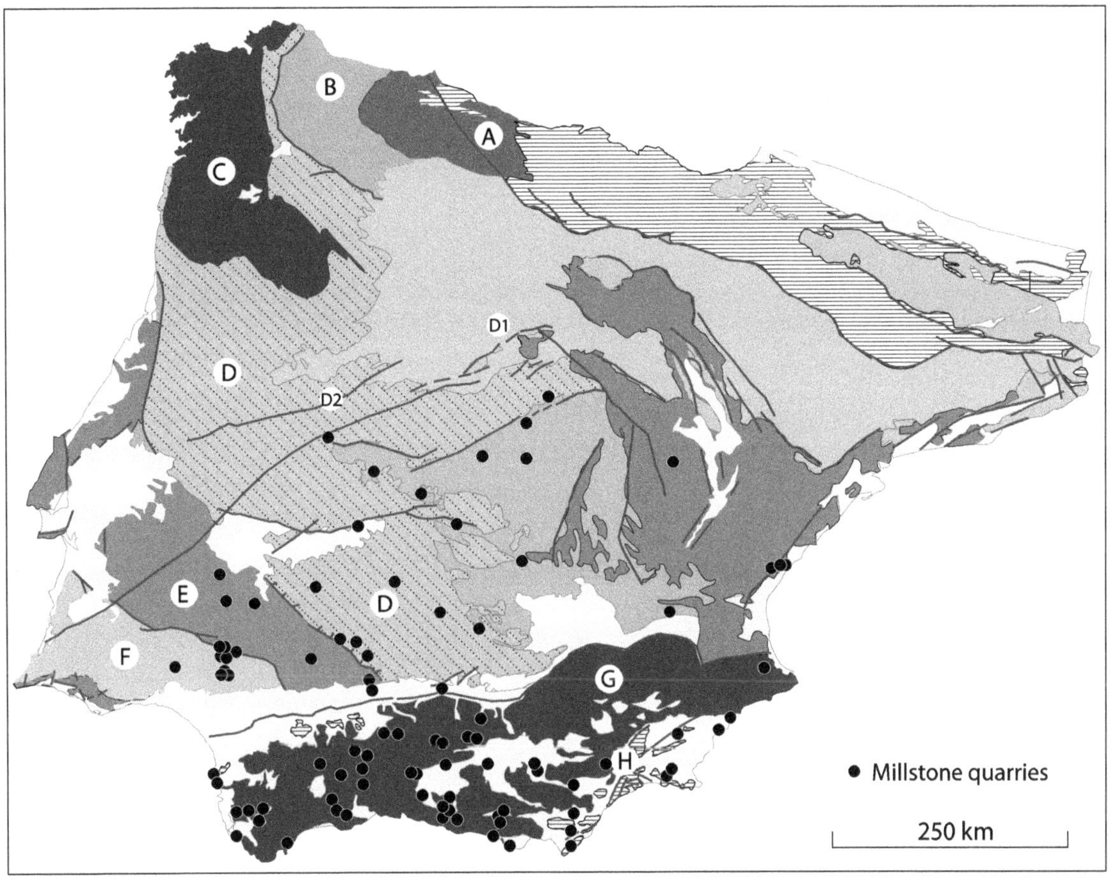

Figure 3. Map of the spread of the millstone quarries (black dots) on the backdrop of the principal geological zones of Iberia. Letters, also mentioned in the text, refer to the following regions. A. Cantabrian Zone; B. Western-Asturian Leonian Zone; C. Galicia Tras-Os-Montes Zone; D. Central Iberian Zone, D1. 'Ollo de Sapo' domain, D2. Greywacke-Schist Complex domain; E. Ossa Morena Zone; F. South Portuguese Zone; G. Bétic range External Zone; H. Bétic range Internal Zone (after Farias et al. 1987).

Ebro river and, broadly, to the south by the Guadalquivir valley. Lying beyond these hydrographic confines are the Cantabrian Cordillera in the north, the Pyrenees in the east and the Bétic Range in the south.

This study of millstone quarries basically concerns the region to the south of the River Tagus that flows east to west, cutting the Peninsula into two roughly equal parts. Thus, rocks exploited for the millstone production considered here include: the Precambrian to, predominantly, Palaeozoic granite-gneiss-schist igneous and metamorphic crystalline rocks of the three most southerly zones of the Iberian Massif (Central Iberian, Ossa-Morena, and South Portuguese); the Mesozoic-to-Cenozoic limestone-sandstone-conglomerate sedimentary rocks of the southerly basins and the Bétic Range; and the Cenozoic volcanic rocks of the recent meridional magmatic activity.

The Central Iberian Zone of the Iberian Massif, encompassing large areas of Extremadura, Castilla-La Mancha and Madrid (Fig. 3, sector D), comprises Proterozoic and early Palaeozoic metasediments and orthogneisses, extensive Lower Ordovician 'Armorican' quartzites and widespread Carboniferous granitoids. The Zone is divided into two domains mainly on the basis of the rocks underlying the quartzite. In the north, the 'Ollo de Sapo' domain pre-Ordovician rocks are potash feldspar augen orthogneisses, high grade regional metamorphic rocks and Variscan syn-tectonic granites (Fig. 3, sector D1). In the south, the Greywacke-Schist Complex domain pre-Ordovician rocks are schists and greywackes, the metamorphic rocks are low grade and the granitoids are syn- and post-orogenic (Fig. 3, sector D2).

The Ossa-Morena Zone occupying parts of the provinces of Huelva and Seville (Fig. 3, sector E) is made up of Upper Proterozoic to Carboniferous rocks. Oceanic basic igneous rocks crop out at the northern contact with the Central Iberian Zone and the southern contact with the South Portuguese Zone. The main lithologies in this region are low grade metamorphic rocks and granitoids of varying ages including Vendian, Cambrian and Carboniferous syn- and post-orogenic intrusions.

The South Portuguese Zone (Fig. 3, sector F), the most meridional division of the Iberian Massif (in the Huelva province), is composed of Devonian to Carboniferous volcano-sedimentary, igneous and low-grade metamorphic rocks that, in places, are overlain by Permian sediments.

The Bétic range in the south and southeast of the Iberian Peninsula (Fig. 4), from Huelva all the way to Alicante, forms the most westerly sector of the Alpine orogen that borders the northern and southern margins of the Mediterranean. The Bétic cordillera is divided into two main zones, which were two separate microplates through the Mesozoic and Cenozoic: the External Zone (South Iberian paleomargin) and the Internal Zone (Alborán Domain fragment of the Mesomediterranean plate). The External Zone, to the north (Fig. 3, sector G), comprises Triassic to Miocene continental margin sedimentary rocks. This zone is further subdivided into the northerly Prebétic relatively undeformed shallow marine sediments and the more strongly deformed southerly Subbétic lower Jurassic pelagic sediments and submarine volcanic rocks. The Internal Zone, to the south (Fig. 3, sector H), is made up of a stack of tectonic units, from base to top: the Nevado-Filábride Complex, the Alpujárride Complex and the Malággide Complex. In contrast to the External Zone, the pre-Mesozoic basement of this region was displaced together with the Triassic to lower Miocene cover rocks. The basement part of the two lower units preserves traces of pre-Alpine deformation, magmatism and metamorphism, which in the Alpujárride Complex includes continental crust and subcontinental upper mantle. Both the basement and cover of all three units were affected by complex Alpine metamorphism.

Throughout part of the Mesozoic through to the mid-Paleogene, turbidites were deposited in a narrow ocean basin that existed between the External Iberian plate and the Internal zone Mesomediterranean plate. These marine sediments are preserved in the Campo de Gibraltar Complex.

Cenozoic volcanic rocks crop out in four regions throughout the Peninsula: Gerona, Gulf of Valencia, Calatrava and Almería-Murcia (Fig. 5). The last two of these are of interest in the current work and include the oldest volcanism, Aquitanian, in Cabo de Gata (Almería) and some of the youngest, Holocene, in the Campos de Calatrava. The Calatrava rocks are basic alkaline and carbonatites. In the Almería-Murcia region, the igneous rocks are more varied including calc-alkaline to alkaline compositions.

4. The millstone quarries

The description of the 88 millstone quarries begins in the Autonomy of Andalusia where roughly two-thirds of the total are spread out over eight provinces. This is followed by the description of the sites in the other autonomies. To facilitate the descriptions we will proceed province by province.

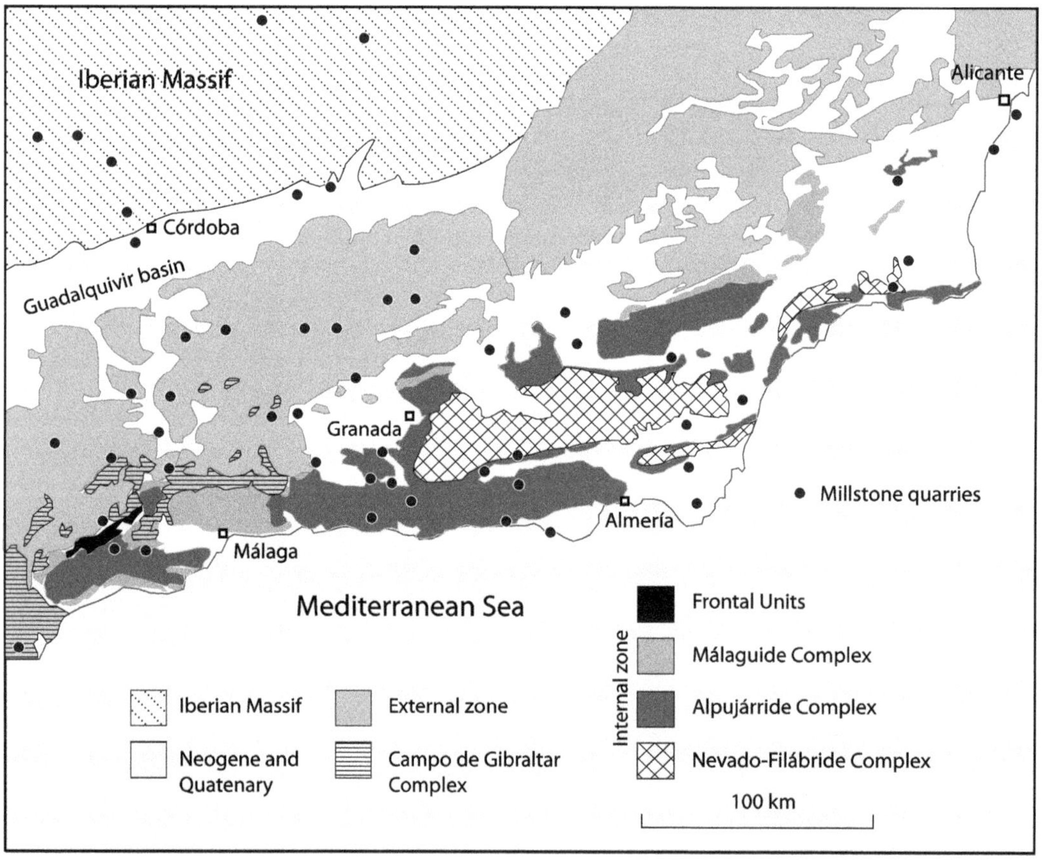

Figure 4. Details of the principal geological formations of the Bétic range (after Martín-Algarra 2004).

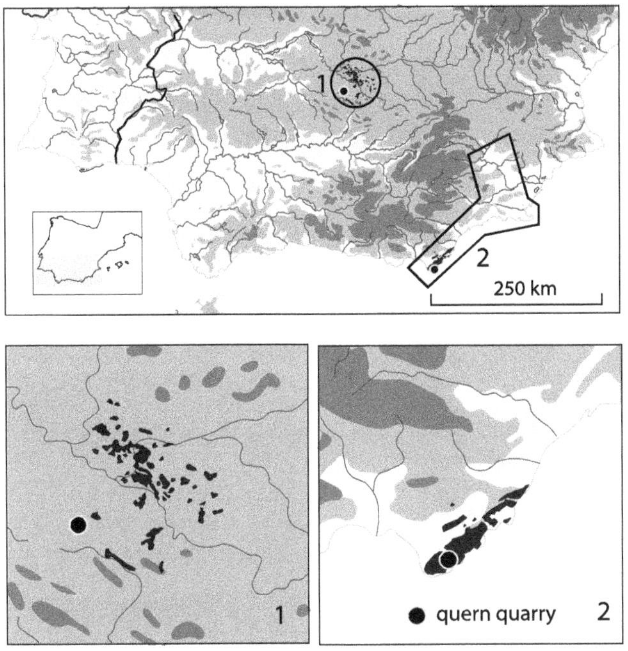

Figure 5. Situation of the two volcanic districts in the southern half of the Iberian Peninsula. 1. Campos de Calatrava and, 2. Cabo de Gata. The dark areas correspond to Neogene Quaternary volcanism (based on the Geological Map of Spain, 1.1.000.000, IGME 1994). Recent millstone quarry finds in each region (grey circles) confirm millstone production dating to Antiquity.

4.1. The millstone quarries in Andalusia

4.1.1. The Province of Granada

The province of Granada possesses, at least for the moment, the highest concentration of millstone quarries. According to the 19th century geographer Madoz (8, 480), the limestone exploitation of Moclín (1) is one of the three most important provincial quarries (Fig. 6). It is located on a promontory to the northeast of the town, beside a Medieval Moorish fortification. The site bears a toponym *Los agujerones* (large holes), possibly related to the circular extractions. In addition to many circular cavities, it is easy to recognize the heaps of working debris on the southern slope. The quarry face is either tiered or tube-shaped where the rock was massive enough, permitting multiple superimposed extractions. On the less weathered areas of the surface, it is possible to distinguish the characteristic diagonal pick marks. It is apparent that the site produced hundreds of millstones. Some of the older residents recall when the quarry was active. Production on this site therefore continued at least into the early 20th century. An author of a study of mills in the Alpujarra region (Rodríguez Monteoliva 1989, 705) states that the Moclín millstones were marketed as far as the south of the province, at least 50 km away, because they yielded white flour.

The second site, Loja, is identified by a proclamation dating to the early 16th century that prohibits the extractions of millstones without a license from the authorities under penalty of having to give up not only the millstones, but the oxen used for their transport (http://www.geocities.com/tdcastros/Historyserver/Fuentes/DocMunic/DOCLOJA.6.htm 1502, enero, 10. Loja, Pregón de la ordenanzas de Loja sobre no sacar piedras de molino. AML, Leg. 49, p. 9. 1 cuartilla + 1).

Furthermore, the 19th century geographer Madoz cites several white or rose *panalizo* exploitations specifically for flour mills (Madoz 10, 360) in the vicinity of the city. *Panalizo* is an old Spanish term that designates a rock that is soft at the moment of extraction but becomes very hard when dry (Ximénez de Guzmán 1756, 9). We have confirmed for the moment two sites near Loja. The first, the Cerro de la Fuente Santa (2), a few kilometres to the west of the town, shows the circular imprints of several dozen cylinders, 1.50 m in diameter, that were extracted directly from the limestone bedrock. There are also signs of roller extraction for the oil industry. The second site, partially destroyed by road construction, consists of a series of sporadic extractions perched on the slope of the mountain to the north of the city at the Camino del Calvario (3). Neither site is very extensive. Nevertheless, since Madoz cites Loja among the three major provincial sites, we think that there are more exploitations in the area, probably in the nearby districts of Chapas, Zagra and Algarinejo.

The site of Fuente de los Morales (4) is in the Sierra de Tejeda mountains outside the city of Alhama de Granada. The name Morales might be a variation of *molares*, a common millstone quarry toponym elsewhere in Spain. Another toponym in the sector, *moladeras*, suggests millstone production. This term however is more often related to whetstones (*piedras de amolar*). The exploitation is about 150 m long with a working face several meters high. There are huge mounds of fine working debris at the foot of the outcrop. The quarrymen apparently carved millstones from previously detached angular blocks of a hard, white layer of limestone less than one meter thick.

Fig. 6. Moclín, Granada. A sector of the quarry with tiered extractions.

Several abandoned cylinders in the area range from 1.10 to 1.50m in diameter. A half cylinder might indicate that the products were not only monolithic but in halves to later be assembled. Among the debris there are also signs of trunco-conical oil rollers. This must be one of the more important quarries of the Granada province considering that the 19th century geographer Madoz reveals that it commercialized its stones all over the region, even as far as the city of Málaga located 50 km on the other side of the mountains (Madoz, 8, 216). Madoz also specifies that these excellent millstones yielded flour for white bread.

In northeastern Granada the site of Pedro Martínez (5) produced both limestone millstones and oil rollers (Madoz 9, 42). Further to the east, in the vicinity of the city of Baza, word of mouth from reliable sources has it that millstones were extracted from stone outcrops along the riverbed (6). Even further east, near the border with Almería, is the site of Las Canteras (7), outside the city of Caniles (Madoz 5, 461). This large site, measuring 300 by 40 m, exploited a meter-thick layer of conglomerate exposed in a dry riverbed. The rock shows rounded pebble inclusions up to 2 cm in a calcite matrix. What remains today is a series of long, parallel cordons of working debris about 2 m high. There are no visible circular cavities in the bedrock. The quarrymen must have carved previously extracted blocks. In our brief visit we identified only one abandoned cylinder measuring 0.90 m in diameter and 0.40 m in thickness. It is noteworthy that Madoz specifies that this quarry produced millstones for *pan moreno* (dark bread). This permits us to possibly associate other similar types of conglomerate quarries with this type of bread.

Los Guillares (8) is in the south of Granada, about 15 km to the east of Padul. It is identified by the municipal website. With its monumental size and imposing extractions, it is surprising that it is not cited in old texts unless it dates to more recent times. Large cylinders (between 1.20 and 1.50 m in diameter) were hewn by pick directly from the bedrock producing diagonal tool marks. This site, along with Moclín, is an excellent example of superimposed extractions producing tube-shaped quarry faces (Fig. 7). An additional facet of Los Guillares - unique for the moment - is the road exiting from the lower sector of the quarry consisting of ruts carved directly into the bedrock. As in the case of other sites elsewhere in Europe, these ruts were meant to facilitate the transport of millstones by wagon.

Further south, about 15 km from the coast, is Vélez Benaudalla (9), a town in a rugged terrain with many natural springs and, consequently, watermills. The quarry, about 1.5 km to the northeast, is about 100 m long and dug into the hillside leaving a working face several meters high. Working debris ranging from large blocks to fine knapped flakes and abandoned millstone drums are strewn about the area (Fig. 8). The material is a white limestone that when weathered takes on a grayish appearance. It seems that both naturally detached blocks and bedrock were exploited.

Figure 7. Los Guillares, Padul, Granada. The superimposition of extractions produce tube-shaped quarry faces.

Figure 8. Vélez de Benaudalla, Granada. View from the top of the quarry of three aborted extractions on three different levels.

Parallel diagonal lines on the quarry face show the use of the pick to cut the circular channels around cylinders measuring between 1.20 and 1.30 m in diameter. Due to the high number of natural fissures, however, the quarrymen could not superimpose their extractions, thus there are no tube-shaped quarry faces. Most of the extractions, as usual, followed the horizontal rock bedding plane. One attempt at a vertical extraction was quickly aborted. It is difficult to estimate the number of extractions. Based on the size and present shape of the outcrop it would have been possible to remove several hundred cylinders. In any case, the 19th century geographer Madoz, places this site with Loja and Moclín in his description of the more important millstone quarries of the province of Granada (Madoz 8, 490; Madoz 10, 464; Madoz 11, 637).

In the vicinity of Vélez Benaudalla, a few kilometres to the west upstream in the valley of the Río de la Toba, are two modest quarries outside the towns of Guajar Alto (10) and Guajar Faraguït (11). We were not able to find the first site and suspect that it was destroyed during the construction of a vacation house. The second, Los Mochos, pointed out in a hiking itinerary on the internet, is hidden away in the fruit plantations adjacent to (and possibly partly covered by) an old farmhouse. There are less than a dozen extractions (some bearing pick marks) of cylinders between 1.10 and 1.20 m in diameter. The small number of extractions and the lack of written sources indicates that these are local village quarries. The nearby site of Otívar (12), also in the Alpujarra mountains near the Mediterranean coast, provided millstones, according to the late 18th century geographer Tomás López, to the watermills of the Jete River (Reyes Mesa 2000, 18). It must have therefore been more than a village exploitation. However, the fact that the later 19th century geographers do not cite it suggests that its production had already halted. Nobody in the town, including the local historian, seems to have knowledge of the site and our attempts to locate it in the rugged terrain have not so far succeeded.

4.1.2. The Province of Almería

The province of Almería, in the southeastern corner of the Iberian peninsula, has a long history of millstone production. The recent identification of the Roman quern quarry of Cerro de los Limones (13), the first of its type certified in the Iberian Peninsula, confirms that the Cabo de Gata volcanic outcrop was exploited massively during Antiquity (Anderson *et al.* forthcoming). This area therefore joins the vast network of volcanic production districts known throughout the Mediterranean basin (Olot in Catalonia, Cape d'Agde and the Massif Central in France, Sardinia, Orvieto in Italy, the Greek island of Milos) as well as the Eifel in Germany. As we will see later in this paper, a second volcanic area in central Spain, the Campos de Calatrava, also produced millstones in Antiquity.

The quarry, with its hundreds of unfinished and aborted roughouts (Fig. 9), is perched on the top of a volcanic dome overlooking the coastal town of San José, a few kilometers to the south. The production consists exclusively of rotary querns approximately 40 cm in diameter hewn from a fractured, reddish-brown, slightly porous rock. The workmen probably detached blocks with levers from the summit of columnar jointing that were then knapped into shape, leaving tons of flakes over the whole of the top of the hill. The proximity of the coastline marked with its small bays probably facilitated the transport of these querns up and down the coast. The extension of the distribution of this product is a project that must be pursued.

Figure 9. Cerro de los Limones, Almería. Detail of a cylindrical blank of a rotary quern (about 40 cm) dating from Roman times.

Turning to the north of the province is the Cantera de la Rambla Honda (14) outside the city of Albox on the edge of a dry riverbed (Martínez López and Granero Gallegos 2005). Here were hewn both small rotary querns and large cylinders (probably destined for windmills or watermills) from an exposed layer of conglomerate bearing relatively large, rounded clasts. The precise date of this site is not certain. The lack of reference in a written source and the presence of small quern extractions, side by side with large extraction, suggests an early date, possibly during the Medieval Moorish domination.

A recent study of the mills of the southern Alpujarra mountains cites interesting information about several recent quarries located in western Almería. The quarry at El Cerro Chispas (or Cerro de la Cantera) (15) is on an isolated hill near Adra (Cara Barrionuevo *et al.* 1999, 153). The authors indicate 32 cavities cut into a conglomerate layer. Four remaining unfinished cylinders measure between 1 and 1.20m in diameter. A second site in this area, also exploiting a conglomerate, is the modest coastal site of Guardias Viejas (16) (Cara Barrionuevo *et al.* 1999, 154) near an 18th century fortification. The quarrymen concentrated their work in a sector of the outcrop containing round pebbles of small and more regular size (up to 2 cm). These inclusions are mostly fine-grained quartzites in a matrix of sand and calcite (petrographical description by Tor Grenne, NGU). In spite of a highly eroded surface, it is possible to observe the diagonal pick marks related to the cutting of the circular trench. The cylinders measure between 1.20 and 1.30 m in diameter. From the small number of circular imprints (8 in number) and abandoned cylinders (2 in number) we can deduce that this site served a very local market.

Cara Barrionuevo also cites several more remote mountain sites such as the calcite exploitation at the Barranco de Palancón (17), near Bayárcal and the conglomerate exploitation at Ugíjar (18) (Ugíjar is technically in the Province of Granada but is cited here because of its proximity to the Almerian sites of Barranco Baena and Bayárcal). However, the more important site in this rugged, isolated area seems to be the conglomerate exploitation at the Barranco Baena (19) between Darrícal and Benínar (Cara Barrionuevo *et al.* 1999, 153). Products from here, according to the author, were transported down a narrow path and marketed all over the Alpujarra. Another recent mill study (Rodríguez Monteoliva 1989, 705) specifies that Barranco Baena millstones yielded dark flour.

To complete the province of Almería we have passing references of the northern sites of Vera (20) (Madoz 15, 670), Loberos (21) (near Sorbas) and Níjar (22) (Torres Montes 1992-1993, 272). These sites probably exploited sedimentary deposits. The only specification we have for the moment is that millstones from Nijar yielded dark flour (Torres Montes 1992-1993, 272).

4.1.3. The Province of Jaén

To date, seven millstone quarries for cereal grinding are known in the province of Jaén. It is, however, not surprising that this province, with its vast olive oil production, has the highest number of quarries dedicated specifically to the extraction of oil rollers (not presented in this study).

The sites of Alcalá la Real (23) and Castillo de Locubín (24) are both found in the Sierra Sur range in the southwest corner of the province. The first, Las Canteras (the quarries), is at the foot of a scarp on the western slope of Mt. Acamuña (Madoz 1, 382). No direct bedrock extractions were observed. It seems that the millstones were hewn from scree blocks. A small quern blank, 50 cm in diameter, suggests an older exploitation of handmills dating to either Medieval, or possibly even Roman, times. Castillo de Locubín, a few kilometres to the north, is in the valley of the river Guadalcotón and is cited by both Miñano (2, 477) and Madoz (6, 190). Miñano specifies that the these products were destined probably for watermills.

Madoz states that around Andújar, there are many abandoned quarries in the valley of the Guadalquivir river in the Sierra Morena mountains that produced millstones for both the bread and olive oil (Madoz 2, 305). He specifies only two granite exploitations, Morales (25) and Pedroso (26), that were active at the time. The sites of Huelma (27), in the southern Sierra Mágina mountains, and Lanchar near Jimena (28), on a hiking trail, also produced millstones and other products.

4.1.4. The Province of Córdoba

A first group of Córdoba millstone quarries is in the northwest of the province in the Sierra Morena, a range that separates the Spanish Meseta from the Guadalquivir river plain. According to Madoz (4, 131 and 9, 39), the site of Belmez (29) in the Albardado river bed furnished piedra basta (rough stone) millstones to both local and regional mills. The term probably refers to a sedimentary deposit. To the southeast is Espiel (30), where Madoz (14, 392) indicates a fault line 30 to 40 km long and 3 km wide that exposes a rough sandstone exploited for millstones. The third of the group is Fuente Obejuna (31), located to the west of Belmez where Madoz (8, 230) cites two ample quarries near the rivers Guadiato and Suja that provided stones for both watermills and animal-driven mills (*tahonas*). A fourth site is Santa María de Trassierra (32), near Arenales, a few kilometres from the city of Córdoba (Madoz 15, 136). There is no mention of the type of rock, but with the name Arenales, meaning sandy terrain, we suppose it is of detritic origin.

A recent article in the proceedings of a molinological colloquium is one of the rare publications in our study area specifically dedicated to the subject of the origin of millstones (Montero 2008). It mentions a quarry in the surroundings of Albaida (33) that probably supplied the stones to the early Medieval watermills of the city of Córdoba. The rock, a carbonate sandstone from the marine Miocene (late Tertiary) rich in fossils, was abundant and easy to carve (Montero 2008, 166). The source for the more recent mills, however, is found in Cabra (34) and Carcabuey (35), twin quarries about 50 km to the southeast of the provincial capital (Montero 2008). It consists of a reddish sandstone from the Late Jurassic bearing ammonite fossil inclusions. These quarries are cited in the 19th century translation and adaptation of Charles Lyell's *Elements of Geology* to Spain. The author reports a vast quarry at Cabra that produced stones for both flour and oil mills (Ezquerra del Bayo 1856, 385). At the time of the report, the quarry had already existed for a long time and benefited from a wide regional market that even extended to the coastal city of Málaga about 80 km to the south. Esquerra del Bayo goes on to say that a white limestone (actually a breccia, in his words) was hewn for flour mills that were sold at a price of between 60 and 100 *reales*. The oil rollers, on the other hand, were hewn from a more compact, reddish siliceous breccia and were sold for between 800 and 1000 *reales*.

4.1.5. The Province of Málaga

A source from the end of the 18th century describes three piedra *tosca* (variety of limestone) quarries in the province of Málaga (García Lena 1789, 106). The author indicates that the better millstones of the three came from El Torcal (36), near Antequera, which is a national park that protects karst formations sculpted by nature over thousands of years. From our observations, the northern perimeter of the park is spotted with abandoned, rose-colored cylinders, about 1 m in diameter, in different stages of work (Fig. 10). Some thicker drums might have been destined for the olive oil industry. The second site identified by García Lena is in the southern Sierra Gorda, outside of Alhaurín el Grande (37), not far from the coast (García Lena 1789, 106). A half century later, Madoz specified that there were several granite exploitations to the south of the town. Since granite is not present in this area, this is most certainly an error on the part of Madoz. The third site mentioned by García Lena, Canteras (quarries), is near Jabonero (38) and apparently produced millstones for dark bread. Although we have not been able to identify the exact location of this site, we suspect that it is several kilometres to the northwest of Antequera.

Figure 10. El Torcal, Málaga. Abandoned cylinder.

Figure 11. Trafalgar Bay, Cádiz. View of the eastern rotary quern exploitation.

Two other sites in the south of the province, not far from Alhaurín el Grande, are Monte del Señor (39), outside the town of Guaro, and Alozaina (40). Although the petrography of the first site is not indicated, Madoz (9, 56) specifies, and Bisso confirms (1869, 20), that it produces stones for flour mills. The quarry of Alozaina, about 10 km to the north of Guaro, was of 'particular merit' (Madoz 2, 186) and was exploited for its *almendrilla* (Bisso 1869, 21), a conglomerate or puddingstone.

The last site identified in the Málaga province is Teba (41) cited by two 19th century sources. The first indicates that its millstones were marketed in the nearby towns (Bergnes y de las Casa, 1831-1833, 9, 608). The second remarks that the town has nothing notable except for the church and the millstone quarries (Miguel Espinosa 1848, 265).

4.1.6. The Province of Cádiz

The province of Cádiz in southeast Andalusia has both an Atlantic and a Mediterranean coastline separated by the Straits of Gibraltar. It has a rich tradition of quarries starting with a quern and millstone quarry on the Atlantic coast of the Bay of Trafalgar (42), site of the famous naval battle of 1805. The *ostionera* rock is a porous, shell-rich sandstone. The eastern sector of the site, with a surface of about 150 square meters, reveals multiple small circular cavities placed side by side (Fig. 11). We estimate that the cylinders extracted were querns about 40 cm in diameter. The rest of the site (about 100 x 20 m), partially under water today, shows numerous extractions ranging between 0.80 and 1.10 m in diameter. Due to a constant erosion by wind and water, they are smoothed over and show no tool marks. The date of the different sectors is impossible to determine with confidence. We have, however, observed small rotary querns of similar dimensions at the coastal Roman city of *Baelo Claudia*, a sign that the quern exploitation could date as far back as Roman times. A Medieval or even more recent date, nevertheless, cannot be discounted. As for the rest of the quarry, the relatively modest diameters (0.80 to 1.10 m) suggest an exploitation

Figure 12. Cerro del Berrueco, Cádiz. General view of the mountain that now has been practically leveled.

of watermills, probably dating to the Moorish domination. We have, in fact, witnessed cylinders of ostionera stone in the museum of the Moorish castle of Almuñecar on the coast of Granada. But, once again, the dating is not certain. In any case, we have not encountered any textual reference to the site, a sign of an early exploitation. Recent fieldwork on the Cádiz coastline has confirmed *ostionera* quarries for both querns and millstones at Rota (43) on the Bay of Cádiz and at Chipiona (44).

For the more recent periods with large extractions there are four sites. For Guadalquitón (45), near the Mediterranean coast (Beneroso Santos 2007, 15), we have very little data. Peña Arpada (46) is an outcrop perched on the summit of a hill outside of the town of Paterna de Ribera that produced flour millstones (Madoz 1, 380). During our visit, we observed abandoned cylinders measuring about 1.30 m in diameter by 0.50 m thick carved from white limestone. The circular negatives in the bedrock are indicative of true extractive processes. Tool marks, however, are too eroded to be observed. A very thick cylinder (around 0.90 m) might be an abandoned oil roller.

The most important site in the region is the massive exploitation of Cerro del Berrueco (47), near the town of Medina Sidonia (Fig. 12). The place name *Berrueco*, generally associated with granite outcrops, is a misnomer. The stone is actually a white limestone that was also used for construction. Today, apart from the huge crater in the middle of the mountain, there is no evidence of millstone extraction. The old quarries were probably destroyed by a modern exploitation for construction. Three different 19th century sources provide unique insight into interesting aspects of the site that are usually ignored by the early geographers.

The earliest source (Cruz y Bahamonde 1813, 13, 91) states that 50 men toiled at various points at the site and that the site was owned by the Duke of Medina, who received 18 millstones yearly for the concession. Madoz, about a half century later, stated that there are five millstone exploitations at the site worked by 23 men. The men resided on the site and they produced the best millstones of the province. Furthermore, the men had the right to bear arms as a consequence of the ooccasional attacks of bandits. Production consisted of 480 *granos*, that is, millstones for animal driven mills (*tahonas*) and 64 millstones for either watermills or windmills (Madoz 2, 290). This last information implies that although stones for watermills and windmills are similar, there must have been a typological difference between them and the animal-driven stones. Unfortunately, this difference remains obscure.

The last written reference to the Berrueco site, 62 years after the first, signals that its millstones for both animal and other mills continued to be distributed throughout the region (Martínez and Delgado 1875, 129). The author adds that these products provided the workers, residents of Medina Sidonia, with advantageous salaries.

The last identified quarry in the province of Cádiz is of La Pila de Casáres (48), about 5 km outside the town of Medina Sidonia. According to Martínez y Delgado (1875, 129), this site also produced millstones for water and animal mills. The author alleges its stones were not as white as that of Berrueco, but were better grinders and, in equal proportions, yielded a higher amount of flour. Today la Pila is a vast factory producing quicklime.

4.1.7. The Province of Sevilla

The site of Cerro Bellido (49) is located in the eastern point of the Sevilla province near the border of the provinces of Córdoba and Málaga. According to the itinerary on the internet retracing the steps of the famous 19th century bandit El Tempranillo, there is a millstone quarry located to the south of the town of Casariche, on the top of the Cerro Bellido hill. The authors indicate that it was exploited during both Roman and Moorish times. These chronological indications must be taken with much precaution. The site of El Hacho (50) in the hills outside of Estepa is about 10 kilometers to the east of Cerro Bellido. According to Madoz (7, 609), a white rock was exploited specifically for flour millstones. The site of Villanueva de San Juan (51), also in eastern Sevilla, is unique in the sense that although not in use at the time, Madoz cites it. He explains that its disuse is due to the lack of access to the site (Madoz 16, 207).

From old texts we can deduce that near the town of Alanis (52) in the Sierra Morena in the north of the province, both whetstones and millstones were exploited. Bergnes (1831, 126) mentions specifically a famous quarry for stones to sharpen barber razors. We suppose that this must be a very fine sandstone. Madoz confirms the whetstone site and goes on to mention that millstones were also exploited in the area (Madoz 1, 190).

Finally, also to the north of the Province, near Almadén de la Plata, a sandstone quern quarry (not on the distribution map) with numerous extractions measuring 50 cm in diameter has been discovered very recently by Miguel Angel Vargas in a dry streambed. Based on the size of the extractions, the site might date to late Antiquity or to the Moorish domination.

4.1.8. The Province of Huelva

The principal data concerning the province of Huelva, on the western side of the peninsula on the Atlantic coast bordering Portugal, comes from the study of the watermills on the River Odiel (Ruiz Gómez 2003). This author cites eight quarries and briefly describes their petrography. Three of them (Campillo Viejo, El Capillo II and El Campillo III) (53-55), of porous limestone tuff, are found in the Umbrías del Río Tintillo region in the east of the province. This type of limestone presumably outfitted the older watermills on the Odiel River that might date as early as the 15th century (Ruiz Gómez 2003, 85). The four other extraction sites are near the town of Alájar about 20 km to the northeast (El Prao del Abad I-II, Las Malenas I-II and La Obra Pía) (56-59). They are of different varieties of biotite granite. According to the author, these products date most likely to the 19th century. To complete this province, we have passing references to sites in Zalamea (60) (Madoz 16, 450) and Puebla de Guzmán (61) (Garrido Palacios 2001, 164) near the Portuguese border. For this last site, the author alleges that the local mills acquired some of their stones from Medina Sidonia (Cádiz), almost 200 km to the southeast, as well as from the cragged outcrop perched above the town.

5. Millstone quarries in the other regions of the southern half of the Iberian Peninsula

5.1. Valencia (Provinces of Castellón, Valencia and Alicante)

Data about millstone quarries in the Autonomy of Valencia are scanty. In the northern province of Castellón, we have information about three sites. Nulles (62), near the coast, at the time of Madoz (12, 195) was a new, little-known exploitation of flour millstones. The second is referred to in the description of the landscape around the town of Soneja (63), about 20 km inland from the coast. In between the two sites is the Dehesa de las Escales (64), near Vall d'Uxo in the L'Alt Palancia mountains, known by means of a hiking itinerary on the internet. In the central Province of Valencia is Les Moles (65), outside Canals, at the source of the River Cànyoles, according to a brief description of the local landscape. In the southern Province of Alicante, two possible quarries are identified due to citations in geological reports. The Illeta Dels Banyets (66) is a small, highly eroded, limestone natural jetty on the coast several kilometres to the north of the city of Alicante. The report cites numerous extractions of millstones and ashlars (Rosselló Verger 1991, 48). The authors of a geological report of Serra del Molar (67), a mountain just to the east of the town of La Marina, suggest that the toponym is related to an unidentified millstone exploitation (Gonzálvez Pérez and Rosselló Verger 1978, 107, note 1).

5.2. Murcia

There are three potential millstone quarries in the small autonomy of Murcia. The first, based on a hiking itinerary on the internet, is at the Rambla del Puerto (68), a mountain pass to the south of the city of Murcia. The source states

Figure 13. Sisapo (La Bienvenida), Ciudad Real. View of one of the three volcanic domes outside the Roman city of Sisapo that was exploited for querns and small millstones. This site is located in the southeastern corner of the Campos de Calatrava.

Figure 14. Sisapo (La Bienvenida), Ciudad Real. Detail of a rotary quern extraction. Diameter 40 cm.

that there are thousands of circular extractions about 1.60 m in diameter that date to the Moorish domination, assertions that would need to be verified. The second, also identified through a tourism brochure, is in the Sierra de Atahona (69), about 15 km to the southeast of Murcia. The name Atahona, without a doubt related to the Arabic term *tahona*, meaning mill, shows a probable link to the period of the Moorish domination. The last site is Fortuna (70), in the north of the province (Matilla Séiquer 2001), which exploited both conglomerate and possibly basalt. Although its main product consisted of rollers for *almanzaras* (or waterpowered oil mills), cylinders for flour mills were also produced (Matilla Séiquer 2001, 270).

5.3. Castilla La Mancha (Provinces of Ciudad Real, Albacete, Toledo and Cuenca)

The Campos de Calatrava volcanic district in the Province of Ciudad Real, covers about 5000 square kilometres. The confirmation of production of querns and millstones in this district, like that of the Cabo de Gata in Almería, is very recent. As we stated in the introduction, we will limit ourselves to only a brief description of these new finds. The first is a small exploitation 500 m to the northwest of the Roman city of Sisapo (71) beside the hamlet of La Bienvenida. The quarry, in the far southeastern sector of the Campos de Calatrava, consists of three small volcanic cones (Fig. 13) spotted with both quern (Fig. 14) and small millstone extractions carved directly into the rock leaving circular imprints. Several cases of single, large wedge holes suggests the use of wooden wedges, drowned in water, to split the cylinder. The rock is a dark grey, almost black, vesicular basaltic lava. The quern cylinders extracted measure 0.40 m in diameter while the diameter of the larger extractions ranges between 0.70 and 0.90 m. The chronology of the site, as in the case of most quarries, cannot be determined with precision. The sizes of the extractions, the proximity of the Roman city and the propensity of the Romans for mills of volcanic rock, nevertheless point to Roman times. The relative thickness of the larger extractions (in proportion to their diameter) suggests that they might have been hewn into bell-shaped or conical volcanic millstones similar to those documented on the watermill sites of Barbegal in southern France and Avenches-En-Chaplix in western Switzerland (Castella 1994).

The whereabouts of a second millstone quarry exploiting a black basaltic lava is based on several unfinished querns in the Roman city of Oterum, in the heart of the Campos de Calatrava and on the banks of the River Jabalón. These finds are aborted querns. Other mills of different types (smaller variations of the Pompeian type and possibly a conical watermill lower stone) are apparently hewn from the same source, which contrasts with the Sisapo rock due to numerous white inclusions. The exact location of this second quarry is probably found in one of the many regional Strombolian domes. These domes, however, are still exploited on a large scale for concrete and other construction materials, and it is possible that the older exploitations have been destroyed. More fieldwork could answer this question.

Non-volcanic rocks were also exploited inside the Campos de Calatrava district. This is the case at Las Canteras (72) near Granatula de Calatrava on the Jabalón riverbank opposite to the Roman ruins of Oretum. During a recent visit, it was covered by floodwaters, so collecting data about extraction size and petrography was not possible. The site is visible, however, on a television documentary (Labordeta, Un país con la mochila). According to this work, it has large cylindrical imprints carved into a sedimentary (conglomerate?) outcrop. From the sizes of the cylinders, it appears to be recent.

Two other sites in the province outside the Campos de Calatrava volcanic fields are Pedrizas de Piédrola (73), in the northeast near Alcázar de San Juan and Chillón (74) in the southwest. We have practically no data about these sites, except that granite was worked at Chillón (Madoz 7, 327).

The quarry near Abengibre (75) in Albacete province has retained the toponym El Molar (Madoz 1, 256). Madoz comments that the grains of this stone are not very compact thus it is less appreciated than the millstones of Barcelona (Madoz 6, 52). This is a direct reference to the most celebrated Iberian millstone quarry of Monjuïc, outside Barcelona, that produced and exported excellent millstones over long distances, even to France and to Italy. The simple comparison with Monjuïc, approximately 500 km away, suggests that Abengibre produced an excellent product. The depiction of a millstone on the heraldic shield of the municipality indicates the important role this industry must have played in the community.

To date we have identified only two millstone quarries in the province of Toledo. The first is Torrecilla de la Jarra (76), referred to in a letter of 1782 responding to a series of queries about the natural resources of the town. Don Lucas Fernández de la Sierra responds that there are no exploitations except for a few cantos (surface blocks) that are carved into millstones for flour mills.

The second site, Las Ventas con Peña Aguilera (77), is found along its southern border. According to the early 19th century geographer Miñano (9, 286), many *berroqueño* (granite) millstones were produced and sold to regional mills as far as 30 leagues away (approximately 120 km). Madoz (15, 662), referring to the town a generation later, cites an 'infinite' number of granite quarries without specifically indicating a millstone production.

The only site we have identified to date in the Province of Cuenca is that of Los Molares (78) outside of the town of Portilla at an altitude of 1200 m. This, for the moment, is the site of our survey found at the highest altitude. Its identification is due to a hiking itinerary organized by the municipality.

5.4. Autonomous region of Madrid

In the relatively small Autonomy of Madrid we have identified four millstones quarries. El Berrueco (79) is about 50 km to the north of the capital and is a stop on a geological itinerary (Díaz Martínez 2005, 17) and has an open air museum opened in 2001 (Museo de la Cantería) dedicated to recovering the stone work tradition. Both conical oil and cylindrical flour millstones were carved from a vast granite outcrop at the foot of the Sierra de Guadarrama mountains. Two other similar sites in the heart of a vast granite outcrop are reported by an old text. The first is Colmenar Viejo (80), about halfway between the capital and Berruecos, with many granite exploitations that produced millstones for four mills (Madoz 6, 530). The second is Chapinería (81), 45 km to the west of Madrid, that had 'good' quarries for both millstones and building stones (Miñano 3, 83).

The last site is Colmenar Oreja (82), about 40 km to the southeast of the capital city. This site is located in a different geological zone, dominated by Tertiary sedimentary deposits (Díaz Martínez 2005). It is cited by three 19th century references. The first (Vallejo 1833, 387) indicates its millstones were traded to Madrid and measured one *vara* in diameter (i.e. 0.84 m). The author goes on to specify that they can grind for 6 straight hours until they have to rest and be dressed, a two hour process. Furthermore, if necessary, these millstones could grind all day and night, provided they were dressed every 6 hours. Toward the middle of the century Madoz alleges that the white stone was used for the most part for millstones for the *tahonas* (animal flour mills) of Madrid (6, 525). Finally, an anonymous report in a Journal of Public Works published 50 years later indicates that the millstone exploitation for animal driven mills was concentrated on the third layer, the *sobrebanco* (0.56 to 0.84 m thick) of a compact limestone.

5.5. Extremadura (Provinces of Badajoz and Cáceres)

Six millstone quarries have been identified in Extremadura. The three in the Province of Badajoz are in the southeast in the Sierra Morena. The first is near Alconera (83) and is mentioned in a text dating to the end of the 18th century (Marin 1791) that makes reference to a large millstone factory. A half a century later Madoz specifies that the exploitation is of a spongy tuff stone (Madoz 16, 435). This does not refer to a volcanic tuff, but rather a highly porous limestone that we have seen used for millstones elsewhere. The second site is just outside the town of Jerez de los Caballeros in the area called Carrascosa (84). Here Madoz identifies quarries of both dark and white rocks for flour mills (Madoz 9, 627). The third site is in Salvaleón (85), about 20 km to the north, where there is, once again, an exploitation of white stones (Madoz 13, 711).

In the province of Cáceres, the first site is Villar de Plasencia (86), where Miñano alleges that there are many *canchales* (screes) in the Sierra de Béjar to the northeast of the town that were exploited for millstones. These were reputedly transported distances from 4 to 6 leagues (20 to 30 km) (Miñano 9, 433). The remaining two sites are on the eastern border near the towns of Bohonal-de-Ibor and Logrosan. At Bohonal-de-Ibor (87) Madoz alludes to *berrocales*-- usually granites -- and to other quarries of very hard stones (Madoz 4, 375-376). The site of Logrosan (88), to the southwest of the town on the Cerro de San Cristobal (Madoz 10, 355), worked a *piedra de grano* (probably a granite) for both millstones and other products related to the olive oil and wine industries.

6. Conclusions and perspectives

6.1. Sources and place names

Old texts, in particular the work of the geographer P. Madoz, are by far the most effective means of identifying millstone quarries. This is followed by recent local and regional molinological studies and geological reports. Hiking itineraries and other sources related to tourism complete the list. The new techniques of searching and consulting these sources through the internet facilitate this procedure. All of these diverse sources, however, require verification in the field because they only rarely provide data about the type of product, the techniques of extraction and the larger question of distribution. We are, therefore, only scratching the surface of what is a vast, largely untapped, subject of research.

Although we have not undertaken a systematic study of place names, there are some recurring toponyms that are worth noting. These include derivatives of the latin term *mola* such as *Molares*, *Molar*, *Amolar* and possibly, *Morales*. Other names are *Berrueco*, presumably of pre-Roman origin associated with granite outcrops and simply *cantera*, meaning quarry.

6.2. Chronology

The majority of the identified sites, as would be expected, are very recent, dating to the last few centuries. There are very few that can be attributed with any degree of certainty to earlier periods. The oldest sites like Cerrro de los Limones and *Sisapo*, of volcanic rock, date presumably to Roman times. The earlier Iron Age Iberian Culture, however, is not represented. Neither do we have any unquestionable sites from the early Middle Ages or the Moorish domination. It is possible that some of the modern sites like Alacalá la Real and Moclín, with signs of rotary quern extractions, hide or have destroyed older phases of exploitation.

6.3. Petrography

The general tendency of the petrographical determinations of the 18th and 19th century texts, as well as our observations of the visited sites, seems to correspond to the determinations defined in recent geological research. There are exceptions like the 'granite' of Alhaurin El Grande (Málaga), a sector of sedimentary deposits devoid of granite, that are obviously mistakes on the part of the early geographers. In general, there are very few sites that have benefited from geological analyses. A systematic sampling of the sites for petrographical determination and hardness of the rock is a project that is far beyond our means.

6.4. Extraction techniques

In what concerns extraction techniques, there is very little data available. The only tool marks that we can observe consistently, when the surfaces are not too weathered, are multiple diagonal lines on quarry faces, typical of cutting circular channels with the pick (Fig. 18). The width of certain circular trenches, surrounding large cylinders, suggests the use of multiples 'passes' with the pick. This term is borrowed from the building stone quarry specialist (Bessac 1996). In certain cases we have observed regularly spaced triangular or trapezoidal holes at the base of cylinders used to detach the cylinder, probably with iron wedges. The few cases of one large wedge slot might indicate the use of water covered wooden wedges. It is interesting to note the contrasting techniques for the two volcanic Roman sites. At the Cerro de los Limones, angular blocks were probably detached using levers, leaving no visible tool marks, while at Sisapo querns and millstones were carved directly into the volcanic mass leaving circular imprints.

6.5. Millstone quarry typology

From the small number of sites that we have been able to observe in the field we can propose a first elementary typology of Iberian millstone quarries.

The exploitation of surface boulders (screes, riverbeds or erratic blocks) for one or a few millstones, is the most simple type of exploitation. Although not confirmed in our study area, this type is known in the north of Spain and elsewhere in Europe. A variation of this quarry-type, also not confirmed, is the exploitation of isolated volcanic 'bombs' in the volcanic districts.

The second quarry-type is more sophisticated and delivered multiple products. It employed several people with specialized skills, yielded standardized products and benefited from a commercial network. This type can be subdivided into several classes.The first is the exploitation of screes such as that of Alcalá la Real (Jaén). This type of quarry is difficult to identify and the working debris is easy to confuse with natural debris.

The second class of quarry is the true extractive exploitation that detached millstones from the bedrock. The multiplication of extractions over time creates a vertical or lateral cavity in the terrain. This type has several categories.

The vast Roman quarry of Cerro de los Limones, exploiting what appears to be the summit of columnar jointed blocks, would also fall into a first category. Since the direction of work is vertical, there is no real quarry face. There is, however, a slight depression in the terrain. This particular site also shows a large amount of working debris.

The second category is that of Fuente de los Morales and Caniles (Granada), where a specific stone layer was sought from which angular blocks were detached and later carved into cylinders. This resulted in a long, linear, quarry face and huge mounds of working debris. Apart from an occasional abandoned cylinder, this type does not leave any material indication of its nature.

A third category of this quarry-class is the true extractive exploitation where cylindrical 'drums' are carved directly into the bedrock leaving circular imprints. In these cases the use of the stone is more economical and there is less debris. A simple version of this type is that of El Torcal (Málaga) and Camino del Calvario (Granada) where the small sizes of the outcrops did not permit side-by-side extractions. On the other hand, there is the case of Trafalgar (Cádiz) where the extension of rock layer permitted multiple, side-by-side extractions following the natural sedimentation.

A more complex variation of this category, in the cases where the rock is massive and homogeneous, permitted the quarrymen to superimpose side-by-side extractions. Over time, due to a lateral direction of work, this type of quarry becomes concave-shaped. This technique produced 'tube-like' quarry faces such as those that can be seen at Moclín and Los Guillares (Granada). This final type, obviously, is the easiest to identify in the field.

There are, of course, quarries of mixed type, like Vélez de Benaudalla (Granada), where both scree and bedrock were worked. All of the sites we have identified are open-air quarries. At the moment no subterranean exploitation has been identified. We suppose this is a question of the state of research, since underground exploitations, although not frequent, do exist in other countries, notably the Cave of Hercules in Tangiers, Morocco, the Mont Vouan (Haute Savoie) in France and the kilometers of galleries in the volcanic district of the Eifel in Germany.

6.6. Production and distribution

The models of production and distribution of the Iberian millstone quarries is as elsewhere in Europe. There appear to be local 'village' quarries with a small number of extractions serving the nearby mills. Such is the case of Guajar Faraguït (Granada) and Guardias Viejas (Almería). The larger regional quarries producing higher-grade products, such as Moclín and Fuente de los Morales (Granada), are identified as regional quarries by written sources. The only quarries that might have had a wider distribution, surpassing the regional scale, are the Roman volcanic quern exploitations of Cabo de Gata and Campos de Calatrava.

6.7. Dark bread, white bread

As we have seen throughout this article, based on old texts, some millstones yield flour for white bread and others for dark bread. This dichotomy is also found in France where this line of research is led by A. Belmont (Belmont 2006, for example). There is certainly a relation, due to the fine stone particles released from the rock that end up in the flour, between the hue of the millstone and the color of the bread. The animal driven mills of Madrid, equipped with the hard, white limestone millstones from Colmenar Oreja that required dressing every 6 hours, must surely have released much powder into the flour. These white particles, although certainly a health hazard to the dental health, might not have affected the color of the bread.

Reyes Mesa, in an article about the old mills in the region of Granada, provides an interesting clue to this question. He states that there is confusion about how the term *pan baxo* is interpreted by some 16th and 17th century documents as 'flat bread' (Reyes Mesa 2000, 18). He goes on that this is a misinterpretation and that *pan baxo* is not flat bread, but rather bread made of *harina baxa* (flour) produced by piedras *baxas*, that is millstones with a rugged texture. These millstones did not separate the bran completely from the flour, yielding a dark bread. This darker bread - of color similar to that of the stones - thus received its name from the millstones (Reyes Mesa 2000, 17).

Therefore, based on the old texts and our field observations, the quarries that produce white bread show the tendency to be of varieties of white limestone (for example, Fuente de los Morales in Granada and Berrueco in Cádiz) while those that yield dark flour are conglomerates such as Caniles in Granada. There is, however, without a doubt still much to be written on this subject.

6.8. Perspectives

As we have noted previously, research on the subject of millstone quarries in the Iberian Peninsula is in an embryonic stage. The study of geographical sources is just an initial phase of a much more complex future study. There are certainly other old texts, such as commercial transactions of notaries that might shed light on the subject. There are also certainly texts in arabic dating to the 700 years of domination that have not been exploited. There is certainly still much to do in the sphere of petrographical and morpho-metric analyses of both individual millstones and the extraction sites to start filling in the many typological and chronological gaps. On a final note, contrary to France, Switzerland and Germany, no millstone quarry in the Iberian Peninsula has yet benefited from a modern archaeological excavation that would surely advance the state of research.

Acknowledgements

TA would like to thank Gurli Meyer, Tom Heldal and Tor Grenne of the Norwegian Geological Survey (NGU) for their support and assistance in the identification of the quarries of El Torcal, Guardias Viejas, and the volcanic sites of Cabo de Gata and Campos de Calatrava; Loïc Martinez for his knowledge of the Cabo de Gata sites; Miguel Angel Vargas, for information about the site of Almadén de la Plata; Aurora Pulido for her artwork, council and patience; and Jeff Hodges for proof reading. JHS was financially supported by the Spanish grant CLG2008-02864 and the Andalusian grant RNM1595.

References

Alonso, N., 1999. *De la llavor a la farina, Els processos agricoles protohistorics a la Catalunya Occidental. Monographies d'archéologie méditerranéenne*, 4, Lattes.

Alonso, N., and Anderson, T., Forthcoming. The current state of research on rotary querns in the Iberian Peninsula. *Proceedings of the Electronic Publications and Knowledge bases*. The grinding materials as a case study, Workshop, Roux, V. & H. Procopiou (coords.). Paris, INHA January 28-29, 2010.

Anonymous 1896. Canteras Españolas, Caliza de Colmenar. *Revista de obras públicas, Boletín*, 4, 145.

Anderson, T., Forthcoming. Un premier bilan sur la production de meules dans la péninsule ibérique de la Protohistoire à la période médiévale. In O. Buchsenschutz, L. Jaccottey and F. Jodry (coords.), *Evolution typologique et technique des meules du Néolithique à l'an mille sur le territoire français*, Actes de la Table Ronde à St. Julien-sur-Garonne, 2 au 4 octobre 2009.

Anderson, T., Fernández Soler, J. M., and Grenne, T., Forthcoming. A preliminary sketch of quern and millstone production in the Spanish volcanic regions of Cabo de Gata (Almería) and Campos de Calatrava (Ciudad Real). In Proceedings of the *ASMOSIA IX* Conference, June 8-13, 2009. Tarragona, Spain.

Belmont, A., 2006. *La Pierre à pain, Les arrières de meules de moulins en France, du Moyen Âge à la révolution industrielle*, 2. Grenoble.

Beneroso Santos, J., 2007. Una aproximación a la toponimia medieval musulmana en el término de San Roque, Alameda. *Nuestra Historia* 13.

Benítez de Lugo Enrich, L., 2001. El registro arqueológico en Alhambra (Ciudad Real). *Cuadernos del Instituto de Estudios Manchegos-C.S.I.C*, 23-24, 9-25.

Bergnes y de las Casa, A., (ed.), 1831-1833, *Diccionario geográfico universal, por una sociedad de literatos (S.B.M.F.C.L.D.)* Barcelona.

Bessac, J.C., 1996. *La pierre en Gaule Narbonnaise et les carrières du Bois des Lens (Nîmes). Histoire, archéologie, ethnographie et techniques* (Journal of Roman archaeology, supplementary series16). Ann Arbor.

Bisso, J., *et al.*, 1869. *Crónica General de España, ó sea historia ilustrada y descriptiva de sus provincias.* Málaga.

Borges, N.C., 1978. Mós Manuais de Conimbriga, *Conimbriga*, 17, 113-32.

Cara Barrionuevo, L., García López, J. L., Lentisco Puche, J. D., Ortiz Soler, D., 1999. Los molinos hidráulicos tradicionales de la Apujarra (Almería). *Instituto de Estudios Almerienses*, 32. Almería.

Castella, D., 1994. L*e moulin hydraulique gallo-romain d'Avenches En Chaplix, Fouilles 1990-1991. Cahiers d'Archéologie Romande*, 62.

Checa, A., Jimeno, A., Tresserras J.J., Benito J.P., and Sanz, A., 1999. Molienda y economía doméstica en Numancia. In Burilllo (coord.), *IV Simposio sobre los Celtíberos, Economia, Homenaje a José Luis Argente Oliver, Daroca 1997*. Zaragoza, 63-8.

de la Cruz y Bahamonde, N., 1813. *Viaje por España, Francia è Italia*. Cádiz.

Díaz Martínez, E., 2005. Itinerario geológico por el norte de la Comunidad de Madrid. Paseos por la Geología Madrileña, 1. Archivo PDF para la página web de la Comunidad de Madrid.

Espinosa, M., 1848. Tera. In M. Fernández de los Ríos (éd.), *Seminario pintoresco español*. Málaga, 265.

Ezquerra Del Bayo, J., 1856. Ensayo de una descripción general de la estructura geológica del terreno de España en la Península. In *Memorias de la Real Academia de Ciencia Exactas, físicas y naturales de Madrid*, 351-99.

Farias, P., Gallastegui, G., González Lodeiro, F., Marquínez, J., Martín-Parra, L.M., Martínez Catalán, J.R., de Pablo Maciá, J.G., Rodríguez Fernández, L.R., 1987. *Aportaciones al conocimiento de la litoestratigrafía y estructura de Galicia Central.* Memorias do Museo e Laboratorio Mineralogico e Geologia, Faculdade de Ciências da Universidade do Porto 1, 411–31.

García de la Leña, C., 1789. *Conversaciones históricas malagueñas ó Materiales de noticias seguras para formar la historia civil, natural y eclesiastica de la M.I. ciudad de Málaga*. Málaga.

Garrido Palacios, M., 2001. Julio Caro Baroja y los molinos de Puebla de Guzman. *Revista de Folklore*, 21a, 164-7.

Gómez de Bedoya, P., 1765. *Historia Universal de la fuentes Minerales en España, T2*. Santiago.

Gomez Ruiz, R., 2003. *Molinos en el Río Odiel, Un estudio de arqueología industrial en los límites de El Andévalo.*

Gozálvez Pérez, V., and Rosselló Verger, V. M., 1978. La Serra del Molar y sus yacimientos pleistocenes. *Cuadernos de Geografía*, 23, Valencia, 107-22. lnstituto Geological y Minero de Espana (IGME), 1994. *Mapa geologico nacional,* Escala 1.1.000.000.

López Cordero, J.A., and Cabrera Espinosa, M., 2004. *Patrimonio histórico-cultural de Arbuniel, Sumantán* 20, 185-218.

Madoz Ibánez, P., 1845-1850. *Diccionario Geográfico, estadístico, histórico de España y des sus possessiones de Ultramar*. Madrid, 1-16.

Maier, J., 1999. Jorge Bonsor (1855-1930). *Un académico correspondiente de la Real Academia de la Historia y la Arqueología Española*. Madrid.

Marín, P., 1791. *Interrogatorio formado de orden del Consejo, para la visita de la provincia de Extremadura que deben hacer el regente y ministros de la Real Audiencia.*

Martín-Algarra, A., (coord.) 2004. Zonas Internas Béticas. In. J.A. Vera (ed.) *Geología de España*, SGE-IGME. Madrid, 395-437.

Martínez y Delgado, F., 1875. *Historia de la ciudad de Medina Sidonia*. Cádiz.

Martínez López, F., and Granero Gallego, A., 2005. Cantera de piedras de moler. In *Actas del IV Congreso Internacional de Molinología, Mallorca*, 1-3 de mayo, 2003, Vol. 2. Palma de Mallorca, 125-42.

Matilla Séiquer, G., 2001. Canteras de piedras de almanzara en Fortuna. *Revista Murciana de antropología*, 7, 267-77.

Mejía Asensio, A., Salgado Olmeda, F., Rubio Fuentes, M., Martín Galán, M., 2007. *Historia moderna de la provincia de Guadalajara*. Ann Arbor.

de Miñano y Bedoya, S., 1826-1829. *Diccionario Geográfico-estadístico de España y Portugal*, 1 -10. Madrid.

Montero Bastarreche, A., 2008. Origen de la piedras de moler de Córdoba. In J.M. Garrido Aranda, M. F. Moreno Pérez, and J. Roldán Caña. *Actas del 6o congreso internacional d Molinología*. Córdoba, 161-74.

Portillo Ramírez, M., 2006. *La mòlta i triturat d'aliments vegetals durant la protohistòria a la Catalunya Oriental*. Thesis, Universidat de Barcelona.

Reyes Mesa, J.M., 2000. Tecnología y arquitectura popular. Los molinos hidráulicos en la provincia de Granada. *Gazeta de Antropología*, 16, 16-21.

Rodríguez Monteoliva, F., 1989. Los molinos de harina en la Alpujarra de Granada, durante los siglos XVI al XVIII. Léxico, etnografía et historia. In *Coloquio de historia y medio físico, I, Instituto de Estudios Almerienses*, Departamento de Historia, 683-709.

Rosselló Verger, V., 1991. *Valoración científica del litoral*

Alicantino. Investigaciones Geográficas, 47-54.

Torres Montes, F., 1992-1993. Los antiguos molinos de agua de la ribera de Huebro (Estudio etnográfico-lingüístico). *Boletín del Instituto de Estudios Almerienses*, 11-12, 255-88.

Williams-Thorpe, O., and Thorpe, R.S., 1987. Els Origens Geologics dels Molins Romans de Pedra del Nord-Est de Catalunya. *Vitrina*, 50-8.

Vallejo, J.M., 1833. *Tratado sobre el ovimiento y las aplicaciones de las Aguas*, 2. Madrid.

Vera, J.A., (ed.) 2004. *Geología de España*. Edited by Sociedad Geológica de España (SGE)/Instituto Geológico y Minero de España (IGME). Facultad de Ciencias, Universidad de Salamanca.

The quern and millstone quarry of the Rambla Honda, Almería, Spain

Francisco José Martínez López, Timothy Anderson and Antonio Granero Gallegos

1. Introduction

The quern and millstone quarry of the Rambla Honda (Fig. 1) is located in the municipality of Albox in the heart of the arid Almanzora Valley of the Province of Almería in the south-eastern corner of Andalusia (Fig. 2). This valley is a natural passageway between the Mediterranean coast and the hinterland and has a long tradition of human occupation, notably the site of El Argar, that has lent its name to the brilliant early Bronze Age Culture of the southeastern Iberian Peninsula. It is, nonetheless, the Moorish domination, between the 8th and the early 15th centuries AD, with its settlements and fortifications, that has left the most visible marks on the landscape.

The name of the site derives from the deep (*honda*) dry river-bed or gulch (*rambla*) that winds through the landscape. Its discovery dates to October 1973 when a disastrous torrent, surpassing 300 litres a square metre, washed away the deposits concealing the site (León 2008). The flood also probably displaced some of the blocks toward the river-bed and might even have buried other sections of the site. The nearby construction of the Negratín-Almanzora aqueduct thirty years later probably also affected part of the site.

This millstone quarry, after that of Fortuna in the Province of Murcia (Séiquer 2001), is the second site of this type in

Figure 1. View of part of Zone 1 of the millstone quarry of Rambla Honda.

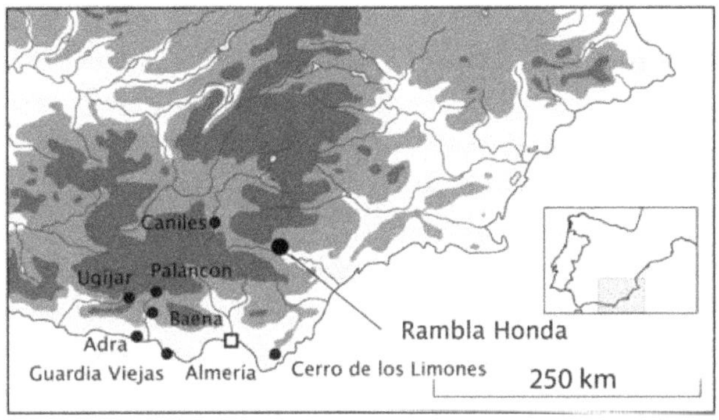

Figure 2. Location of the Rambla Honda millstone quarry in southeastern Spain as well as other millstone production sites cited. The three shades of the topographical map correspond respectively to intervals of 500 metres.

the southern half of the Iberian Peninsula to benefit from a scientific publication (Martínez and Granero 2005). The working hypothesis of the initial article of 2005 speculated that the products of this quarry were rollers for the regional *almazaras* (olive oil mills). In this current study we revise this initial interpretation. Olive oil rollers such as those at Fortuna (Séiquer 2001, 273) are either conical or truncated cones, while the extractions from this quarry are discoidal, with the ratio of diameter about three times that of the width. The cylinders from the Rambla Honda are therefore most likely grinding stones for cereal mills. Moreover, the rock, a rough conglomerate, is well known to have been traditionally exploited for flour mills in both southern Spain and elsewhere in Europe.

2. Petrography

A general sketch of the geology of the southeast of the Iberian Peninsula is described by Jane H. Scarrow (Department of Mineralogy and Petrology, University of Granada) in a related article in this volume (cf. Millstone quarries in Southern Spain) based on the monograph *La Geología de España* (Vera 2004). More specifically, the Rambla Honda quarry is located between the Alpujárride and Nevado-Filábride Complexes of the Internal Zone (Fig. 3) of the Penibetic range. This sector is especially known for its exploitations of iron ore and, at Macael, for a fine white marble that probably has over the years rivaled

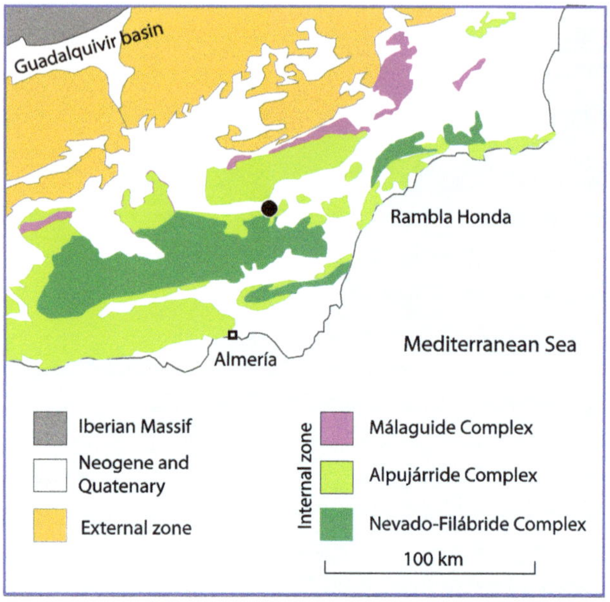

Figure 3. The location of the Rambla Honda quarry in relation to the principal geological formations of the Bétic range (after Martín-Algarra 2004).

the celebrated Carrara marbles of northern Italy.

The more mundane conglomerate layer exploited for millstones, between 1.30 and 1.50 m thick, probably corresponds to post-Alpine sedimentary deposits placed above the original metamorphic basements. One sample of the rock has benefited from a petrographical portrait by J. I. Soto of the Department of Geo-dynamics of the University of Granada. The following is his description:

'The sample is an unsorted detritic conglomerate formed by clasts of different size and lithology, cemented probably by calcite (Fig. 4). There is also a perceptible matrix and all the clasts are tightly glued by the carbonate cement. The clasts vary in size from 2.5 - 3 cm to 0.5 cm, and the smaller fragments form part of the matrix with a constant size under 2 - 3 mm. They are of different compositions and indicate a polymictic origin. The majority have a carbonate composition, corresponding to light-colored limestones (light yellow and grey), although there are also a few dark-colored planar limestone fragments. Although the larger fragments are generally rounded and elongated, there are some cases presenting angular edges. These cases, less than 10-15% of the rock fragments in the sample, correspond mostly with zoned chert (light-grey and white) and occasionally quartzite (light-brown). Some of these more rare fragments present a cubic shape with angular cutting edges. From the composition of the different clasts it can be established that the hardness of this rock varies strongly from the calcite-rich (and more abundant) clasts to the quartz-rich rock fragments (quartzite and zoned chert). The fragments of the first type have a Mohs hardness of about 3, whereas the quartz-rich ones have a hardness of 7 (the maximum hardness in the Mohs scale is 10 and corresponds to diamond).'

Figure 4. Macroscopic view of the surface of the conglomerate. The clasts are generally between 2.5 and 3 cm.

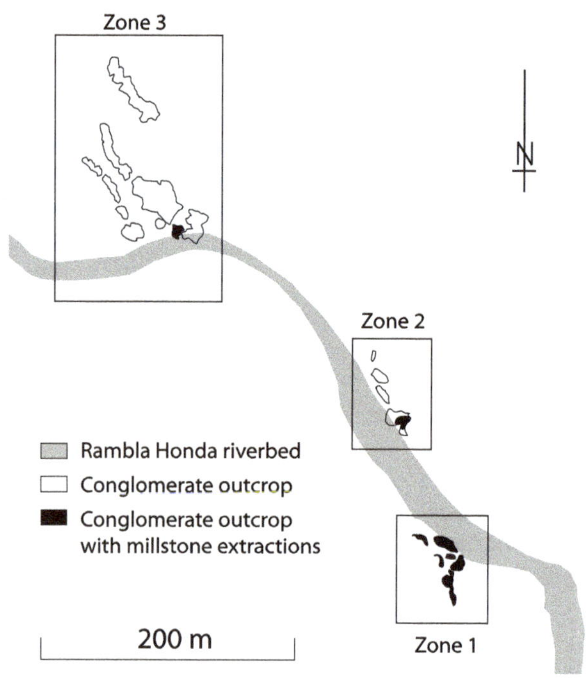

Figure 5. Schematic map of the winding Rambla Honda river-bed and three extraction zones.

The contrast of hardness between the various types of inclusions and the matrix is most likely the factor that guaranteed the 'bite' of the rock in the grinding process. It was probably this property that led the quarrymen to exploit this rock.

3. The Quarry

The whole of the quarry is spread out over a surface of roughly 500 m long between two bends in the river-bed. It is divided into three zones (zones 1, 2, and 3; Fig. 5). Zone 1, on the west bank, is the main extraction area with the highest concentration of circular imprints. Zone 2, slightly to the north on the eastern bank, has only one millstone

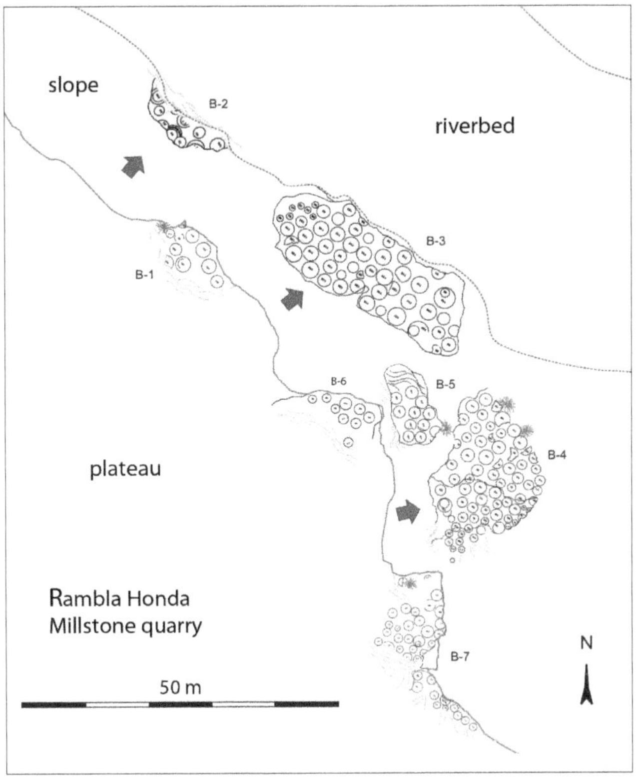

Figure 6. Schema of Zone 1 of the Rambla Honda quarry. Blocks 1, 6 and 7 are in their original position bordering above the river-bed. Blocks 2, 4 and 5 are displaced on the slope above the river-bed. Blocks 2 and 3 are steeply inclined and partially buried in the river-bed.

extraction. Zone 3, still further to the north on the eastern bank, is dedicated principally to extraction of ashlars. There are, however, indications that in this last sector both medium-sized and large millstones were previously exploited.

3.1. Zone 1: extraction of querns and millstones

Zone 1, composed of 7 major blocks, is the southern-most sector of the quarry and shows the highest density of quern and millstone extractions. After the flood of 1973 it was known as the 'place of the holes'. It is located on the west bank of the riverbank (Fig. 6) at a slight bend, a feature possibly provoked by the presence of the massive bedrock.

Figure 7. View from the southeast of part of the quarry. Block 3 is in the foreground and Block 1 is perched above it on the edge of the plateau.

The proximity of the conglomerate layer to the surface, and its exposure on the edge of the river-bed, made it readily accessible to the quarrymen. A soft deposit below it, easily washed out by erosion, explains why several of the blocks were detached from their original position and displaced during periodic flooding.

Three of the blocks (B-1, B-6 and B-7) are in their original horizontal position on the top of the plateau, perched about 10 meters above the river-bed. One of them (B-1) hangs dangerously in the air and will probably soon fracture and plunge below under its own weight (Fig.7). Blocks 4 and 5 are also detached and now on the slope bordering the river-bed. Their base is slightly buried in the sand and gravel. Blocks 3 (Fig. 8) and 2 (Fig. 9) travelled even further and are largely buried in the river-bed and present sharp inclinations of about 50°.

The circular imprints of cylinder extractions are generally placed side by side (Fig. 10), indicating an economical use of the raw material. Only one or two levels of the rock were exploited. There are relatively few superimposed extractions and in no case has the total width of the rock been exploited. We do not know why the quarrymen limited their extractions to the upper levels of the rock layer avoiding superimposed extractions. They may have considered that they had enough raw material available

Figure 8. Panoramic view from the northeast of Block 3.

Figure 9. View from the southeast of Block 2 (displaced by the flash flood).

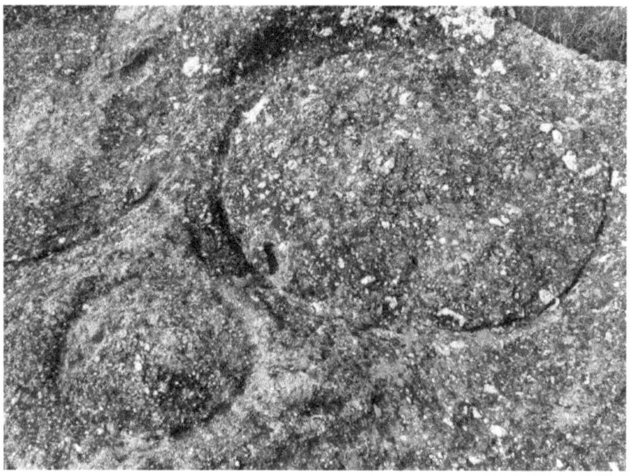

Figure 10. Detail showing both quern (approx. 40 cm) and millstone extractions.

close to the surface, permitting them to progress in a horizontal sense and avoid the complications inherent with vertical superimposed extractions. Another explanation is that the quarryman may have decided that the lower half of the rock was not apt for millstones. We, however, have not observed any notable difference in the stratification of the rock.

There are no visible heaps of spoil or working debris. It appears that the flooding also removed these features that would have most likely accumulated in the river-bed.

For the most part, the extraction planes follow the layers of sedimentation of the rock, resulting in horizontal extractions. In the cases where the blocks have been naturally displaced and are now on an inclined plane, the extractions are also inclined revealing that they took place before the block was detached. In the case of Block 3, however, there is a series of horizontal extractions on the top of the block (Fig. 11) that suggest that millstones were also hewn after the block was detached from its original position.

Figure 11. Detail of Block 3 located in the river-bed showing extractions on an inclined plane (right) and extractions on a horizontal plane (left) that might represent two phases of work, before and after the block was displaced.

An intense weathering of the surface of the rock has erased most of the original tool marks hindering the observation of extraction techniques. Circular channels surrounding the cylinder are traditionally cut with an iron pick, leaving multiple diagonal marks on the working faces. We do not have any indication that this traditional technique has not been put to use, but, once again, the weathering of the rock probably erased these marks. In what concerns the means of splitting the cylinder from the bedrock, we do not have much more information. Small, relatively deep, inclined wedge holes, are clearly visible in the case of a few large extractions (Fig. 12). They probably correspond to iron

Figure 12. View of a series of wedge holes used to split the cylinder from the bedrock.

wedges, although the option of wooden wedges or pegs (or other means of splitting such as levers) cannot be ruled out.

A total of 203 circular imprints have been identified in this area. This is, of course, a minimum count since an unknown number have probably disappeared. Of the 203, we have been able to estimate the diameter of 158, about two thirds of the lot. The dimensions of the others cannot be measured precisely due to the weathering.

The result of these measurements indicates that the diameters of the cylinders extracted range from 40 to 140 cm. A simple graph displaying the total number of extractions with regards to the diameters (in increments of 5 cm) clearly reveals four categories of sizes of circular extractions (Fig. 13). To the bottom left of the graph appear a total of 11 small imprints between 40 and 50 cm. The next category, after a notable hiatus of 25 cm, corresponds to cylinders between 75 and 100 cm. The next category, between 105 and 120 cm, with over 80 extractions, is the class that prevails by far. Finally, we have a small group of about 20 very large cylinders ranging from 125 to 140 cm.

The nature of the first category, between 40 and 50 cm in diameter, is quite obvious. These are rotary querns that can be driven by one person. The nature of the larger three categories of cylinders is not secure. It would be tempting to speculate that the smaller millstones between 75 and 100 cm were destined for a specific type of mill such as animal driven installations, while the other two larger categories (between 105/120 and 125/140 cm) were meant for either windmills or watermills. Due to the absence of morphometric studies of querns and millstones in this area, however, these ideas remain in the realm of conjecture.

The spatial distribution of the four categories of extractions (Fig. 14) is not easy to interpret. If we accept the idea that the rotary querns are the older production, then the group of seven extractions on the northeastern corner of Block 3 (Fig. 15) correspond to the first phase of exploitation of the site. There are, however, several other small rotary quern extractions elsewhere on Blocks 4, 6 and 7 intermingled with larger extractions that would seem to refute this notion. In addition, we know that the rotary quern did not die out in Antiquity but lasted even until very recent times. All of the other larger extractions do not show any particular clustering, suggesting that millstones of different

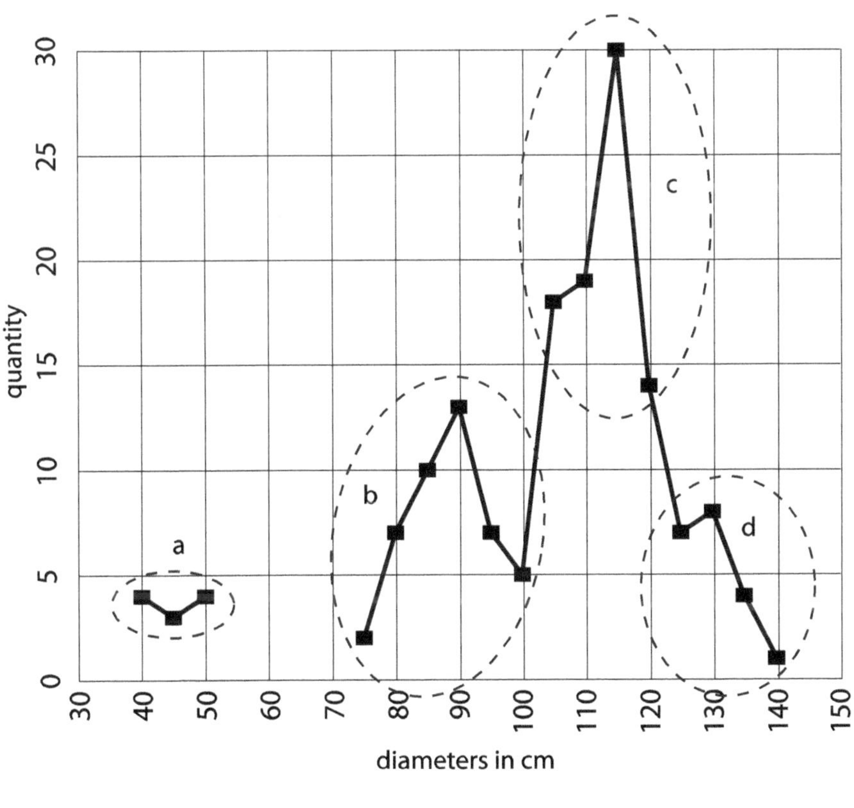

Figure 13. Line graph revealing the four categories of extractions. a) 40-50 cm. rotary querns, b) 75-100 cm. small millstones, c) 105-120. medium-sized millstones, d) 125-140 cm. large millstones.

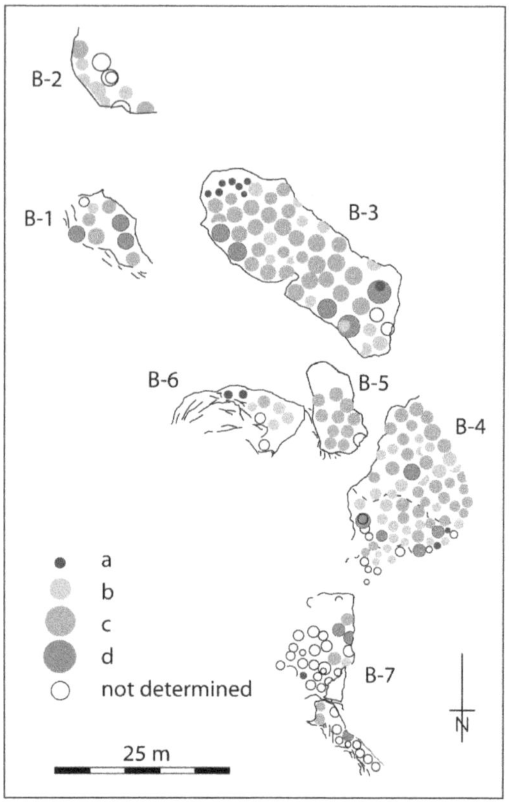

Figure 14. Spatial distribution of the four categories of millstones based on the measurements (in cm) of the extractions.

Figure 15. Detail of the northern corner of Block 3 with the group of smaller rotary quern extractions.

dimensions were extracted simultaneously. We therefore do not have the impression that the different categories are extracted in any particular chronological order.

3.2. Zones 2 and 3: millstones and ashlars

Zone 2, on the opposite bank of the river-bed about 100 m to the north, presents only one isolated cylindrical imprint measuring about 0.80 m in diameter. We do not know why the large number of easily accessible blocks in this area where not exploited.

Figure 16. Millstone extraction in Zone 3 of the Rambla Honda quarry. This area was for the most part exploited in recent times for ashlars.

Zone 3, further to the north, is a much larger zone composed of about 20 blocks arranged in three parallel lines. The extractions now visible here corresponds to rectangular ashlars dating, according to oral information, from the middle of the 20th century. There are also remnants indicating the extraction of both medium and large millstones (Fig. 16).

4. Conclusions

The chronology of the site is far from secure. The older residents of the area do not recall the quern and millstone extractions, so a very recent date can be discarded. There are no text citations, for example, in the geographical dictionaries of the 19th century of Pascual Madoz and Sebastian Miñano. This absence, however, might simply be due to an omission or that the site was not deemed important enough to be mentioned. The exploitation of small rotary querns, in combination with the other larger millstones, suggests an early date, possibly in the Middle Ages, during the Moorish domination. On the other hand, the very large millstone extractions between 1.20 and 1.40 m in diameter, destined for larger more complex mills, are probably not more than a few hundred years old.

The precise magnitude of the quarry also escapes us. Portions of it could be hidden, either under the alluvial deposits of the river-bed or under the topsoil on the upper plateau in the almond orchards. The elevated number of confirmed extractions in Zone 1 (203), however, tends to tilt the balance toward a quarry that marketed its products on more than just a local scale. It was therefore not simply the village quarry that traded its products to the surrounding towns and villages. There are no major topographical obstacles hindering access to even larger markets.

Conglomerate millstones for cereal mills are known over large portions of Europe. As we stated before, the variable hardness of the pebble inclusions rendered it particularly apt for grinding cereal. In the nearby site outside of Caniles in the Province of Granada, about 45 km to the west, a rock with very similar characteristics was also exploited for millstones. In a reference to Caniles, Madoz in the middle of the 19th century indicates that its millstones yielded flour for dark bread (Madoz 1845-1850, 5, 461). Due to the similarity of the rock, we suppose the Rambla Honda millstones also produced dark bread flour.

H. Amouric has noted an interesting development in southeastern France somewhere in the period of transition from the end of Antiquity to the Middle Ages. The basaltic lavas of the volcanic districts (for example Agde and probably certain regions of the Massif Central), as well as the imports from elsewhere in the Mediterranean, were at some point abandoned in favour of other types of stones, notably conglomerates (Amouric 1991, 444). This shift away from volcanic rocks must be related to the quality of grinding and possibly a more economical raw material. This model seems to apply equally to the southeastern corner of Andalusia. The antique exploitations of volcanic outcrops, notably the quern quarry of the Cerro de los Limones in the Cabo de Gata area (Anderson forthcoming), were also abandoned at some point in time in favor of other rock types, notably conglomerates. In the register of extraction sites in southeastern Andalusia there are, besides the Rambla Honda, conglomerate exploitations known (cf. Fig. 2) at Caniles (Madoz 1845-50, 5, 461), Adra, Guardias Viejas, and Ugíjar (Cara Barrionuevo *et al.* 1999, 153).

On a final note we lament that the site of the Rambla Honda, like the multitude of other quern and millstone exploitations in Spain, is not classified as an archaeological monument. Hopefully this paper will be a first step to create an awareness of the existence of this type of rare and spectacular monument.

Acknowledgements

We would like to thank both Jane H. Scarrow and Juan Ignacio Soto of the Geology department of the University of Granada for their assistance in defining the portrait of the rock. We also thank Jeff Hodges for proofreading the text.

References

Amouric, H., 1991. Carrières de meules et approvisionnement de la Provence au Moyen Age et à l'époque moderne. In J. Lorenz and P. Benoit (eds.), *Carrières et constructions en France et dans les pays limitrophes, Actes du 115e Congrès National des Sociétés Savantes, Avignon, 9-12 avril 1990*. Paris.

Anderson, T., forthcoming. Un premier bilan sur la production de meules dans la Péninsule Ibérique depuis la Protohistoire à la période médiévale. In O. Buchsenschutz, , L. Jaccottey and F. Jodry (eds.), *Evolution typologique et technique des meules du Néolithique à l'an mille sur le territoire français, Actes de la Table Ronde de St. Julien-sur-Garonne*, 2 au 4 octobre 2009.

Cara Barrionuevo, L., García López, J. L., Lentisco Puche, J. D. and Ortiz Soler, D., 1999. *Los molinos hidráulicos tradicionales de la Apujarra (Almería).* Instituto de Estudios Almerienses, 32. Almería.

León, M., 2008. Cuando rugió la marabunta. *La Voz de Almería* (newspaper).

Madoz Ibánez, P., 1845-1850. *Diccionario Geográfico, estadístico, histórico de España y des sus possessiones de Ultramar*. Madrid, 1-16.

Martínez López, F. J. and Granero Gallego, A., 2003. Cantera de piedras de moler. *IV Internacional Molinology Conference*, 2,125-142.

Matilla Séiquer, G., 2001. Canteras de piedras de almazara en Fortuna. *Revista Murciana de antropología*, 7, 267-277.

Martín-Algarra, A., (coord.) 2004. Zonas Internas Béticas. In J. A.Vera (ed.) *Geología de España*, SGE-IGME. Madrid, 395-437.

Vera, J. A., (ed.) 2004. *La Geología de España*. SGE-IGME. Madrid.

Quern and millstone quarries in the north of Spain

Pilar Pascual Mayoral and Pedro García Ruiz

Translated by Timothy Anderson

1. Introduction

Our interest in the process of millstone manufacture (Fig. 1) dates to the discovery of several rotary quern roughouts a few years back. This led us to the idea that archaeologists have accepted the presence of rotary querns on archaeological sites without wondering about their origin (Pascual and García 2001, 241).

Moreover, during consultation of bibliographical sources on the Jubera Valley in the region of La Rioja, we came across a quote from Pascual Madoz that would prove essential for the study of this industry. The 19th century geographer states that 'in the municipality of Robres del Castillo, there is a very beautiful stone quarry that produces flour and oil millstones' (Madoz 1849, T. 13, 529). And indeed, during our survey of the hills of Robres del Castillo, we discovered that the millstone quarry surpassed by far the account of Madoz. In the light of this find, we expanded our area of investigation of millstone quarries to the other valleys of La Rioja and published a first distribution map of millstone quarries for the whole of the La Rioja Autonomy (Pascual and García 2003, 136). Four years later, we were invited to collaborate with the research group related to the European millstone quarries database. At this moment, we began to survey other regions of northern Spain.

Figure 1. Millstone makers (moleros) from the region of Palencia. The date of the photograph is unknown. We thank the authorities of Barruelo for permission to publish the photograph.

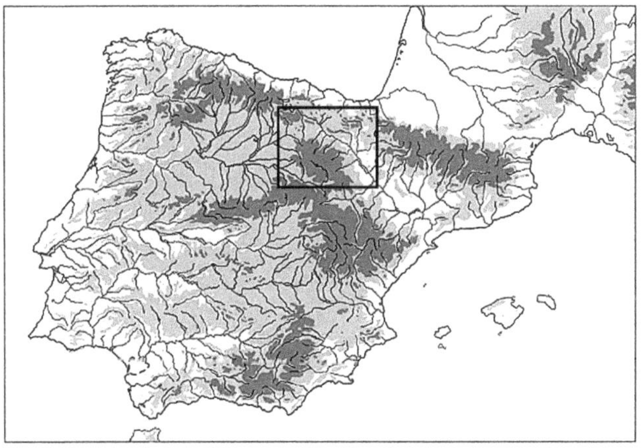

Figure 2. Survey area.

2. Study area and geology

The methodology adopted has varied in our two main study areas. In the mountain range of the *Sistema Ibérico*, we performed a systematic study, based primarily on geology and place names. In northern Burgos, Navarra, and the Basque Country, our research was based exclusively on the study of texts. From a quantitative standpoint, the results are striking in the *Sistema Ibérico* while in the second area, the documents consulted allowed us to approach other important issues, such as trade routes (Fig. 2).

The *Sistema Ibérico* range was formed forty million years ago during the Pyrenean phase of the Great Alpine Orogeny. Outcrops of conglomerate are visible at altitudes ranging between 700 and 1400 m, while limestone formations appear down to 500 m.

3. Millstone production in the *Sistema Ibérico* mountains

3.1. Toponymy as a working tool

The study of place names has been an extremely valuable means of locating millstone quarries. The author that stated that the dictionary of toponymy of La Rioja is a 'huge silent quarry waiting to be exploited' (González Blanco 1987, 35) was right. The name '*Molares*', deriving from

Municipality	Valley	Production centre	Rock	Millstone Atlas
Aguilar	Alhama	Contre. Leucade	Limestone	530
Almarza	Iregua	Tardiego	Conglomerate	222
Almarza	Iregua	P. San Agustin	Conglomerate	251
Arnedillo	Cidacos	Los Molares	Conglomerate	194
Grávalos	Linares	Los Molares	Micro – conglom.	203
Hornillos	Leza	Peña el Zorro	Conglomerate	363
Igea	Linares	Los Molares	Micro- conglom.	200
Islallana	Iregua	Las Planas	River boulder	346
Jubera	Jubera	La Valleja	Micro – conglom.	192
Jubera	Jubera	Peña Tejero	Conglomerate	198
Luezas	Leza	Los Molares	Conglomerate	212
Muro de Aguas	Linares	La Lobera	Conglomerate	204
Matute	Río Najerilla	Crispanas	River boulder	545
Robres Castillo	Jubera	Los Molares	Conglomerate	195
Robres Castillo	Jubera	La Viñaza	Conglomerate	199

Figure 3. Production centres in the Autonomy of La Rioja.

the latin *mola* (millstone), according *El Diccionario de la toponímia actual de La Rioja,* is indicative of millstone extraction (1849, T. 13, 529). This was rapidly confirmed for the case of the quarry of Robres del Castillo in La Rioja cited by the 19th century geographer Pascual Madoz.

We therefore took the term '*Molares*' and surveyed all the towns that still retained this name. The results were positive in five valleys of La Rioja: Linares, Cidacos, Iregua, Leza, and Jubera. Furthermore, subsequent surveys conducted in the Province of Soria of the Autonomy of Castilla y Léon proved that the name '*La Cuerda*', related to certain types of geological formations that coincided with millstone extraction, shared the stage with '*Molares*'.

3.2. Millstone production in La Rioja

La Rioja is the most intensively studied area in northern Spain due to its small size (5045 km^2). The millstone production centres identified are located in Fig. 3.

Specific information concerning each site (pictures, geographical co-ordinates and bibliography) can be consulted in the online database of the European Millstone Atlas.

3.3. Millstone production in the Province of Soria (Castilla *y* León)

In the province of Soria, we have identified 25 millstone production centres. In the region of Pinares and in the mountain ranges of Madero, the Almuerzo and Carcaña conglomerates were exploited, while limestone outcrops were exploited in the municipality of Fuentelárbol (Fig. 4).

The site of Fuenteárbol is particularly interesting. In this town, located about 45 km from Soria, there is a very curious feature directly related to the millstone quarry. It is an alignment of about 225 millstones (Fig. 5) placed by the residents as a type of tribute to the authorities on the outskirts of the town. Oral tradition states that the young people who married and decided to stay in the town had the right to exploit a parcel of the quarry on the condition that the first millstone they produced would be placed in the alignment.

4. Production of rotary querns in Antiquity

4.1. The *Sistema Ibérico* mountain area during Antiquity

During Antiquity, this area formed part of *Hispania Citerior*. To the north passed the *Via* 1, '*De Italia in Hispanias*' and to the south passed *Via* 27, *Item ab Astúrica per Cantabria Caesaraugusta* of the Itinerary of Antoninus (Roldán Hervás 1973. 35-36). The enclaves of *Graccurris* (Alfaro), *Calagurris* (Calahorra), *Vareia* (Varea- Logroño), *Tritium Magallum* (Tricio) and *Libia* (Herramélluri) were located along Via 1 while to the south, along the Via 27, were the sites of *Turiasone* (Tarazona), *Augustóbrig*a (Muro de Agreda) and *Numantia* (Garray). These political, cultural, and commercial centres favoured the construction of many smaller settlements (*villae*) in their rural orbits.

The demographic vitality of this region is reflected in

Municipality	Valley	Production centre	Rock	Millstone Atlas
Calderuela	Almuerzo	La Hoya	Conglomerate	284
Canos	Almuerzo	La Cuerda	Conglomerate	304
Canos	Almuerzo	La Cuerda	Conglomerate	306
Canredondo	Carcaña	Cuerda Larga	Conglomerate	431
Canredondo	Carcaña	El Carrascal	Conglomerate	259
Cortos	Almuerzo	El Monte	Conglomerate	297
Cortos	Almuerzo	La Soledad	Conglomerate	295
El Espino	Madero	Las Peñas	Conglomerate	254
Fuentelárbol	C. Villa y Tierra	Pueblo	Limestone	196
Fuentelárbol	C. Villa y Tierra	Las Canteras	Limestone	207
Matalebreras	Madero	Pizarrales	Conglomerate	534
Muro de Ágreda	Madero	Cumbres	Conglomerate	310
Muro de Ágreda	Madero	Monte Oncillos	Conglomerate	312
Muro de Ágreda	Madero	Cº de Ágreda	Conglomerate	313
San Felices	Madero	Los Molares	Conglomerate	248
Trébago	Madero	Cerro Balcones	Conglomerate	260
Trébago	Madero	Peña el Mirón	Conglomerate	253
Valdelagua	Madero	El Sardón	Conglomerate	321
Valdegeña	Madero	Las Matas	Conglomerate	294
Villar de Campo	Madero	Castellanos	Conglomerate	257
Villar de Campo	Madero	Los Molares	Conglomerate	255
Vinuesa	C. Pinares	Las Majadillas	Conglomerate	460
Vinuesa	C. Pinares	La Muedra	Conglomerate	462
Vilviestre Nabos	Carcaña	El Rebollo	Conglomerate	448

Figure 4. Table of production centres in the Autonomy of Castilla y León.

Figure 5. Alignement of 225 millstones in Fuentelárbol.

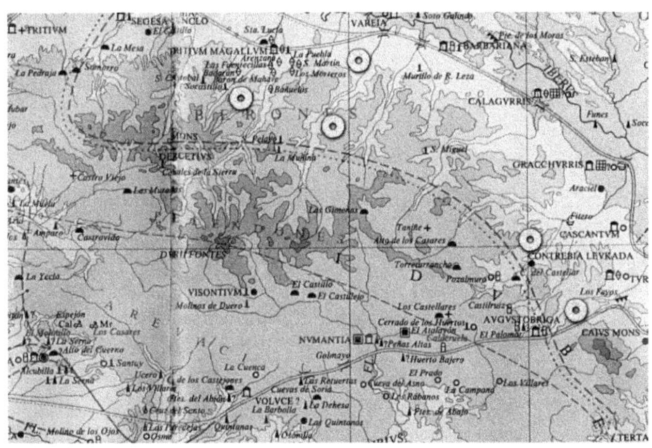

Figure 6. Centres of production of rotary querns (Hoja K-30, TIR).

the high number of sites, as seen in the figure reflecting the *Tabula Imperii Romanii* (Fig. 6). The archaeological record of millstones is, nevertheless, not proportional to the number of settlements, probably due to the logistical difficulties caused by the transport and storage of millstones recoverd on excavations. As for the origin of the millstones, the finds in this area point to local exploitations using two production techniques: the collection of surface boulders and the use of outcrops in true extractive quarries. We have not, at least for the moment, identified long-distance imports, for example, of volcanic rocks.

4.2. Manufacture of rotary querns from small boulders

Our understanding of the process of knapping rotary querns from small boulders is based on six unfinished rotary querns discovered at the Roman *villa* of Crispano (La Rioja). This first stage of the process is the collection of boulders in river beds or on alluvial terraces (Fig. 7). The stones present a size of about 40 cm in diameter. The average thickness is 30 cm for lower stones (*metae*) and 15 cm for upper stones (*catilli*). The manufacture of the lower stone begins with the choice of the boulder, which

Figure 7. Small surface boulder selected for a lower stone (meta).

Figure 10. Lower stone in an advanced state of carving (small surface boulder).

Figure 8. Roughly knapped small surface boulder (meta).

Figure 11. Unfinished upper stone knapped from a surface boulder.

Figure 9. Lower stone roughout (small surface boulder).

is then roughly carved into a cylindrical shape similar to those found in quarries (Fig. 8). The next step consists of pecking a small hole in the centre (Fig. 9), which will serve to trace the circumference with the compass (Fig. 10) for the subsequent carving. The carving of the upper stone also begins by roughly rounding the stone and ends with careful perforation of the eye. To avoid breaking, the eye was carved from both sides (Figs. 11-12).

4.3. Production of rotary querns in quarries

The finds of several unfinished rotary querns confirm their production in quarries. In the quarry of Tardiego (La Rioja), we have located several unfinished querns among the rubble of more recent exploitations. In Muro de Agreda (Soria), presumably the Roman city of *Augustóbriga*, we find the same situation although the querns are now stored in private homes.

In Valdegeña (Soria) a wide variety of finished and unfinished rotary querns (Figs. 13-14), probably collected from the Roman settlement of 'Los Villares', decorate the patios and façades of houses. Their cylindrical shape - without rounded edges - suggest they derive from a nearby extractive quarry and not from riverbed surface boulders like the querns found near the Roman villa at Crispano.

Archaeological excavations at the site of Contrebia, Leukade (La Rioja) uncovered holes carved into the bedrock that were interpreted as supports to brace large storage vessels (Hernández Vera 2007, 54). This hypothesis should be reviewed because these extractions appear actually to be negatives of quern extractions (Fig. 15).

Figure 12. Unfinished upper stone knapped from a surface boulder.

Figure 14. Unfinished lower stones from the Roman settlement of "Los Villares".

Figure 13. Cylinder from the Roman settlement of "Los Villares".

Figure 15. Circular extractions carved into the bedrock. Contrebia Leucade (La Rioja).

5. Production of millstones in the *Sistema Ibérico* mountains

5.1. The transition to the Medieval and Modern periods

The Roman cities and *mansios* distributed at the foot of the *Sistema Ibérico* mountain range faced invasion and destruction in the 5th century AD. The Suebis, Vandals and Alans tribes razed everything in their way. In the 10th century, there was once again conflict during the Muslim domination. The winds of war probably affected the development of the millstone industry and prolonged the use of the rotary quern. The first reports of hydraulic mills in La Rioja appear in monastic documents towards the middle of the 10th century. It is nonetheless in the 11th century that there is multiplication in the number of documents of purchases, sales, and donations to monasteries. By the 16th century, the stones that outfit the mills of Castile, Navarra, and Aragon can be counted by the hundreds.

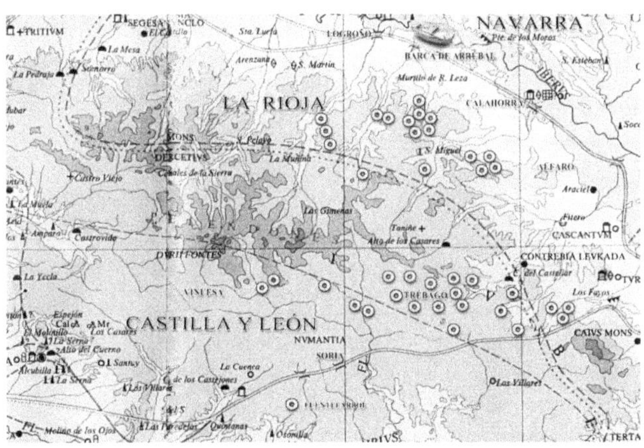

Figure 16. Medieval and modern millstone production centres.

5.2. Millstone production in the Modern period

Medieval and Modern quarries in the north of the peninsula produced stones basically for both flour mills, with diameters ranging from 0.80 to 1.80 m, and oil rollers, on the average 1.00 m in diameter and 0.40 m thick (Fig. 16). There was to a lesser extent demand for stones for other

Figure 17. Trenching.

Figure 18. Extraction.

Figure 19. Circular negative.

Figure 20. Carving surface A.

Figure 21. Carving surface B.

Figure 22. Cutting of the eye.

industries, such as pulp mills (which began to decline in Spain around the 16th century).

The production centres of Igea, Grávalos, and Villarroya in La Rioja, with their millstones measuring 0.90 m in diameter and 0.10 m thick, were an exception. The extraction techniques also differed in this area. Blocks were extracted from the outcrop, roughly knapped into shape with a sledgehammer, and finally finished with hammer and chisel. The modest diameter and thickness of some millstones point to their use as sharpening stones and the tendency toward an oval shape of other millstones reveals their possible use in either the oil or the wine industry (Pascual and Arrastia 1980, 199-210).

The permanent abandon of these exploitations probably began in the second half of the 19th century with the arrival of the railroad. This new means of transport most likely introduced higher quality millstones from further away and eclipsed the traditional means of transport from the local traditional quarries. This explains the high number of abandoned millstones in the quarries that illustrate the different stages of production and constitute such a rich heritage (Figs. 17-22) that could be exploited for tourism.

6. Quarries and municipal boundaries

Through the study of the distribution of millstone production sites, we have observed some coincidences that are noteworthy. Although the quarries of Villarroya, Grávalos, and Igea are all located in an area called '*Los Molares*', they are separated by municipality boundaries. We find the same demarcations in the valley of Jubera. The site of Robres del Castillo was the main producer of millstones in the region. Its industrial border ends where the exploitations of Jubera and San Vicente de Robres begin. Furthermore, in the valley of Iregua, the quarries of Almarza de Cameros and Torrecilla en Cameros are cut by a line that distinguishes two jurisdictions.

This peculiar phenomenon is also found in the province of Soria (Castilla y León). At the foot of the Madero mountains, there are several villages with a long millstone extraction tradition, notably, Trébago, Matalebreras, Fuentestrún, Valdegaña, and Villar Field Valdelagua, whose jurisdictional limits scrupulously separate millstone production centres.

The curious concordance of industrial and municipal boundaries has also been detected in France, south of Grenoble, where the number of quarries on Mount Sénépy matches the number of towns (Belmont 2006, 82). An agreement reached in the 18th century between the councils of the municipalities of Valle Redondo, Brañosera, Celada of Roblecedo, Salcedillo, and Herreruela might bring light to this subject (Basterra 2003, 149-268).

This legislation was intended to control the quality and quantity of millstone production to avoid deterioration of the market, and also to maintain a balance of benefits and avoid potential conflicts between producers. The sizes of the millstones were perfectly regulated. In addition, the legislation included penalties that could be as high as 300 *reales* for fraudulent millstones. This sum represented a fortune at the time.

7. Commercial routes

Trade routes played an essential role in the millstone industry. An excellent example comes from the commendable work of the ethnographer Inés Sainz Azuelo on the well-preserved archives of the municipality of Mendavia (Navarra). According to accounts between 1690 and 1805, the municipality purchased 19 millstones from eight different production centres. Until the middle of the 18th century, the mill of Mendavia purchased its stones from the La Rioja quarries (Robres del Castillo, San Vicente de Robres and Jubera), located about 35 km away. However, in the second half of the century, the quarries of La Rioja are overshadowed by the quarry of Trébago (Soria), located 80 km away.

Between 1764 and 1783, the authorities of Mendavia ordered two millstones from Estella (Atlas, entry 323) and made small purchases from the quarries of Arbaitza (Atlas entry 345), but in the 19th century, the only known supplier of millstones was in the area of the Condado de Treviño.

The commercial dynamism observed in Mendavia is also seen elsewhere in the north of the peninsula. Toward the middle of the 19th century, Pascual Madoz states that 'the Condado de Treviño is famous for its stone quarries located in the villages of Arana, Dórdoniz, Armentera, Pedruzo. and Torre, from which are hewn millstones exported to the regions of Aragón, Vizcaya and Burgos' (Madoz 1849,T. 15, 153-154; Atlas entries 236, 237, 242, 359, 356, 361, 408 and 409).

In his study of the production centres of the mountains of Palencia, Basterra Adán states that the Palencia productions rivalled those of La Rioja, Segovia, and Navarra, while the quarries of the Gorbea mountains sold their millstones to millers of Alava and surrounding areas of Vizcaya, Burgos, and La Rioja (Iturrate 2001, 129).

These documents present us with an image of the last few centuries of the commercial map of millstones in northern Spain. An impoverished network of roads nonetheless hindered this commerce. Larruga describes it as follows: 'Another very serious calamity is the pitiful state of the roads of La Rioja, suffice to say that from November to May cannot travel without serious dangers and delays, and that even in Summer very few coaches transit through the country road. This is a county you can only reach by air!' (Madrazo 1984, 260).

Nevertheless, commercial necessity by far surpassed the inconveniences. An example is the transport of a millstone measuring two *varas* (approximately 167 cm) by Santiago Garcia in 1766 between Trébago in the Sierra del Madero and Mendavia (Navarra). The first obstacle was crossing the Iberian mountains with passes exceeding 1000 m. Then his team of oxen had to cross the La Rioja territory to reach the right bank of the Ebro River. From the edge of the La Rioja region, with sights into Navarra, the quarryman had to make one last effort to load oxen, cart, millstone into a ferry at Arrúbal (cf. Fig 16.) and cross the river with the strongest-flow current in Spain before arriving a few miles later at the mill Mendavia.

We assume that Santiago García would remember the details of this extraordinary adventure again and again while returning to Trébago. The return was more bearable given that the commercial transaction of an upper stone was worth nothing less than the price of 704 *reales* paid by the authorities of Mendavia, a large sum at the time.

8. Conclusions

In this article, we have attempted to summarise our research on millstone-producing centres in northern Spain. For the moment, our research has only been published in local reviews and in specialised journals. When we started our research in 2000, not a single line about millstone quarries in La Rioja had been written since the work of Pascual Madoz in 1850. However, much work remains to be undertaken to understand the scale of this specific industrial world. We conclude with two final reflections.

There is a need to focus research on the early quern and millstone production sites dating to Protohistory and to Roman times. This would help correct the imbalance between the finds from settlements and from quarries, given that every residence counted one or two rotary querns among their possessions.

Trade routes also offer an interesting field of study if we are able to analyse them beyond simply the geographical distance separating the production centres from the settlements where the millstones were used. For example, the transport of commercial goods between Castilla and Navarra required crossing the Ebro River. This condition certainly did not escape the insatiable royal and noble tax collectors. The merchants and carriers had no choice but to increase the price of their products to offset the cost of taxes and tolls at bridges and barges.

Finally, an enormous amount of information remains to be recovered in archives and in field surveys to assists us to piece together the daily life of the past and to bring the reader closer to this aspect of molinology.

References

Basterra, A. and Vicente, M., 2003. Convenio y concierto subscrito en el siglo XVIII entre los concejos del Valle de Redondo, Brañosera, Celada de Roblecedo, Salcedillo y Herreruela para la fabricación de muelas de molino. *PITTM* 74. Palencia, 149-268.

Belmont, A., 2006. Les meulières médiévales. In A. Belmont and F. Mangartz, (dir.), *Les Meulières. Recherche, protection et valorisation d'un patrinoine industriel européen*, Colloque international Grenoble -22 au 25 septembre 2005. Mainz, 3-4.

Catalogo Monumental, 1968. *Diócesis de Vitoria. Arciprestazgo de Treviño y Campezo*. Tomo II, Vitoria.

González Blanco, A., 1987. *Diccionario de Toponimia actual de La Rioja*. Murcia.

Hernández Vera, J. A., 2007. *Contrebia Leucade. Guía arqueológica*. Logroño.

Iturrate, J., 2001. Notas etnohistóricas sobre la fabricación de piedras de molino en las canteras del macizo de Gorbea (Álava). *Kobie* 10, Bilbao, 129-141.

Madoz, P., 1845-1850. *Diccionario geográfico - estadístico- histórico de España y sus posesiones de ultramar*. (reprint Logroño 1985).

Madrazo, S., 1984. *El sistema de transportes en* España, *1750-1850*. Vol 1. *La red viaria*. Madrid.

Moreno, M., 1985. *Memorial de Soria I. Relatos*, Soria.

Pacual Mayoral, P. and García Ruiz, P., 2001. Canteras y tecnología molinar en el río Jubera (La Rioja). *Revista Murciana de Antropología*, 7, Murcia, 237-266.

Pacual Mayoral, P. and García Ruiz, P., 2003. Las canteras de piedras de molino. una industria riojana desconocida. *Altza, Hautsa Kenuz,* 7, Donostia - San Sebastián, 135-146.

Pacual Mayoral, P. and Moreno Arrastio, F. J., 1980. Prensas de aceite romanas en La Rioja. *Archivo Español de Arqueología*, v 53, nº 141-142, 199-210.

Porres Marijuán, M. R., 1983. *Un ejemplo de economía rural de Antiguo Régimen. El condado de Treviño (1650 – 1800)*. Vitoria.

Roldán Hervás, J. M., 1975. *Itineraria Hispana*. Vitoria.

Sainz Azuelo, M. I., 2007. Etnografía histórica de Mendavia. Panadería, molino y tiendas. *Cuadernos de Etnografía y Etnología de Navarra*, nº 82. Navarra, 163-199.

Seven thousand years of millstone production in the Serre Mountain range of the French Jura

Luc Jaccottey

Translated by Timothy Anderson

The grinding of cereals into meal or flour to render it suitable for human consumption is an activity known since the beginning of agriculture. Milling devices such as saddle querns and rotary querns (for domestic use), and later millstones driven either by animal, water or wind (for collective use) therefore played an essential role in early societies. Accordingly the search for a rock with excellent milling properties has preoccupied mankind since prehistory.

The Serre Mountain Range (*Massif de la Serre*) in the French Jura is a unique geological feature that very early attracted people. It is the only 'island' of granite and sandstone in the heart of the Jura limestone formation. It is also a natural passage from the Vosges Mountains to the east and the Morvan Range to the west, each about 100 km away. The Serre Range provided a raw material that was exploited up until recent times. Due to its relatively modest size (3000 hectares), the whole of the region was surveyed, therefore yielding an accurate picture of different exploitation over time.

Querns and millstones coming from these mountains, dating from the Neolithic to Antiquity, are present in regional archaeological collections. But production did not stop there as this region is also known to have been a source of millstones up to very recent times.

Since 2005, research on this subject has followed three different lines. The first is to describe and characterize the geological formations of the Serre Massif. The second is to search for and define the different systems of extraction. Finally, the typology and distribution of the products are studied with respect to different chronological phases starting with the Neolithic. This research permits us to examine numerous aspects of production and millstone production economy from late Prehistory to very recent times. It also allows us to compare the way extraction evolved with other districts in France and neighbouring countries.

While information linked to millstone quarries in Europe is constantly growing, there is little data concerning one single geographical area over a long period of time. In France, at present, there is only one other site that produced millstones over an extended period of time: of Claix in the Charente where extraction persisted from late Antiquity to the 19th century (Belmont forthcoming). Outside France, the lava flows of the Eifel in Germany produced grinding stones from the Neolithic to the 20th century (Hörter 1994; Harms and Mangartz 2002). The Serre, with production from Neolithic to contemporary times is therefore one of the few sites that permits a view of production over an extended period of time.

1. Geographical and petrographical portrait of the Massif de la Serre

The Serre Mountain Range is a small metamorphic horst located in the northern part of the Jura Department between the cities of Dole, Besançon and Dijon. It is between 150 and 200 metres higher than the Doubs, Saône and Ognon rivers which cut through it in the north-west sector of the Jura. The outcrop, 15 km long and 4 km wide, is composed of granite formations organised in parallel bands and covered by Triassic sandstone.

The granite of the Serre Massif is one of the oldest outcrops of the region and corresponds to the basement platform. It consists of biotite, muscovite, quartz and feldspar (both alkali and plagioclase). Also many myrmekites originating from the syn-crystallisation of the two types of feldspars are observed in thin sections. To the naked eye, the granite has a light, rosy tint. Metamorphism has transformed it into several types of gneiss. Where the rock platform is visible on the surface both, metagranite, and mylonite can be seen. In the field its surface is most often altered and crumbly.

Capping this geological platform are various Early Triassic Bunter sandstone layers alternating between fine-grained and rose-coloured to coarse-grained (Zogo 1981). This last lithostratigraphic type consists principally of quartz and microcline feldspar. The matrix is siliceous, with syntaxial overgrowths on the quartz grains cementing the rock and increasing its induration.

The Serre sandstone, like all sandstones, has silicious grain inclusions (Boyer 2008, 53) that make for an abrasive and coarse texture. The tenor of silica, comparable to that of the Vosges (94% at Moissey and Mont-Saint-Odile), is

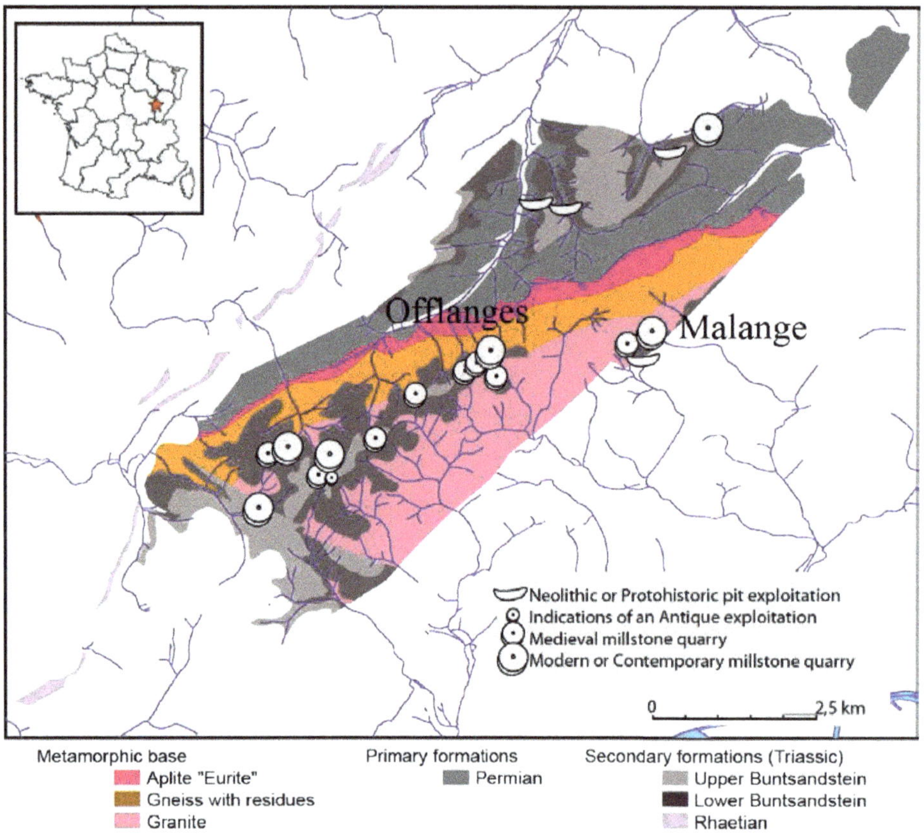

Figure 1. Geological map of the Serre Mountains with the main quern and millstone extraction sites. CAD by L. Jaccottey, INRAP.

one of the highest found in French millstones. Only the sandstone of Bézu in the Aude, with 99%, has a higher quantity (Fabre *et al.* 2006, 94). Compared to other silicious sandstones, the Serre stone, at about 8.5%, has a low level of porosity, compared to the usual 12 to 20%. In addition, it has a high resistance to mechanical compression: 70, while other sandstones do not surpass 59 megapascals (Fabre *et al.* 2006, 95). These different qualities explain why the Serre sandstone is one of the most sought-after rocks in the region for millstones.

In the study of querns and millstones stored in the museum collections, the different categories of stones are determined macroscopically. The term 'Serre type' is therefore employed in a global manner to define the different sandstones and granites (Fig. 1).

2. The production of Prehistoric and Protohistoric saddle querns

During the Neolithic, cereals were ground by means of a to-and-fro motion, with a mill consisting of a stationary lower stone (saddle quern) and an active upper stone (rubber or *mano*). Based on the study of querns in the local museums, Serre saddle querns from the early phases of the Neolithic are almost exclusively of Triassic sandstone. Granite querns are rare and their diffusion is restricted in space. The spread of sandstone querns, however, is wider (Fig. 2). The geographical diffusion of sandstone querns followed natural passageways through the Doubs, Ognon

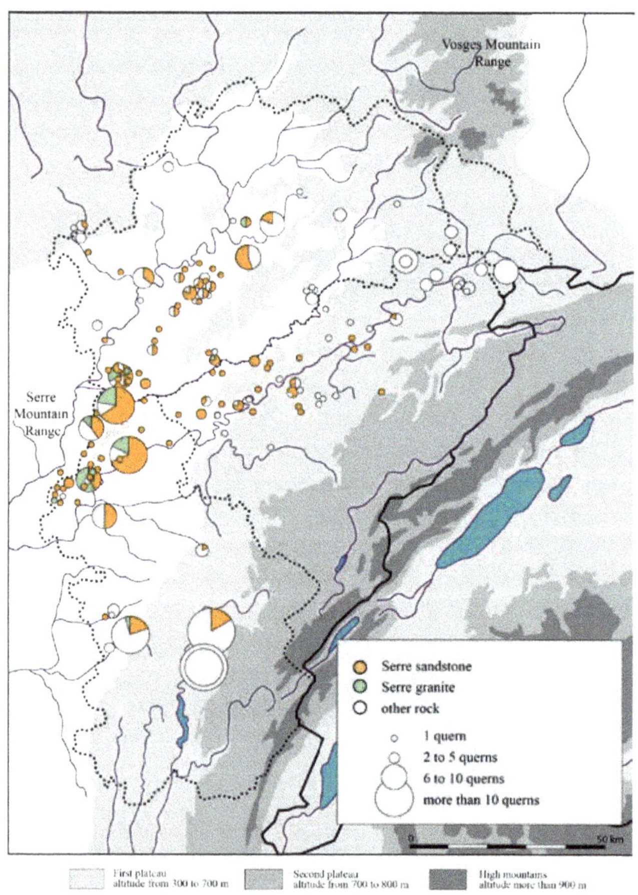

Figure 2. The distribution of Serre granite and sandstone querns in the Neolithic and in Protohistory.

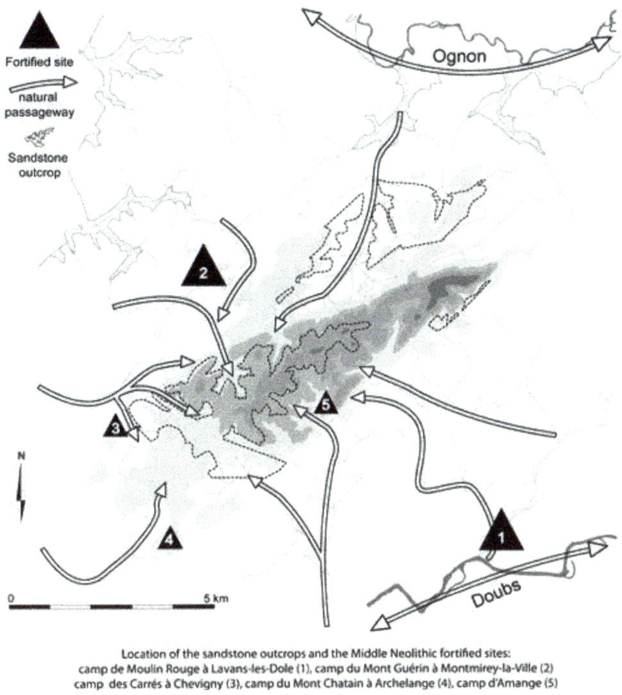

Figure 3. The spread of fortified settlements in and around the Serre Mountain Range dating to the middle Neolithic.

and Saône River valleys and did not exceed a distance of 40 to 60 km (Jaccottey and Milleville 2007b). Beyond this sphere of distribution, the sandstone querns of the Serre met the competition of another sandstone product originating in the Vosges. This second production centre took advantage of a pre-established trade network of pelitic-quartz stone axes (Pétrequin and Jeunesse 1995) to market its product. At this time, the Serre region was dotted by a series of fortifications that are in fact quern production centres (Jaccottey *et al.* 2007). Other areas with a high density of fortified settlements, generally near salt sources, are known in the region and produced pelite axes (Pétrequin and Jeunesse 1995) (Fig. 3).

Toward the end of the Neolithic the decline of the production of pelitic-quartz axes at Plancher-les-Mines, and the subsequent decline of the axe trade network, benefited the Serre production. Consequently, there were fewer Vosges querns circulating, while there was an increase in the level of Serre production (Fig. 4). Later, in protohistory, the main production materials for Serre querns shifted from sandstones to granites. Systematic field surveys have not revealed any of these ancient granite exploitations. At this time granite is rarely seen in the regional collections, but it was most likely gathered in screes which would leave no perceptible marks in the field (Fig. 5). Several sandstone exploitations, nonetheless, have been identified in the field. These sites consist of oval-shaped pits dug into screes at the base of the slope (Fig. 6).

The results of a recent excavation of one of these production sites (Jaccottey and Milleville 2007), complemented with data collected from modern reconstructions of techniques of digging pits to search for the raw material as well as experimental reconstructions of knapping querns, reveal

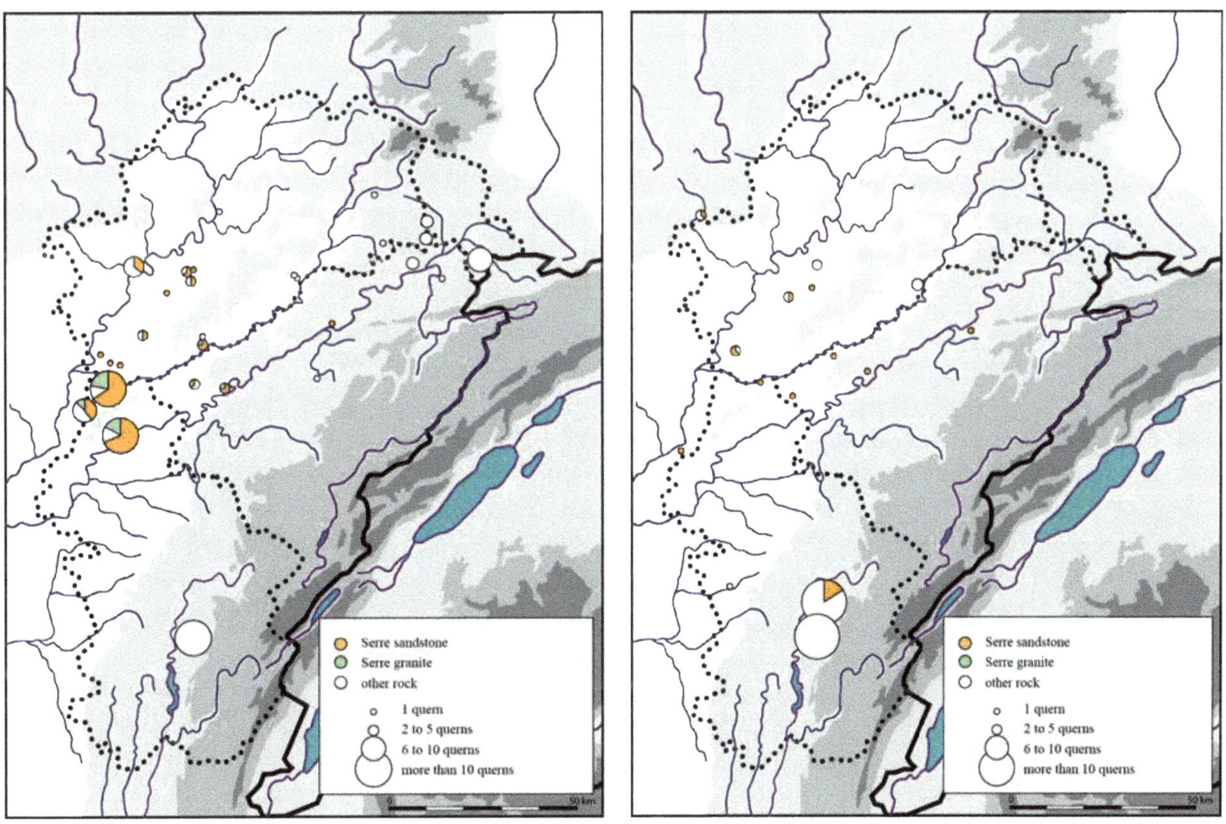

Figure 4. Distribution of Serre sandstone and granite querns in the middle and late Neolithic.

Figure 5. Reconstruction of the Neolithic and Protohistoric technique of quern production exploiting screes found at the base of either granite or sandstone low cliffs. Drawing by A. Pulido.

fresh data concerning the production sequence (*chaîne opératoire*) of querns from the Neolithic and the Bronze Age.

The pits were dug at the base of the scree with picks with short handles (approx. 50 cm). Picks with long handles (longer than 80 cm), used while standing, do not seem to be effective. These picks are either of wood or of antler. Antler, however, seems to be much better adapted to this type of work. Spoil was removed with small hand shovels fashioned from either pig or sheep shoulder blades. The larger shovels made from oxen shoulder blades were not effective, especially between the stones and tree roots. The debris was transported by means of either a cloth, a basket or an animal skin (Fig. 7).

According to our experiments, between 0.6 and 0.8 m3 of earth can be removed by one person in one hour. This quantity is certainly less than an early quarrymen, with experience and skill, could excavate. If we project this onto the volume of ancient pits, we obtain an estimate of 14 to 18 hours of work for a smaller pit (pit 2b) and 37.5 to 50 hours for the largest (pit 1). Since the early quarrymen probably did not work alone, the total time spent to dig a pit in search of appropriate stones for querns was at most a few days (Fig. 8).

After the appropriate sandstone blocks for both rubbers and lower stones were selected, the knapping process could begin. Finds of heaps of working debris consisting of stone flakes, as well as quartz hammerstones (*percuteurs*) and quern blanks and roughouts, show that the querns were

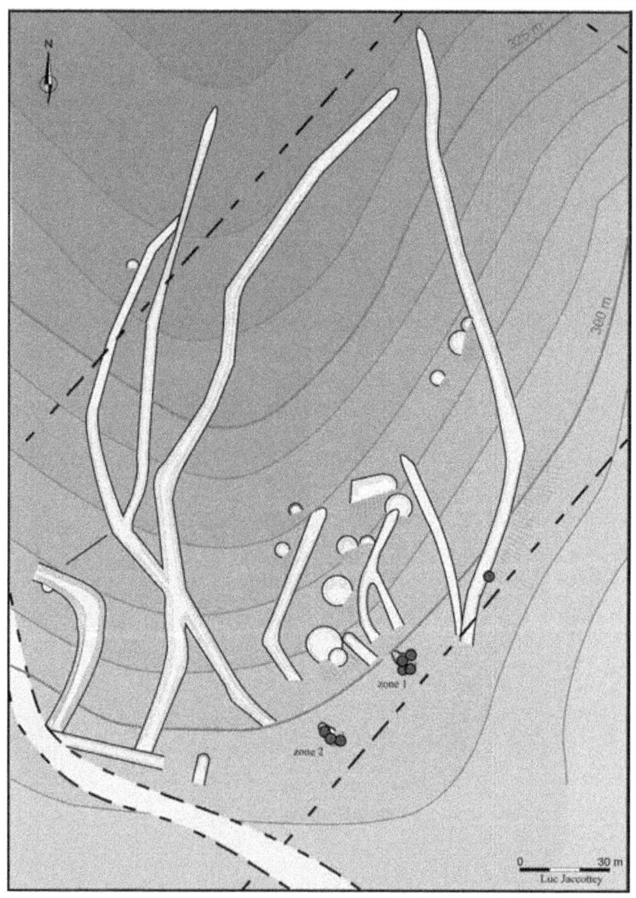

Figure 6. Site of Malange with in red the location of pit exploitation (Zone 1. 18 x 13 m, 1.2 deep; Zone 2a. 8 x 7 m, 1 m deep; Zone 2b. 5 x 5 m, 0.9 m deep).

knapped at least roughly into shape at the quarry itself (Fig. 9).

The experiments related to knapping rubbers and saddle stones suggests that small hammerstones of around 800 grams are effective only for making small rubbers. Larger querns require larger, heavier hammerstones involving the use of two hands (Fig. 10). The technique consists first of roughly knapping the block into shape, removing uneven areas and irregularities with rough knapping strokes (Fig. 11). The future surfaces are then fashioned with finer work to obtain the roughout. Finally, the external surfaces are adjusted and the grinding surfaces of either the hand stone or the lower stone are dressed. In the German Eifel, similar scree exploitations with numerous discarded quern roughouts have been identified at Rossbüsch (Hörter 1994, 14,107).

These newly identified quarries in the Serre Mountains, like that of Malange, consist of small surfaces and cannot themselves account for the many querns in the regional archaeological collections. It is probable that a large part of these very early extraction sites were destroyed by later quarry work. Other extraction techniques, difficult to detect in the field, might have been put to use. The technique of thermal shock, consisting of placing a fire either directly on the rock outcrop or beside small embankments, might have been used. Extraction by fire produces large naturally curved flakes that split on curves perpendicular to the natural layering of the stone. Although querns in museum collections seem to verify this technique, surface surveys, for the moment, have not identified any sites of this type (Fig. 12).

Figure 8. Reconstruction of extraction of blocks for quern production by means of a pick fashioned from a deer antler. Drawing by A. Pulido.

The Serre exploitation of scree, in pit or by fire, involved detaching blocks. A variation of this type, the simple exploitation of small surface blocks, is well documented,

Figure 7. Reconstruction of quern extraction in pits dating to the Neolithic and to protohistory. Drawing by A. Pulido.

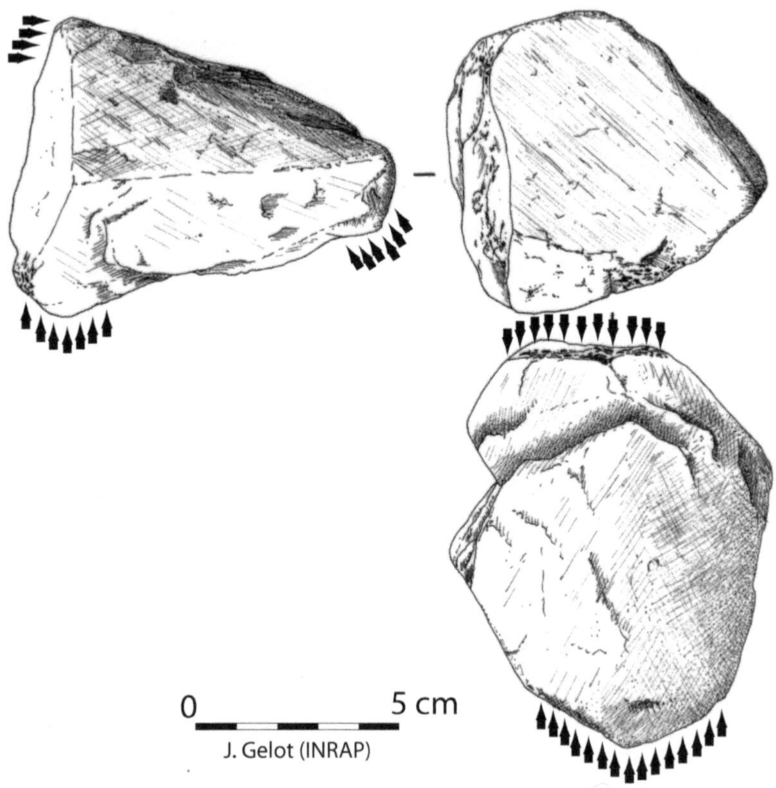

Figure 9. Quartz percuteur with working marks. Drawing by J. Gelot, INRAP.

for example, in the Parisian Basin during the early Neolithic (Hamon 2006). The exploitation of blocks in moraines is also known during the middle Neolithic in the Vosges region (Jaccotey and Milleville 2007b), as well as along the shores of the lakes of Switzerland during both the Neolithic and the Bronze Age (Ribeau 1986; Leuvrey 1999).

3. Millstone production in Antiquity

During the second half of the La Tène period, the traditional saddle quern was replaced by the circular rotary quern. This innovation is documented in the Iberian Peninsula as early as the 5th century BC (Alonso 2002) and in the 4th century BC in the south-east of France. Its introduction into the Serre area does not occur until later in the 2nd century BC. These new millstones consist of a stationary lower stone, the *meta*, mounted by the runner, the *catillus*. Although granite versions of this quern-type are found within a radius of 40 to 60 km around the Serre outcrop, no granite quarry from this period has been detected in the field.

During the Roman period, sandstone rotary querns have a distribution similar to that of querns of the preceding late Iron Age (Jaccottey *et al.* 2007). One Roman quarry was unfortunately uncovered and destroyed during millstone quarry work in the 19th century in the western area of the Serre Range (Feuvrier 1920). This explains the find of a rotary quern blank discovered in the debris of the more recent exploitation (Fig. 13). It is most probable that the

Figure 10. Experimental knapping with a large hammerstone for two hands.

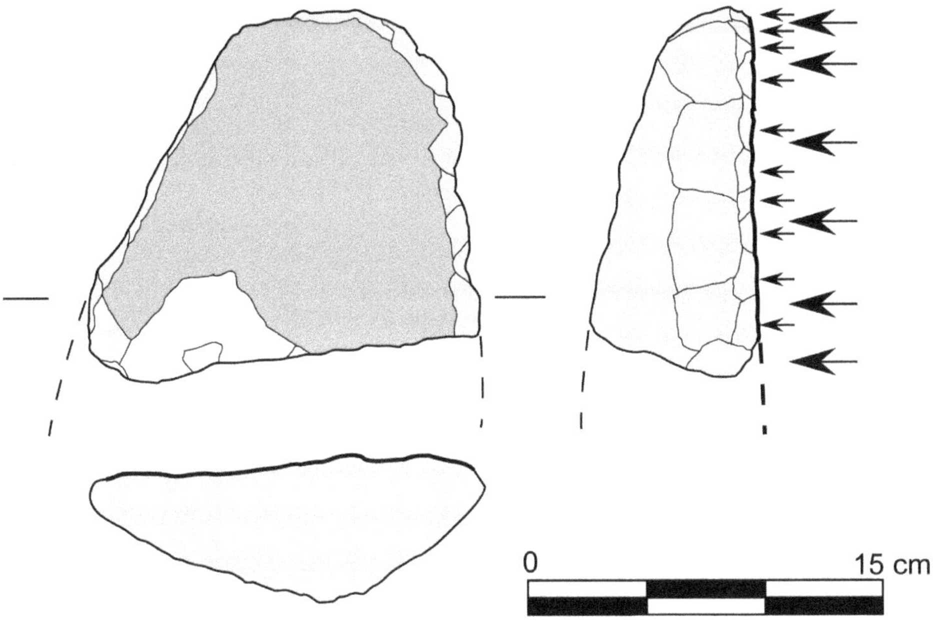

Figure 11. Example of a sandstone rubber (mano) roughout with scars of detached flakes. Drawing by L. Jaccottey, INRAP.

even more recent quarry work here has also obliterated the older rotary quern extractions. In spite of the small number of vestiges, it is possible to reconstruct part of the Roman production sequence. A circular trench was cut, producing a cylinder slightly larger in diameter than the future quern. The cylinder was then split from the bedrock by means of a pick or an iron bar acting as a lever. This is inferred from the marks left on the lower face of the blank (Fig. 14).

These extraction techniques are very similar to those seen in the quern quarries of Châbles and Chavannes-le-Chêne in western Switzerland (Anderson *et al.* 2003) and Wurenlös near Zürich (Doswald 1994). In France, they have been observed at Saint-Quentin-la-Poterie (Longepierre 2006).

Figure 12. Hypothetical reconstruction of a quern exploitation using thermal shock. Drawing by A. Pulido.

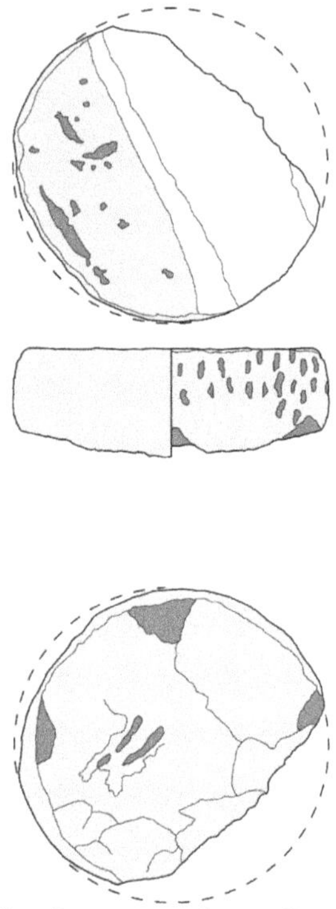

Figure 13. Sandstone roughout. On its base are the marks of iron tools used to split it from the bedrock. Drawing by L. Jaccottey, INRAP.

In the Eifel of Germany, the basaltic lava columns were sectioned for rotary querns (Harms and Mangartz 2002).

In Antiquity, rectangular blocks were also knapped into cylindrical querns. This different technique is known at Avrilly in Normandy (Guillier *et al.* 2005), Vendresse and Beaulne in the Aisne Department (Naze *et al.* forthcoming), Vauxrezis (Robert and Landreat 2005), La Salle in the Vosges (Farget and Fronteau forthcoming), Le Portus in Bourgogne and Schweigmatt in the Black Forest in Germany (Joss 1975) and Marèze in the south-west of France (Servelle 2006).

Very few millstone productions are dated securely. Châbles, which ran from 80 to 120 AD (Anderson *et al.* 2003), and Saint-Quentin-la-Poterie dated to the start of the 4th century AD (Longepierre 2006) are exceptions. The workshop of Avrilly dates to the early Roman period (Guillier *et al.* 2005). The site of La Salle has quern roughouts dating from the end of the Celtic Iron Age (Farget and Fronteau forthcoming). Finally, the exploitation of Schweigmatt (Joss 1975), Vendresse and Beaulne (Naze *et al.* forthcoming), as well as that of Marèze (Servelle 2006), all date to the end of the La Tène or the beginning of the Roman period.

Figure 14. Reconstruction of the techniques of rotary quern sandstone production during Antiquity. Drawing by A. Pulido.

In spite of the small number of quarries studied, it is possible to sketch a general chronological trend between the sites earlier and later than the middle of the first century AD. It appears that the earlier sites dating between the Iron Age and the beginning of the Roman period followed the Neolithic tradition. A shift then took place after the Roman conquest when querns and millstones were extracted as cylinders directly from bedrock a technique which endured until modern times. This model of transition from one technique to another would need to be verified as the number of quarries grows.

The use of hydraulic force to drive larger millstones has been known known since Roman times through written sources (Vitruvius, *De architectura X*, 5). Archaeological finds of this type of more complex mill include the famous site of Barbegal in southern France (Benoit 1940) and the more modest watermill of Avenches in Switzerland (Castella 1994). We cannot confirm through our field survey the use of the Serre sandstone for the larger Roman water millstones. There must however be sites to be found based on the millstones of this type in the archaeological collections of Franche-Compté and Bourgignon.

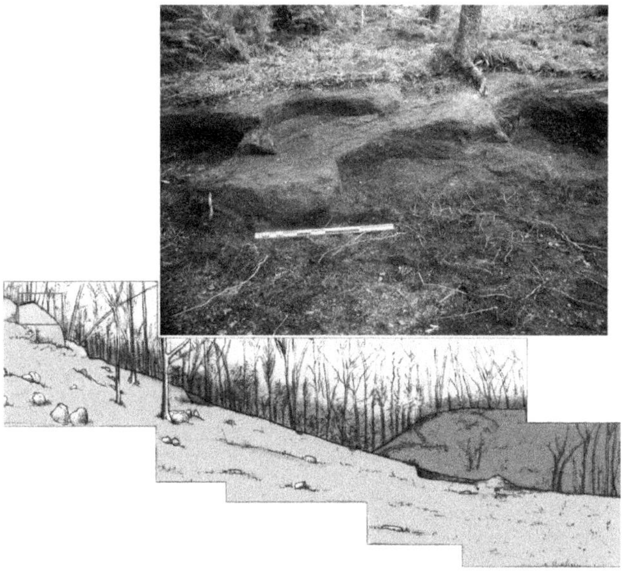

Figure 15. View of a Medieval millstone exploitation beside the sandstone plateau. Photograph by L. Jaccottey, INRAP and drawing by A. Pulido.

4. Medieval millstone production

This production concerns for the most part millstones for watermills, installations that are ubiquitous and found along the waterways of Medieval France. The channelled hydraulic force was sufficient to drive a large flat millstone measuring about 1 m in diameter.

Medieval written sources cite several millstone quarries in the Serre region between 1371 and 1475 (Theurot 1998,

Figure 16. Map of the Medieval millstone quarry. Drawing by L. Jaccottey, INRAP.

Figure 17. Reconstruction of the technique of tracing the circumference of a future Medieval millstone using a thin iron or a piece of charcoal attached to a string. Drawing by A. Pulido.

214-5). The study of these old texts also throws light on the sales of Serre millstones during Medieval times from the quarries to Salins-les-Bains and Poligny, about 40 km away. There is also mention of products reaching Pontarlier in the upper range of the Jura, about 80 kms away (Theurot 2006). The recent discovery of two Medieval watermills dating respectively from the 10th and 12th centuries near the quarries in the Commune of Thervary (excavation by G. Rollier, INRAP) allows comparison directly between the millstones of the watermill with the roughouts and circular extraction holes in the quarries.

A few of these millstone quarries have been identified in the field and one was excavated. It is estimated that this small quarry yielded not more than 20 millstones. The quarry progressed essentially in a horizontal direction leaving few superimposed extractions. As in the case of the other small sites, this quarry is located on the edge of a sandstone plateau overlooking the Serre River. It is an example of a modest, short-lived quarry, contrary to other much larger ones that endured over long periods of time (Figs. 15-16).

The Medieval extraction technique begins by tracing the circumference of the future cylinder with either a compass (or possibly an even more simple system consisting of a piece of charcoal or an iron shank at the end of a string) serving as a guide for cutting a circular trench with a pick. The circular trench with vertical faces is carved in several stages. A first segment is cut in one direction while the second segment, which joins the first, is cut in the opposite direction. This change of direction allows the quarryman to benefit from more working space. The tool marks visible on the stone suggest that the quarryman employed the technique of successive 'passes' (French '*passe*') with the

Figure 18. Reconstruction of the technique of cutting a trench with 3 'passes' around a Medieval millstone. Drawing by A. Pulido.

pick, producing linear tool marks measuring between 0.5 and 1 cm. Each 'pass' consists of multiple aligned strokes of the point of the pick. The impact of the point of the pick, crushing the crystals of the stone, resulting in a clearly visible mark. To cut the trench, the quarryman used three passes: two on each side and one in the middle to remove the residual stone (Fig. 18). This technique is similar to that observed in the Roman quern quarry of Châbles in western Switzerland (Anderson *et al.* 2003), as well as in the modern millstone quarries of Quaix-en-Chartreuse (Belmont 2001) (Fig. 17).

Once the circular trench was completed, the cylinders were split from the bedrock, either by long iron bars (*pinces*) or picks acting as levers. The marks observed at the base of cylindrical roughouts clearly indicate regularly spaced extraction cavities. These cavities are similar to those resulting from the technique observed on the Roman site of Châbles in Switzerland, where the cylinder was split by repeated, regularly spaced strokes at the base of the cylinder (Anderson *et al.* 2003, 51). Often, the resulting shape of the lower face of the cylinder in contact with the bedrock therefore is not flat but convex-shape. In any case, the extraction and splitting techniques match those of Antiquity. The use of iron bars or picks for splitting is identified at the Medieval site of Ecouges in the French Alps dating from the end of the 13th to the beginning of the 15th century (Belmont 2005, 86).

At the site of Molères in the Commune of Boulu in the Pyrénées Atlantiques departement, the technique of splitting with the pick following the bedding of the stone is

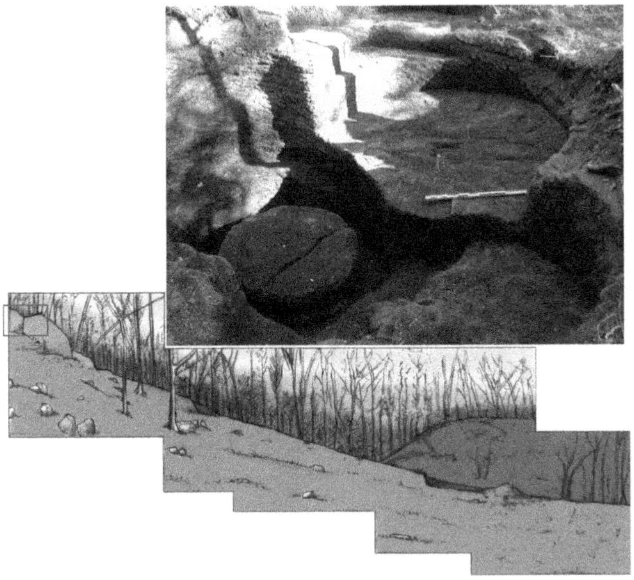

Figure 19. View of the contemporary millstone exploitation. Photograph by L. Jaccottey, INRAP and drawing by A. Pulido.

known for millstones that measure about 1 m in diameter and date from Medieval times (Martzluff *et al.* 2008; Martzluff 2009). Further away, at the production centre of Hyllestad in Norway, the extraction technique consists of first cutting a circular trench with the pick and then splitting the cylinder by means of pointed chisels following the cleavage plane of the schist. The earlier exploitations progressed horizontally. This technique then switched at the beginning of the 12th century to vertical successive, superimposed extractions (Grenne *et al.* 2008).

5. Modern and contemporary millstone production

The contemporary millstones produced in the Serre Massif differ little from their earlier counterparts except in size. Their diameter ranges between 1.10 and 1.25 m. In the 18th and 19th centuries, there were many quarries exploiting the sandstone outcrops of the Serre Massif (Figs. 19-20). The production of millstones from the Ferté-sous-Jouarre in the Paris basin began to invade progressively the regional markets and finally eclipsed the modest productions of the

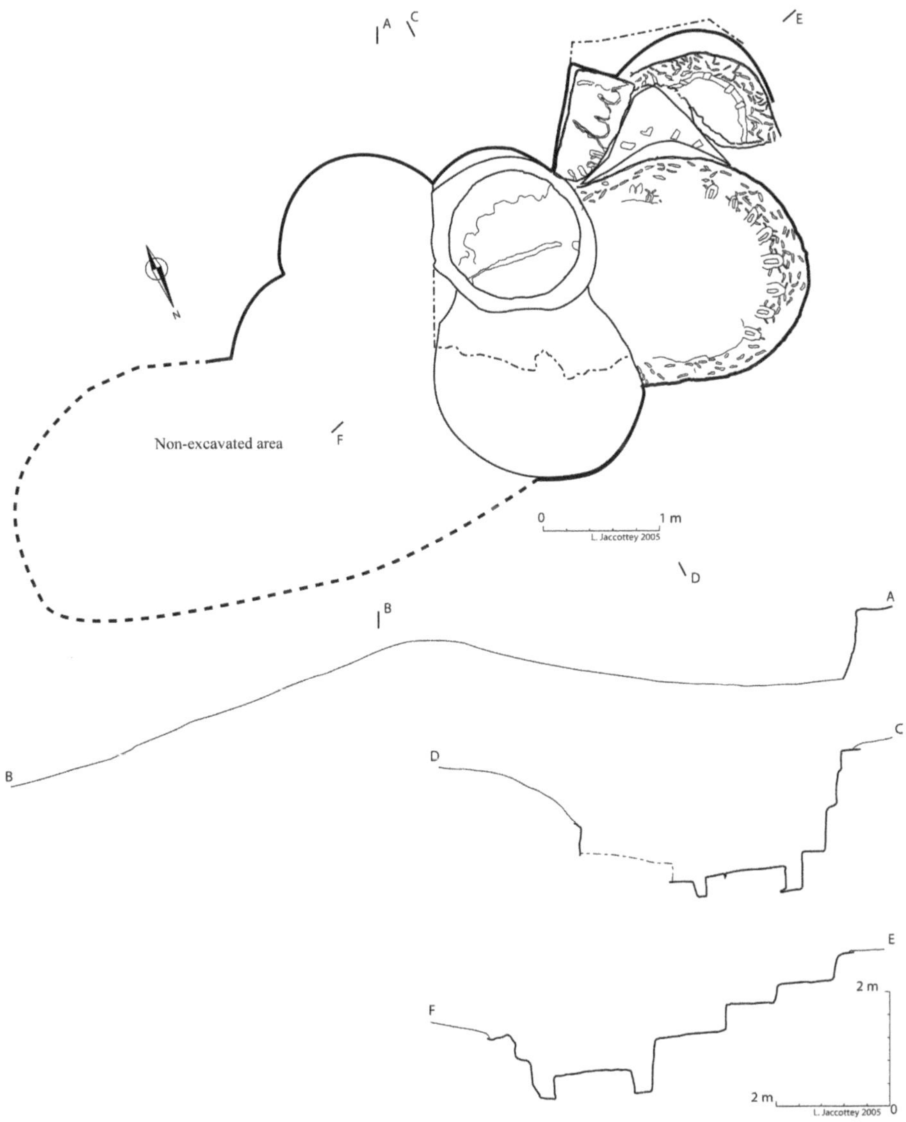

Figure 20. Map of the contemporary millstone quarry. Drawing by L. Jaccottey, INRAP.

Serre Moutains (Belmont 2006b; Jaccottey 2009). Joseph Maillot, for example, closed his millstone quarry in 1909 (Fig. 21).

The extraction technique is similar to that of preceding times. The circular trench was dug by an iron pick with a long handle (Fig. 22). The extremity of the pick (French *escoude*) used at this time had one point. The vertical trench was about 20 cm wide, allowing the quarryman to stand in it. As in the case of earlier extractions, work progressed alternatively, first in one direction and then in another.

The cylinder was split by a series of slots dug along part of the base of the circumference. These cavities were destined for either iron or wood wedges. According to Diderot 'Wooden wedges were inserted in the holes and then were filled with water making the wedges swell and splitting the millstone on a horizontal plane' (Diderot and d'Alembert, 1757, vol. II, Tome X, 880). The discovery of iron wedges at the site confirms the use of iron wedges that were placed in slots and hammered alternately until splitting occurred (Fig. 23). Placing the wedges around the whole of the cylinder produced a more regular, flat lower surface (Fig. 24). In some cases, deep spout-shaped slots were cut into the edges of the extractions to enable sliding an iron bar under the millstone to lift it once it was split from the bedrock. These could also have been used

Figure 22. Reconstruction of the technique of cutting a circular trench around a contemporary millstone. Drawing by A. Pulido and J. I. Motoro.

Figure 21. Reconstruction of the technique of tracing the circumference of a contemporary millstone with a compass. Drawing by A. Pulido and J. I. Motoro.

to place wooden beams for sliding the millstone out of the quarry.

The chronology of these extractions, between the end of the 18th century and throughout the 19th century, is certified by the archaeological finds on the site. Similar circular extractions dating to the 17th or 18th centuries were found at the site of Vic-le-Comte at the foot of the Massif Central (Cabezuelo 2006).

The final phases of millstone carving took place near the quarry site. The edges and the surfaces of the future millstone were regularized, and the central eye was perforated. During the process of eye perforation, many millstones fractured. The number of broken millstones corresponds to about half of the total number of roughouts seen on the site. It is therefore logical that this final phase of work, resulting in many discarded millstones, took place near the original extraction site (Fig. 25).

The use of iron or wooden wedges, besides permitting an increase in the diameter, seems to be the principal innovation of modern and contemporary times. This technique is observed in the quarries of Offlanges and was also well known later in the second half of the 12th century at Boulou (Martzluff *et al.* 2008; Martzluff 2009). This technique is also known in the more recent quarry

Figure 23. Square-shaped holes for iron wedges. Photograph by L. Jaccottey, INRAP.

of Quaix-en-Chartreuse, dating from the 16th to the 18th centuries (Belmont 2001). In Norway, this technique seems to date to the 16th century at the quarries of Selbu (Grenne *et al.* 2008). At these sites, as in the case of Hyllestad toward the beginning of the 12th century, extraction began to change from a horizontal progression to superimposed extractions leaving vertical columns. The same can be said of Offlanges.

Figure 24. Reconstruction of the technique of splitting millstones with iron wedges. Drawing by A. Pulido.

Figure 25. Abandoned millstone near the quarry. Photograph by L. Jaccottey, INRAP.

Conclusion

The unique geological characteristics of this small metamorphic mountain range located in the heart of the Jurassic limestone formation, combined with specific grinding qualities, resulted in continuous exploitation from the beginning of the Neolithic (c. 5200 BC) until the 20th century. We can therefore follow the evolution of extraction techniques over seven thousand years from the earliest exploitation of surface blocks in screes in the Neolithic and protohistory to the true extractive sandstone quarries located on the edge of the plateau in Antiquity. The type of rock sought also changed over the years. In the Neolithic, only sandstone was used. Little by little granite querns appear in archaeological collections and assume an important role from the Bronze Age and into the La Tène period when about half of the querns in the collections are of granite. With the Roman conquest, granites are once again relegated to a lesser role, giving place to local sandstones and basaltic lava querns imported from long distances. During the Medieval period, there is a return exclusively to sandstones, which in turn are replaced from the 16th century by the importation of chert millstones from La Ferté-sous-Jouarre, near Paris. The rotary querns of Antiquity and the watermills of Medieval times, although of different typology, seem to have been extracted using similar techniques. However, extraction techniques, in particular the means of splitting the cylinder from the bedrock, change radically between the Middle Ages and modern times.

The evolution of extraction techniques over the last few thousand years in the Serre region appears to be similar to that observed in other areas of France. Furthermore, the spread of La Serre millstones has varied over time.

This study brings forward new data on the question of millstone production and economy. This new data must, nevertheless, be confronted with other analogous work. Research in France on this subject is currently in a dynamic state, with new excavations and investigations of querns and millstones in museum collections. This new research should allow us soon to evaluate the different French production centres and compare the evolution of techniques and compare observations from historical and social spheres, following the example of recent research in other European countries (Grenne *et al.* 2008).

Acknowledgments

The translation of this text was the work of Timothy Anderson. Aurora Pulido is responsible for the graphic reconstructions.

References

Alonso, N., 2002. Le moulin rotatif manuel au nord-est de la Péninsule Ibérique: une innovation technique dans le contexte de la mouture des céréales. In H. Procopiou and R. Treuil (dir.), *Moudre et broyer*, Comité des Travaux Historiques et Scientifiques, 111-27.

Anderson, T., Augustoni, C., Duvauchelle, A., Serneels, V. and Castella, D., 2003. *Des artisans à la campagne. Carrière de meules, forge et voie gallo-romaines à Châbles (FR), Archéologie Fribourgeoise*, 19, Fribourg.

Belmont, A., 2001. Les carrières de meules de Quaix-en-Chartreuse (XVIe- XVIII^{e}siècle). *Histoire et Sociétés Rurales*, 16, 45-79.

Belmont, A., 2005. Un chantier du Laboratoire de Recherche Historique Rhône-Alpes: les fouilles archéologiques de la carrière de meules de moulins des Ecouges (Isère), publication en ligne, site internet du LARHRA, *http://www.msh-alpes.prd.fr.*

Belmont A. 2006a. *La pierre à Pain, les carrières de meules de moulins en France du Moyen Age à la révolution industrielle, Tome I*, Grenoble.

Belmont A. 2006b. *La pierre à Pain, les carrières de meules de moulins en France du Moyen Age à la révolution industrielle, Tome II*, Grenoble.

Belmont, A., and Mangartz, F., 2006. *Les meulières. Recherche, Protection et Valorisation d'un Patrimoine Industriel Européen, actes du colloque international, Maison des Sciences de l'Homme-Alpes, Grenoble, septembre 2005*. Mainz.

Benoit, F., 1940. L'usine de meunerie hydraulique de Barbegal (Arles). *Revue Archéologique*, 15, 19-80.

Boyer, F., 2008. Roches meulières : de la classification pétrographiques à la classification texturale 'mécanique'. In *Evolution typologique et techniques des meules du Néolithique à l'an mille sur le territoire français, rapport de Programme Collectif de Recherche 2008-2010, année 2008*, CNRS, Ecole Normale Supérieure, 52-3

Boyer, F., Buchsenschutz, O., Hamon, C., Jaccottey, L., Lagadec, J.-P., Milleville, A., Thomas, E., and Triboulot, B. 2006. Production et diffusion des meules du Néolithique à l'Antiquité: quelques exemples français. In Belmont and Margartz 2006, 5-13.

Cabezuelo, U., 2006. Le site de la ZAC des 'Meules' à Vic-le-Comte (Puy-de-Dôme). In A. Belmont and F. Mangartz (coord.) *Les meulières. Recherche, Protection et Valorisation d'un Patrimoine Industriel Européen,* actes du colloque international, Maison des Sciences de l'Homme-Alpes, Grenoble, septembre 2005, Verlag des Römisch-Germanischen Zentralmuseums, Mainz, 109-14.

Castella, D., 1994. *Le moulin hydraulique gallo-romain d'Avenches 'En Chaplix'. Cahiers d'Archéologie Romande,* 62.

Diderot, D., and D'Alembert, J., 1757. *Encyclopédie, ou dictionnaire raisonné des sciences, des arts et des métiers*, société de gens de Lettres, vol II, Tome X.

Doswald, C., 1994. Herkunft und Verbreitung der römerzeitlichen Mülhlsteine im Kanton Aargau. *Minaria Helvetica,* 14a, 22 -33.

Fabre, D., Carrio, E., Orengo, Y., and Malacour, C., 2006. Analyses pétrographiques et mécaniques d'un ensemble de roches meulières françaises. In Belmont and Mangartz 2006, 91-7.

Feuvrier, J., 1920. Une vieille industrie comtoise éteinte. *Franche-Comté et Monts Jura, Revue Régionale Mensuelle*, 10, avril 1920.

Grenne, T., Heldal, T., Meyer, G.B., and Bloxam, E.G., 2008. From Hyllestad to Selbu: Norwegian millstone quarrying through 1300 years. In T. Slagstad (ed.), *Geology for Society*. Geological Survey of Norway Special Publication, 11, 44-66.

Guillier, G., Biard, M., and Chérel, A.-F., 2005. Un atelier augustéen de taille de meules en poudingue au 'Clos des Forges' à Avrilly (Eure), *Revue Archéologique de l'Ouest*, 22, 199-220.

Hamon, C., 2006. *Broyage et abrasion au Néolithique ancien. Caractérisation technique et fonctionnelle des outillages en grès du Bassin parisien,* BAR S1551, Oxford, 342.

Harms, E., and Mangartz, F., 2002. *Vom Magma zum Mühlstein, Eine Zeitreise durch die Lavaströme des Bellerberg-Vulkans*, Vulkanpark-Forschungen 5, Mainz.

Hörter, F., 1994. *Getreidereiben und Mühlsteine uas der Eifel.* Mayen.

Jaccottey, L., Anderson, T., and Jodry, F., 2007. *La circulation du matériel de mouture de part et d'autre de l'arc jurassien durant l'Antiquité, Première journée archéologique frontalières de l'Arc jurassien (JAFAJ), Delle, Territoire de*

Belfort (F), Boncourt, Jura (CH), octobre 2005, 223-228.

Jaccottey, L., 2009. Les Carrières de meules de la Serre. Révision des données sur l'extinction d'une vieille industrie comtoise, in *Les recherches de Julien Feuvrier (1851-1936), historien, archéologue, archiviste et conservateur du musée. Dole et sa région, de la préhistoire au 18e siècle,* 180-1.

Jaccottey, L., Milleville, A., and Pétrequin, P., 2007. Des meules et des villages fortifiés : exploitation des matières premières au Néolithique dans le Jura et organisation territoriale. In M. Besse (ed.), Sociétés néolithiques. Des faits archéologiques aux fonctionnements socioéconomiques, actes du 27ème colloque INTERNEO, Neuchâtel (Suisse), octobre 2005. *Cahiers d'archéologie romande*, 108, Lausanne43-50.

Jaccottey, L., and Milleville, A., Forthcoming. Schéma d'occupation interrégional: l'exploitation du massif de la Serre (Jura, France) et la diffusion du matériel de mouture au Néolithique. *Archeologia Mosellana.*

Jaccottey, L., and Milleville, A., 2007. Première découverte de zones d'extraction de moulins de type 'va-et-vient', Jura. *Société Préhistorique Française*, 104, 827-831.

Joos, M., 1975. Eine permische Brekzie aus dem Südschwartwald und ihre Verbreitung als Mühlstein im Spätlatène und frührömischer Zeit. *Archäologisches Korrespondezblatt*, 5, 197-9.

Leuvrey, J.M., 1999. Hauterive-Champréveyres, l'industrie lithique du Bronze final, étude typo-technologique. *Archéologie Neuchâteloise*, 12, 97-9.

Longepierre, S., 2006. Aux environs de Saint-Quentin-La-Poterie (Gard) durant l'antiquité tardive : une microrégion très impliquée dans l'activité meulière. In Belmont and Mangartz 2006, 47-54.

Martzluff, M., 2009. Au temps des pierres amoureuses. Typologie du débitage des roches monumentales depuis l'an mil dans les Pyrénées catalanes, in *De Méditerranée et d'Ailleurs... Mélanges offerts à Jean Guilaine, Archives d'Ecologie Préhistorique*, Toulouse 2009, 483-508.

Martzluff, M., Aloisi, J., Passarrius, O., and Catafau, A., 2009. Meules et moulins de Vilarnau, in Passarrius O. Donat R. et Catafau A. (coord) 2009 : Vilarnau. Un village du Moyen Age en Rousillon, Conseil Général des Pyrénées-Orientales et Trabucaïre , 314-87.

Naze, Y., Fronteau, G., and Robert, B., forthcoming. L'atelier de meules rotatives en calcaire à cérithes de Vendresse-Beaulne (Aisne). Note à propos des outils de mouture en calcaire Lutétien, in *Évolution typologique et technique des meules du Néolithique à l'an mille sur le territoire français.* Chronique sur la Table ronde de Saint-Julien-sur-Garonne (F) du 2 au 4 octobre 2009, supplément Aquitania.

Pétrequin, P., and Jeunesse, C., 1995. *La hache de pierre. Carrières vosgiennes et* échange *de lames polies pendant le Néolithique (5400-2100 av. J.C.).* Dijon-Quétigny.

Ribaux, P., 1986. Cortaillod-Est, un village du Bronze final, Tome 3. L'homme et la pierre. *Archéologie Neuchâteloise*, 3, 140.

Robert, B., and Landreat, J.L., 2005. Les meules rotatives en calcaire à glauconie grossière et l'atelier de Vauxrezis (Aisne). Un état de la question, in *Revue Archéologique de Picardie*, n° spécial 22, 105-14.

Servelle, C., 2006. Les meulières antiques de la Marèze (Saint-Martin-Laguépie et le Riols, Tarn, France), géologie, géomorphologie, techniques d'exploitation et de façonnage. In Belmont and Mangartz 2006, 61-9

Theurot, J., 1998. *Dole, genèse d'une capitale provinciale, des origines à la fin du XVe siècle*, Cahier Dolois 15, 2, Dole.

Theurot, J., 2006. La pierre et le Grain, à propos d'une récente découverte archéologique. *Le Jura français*, 12-14.

Zogo, J., 1981. *Le permien et le trias du massif de la Serre (Jura) étude sédimentologique*. Doctoral thesis, Université de Dijon, Institut des Sciences de la Terre.

The widening use of Lodsworth Stone: Neolithic to Romano-British quern distribution

Ruth Shaffrey and Fiona Roe

Introduction

In 1987 David Peacock published a seminal paper in the *Antiquaries Journal*, in which he described a distinctive rock type, the Lodsworth Greensand, which was used for saddle and rotary querns across the south of England (Peacock 1987). The paper located the area where the stone was quarried from Lower Greensand Hythe Beds in the parish of Lodsworth, between Petworth and Midhurst in West Sussex, and described the sandstone thus:

'*a hard, medium-grained, greenish-grey or brownish-grey, silicified, glauconitic, quartz sandstone, with characteristic swirls and stringers of dark cherty material, rich in glauconite*' (Peacock 1987, 62).

During the course of some fifteen years of research, Peacock visited 59 museums. In the resulting publication he investigated all aspects of the use of Lodsworth stone for querns, including the production technology, the typology of the querns (including the transition from saddle to rotary varieties) and the periods during which they were used. He then went on to plot and discuss the distribution of both saddle and rotary querns, which were listed in an Appendix. This publication became and has remained the reference source whenever querns of Lodsworth stone are found. The paper presented here aims to build upon the foundations laid by Peacock with data accumulated by both authors during the course of their work across the south of England. As with the original paper, it is not intended to be an exhaustive survey of the use of Lodsworth stone, but includes data collected from a combination of museum visits, independent research and reports for professional archaeological bodies carried out since 1985. As the authors' work has been concentrated in the Thames valley, a noticeable bias towards this area may be observed; however, the additional sites are real finds and make a significant contribution to our understanding of Lodsworth Greensand quern distribution. It is worth noting that neither of the authors has had much involvement in the Greater London area so that the lack of new discoveries here may be considered something still to be investigated.

This survey adds 69 new sites to the original gazetteer, amounting, with some overlap, to 21 saddle quern sites, 11 Iron Age rotary quern sites and 51 late Iron Age / Romano-British and Romano-British rotary quern sites. Nearly all the data included in the main analysis has been personally recorded by one or both authors, or else has reliable identifications. Bibliographical references to the sites are to be found in the gazetteer rather than in the text. All new sites are mapped, as are any sites listed by Peacock which have now produced finds with a wider chronological range; those with additional finds of the same dates as in the original publication are listed in the gazetteer but are not mapped as new sites. Sites with querns of possible Lodsworth stone noted in publication but not seen by the authors have been included in the gazetteer but are not mapped.

The paper is subdivided into the Prehistoric use of Lodsworth stone for saddle querns and the Iron Age and Roman use for rotary querns followed by a brief discussion of typology.

Saddle querns (Neolithic to middle Iron Age)

In 1987 there were only 14 known sites with saddle querns of Lodsworth stone (Peacock 1987, fig. 6). The earliest saddle querns then recorded were from the late Bronze Age site of Green Lane, Farnham, some 25 km (15.5 miles) north of the quarry area. Two further saddle querns of late Bronze/early Iron Age date, from Harting Beacon and Park Brow were both from West Sussex, and so also fairly local to Lodsworth. The distribution of the originally recorded saddle quern sites barely stretched beyond the limits of Sussex and Hampshire with most finds occurring westwards from the quarry area. It is now possible to add a further 21 definite sites to the original list, of which 12 occur outside the original distribution area (Fig.1 and Table 1).

In 2008 the first saddle quern from a Neolithic context came to light, the result of an evaluation by Cotswold Archaeology at Lavant Quarry, near Chichester, 16 km (10 miles) south of Lodsworth. The associated pottery is provisionally identified as early Neolithic Plain Bowl, while the quern is made from a boulder, which suggests utilisation of surface deposits at the quarry area. Numbers

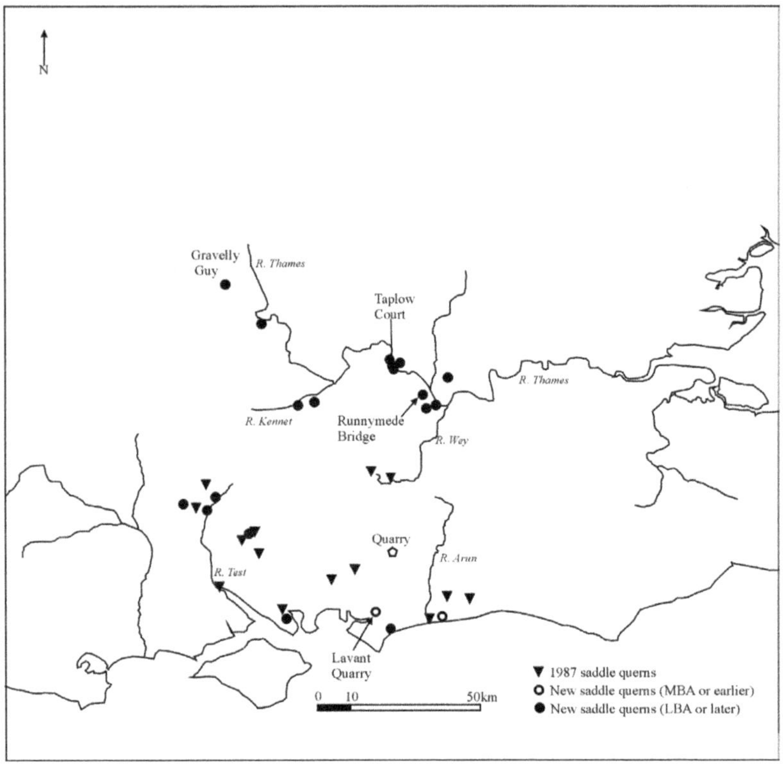

Figure 1. Map of saddle quern distribution.

of Lodsworth querns continue to be low in subsequent phases, with no finds from early Bronze Age contexts, and only one from a middle Bronze Age site at Angmering, 24 km (15 miles) from the quarry area. Both of these early finds are within the original distribution area.

There has been a proliferation, however, of examples dated from the middle Bronze - late Bronze/ early Iron Age transition onwards. The most striking feature of these is that many are not local to the quarry area (Fig. 1 and Table 1). Both the middle/late Bronze Age sites known to us at present are on or near the River Thames, at Taplow Motorway Division, and Imperial College Sports Ground, Harlington. These sites are some 57 km (36 – 37 miles) north from Lodsworth. By the late Bronze Age the number of certain sites with querns of Lodsworth stone now identified amounts to seven, only two of which, Green Lane, Farnham (Peacock 1987, 77) and Bognor Regis, are local to the quarry area. There are four new sites located in the middle Thames region: Cippenham, Lot's Hole (Dorney), Home Farm (Laleham) and Runnymede Bridge. Two further Bronze Age sites, less well documented, are on the River Kennet, at Pingewood and Field Farm, Burghfield, both near Reading.

With chronological progression towards the Iron Age, there are two late Bronze/early Iron Age sites with Lodsworth stone saddle querns in the middle Thames area, Taplow Court and Imperial College Sports Ground, Harlington (the latter site already noted with middle to late Bronze Age fragments). It seems likely that the observed distribution area could now be extended into the upper Thames Valley since excavations at Appleford, Oxfordshire produced a possible Lodsworth stone saddle quern of late Bronze/ early Iron Age date. This quern has not been recorded by the authors (and is therefore not on the map) but it is given the same published lithological description as a rotary quern that has been identified by us (FR) as Lodsworth stone.

By the early Iron Age it appears that the middle Thames Valley was less densely occupied, although fragments of Lodsworth stone were found in a probable hillfort enclosure at St. Ann's Hill, Chertsey. Despite this, there is evidence that Lodsworth stone was being taken further up the Thames than before, with finds at Gravelly Guy, Oxfordshire from both early and middle Iron Age contexts. The minimum distance from here to the quarry site is about 100 km (62 miles) but the actual journey would no doubt have been somewhat longer. Peacock's work demonstrated an emphasis on Hampshire during the Iron Age, where he recorded a number of sites with Lodsworth stone saddle querns of both early and middle Iron Age date. The most distant of these sites from Lodsworth was probably Danebury hillfort, near Middle Wallop (Cunliffe and Poole 1991, 396), which would have required a journey of more than 66 km (41 miles) to transport the numerous finds of Lodsworth stone querns to the hillfort. Further Lodsworth stone saddle querns of early and middle Iron Age date have subsequently been recorded from the Danebury Environs project.

Discussion of saddle querns

Sussex is well provided with earlier prehistoric sites and examination of quern fragments from some of these could well yield further examples of the early use of Lodsworth

Time Period	Local	Thames and Kennet Valleys	Coast	Hampshire
NEO	Lavant Quarry, W Sussex			
	(The Trundle, W Sussex?)			
MBA			Angmering, W Sussex	
M/LBA		Taplow Motorway Division, Bucks		
		Harlington, Greater London		
		(Pingewood, Berks)		
LBA	*Green Lane, Farnham, Surrey*	Cippenham, Berks	Gosport, Hants	
		Dorney, Lot's Hole, Bucks	Bognor Regis, W Sussex	
		Laleham, Surrey		
		Runnymede, Surrey		
		(Burghfield, Field Farm, Berks)		
		(Hurst Park, Surrey?)		
LB/EIA	*Harting Beacon, W Sussex*	Taplow Court, Bucks	*Park Brow, W Sussex*	
		Harlington, Greater London		
		(Appleford, Oxon)		
EIA		Gravelly Guy, Oxon	*Southampton, Hants*	*Balkesbury*
		St Ann's Hill, Surrey		Danebury Environs
				Winchester, Staple Gardens
MIA		Gravelly Guy, Oxon		*Balkesbury*
				Danebury
				Danebury Environs
				Owslebury
				Winnall Down
			Littlehampton, W Sussex	
			Fareham, Hants	

Table 1. List of sites with saddle querns of Lodsworth Greensand. Peacock's sites in italics, sites of less certain date in brackets, uncertain identifications with question mark.

stone. One such Neolithic candidate would be a quern found in 1928 at The Trundle near Goodwood (not in gazetteer), which is described as being made from a fine siliceous or glauconitic sandstone from the Lower Greensand Hythe Beds (Curwen 1929, 63). It would be no surprise if further Neolithic querns of Lodsworth stone were to be recorded, since in other areas it has become clear that the best quern materials were identified during the Neolithic period, and then continued in use for thousands of years (Roe 2009, 30). Further identification of Bronze Age sites with querns made from Lodsworth stone could also be expected, both in Sussex and elsewhere. One possibility is at Hurst Park, East Molesey in the middle Thames area where five Greensand fragments (not listed in the gazetteer) were found in late Bronze Age contexts (Andrews and Crockett 1996, 80).

From the Bronze Age onwards, the distribution of Lodsworth stone saddle querns can be divided into four groups (Table 1). Firstly, there are the core sites local to the quarry area, within about one day's walking distance, which might be up to some 32 km (20 miles) away. Then there are the sites along the middle Thames Valley, with two prominent clusters around Taplow and Runnymede Bridge, the latter near the confluence of the River Wey with the Thames. The discovery of these sites owes much to intensive development in the areas around Windsor and Staines, but their existence may be connected to the important role that the River Wey must have played in transporting the querns up from Sussex. Other sites with Lodsworth stone querns are situated on the coast. For these the River Arun may have provided a means of transport to the Sussex coast, leading to Littlehampton and Angmering, and then westwards along the coast to Bognor Regis, Gosport, Fareham and Southampton (Peacock 1987, 77). Some coastal distribution could also partly explain the numbers of Lodsworth stone querns recorded in Hampshire from the early Iron Age onwards, since the River Test could have led from Southampton to the sites at and around Danebury, near Andover, and the River Itchen could have led to the sites clustered round Winchester.

There have been a number of finds of log boats from the Rivers Wey and Arun and one from the River Test (McGrail 1978, fig. 207); none is dated to the prehistoric period but a long tradition of usage of log boats on these rivers could be assumed. There is an early Bronze Age example of a dug-out canoe from Locharbriggs, Dumfries and Galloway (McGrail 1998, 85), while in Derbyshire a log boat found near Shardlow has been radiocarbon dated to the middle Bronze Age (Brown 2009, 14). This boat had been carrying blocks of stone and the technology for transporting querns by this method was undoubtedly there.

The geology of the areas where the earliest Lodsworth stone querns have occurred also helps to explain their presence at sites some distance from the quarry area. The two main groups of Hampshire sites near Andover and Winchester are on the chalk, with only limited availability of sarsen and other hard stone for querns. In the Thames Valley there was a similar shortage of good, hard rocks that would grind corn well and in both regions an incentive thus existed to

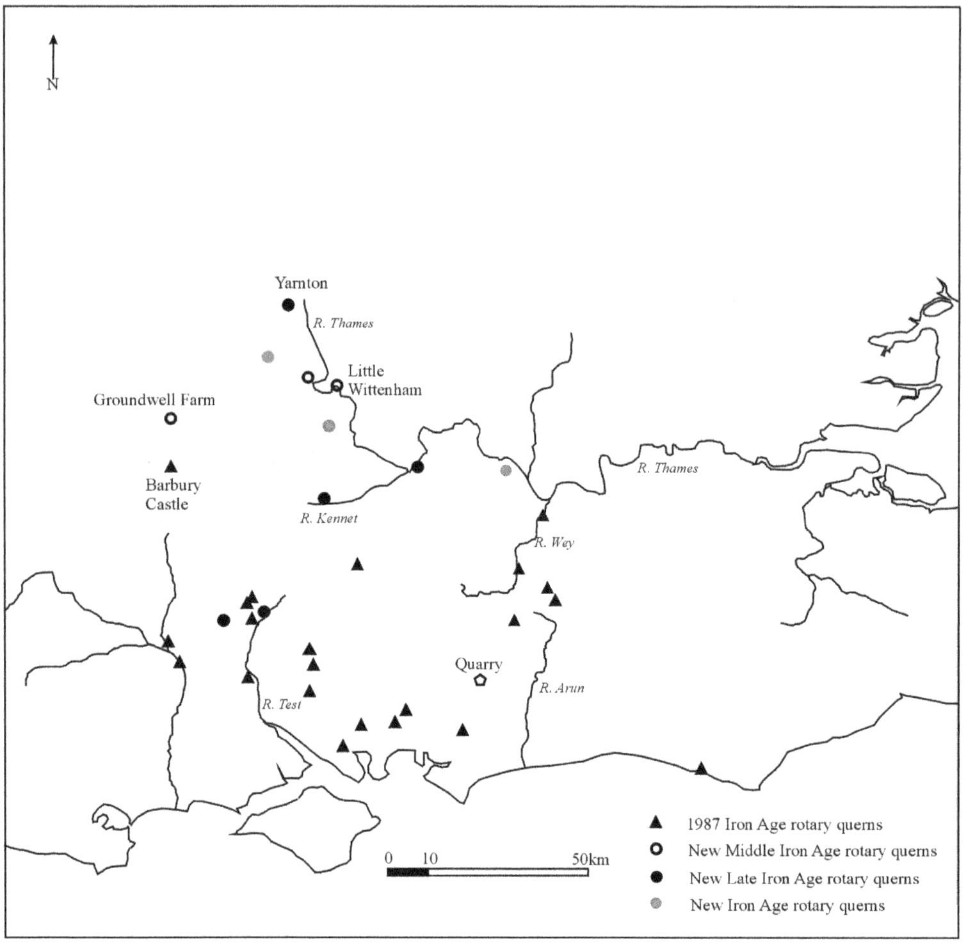

Figure 2. Map of Iron Age rotary quern distribution.

obtain the good quality Lodsworth stone querns. Similar initiative was applied in other areas, as in Gloucestershire, where the Jurassic limestone was unsuitable for corn grinding. Here, saddle querns of a hard Silurian gritstone obtained on May Hill were transported across the Cotswolds to sites that even in Neolithic times might be up to 43 km (27 miles) distant from the source (Roe 2009, 27). Those travelling to the Thames from Sussex may have hoped to acquire bronze tools and weapons in return for their querns, while the number of sites along the Thames Valley where flax seeds have been recorded (Lambrick and Robinson 2009, 253) indicates that it may also have been possible to acquire linen garments. Other commodities were no doubt exchanged as well and these could have included salt from the south coast and pottery. It has been suggested that the River Wey played an important role in the transport of Neolithic stone axes to Surrey, where finds are concentrated by the old ford at Weybridge and also in the Thames and its tributaries (Field and Woolley 1984, 101). This route may long have been a traditional one, as may other proposed river journeys.

The recently accumulated evidence indicates that not only were saddle querns of Lodsworth stone distributed further than was initially appreciated, but that an expansion in this distribution began sometime during the middle to late Bronze Age. The increased availability of bronze implements in the Thames Valley at this time could have provided initial motivation for the effort required to transport the querns northwards over increasingly long distances.

Rotary querns (middle Iron Age to Roman)

In 1987 there were 25 known sites with Iron Age rotary querns, 31 with Roman rotary querns and a further 33 of uncertain date. A further 51 sites dated to the Roman period (including any finds dated to the late Iron Age/ Roman period) and 11 dated to the middle or late Iron Age can now be added. Those considered here as Iron Age are specifically noted as being pre-Roman and of them, three are dated to the middle Iron Age, with further finds of comparable date from the Danebury Environs. It may be noted that a few sites have rotary querns dating to both periods and are thus considered twice.

Iron Age rotary querns

Information on the earliest appearance of rotary querns in the UK remains limited (Buckley 1979, 89; Heslop 1987, 111; Shaffrey 2007a, 89; but see Peacock and Cutler, this volume), although introduction of the new technology during the fifth or fourth centuries BC seems likely. At Danebury a few rotary querns from ceramic phase 5 have been recorded (Brown 1984) suggesting they were in use here by the middle of the fourth century BC, and at least one of these was made from Lodsworth stone (Peacock 1987, 69). There is a growing number of rotary querns from middle Iron Age contexts, but these querns tend to be fragmentary and dating evidence is usually only approximate.

Peacock was working at a disadvantage, because many of the Lodsworth stone querns he recorded were from older excavations and so not well dated. His research revealed a greater number of rotary than of saddle querns with a clear density of find spots in Sussex and Hampshire, many within 50 km of the quarry site (Peacock 1987, fig. 7). There was one outlier of probable Iron Age date at Barbury Castle, Wiltshire. In addition, he had two find spots north of the River Thames for querns of possible Iron Age or early Roman date, at Odell in Bedfordshire and Hailes in Gloucestershire. Published individual dates for these querns suggest they are all of Roman or LIA/ER transition date (King 1980, 80) and are thus shown here only on Fig. 3.

Recent work has shown that during the middle Iron Age, rotary querns of Lodsworth stone were distributed further from the quarry than the most far travelled of the Bronze Age saddle querns (Fig. 2 and Table 2). One new site has been recorded in Wiltshire, at Groundwell Farm, Blunsdon St Andrew, while in the upper Thames Valley rotary querns were found at Abingdon Vineyard, and Hill Farm, Little Wittenham. At the latter site fragmentary but matching upper and lower stones were associated in a pit with middle Iron Age pottery but with a relatively late C14 date of 200 BC – 1 AD at 95% confidence (Allen *et al,* 2010, 215). There are also probable rotary querns of Lodsworth stone from Blewburton Hill, Oxfordshire, which may belong in the middle Iron Age, and a quern found during fieldwalking at Datchet, Berkshire, should on typological grounds be of mid rather than later Iron Age date.

The growth in the distances over which querns were distributed is more obvious during the late Iron Age, when Lodsworth stone was transported further up the Thames into Oxfordshire, as for example to Yarnton. There have been other finds from around Oxford that may be of this date, and could relate to the "Big Ring" at Cassington, since querns in the Ashmolean Museum collection include a lower stone of Iron Age style, (probably 1943.24) from Cassington but not in the Gazetteer. There is also a complete Iron Age style rotary quern of Lodsworth stone which is unregistered but likely to have been found locally. A short way up the river Windrush, an undated rotary quern found at Eagle Farm, Standlake is of Iron Age form.

During this period, there appear to be relatively few sites in the middle Thames and Kennet Valleys, though Lodsworth stone querns were found at Thames Valley Park, Reading and also within reach of the Kennet at Upper Bucklebury. By contrast, throughout the Iron Age there were numerous sites in Hampshire at which Lodsworth stone querns were in use. This may partly reflect the amount of excavation carried out but does also seem to indicate a certain density

Time Period	Local	Thames and Kennet valleys	Coast	Hampshire	Dispersed
MIA	*Holmbury, Surrey*	Abingdon Vineyard, Oxon		Danebury Environs	Groundwell Farm, Wilts
		Little Wittenham, Oxon		*Danebury*	
				Owslebury	
				Winklebury	
				Winnall Down	
M/LIA				Danebury Environs	
				Balkesbury	
LIA	*Holmbury, Surrey*	Reading, Thames Valley Park, Berks	*Horndean, Hants*	Danebury Environs	
	Hascombe, Surrey	Upper Bucklebury, Berks	*Bishopstone, E Sussex*	*Balkesbury*	
		Yarnton, Oxon		*Bury Hill?*	
		(Cassington?, Oxon)		*Danebury*	
				Owslebury	
				Winklebury	
				Winnall Down	
IA	*Burpham, Surrey*		*Chalton, Hants*	*Twyford Down, Hants*	
	Shere-Peaslake, Surrey		*Fareham, Furzehall Farm, Hants*		Blewburton Hill, Oxon
	The Trundle, W. Sussex		*Fareham, Wallington Military Road, Hants*		
IA prob		Datchet, Berks		*Timsbury, Hants*	*Barbury Castle, Wilts*
		Standlake, Eagle Farm, Oxon			*Old Sarum, Wilts*
		Weybridge, Surrey			

Table 2. List of Iron Age rotary quern sites. Peacock's sites in italics, sites of less certain date in brackets.

of population, while the apparent lessening of prehistoric activity in the middle Thames valley remains unexplained.

This widening of the distribution of Lodsworth stone rotary querns may be related in some areas to a general change in quern production. Some materials were either found to be less suitable for rotary quern production or their limited resources were not able to cope with increased demand and so ceased to be utilised. In the upper Thames Valley a significant change of quern materials occurred with the introduction of rotary querns, probably between the later middle Iron Age and the late Iron Age (Roe 1999, 46) but varying from site to site depending on status and size. Lodsworth stone was one of a range of stone types that were then imported to this region specifically for rotary querns, along with Millstone Grit, Old Red Sandstone and later, Niedermendig lava. Between them they replaced the mostly local stones such as Culham Greensand and Lower Calcareous Grit which had long been used for saddle querns, for example at Gravelly Guy and the Wittenhams (Bradley *et al.* 2004, 373 and Roe 2010, 172). As a result the best available rotary quern materials were distributed over greater distances and it is often around the time of the transition from saddle to rotary quern that a change in the lithologies occurred.

Lodsworth stone is one of the lithologies suitable for manufacture into both saddle and rotary querns. In Hampshire, where it had been much used for saddle querns, it continued to be imported in the form of rotary querns. There are also a number of coastal sites where rotary querns of Iron Age date were recorded by Peacock (Table 2). The site of Bishopstone, in East Sussex, seems to presage further transport of querns in Roman times eastwards along this part of the Sussex coast, while other coastal sites, such as Chalton, Horndean, and Fareham could indicate routes being taken westwards towards other Hampshire sites. Finds of querns at other sites on or near the River Wey, such as Burpham, Hascombe, Holmbury and Weybridge suggest that this river continued to provide an important routeway to the Thames. Taken together, these coastal and riverine sites may indicate that, as in the Bronze Age, transport of bulky items such as querns was best achieved by boat, following routes that by then had long been traditional.

It is likely that in addition to the querns, a variety of other goods were moved around during the Iron Age. There is evidence from Gloucestershire and Worcestershire of flourishing Iron Age trade, with finds from many sites both of Droitwich briquetage and high status Malvernian pottery, which appear to have been transported alongside querns of May Hill sandstone, all of which also reached Oxfordshire (Lambrick and Robinson 2009, 209). A similar trading pattern might be expected for the areas under consideration here, with the all important supplies of salt coming from the south coast, while the distribution of pottery may also have been of relevance. At the high status settlement of Abingdon Vineyard, imports additional to the querns of Lodsworth stone included pottery and brooches from Sussex (Tim Allen pers. comm.) At Danebury imported querns of Lodsworth Greensand and briquetage were both of importance, while it has been suggested that surplus produce for possible export would have included wool and grain (Cunliffe 2003, 142). Iron from the Weald must have been another significant commodity that needed to be carried as efficiently as possible around the region.

Late Iron Age - early Roman and Romano-British

A total of 51 new sites have been added to the 31 definite Romano-British finds spots known in 1987 and the outer limits of distribution pushed eastwards from Surrey into Kent and north-westwards from Northamptonshire into Warwickshire and Worcestershire. The revised map reveals a significant change in the distribution of late Iron Age and Roman rotary querns since 1987 and thus a marked difference in area reached in comparison to early rotary querns (Fig. 3). Because of the much larger number of sites, no table of Roman finds has been included.

To the north, the periphery of distribution now has the addition of Brigstock as a probable Roman site in Northamptonshire, with Desborough and Clay Coton remaining as outliers in the area. Significantly, however, this 'edge' has been further defined westwards by finds at Tiddington, Beckford and Bourton-on-the-Water, each of which produced a single quern of Lodsworth stone. The single item found at each site, and in particular the example at Tiddington, which is from an assemblage of approximately 300 quern fragments, suggests that these sites indicate a 'real' periphery to Lodsworth stone quern distribution and not one which will be markedly changed through future finds. Querns of Lodsworth stone also remain relatively scarce in the south-east Midlands (Buckinghamshire, Hertfordshire and Cambridgeshire), with only two single finds added to the cluster around Odell in the east Midlands, at Great Barford and Broughton.

In the east, the presence of Lodsworth stone was unknown until very recently but has now been recorded in Kent at Westhawk Farm (Ashford) and Springhead Roman town. Multiple querns were found at each site making their presence perhaps more significant than the 'one-off' events at the northern periphery. In both cases the sites produced large assemblages comprising a wide range of quern lithologies, but nevertheless, the presence of Lodsworth stone is significant as the county's quern supply was strongly dominated by Millstone Grit and Niedermendig lava. In Kent, early rotary querns appear to have relied entirely upon local materials such as the well-known Folkestone Beds Greensand (Keller 1988), perhaps because of strong tribal boundaries in that region restricting imports. This dependence appears to have lessened with

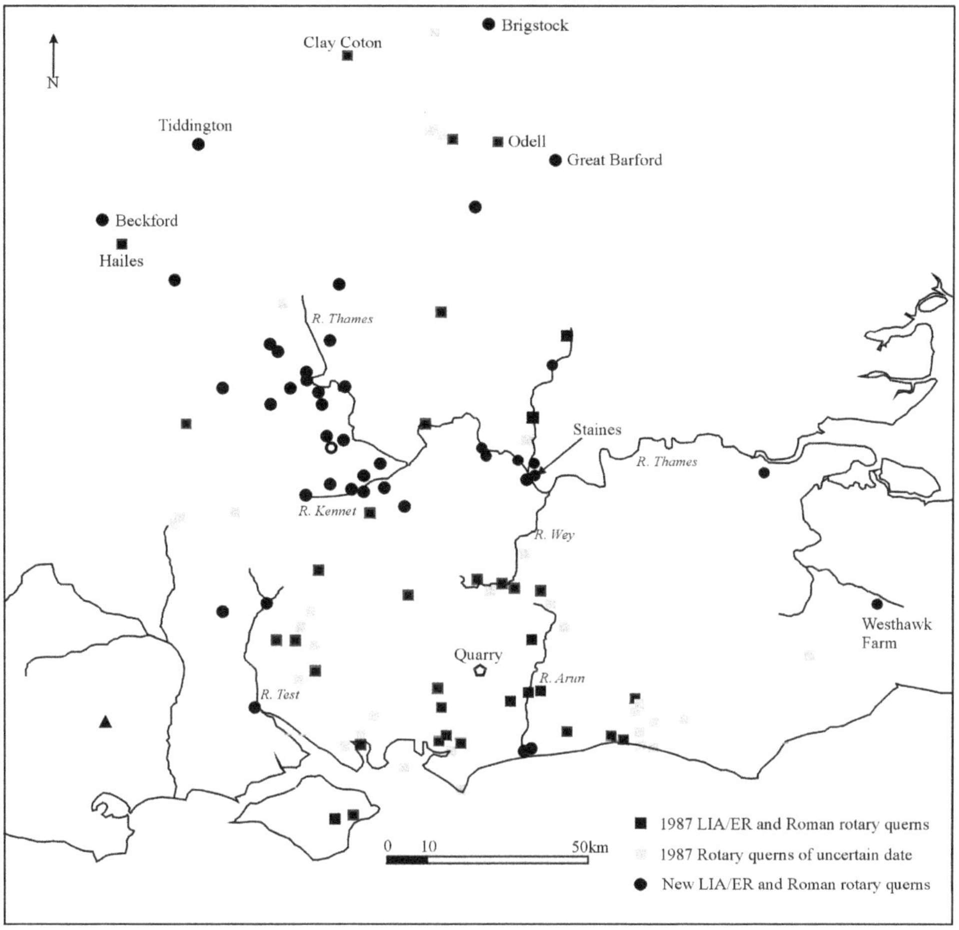

Figure 3. Map of late Iron Age/Roman and Roman quern distribution.

the introduction of imported stone such as Lodsworth Greensand and more significantly Niedermendig lava and Millstone Grit (Ingle 1990, 270). Although this suggests that the change to imported materials was entirely post-conquest, unlike the upper Thames Valley, in Kent there is actually little evidence for the use of rotary querns at all before the very late Iron Age (Blanning 2006, 24). These finds could have reached Kent either by transport along the south coast (Roe 2008, 191) or else down the River Thames to Springhead and on down the Medway to Westhawk Farm. Perhaps it was both: it may be noted that the two groups actually represent no more than one or two journeys in total.

In addition to the extension of the area known to be supplied by querns of Lodsworth stone, some prominent gaps in the 1987 map (Fig. 3) have also now been filled. The most substantial of these is the upper Thames Valley (including most of Berkshire and Oxfordshire) from which Lodsworth stone was thought to be almost entirely absent but which now contains numerous sites. This density of find spots may be in part related to the amount of work the authors have carried out in that region, reflecting the examination of both new finds and older ones in museum collections. The discovery of so many sites in the upper Thames Valley region is nevertheless of interest for it alters our perception of Lodsworth stone quern distribution. It is now clear that there is a much more continuous spread of finds from the quarry site northwards.

Discussion of rotary querns

It is clear that during the late Iron Age/1st century AD and subsequently during the Roman period, querns of Lodsworth stone were transported long distances with much more frequency than during the prehistoric period. The main distribution area is greater and the edges are extended with an increasing number of sites some significant distance from source. Prominent concentrations of sites along major waterways such as the Rivers Arun, Test, Wey, Thames and Kennet, are testament to their continued use as a mechanism for quern distribution. Sites such as Binscombe, Wagdon Common and Bramley (Peacock 1987) indicate an overland link between the Rivers Arun and Wey, although how far to the end of these routes the rivers would have been navigable is debatable.

Not all the sites with rotary querns of Lodsworth stone could have obtained these items solely via the river network. There are no prominent concentrations of find spots along Roman roads but the scatter of sites distant from the main river routes probably reflects the increasing availability of road transport during the Roman period. A considerable effort and time commitment would have been involved in transporting any quantity of querns from one riverine connection to another (Shaffrey 2006b, 68) and led to a natural restriction in distances covered during the prehistoric period. Alternative routes were more easily available following the conquest and in some instances a single road journey may even have been preferable to a combination of roads and rivers, depending on the quantity of material being moved (*ibid.*).

The greater distances involved, however, make it increasingly less likely that querns, especially those on the periphery, would have reached their final destination in a single journey. One possible explanation for long journeys of single items is that they were being brought into the area as high cost gifts or part of dowries. But even high cost gifts could have been acquired from an intermediate point. Could those making the querns really have known how far their products were headed? One can speculate that it is more likely that they were obtained nearer to their final home, perhaps at a site such as Odell in the east Midlands with its unusually large number of querns, including Lodsworth Greensand.

The increased distribution will also have been partly a result of the greater quantity of querns being produced from the late Iron Age / early Roman transition onwards. Although the lack of closely dated finds hinders any attempt to analyse the changing use of Lodsworth stone throughout the Roman period, the evidence increasingly appears to support the idea that its main focus was during the early Roman period (Peacock 1987, 73). Sites such as Staines, where all the early Roman querns were of Lodsworth Greensand, provide strong evidence of a predominantly early use of the stone. At Silchester meanwhile, not only is there strong evidence for early Roman use (Shaffrey 2003, 161-2) but there is extremely scant evidence for late Roman use (Shaffrey 2006a, 133-134). In addition, there are no finds of querns in the peripheral areas (such as Kent, Northamptonshire, Warwickshire and Worcestershire) dated to the later Roman period, suggesting that where there were less querns of a type in use, they ceased to be functional sooner. Indeed, many of these examples actually date to the late Iron Age/early Roman period.

In 1987, the evidence indicated a clear division between saddle and rotary quern distribution and Peacock interpreted this as showing that saddle querns were made by locals with direct access to the quarry and rotary querns by specialists with access to wider distribution networks (1987, 76). He suggested that the overall distances reached by the earliest rotary querns showed a greater similarity with their Romano-British counterparts than with the earlier saddle quern distribution. The new data has made a significant difference to our understanding of quern use because it is now clear that saddle querns reached the middle Thames Valley, much further than was initially understood. Knowing this also affects our interpretation of the prehistoric rotary quern data, which, except for a few outliers, reached much the same area as Iron Age saddle querns. This would suggest some continuity in the production and distribution between the two artefact types and that the people responsible for making saddle querns were also responsible for the earliest rotary querns.

The biggest change in quern distribution occurred either during the very late Iron Age or Roman period, probably very soon after the conquest. The frequency with which querns of Lodsworth stone were transported long distances during this phase increased significantly with many more finds spots in the Thames Valley and an increased number of outliers. There is no evidence for any further expansion and although there are insufficient dated examples to define an end period, Lodsworth Greensand does not seem to have been used in the post-Roman period. Some later Roman finds may be re-deposited, as may the rare examples from Saxon or medieval contexts and the source is likely to have been worked out at some time during the Roman period.

Typology

Saddle querns and Mortars

The relative scarcity of complete or near complete saddle querns of Lodsworth stone has not permitted a detailed analysis of their forms. Peacock found them to be 'generally poorly formed' and of varying levels of quality with some stones only roughly shaped and others pecked into shape and worked all over or 'formed' (Peacock 1987, 67). Findings of Lodsworth stone saddle querns continue to include examples of both well-dressed and crudely used types although recent evidence includes sites where most or all the saddle querns are of the 'formed' type, for example at Balksbury. The existence of two clear size or types here and the absence of debris suggests importation in the finished state (Buckley 1995, 42) and more organisation to production than initially thought, although the numbers are still low.

Amongst the new examples of Lodsworth stone artefacts are some that do not sit easily with the label of saddle quern. Some are, or are likely to be, reused rotary querns, including a fragment from Overton, Pilgrims Field said to be reused as a mortarium (Peacock 1987, 79). Another example from Rustington (Gilkes 2000, 24; here Fig. 4. 8) is a partially reshaped rotary quern reused as a mortar while some are obviously rotary querns with one surface simply used for secondary grinding (Gilkes 1993, 13-15 and here Figs. 4. 4 and 5.). These indicate casual reuse of broken or obsolete querns.

Some recent finds hint at the possibility of a different class of artefact, which can be distinguished from typical saddle querns by their well-defined rectangular edges or distinct rim. Examples include one found in/on a 2nd century AD possible trackway along the Newbury reinforcement

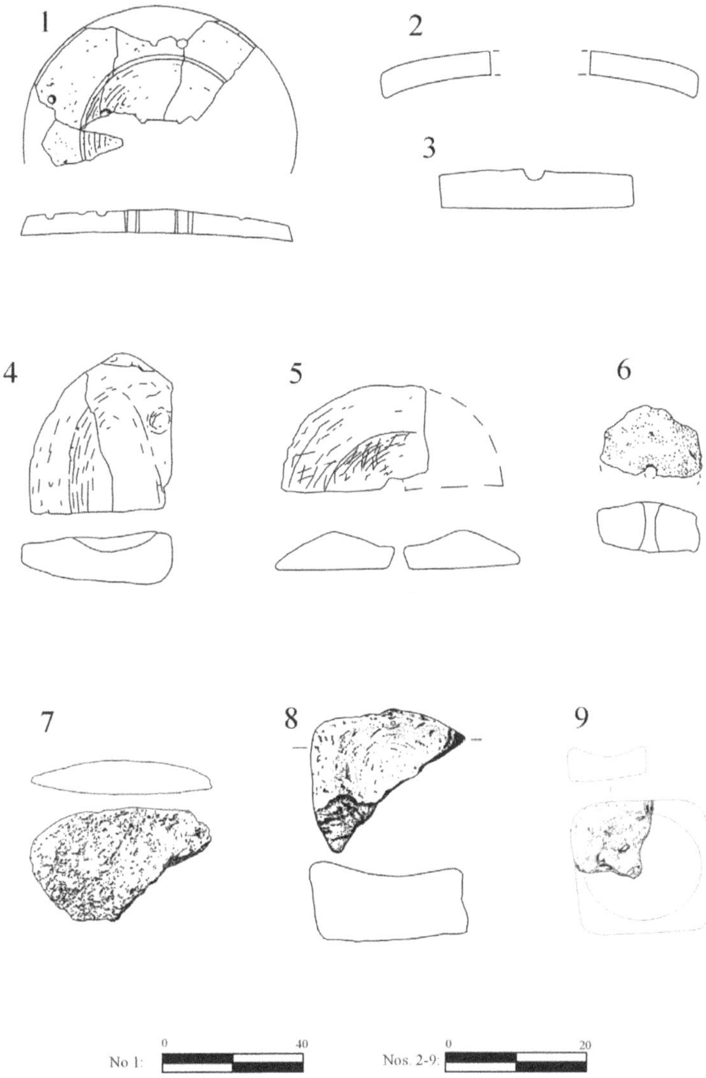

Figure 4. Lodsworth artefacts:
1: millstone from Littlehampton (after Gilkes 1993, 13); 2 and 3: disc type querns from Alchester (after Roe 2002, 248) and Faringdon (after Shaffrey 2004, 241); 4 and 5: rotary querns from Littlehampton reused as saddle querns or mortars (after Gilkes 1993 Fig. 7); 6: rotary quern from Tiddington reused as a weight; 7: rubber from Runnymede Bridge (after Freestone 1991, Fig. 57); 8 and 9: mortars from Rustington (after Gilkes 2000, 24) and Newbury Pipeline (after Shaffrey 2005, 269).

pipeline (Shaffrey 2005, 269 and here Fig. 4.9) and a further rectangular example from nearby Wanborough (Swindon Museum: Acc B1989.1; not illustrated). The design of both is different to anything previously illustrated for Lodsworth stone (Peacock 1987, 68). In addition, they are mainly recovered from Roman contexts alongside rotary querns of Lodsworth stone, suggesting they were used for processing something other than grain for flour.

Rotary querns are largely assumed to have come as a pair because of the necessity of marrying the upper and lower stones. One exception was found at Eton Area 3, where a pair of rotary stones are made from different materials, with just the lower stone of Lodsworth greensand and a re-used upper stone made from another variety of greensand. Determining whether the upper and lower stones of a saddle quern were made and transported together is more problematic. Both rubbers and saddle querns often survive only as small fragments with the result that in the archaeological record they are difficult to distinguish and often catalogued as 'rubber or saddle quern fragment'. It is clear that the two stones needed to be of similar hardness; however many of the available sandstones would have been compatible with one another and the smaller upper stone was potentially easier to obtain, possibly just collected locally. At Eton a pair of stones are made from different materials but at Runnymede Bridge there are matching upper and lower stones of Lodsworth stone; the rubber in question is a particularly well shaped and pecked example (Freestone 1991, fig. 57 and here Fig. 4.7). This pair suggests that in some instances, rubbers may have had similar status or desirability as the saddle quern and also that they were transported together. Unfortunately this is the only complete rubber of Lodsworth stone on record with only one other likely fragment (also from Runnymede Bridge: Higbee 1997, 165).

Rotary querns and millstones

The addition of a sizeable amount of new data has not altered our understanding of the typology of Lodsworth stone querns, which are of a very limited range, generally falling within the 'Sussex' type (Peacock 1987, 69). Recent findings add only a small number of variations. These include Curwen's disc type quern (1937), examples of which have been found at Faringdon (here Fig. 4.3 and

Shaffrey 2004, 241) and Great Barford. Two further querns of the same design were recovered from specifically late Roman contexts at Alchester (here Fig. 4.2 and Roe 2001, 248) and Littlehampton (here Fig. 4.1 and Gilkes 1993, 13). Very few querns of other designs have been recorded although a quern of probable Oxfordshire provenance (Ashmolean Museum) is of typically small Lodsworth stone dimensions (265 mm diameter) but with a keyhole shaped fitting in the centre.

Lodsworth stone rotary querns are on the whole fairly small in size with an observable range in this study of between 260 and 760 mm. The majority of querns measure between 300 and 450 mm and only four examples measure more than 450 mm. This range is similar to, but much more confined than that observed for querns of Old Red Sandstone, where the number of querns falls off markedly only after 550 mm diameter (Shaffrey 2006b, fig. 4.3). Querns of altogether smaller diameters (less than 300 mm) also seem to be more commonly made of Lodsworth stone than of other materials, with the exception of Hertfordshire Puddingstone, an altogether much harder stone to work (Green this volume).

Despite the usually small size of Lodsworth stone rotary querns, there are a few that were likely to have been mechanically operated. Peacock observed only one example of a millstone from Bramley in Surrey (1987, 80) and now three or four more examples can be added, all dating from the 2nd century AD onwards. The most significant is an example from a 2nd century or later context at Eton Rowing Course, Buckinghamshire, measuring 760 mm in diameter. A further example from Winchester, Staple Gardens (recovered from a medieval context) measures approximately 600 mm diameter while two smaller examples from Littlehampton in Sussex may also be from millstones. One has a diameter of 500 mm - larger than the consistently small typical diameters of Lodsworth stone (almost always 450 mm or less: see above). A fourth example has the sort of complex fittings usually associated only with mechanically operated mills but it has no recorded diameter and should perhaps be discounted on the evidence of a quern of small diameter from Oxfordshire with keyhole-shaped fitting (see above). The apparent rarity of millstones of Lodsworth stone may be chronological as millstones are much more common during later Roman phases and for some materials occur largely from the 3rd century onwards (Shaffrey 2006b, 30).

In addition to the millstones, we know that on occasion Lodsworth stone was used to make other items. Peacock recorded an end-runner used for crushing from Crookhorn, Portsdown of uncertain date (Peacock 1987, 78). A single piece of Lodsworth stone from a Bronze Age context at Runnymede Bridge (Freestone 1991, 133 and fig. 56) has been identified as a possible anvil.

There are several possible explanations for the existence of unusual querns, millstones and other artefacts. They may reflect a later use of the quarry, when it was less rigidly managed and a broader range of artefact class and styles being produced, although there seems to be no particular pattern to the millstones, mortars and other artefacts to support this theory. The possibility also remains that there was another smaller source of the stone being exploited on a more haphazard basis. However, it is clear that some items (the mortars and possibly the anvil) represent reuse and were therefore created away from the site. Other atypical items such as the millstones were probably produced 'to order' either at the quarry or elsewhere; there is evidence for the transportation of other quern materials (e.g. Hertfordshire Puddingstone) as raw material, (Green this volume). The production of the occasional unusual artefact away from the quarry would leave very little evidence in the archaeological record and thus cannot be entirely discounted.

Conclusion

The purpose of this paper was to enhance Peacock's 1987 report by incorporating a growing body of data into the foundations of his original work. The additional information accumulated in the last 25 years has extended our understanding of the distribution of querns of Lodsworth stone. We have been able to show that saddle querns of Lodsworth Greensand were first transported out of Sussex during the Bronze Age and that querns of this date are being found in increasing numbers along the middle Thames Valley. We also now know that large numbers of Lodsworth stone rotary querns were being utilised in the Upper Thames Valley during the Roman period. Analysis of the data suggests a greater continuum between the distribution of saddle querns, middle Iron Age rotary querns and Romano-British rotary querns than was at first envisaged. It has also clearly shown that the distribution of the querns is more widespread from the Bronze Age onwards and was more consistently utilised across this area. The data has also revealed that while rotary querns of Lodsworth stone do, on the whole, fall within a narrow range of designs, a greater variety of object forms and types exist than had previously been recognised. It has also shown that the main expansion in distribution was from the late Iron Age / early Roman transition rather than from the introduction of the rotary quern. More detailed analysis of other varieties of quernstone might reveal similar patterns of prehistoric and Roman transportation, casting further light on the cultural links built up by the communities using these essential artefacts.

Gazetteer

69 sites in total (some counted together if part of same project or immediately adjacent).

Saddle querns (Fig. 1)

Berkshire

Burghfield, Field Farm. SU 671 702. Single quern fragment, of unknown date but likely to be Bronze Age (Butterworth and Lobb 1992 and Reading Museum Acc: 1996.11).
Cippenham, Old Way Lane. Approx. SU 941 810. Single fragment from LBA pit or posthole (Williams 2003a, 138).
Pingewood. SU 6869. One fragment of saddle quern, undated but likely to be Bronze Age. Also three other worked and three unworked fragments, Bronze Age or Roman (Johnston 1985 and Reading Museum Acc: 1978.1).

Buckinghamshire

Dorney, Lot's Hole. SU 922 797. Single saddle quern fragment (Roe in prep a), LBA.
Taplow Court. SU 907 823. Two saddle quern fragments (Shaffrey 2009), LB/EIA.
Taplow Motorway Division/Marsh Lane East Site 1. SU 917 801. Single saddle quern fragment (Roe in prep a), M/LBA.

Greater London

Harlington, Imperial Hospital College Sports Ground. TQ 088 783. M/LBA fragments and LB/EIA saddle quern, Wessex Archaeology, in prep.

Hampshire

Danebury Environs
1. *Nettlebank Copse, Wherwell.* SU 389 409. At least one Lodsworth saddle quern, cp 3 (Cunliffe and Poole 2000c, 90 & fiche 10). Earlier Iron Age.
2. *New Buildings, Longstock.* SU 359 373 Three saddle querns and four small fragments (Cunliffe and Poole 2000b, 69 & fiche 8). Probably from earlier Iron Age.
3. *Suddern Farm, Over Wallop.* SU 284 383. At least 11 saddle quern fragments from this large assemblage, dating from cp 3 (Cunliffe and Poole 2000a, 123 and fiche 5). Earlier Iron Age.
Gosport, Grange Road. Approx SU 588 003. Five saddle quern fragments and seven other fragments (Williams 1995, 29). LBA.
Winchester, Staple Gardens. Approx. SU 4829. Single saddle quern fragment, EIA (Shaffrey in prep a).

Oxfordshire

Appleford. SU 526 937. Part of saddle quern, No 5 in report, not seen but same lithology as two Lodsworth rotary querns (Hinchcliffe and Thomas 1980, 61). LB/EIA.
Stanton Harcourt, Gravelly Guy. SP 403 053. Two indeterminate quern fragments, EIA, also one saddle quern (MIA but reused) (Bradley et al 2004, 339-340).

Surrey

Chertsey, St Ann's Hill. TQ 026 676. Two fragments (Jones in prep), EIA.
Laleham, Home Farm. Approx. TQ 0568. A total of 18 fragments including one of a rubber (Hayman 2002). LBA.
Runnymede Bridge. TQ 018718. Various items including an anvil or quern, a possible saddle quern, a flake and a fragment (Freestone 1991, 138; figs. 56-7). Also eight fragments including two worked pieces (Higbee 1997, 165), LBA.

Sussex, West

Angmering By-Pass. TQ 079 034. Single saddle quern, almost complete (Shaffrey in prep b), MBA.
Bognor Regis. SU 912 008. Six saddle quern fragments, four of them fitting. Also a complete rubber (Cotswold Archaeology, in prep). LBA. Destined for Chichester District Museum, CHCDM 2009.8.
Lavant Quarry, near Chichester. SU 858 083 Part of saddle quern found with early Neolithic pottery (Cotswold Archaeology, evaluation). Neolithic
Littlehampton. TQ 039 026. One saddle quern. An addition to those recorded by Peacock, *so not shown on map* (Gilkes 1993, 13-15). Iron Age.

Iron Age rotary querns (Fig.2)

Berkshire

Datchet, South Lea Farm. SU 988 759. Two rotary querns, finds from fieldwalking (Datchet Archaeological Society). Unstratified, probably Iron Age.
Reading, Thames Valley Business Park. Approx. SU 7474. A total of 27 fragments from at least three individual rotary querns (Barnes et al 1997, 45-8 & Reading Museum Acc.1996.16), LIA.
Upper Bucklebury, Hartshill Copse. SU 531 686. One

rotary quern, LIA and one fragment, undated (Roe in prep b).

Hampshire

Danebury Environs
1. *Nettlebank Copse*. SU 389 409. Five rotary quern fragments from cp 8-9 (Cunliffe and Poole 2000c, fiche 10). Later Iron Age.
2. *Suddern Farm, Over Wallop*. SU 284 383. Numerous rotary querns, from cp 8-9 (Cunliffe and Poole 2000a, 123 and fiche 5). Some probably later Iron Age.

Oxfordshire

Abingdon Vineyard. Approx. SU 496 973. One rotary quern and about nine indeterminate fragments (Roe in prep c). Middle Iron Age.
Blewburton Hill. SU 547 862. Rotary quern fragments from two contexts, possibly Lodsworth Greensand (Collins 1947, 21) and at least one fragment that has been positively identified. Iron Age.
Little Wittenham, Hill Farm. SU 564 926. Fragments of matching upper and lower rotary querns in a MIA pit (Roe 2010).
Standlake, Eagle Farm. ST 3903. Single rotary quern fragment of thick Iron Age style (Allen & Moore 1987).
Yarnton, Worton Rectory Farm. SP 4416. One LIA rotary quern (Roe in prep d).

Wiltshire

Blunsdon St. Andrew, Groundwell Farm. SU 157 889. Single lower rotary quern fragment (Gingell 1981 and Swindon Museum), MIA.

Late Iron Age/Romano-British and Romano-British rotary querns (Figure 3)

Bedfordshire

Great Barford Road Scheme, Site 8. TL 106 516. Single rotary quern probably Roman on the basis of its design, although recovered from an undated fill in the top of a MIA pit (Shaffrey 2007b, 284).

Berkshire

Aldermaston Wharf. SU 605 681. Four unregistered fragments of probable rotary querns not mentioned in report (Cowell et al 1978) but probably 1st century AD in date (Reading Museum).
Aldworth (Newbury Pipeline). SU 5579. Single mortar or saddle quern (Shaffrey 2005, 269), ERB.
Cippenham:
1. *Wood Lane Site*. Approx. SU 941 810. Fragments from at least 11 contexts, including 3 upper rotary quern fragments (Williams 2003a, 88-9), Roman.
2. *Brook Farm trenches* ? Single quern fragment with tooled surface (Williams 2003a, 27), Unstratified.
Horton, Lower Horton Channel. Approx. TQ012 760. One upper rotary quern (Williams 2003b, 84), LIA/RB.
Newbury Hospital. Approx SU 4967. Single rotary quern (Shaffrey 2008b), undated but probably Roman.
North Street (Theale to Bradfield pipeline). SU 636 725. Two fragments, one with surviving grinding surface (Williams 1997, 53), late 1st century AD.
Pingewood. SU 6869. Seven fragments of rotary querns (from three contexts), Roman, possibly early; also three other worked and three unworked fragments, undated (Johnston 1985 and Reading Museum Acc: 1978.1).
Reading, Cockney Hill. Approx. SU 6775. Single rotary quern fragment (Reading Museum Acc:160.80). Unphased but presumed to be Roman.
Streatley. SU589 810. Single upper rotary quern found during fieldwalking (Allen et al 1993). Unstratified but probably LIA/ER
Swallowfield, Riseley Farm. SU730 638. Five fragments including at least two definite rotary querns (Lobb and Morris 1993, 50). Found with Terra Rubra vessel so probably LPRIA.

Buckinghamshire

Broughton Manor Farm. SP 902393. Single rotary quern quarter (Shaffrey in prep c). Early Roman.
Dorney: (shown as one dot)
1. *Lake End Road West*. SU 929 796. Four rotary quern fragments from three contexts (Roe in prep a). Early Roman.
2. Lot's Hole. SU 929 796. Three probable Roman rotary quern fragments from medieval/Saxon contexts but probably of Roman origin (Roe in prep a).
Eton Rowing Course: (shown as one dot)
1. *Area 3*. SU 933 775. Single rotary quern (Roe in prep a). Unstratified but probably Roman.
2. *Area 10*. SU 932 777. Single rotary quern (Roe in prep a), LIA-ER.
3. *Area 16*. SU 919787 to SU 937774. Single rotary quern fragment (Roe in prep a), early Roman.
4. *Area 16D*. SU 929 778. Adjacent fragments of millstone (Roe in prep a), late Roman.

Gloucestershire

Bourton-on-the-Water, Bourton School. SP 166 212. Single rotary quern fragment (Roe in prep e). Unstratified but probably LIA or Roman.

Greater London

Heathrow, Terminal 5. TQ052761. Three fragments, two from rotary quern, part of rotary quern and 6 fragments, Roman; four millstone fragments from medieval well but probably Roman (Roe in prep f). LIA / Roman.

Hampshire

Nursling, Dairy Lane, Southampton. SU 366 161. Two rotary quern fragments (Seager Smith 1997, 41), probably early Roman.
Winchester, Staple Gardens. Peacock included finds from Winchester, but recent excavation produced a millstone, 600mm diameter, included here as millstones are so infrequent. Presumed to be Roman, although recovered from a medieval context (Shaffrey in prep a).
Danebury Environs:
1. *Fullerton, Danebury*. Approx. SU 374 392. Single rotary quern (Shaffrey 2008a), Late Roman.
2. *Suddern Farm, Over Wallop*. SU 284 383. Numerous rotary querns, from cp 8-9 (Cunliffe and Poole 2000a, 123 and fiche 5). Some probably Roman.

Hertfordshire

Leavesden aerodrome. TL 097 003. Two rotary querns and one chip (Roe in prep g), early Roman.

Kent

Ashford, Westhawk Farm. TQ 9040. Three rotary quern fragments (Roe 2008), early Roman.
Springhead Roman town. TQ 617 713. Two rotary quern fragments (Shaffrey in prep d), early Roman.

Northamptonshire

Brigstock. SP 9485. Single quern (Ingle 1990, 105), probably Roman.

Oxfordshire

Abingdon Reservoir. SU 4594. Single rotary quern (Hearne 2000, 9), Roman.
Abingdon Vineyard. Approx. SU 496 973. Six rotary querns and about a dozen indeterminate fragments (Roe in prep c). Late Iron Age/early Roman and Roman.
Abingdon, West St. Helen's Street. SU 497 969. Single rotary quern (Tim Allen pers comm), LIA/ER.
Alchester. SP 571 209. Single quern (Roe 2001, 248). Roman (between AD 240/50 & 300/20).
Appleford 1973. SU 526 937. Two rotary querns, Roman (Hinchcliffe & Thomas 1980, 82 and Ashmolean Museum).
Barton Court Farm. SU 4090. Single rotary quern (Miles 1984 and Oxfordshire County Museums Store Acc: 1984. 194), Roman.
Berinsfield, Mount Farm. SU 5090. Three rotary querns and one fragment (Roe and Brown 2010), Roman.
Dorchester-on-Thames. SU 595 945. Single rotary quern fragment (Frere 1962 and Ashmolean Museum, DOR 62). Roman.
Faringdon, Coxwell Road. SU 281 946. Two rotary quern fragments (Shaffrey 2004, 241), Roman. Two further fragments from a mainly EIA site (*SU 280 945)* but these are not on the map as date of querns not specified in publication (Weaver and Ford 2004).
Lowbury Hill, Aston Upthorpe. SU 540 823. Single quern fragment (Fulford & Rippon 1994, 179; Oxford County Museum Store, 1995.75). Probably Roman.
Northmoor, Watkins Farm. SP 427 034. Single rotary quern, Roman (Allen 1990 - but not identified as Lodsworth stone in the report - and Oxfordshire County Museums Store Acc: 1985.48).
Oxford, Churchill Hospital. SP 543 060. Two querns, Roman (unpublished, Oxford Archaeology).
Stanton Harcourt, Gravelly Guy. SP 403 053. Three rotary quern fragments from LIA-ER contexts (Bradley et al 2004, 398).

Surrey

Staines:
1. *Elmsleigh Centre 1975 - 78*. Approx. TQ 0371. Fragments from 35 contexts including 14 definite rotary quern fragments (Jones 2010), Roman.
2. *Hengrove Farm.* TQ 052 722 Fragments of rotary querns from 3 contexts (Jones 2010), Roman.
3. *Johnson and Clarks.* Approx TQ 0371. Four rotary quern fragments (Jones 2010), Roman.
4. *Percy Harrison's.* Approx TQ 0371. Three fragments including one definite rotary quern fragment (Jones 2010), Roman.
5. *Prudential.* TQ 0371. Ten fragments (Jones 2010), Roman.

Sussex, West

Littlehampton. TQ 039 026. At least seven rotary querns, Roman (Gilkes 1993, 13-15). Peacock recorded saddle querns only (Peacock 1987, 78).
Rustington Bypass. TQ 046 038. Two rotary querns fragments, both reused as saddle querns (Gilkes 2000, 24), early Roman.

Warwickshire

Tiddington. SP 2255. Single rotary quern (Warwickshire

Museums Service), Roman.

Worcestershire

Beckford. SO 980 361. Single fragment of beehive style rotary quern (Roe in prep h), LIA/Roman.

References

Allen, S., Allen, J.R.L. and Fulford, M.G., 1993. A Late Iron Age to Early Roman site at Streatley. *Berkshire Archaeological Journal,* 74, 145-6.

Allen, T.G., 1990. *An Iron Age and Romano-British Enclosed Settlement at Watkins Farm, Northmoor, Oxon.* Oxford.

Allen, T.G. and Moore, J., 1987. Standlake: Eagle Farm, *South Midlands Archaeology,* 17, 96-7.

Allen T., Cramp K., Lamdin-Whymark H. and Webley L., 2010. *Castle Hill and its Landscape; Archaeological Investigations at the Wittenhams, Oxfordshire.* Oxford Archaeology Monograph No 9. Oxford.

Andrews, P. and Crockett, A., 1996. *Three Excavations along the Thames and its Tributaries, 1994: Neolithic to Saxon Settlement and Burial in the Thames, Coln and Kennet Valleys.* Wessex Archaeology Report 10. Salisbury.

Barnes, I., Butterworth, C.A., Hawkes, J.W. and Smith, L., 1997. *Excavations at Thames Valley Park, Reading, 1986 – 88. Prehistoric and Romano-British Occupation of the Floodplain and a Terrace of the River Thames.* Wessex Archaeology Report 14. Salisbury.

Blanning, E., 2006. *The Archaeology of Beehive querns in Kent.* Unpublished undergraduate Dissertation. University of Canterbury.

Bradley, P., Roe, F. and Wait, G.A., 2004. Stone. In G. Lambrick and T. Allen, *Gravelly Guy, Stanton Harcourt, Oxfordshire: the development of a Prehistoric and Romano-British Community,* Oxford Archaeology, Thames Valley Landscapes Monograph 21. Oxford, 368-76 and 398-400.

Brown, L., 1984. Objects of stone. In B. Cunliffe, *Danebury: an Iron Age Hillfort in Hampshire, Volume 2. The Excavations 1969-78: the finds.* London, 407-456.

Brown, T., 2009. The Bronze Age climate and environment of Britain. *Bronze Age Review,* 1, 7-22. Online journal, British Museum.

Buckley, D.G., 1979. The Stone. In G.J. Wainwright, *Gussage All Saints An Iron Age Settlement in Dorset.* Department of the Environment Archaeology Report 10. London, 89-97.

Buckley, D.G., 1995. Stone. In G.J. Wainwright and S.M. Davies, *Balksbury Camp, Hampshire.* English Heritage Archaeological Report 4. London, 40-49.

Butterworth, C.A. and Lobb, S.J., 1992. Excavations at Field Farm, Burghfield, Berkshire, in C.A. Butterfield and S.J. Lobb (eds), *Excavations in the Burghfield Area, Berkshire: Developments in the Bronze Age and Saxon Landscapes,* Wessex Archaeology Report 1. Salisbury, 5-72.

Collins, A.E.P., 1947. Excavations on Blewburton Hill, 1947. *Berkshire Archaeological Journal,* 50, 4-30.

Cowell, R.W., Fulford, M.G. and Lobb, S., 1978. Excavations of Prehistoric and Roman Settlement at Aldermaston Wharf, 1978 – 77. *Berkshire Archaeological Journal,* 69, 1-35.

Cotswold Archaeology, in prep. Report on Excavations at Bognor Regis Eco Quarter, West Sussex.

Cunliffe, B., 2003. *Danebury Hillfort.* Stroud, Tempus.

Cunliffe, B. and Poole, C., 1991. *Danebury: An Iron Age Hillfort in Hampshire: Vol 5, The excavations 1979 – 1988: the finds.* CBA Research Report 73. London.

Cunliffe, B. and Poole, C., 2000a. *The Danebury Environs Programme, The Prehistory of a Landscape. Vol 2, Part 3, Suddern Farm, Middle Wallop, Hants, 1991 and 1996.* Oxford.

Cunliffe, B. and Poole, C., 2000b. *The Danebury Environs Programme, The Prehistory of a Landscape. Vol 2, Part 4, New Buildings, Longstock, Hants 1992 and Fiveways, Longstock, Hants.* Oxford.

Cunliffe, B. and Poole, C., 2000c. *The Danebury Environs Programme. The Prehistory of a Wessex Landscape. Vol 2, Part 5, Nettlebank Copse, Wherwell, Hants, 1993.* Oxford.

Curwen, E.C., 1929. Excavations in the Trundle, Goodwood, 1928. *Sussex Archaeological Collections,* 70, 33-85.

Field, D. and Woolley, A.R., 1984. Neolithic and Bronze Age ground stone implements from Surrey: morphology, petrology and distribution. *Surrey Archaeological Collections,* 75, 85-109.

Freestone, I.C.,1991. The petrography of the stone materials. In S. Needham, *Excavation and Salvage at Runnymede Bridge, 1978: The Late Bronze Age Waterfront Site,* BM press, London, 138-9.

Frere, S., 1962. Excavations at Dorchester-on-Thames, 1962. *Archaeological Journal,* 119, 114-149.

Fulford, M.G. and Rippon, S.J., 1994. Lowbury Hill, Oxon: a Re-Assessment of the Probable Romano-Celtic Temple and the Anglo-Saxon Barrow. *Archaeological Journal,* 151, 158-211.

Gilkes, O., 1993. Iron Age and Roman Littlehampton. *Sussex Archaeological Collections,* 131, 1-21.

Gilkes, O., 2000. The Quernstones. In D. Rudling and O. Gilkes, Important archaeological discoveries made during the construction of the A259 Rustington Bypass, 1990. *Sussex Archaeological Collections,* 138, 15-29.

Gingell, C., 1981. Excavation of an Iron Age Enclosure at Groundwell Farm, Blunsdon St Andrew, 1976-7. *Wiltshire Archaeological Magazine,* 76, 33-75.

Green, C., 2011 Hertfordshire Puddingstone. Working with a difficult rock. This volume.

Hearne, C.M., 2000. Archaeological Evaluation in the Vale of the White Horse, near Abingdon, 1992-99. *Oxoniensia,* 65, 7-12.

Hayman, G.M., 2002. *Archaeological discoveries, principally of Neolithic and Bronze Age date, within the Home Farm, Laleham, mineral extraction site 1991-1999.* Client report, Surrey County Archaeological Unit. Woking.

Heslop, D.H., 1987. *The Excavation of an Iron Age Settlement at Thorpe Thewles, Cleveland, 1980 – 1982.* Council for British Archaeology Research Report 65, London.

Higbee, L., 1997. Imported stone: morphology and utilisation. In S. Needham and T. Spence. *Refuse and Disposal at Area 16 East Runnymede.* Runnymede Bridge Research Excavations Volume 2. British Museum Press. London, 165-8.

Hinchcliffe, J. and Thomas, R., 1980. Archaeological investigations at Appleford. *Oxoniensia,* 45, 9-111.

Ingle, C.J., 1990. *Characterisation and Distribution of Beehive Querns in Eastern England.* Unpublished Ph. D. Thesis, University of Southampton.

Jones, P., 2010. *The Roman and Medieval town of Staines.* Spoilheap Publications Monograph 2, Woking.

Jones, P., in prep. Excavations at St Ann's Hillfort, Chertsey. In G. Hayman, G. Jones and R. Poulton, *Settlement sites and sacred offerings: prehistoric and later archaeology in the Thames Valley, near Chertsey.* Spoilheap Publications, Woking.

Johnston, J., 1985. Excavations at Pingewood. *Berkshire Archaeological Journal,* 72, 17-53.

Keller, P.T., 1988. Quern Production at Folkestone, South East Kent: An Interim Report. *Britannia,* 20, 193-201.

King, D., 1980. Petrology, Dating and Distribution of Querns and Millstones. The Results of Research in Bedfordshire, Buckinghamshire, Hertfordshire and Middlesex. *University of London Institute of Archaeology Bulletin,* 23, 65-126.

Lambrick, G., and Robinson, M., 2009. *Thames though Time. The Archaeology of the Gravel Terraces of the Upper and Middle Thames. Late Prehistory:1500 BC – AD 50,* Oxford Archaeology, Thames Valley Landscapes Monograph 29. Oxford.

Lobb, S.J. and Morris, E.L., 1993. Investigations of Bronze Age and Iron Age features at Riseley Farm, Swallowfield. *Berkshire Archaeological Journal,* 74, 37-69.

McGrail, S., 1978. *Logboats of England and Wales, with comparative material from European and other countries.* British Archaeological Reports 51, Oxford.

McGrail, S., 1998. *Ancient Boats in North-West Europe: The Archaeology of Water Transport to AD 1500.* Longman Archaeological Series, London.

Miles, D., (ed.) 1984. *Archaeology at Barton Court Farm, Abingdon, Oxon: An investigation of late Neolithic, Iron Age, Romano-British and Saxon settlements.* CBA Research Report 50. Oxford.

Peacock, D.P.S., 1987. Iron Age and Roman Quern Production at Lodsworth, West Sussex. *The Antiquaries Journal,* 67, 61-85.

Roe, F., 1999. The Worked Stone. In J. Muir and M. R. Roberts, *Excavations at Wyndyke Furlong, Abingdon, Oxfordshire, 1994.* Oxford Archaeological Unit, Thames Valley Landscapes Monograph 12. Oxford, 44-6.

Roe, F., 2001. Worked stone. In P. M. Booth, J. Evans and J. Hiller, *Excavations in the Extramural Settlement of Roman Alchester, Oxfordshire, 1991: A41 (formerly A 421) Wendlebury – Bicester dualling,* Oxford Archaeology Monograph 1. Oxford, 248-53.

Roe, F., 2008. The Worked Stone. In P. Booth, A. Bingham and S. Lawrence, *The Roman Roadside settlement at Westhawk Farm, Ashford, Kent: excavations 1998-9,* OA Monograph 2. Oxford, 188-194.

Roe, F., 2009. Corn grinding in southern England: what can the querns tell us? In K. Brophy and G. Barclay (eds), *Defining a Regional Neolithic: The Evidence from Britain and Ireland,* Neolithic Studies Group Seminar Papers 9, Oxford, 26-34.

Roe, F., 2010. Worked Stone. In T. Allen, K Cramp, H. Lamdin-Whymark and L. Webley, Castle Hill and its Landscape; Archaeological Investigations at the Wittenhams, Oxfordshire. Oxford Archaeology Monograph No 9, Oxford, 170-175.

Roe, F., 2011. The Worked Stone. In J. Lewis, M. Leivers, L. Brown, A. Smith, K. Cramp, L. Mepham and C. Phillpotts, *Landscape Evolution in the Middle Thames Valley: Heathrow Terminal 5 Excavations Volume 2.* Framework Archaeology.

Roe, F., in prep a. The Worked Stone. In T. Allen, A. Barclay and P. Bradley, *Bridging the river, dividing the land: the Archaeology of a Middle Thames landscape: Middle Bronze Age to Roman.* Oxford Archaeology Thames Valley Landscapes Monograph.

Roe, F., in prep b The Worked Stone. In *Report on Excavations at Hartshill Quarry, Upper Bucklebury, Berkshire,* Cotswold Archaeology.

Roe, F., in prep c. The Worked Stone. In *Abingdon Vineyard, Oxon.* OA Assessment Report.

Roe, F., in prep d. The Worked Stone. In G. Hey and J. Timby, Yarnton Iron Age and Roman settlement and landscape, Thames Valley Landscapes Monograph, Oxford.

Roe, F., in prep e. The Worked Stone. In *Bourton School,*

Bourton-on-the-Water, Gloucestershire County Council.
Roe, F., in prep g. The worked Stone. In Leavesden Aerodrome, Herts, *Hertfordshire Archaeological Journal.*
Roe, F., in prep h. The Worked Stone. In J. Wills, *Excavations at Beckford, Worcestershire, 1972 –* 79, CBA Research Report. York.
Roe, F. and Brown, S., 2010. Querns. In G. Lambrick, *Neolithic to Saxon social and environmental change at Mount Farm, Berinsfield, Dorchester-on-Thames*, OA Occasional Paper 19. Oxford. Appendix 13.
Seager Smith, R., 1997. Other Artefacts. In N.J. Adam, R, Seager Smith and R. J. C. Smith, An early Romano-British settlement and Prehistoric field boundaries at Dairy Lane, Nursling, Southampton, *Proceedings of the Hampshire Field Club and Archaeological Society,* 52, 41.
Shaffrey, R., 2003. The Rotary Querns from the Society of Antiquaries' excavations at Silchester, 1890-1909. *Britannia,* 34, 143-174.
Shaffrey, R., 2004. The Worked Stone Objects. In J. Cook, E. B. A. Guttman and A. Mudd, Excavations of an Iron Age Site at Coxwell Road, Faringdon. *Oxoniensia,* 69, 242-9.
Shaffrey, R., 2005. Worked Stone. In J. Timby, D. Stansbie, A. Norton and K. Welsh, Excavations along the Newbury Reinforcement Pipeline: Iron Age-Roman activity and a Neolithic pit group. *Oxoniensia,* 70, 268-274.
Shaffrey, R., 2006a. The Worked Stone. In M. Fulford, A. Clarke and H. Eckardt, *Life and Labour in Late Roman Silchester. Excavations in Insula IX since 1997.* Britannia Monograph Series 22, Society for the Promotion of Roman Studies. London, 133-134.
Shaffrey, R. 2006b. *Grinding and Milling. Romano-British Rotary Querns made from Old Red Sandstone.* BAR British Series 409. Oxford.
Shaffrey, R., 2007a. Worked stone. In L. Webley, J. Timby and M. Wilson, *Fairfield Park, Stotfold, Bedfordshire: Later Prehistoric Settlement in the Eastern Chilterns.* Bedfordshire Archaeology Monograph 7, 86-92.
Shaffrey, R., 2007b. Worked and Utilised Stone. In J. Timby, R. Brown, A. Hardy, S. Leech, C. Poole and L. Webley, *Settlement on the claylands: Archaeology along the A421 Great Barford Bypass, Bedfordshire.* Bedfordshire Archaeology Monograph 8, 279-284.
Shaffrey, R., 2008a. The Millstones. In B. Cunliffe and C. Poole, *The Danebury Environs Roman Programme. A Wessex landscape during the Roman era. Vol 2, part 3, Fullerton, Hants, 2000 and 2001.* English Heritage and Oxford University School of Archaeology Monograph 71, 124-130.
Shaffrey, R., 2008b. The Worked Stone. In A. Simmonds, The excavation of a 1st century AD field system and associated cremation burials at the Community Hospital, Newbury, West Berkshire. *Berkshire Archaeological Journal,* 77, 27.
Shaffrey, R., 2009. The Worked Stone. In T. Allen, C. Hayden and H. Lamdin-Whymark, *From Bronze Age enclosure to Saxon settlement: new light on Taplow Hillfort, Buckinghamshire.* Thames Valley Landscapes Monograph No 30, Oxford Archaeology. Oxford, 145-146.
Shaffrey, R., in prep a. The Worked Stone. In S. Teague, E. Biddulph, L. Brown, A. Hardy and B. Ford, *Excavations at Staple Gardens, Winchester 2004-07.* OA monograph.
Shaffrey, R., in prep b. The Worked Stone. In *Angmering By-Pass, W Sussex.* Oxford Archaeological Unit Assessment Report.
Shaffrey, R., in prep c. The Worked Stone. In R. Atkins and G. Rees *Broughton Manor Farm, Milton Keynes, A middle Iron Age to Late Roman Settlement.* Oxford Archaeology Monograph.
Shaffrey, R., in prep d. The Worked Stone. In P. Andrews, E. Biddulph and A. Hardy, *Settling the Ebbsfleet Valley. CTRL Excavations at Springhead and Northfleet, Kent. The Late Iron Age, Roman, Saxon, and Medieval Landscape.* Oxford Wessex Archaeology Monograph.
Weaver, S.D.G. and Ford, S., 2004. An Early Iron Age Occupation Site, a Roman Shrine and Other Prehistoric Activity at Coxwell Road, Faringdon, Oxfordshire. *Oxoniensia,* 69, 119-180.
Wessex Archaeology, in prep. Report on Excavations at Imperial College Sports Ground, Harlington, Greater London.
Williams, D.F., 1995. Saddle Querns. In M. Hall and S. Ford, Archaeological Excavations at Grange Road, Gosport, 1992. *Proceedings of the Hampshire Field Club and Archaeological Society,* 50, 29-30.
Williams, D.F., 1997. The Stone. In F. Raymond, The Investigation of Roman and Medieval Settlements found during the construction of the Theale to Bradfield Pipeline, *Berkshire Archaeological Journal,* 75, 53.
Williams, D.F., 2003a. Stone. In S. Ford with R. Entwhistle and K. Taylor: *Excavations at Cippenham, Slough, Berks, 1995-7,* Thames Valley Archaeological Services Monograph 3, Reading, 27 (Brook Farm), 88 (Wood Lane) and 138 (Old Way Lane).
Williams, D. F., 2003b. Stone. In S. Preston (ed), *Prehistoric, Roman and Saxon Sites in Eastern Berkshire: Excavations 1989 – 1997.* Thames Valley Archaeological Services Monograph 2, Reading, 84.

The rise and fall of the Hyllestad millstone quarry landscape, western Norway

Tom Heldal and Gurli Birgitte Meyer

Introduction

For more than one and a half millennia, good raw materials for the manufacturing of rotating millstones have been sought after and quarried in Norway, leaving traces in the landscape of one of the longest lasting extractive industries in the country. Certain mica-schists that had a knobbly surface due to garnet or staurolite porphyroblasts proved to be of particularly good quality and resulted in huge quarry landscapes, each of them active for many centuries. Five such quarry landscapes stand out as particularly important production sites: Saltdal (Titland 1993; Monssen 1997; Helberg 2007), Brønnøy (Sognnes 1980; Krokvik 1999), Selbu (Rolseth 1947; Grenne *et al.* 2008), Vågå (Brekken 1980) and Hyllestad, which is the subject of the present paper (Fig. 1), building on an unpublished survey report (Heldal and Bloxam 2007). Parts of the data presented and some figures have previously been published by Grenne *et al.* (2008).

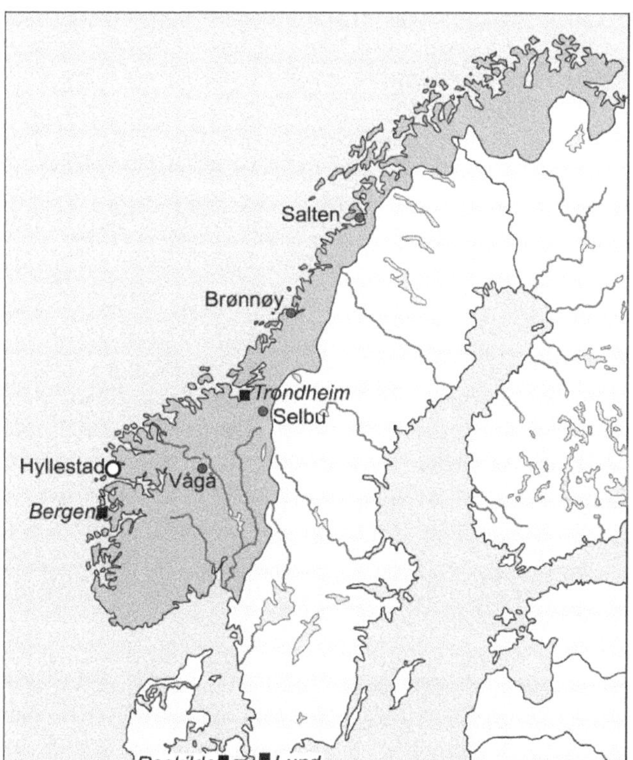

Figure 1. Location of Hyllestad and the four other most important millstone landscapes in Norway.

Figure 2. a) Remains from millstone production are found in many places in Hyllestad. The Lihesten mountain stands up as a landmark in the background, b) quarries along the shore display the typical spherical marks from extraction of millstones, c) quarries in the hillsides are often covered with vegetation and soil, leaving a terrain of overlapping quarry pits, d) 'fresh' quarry marks uncovered beneath the vegetation, e) built-up quarry road, f) unfinished water-millstone still attached to the rock.

The quarry landscape is named after the municipality of Hyllestad in the western part of Sogn og Fjordane county, set in a typical western Norwegian fjord landscape. The communal centre is situated in the inner part of the Åfjorden fjord and most of the population live along or close to the shoreline. Most of the quarries are found along the fjord and in the slopes above, less than 1 km from the sea (Fig. 2 and 3) (Heldal and Bloxam 2007). The highest concentrations of quarries are found in the southern part of the area, in particular in the quarry areas named Myklebust-Hyllestad and Rønset-Berge (Fig. 3). In these areas, numerous quarries occur side-by-side and even on top of each other. They are connected by roads and tracks to several natural harbours where the millstones were loaded onto ships. Outside these core areas, only small and scattered millstone quarries are found (Sæsol-Stigedalen and other areas named in Fig. 3).

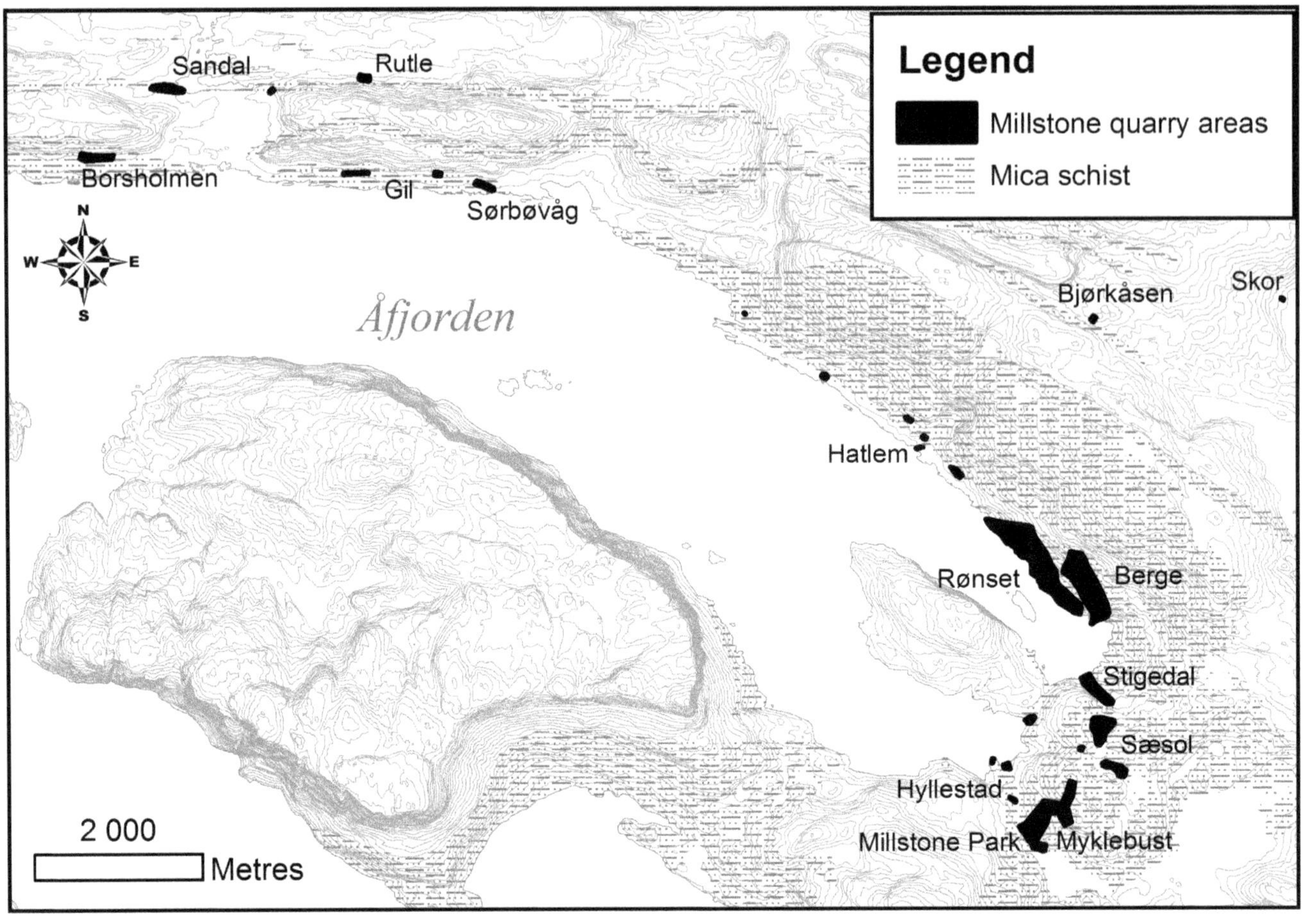

Figure 3. Distribution of garnet-mica-schist and millstone quarries in Hyllestad.

Millstone quarrying has made a significant impact on the landscape in Hyllestad and must have been an important part of the community's identity over hundreds of years. Nevertheless, this activity became absent in the collective memory of the community in an amazingly short period of time. Ottar Rønneseth (1968; 1977; 1988) 're-discovered' the millstone quarries and understood their significance in the 1960s. Marine archaeology was carried out along Åfjorden in Hyllestad after shipwrecks with cargoes of millstones from Hyllestad had been localised along the west coast of Norway (Hansen 1991). This led to the discovery of several harbours where the stones had been loaded onto boats (Hansen 1997).

A provenance study by Carelli and Kresten (1997) demonstrated that Hyllestad millstones were widely distributed in settlements from the late Viking Age and early Medieval times in Denmark and southern Sweden. Specifically, they were predominant among the early Medieval millstones used in the Swedish town of Lund. Excavations in Hyllestad suggest that the earliest quarrying dates go back to at least the 8th Century AD, with a peak between the 12th and 14th centuries and only minor extractions thereafter (Baug 2002). At some stage after the Reformation (AD 1537), the production techniques changed from cutting the millstones directly from the bedrock with pick-axes, to using wedges and later gunpowder for the primary extraction of blocks. Quarries near Rønset (Fig. 3), where gunpowder was evidently used, were abandoned in 1750 (Rønneseth 1968), indicating that blasting techniques were already introduced in the first half of the 18th century. Quarrying with powder continued in Hyllestad until 1930, when the last millstone was made.

One of the important milestones in Hyllestad millstone production was the introduction of water mills in Norway. Baug (2002) suggests that the first production of water-millstones (60 to 120 cm in diameter) took place around AD 1100 or even earlier. Before that period only hand-millstones (35 to 60 cm) were produced (Fig. 4). However, production of hand-millstones continued also after water mills were introduced and probably remained an important product for a long time. Not only millstones were produced at Hyllestad in the Middle Ages, but also several of the large stone crosses from the 11th century found at various locations along the west coast of Norway (Baug 2002; Baug 2008; Baug and Løland 2011) (Fig. 5). The schist was also used for grave-slabs and a range of domestic purposes (Fig. 4).

Since 1995, efforts in the local community along with multi-disciplinary research (see Heldal and Bloxam 2007 and reports from the annual seminar at Hyllestad, see www.kvernstein.no) have contributed in re-establishing Hyllestad as a 'millstone community' and an outdoor museum ('The Millstone Park'), and several successful

Figure 4. Products from garnet-mica-schist. a) hand-millstones, b) water-millstone, c) water-millstone put together from several parts, d and e) vessels, f) chimney-stone.

educational and promotional activities have put the site on the tourist map. The area is also an important case study in an ongoing multidisciplinary research project ('Millstone - the Norwegian millstone landscape') granted by the Norwegian Research Council. The project seeks to investigate several such quarry landscapes in Norway from different periods, the distribution and trade of millstones and how significance and landscape values can be articulated for such sites.

Geology and resources

At Hyllestad, millstone quarrying has targeted a specific variety of garnet-kyanite-mica-schist within the *Hyllestad complex*, a unit of high-pressure metamorphic rocks (Chauvet *et al.* 1992; Chauvet and Dallmeyer 1992; Tillung 1999). The main metamorphic minerals (garnet, muscovite, kyanite, staurolite) and the cleavage of the rocks were formed at depths of ca. 50 km (Hacker *et al.* 2003; Hacker and Gans 2005). Subsequently, the complex underwent deformation and formation of new metamorphic minerals (chloritoid, chlorite) at lower pressures and temperatures during late-orogenic uplift and extension along the Nordfjord-Sogn Detachment Zone (Hacker *et al.* 2003). The topography of the area is controlled by the bedrock geology, where weathering-resistant quartzo-feldspathic gneiss stands up as hills and ridges in the terrain. The mica-schist unit occurs in several zones striking parallel to the fjord, of which the south-western one (outcropping in

Figure 5. Stone crosses. Left: 11th century 'landmark' cross, right: Medieval grave-cross.

the hillside along the fjord) is the most heavily exploited for millstones (Fig. 3). Thus, most of the quarry areas are situated close to the sea and natural harbours. The millstone quality mica-schist is characterized by having hard minerals, predominantly garnet and kyanite (more rarely also chloritoid and staurolite), occurring as porphyroblasts (large grains) set in a matrix with softer and/or more fine-grained minerals (Fig. 6, Table 1). The matrix minerals display a fine banding between layers containing the large muscovite flakes, layers dominated by very fine-grained mica (sericite) and quartz layers (Fig. 7). Along with the occurrence of kyanite, these are some of the diagnostic features of the Hyllestad schist, separating it from other Norwegian deposits of mica-schist used for millstones.

The size and distribution of the garnet has been an important measure of quality, and there are many examples of quarrying that ended when it reached garnet-poor zones or zones with garnet that were too large for serving the purpose. Roughly, two subtypes of good-quality mica-schist can be recognised. In most of the area the schist is highly micaceous and 'soft', containing large flakes of mica in a quartz-poor matrix (Type 1) (Fig. 7a). This subtype was the target for quarrying when millstones were cut directly from the bedrock (see below). The garnet porphyroblasts are most commonly rounded and broken up, enveloped by muscovite flakes, often with chlorite filling the cracks.

Figure 6. Close-up photo of the Hyllestad mica-schist. Round, dark grains are garnet, light grey tabular grains are kyanite. White mica in between.

The other type (Type 2) is found in the eastern part of the mica-schist zone (south-east Sæsol area and Berge; Fig. 7b). It differs from Type 1 in being less micaceous (more quartz in the matrix) and generally harder. The garnets appear more euhedral in shape, are less cracked and are better attached to the matrix. Type 2 was mainly

Occurrence	Minerals
Porphyroblasts	Garnet
	Kyanite
	(Chloritoid)
Main matrix minerals	Muscovite
	Quartz
	Chlorite
Subordinate matrix minerals	Plagioclase
	Pottasium feldspar
	Staurolite
Accessory minerals	Apatite
	Rutile
	Clinozoisite
	Zoisite
	Tourmaline
	Sphene
Opaque minerals	Sulphide (Chalcopyrite, pyrite)

Table 1. Mineral composition of the Hyllestad micaschist

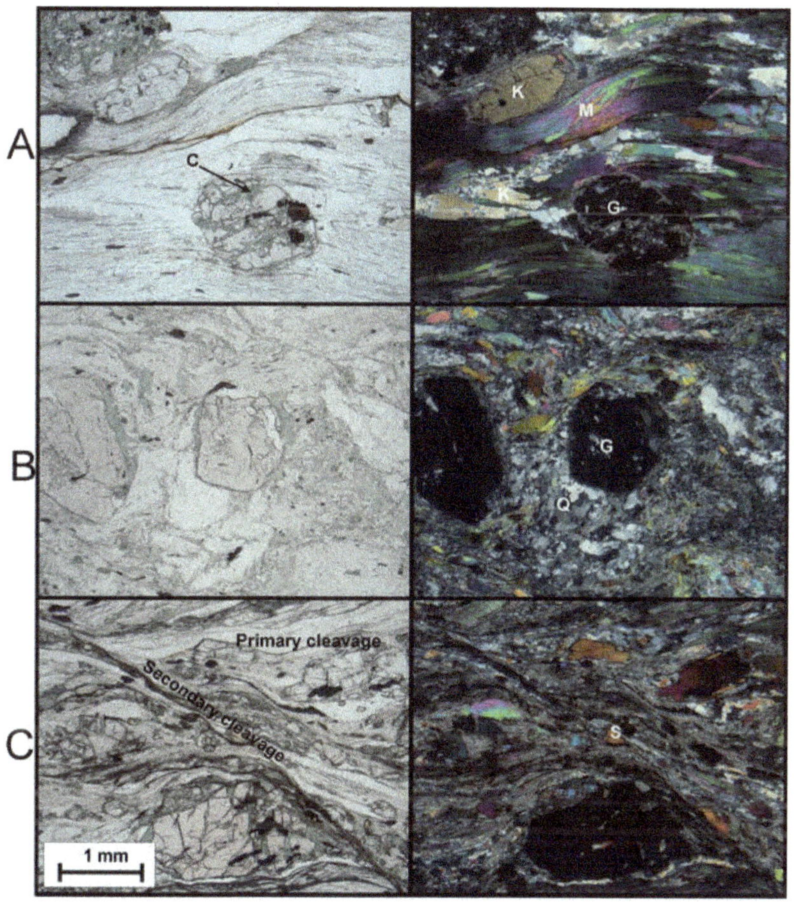

Figure 7. Photomicrographs of thin sections of the Hyllestad mica-schist. Left: plane-polarized light, right: crossed Nichols: a) Type 1 mica-schist, b) Type 2 mica-schist, c) mica-schist with secondary cleavage (shear zone). G=garnet, K=kyanite, M=muscovite, S=staurolite, Q=quartz, C=chlorite.

exploited in the later periods of quarrying, when wedging and blasting were the primary extraction methods. The difference between the two types may have been significant for the grinding properties and durability of the millstones, because the garnets of Type 1 would have been more easily detached from the rock during use. Yet, the 'soft' mica-rich schist must have been the easiest one of the two to cut with a pick axe. This could have been the reason for the preference for this type until the introduction of gunpowder in the millstone quarrying facilitated extraction of the 'hard' variety.

In addition to the mineralogy, structural features in the schist had an influence on quality in production and use, and consequently, on the layout and productivity of the quarries. The cleavage of the schist was the main controlling feature, since the millstones essentially were split from the rock along this plane of weakness. The dip (angle of inclination) of the main cleavage varies through the area, from almost horizontal to 65 degrees. The dip must have played a role in quarrying, since it would be more difficult to work on a steep cleavage than a flat-lying one, and more challenging to work in areas were the dip and strike of the cleavage vary a lot than in areas with a stable orientation.

In addition to the variation of the main cleavage, other structural features were evidently important: some places the cleavage is strongly folded (bended). Such areas have generally been avoided in quarrying. Shear zones form

Figure 8. Examples of trial extractions of millstones from the Rønset area.

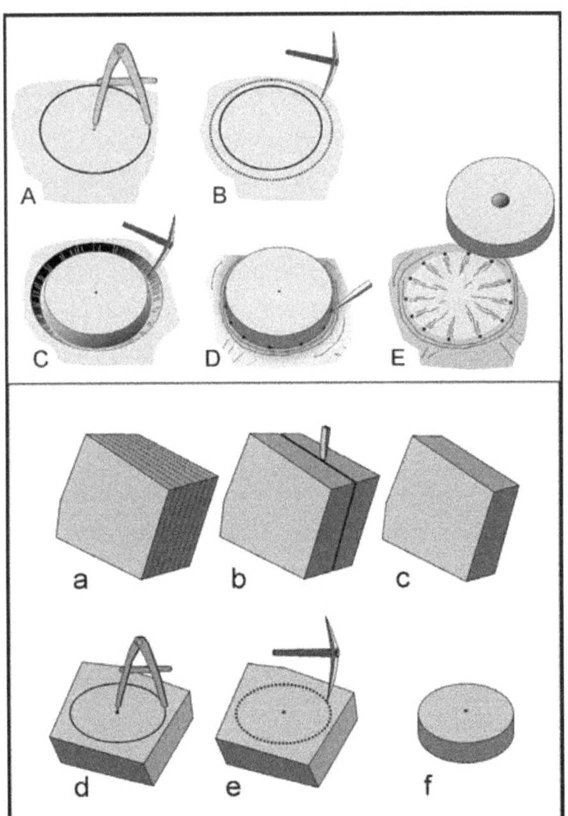

Figure 9. Cutting techniques. From bedrock (top) and from loose blocks extracted by wedging or blasting (bottom). Drawings by Tor Grenne and Tom Heldal.

locally a secondary cleavage in the schist (Figs. 7c), thus causing two possible splitting directions. This caused a lower yield in some quarries, since many of the millstones broke along the cleavage not defining the splitting surface. The shear zones are most likely formed during late-orogenic extension (Chauvet *et al.* 1992; Hacker *et al.* 2003), and associated with them is also a general crushing of the minerals (such as garnet) and growth of new minerals such as chlorite. The shear zones thus contributed to a general decrease of production quality and usability.

Brittle fractures and faults were important. If the spacing of such features was dense the yield was relative low. On the other hand, vertical fractures defined natural 'walls' in the quarries and thus facilitated the extraction. Thus, quarrying primarily targeted areas where the combined geological conditions were most favourable for extraction. This can partly explain the large concentration of quarrying activity at Rønset and Myklebust. However, the near absence of exploitation in qualitatively similar areas further north (Heldal and Bloxam 2007) also points at other mechanisms, i.e., land use and ownership, as the driving force for quarrying (Baug 2005).

Quarries and other features related to millstone quarrying

The survey at Hyllestad is aimed at recording quarries and other features related to the production of millstones. The main categories of recordings were quarries (including spoil heaps and work areas), single extraction sites (predominantly trial extractions) (Fig. 8), artefacts (i.e. loose millstones), quarry roads and man-made features (i.e., dry-stone walls, shelters). The quarries were divided into main groups based on the technology involved in the extraction: cut directly from bedrock (the majority of the quarries; Fig. 9), cut from large boulders in rock-fall areas, cut from smaller blocks in scree deposits, cut from blocks extracted by wedging techniques (only one example) and finally, cut from blocks extracted by the use of gunpowder (Fig. 10). Two other groups are recorded as 'depression, possibly quarry' (diffuse pit in the terrain) and 'depression, unknown quarrying method' (pit in the terrain that is clearly a quarry, but too overgrown to interpret the quarrying method) (Fig. 2c).

The quarries where the millstones were cut from bedrock were further sub-divided in several groups based on their morphology. *Shallow quarries* display a production technique exploiting layer by layer of mica-schist. The millstones were carved out side by side along the cleavage plane, so that the resulting morphology of circular depressions from the extraction of millstones resembles 'pastry-cutting' (Fig. 11). Another group – *deep quarries* - reflects the cutting of slices perpendicular to the cleavage, one millstone on top of the other, resembling 'piles of coins'. This resulted in deep quarries with tall,

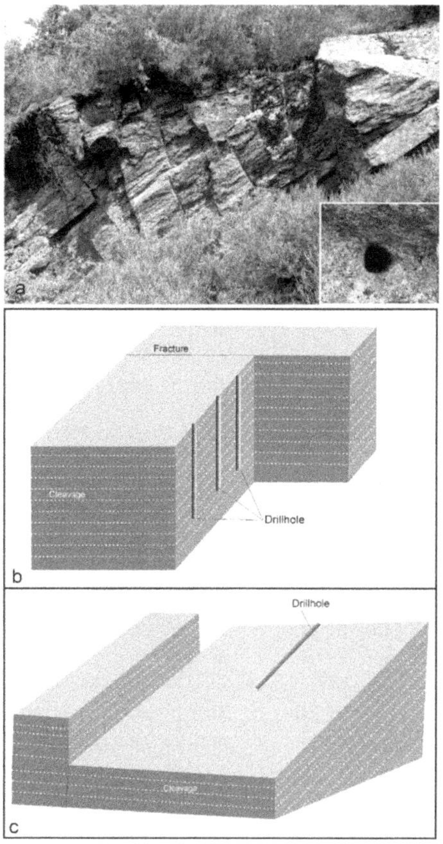

Figure 10. Quarrying by drilling/gunpowder: a) quarry face with drillholes set perpendicular to the cleavage, b) drawing of the situation seen in a), c) drawing of a situation where drillholes are made along the cleavage.

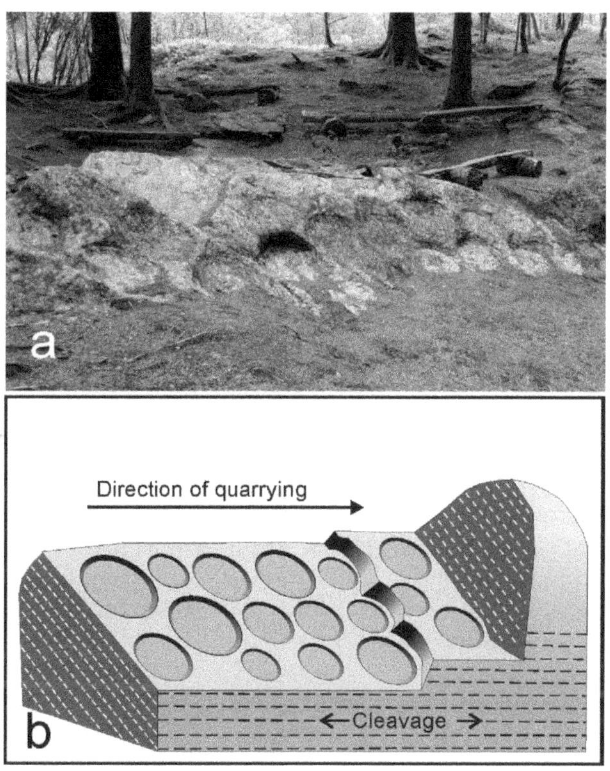

Figure. 11. Shallow quarries where millstones are cut along layers parallel to the cleavage. Drawing by Tor Grenne and Tom Heldal.

Figure 12. Deep quarries where millstones were cut from slices perpendicular to the cleavage ('coin piles'). Drawing by Tor Grenne and Tom Heldal.

steep walls displaying marks from cutting (Fig. 12). These two subtypes of quarries seem to reflect 'extreme principles of quarrying' that may represent change of production techniques in different periods. A large part of the quarries, however, display *combinations* of the two, indicating several episodes of production and/or that the two lived side by side. Yet others are difficult to put into one of the three former groups. Production techniques are discussed in more detail by Grenne *et al.* (2008) and Baug and Løland (2011). The distribution of quarry types in the different quarry areas is shown in Fig. 13.

In addition to such morphological and technological differences, the size of the millstones (both of the discarded ones found in the spoil heaps and the negative imprints from extraction) was recorded. Predominantly, this included small sizes used for hand querns and larger water-millstones. In some quarries there is also evidence of the extraction of slabs meant for other purposes than millstones (grave slabs, stone crosses). In Fig. 14 the distribution of products found in different quarry areas is shown. Overall, signs of hand-millstone production only is found in about one third of the quarries. The proportion of quarries only producing hand-millstones is largest in

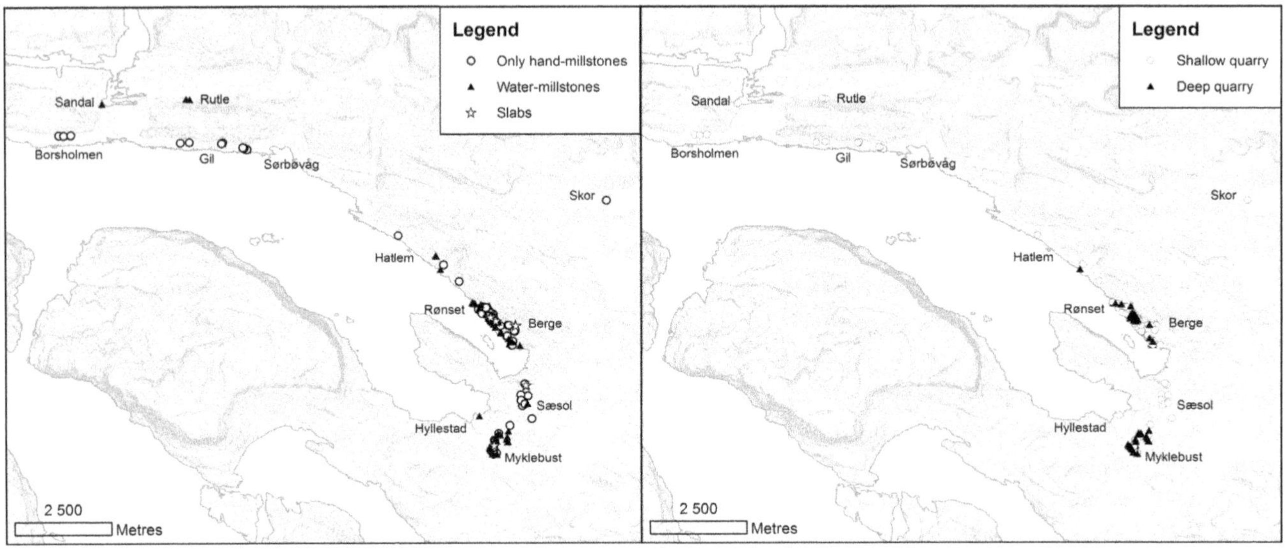

Figure 13. Distribution of observed shallow and deep quarries (right) and observed millstone type (quarries where only hand-millstones are found, quarries where water-millstones are found and quarries where slabs have been extracted).

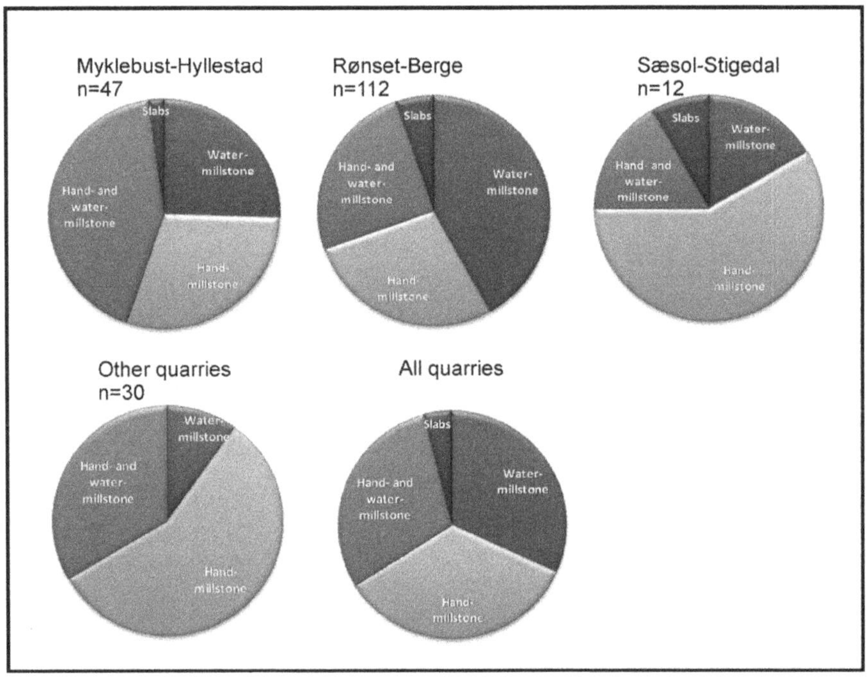

Figure 14. Production in the quarries with observations of hand-millstones, water-millstones and slabs in the different quarry areas.

the Sæsol-Stigedal area and the marginal quarries, whilst in the two big production areas (Myklebust-Hyllestad and Rønset-Berge) the proportion of hand-millstone quarries only is relatively lower. This may give us some input to the interpretation of the evolution of millstone quarrying in Hyllestad, as we will address below.

The quarry roads are rather modest features including paths made from wearing, cleared tracks, hollow roads (almost cylindrical depressions in the terrain made from repeated wearing) and in rare cases, roads built up with dry-wall stone for evening out the terrain. Only fragments of the roads are still visible, due to the dynamics of quarrying (frequent changes of extraction area and spoil deposition), overgrowth and later re-use by the farms in the area. The road fragments do, however, lead from the quarries down to the closest harbour from which the millstones were shipped. The harbours themselves are found where natural features (cliffs, water-depths) were suitable for mooring ships. Mooring sockets (cut in the bedrock and/or large blocks) are found in such places (Fig. 15), and dropped millstones have been located on the bottom of the sea near them (Baug 2002; Hansen 1997).

Man-made features, such as dry-stone walls and shelters, are found in the area. It is difficult to know if these had a direct connection to millstone quarrying or to farming, and more detailed archaeological investigations are needed before one can draw more conclusions regarding such structures.

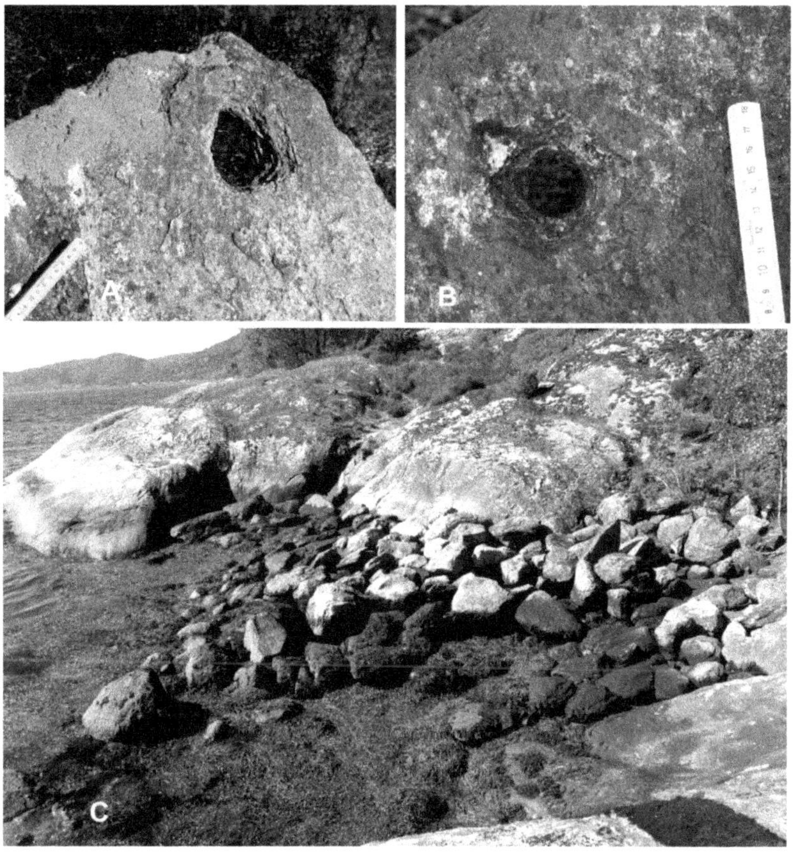

Figure 15. Harbour features: a) mooring socket, cut by pick, b) mooring socket drilled with hand-drill, c) demolished quay.

Table 2. Number of quarries and single extractions (trial quarrying) recorded in the different quarry areas.

	Quarries	Single extractions
Rønset-Berge	176	52
Myklebust-Hyllestad	75	5
Sæsol-Stigedal	12	4
Sæsol-Stigedal	32	11

The quarry areas

Two of the quarry areas contain the largest concentration of quarries in Hyllestad, Rønset-Berge and Myklebust-Hyllestad. In these areas, the landscape is completely transformed by quarrying, leaving numerous pits, quarry faces and piles of spoil from quarrying overlapping each other. The other quarry areas, scattered over a large part of the municipality, are insignificant in size, although there may be more quarries hidden beneath farmland. Table 2 summarizes the number of recorded quarries within the different areas.

The *Myklebust-Hyllestad area* is the southernmost production site (Fig. 16), and includes the Millstone Park, an outdoor museum built up around some of the most spectacular quarries. Large quarries are also found in the hillside to the north of the park. The schist is of the first type, rather micaceous and easy to cut with picks. Shear zones are quite abundant, particularly in the south-eastern part of the Millstone Park where the secondary cleavage in parts completely overrules the main one, and there are several examples showing that millstones were split along the secondary cleavage. The dip of the main cleavage varies from 10 to 50 degrees and some of the deepest quarries are situated in schist with low dip angles.

All the quarries display cutting of millstones directly from the bedrock. In the lowland around the Millstone Park there are many quarries worked only for hand-millstones (shallow type), together with deeper quarries that were exploiting both hand-millstones and water-millstones ('coin pile' quarries or combined). Similar quarry landscapes are probably buried under recent buildings and constructions of the Hyllestad community. Radiocarbon dating of charcoal from a quarry in the Millstone Park indicates that production took place from the Viking Age to the late Middle Ages (Baug 2002). In 2008, excavations in one quarry just to the north of the Millstone Park revealed several unfinished and broken stone crosses (Baug 2008; Baug and Løland 2011), a significant discovery that without doubt linked the late Viking Age/early Medieval stone crosses to the millstone production sites.

Just to the north of the Myklebust area, a group of small and scattered quarries are situated on top of a hilly plateau near the Sæsol farm (the *Sæsol-Stigedal area*, Fig. 16). The schist is here strongly folded, which may explain the

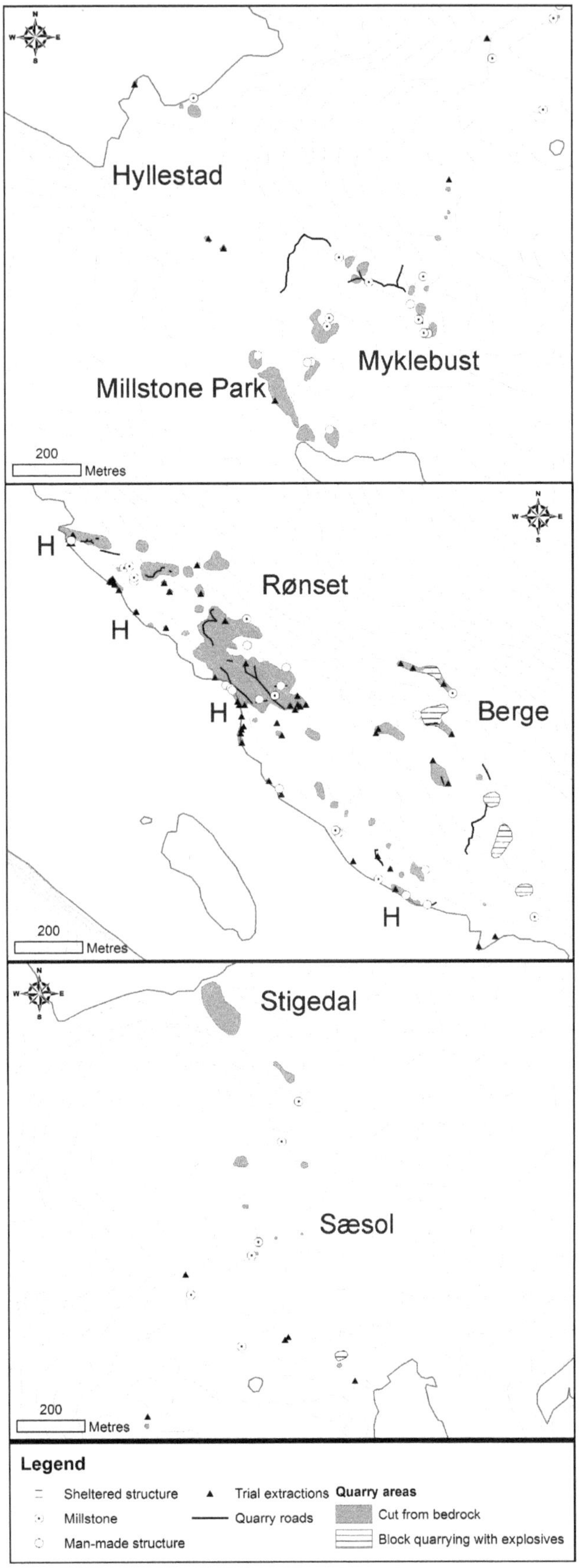

Figure 16. The most important production areas. Myklebust-Hyllestad (top), Berge-Rønset (middle) and Sæsol-Stigedal (bottom).

distribution of quarries; most of the area is not of good enough quality for millstone production. Most of these quarries are typically of the shallow type and produced only hand-millstones. One of them gave a calibrated radiocarbon date of AD 715–890, thus being one of the oldest recorded quarries in Hyllestad (Baug 2002). A few attempts at millstone production later with gunpowder blasting never developed into anything sizeable. In the northern part of the Sæsol area, along the Stigedalen valley, the mica-schists and quarries can be followed down towards the fjord where ancient harbour facilities have been found. However, few quarries or spoil heaps are visible at the present time due to modern road construction.

The *Rønset-Berge area*, north of Sæsol, contains the largest number of quarries in Hyllestad (Fig. 16). The schist is predominantly of Type 1, but contains less shear zones than at Myklebust-Hyllestad. However, in the eastern part (Berge), a zone of harder Type 2 schist occurs that has been exploited in late periods when explosives were used in quarrying. In fact, most of the modern period quarries are found there. The orientation of the main cleavage is essentially favourable in the area, dipping parallel to the hillside. The number of quarries increases towards the north in the Rønset area, which is the largest and most significant quarry area. Numerous quarries are partly overlapping each other and display intensive quarrying during several periods, particularly the Viking Age and Middle Ages (Baug 2002; Heldal and Bloxam 2007). There are also some quarry pits in scree deposits. These may represent remains of a very early phase of quarrying, exploiting single blocks of suitable size for the production of hand-millstones only.

Several fragments of the road network leading from the quarries to the shore still remains visible in the area and along the shore there are numerous trial extractions. At least four sites along the shore have been identified as harbours, of which the Otringsneset in the north is the most well-known. Although many quarries have been recorded in the area, many more are probably hidden beneath farmland and vegetation.

North of Rønset, there are several small quarry areas scattered along the fjord (Fig. 3). Except for one at Hatlem, they all display cutting directly from bedrock, and the majority of them contain only traces of hand-millstone production. Most significant are the *Sørbøvåg* quarries, the majority of which are covered by agricultural land. Several small and more remote quarries are situated at *Gil*, *Borsholmen*, *Sandal* and *Rutle*, all of them probably dating from the Middle Ages (Heldal and Bloxam 2007). Some kilometres to the west of the village of Skor, a very small quarry site is located on the top of a hill. This may be the last millstone production site in Hyllestad, where one pair of water-millstones was produced in the winter of 1930.

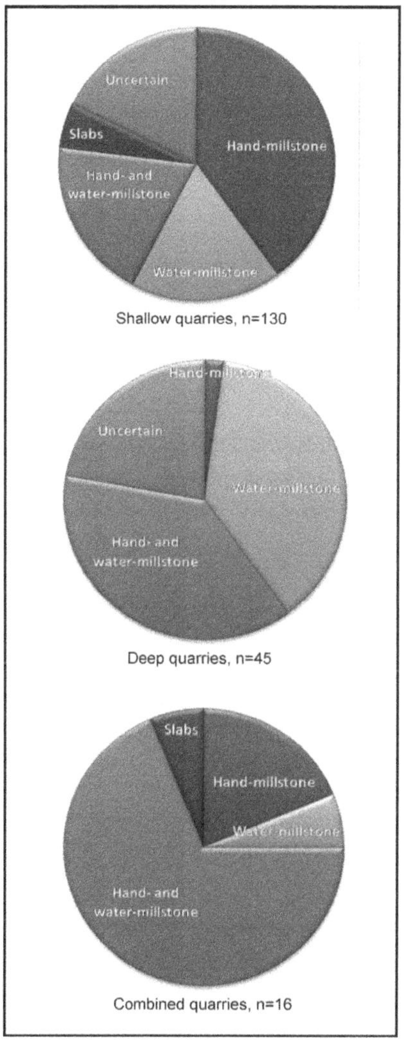

Figure 17. Observed products in different quarry types, showing that most of the quarries where only hand-millstones are found are of the shallow type.

The evolution of the Hyllestad quarry landscape in time and space

From the archaeological excavations of some of the Hyllestad millstone quarries, Baug (2002) and Baug and Løland (2010) have concluded that the peak of production in the quarries seem to have been in the Middle Ages between AD 1100 and 1350. The data suggests the following:

1. The oldest quarry (at Sæsol) dates to the pre- or early Viking period. This particular quarry and other ones in the same area involved extraction of hand-millstones only, and the cutting took place 'side by side' along the cleavage (shallow quarry).
2. Quarries (cut in bedrock) in the Millstone Park and at Rønset suggest multiple periods of quarrying, from the Viking period up to as late as the 17th century. The quarries are either of the deep 'coin pile' type or display combined extraction methods.

As shown in Fig. 17, most water-millstones are found in deep quarries, whilst about half of the shallow quarries seem to have involved production of hand-millstones only. The quarries where both of the methods are found ('combined quarries') contain remains from the production of both hand- and water-millstones. Given that hand-millstones must have been the only product made in the Hyllestad quarries at an early stage of production (until at least the late Viking period) it seems likely that the typical early quarry was of the shallow type.

Although we do not know exactly when the 'coin pile' type of quarrying was initiated, it seems to be strongly linked to the introduction of the water mill. Baug (2002) suggests that this happened sometime before AD 1100, perhaps already in the late Viking period. The change in quarry layout may be linked to a need for more efficient methods for extracting larger millstones and/or a change in organisation of the quarrying activity. Deeper quarries involved more production per area, and would trigger a need of larger teams of quarrymen and more specialisation of tasks. We know that Munkeliv Abbey took over land in Hyllestad in about the early 12th century (Baug 2002), which included some of the most important millstone-quarry areas. Could the change in production methods be linked to such change in landownership? We also know that the monasteries were deeply involved in the quarrying of other stone for buildings and other purposes (for instance soapstone), and it may be that here we see signs of transferring quarrying methods from other sectors (Storemyr and Heldal 2002). What we do see from the survey data, is that such quarries have a much more concentrated distribution than the shallow ones, namely at Rønset and Myklebust.

Thus, we may see a significant pattern of change in millstone quarrying due to the introduction of the water mill with large and efficient quarries concentrated in small areas. Before that, quarrying may have taken place in many small quarries spread out through the community. This may of course also have continued later (in the 'old fashion way'), but clearly the main production output must have come from these concentrated areas. Such ideas are also supported by some quarries displaying both types of production, where the 'coin pile' type clearly postdates the 'pastry cutting' one, and where charcoal dating indicates two episodes of quarrying (Baug 2002).

At Rønset, a few quarries are found in scree deposits close to the shore. Small blocks of schist, large enough for production of a hand-millstone, have been selected from the scree and worked, leaving circular depressions in the scree deposit. We have no information about the age of such quarries, but it seems strange that this kind of production could have taken place alongside more efficient quarrying of millstones from the bedrock. Thus, it is possible that these quarries are the remains of the earliest production of millstones in Hyllestad, pre-dating bedrock quarrying. We may assume that around AD 1700, the quarrying technique changed from cutting

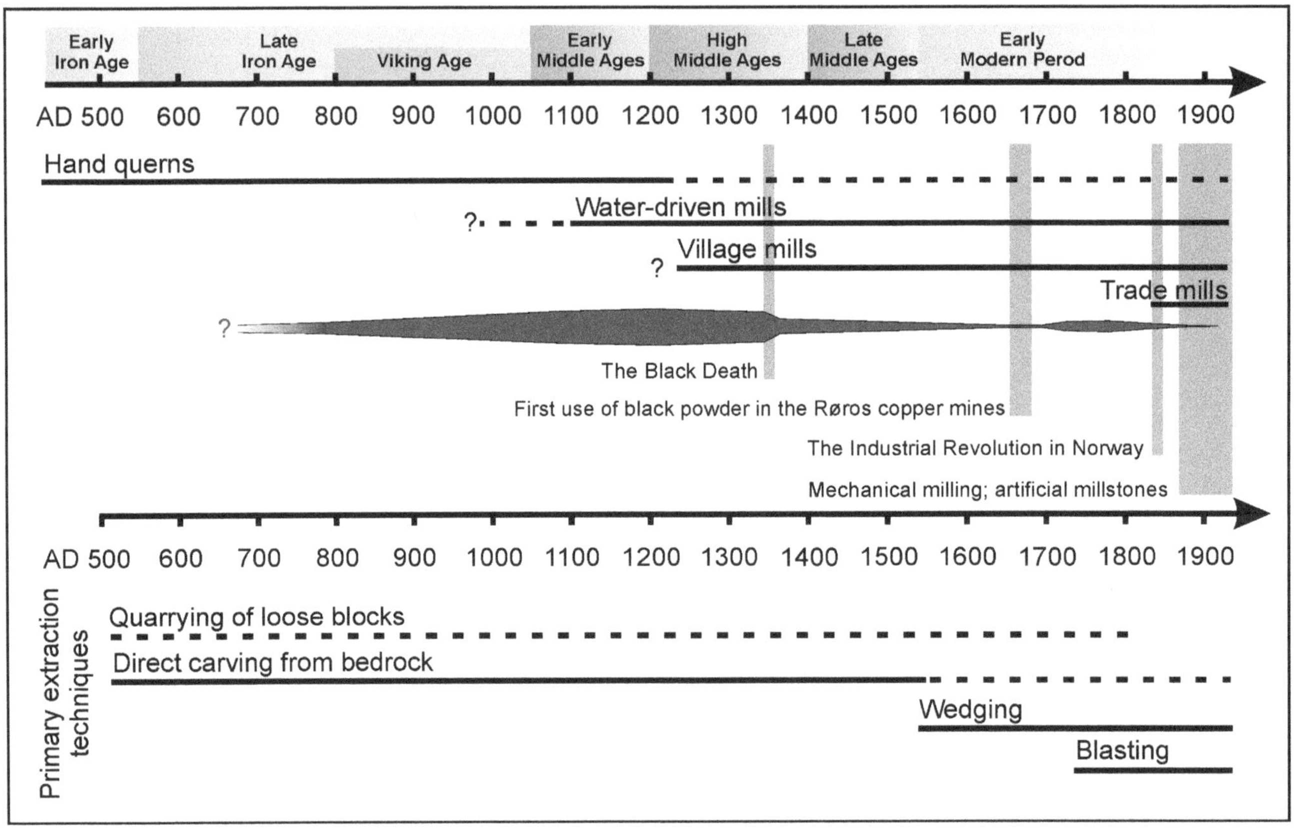

Figure 18. Assumed historical evolution of millstone production in Hyllestad with connection to changes in society and technological innovation (modified from Grenne et al. 2008).

directly in bedrock to extracting blocks with explosives. Simultaneously, there was a shift in raw material from 'soft' to 'hard' mica-schist. This type of quarrying involved hand-drilling of holes either perpendicular to or parallel to the main cleavage. Gunpowder was detonated in the holes, and the most suitable of the irregular blocks resulting from the blasting were carved into millstones (exclusively water-millstones). The use of explosives continued until the production stopped around 1930. It is interesting that the number of such quarries is low in Hyllestad, and compared with the contemporary Selbu quarries (Grenne *et al.* 2008), insignificant. This means that Hyllestad never got back to its Medieval hay-days, and had to live in the shadow of other millstone quarry areas throughout the modern period.

An attempt to illustrate the 'rise and fall' of the Hyllestad millstone quarrying is given in Fig. 18. Fig. 19 shows how this may have looked in the main production areas:

1. In the pre-Viking period small-scaled production of hand-millstones from scree deposits was practiced.
2. In the early Viking period, production of hand-millstones in an increasing number of bedrock quarries, spread throughout the area. They exported to other regions and later also to other countries.
3. In the late Viking to early Medieval period the production of hand-millstones and water-millstones was initiated in large and efficient quarries at Rønset and Myklebust. Exports grew and the Hyllestad quarries became the most important in Norway. Perhaps a continuation of 'artisan' production remained in other parts of Hyllestad. Production decreased after the Black Death.
4. Sometime after the Reformation quarrying methods changed, first to wedging and later to the use of explosives, but the quarrying activity remained on a low scale until it ceased around 1930.

Two perspectives of rise and fall

Why did the Hyllestad quarries grow to great heights during the Viking and Medieval periods? And why did they decline to insignificance in the modern period? There are numerous possible explanations to this, and some may be found in other areas (Grenne *et al.* 2008). We will probably never find the 'true' answers to these questions. However, two perspectives may get us further on the road to the truth. One is the workability and availability of the Hyllestad schist, how easy (or difficult) the rock was to extract and its proximity to the coast, in short, the production quality of the deposit. During the time when the millstones were cut directly from the outcrops, the softer variety was preferred. This may be due to this type being most abundant close to the shore, but also it would be the easiest to work with a pick axe.

The focus changed towards the harder variety when gunpowder came into use. A good reason for this could be that it was better suited for blasting: the softer variety

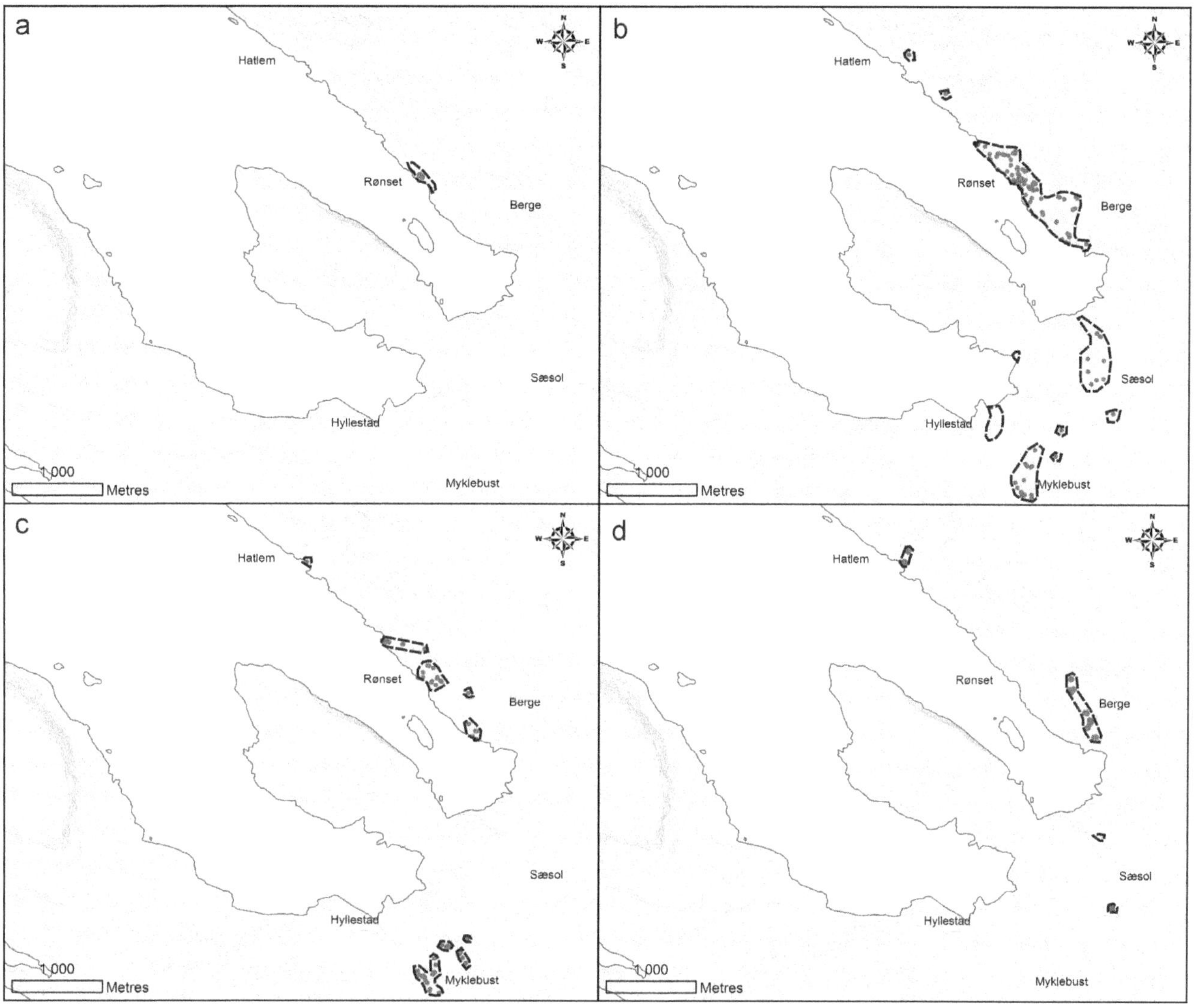

Figure 19. Proposed evolution of the central part of the Hyllestad quarry landscape (quarries=dots, quarry areas=dotted lines): a) early pre-Viking sporadic quarrying of hand-millstones from scree deposits, b) production of hand-millstones in shallow quarries in the Viking period, c) production of hand- and water-millstones in deep quarries in the Medieval period, d) production with explosives in the late 19th century.

would more easily be damaged by explosives, but it may also be triggered by a need for more durable millstones. These would be needed for meeting the competition from other quarries and for adapting to quality demands when the size and weight of millstones grew.

This leads to the second perspective. The softer variety of schist may have been of sufficient quality for hand querns and small water-millstones. But the garnets, being broken, rounded and rather loosely tied to the matrix, would not last long before they fell out, and thus the quality could have been just too poor for use in the modern period. The shift towards quarrying the harder one may have helped some, but obviously not enough to ensure a sizeable production throughout the period. In particular, the millstones from Selbu had properties which made them much more appreciated in the market (Grenne *et al.* 2008).

Although the above may explain some important limitations of the resource, several other questions need to be considered. It may be, for example, that the plague in 1349 ('Black Death') and the succeeding ones, caused a complete re-setting of demographic and production/trade patterns, in addition to literally killing half the market. Moreover, changes in land-ownership (i.e., the Reformation), trade routes and cereal production patterns may have played important roles in the millstone production history.

The Hyllestad quarry landscape as projections of the past

When viewing the Hyllestad quarry landscape with modern eyes, it is fortunate that production never grew large in the modern period, thus leaving a well preserved, 'industrial' landscape from the Viking and Medieval periods. The quarries themselves display evidence of production techniques that are long since forgotten. Millstones in museum collections in Norway and other countries witness how far the millstones from Hyllestad were traded over centuries, and sunken ships loaded with millstones remind us about the dangers involved in transport.

The Hyllestad quarry landscape remains the most long-lived so far discovered in Norway, the only one where the whole history of standardised millstone production is exhibited. New data from ongoing research will contribute to shed more light on the millstone production in Hyllestad, and our knowledge of the people behind it and trade patterns will increase in the years to come. Numerous stories are still hidden in the landscape, and hopefully we will gradually be able to view more of the greater story of Hyllestad and how people interacted with a specific geological resource over almost one and a half millennium.

Acknowledgements

The present paper is a result of multidisciplinary research in Hyllestad, funded by NGU, the municipality of Hyllestad and Sogn og Fjordane county. We are very grateful to Irene Baug, Torbjørn Løland, Finn B. Førsund and Bjarne Akse for sharing their knowledge of the Hyllestad area, and to Elizabeth Bloxam for contributions during field work. The enthusiasm and interest of the Hyllestad municipality have been crucial for the implementation of the studies and is greatly acknowledged. Thanks are also extended to Astrid Waage (Hyllestad) and Øystein Jansen (Bergen Museum). Last, but not least, thanks to the Norwegian Research Council who saw the great value in supporting the project 'Millstone'.

References

Baug, I., 2002. *Kvernsteinsbrota i Hyllestad. Arkeologiske punktundersøkingar i steinbrotsområdet i Hyllestad, Sogn og Fjordane. Bergverksmuseet. Skrift 22.* Kongsberg.

Baug, I., 2005. Who owned the products? Production and exchange of quernstones, Hyllestad in Sogn, Western Norway. In I. Holm, S. Innselset and I. Øye. (eds.), *UBAS International 1. University of Bergen Archaeological Series. "Utmark". The Outfield as Industry and Ideology in the Iron Age and the Middle Ages.* Bergen, 99-108.

Baug, I., 2008. Steinkrossbrotet i Hyllestad. In F.B. Førsund and H. Haukøy (eds.), *Sogeskrift frå Hyllestad 2008.* Førde, 8–13.

Baug, I. and Løland, T., 2011. The millstone quarries in Hyllestad – an arena of research and education. *This volume.*

Brekken, J., 1980. Tolstadkvernberget og andre kvernsteinbrot i Vågå og Sel. Årbok *for Gudbrandsdalen 1980*, 78-83.

Carelli, P. and Kresten, P., 1997. Give us this day our daily bread. A study of Late Viking Age and Medieval Quernstones in South Scandinavia, *Acta Archaeologica*, 68, 109-137.

Chauvet, A. and Dallmeyer, R.D. 1992. 40Ar/39Ar mineral dates related to the Devonian extension in the southwestern Scandinavian Caledonides. *Tectonophysics*, 210, 155-177

Chauvet, A., Kienast, J.R., Pinardon, J.L. and Brunel, M., 1992. Petrological constraints and PT path of Devonian collapse tectonics within the Scandinavian mountain belt (Western Gneiss Region, Norway). *Journal of the Geological Society,* 149, 383-400.

Grenne, T., Heldal, T., Meyer, G.B. and Bloxam, E., 2008. From Hyllestad to Selbu: Norwegian millstone quarrying through 1300 years. In T. Slagstad (ed.), *Geology for Society. Geological Survey of Norway Special Publication 11.* Trondheim, 47-66.

Hacker, B.R. and Gans, B.G., 2005. Continental collisions and the creation of ultrahigh-pressure terranes: Petrology and thermochronology of nappes in the central Scandinavian Caledonides. *GSA Bulletin,* 117, 1-2, 117-134.

Hacker, B.R., Andersen, T.B., Root, D.B, Mehl, L., Mattinson, J.M. and Wooden, J.L., 2003. Exhumation of high-pressure rocks beneath the Solund Basin, Western Gneiss Region of Norway. *Journal of Metamorphic Geology,* 21, 613-629.

Hansen, A.M., 1991. Kverna som maler på havets bunn. Et kvernsteinsfunn i Alverstraumen, Lindås i Hordaland. *Sjøfartshistorisk årbok 1991*, 195-215.

Hansen, A.M., 1997. Maritime perspektiv på kvernsteinsproduksjonen i Hyllestad. In H. Sørheim (ed.), *Arkeologi og kystkultur, Sunnmøre Museum*. Ålesund, 58-63.

Helberg, B.H., 2007. Rapport vedrørende overvåking av inngrep i kvernsteinsbrudd i damområdet for Saksenvik kraftverk, Saksenvik i Saltdal Kommune, Nordland. *Unpublished report.* Tromsø.

Heldal, T. and Bloxam E. G., 2007. *Kartlegging og karakterisering av kvernsteinsbruddene i Hyllestad.* Geological Survey of Norway Report, 2007.079. Trondheim

Krokvik, O., 1999. Kvernsteindrift. *Blad av Brønnøy Historie 1999*, 26-30.

Monssen, J.D., 1997. Kvernfloget, *Saltdalsboka.* Saltdal kulturstyre.

Rolseth, P.O., 1947. *Kvernfjellet.* Selbu og Tydals Historielag.

Rønneseth, O., 1968. Das Zentrum der ältesten Mühlsteinindustrie in Norwegen. In C.W. Haarnagel and K. Raddatz (eds.), *Studien zur europäischen Vor- und Frühgeschichte.* Neumünster, 241-252.

Rønneseth, O., 1977. Kvernsteinsbrota ved Åfjorden.

Sogeskrift frå Hyllestad, 2.
Rønneseth, O. 1988. Kvernsteisnbrota ved Åfjorden. *Årbok for Sogn* 1988.
Sognnes, K. 1980. Kvernsteinsdrifta i Monsholet. *Årbok for Helgeland* 1980, 171-176.
Storemyr, P. and Heldal. T., 2002. Soapstone production through Norwegian history: geology, properties, quarrying and use. In J.J. Herrman, N. Herz and R. Newman (eds), *Asmosia 5 – Interdisciplinary studies on ancient stone. Proceedings of the fifth international conference of the Association for the study of marble and other stones in antiquity*. Museum of Fine Arts, Boston. 359 – 369.
Tillung, M., 1999. *Structural and metamorphic development of the Hyllestad-Lifjorden Area, Western Norway*. Unpublished Cand. Scient. Thesis, University of Bergen
Titland, H.P., 2003. Kvernsteinsbruddene i Saltdal. *Saltdalsboka 2002/2003*. Saltdal kulturstyre.

The function of querns

Susan Watts

The quern, one of the oldest and longest used of all craft tools, is traditionally seen as a simple, practical implement, synonymous with utility and grinding grain. This article, which is based on a paper given at the Millstone Colloquium in Rome in 2009 and also on that presented at the Theoretical Archaeology Group Conference at Exeter University in 2006, aims to break down this commonly held perception, suggesting that querns are actually multi-functional tools, of which utility is but one aspect, and that the modes of action by which they fulfil their purpose is varied and changeable and can be perceived in both physical and abstract terms. To tease out these different functions one needs to better understand how querns operated in the societies that used them and in order to do this one needs to explore their life history or object biography. Object biography encompasses more than the practical and technical aspects of manufacture, use and discard. It also includes the social, economic and symbolic aspects of those processes and, in doing so, considers the changing meaning of the object in dynamic interaction with human lives (Kopytoff 1986; Gosden and Marshall 1999; Hurcombe 2007, 38-42).

Querns (the term is used generically to mean all forms of hand-operated tools used for milling grain and other products, such as saddle querns, rotary querns, grinding slabs and metates) actually have rather long and involved life cycles. On the one hand the task of milling is seen as one of the 'less pleasant household tasks' (Moritz 1958, xxv) and in the Old Testament there are a number of references indicating that the grinding of flour was considered a lowly chore, fit only for slaves and prisoners (see for example Exodus 11.5; Isaiah 47.2; Judges 16.12). Yet the laws of Moses stated that one should not take a millstone as a pledge for that would be the equivalent of taking a man's life (Deuteronomy 24.6) and, in concurrence with this, Crawford describes querns as 'an excellent instance of necessary things for they grind the corn which for an agricultural people is the chief basis of life' (Crawford 1953, 98-99).

Querns can be long-lived artefacts, being passed down in some societies from mother to daughter (see for example Anon 1850, 394) and they can also potentially see several phases of secondary use, as building material for example. Thus many years may pass between the date of a quern's manufacture and the date of its final deposition in the archaeological record (Gwilt and Heslop 1995, 40). And throughout their life history querns are enmeshed in a network of associations, interactions and relationships between people and other elements of material culture both spatially and temporally, physically and mentally. The function, importance and meaning of a particular quern within the society in which it is used is dependent upon the uses to which that quern is put and who uses it and this may alter as it moves from one stage of its life history to the next (after Barrett *et al.* 2000, 45; Walker and Lucero 2000, 133). Querns can be seen therefore as 'social objects with multiple values' (Lidström Holmberg 2004, 202). These values may be related to the physical, practical aspects of grinding and/or to the symbolic properties associated with the transformation of raw material into a usable product and which it is suggested had a bearing on the querns' final resting place in the archaeological record (Hill 1995, 108; Lidström Holmberg 2004, 226; Watts 2008; Watts in preparation).

Each quern, of course, like each person, has its own biography and not all querns follow the same unilinear path of manufacture, use and discard (after Schiffer 1995, 27). The following is, therefore, generic, drawing on historical, ethnographical and archaeological evidence. It is not to say this is how it is or was but to highlight the possibilities of what could be (after Parker Pearson and Ramilisonina 1998, 309).

The life history of a quern begins with a decision that this particular tool is required and the reasoning behind that requirement begins to set the meaning and value of that quern. Thus, although the quern may have been intended as a milling tool, we can already see it potentially performing several different functions – as a commodity for trade or exchange, as a gift, as a replacement. Its manufacture may be an individual act but it is also part of a wider social and economic network, providing a framework in which technical skills can be learnt and improvements and innovations made, although those same networks can also inhibit developments. It can also be a symbolic process as a quern stone emerges from the rock (after Pfaffenberger 1988, 249; Jones 2000, 127, 133; Knappett and Leeuw forthcoming).

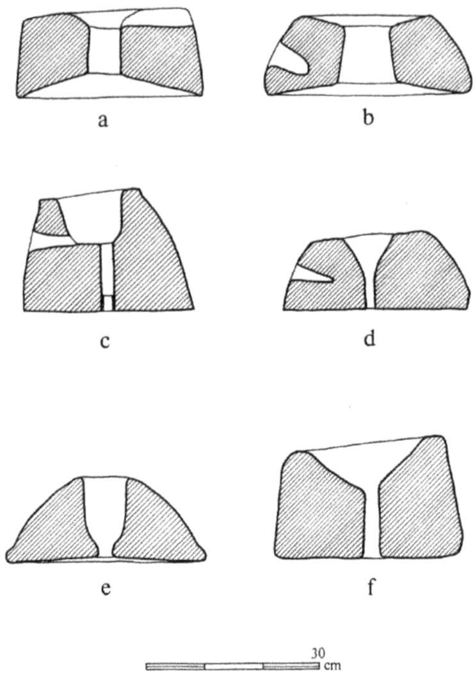

Figure 1. Iron Age beehive quern upper stones: a) Sussex Type, b) Wessex type, c) Hunsbury type, d) Yorkshire (northern) type, e) Puddingstone (East Anglian) type, f) Folkestone type (Drawing: M. Watts, 2002, fig. 9).

The finished quern can itself be a means of expressing identity as seen in the different types of British Iron Age beehive quern, their forms indicating which part of the country they originate from (Fig. 1). It is noticeable that beehive querns from southern England have sloping grinding surfaces and wide central feed pipes whereas those from the midlands and north of England, southern Scotland and Ireland have flat grinding surfaces and narrow feed pipes. There are also differences within each basic form, mainly in the positioning of the handle hole or holes. It is notable that regional pottery styles were also developing at this time and these differences may be due, therefore, to ethnic choices (Curwen 1937; Curwen 1941; Watts 2002, 31-33; Cunliffe 2005, 120, 122). An interesting exception to the north-south divide, however, is found in the querns produced at a quarry at East Wear Bay, Folkestone in Kent in the first century BC and AD which are similar in form to those from the midlands and north (Fig. 1f.) (Keller 1989; Watts 2002, 33).

Decoration too can be a means of expressing identity. A number of Roman military querns, for example, are inscribed with the name of the century or contubernium to which they belonged, such as that in the museum of the fort at Saalburg on the Roman frontier in Germany which bears the inscription 'CON[TUBERNIUM] BRITTONIS'. On another, from Greatchesters on Hadrian's Wall, the name of the centurion is followed by 'MOLAVII[', while one from Vindolanda, is inscribed with a backwards 'C' for century, followed by 'AD' (Birley 1932, 219). Most decoration, however, appears more symbolic in its application. Several quern fragments, also of Roman date, have been found bearing the carving of a phallus, seen as a fertility symbol and protection against evil (Henig 1984, 167, 185-186). One such example was found in the extra mural area of the fort at Rocester, Staffordshire (Frere *et al.* 1983, 302). The sign of the cross found on some medieval querns such as that from Dunadd, Argyll, Scotland (Campbell 1987), may also have been to ward off evil or to have been a blessing, to lighten the load of milling, or perhaps indicated that the quern was to be used solely for grinding flour for communion bread. It is also not uncommon for the spouts of medieval pot querns to be carved with the likeness of a face, the ground meal being spewed through the mouth (Watts 2002, 41). Is this an example of medieval humour or is there another deeper, symbolic meaning behind it? Likewise, there is presumably a meaning behind the abstract decoration found on a number of Iron Age beehive querns from Ireland, Wales and Cumbria, some of which is similar to that found on contemporary metalwork (see for example Griffiths 1951; Caulfield 1977, 121-123; Ingle 1987, 13) and also to the later so-called 'sleeping fox querns' on which the upper stone is carved into the semblance of a curled-up fox (Fig. 2). Decoration communicates information, enabling the quern to function in other ways. Even if one may not know today specifically what that decoration signifies it can be seen to raise the quern out of the ordinary, out of the domestic state (Bradley 2005, 100-101).

Figure 2. An elegant and sophisticated example of a 'sleeping fox' quern from Merona, northern Italy. Used for grinding salt, the quern is thought to date from the 17th century (G. Clausse).

Figure 3. A social scene showing a group of women shelling lentils in Tille village in Adıyaman province, on the Euphrates in South-East Turkey (Photo: S. Blaylock).

Once finished, unless it was made by the person who intended to use it, the quern became a commodity. A quern could thus come into a community by a number of means: it could have arrived through trading networks or have been presented as a gift or form part of a dowry, or it could have been a replacement tool. Although intended as a milling tool each method and reason for arrival would have conveyed different meanings and associations. Of course, it may not be possible to fully interpret the circumstances that led to the presence of querns on archaeological sites but what can be seen are querns functioning as indicators of relationships and contacts between peoples whether indirectly through the movement of resources or directly through the movement of peoples (Moore 2007, 93-4). A greater significance, however, may be attached to exotic stones such as the broken saddle quern of puddingstone from central Normandy found in a Neolithic pit at Maiden Castle, or the fragments of German lava saddle quern found at the Sanctuary near Avebury, Wiltshire (Peacock and Cutler 2010; Cunnington 1930-1932, 332). The rock from which the quern is made thus also functions as an expression of identity, through which the quarry or area from which the quern originally derived can be identified.

Although primarily associated with grinding cereals there is much ethnological, historical and archaeological evidence to illustrate the use of querns for grinding many other vegetable, mineral and animal products such as salt, lentils, rice, ore, pottery temper and snuff. Undoubtedly, the quern, in its various forms, functions as a very capable milling tool. However, although the physical action of using a quern is the same for each product, be it the forward and back motion of the saddle quern or the circular motion of the rotary quern, the social significance and meaning behind that action is determined by that product. Compare, for example, grinding gold ore to grinding cereals for an everyday meal, to grinding roots for medicinal use. But in all these activities the quern functions, most importantly, as a transformer, turning raw product into usable material (Hill 1995, 108; Lidström Holmberg 2004, 226).

A quern thus functions as part of a particular social setting, reinforcing social identity, ties and obligations (Fig. 3). It is central to the task of milling, functioning as a primary interactor in the life history of the product being ground and as such is linked with other interactors necessary for the advancement of that products' life history (after Schiffer 1999, 25-27). As such a quern is associated not only with the person using it but also with a set of artefacts that play a less direct but nevertheless necessary role such as a container for the raw material and something to collect the ground product in. It is also indirectly associated with those people, artefacts and other items of material culture that are not related to the task of milling but which form part of the social setting in which the quern is used.

Although there are significant exceptions, such as the all male enclaves of armies and monasteries, ethnographic evidence indicates that for the preparation of food querns are very much associated with women. In some societies

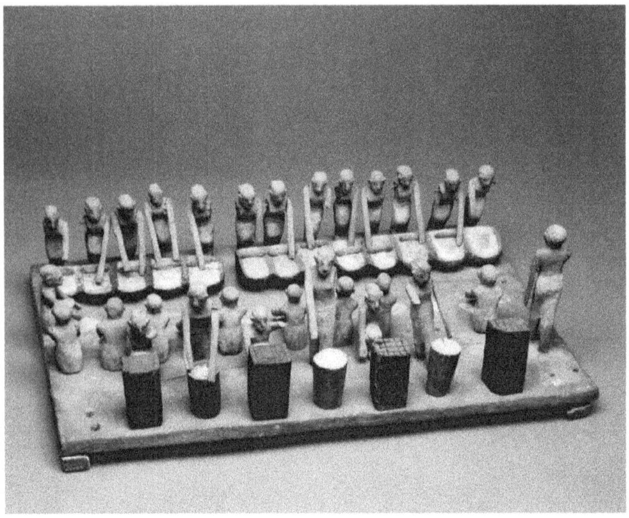

Figure 4. Model from the tomb of King Nebhepetre Mentuhotep II (2055-2004BC). The model is schematic but is illustrative of the multiple milling stations found in state run bakeries of the period, showing a line of women working in unison. The figure at the right hand end is presumably the foreman (Photo © Trustees of the British Museum).

they are still considered an important part of the household equipment necessary for a newly married woman and frequently form part of the dowry or bridal gift (see for example Cook 1970, 779; Thompson 1979, 318 n94; Hamon pers. comm.). In this respect querns can be seen as symbolic of the female state, associated with the home and the provision of food. In early 20th century rural Mexico, for example, the grinding stone together with the hearth, griddle and pot were considered to be the four items necessary for the house (Brumfield 1991, 237). As the quern is used within the daily domestic setting so it also provides a context for learning as a young girl finds out by observation and experience how best to hold the rubber of a saddle quern for example, how much pressure to exert and the most comfortable position in which to operate the quern. In some communities a proven ability with a quern makes a girl more desirable as a marriage partner (Katz 2003, 46).

This domestic scene can be contrasted with the milling rooms of the prisons and state-run or royal milling establishments found in Egypt, Mesopotamia and Syria in the 3rd and 2nd millennium BC (Fig. 4) (Curtis 2001, 202-203; Englund 1991; see also Judges 16.21). The quern is still central to the scene and directly associated with a similar set of artefacts but the values and meanings attached to the quern would have been quite different. The person labouring at the quern would not have been its owner but simply assigned to it and there would have been a foreman watching over to ensure that the work was completed on time. The hours spent at the quern are likely to have been longer, particularly in prisons, where the task of milling would have been a burdensome chore, the quern functioning as an instrument of punishment and even torture.

The physical location of a quern in situ in its primary working position is also important, potentially operating on both a practical and symbolic level. In a domestic situation, for example, the location of a quern may denote a female activity area. However, for the Batammaliba of West Africa, for whom the house functions as a metaphor for the human body, the location of the quern plays a specific part in their cosmology. The house is made of earth (flesh), water (blood) and pebbles (bone); the door is the mouth and just inside the door lies the quern, the teeth, for grinding grain (Tilley 1999, 44-45).

Although, as mentioned above, querns can be used to grind a wide variety of materials their significance in grinding staple foods should not be underestimated. This may be a utilitarian, practical function but the provision of food is a necessary, life-giving task. It is perhaps not surprising, therefore, that in the Bible the absence of the sound of millstones is used as a sign of desolation, symbolic of a place that is forsaken (Deuteronomy 24.6; Jeremiah 25.10; Revelation 11.22). Even the rasping whir of the quern as it turns performs an important function. It would have been a familiar noise, coupled perhaps with the sound of women singing as they worked both to keep the rhythm of milling going and offset the tediousness of the task. There would have been a subconscious link between the sound of milling and the provision of food (Thomson 1877, 526; Watts 2008, 95). The sound can also incidentally function as an early morning alarm call. In an article on Bread in the East published in *The Penny Magazine*, a 19th century English periodical, it was recorded that 'this labour is generally performed in the early morning by the women of the household' and 'as the upper stone is whirled round, the women beguile their labours by singing, at the top of their voices, certain songs which seem almost appropriated to this service. The simultaneous noise of grinding and singing in an Oriental city warns the indolent that it is time to rise' (Anon 1834, 3).

Milling is thus a vital, socially meaningful act that has pragmatic and emotional values through which a symbolic link can be seen between querns and life, death, transformation and regeneration. These values can be applied to querns whatever the product they grind but are most potent in their relation to grinding cereals. During milling whole, inedible grains are reduced to edible meal but are killed during that process. Yet although they cannot be sown and grown they nevertheless give life to those they feed (Watts 2008, 100). The killing of the grain is echoed in an old English Folksong, John Barleycorn, which is thought to be the survival of a myth relating to the slaying of a corn god. The song tells of three men who vowed to kill John Barleycorn. They planted him and were sure he was dead but he grew. So they cut him down in his prime, bound and beat him. 'The miller', however,

Figure 5. The upper stone of a rotary quern reused as a gravestone from Glendalough, Eire. The inscription includes a cross and the name of the deceased, 'Sechnasach' (Photo: G. Clausse).

'served him worse than that, for he ground him between two stones' (Vaughan Williams and Lloyd 1959, 57, 116). This story is reminiscent of that surrounding the death of Mot, a Canaanite and Phoenician god of death, who in retribution for him trapping Baal, vegetation god, beneath the earth is cleaved, winnowed, burnt and ground by Anat, twin sister of Baal. Her actions brought Baal back to life (Jordan 2002, 167).

It is the role of querns in grinding cereals and its symbolical link to life, death and regeneration as witnessed in the agricultural year that is thought to be manifested in the structured deposition of querns during the prehistoric period, that is they were placed with thought and meaning in pits, ditches and other places rather than being unthinkingly or unwontedly discarded (Hill 1995, 55; Heslop 2008, 73-80; Graefe *et al.* 2009; Watts in preparation). Broken querns in particular are seen as symbolic of death (Campbell 1987, 112; Brück 1999, 155). Indeed it is the fact that such querns are often apparently deliberately broken, or placed upside down that implies that querns function on more than a simple, utilitarian scale. This symbolical link can be traced back to the Neolithic period and the beginnings of agriculture. Residue and use-wear analysis indicates a strong link between saddle querns and cereal cultivation and it has also been suggested that an increased use of flour-based foods led to the increasing use of the saddle quern in the Neolithic Near East (Wright 2000, 98; Dubreuil 2004; Hamon 2008). The use of querns as structured deposits can be found across Neolithic Europe and may, therefore, have been introduced to Britain as part of a pre-existing belief system. The placement of a broken saddle quern from central Normandy at the bottom of a Neolithic pit at Maiden Castle lends credence to this suggestion (Watts 2008, 100; Graefe *et al.* 2009; Peacock and Cutler 2010). However, such depositions may also have had more personal meanings, functioning, for example, as tokens of remembrance or as offerings.

It has also been suggested that apparently still usable roughouts found at prehistoric quern quarries were not simply abandoned but deliberately left on site as some form of structured deposit (Heslop 2008, 45). Are such deposits related to agriculture and milling or are they more closely related to the actual process of extracting the quern from the rock, giving back to the earth something of what had been taken, a thank offering perhaps? Although one may not understand the meaning behind particular depositions of querns, the nature and context of those deposits may help to distinguish differences in levels or category of meaning (after Bradley 1987, 352; Needham and Spence 1997, 86-87).

Querns can also fulfil secondary functions, that is functions for which they were not originally intended but which their physical characteristics lend them to. A rubber, for example, makes a handy weapon, as recorded in the story of the Israelite King Abimelech who was fatally wounded by such a stone while attempting to burn down a fortified tower filled with people at Thebez c. 1149 BC (Judges 9.53). On a more everyday level, querns can also make useful work surfaces or they can be reused as hone stones or turned into mould stones for casting metal. They are also frequently found reused as building stones and incorporated within the construction of hearths and kilns (see for example Gray and Bulleid 1953, 130, 180, 183; Frere 1972, 78; Gwilt and Heslop 1995, 40; Watts 2004, 221; Pitts 2007, 6). Now while these latter uses can certainly be seen to fulfil a pragmatic function it can also be argued that these too are aspects of structured deposition. If querns can be utilised as foundation deposits, such as in the Neolithic tomb of La Hougue Bie on Jersey, where a worn saddle quern was placed beneath the large granite slab that separated the terminal cell from the 'sanctuary', or play an important part in the abandonment procedures witnessed at a number of Bronze Age sites in south west England, such as at Trethellan Farm and Callestick in Cornwall, then why should not a fragment higher up in a wall or in a hearth be equally socially significant, linked perhaps to events within the lifecycle of the structure? (Société Jersiaise 1977, 12; Nowakowski 1991, 25, 73; Jones 1998-9, 14-15; Brück 1999, 154; O'Sullivan and Kenny 2008). Querns have also been reused in burial cists and as grave markers (Fig. 5). Although there could be a connection between the quern and the person buried, the use of querns in this respect forges a particularly potent link between life and death (Bennett and Elton 1898, 144-5; Henshell 1955-56, 261; Campbell 1987, 113).

A quern's life history does not end with its deposition in the archaeological record, however. New relationships are created and a new series of interactions begin to take place as, when and how that quern is retrieved, through excavation or as a chance find for example. The quern can then take on a new range of functions, such as an item for analysis and research, as part of a museum display, as an experimental or educational tool, or as a decorative building or garden feature (Fig. 6) (after Leone 1981, 5; Hurcombe 2007, 42).

Figure 6. The lower stone of a rotary quern reused as a decorative feature in the wall of a building near Bristol (Photo: Author).

Through the exploration of the life history or object biography of querns it can be shown that they are indeed multi-functional objects, operating on several co-existing, interrelated levels within which their physical and symbolic properties come together in relationship to specific social and cultural contexts. These different functions are both determined and changed by the series of interactions that take place between a quern, people and other aspects of material culture throughout its life history. Querns can be used for grinding a wide variety of products but it is their role in grinding staple foods that can be seen to engender symbolic meanings associated with the agricultural year, with life, death and transformation, meanings it is believed are witnessed in the structured deposition of querns during the prehistoric period. A quern's biography, however, continues beyond its deposition in the archaeological record. New connections are created and new functions emerge as archaeologists, museum curators and members of the public, for example, engage with it. These new relationships and functions provide interesting contrasts with those that have gone before. The quern may no longer function as a key part of the process that transforms raw material into usable product but it still nevertheless has an important task to perform, that of informer on past behavioural practises and technologies.

Acknowledgements

The author is grateful to S. Blaylock, G. Clausse, M. Watts and The Trustees of the British Museum for figures 1-5.

References

Anon, 1834. Bread in the East. *The Penny Magazine*, 4th January, 2-4.

Anon, 1850. Proceedings at the Meeting of the Archaeological Institute, December 6, 1850. *Archaeological Journal*, 7, 391-398.

Barrett, J.C., Freeman, P.W.M., and Woodward, A., 2000. *Cadbury Castle Somerset. The Later Prehistoric and Early Historic Archaeology*, English Heritage Archaeological Report, 20.

Bennett, R., and Elton, J., 1898. *History of Corn Milling, 1, Handstones, Slave & Cattle Mills*. London.

Birley, E., 1932. Excavations at Chesterholm-Vindolanda. *Archaeologia Aeliana*, Series 4, 9, 219-221.

Bradley, R., 1987. Stages in the Chronological Development of Hoards and Votive Deposits. *Proceedings of the Prehistoric Society*, 53, 351-362.

Bradley, R., 2005. *Ritual and Domestic Life in Prehistoric Europe*. London.

Brück, J., 1999. Houses, Lifecycles and Deposition on Middle Bronze Age Settlements in Southern England. *Proceedings of the Prehistoric Society*, 65, 145-166.

Brumfield, E.M., 1991. Weaving and Cooking. Women's Production in Aztec Mexico. In J.M. Gero and M.W. Conkey (ed.), *Engendering Archaeology. Women and Prehistory*. Oxford.

Campbell, E. 1987. A Cross-marked Quern from Dunadd and Other Evidence for Relations Between Dunadd and Iona. *Proceedings of the Society of Antiquaries of Scotland,* 117, 105-117.

Caulfield, S., 1977. The Beehive Quern in Ireland. *Journal of the Royal Society of Antiquaries of Ireland*, 107, 104-138.

Cook, S., 1970. Price and Output. Variability in a Peasant-Artisan Stoneworking Industry in Oaxaca, Mexico: An Analytical Essay in Economic Anthropology. *American Anthropologist*, 72.4, 776-801.

Crawford, O.G.S., 1953. *Archaeology in the Field.* London.

Cunliffe, B., 2005. *Iron Age Communities in Britain.* London.

Cunnington, M.E., 1930-1932. The 'Sanctuary' on Overton Hill Near Avebury. *Wiltshire Archaeological Magazine*, 45, 300-335.

Curtis, R.L., 2001. *Ancient Food Technology*. Leiden.

Curwen, E.C., 1937. Querns. *Antiquity*, 11, 133-151.

Curwen, E.C., 1941. More about Querns. *Antiquity*, 15, 15-32.

Dubreuil, L., 2004. Long-term Trends in Natufian Subsistence. A Use Wear Analysis of Ground Stone Tools. *Journal of Archaeological Science*, 31, 1613-1629.

Englund, R.K., 1991. Hard Work Where Will it Get You? Labor Management in Ur III Mesopotamia. *Journal of Near Eastern Studies,* 50.4, 255-280.

Frere, S., 1972. *Verulamium Excavations*, 1. Reports of the Research Committee of the Society of Antiquaries of London, 28.

Frere, S.S., Hassall, M.W.C., and Tomlin, R.S.O, 1983, Roman Britain in 1982. *Britannia*, 14, 280-356.

Gosden, C., and Marshall, Y., 1999. The Cultural Biography of Objects. *World Archaeology*, 31.2, 169-178.

Graefe, J., Hamon, C., Lidström Holmberg, C., Tsoraki, C., and Watts, S., 2009. Subsistence, Social and Ritual Practices. Quern Deposits in the Neolithic Societies of Europe. In S. Bonnardin, C. Hamon, M. Lauwers and B. Quilliec (ed.), *Du Matériel au Spirituel. Réalités Archéologiques et Historiques des Dépôts de la Préhistoire à nos Jours XXIX*. Antibes.

Gray, H. St. George, and Bulleid, A., 1953. *The Meare Lake Village*, 2. Taunton.

Griffiths, W.E., 1951. Decorated Rotary Querns from Wales and Ireland. *Ulster Journal of Archaeology*, 14, 49-61.

Gwilt, A., and Heslop, D., 1995. Iron Age and Roman Querns from the Tees Valley. In B. Vyner (ed.), *Moorland Monuments. Studies in the Archaeology of North-East Yorkshire in Honour of Raymond Hayes and Don Spratt*, CBA Research Report, 101, 38-45.

Hamon, C., 2008. The Symbolic Value of Grindingstones Hoards. Technical Properties of Neolithic Examples. In C. Hamon and B. Quilliec (ed.), *Hoards from the Neolithic to the Metal Ages*, BAR International Series. Oxford, 19-28.

Heslop, D.H., 2008. *A Corpus of Beehive Querns from Northern Yorkshire and Southern Durham*, Yorkshire Archaeological Society Occasional Paper, 5.

Henig, M., 1984. *Religion in Roman Britain.* London.

Henshall, A.S., 1955-56. Long Cist Cemetery at Parkburn Sandpit, Lasswade, Midlothian. *Proceedings of the Society of Antiquaries of Scotland*, 89, 252-283.

Hill, J.D., 1995. *Ritual and Rubbish in the Iron Age of Wessex*. BAR British Series, 242. Oxford.

Hurcombe, L.M., 2007. *Archaeological Artefacts as Material Culture*. London.

Ingle, C.J., 1987. The Production and Distribution of Beehive Querns in Cumbria. *Transactions of the Cumberland and Westmoreland Archaeological and Antiquarian Society*, 87, 11-17.

Jones, A.M., 1998-99. The Excavation of a Later Bronze Age Structure at Callestick. *Cornish Archaeology*, 37-8, 5-55.

Jones, A., 2000. Life After Death. Monuments, Material Culture and Social Change in Neolithic Orkney. In A. Ritchie (ed.), *Neolithic Orkney in its European Context*. Cambridge, 127-138.

Jordan, M., 2002. *Encyclopedia of Gods*. London.

Katz, E., 2003. Le Metate, Meule Dormante du Mexique. In M. Barboff, F. Sigaut, C. Griffin-Kremer and R. Kremer (ed.), *Meules à Grains. Actes due Colloque International La Ferté-sous-Jouarre 16-19 Mai 2002*. Paris, 32-50.

Keller, P.T., 1989. Quern Production at Folkestone, South-East Kent. An Interim Note. *Britannia*, 20, 193-200.

Knappett, C., and van der Leeuw, S., forthcoming. The Space of Innovation. Cognitive, Social and Physical Considerations, for submission to *Current Anthropology*.

Kopytoff, I., 1986. The Cultural Biography of Things. Commoditization at Process. In E. Appadurai (ed.), *The Social Life of Things, Commodities in Cultural Perspective.* Cambridge.

Leone, M.P., 1981. Archaeology's Relationship to the Present and the Past. In R.A. Gould and M.B. Schiffer, *Modern Material Culture. The Archaeology of Us*. London, 1-14.

Lidström Holmberg, C., 2004. Saddle Querns and

Gendered Dynamics of the Early Neolithic in Mid Central Sweden. *Arrival. Coast to Coast*, 10, 199-213.

Merrifield, R., 1987. *The Archaeology of Ritual and Magic*. London.

Moore, T., 2007. Perceiving Communities. Exchange Landscapes and Social Networks in the Later Iron Age of Western Britain. *Oxford Journal of Archaeology*, 26.1, 79-102.

Moritz, L.A., 1958. *Grain-Mills and Flour in Classical Antiquity*. Oxford.

Needham, S., and Spence, T., 1997. Refuse and the Formation of Middens. *Antiquity*, 71, 77-90.

Nowakowski, J.A., 1991. Trethellan Farm, Newquay: The Excavation of a Lowland Bronze Age Settlement and Iron Age Cemetery. *Cornish Archaeology*, 30, 5-242.

O'Sullivan, A. and Kenny, N., 2008. A Matter of Life and Death?, *Archaeology Ireland*, 22.4 Issue 86, 8-11.

Parker Pearson, M., and Ramilisonina, 1998. Stonehenge for the Ancestors. The Stones Pass on the Message. *Antiquity*, 72, 308-326.

Peacock, D.P.S., and Cutler, L., 2010. A Neolithic voyage. *International Journal of Nautical Archaeology*, 39, 116-24

Pfaffenberger, B., 1988. Fetishised Objects and Humanised Nature. Towards an Anthropology of Technology. *Man*, NS 23.2, 236-252.

Pitts, M., 2007. Ancient Trading Power near Inverness. *British Archaeology*, 92, 6.

Raftery, B., 1984. *La Tene in Ireland*. Marburg.

Schiffer, M.B., 1995. *Behavioral Archaeology*. Salt Lake City.

Schiffer, M.B., 1999. *The Material Life of Human Beings*. London.

Société Jersiaise 1977. *La Hougue Bie, Jersey, Channel Islands. Prehistoric Tomb and Ancient Chapels*. St. Helier.

Thompson, F.H., 1979. Three Surrey Hillforts. Excavations at Anstiebury, Holmbury, and Hascombe, 1972-1977. *Antiquaries Journal*, 59, 245-318.

Thomson, W.M., 1877 (new impression 1985). *The Land and the Book*. London.

Tilley, C., 1999. *Metaphor and Material Culture*. Oxford.

Vaughan Williams, R., and Lloyd, A.L., 1959. *The Penguin Book of English Folk Songs*. Harmondsworth.

Walker, W.H., and Lucero, L.J., 2000. The Depositional History of Ritual and Power. In M Dobres and J. Robb (ed.), *Agency in Archaeology*. London. 130-147.

Watts, M., 2002. *The Archaeology of Mills & Milling*. Stroud.

Watts, S., 2004. Querns. In P.A. Rahtz and L. Watts *Wharram, A Study of Settlement on the Yorkshire Wolds, IX, The North Manor Area and North-West Enclosure*. York University Archaeological Publications, 11, 219-224.

Watts, S., 2008. Object Biography and its Importance in Furthering our Understanding of the Structured Deposition of Querns in Neolithic Britain. In C. Hamon and J. Graefe (eds.), *New Perspectives on Querns in Neolithic Societies*, Archäologische Berichte, 23, Bonn, 93-102.

Watts, S., in preparation. *Contexts of Deposition. The Structured Deposition of Querns in the South West of England from the Neolithic to the Iron Age*. Unpublished PhD thesis. University of Exeter.

Wright, K.I., 2000. The Social Origins of Cooking and Dining in Early Villages of Western Asia. *Proceedings of the Prehistoric Society*, 66, 89-121.

The millstone quarries in Hyllestad: an arena of research and education

Irene Baug and Torbjørn Løland

One of the largest and most long-lasting production areas for millstones in the country is located at Hyllestad on the west coast of Norway (Fig.1). The natural source for the production was a specific variety of rock type, a garnet-kyanite mica-schist, which is found along Åfjorden. The activity in Hyllestad dates back a long way and the quarries are the remains from a long-abandoned industrial landscape that has developed over hundreds of years, leaving quarries, spoil-heaps, remnants of roads, harbours etc. as visible traces.

A few scholarly works on the production and distribution were published in the latter half of the 20th century (Rønneseth 1968; Hansen 1997; Carelli and Kresten 1997), and in the last years an interdisciplinary research along with a strong local engagement has put the quarry landscape in Hyllestad back on the map. From the 1990s a renewed local interest in the quarries was awakened, which developed into a local engagement with the aim of passing on the historical significance of the quarries. This resulted in an establishment of the 'Millstone-guild', a local informal association where craft and trade are important aspects. The members of the guild enjoyed festivals and chronicle plays and brought with them stones from Hyllestad, so they could demonstrate possible extraction techniques from the quarries. In this way they managed to present the quarrying in Hyllestad to others in a fascinating and charming way.

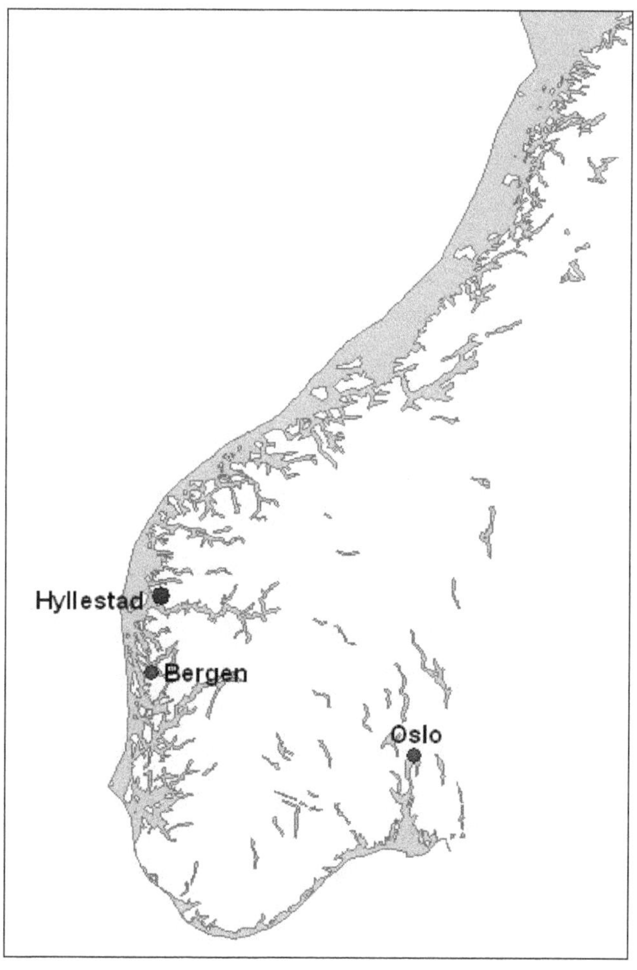

Figure 1. Map of the location of Hyllestad.

However, behind this somewhat informal guild was also a strong desire to learn more about the quarries and their significance. Millstones were vital tools for the daily bread. The large-scale production of these products is, however, only rarely documented in written sources and has until recently gained little attention in research projects. More knowledge was needed, and different research institutions were contacted. The result is an interdisciplinary research programme involving archaeology, geology and the study of craft techniques (see e.g. Baug 2001; 2002; in prep.; Heldal and Bloxam 2007; Løland in prep.).

The first part of this paper will look into the three different research projects that have been conducted in Hyllestad during the last few years, and what knowledge we have gained from this so far. Secondly, we want to shed light on the educational project that has developed in connection with the quarries, partly because of the research conducted. The results from both geological and archaeological investigations, as well as the study of craft techniques, are being integrated in the teaching program in the primary school in Hyllestad in the most unique way.

The quarry landscape

A survey in the quarry landscape conducted by the Geological Survey of Norway (NGU) has identified and mapped different quarry types and geological conditions, and so far about 370 quarries have been identified (Heldal and Bloxam 2007). The result from this investigation is being presented by Tom Heldal (this volume), and we will only summarise here certain points relevant to the archaeological investigations of the quarries. The survey provides a great understanding of the production landscape as a whole and has been important for the choice of investigation sites for archaeological excavations.

Two different extraction techniques were used for production in Hyllestad; with and without gunpowder. Gunpowder was used in the most recent production, and continued up until the 1930's before it ceased (Rønneseth 1977, 61). With this technique the stones were cut out of blocks that had already been blasted out of the rock. Production based on gunpowder was, however, of a much smaller scale than the older manual extraction.

Only the older quarry types, without use of gunpowder, are included in this paper: an extraction where millstones were normally shaped and cut directly from the bedrock (Fig. 2). However, these quarries can be further divided into subtypes that represent different technologies and ways of organising the extraction (Heldal and Bloxam 2007). It has been important to include all types of quarries in the archaeological investigation, and the three main types; shallow quarries, deep quarries and combination quarries have been excavated.

Shallow quarries are quarries where millstones were cut along the cleavage plane, leaving traces in the form of circles from the extracted millstones. In the *deep quarries* millstones were quarried in piles, one under another, leaving tall, carved walls, sometimes with a step-like shape. Heldal has suggested that the shallow quarries represent the oldest production phase in Hyllestad, while the deep quarries are the remains from a somewhat younger and more intensive activity (Heldal and Bloxam 2007, 48–54). This can only be investigated through an archaeological survey.
A third common quarry type in Hyllestad is a combination of a shallow quarry and a deep quarry. Several phases in the extraction may explain this quarry type. A quarry may have started as a shallow quarry in the first period and in a later period the production continued as a deep quarry (Heldal and Bloxam 2007, 54). The shift in quarry types, and thus quarrying techniques, may have to do with an increased efficiency. However, the possibility that the deep quarries coincide with a large scale production of building stones at other Norwegian quarries from the Middle Ages onwards should also be considered (Grenne *et al.* 2008, 63–64). In order to assess such problems, it is necessary to know the age of the quarries.

The survey conducted by NGU shows a much more diverse production landscape in Hyllestad with a variety of different quarry types than what was known before. A certain chronology is also indicated, as well as possible explanations for changes in the extraction techniques and the location of the different quarry types. As a result, the production landscape that previously appeared as complex and confusing, is now much more understandable. This is a great advantage for further analysis at the quarry sites and the survey conducted by NGU thus forms an important basis for archaeological investigations.

Archaeological investigations

The production landscape in Hyllestad comprises remains of a near-industrial activity. Important goals would be to look at quarries, production and distribution in a larger

Figure 2. Millstone quarry with direct carving from bedrock.

cultural and social context and to provide an insight into the organisation of the activities. The archaeological investigations of the production sites will constitute an essential basis for the study. They are important in order to date the activity and hence place it in a temporal and social context. Small scale archaeological investigations were carried out at selected sites within the production landscape. The main goal has been to date the extraction and products in different quarries and if possible to clarify both the beginning and the end of the activities. The considerable size of the production landscape makes it possible only to investigate selected sites within this area and to date nine quarries have been excavated. This was done by digging search trenches into the spoil-heaps in the selected quarries (Baug 2001; 2002; in prep.) (Fig. 3).

Dating of the spoil heaps: What does the presence of charcoal mean?

The main method for dating the quarries is stratigraphical analysis combined with radiocarbon dating of organic material, mainly charcoal. However, a radiocarbon date is not absolute but given within parameters, making it difficult to assess an exact chronology for the quarries.

Charcoal is found either as small and scattered pieces in the trenches or as rather large collections of charcoal and ashes, most likely representing remains from fires. The amount and scattering of the charcoal indicate that it stems from activities related to the investigated quarries, and was most likely deposited at the same time as the masses of debris within which it is found. Larger amounts of charcoal and ashes most likely represent deliberate actions of the workers in the quarries – dumping of deposits from nearby fires, while small and more scattered pieces of charcoal may have come into the spoil-heap by coincidence. Charcoal may also have been transported by wind from fires at the quarry site, or by people who worked in the quarries.

A repeated and important activity in connection with the quarrying must have been sharpening of tools - a needed daily activity in the production. The presence of smithies near or at the quarry sites is therefore also likely. In one of the excavated trenches a piece of slag from a smithy has been found, confirming such a hypothesis (Baug in prep.). The possibility that there were fires at the quarry sites for domestic use, such as preparing of food, should be considered also.

The age of the quarries

So far, dates from four of the nine investigated quarries are available. The production in Hyllestad most likely started in the early Viking Period – maybe as early as the 8th century AD. This seems to be the case for a shallow quarry at the Sæsol farm and a combination quarry at the Rønset farm – both with the extraction solely of millstones for

Figure 3. Excavated trench in a spoil-heap, with several fragments of millstones.

hand-querns (Baug 2001; 2002, 50, 57) (Fig. 4). However, the spoil-heaps where the excavations took place cover older quarries, meaning that the oldest production in Hyllestad is older than the oldest dating. Also, deep quarries seem to date to the Viking Period, although not the very beginning of the period, and they continue throughout the Middle Ages. There are also indications that the extraction of millstones for watermills in a deep quarry dates to the Viking Period – or at least to the early Middle Ages (Baug 2001; 2002, 38–42, 53). These dates demonstrate activities in several of the quarries at the same time. The investigation is still not finished, but the period between the eleventh and the fourteenth century is, so far, the period with the most activity in the investigated areas (Fig. 4) (Baug 2001; 2002, 59).

A different production

The Hyllestad-quarries produced more than just millstones (Baug 2001; 2002, 67, 88), and the survey conducted by NGU identified quarries where the extraction of slabs rather than millstones seems to have taken place (Heldal and Bloxam 2007, 39–45). Two of these quarries have been investigated archaeologically – one of them with a special result. The quarry in question is a deep quarry with quite straight carved walls and a rounded inner corner. The

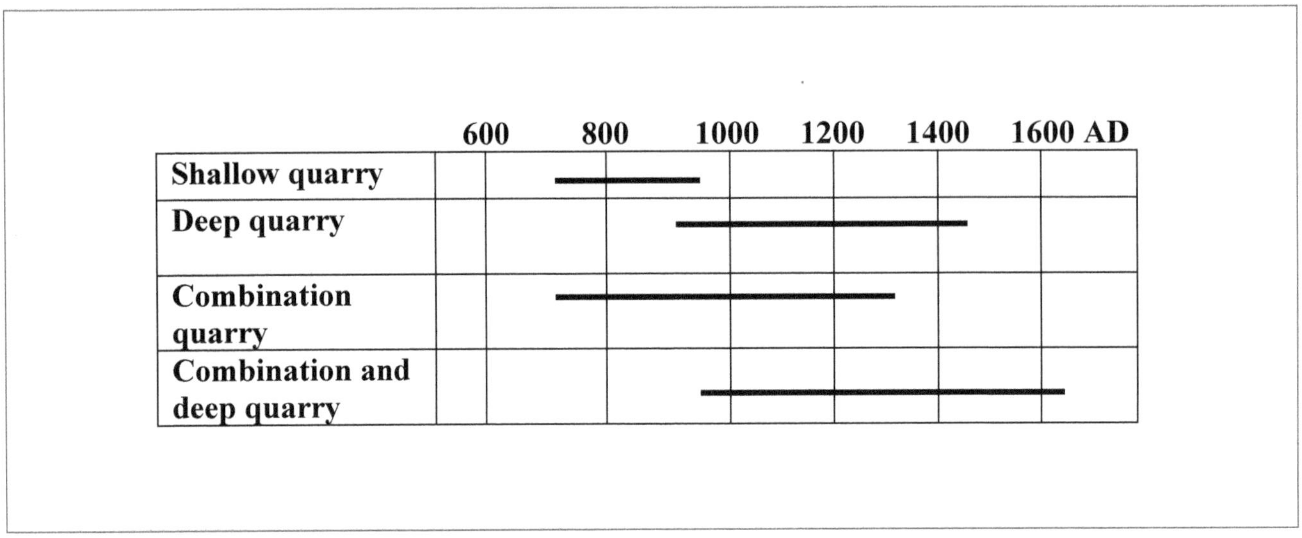

Figure 4. Dating of the extraction in four of the investigated quarries.

Figure 5. Photo of a cross-fragment during excavation.

shape of the carved walls indicated an extraction of slabs, and it was thus considered an important investigation site.

During excavation it became clear that a large quantity of millstones for hand-querns was produced here. This was not very surprising, but reflected what could be seen on the surface where many querns are located. However, on the first day of excavation another product was found – a stone cross (Fig. 5). This was not at all what we had expected, and during six weeks of digging the remains of seven crosses were found in the quarry (Baug 2008; in prep.).

The crosses vary in shape and size. The smallest cross is only 26 cm, while the largest one is 1.2 m. All of the crosses are broken, and an estimation of the intended size is difficult to give. The investigated quarry clearly demonstrates that millstones and crosses were extracted in the same quarries – maybe also by the same stonecutters.

A serial production aimed at a larger market?

An investigation of the spatial distribution of different products at a local, regional and over-regional level has shed light on networks and commercial contacts related to the quarries, and given an indication of the degree of exchange/trade with different areas and the scale and importance of the activity in the quarry landscapes.

As noted, the investigations show that large-scale production took place in Hyllestad and the extraction of millstones may be described as a serial production. This pre-supposes a considerable number of customers and a large market (Skre 2007, 450). The millstones are found in towns and market places from the Viking Period onwards, confirming such a hypothesis. Stones of garnet-muscovite-schist have been found in great quantities in most parts of Norway. However, closer investigations of many of the stones are necessary before the provenance can be decided for certain. Millstones from Hyllestad have also been identified in large parts of Denmark and Sweden (Carelli and Kresten 1997), and the stones also seem to have been distributed to northern Germany (Schön 1995) and probably to the North Atlantic region (Eldjárn 1964, 541; Arge 1989, 119). More investigation is, however, needed and at the moment an ongoing study on millstone distribution and trade networks in the Nordic region is being conducted (see www.millstone.no; Baug in prep.).

To what degree the distribution of millstones in different areas of Northern Europe may be explained by exchange between farmers and fishermen or as the result of an organised trade should be discussed. The quantity of millstones dated to the Middle Ages in Norwegian towns indicate an organised trade in this period (Baug in prep.), and the quantity of Hyllestad-stones in Sweden and Denmark indicates an organised trade/exchange network already existing in the Viking period (Carelli and Kresten 1997, 120–121).

The production and distribution of the crosses was probably of another character, even though they occur in the same quarry as millstones. This was most likely a production on demand where the crosses were transported directly to the receivers and not traded at market places and in towns. Additionally, the cross-production was undertaken during a much more limited period than the millstones.

Who owned the quarries?

An important part of the archaeological project is to investigate what role social and socio-political aspects like ownership, disposal, control and power played in the exploitation of the resources. The character of the production and the scale of the distribution indicate an intense and well-organized activity, and the quarry landscape may in this way illuminate aspects concerning professionalisation and social and socio-political relations within this period.

As noted, the quarries in Hyllestad are not mentioned in any documentary sources from the Middle Ages or early modern period and no information regarding disposal of quarries and production is given. A study of the local community where the quarries are located, including both archaeological as well as documentary sources, has therefore been carried out in order to gain insight into the landholding and property structures in the period of the quarrying. The aim is to shed light on aspects like disposal and control over the mineral resources.

According to Medieval written sources, all the farms with quarries in Hyllestad were parts of various larger estates in the Middle Ages. Most of the farms were owned by the clergy, divided between the local churches of Øn and Hyllestad, and the ecclesiastical institutions of Munkeliv monastery in Bergen and the Bishop. Neither before nor after the Reformation were the farms with quarries within their boundaries owned by the peasants themselves (Baug 2001; 2002, 90; Baug 2005). Whether or not these institutions were involved in the quarrying needs to be discussed.

Studies of craft and carving techniques

The quarries in Hyllestad are protected cultural monuments, but a small area has been released in order to conduct experiments on the carving techniques. Norwegian Crafts Development is engaged in this project and Tørbjørn Løland is the consultant on craft techniques and the one who carries out the experiments. Carving traces in the old quarries are being documented and the techniques and choice of tools reconstructed. The experiments conducted are also used to assess estimated time of production and difficulty of the procurement (Løland in prep).

Experimental archaeology can be especially useful for creating and testing hypotheses about past production. The experiments in Hyllestad shed light on the technical and functional aspects of the production, and they may be used as a way to understand how the products were extracted in the Viking Period and the Middle Ages.

As noted, in a majority of the quarries the millstones were extracted directly from the bedrock. This was done by making a circular groove outlining the size of the millstone. Then a channel was carved by using a pick or a pointed chisel. The depth of the channel decided the thickness of the millstone. The millstone was loosened along its base which had been thickly set by holes and by repeatedly striking against these holes the stone loosened (Løland in prep.).

Along with this experimental project is also an ongoing documentation and comparison of carving traces and interpreting of techniques in five of the largest millstone quarries in Norway, including Hyllestad (see www.millstone.no; Løland in prep.). The knowledge gained through both of these projects can be transferred to other quarries as there are common features independent of the product types. This will increase our knowledge of production techniques and the use of tools from the Viking Period and far into recent times.

Several of the oldest Norwegian millstone quarries seem to have had common carving techniques. However, the extraction of millstones has not necessarily been strictly controlled. A few basic skills were most likely needed, but the quality of the work may have varied in different periods as well as in different areas. The experiments conducted in Hyllestad will shed light on the basic common features that seem to occur both in Norwegian and foreign quarries (Løland in prep.).

The millstone history as an educational arena

The increased knowledge concerning the quarries in Hyllestad gained from the different researches has lead to some unique educational projects where the primary school in Hyllestad and the schoolchildren are being involved in the guiding and presentation of the quarries. The school has in this way integrated a local cultural monument in the daily school activities, and the results from the different research projects are being incorporated in the education. Young students in Hyllestad know that when they are 8th graders (13 years old), some of their education will be about a part of their cultural heritage and local history, namely the millstone quarries. The pupils are also aware that this year they will be the local presenters of this heritage: they are the ones who are going to show and tell others about the more than 1200 years old millstone production. This is a big assignment for a group of young people.

The youngsters are taught to be presenters of the millstone history. They learn the history in the way that archaeologists, historians and geologists describe it. Furthermore they learn, both theoretically and practically, how everyday life might have been in this area in the Viking period and the Middle Ages, when the large scale production of millstones had its glory days. The experience so far shows that the youngsters very quickly develop into good and credible presenters. The visitors highly appreciate learning the history from these young people and the youngsters themselves have great fun conducting this task.

The teaching of history to the pupils mainly takes place outside the school building – in the immediate vicinity of the cultural heritage site. Close to the quarries, a millstone park has been established with different buildings as well as a smithy. The main building, a timber building of 15 x 8 metres in size with a turf roof, creates the basis for education and presentations. The building has a fireplace in the middle of the room, and several long, wooden tables and benches makes it suitable for receiving groups of up to 50 persons. Inside the building there are also different tools and equipment used in different types of crafts typical for the Viking period and the Middle Ages. Of special importance is a hand-quern carved in Hyllestad.

From the buildings the visitor steps directly into the millstone landscape, where they sense the more than 1200 years old history. This landscape is the classroom for the youngsters – 3 lessons per week during the whole school year. In reality, however, it is used far more as the pupils train for and perform different presentations and guiding assignments. A lot of time is also spent on field studies in the quarries, where the students practice recognising different carving traces as well as formations in the landscape that give information about the production. Even mathematics has been included in the project, where the students do measurements, including the weight of different millstones.

In the beginning of the school year the youngsters immediately start to study theoretical and practical themes connected to the past, and selected craft techniques are learned. Most important are the carving of millstones and the work in the smithy (Fig. 6). However, the most demanding task for the schoolchildren is the guiding of visitors through the quarries. During the first month of the school year, the teaching is organised in several teaching units. An important part of the project is pupils supervising and teaching other pupils. The supervisors are specially selected 9th graders, who teach the youngsters in the 8th grade about the millstone history and all the activities connected to the project.

The school children as presenters of the millstone history

During the whole period of learning, the pupils know that one of their most important tasks is to present what they learn to others. The first assignment is normally in November. The audience may be all kinds of people, from pensioners on day trips to younger school classes from Hyllestad or other places. Every spring an international group of students from UWC (United World College) in Fjaler comes for a visit. This is an extra challenge, because all the oral communication must be in English. In this way language lessons are brought into the project as well, and a few weeks ahead of this event the millstone history is the main theme in the English lessons at school.

Another annual event is the 'Hyllestad-seminar', where both professionals and others with a historical interest join in. The topics at the seminars are most often related to the millstone history. In this seminar the pupils from Hyllestad have come to play an important role. They are the ones who guide the visitors through the quarry landscape, presenting the history of the production in the most fascinating way, and they are the ones who demonstrate and teach the visitors of long forgotten crafts and techniques (Fig.7).

Our experience shows that the encounter between audience and youngsters is in itself an event to remember. These young people give the visit to the cultural heritage an extra dimension, and after several of the seminars the school in Hyllestad has received letters from satisfied visitors who want to give their praise to the pupils and the work they conducted during the seminar. For many, the peak of the seminar is the guided tour conducted by the schoolchildren.

Figure 6. Schoolchildren are carving millstones.

Figure 7. Guiding in the Mill Stone Park.

The Learning Pyramid as guiding principle

Currently, the Norwegian primary school has a very theoretical approach to the learning process. As a consequence, some students have problems adjusting to the everyday activities at school, and thus need other forms of adapted teaching. The experience from the educational project in Hyllestad indicates that it is an advantage, not only for students that are struggling with the theoretical approach, but for all students to have a more practical approach to the syllabus.

Innumerable studies and analyses have shown that learning is a personal process that demands the use of all our senses. David and Roger Johnson's Learning Pyramid shows this in a way which is easy to understand (Fig. 8). It describes the learning effect compared to the choice of learning methods. The pyramid should be evaluated critically, but within several fields and subjects it is a good guiding principle in the planning of lessons. The more senses we use, the better the learning effect is. Very often, however, teaching in the schools is based on theoretical lectures, also on subjects that could easily be taught with a larger scale of student-activity. In a group of youngsters who are not too eager to learn, an oral lecture usually gives a small outcome.

Looking at the pyramid, we find 'learning by doing' as a good method. However, the most successful way to learn something is to teach others. The Hyllestad project where the students have to teach others has thus turned out to be very successful for the youngsters in the instructor-role. This has consequences also for the teacher's role in general. The essential point is not that the teacher teaches, but that the pupils learn. Or as Galileo Galilei once said: '*You cannot teach a human being anything at all. You can only help him to realize it himself.*'

The school project at Hyllestad School tries to use the learning pyramid as a basic principle in teaching. In addition to the pupils' profit, the visitors will experience that they get practically engaged. If you visit Hyllestad and meet these young people as presenters of the history, you will get the chance to practise carving, grinding of corn on the hand-quern and other different craft techniques.

Several benefits

This use of the millstone history can easily be transferred to other kinds of cultural heritage. In each case we must start with the attraction itself, and find out what is suitable in each particular case. It is of course also important to consider how much the actual attraction can stand being used in this way. It is thus necessary and useful to have a close co-operation with the county, the municipal cultural authorities and other specialists.

The principle that young people should be educated as presenters should be the basic starting point. From there on other ideas and solutions will occur, as the project develops. In Hyllestad the experience from the educational project shows us that a project like this has several benefits:

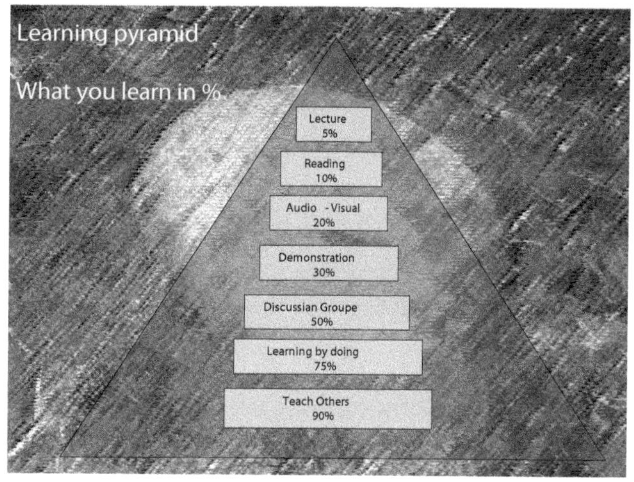

Figure 8. The learning pyramid.

- The local cultural heritage gets focus and status
- Knowledge and skills connected to the attraction are passed on
- Local people feel they are a part of the heritage, they feel pride and the project can strengthen their local identity
- Visitors get an interesting and memorable experience

There are many ways of presenting cultural heritage. Whatever the method, we may say that those who visit a cultural attraction will have a personal experience. In one way or another, our encounter with the cultural heritage will have an influence upon us. Something will, more or less, positive or negative, be different, before and after the encounter with the cultural heritage. We all have different assumptions and needs when we visit a cultural attraction. The project in Hyllestad tries to consider this, when using the millstone history as a pedagogical arena.

Conclusions

The millstone quarries in Hyllestad are an example of the importance and values of a co-operation between different disciplines and the local community. The research conducted indicates that Hyllestad is one of the largest and oldest quarry areas for millstones in Norway. Different quarry types – and thus different technologies – are identified. The activity started in the early Viking period – perhaps a little bit prior to this – and the extraction continued far into recent times, with the early and high Middle Ages as the most intensive period.

Not only millstones were extracted at these quarries, the production of stone crosses was also an important part of

Figure 9. Apprentice and master.

the activity. This production was, however, much smaller compared to the millstone production. Nevertheless, millstones and stone crosses were produced at the same quarries.

Study of carving traces in the quarries and experiments conducted, has given new information about long forgotten craft techniques and the use of tools. The results from the ongoing studies may shed light on, and bring to life, both the millstone extraction as well as techniques used in other types of quarries. In this way we will be able to carry on the practical dimension of handicraft in the cultural monument.

The Hyllestad quarries are an example of how a local community can benefit from a cultural monument. The quarries create good and unique learning-situations, and the educational project in Hyllestad shows the value of a cultural monument as well as how local history can be presented and taught to young people (Fig. 9). The project also shows the importance and benefit of a local engagement in order to take care of a cultural monument and to pass on the knowledge concerning this to others. In order to do so, we need a good relation and a good communication between the different disciplines involved and the local community.

References

Arge, S.V., 1989. Om landnåmet på Færøerne. *Hikuin,* 15, 103-128.

Baug, I., 2002. *Kvernsteinsbrota i Hyllestad. Arkeologiske punktundersøkingar i steinbrotsområdet i Hyllestad, Sogn og Fjordane. Bergverksmuseet. Skrift nr. 22.* Kongsberg.

Baug, I., 2005. Who owned the products? Production and exchange of quernstones, Hyllestad in Sogn, Western Norway. In Ingunn Holm, Sonja Innselset & Ingvild Øye (eds.). *UBAS International 1. University of Bergen Archaeological Series. 'Utmark'. The Outfield as Industry and Ideology in the Iron Age and the Middle Ages.* 99-108. Bergen.

Baug, I., 2008. Steinkrossbrotet i Hyllestad. In F.B. Førsund and H. Håkon (eds.), *Sogeskrift frå Hyllestad 2008.* Forde, 8-13.

Baug, I., in prep. *Quarrying in the late Iron Age and the Middle Ages - social and socio-political aspects in relation to production and distribution.* Unfinished doctoral thesis at AHKR, University of Bergen.

Carelli, P., and Kresten, P., 1997. Give us this day our daily bread. A study of Late Viking Age and Medieval Quernstones in South Scandinavia. *Acta Archaeologica,* 68-1997, 109-137. Munksgaard.

Eldjárn, K., 1964. Kvarn. Island. *Kulturhistorisk leksikon for nordisk middelalder fra vikingtid til reformajsonstid, 9.* 540-541. Oslo.

Grenne, T., Heldal, T., Meyer, G.B., and Bloxam, E.G., 2008. From Hyllestad to Selbu. Norwegian millstone quarrying through 1300 years. *Geology of Society. Geological Survey of Norway Special Publication (Norges geologiske undersøkelse),* 47-66, Geological Survey of Norway. Trondheim.

Hansen, A. M., 1997. Maritime perspektiv på kvernsteinsproduksjonen i Hyllestad. In H. Sørheim (ed.), *'Arkeolog og kystkultur'. Foredrag på Sunnmøre Museum 25-26/10 1997.* 58-63, Ålesund.

Heldal, T., and Bloxam, E. G., 2007. *Kartlegging og karakterisering av kvernsteinsbruddene i Hyllestad. NGU Rapport, 2007.079.* Geological Survey of Norway. Trondheim.

Løland, T., in prep.. *Millstone handcraft through 1000 years.* (Preliminary title of an unfinished report).

Rønneseth, O., 1968. Das Zentrum der ältesten Mühlsteinindustrie in Norwegen. In M. Claus, W. Haarnagel and K. Raddatz (eds.), *Studien zur europäischen Vor- und Frühgeschichte,* 241-252, Neunmünster.

Schön, V., 1995. *Die Mühlsteine von Haithabu und Schleswig. EinBeitragzur Entwicklungsgeschichte des mittelalterlichen Mühlenwesens in Nordwesteuropa.* Neunmünster.

Skre, D., 2007. Towns and Markets, Kings and Central Places in South-western Scandinavia *c.* AD 800-950. In D. Skre (ed.), *Kaupang in Skiringssal.* Kaupang Excavation Project Publication Series, Vol. 1. Norske Oldfunn, 22, 445– 98, Aarhus.

Les meules de l'Esterel (Var, France) un diagnostic par SIG

André Buisson

Le massif de l'Esterel s'étend entre Fréjus (Var) et Cannes (Alpes Maritimes), bordé à l'Est par la Méditerranée, au Nord par le massif du Tanneron et au Sud par les Maures. Sa couleur rouge caractéristique lui a été donnée par l'affleurement presque exclusif de rhyolite, roche volcanique ignimbritique de formation permienne (Bordet 1966; Toutin-Morin *et al,* 1994). Plusieurs études ont déjà mis en évidence la richesse du massif de l'Esterel pour la production de meules à broyer (Palausi 1965; Amann 1976; Brun 2000; Désirat 1980; Buisson et Olive 2009). Un diagnostic effectué depuis quelques années montre que les types de meules extraites dans ce massif sont nombreux (diamètres de 0,35 m. à 1,60 m.), que la durée d'extraction est très importante (de l'époque protohistorique – meules carrées, à va-et-vient – à l'époque moderne) et qu'elle reflète assez exactement l'évolution des méthodes de broyage et des produits à broyer : grains, olives, écorces, par des particuliers – moulins à main – ou des collectivités – moulins à eau, à vent ou à sang –). L'exploitation de ces meulières est multiforme: des traces d'extraction sont découvertes à de nombreux endroits, dans certains cas pour seulement quelques unités, et dans d'autres cas, l'emplacement s'étend sur une surface prouvant l'enlèvement de volumes très importants, certainement supérieurs à plusieurs dizaines de milliers d'unités de chaque type (Buisson et Olive 2009).

Problematique de la recherche

L'intérêt d'une analyse archéogéographique (Chouquer 2008) n'est plus à démontrer. Elle a été largement appliquée à l'étude de l'évolution urbaine (Galinié and Rodier 2002) et à l'évolution géomorphologique fluviale (par exemple, Beck *et al.* 2009). Dans le domaine de l'extraction meulière, la démarche ouvre la voie à la compréhension de l'adaptation des paysages par l'homme au cours de l'exploitation des carrières: aménagement de chemins d'accès, de pentes pour l'évacuation des pièces extraites, la gestion des résidus de taille et l'évolution des carrières après abandon de l'activité.

Deux buts ont été recherchés dans cette perspective : cartographier les centres d'extraction des meules (à travers les restes de taille encore en place, les accumulations de déchets de taille et enfin les centres d'extraction) ; assurer, grâce à cette cartographie, le projet de leur protection (une partie de ces meulières se trouve sur un terrain militaire et des engins passent, effaçant progressivement les traces).

Les méthodes

La méthode de recherche repose tout à la fois sur une prospection du massif et sur une étude de l'aire de diffusion des productions. En matière de prospection, nous avons volontairement effectué un choix en divisant le massif en deux zones principales : la première, qui fait l'objet de toute notre attention, est constituée du cœur du volcan, zone dans laquelle la rhyolite est la plus compacte et la moins diaclasée, et nous avons laissé de côté, dans un premier temps, les espaces éloignés du cœur du volcan permien, considérant à raison que la texture de la roche était trop fragmentaire et diaclasée pour avoir permis des extractions de meules nombreuses, même si des découvertes de meules isolées ont été effectuées dans les zones périphériques au cours d'études précédentes (Brun 2000). Pour autant, la zone étant trop vaste pour être abordée globalement, la prospection a été dans un premier temps limitée à des sites aisément repérables dans l'environnement (certaines zones sont directement visibles sur *Google Earth*) et accessibles, on a donc axé la prospection sur les zones découvertes (la végétation de maquis est très peu accessible, du fait de la présence de plants épineux et de salsepareille très envahissante, sauf après incendie !), et notamment en bordure dominante de talus.

En dehors de l'activité de terrain, nous avons vérifié les données déjà publiées et éliminé les emplacements suspects (toponymes non repérés). La recherche en archives, beaucoup plus longue, nous permettra sans aucun doute, de compléter ce corpus, à travers les actes de vente de meules et le recensement des meuniers.

Le SIG 'Rhyolite'

Le SIG est établi grâce au logiciel ArcGis 9.2.

La base de données chiffrées

Elle est constituée sous la forme d'un tableur. Chaque meule (considérée comme une 'entité') est identifiée par des données recensant les coordonnées Lambert en x et en y du lieu de sa découverte ; on renseigne également les thèmes suivants : les dimensions (par un code identifiant les formes –carrée ou circulaire- et les différents diamètres connus -de 0,35 à 1,60 m-), si elle est encore 'encastrée', si elle est 'libre', si elle est 'brisée', si elle est 'dans un contexte archéologique' (habitat ?) ou 'en remploi utilitaire'. On renseigne également les zones d'extraction repérées lors de la prospection (sous forme d''aires', ou de 'polygones').

La base de données 'images'

Une base de données 'images' est agrafée à ce tableur, chaque meule identifiée par une photographie et un code retrouve ainsi commodément son emplacement géographique lors de l'interrogation de la base.

La base de données cartographique

Le fond de carte est établi grâce à la BD Ortho de l'IGN au 1/50 000 et à la BD géologique du BRGM. Le fond cadastral peut être interrogé si le besoin s'en fait sentir. La base de données images est identifiée par des codes numériques correspondant à des secteurs géographiques (par exemple, 'champ de tir'= 0010 et suivantes, 'les lauriers'= 0020 et suivantes…) ou thématiques (meule, emplacement d'extraction…). Pour la cartographie historique, la carte de Cassini est géoréférencée pour la zone qui nous intéresse, elle permet de mettre en valeur notamment les axes routiers anciens de la région, la voirie ayant évolué au cours des derniers siècles. Les anciennes cartes régionales pourront s'intégrer à la base au fur et à mesure de leur numérisation (la base est actuellement limitée aux communes de Fréjus, Bagnols et Saint-Paul-en-Forêt, mais est susceptible d'être étendue aisément).

L'interrogation des données

L'interrogation peut porter sur le recensement de toutes les meules 'encastrées' (indiquant alors la présence de lieux d'extraction), de toutes les meules 'brisées' + 'libres' (permettant ainsi la visualisation des axes de déplacement, et dans quelques cas, des lieux d'utilisation), voire de toutes les meules de tel ou tel type, etc.

L'analyse des données

Jointes à l'ensemble formé par les cartes géologiques et topographiques, les données chiffrées forment un éparpillement de points révélateurs, en l'absence de fouilles, de plusieurs types d'enseignements : sur les zones d'extraction, sur les voies d'acheminement des meules et sur les 'risques' de casse de ces pierres au cours du transport. Des zones de casse plus fréquentes soulignent le risque 'transport' et les prospections plus fines sur ces lieux pourront mettre en évidence les aménagements de voirie (descenderies) du type restanques ainsi que les remplois de ces meules cassées dans les murets de ces terrasses.

Les perspectives

Une extension de ce système permet également la visualisation de l'aire d'extension de la zone de commercialisation de ces meules, spécialement pour les époques récentes et les meules de fort diamètre (1,60 m), sur les moulins du Var et des départements circonvoisins, dans le cadre de l'enquête sur les moulins. Une base de données recensant tous ces moulins utilisant des meules de rhyolite permet d'évoquer les axes de circulation et de transport de ces objets très pondéreux (0,6 m^3 en moyenne, soit 800 kg environ). L'étude des voies routières et fluviales régionales est très importante pour ce diagnostic et les documents cartographiques, comme la carte de Cassini, géoréférencée, sont très utiles.

Une question doit également être débattue, celle de l'origine de la rhyolite. En effet, plusieurs centres d'extraction ont été actifs simultanément en Méditerranée (des meules ont été découvertes dans plusieurs épaves découvertes le long des côtes, sans connaître la provenance de chacun de ces navires), et il sera nécessaire de délimiter pour chacun d'entre eux les aires diffusion de la production (elle ne pourra être effectuée qu'à partir de l'étude de lames minces).

Bibliographie

Amann, A.H., 1976. Les meules préromaines en rhyolite et en basalte du département duVar. *Annales de la Société des Sciences Naturelles et d'archéologie de Toulon et du Var*. Toulon, 27-37.

Beck, C., Guizard-Duchamp, F., et Heude, J., (dir.) 2009. *Lit mineur, lit majeur, lit voyageur. mémoires et cours d'eau* / 11e[s] Rencontres internationales de Liessies, (24-25 septembre 2008), Villeneuve-d'Ascq, *Revue du Nord* (Collection Art et archéologie, n° 14).

Bordet, P., 1966. *L'Esterel et le massif de Tanneron (histoire géologique, itinéraires géologiques).* Paris. Hermann, 1 carte h.t.

Brun, J.-P., 2000. *Carte archéologique de la Gaule romaine, 83. Le Var*. Paris.

Buisson, A., et Olive, G., 2009. L'exploitation des meules de rhyolite de l'Estérel, problèmes et methods. In J.-R. Gaborit (dir.), *'Circulation des matières premières en* Méditerranée, transferts de savoirs et de techniques (édition électronique)', 128e congrès national des sociétés historiques et scientifiques. Bastia, 2003. Paris, 58-73.

Chouquer, G., 2008. *Traité d'archéogéographie, I, La crise des récits historiques*. Paris.

Desirat, G., 1980. *Bagnols-en-Forêt, contribution à l'étude de la Provence orientale.* Toulon, Riccobono Impression, t.1 (seul paru).

Galinie, H., et Rodier, X., 2002. ToTopI, Topographie de Tours pré-industriel, *Les petits cahiers d'Anatole*, n°11, 9/12/2002 (http.//www.univ-tours.fr/lat/pdf/F2_11.pdf.

Palausi, G., 1965. Les tailleries de meules anciennes dans la région de l'Esterel et leur relation avec la géologie. *Bulletin philologique et historique du Comité des Travaux Scientifiques et Historiques*, 44, 707-714.

Raybaut, P., (P. Castella, A. Compan coll.) 1979. *Les sources régionales du Pays de Nice.* Paris.

Toutin-Morin, N., Bonijoly, D., Brocard, C., Broutin, J., Crevola, G., Dardeau, G., Dubar, M., Feraud, J., Giraud, J.D., Godefroy, P., Laville, P., Meinesz, A., 1994. *Notice explicative, Carte géologique France (1/50 000), feuille Fréjus-Cannes (1024).* Orléans . BRGM.

LE GROUPE MEULE

« Evolution typologique et technique des meules du Néolithique à l'an mille sur le territoire français »

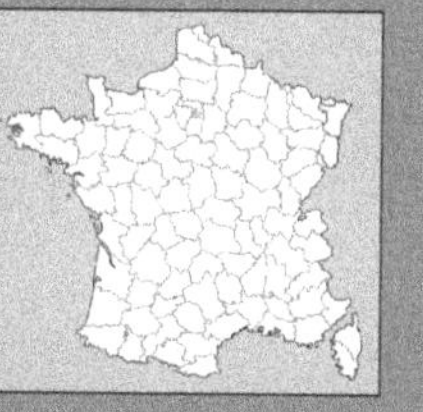

Ce Programme Collectif de Recherche issu d'une dynamique d'échanges et de collaborations entre une vingtaine d'archéologues et géologues est soutenu par le Ministère de la Culture, le CNRS et l'INRAP.

Le projet d'étude des moulins qu'ils soient rotatif ou va-et-vient s'inscrit dans des **thématiques vastes**, qui concernent les meules depuis leur **extraction** dans les carrières jusqu'à leur **rejet**, comprenant les étapes de façonnage, de transport, d'utilisation et d'entretien.
Près d'une dizaine de régions françaises sont couvertes : des Vosges au Massif Armoricain et de la Provence à la région Nord-Picardie

Dans ce cadre, quatre axes de recherche ont été définis :

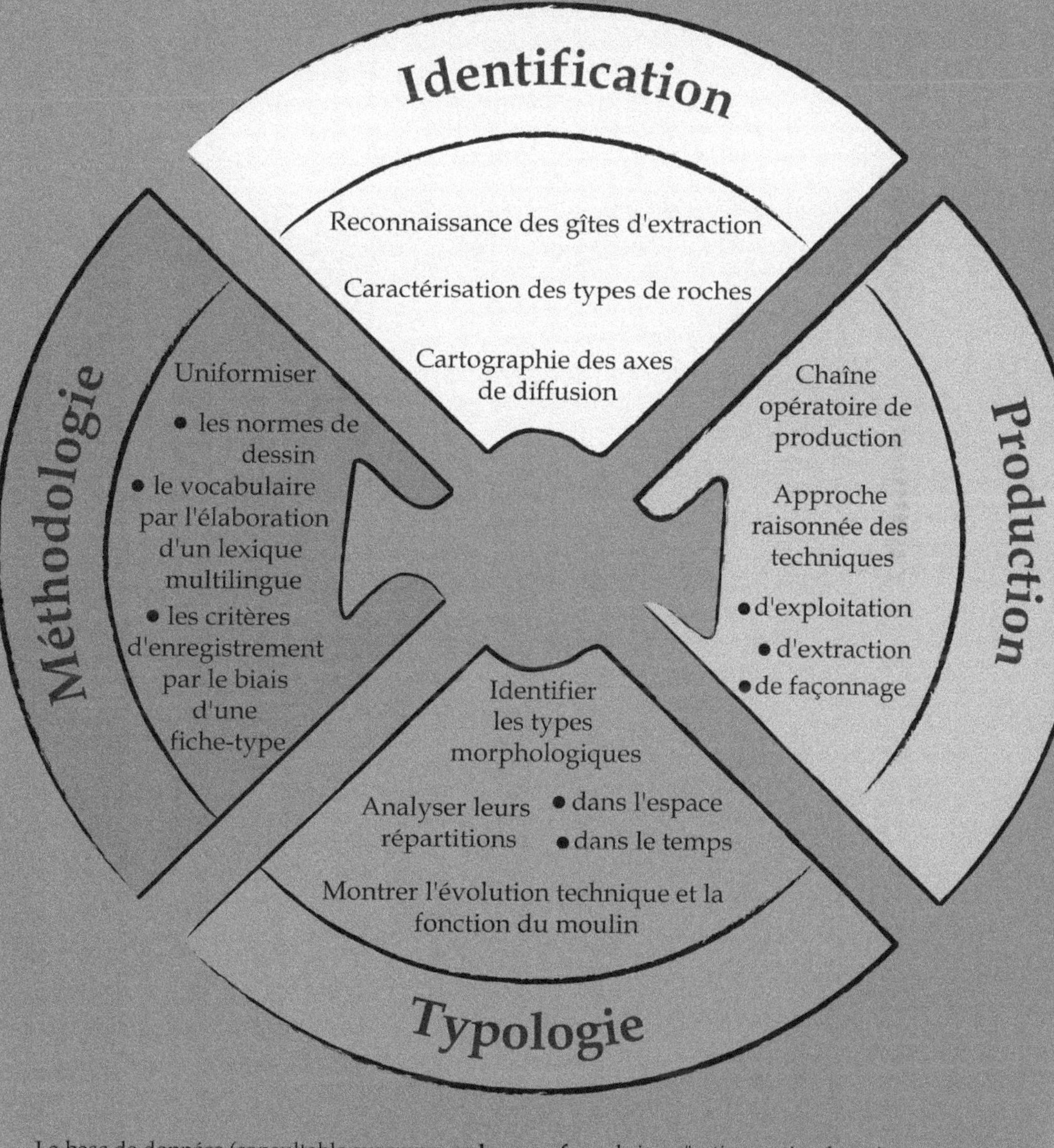

La base de données (consultable sur **www.archeo.ens.fr** - rubrique "*actions nationales et internationales*") est à l'usage de toute personne intéressée par ce domaine de recherche. Elle peut être également alimentée en prenant contact avec le groupe (**groupe-meule@orange.fr**).

www.ingramcontent.com/pod-product-compliance
Ingram Content Group UK Ltd.
Pitfield, Milton Keynes, MK11 3LW, UK
UKHW061844041026
15591UKWH00015B/903

9 781407 308487